The Most Successful Way of JEE Preparation

DPP

Daily Practice Problems

JEE Main & Advanced

Aromatics & Biomolecules

CHEMISTRY Vol-9

Preeti Gupta

AF518223

arihant
ARIHANT PRAKASHAN (SERIES), MEERUT

DPP

Daily Practice Problems

arihant

ARIHANT PRAKASHAN (SERIES), MEERUT

All Rights Reserved

© Publisher

No part of this publication may be re-produced, stored in a retrieval system or by any means, electronic mechanical, photocopying, recording, scanning, web or otherwise without the written permission of the publisher. Arihant has obtained all the information in this book from the sources believed to be reliable and true. However, Arihant or its editors or authors or illustrators don''t take any responsibility for the absolute accuracy of any information published, and the damages or loss suffered thereupon.

All disputes subject to Meerut (UP) jurisdiction only.

Administrative & Production Offices

Corporate Office 'Ramchhaya' 4577/15, Agarwal Road, Darya Ganj, New Delhi -110002
Tele: 011- 47630600, 43518550; Fax: 011- 23280316

Head Office Kalindi, TP Nagar, Meerut (UP) - 250002
Tel: 0121-2401479, 2512970, 4004199; Fax: 0121-2401648

Sales & Support Offices

Agra, Ahmedabad, Bengaluru, Bareilly, Chennai, Delhi, Guwahati, Hyderabad, Jaipur, Jhansi, Kolkata, Lucknow, Meerut, Nagpur & Pune

ISBN : 978-93-13193-48-7

Published by Arihant Publications (I) Ltd.

Production Team

Publishing Managers Mahendra Singh Rawat & Keshav Mohan
Project Head Sorabh Chaudhary
Cover Designer Shanu Mahad
Inner Designer Mazher Chaudhary
Page Layouting Sunil Sharma, Hemant
Proof Reading Salman

For further information about the products from Arihant, log on to www.arihantbooks.com or email to info@arihantbooks.com

/arihantpub /@arihantpub Arihant Publications /arihantpub

ABOUT DPP

Daily Practice Problems

Indian Institutes of Technology better known as IITs have always been the dream destination of each science group students who want to pursue their career as Engineers. These are the institutes which have got very high value not even in India but in the whole world. To get the admission in such institutes one need to face entrance Examinations; JEE Main & JEE Advanced, out of which JEE Main is considered as the preliminary exam & JEE Advanced is the final exam to get into the IITs. Approximately per year 12 lakh students enroll themselves for JEE Main and only the top 2.25 lakh in JEE Main merit can attempt for JEE Advanced, the fact clearly express the degree of toughness of JEE Main & JEE Advanced.

To have success in such an exam one needs to be very planned & systematic in its preparation. A very popular wise saying 'Rome was not built in a day' suggests that to achieve a big target like IITs we need to work continuously. For continuous work, one need to have proper planning that how much work he/she has to do on regular basis, i.e., on daily basis. Each student preparing for JEE has many books but unfortunately none of them gives the students a plan for daily studies, which is the most effective success key for JEE.

Keeping the daily practice needs of the students in mind we have come up with DPP Series (Daily Practice Problems) having approximately 26 books. The primary focus of this series is on achieving success through practice & proper planning. For that we have divided the whole syllabus of JEE Main & JEE Advanced; Physics, Chemistry & Mathematics, into different volumes (around 8 volumes of Mathematics & 9 volumes each of Physics & Chemistry), each volume has Daily Practice Problems Sheets having specific questions on various Topics of the individual chapter of the volume ensuring the complete Practice & Assessment of the topic.

DPP

Daily Practice Problems

In this way we have tried to give you an effective daily planning for JEE keeping in mind the daily hours that a student can devote to his practice, you are just required to practice one sheet of each subject on daily basis. This habit of daily practice will keep you away from exam stress and thoughts like; how much I have prepared/am I preparing properly or not/ do I need to study more on daily basis or what I am doing is ok, etc., which regularly tease your mind and can give you panic at any stage of your preparation.

Some special features that we have tried to incorporate in our Daily Practice Problems (DPP) series are discussed below

1. Micro Level coverage of all the topics of a Chapter/ Unit given in JEE Main & Advanced Syllabus.
2. Each DPP has various types of questions; Subjective Questions, Single/ Multiple Correct Options, Statements type, Comprehension based, Integer Value Answer type, Matching, etc., making you practice & ready for all the question formats of JEE.
3. New & innovative questions in each DPP challenging & improving your problem solving skills.
4. Along with Topical Coverage, Revisal DPPs for JEE Main & JEE Advanced with each Chapter, covering multidisciplinary questions proving comprehensive practice.
5. JEE Main & JEE Advanced Archive (collection of last 13 years' exams questions) with each chapter.
6. Complete Solutions for each DPP.

We hope that DPP series will prove to be very beneficial for the students preparing for JEE, by giving them daily practice in a systematic & planned manner. The suggestions/criticism from the students, teachers & parents about DPP series will be welcomed whole heartedly and we promise to update this series accordingly from time to time.

DPP

Daily Practice Problems

DAILY TASKS...

1 ARYL HALIDES AND PHENOLS

2 AROMATIC ALDEHYDES, KETONES AND ACIDS

DPP
Daily Practice Problems

3 ALIPHATIC NITROGEN CONTAINING COMPOUNDS

4 AROMATIC NITRO COMPOUNDS, ANILINES AND DIAZONIUM SALTS

DPP
Daily Practice Problems

5 BIOMOLECULES

6 POLYMERS

1. Aryl Halides and Phenols

DPP
DAILY PRACTICE PROBLEMS

DPP-1 Methods of Preparation of Aryl Halides

Subjective Questions

Working Space

Direction *(Q. Nos. 1-4) These questions are subjective in nature, need to be solved completely on notebook.*

1 Synthesise the following from the given reactants.

i. *Ortho*-bromotoluene from toluene

ii. *m*-dibromo+benzene from benzene (without involving nitration)

iii. 1, 3, 5-tribromo-2-fluorobenzene from aniline

iv. BHC from chlorine

2 Describe all the steps necessary for the conversion of benzene and/or toluene into the following compounds.

i. *p*-bromobenzylchloride

ii. *p*-chloronitrobenzene

iii. 3, 4-dibromonitrobenzene

iv. *p*-bromobenzene sulphonic acid

v. *p*-bromostyrene

3 i. Give the product of reaction of $\mathrm{Ph\underset{\underset{Ph}{|}}{C}HCH_2OH}$ with HBr and explain its formation.

ii. What product is expected from the reaction?

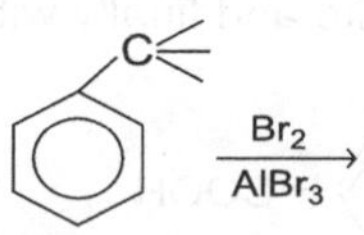

Justify your answer with the mechanism.

4 i. Bromine adds to styrene to give styrene dibromide only, no nucleophilic substitution takes place. Explain.

ii. Both benzene and hexadeuterobenzene (C_6D_6) are brominated at equal rates when treated with $Br_2/FeBr_3$. Explain.

Only One Option Correct Type

Direction *(Q. Nos. 5-10) This section contains 6 multiple choice questions. Each question has four choices (a), (b), (c) and (d), out of which ONLY ONE is correct.*

5 What function does nitric acid do in the reaction of benzene with I_2 to produce iodobenzene?

a. HNO_3 converts I^- to I_2

b. It acts as a catalyst

c. HNO_3 converts I_2 to HI

d. It oxidises I_2 to I^+

Working Space

6 The reaction of biphenyl with HOCl in the presence of a strong acid gives (major)

a. 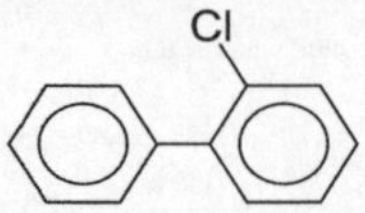

b. 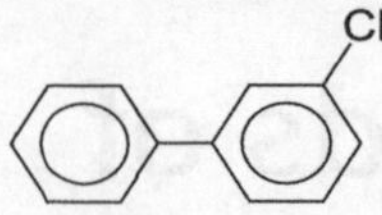

c. 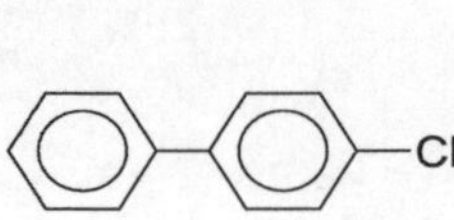

d. 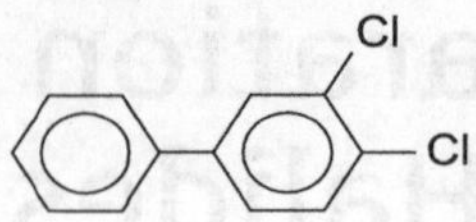

7 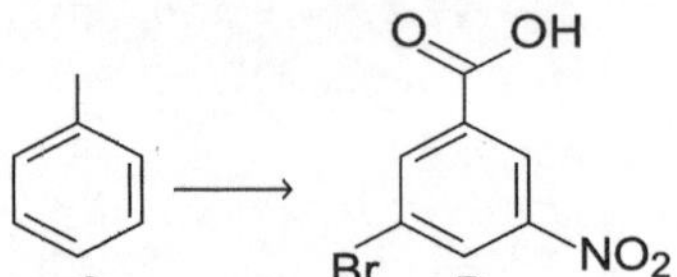

A B

It can be performed by using reagents :

I. $Br_2/FeBr_3$ II. HNO_3/H_2SO_4 III. $K_2Cr_2O_7/H_2SO_4$, Δ

These reagents should be used in following order

a. $A \xrightarrow{I} \xrightarrow{II} \xrightarrow{III} B$

b. $A \xrightarrow{II} \xrightarrow{I} \xrightarrow{III} B$

c. $A \xrightarrow{III} \xrightarrow{II} \xrightarrow{I} B$

d. $A \xrightarrow{II} \xrightarrow{III} \xrightarrow{I} B$

8 Which of the following will best convert nitrobenzene into 3-fluorobromobenzene?

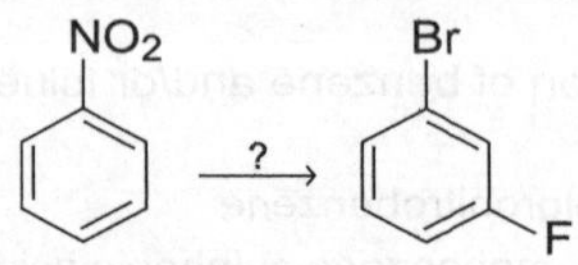

a. $F_2/AlCl_3$, Zn/HCl, $NaNO_2$/HCl-0° C, CuBr
b. $Br_2/FeBr_3$, $SnCl_2$/HCl, $NaNO_2/HBF_4$ -0° C, heat
c. $SnCl_2$/HCl, $NaNO_2/HBF_4$ -0° C, heat, $Br_2/FeBr_3$
d. $SnCl_2$/HCl, $Br_2/FeBr_3$, $NaNO_2/HBF_4$ -0° C, heat

9 What is the final product when benzoic acid is treated with thionyl chloride, then ammonia, followed by reaction with sodium hypobromite and finally with bromine?

a. 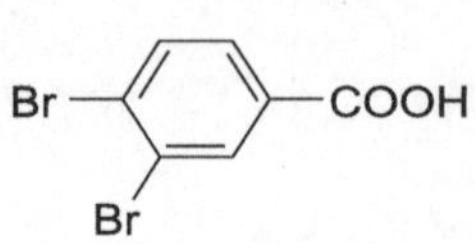

b. 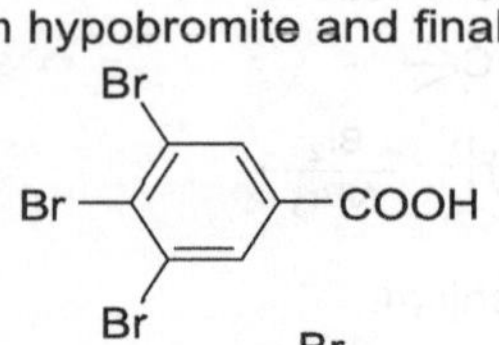

c. 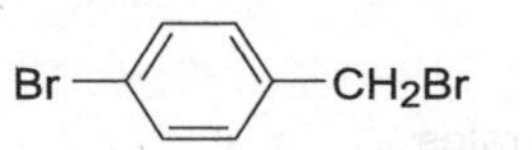

d. 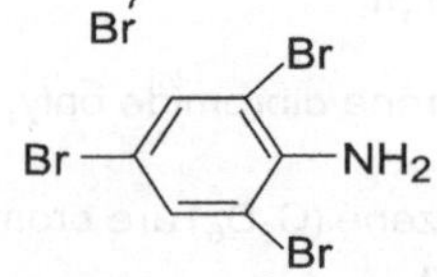

10 A hydrocarbon *A* of molecular formula C_9H_{10} adds Br_2 to give *B* $C_9H_{10}Br_2$. *B* on hydrolysis gives *C*, which on oxidation gives benzoic acid and acetic acid. *B* in this sequence is

a. 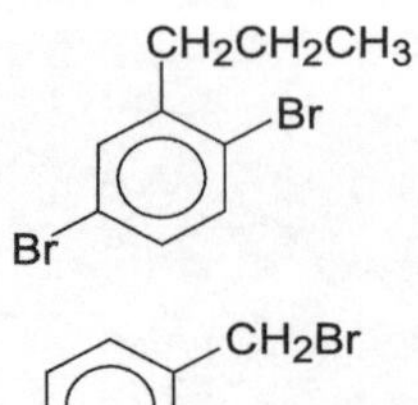

b. 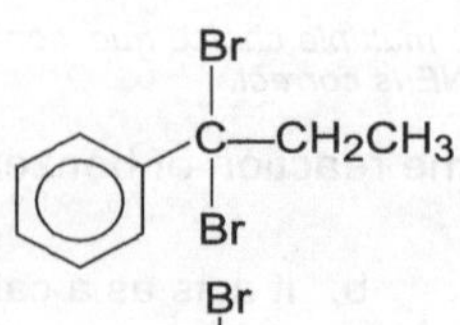

c. 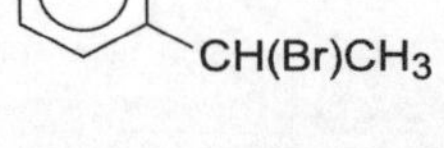

d. 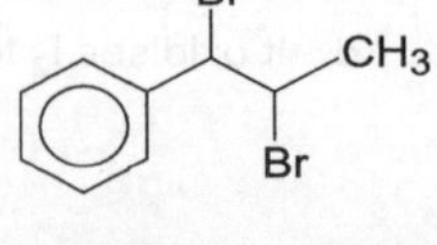

Working Space

Matching List Type

Direction (Q. No. 11) Choices for the correct combination of elements from Column I and Column II are given as options (a), (b), (c) and (d), out of which one is correct.

11 Match the Column I with Column II and choose the correct answer using the codes given below :

	Column I (Reaction)		Column II (Reaction name)
i.	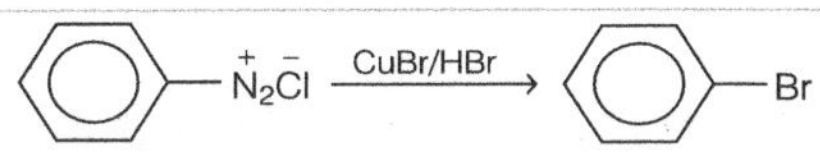	p.	Diazotisation
ii.	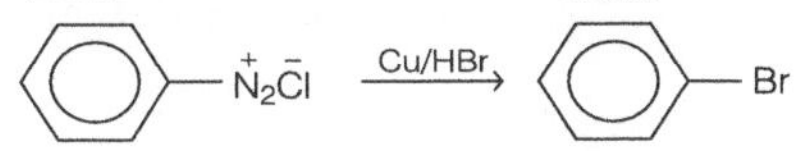	q.	Raschig
iii.	Benzene + HCl + $\frac{1}{2}O_2 \longrightarrow C_6H_5Cl$	r.	Gattermann
iv.	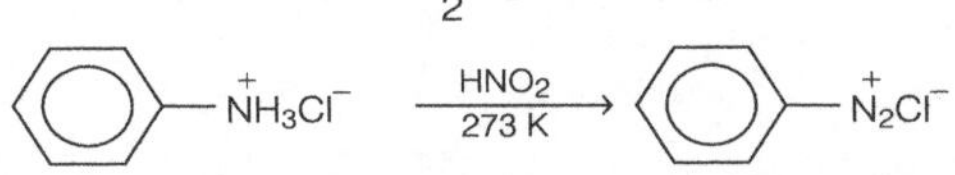	s.	Sandmeyer

Codes

	i	ii	iii	iv			i	ii	iii	iv
a.	s	r	p	q		b.	p	q	r	s
c.	s	r	q	p		d.	r	s	q	p

One or More than One Options Correct Type

Direction (Q. Nos. 12-14) This section contains 3 multiple choice questions. Each question has four choices (a), (b), (c) and (d), out of which ONE or MORE THAN ONE are correct.

12 Which of the following reaction will produce chlorobenzene as a major product?

a. $C_6H_5NH_2 \xrightarrow[HCl;\,0°C]{NaNO_2} \xrightarrow{Heat}$

b. $C_6H_5NH_2 \xrightarrow[HCl;\,0°C]{NaNO_2} \xrightarrow{CuCl}$

c. $C_6H_6 + C_6H_5COCl \xrightarrow{AlCl_3}$

d. $C_6H_6 + Cl_2 \xrightarrow{hv}$

13 Which of the following reaction(s) will give 1,3-dichlorobenzene as a significant organic product?

a. 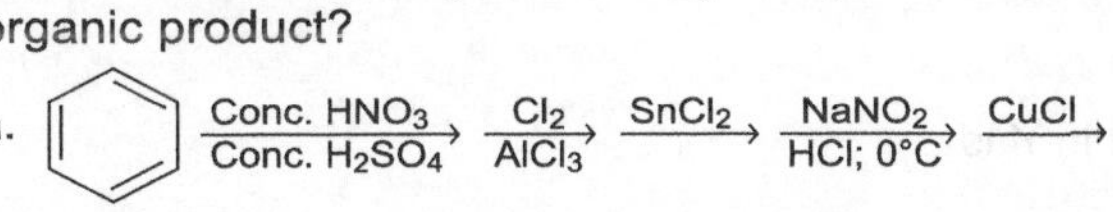

b. 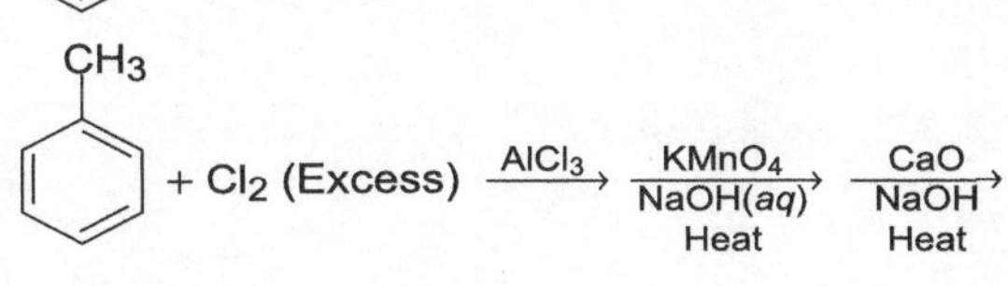

c. 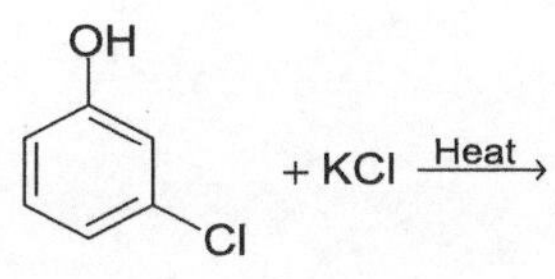

d. 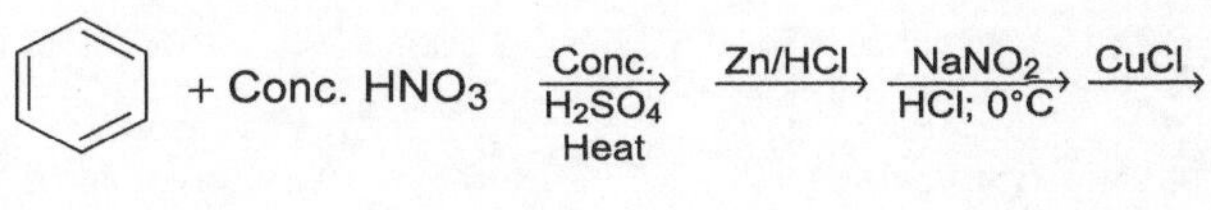

Working Space

14 The correct statement(s) about the following reaction is/are

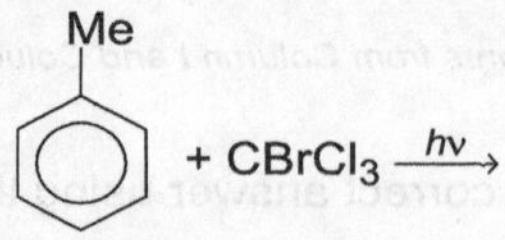

a. $^{\bullet}CCl_3$ intermediate is involved

b. Product is

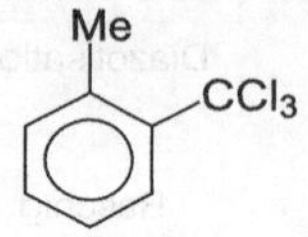

c. Product is $C_6H_5CH_2Br$

d. Formation of $CHCl_3$ is more stable than the formation of HBr

Statement Type

Direction (Q. No. 15) *This section is based on Statement I and Statement II. Select the correct answer from the codes given below.*

a. Both Statement I and Statement II are correct and Statement II is the correct explanation of Statement I

b. Both Statement I and Statement II are correct and Statement II is not the correct explanation of Statement I

c. Statement I is correct but Statement II is incorrect

d. Statement II is correct but Statement I is incorrect

15 **Statement I** 2,4-dihydroxy toluene gives benzyclic substitution product when treated with NBS.

Statement II Presence of —OH group favours electrophilic substitution.

One Integer Value Correct Type

Direction (Q. No. 16) *This section contains a question. When worked out will result in an integer from 0 to 9 (both inclusive).*

16 X is insoluble in water, dilute HCl and aqueous $NaHCO_3$, but dissolves in dil. NaOH. When treated with Zn dust / $KMnO_4$, X gives an acid, sodium salt of which is used as a preservative for food.

$$X \xrightarrow[H_2O]{Br_2} Y$$

The number of types of functional groups present in Y is

DPP-2 Physical Properties and Electrophilic Substitution in Aryl Halides

Subjective Questions

Working Space

Direction (Q. Nos. 1-3) *These questions are subjective in nature, need to be solved completely on notebook.*

1 a. Draw the structure of the major product obtained from each of the following sequence :

i. 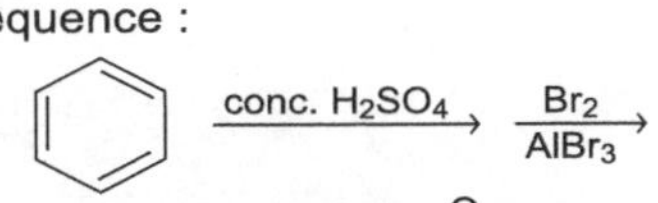

ii. 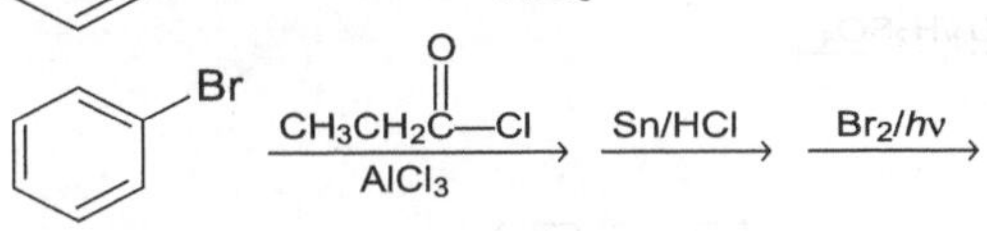

iii. 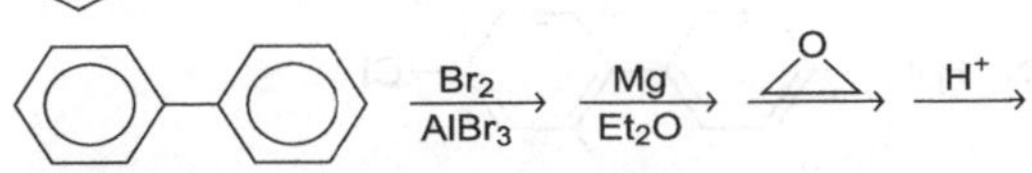

b. Why 1,4-dichlorobenzene has higher melting point than the other two isomers ?

2 Give structures of principle organic product of the following reagents when they react with bromobenzene.

i. Mg/ether, followed by treatment with CH_3COOD.
ii. $CH_3CH_2CH_2Cl$ / $AlCl_3$ / 0° C
iii. Conc. H_2SO_4, Δ
iv. Boiling 10% aqueous NaOH
v. $CH_3C{\equiv}CNa$

3 Arrange the following in decreasing order of $\left(\frac{m}{o+p}\right)$ ratio.

i. I. $PhCH_3$ II. $PhCBr_3$ III. $PhCHBr_2$ IV. $PhCH_2Br$
ii. I. $PhCH_3$ II. $PhCBr_3$ III. $PhCCl_3$ IV. $PhCl_3$

Only One Option Correct Type

Direction (Q. Nos. 4-12) *This section contains 9 multiple choice questions. Each question has four choices (a), (b), (c) and (d), out of which ONLY ONE is correct.*

4 Which one among the following compounds has the highest dipole moment?

a. *o*-dichlorobenzene
b. *o*-dibromobenzene
c. *m*-dichlorobenzene
d. *o*-bromochlorobenzene

5 Partial rate facts are given for the nitration of a particular aromatic compound. On the basis of these data, the most reasonable choice for substituent *X* is

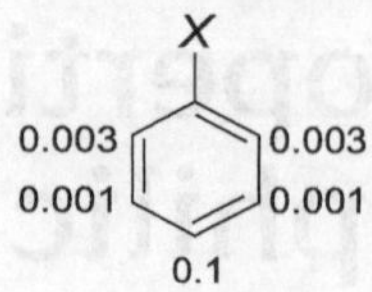

a. —$N(CH_3)_2$ b. —Cl
c. —CN d. —OMe

6 Which of the following substituted benzene derivatives would furnish three isomers when one more substituent is introduced (you are not to consider directive influence rule)?

I. 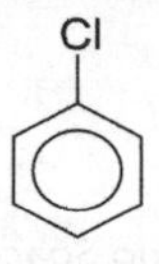 II. 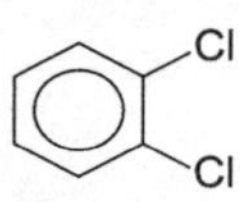 III. 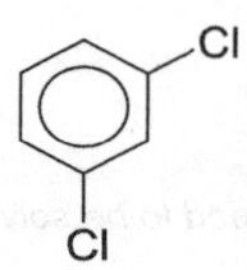 IV. 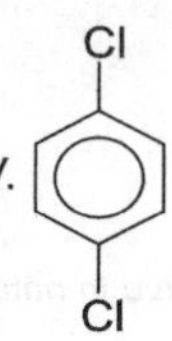

The correct answer is

a. I, II and III b. Only I
c. I and III d. II and IV

7 The reaction

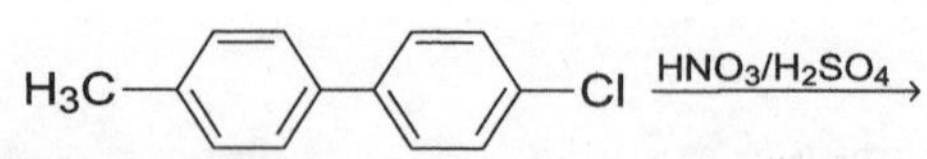

gives the following major product

a. 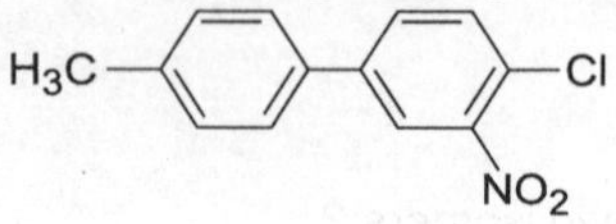 b. 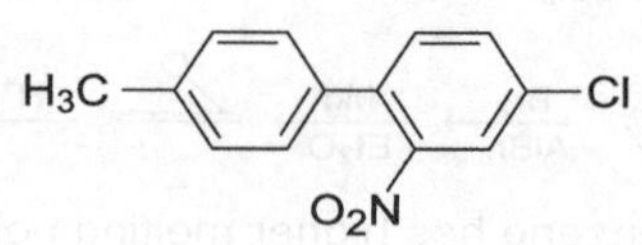

c. 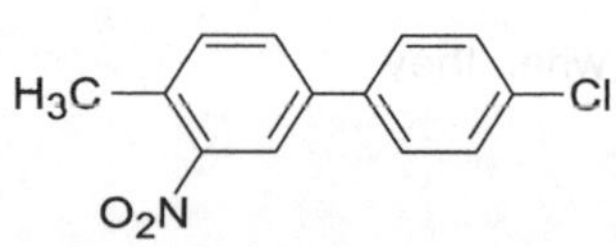 d. 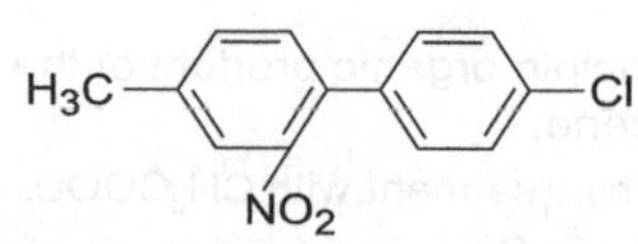

8 Which of the following procedures would be best for achieving the following reaction?

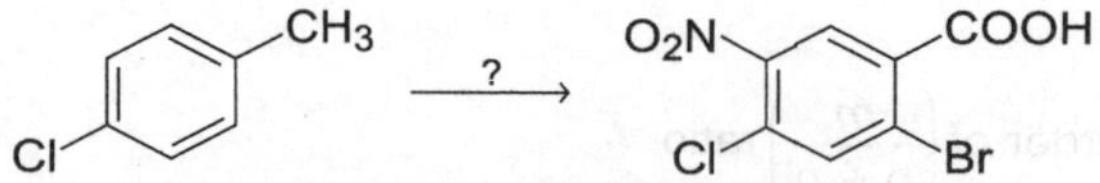

a. i. $Br_2 + FeBr_3$, ii. $KMnO_4$ and heat, iii. HNO_3 and H_2SO_4
b. i. $KMnO_4$ and heat, ii. $Br_2 + FeBr_3$, iii. HNO_3 and H_2SO_4
c. i. NBS in CCl_4 and heat, ii. $KMnO_4$ and heat, iii. HNO_3 and H_2SO_4
d. i. NBS in CCl_4 and heat, ii. $NaNO_2$, iii. $KMnO_4$ and heat

9 Consider the following reaction,

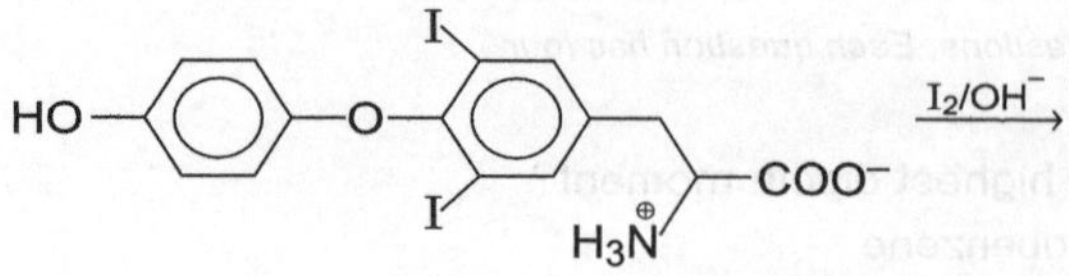

Thyroxine, a thyroid hormone that helps to regulate metabolic rate.

The formula of thyroxine is

a. 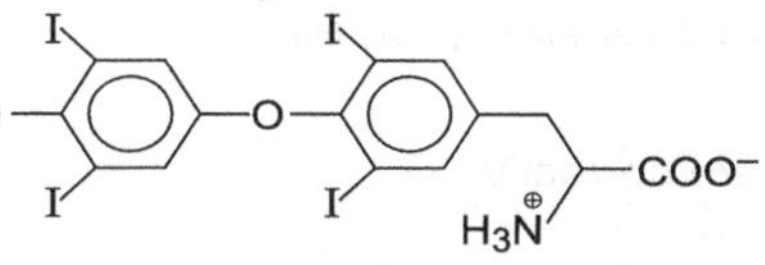 b. 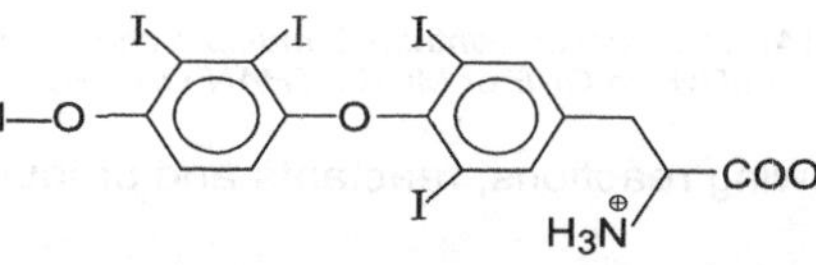

c. 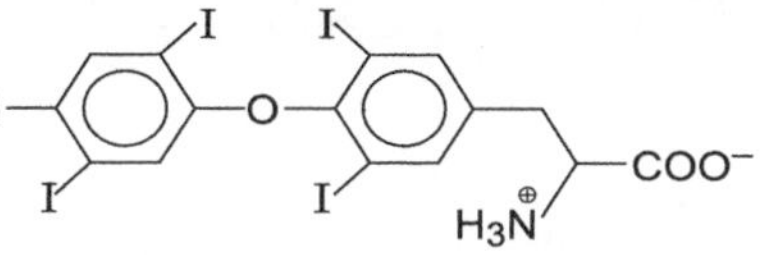 d. 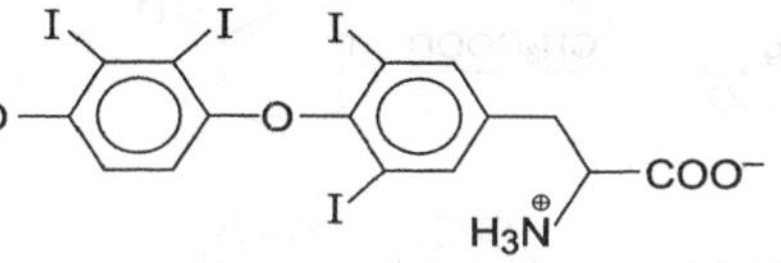

10 X, Y and Z have the value of μ as 1.78, 1.9 and 1.3 D respectively. Which of the following could be X ?

a. *o*-chlorotoluene b. *p*-chlorotoluene c. *m*-chlorotoluene d. Data insufficient

11 The possible product/products of the following reaction is/are

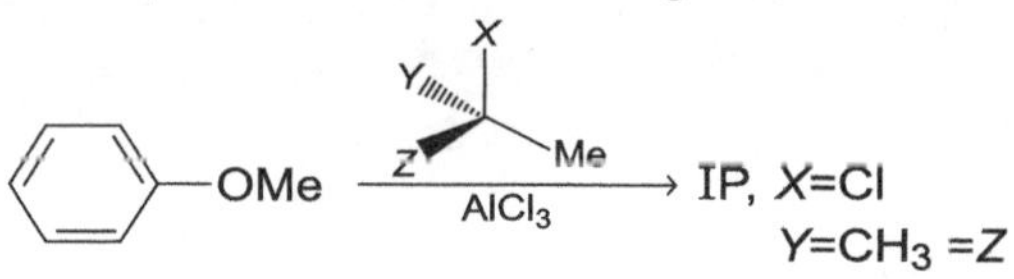

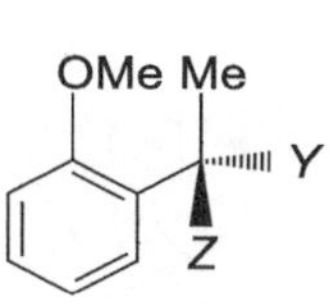
I.

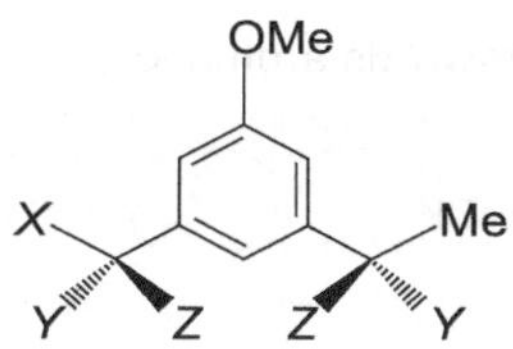
II.

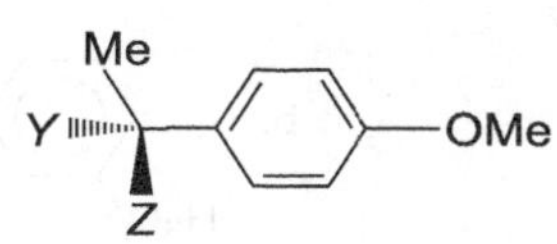
III.

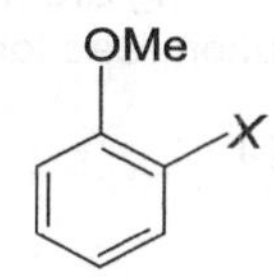
IV.

a. Only I
b. II and III
c. Only III
d. I and IV

12 Treatment of tetrachloro salicylanilide with fuming sulphuric acid will provide which of the following compound as the major product?

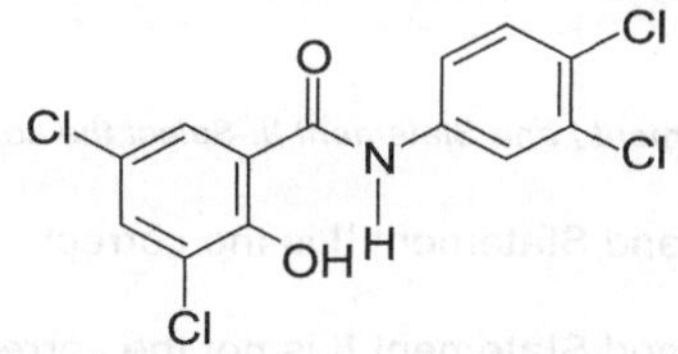

a. 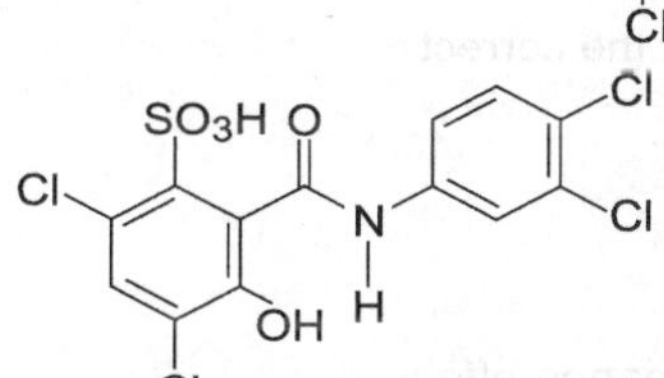

b. 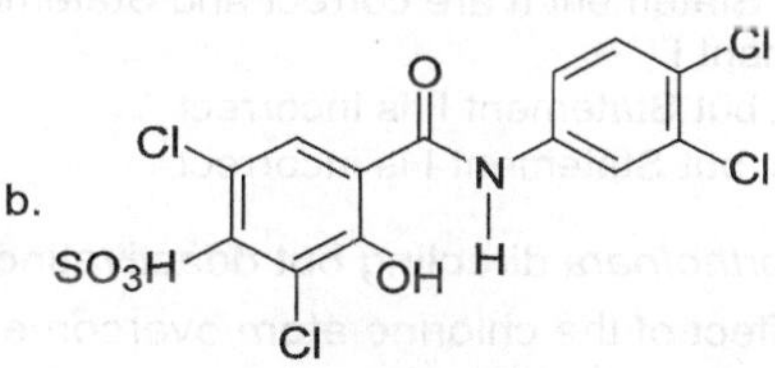

c. 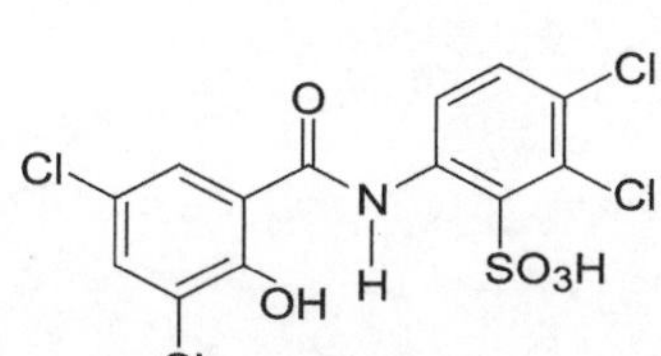

d. 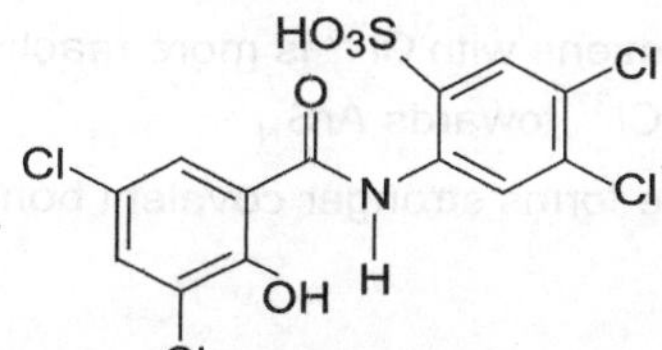

Working Space

Working Space

One or More than One Options Correct Type

Direction (Q. Nos. 13 and 14) This section contains 2 multiple choice questions. Each question has four choices (a), (b), (c) and (d), out of which ONE or MORE THAN ONE are correct.

13 In which of the following reactions, reactants and product(s) are correctly matched?

a. 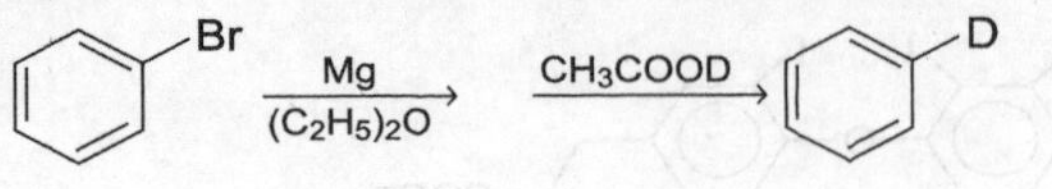

b. 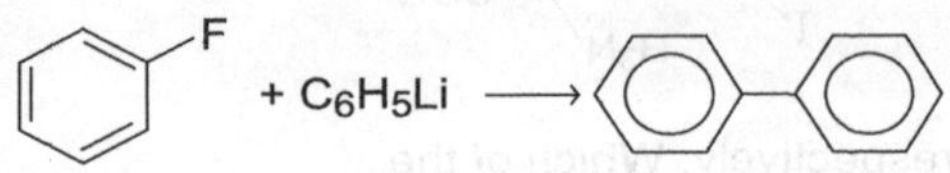

c. 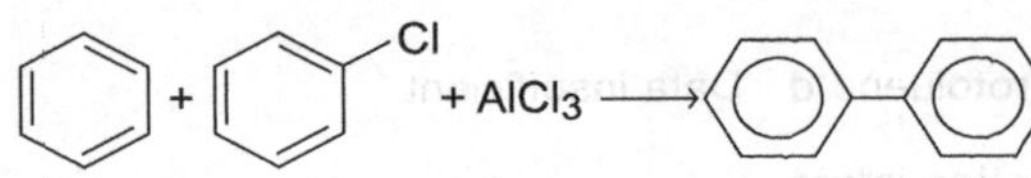

d. 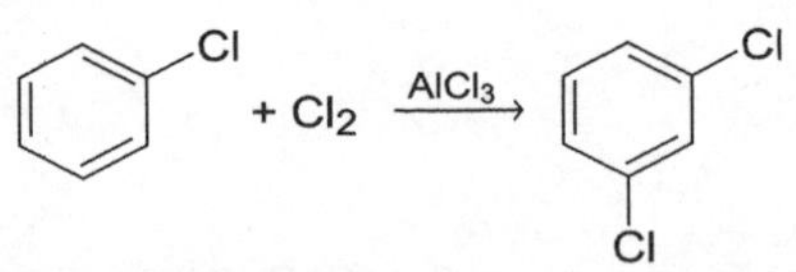

14 Which two of the following are more reactive than other two towards aromatic electrophilic substitution reaction?

a. 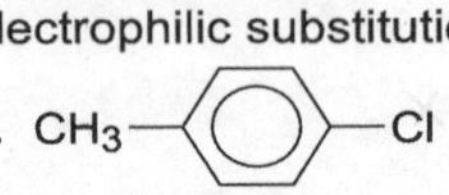

b. 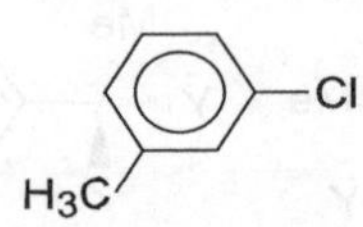

c. 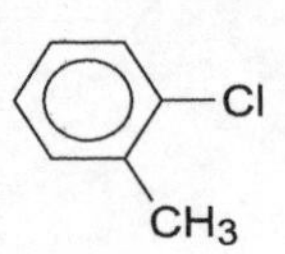

d. 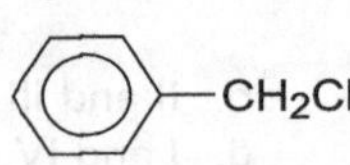

Statement Type

Direction (Q. Nos. 15 and 16) This section is based on Statement I and Statement II. Select the correct answer from the codes given below.

a. Both Statement I and Statement II are correct and Statement II is the correct explanation of Statement I
b. Both Statement I and Statement II are correct and Statement II is not the correct explanation of Statement I
c. Statement I is correct but Statement II is incorrect
d. Statement II is correct but Statement I is incorrect

15 **Statement I** Chlorine is *ortho*/*para* directing but deactivating group.

Statement II Inductive effect of the chlorine atom overcomes the resonance effect.

16 **Statement I** *p*-nitrochlorobenzene with Cl^{35} is more reactive than *p*-nitrochlorobenzene with Cl^{37} towards ArS_N.

Statement II Heavier isotope forms stronger covalent bond.

DPP-3 Chemical Reactions of Aryl Halides (Other than Electrophilic Substitution)

Subjective Questions

Working Space

Direction *(Q. Nos. 1-6) These questions are subjective in nature, need to be solved completely on notebook.*

1 Complete the following reactions :

i. 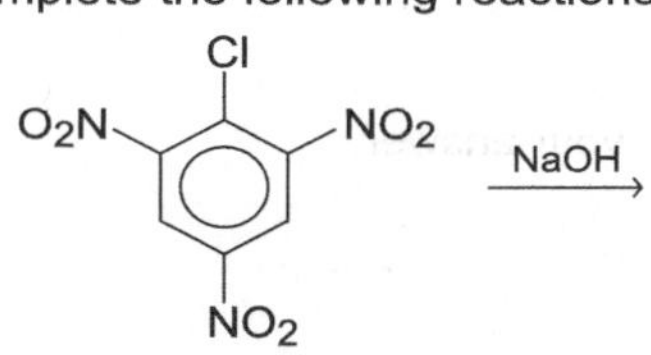

ii. 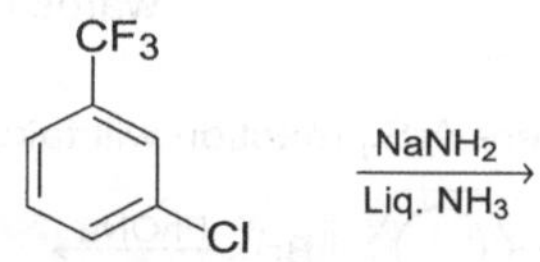

iii. 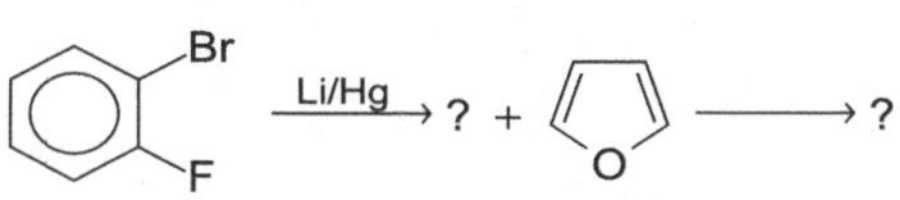

iv. 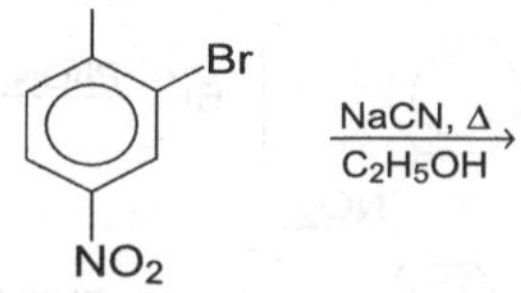

v. 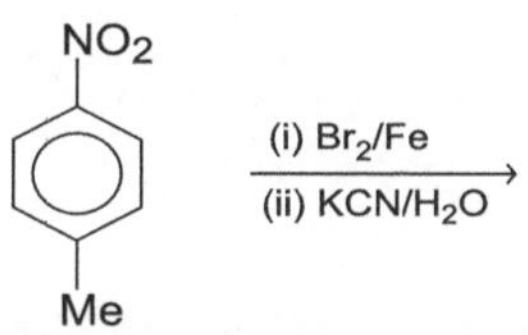

2 i. When a trace of KNH_2 is added to a solution of chlorobenzene and potassium triphenyl methide ($Ph_3C^-K^+$) in liquid ammonia, a rapid reaction takes place to give a product of formula $C_{25}H_{20}$. What is the product ? Also give the role of KNH_2 in this reaction.

ii. In the following reactions :

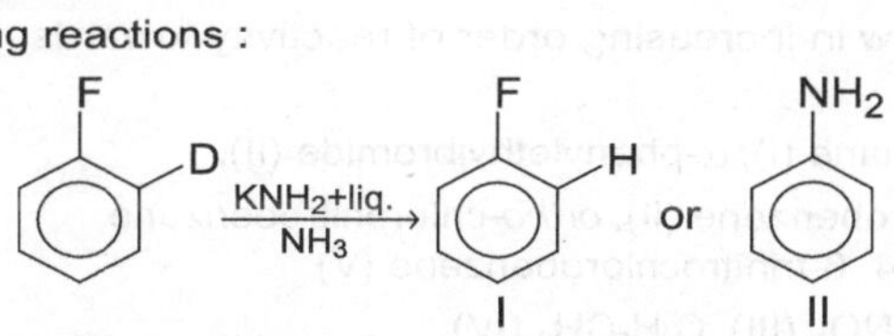

Explain, why the formation of I is faster than that of II?

3 Arrange the following in the decreasing order of aromatic nucleophilic substitution reaction.

i. PhCl, p-NO_2— C_6H_4 —Cl, 2, 4, 6-trinitrochlorobenzene, 2, 4-dinitro chlorobenzene.

ii. PhF, p-HO_3S— C_6H_4F, p-HOOC— C_6H_4F, p-NO_2— C_6H_4— F

iii. PhCl, p-NO_2— C_6H_4Cl, o-$NO_2C_6H_4Cl$, m-$NO_2C_6H_4Cl$

Working Space

4 Arrange the following in order mentioned in bracket.

i.

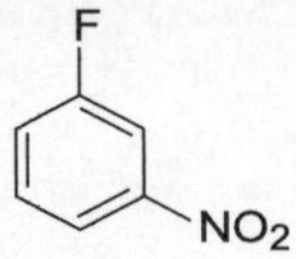

I

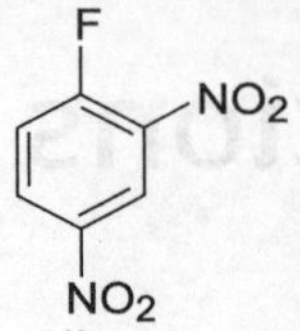

II

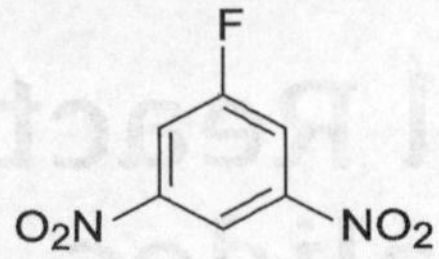

III

(Reactivity with CH_3ONa)

ii.

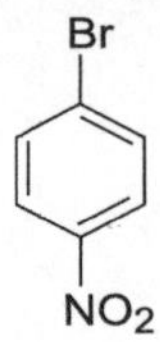

I

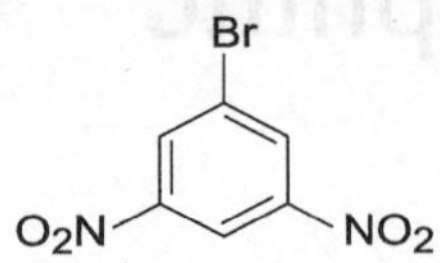

II

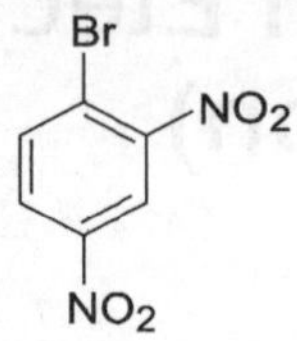

III

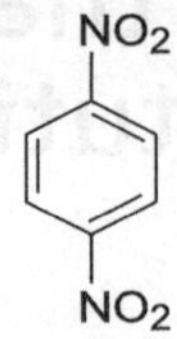

IV

(Reactivity towards $\bar{O}H$)

iii.

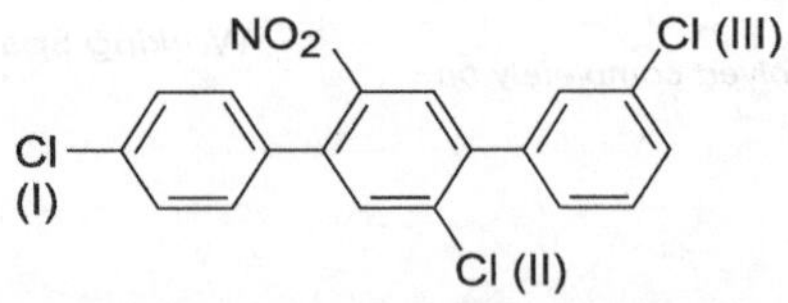

Cl (I), Cl (II) and Cl (III)

(Reactivity towards $NaOCH_3$)

5 i. Indicate the position where ArS_N reaction will take place. Also justify your answer.

a. 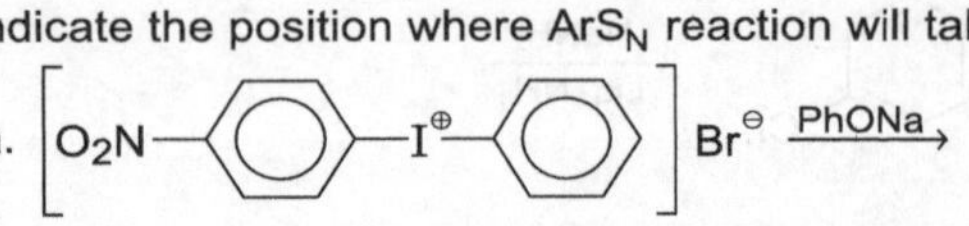

b. 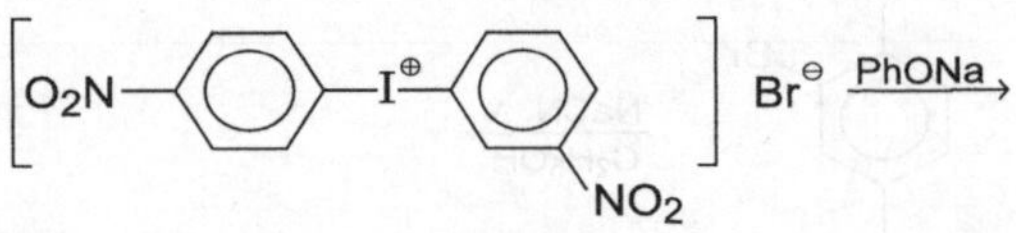

c. 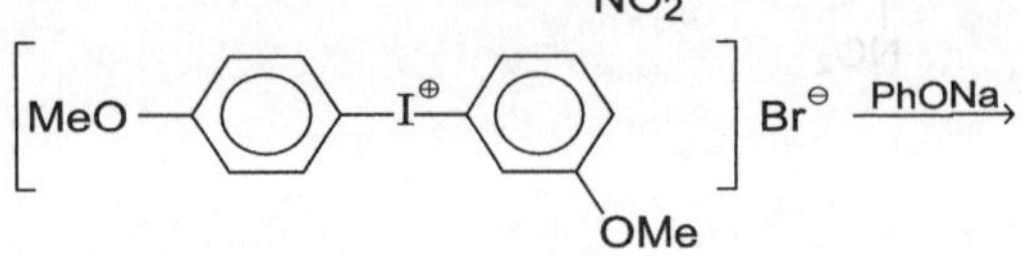

d. 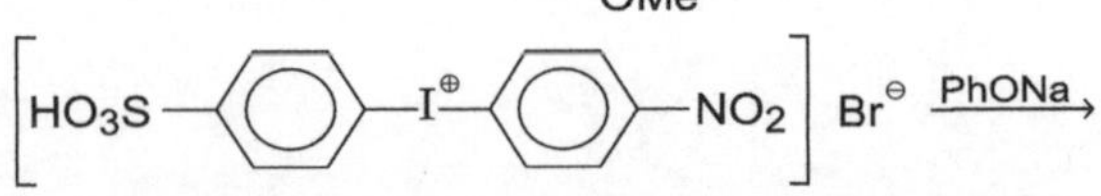

ii. How to distinguish between the following compounds :

a. *m*-iodotoluene and benzyl iodide.

b. *p*-bromoanisole and *p*-nitrobromobenzene.

6 Arrange the compounds in each set below in increasing order of reactivity towards the indicated reagents.

i. Ethanolic $AgNO_3$: 2-bromo-1-phenylethane (I), α-phenylethylbromide (II).

ii. NaOH : Chlorobenzene (I), *m*-chloronitrobenzene (II), *ortho*-chloronitrobenzene (III), 2, 4-dinitrochloro benzene (IV), 2, 4, 6-trinitrochlorobenzene (V)

iii. HNO_3 / H_2SO_4 : C_6H_6(I), C_6H_5Cl (II), $C_6H_5NO_2$ (III), $C_6H_5CH_3$ (IV)

iv. Ethanolic $AgNO_3$: 1-bromo-1-butene (I), 3-bromo-1-butene (II), 4-bromo-2-phenyl-2-butene (III);

v. Conc. H_2SO_4/Heat : C_6H_5Br (I), *p*-bromotoluene (II), *p*-dibromobenzene (III), toluene (IV)

vi. KCN : $C_6H_5CH_2Cl$(I), C_6H_5Cl(II), C_2H_5Cl(III)

Only One Option Correct Type

Working Space

Direction (Q. Nos. 7-14) *This section contains 8 multiple choice questions. Each question has four choices (a), (b), (c) and (d), out of which ONLY ONE option is correct.*

7 What is the major organic product of the reaction?

a. 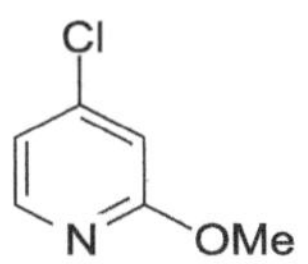

b. 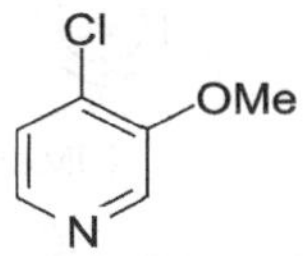

c. 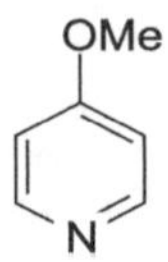

d. 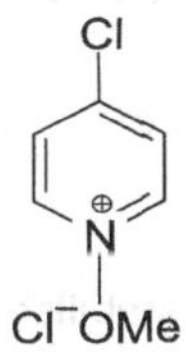

8 Which one of the following occurs by an addition-elimination mechanism?

a. Toluene, Br_2, $FeBr_3$ $\longrightarrow$ *p*-bromotoluene

b. Benzene, HNO_3, H_2SO_4 $\longrightarrow$ nitrobenzene

c. Benzene, CO, HCl, $AlCl_3$, CuCl $\longrightarrow$ benzaldehyde

d. 2, 4-dinitrochlorobenzene, NaOH, heat, then H^+ $\longrightarrow$ 2, 4-dinitrophenol

9 What term describes the reactive intermediate formed in the following reaction?
p-bromotoluene, $NaNH_2$, NH_3, $-33°C$ $\longrightarrow$ $CH_3(C_6H_4)NH_2$ in a ratio 50% *para* isomer, 50% *meta* isomer.

a. Aryl radical
b. Benzo cation
c. Sigma complex
d. Benzyne formation

10 In the following reaction, which of the following steps is wrong?

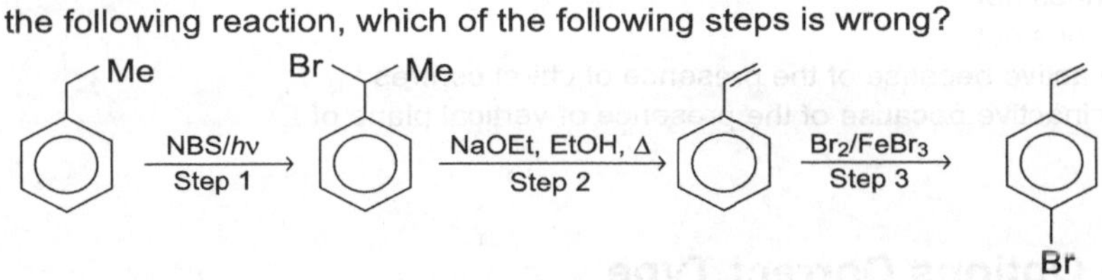

a. step 1
b. step 2
c. step 3
d. None of these

11 The insecticide DDT is prepared by heating chlorobenzene with chloral (CCl_3CHO) in the presence of conc. sulphuric acid. Which of the following compounds is DDT?

a. 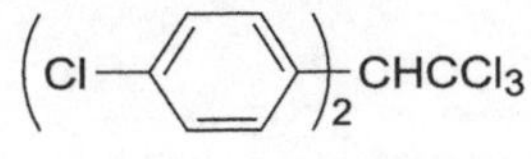

b. 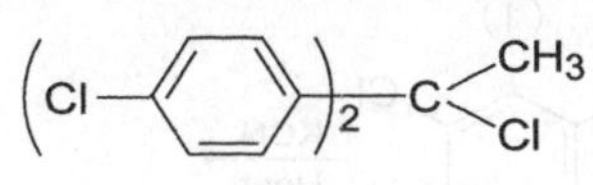

c. 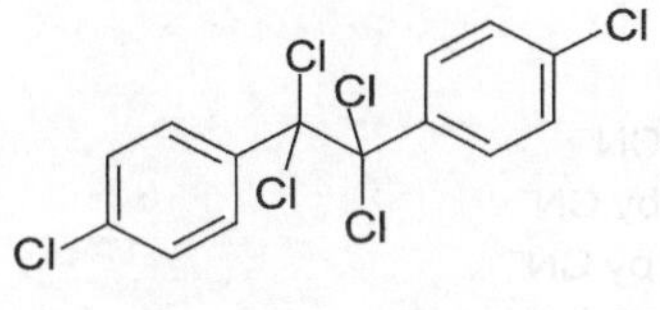

d. 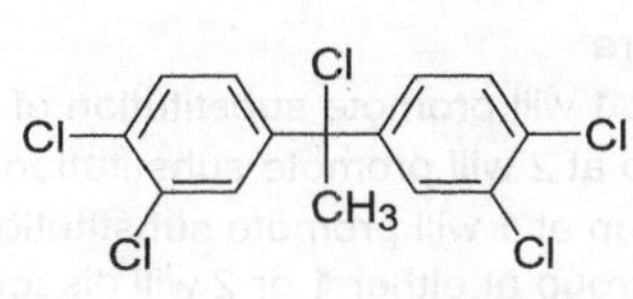

12 The product of the reaction

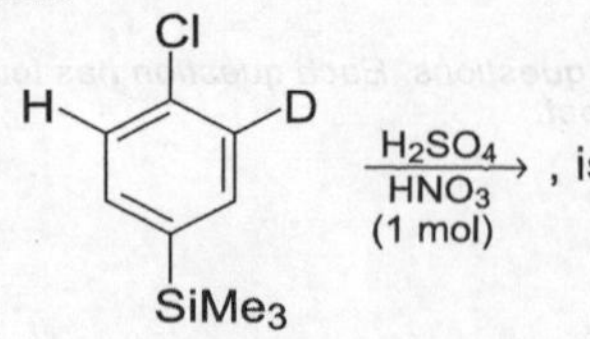

, is

a. 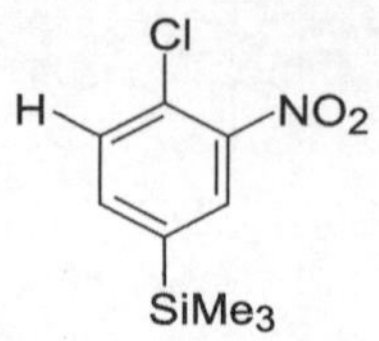 b. 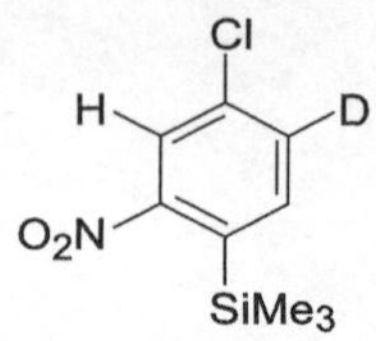 c. 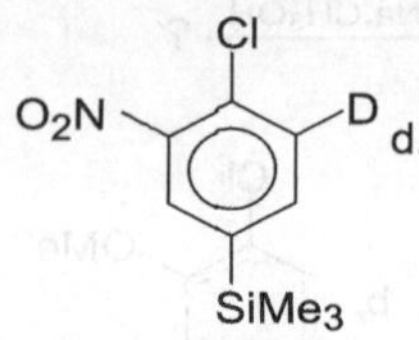 d. 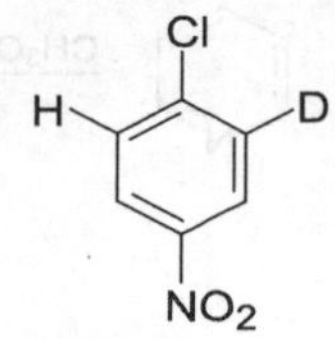

13 In the conversion of *p*-nitrofluoro benzene to *p*-nitroanisole, intermediate *X* is involved.

The true statements about the intermediate *X* is/are

I. the intermediate is aromatic.

II. the intermediate is resonance stabilised anion.

III. electron withdrawing group on the benzene ring stabilise the intermediate.

a. Only I
b. Only II
c. I and III
d. II and III

14 Select the correct statement about *A* and *B*

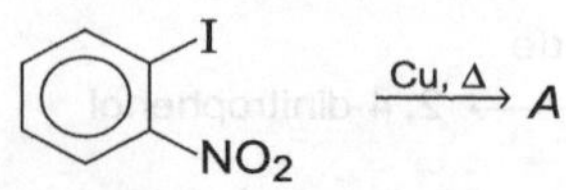

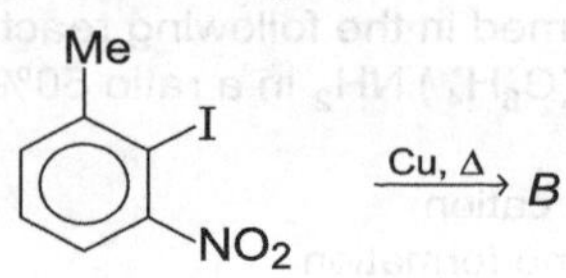

a. *A* is optically active but *B* does not
b. *B* is optically active but *A* does not
c. Both *A* and *B* are optically active because of the presence of chiral centres
d. Both *A* and *B* are optically inactive because of the presence of vertical plane of symmetry.

One or More than One Options Correct Type

Direction *(Q. Nos. 15-17) This section contains 3 multiple choice questions. Each question has four choices (a), (b), (c) and (d), out of which ONE or MORE THAN ONE are correct.*

15 Consider the following chlorofluorobenzene undergoing nucleophilic aromatic substitution (S_NAr) reaction with KCN.

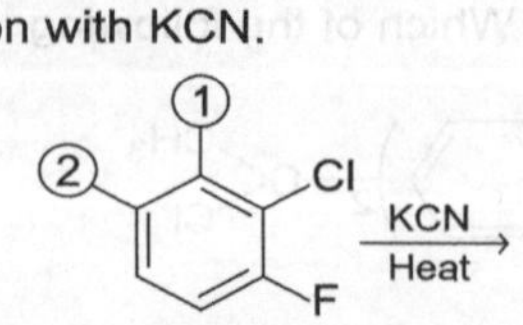

The correct statement(s) is/are

a. substitution by a chlorine at 1 will promote substitution of fluorine by CN^-
b. substitution by a nitro group at 2 will promote substitution of fluorine by CN^-
c. substitution by a cyano group at 1 will promote substitution of chloro by CN^-
d. substitution by a methoxy group at either 1 or 2 will discourage the reaction as a whole

Working Space

16 Which of the following statements are correct for the below reaction?

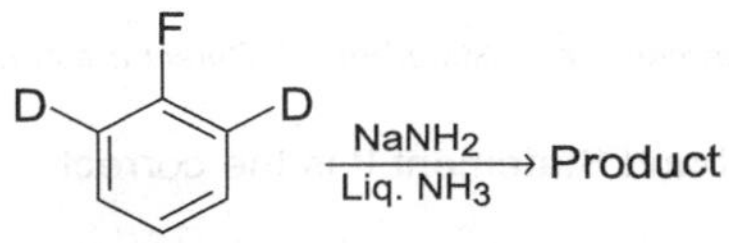

a. The product is a mixture of

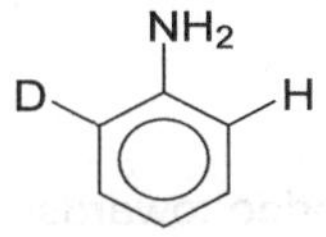

I

and

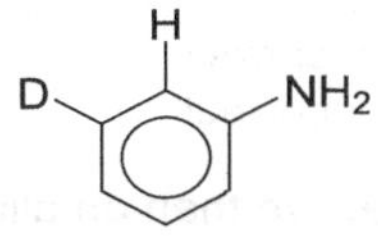

II

b. The product (II) is a cine substitution product
c. The reaction proceeds *via* benzyne intermediate
d. The reaction is ArS_N (addition-elimination)

17 In Dow's process for the manufacture of phenol, PhCl is fused with NaOH at elevated temperature under pressure.

$$\text{PhCl} \xrightarrow[\text{623 K, 300 atm}]{\text{NaOH}} \text{[Intermediate]} \xrightarrow{H_2O} \underset{A}{\text{Phenol}} + \underbrace{B + C}_{\text{Side product}}$$

Which of the following statements are correct ?
a. Phenol is formed *via* the formation of benzyne intermediate
b. *p*-phenylphenol is also formed as a by-product
c. Diphenylether is also formed as a by-product
d. Biphenyl is also formed as a by-product

Matching List Type

Direction *(Q. No. 18) Choices for the correct combination of elements from Column I and Column II are given as options (a), (b), (c) and (d), out of which one is correct.*

18 Match the Column I and Column II and mark the correct option from the codes given below.

	Column I (Reaction)		Column II (Products/mechanism)
i.	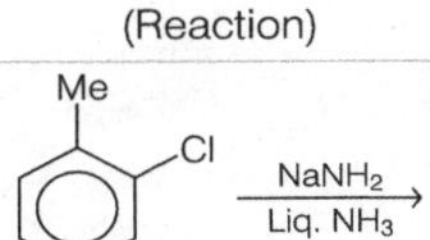	p.	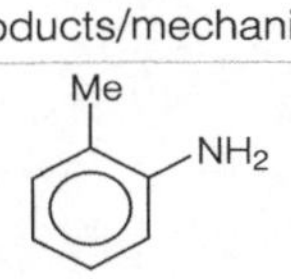
ii.	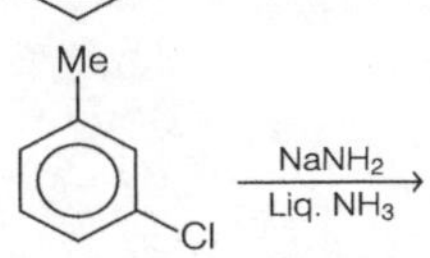	q.	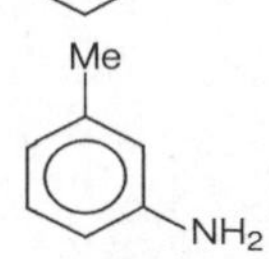
iii.	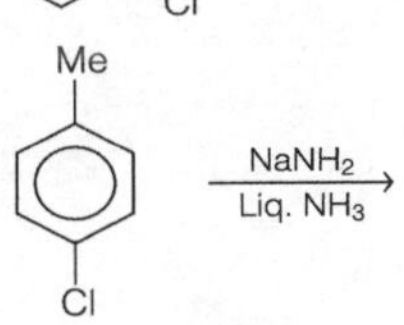	r.	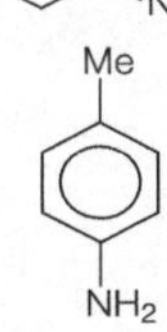
		s.	ArS_N (elimination-addition)

Codes

	i	ii	iii
a.	p,q,s	p,q,r,s	q,r,s
b.	p,q,s	q,s	p,r,s
c.	p	p,s	q,s
d.	p,q,r,s	p	p,q,s

Working Space

Statement Type

Direction (Q. Nos. 19 and 20) *This section is based on Statement I and Statement II. Select the correct answer from the codes given below.*

a. Both Statement I and Statement II are correct and Statement II is the correct explanation of Statement I
b. Both Statement I and Statement II are correct and Statement II is not the correct explanation of Statement I
c. Statement I is correct but Statement II is incorrect
d. Statement II is correct but Statement I is incorrect

19 **Statement I** Chlorobenzene is very less reactive than an aliphatic chloride towards a nucleophilic substitution reaction.

Statement II α-carbon of alkyl halide is sp^3-hybridised while α-carbon of chlorobenzene is sp^2-hybridised.

20 **Statement I** A dichlorobenzene is slightly more reactive than chlorobenzene towards an aromatic nucleophilic substitution reaction.

Statement II Chlorine has electron withdrawing effect on aromatic ring.

One Integer Value Correct Type

Direction (Q. No. 21) *This section contains a question. When worked out will result in an integer from 0 to 9 (both inclusive).*

21 i. How many products are expected from the following reaction of *m*-bromo toluene with KNH_2 in liquid ammonia?

ii. In the following sequence of reactions how many electrophilic substitution takes place?

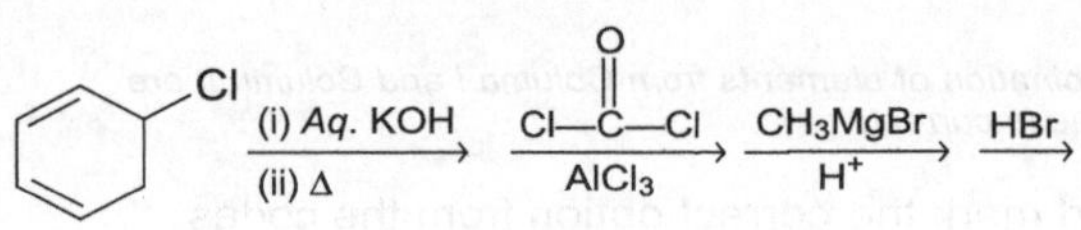

DPP-4 Methods of Preparation of Phenols

Subjective Questions

Direction (Q. No. 1) *This question is subjective in nature, need to be solved completely on notebook.*

Working Space

1 When triphenyl carbinol $[(C_6H_5)_3COH]$ is treated with acidic H_2O_2, a 50 : 50 mixture of benzophenone and phenol is produced.

i. Discuss the mechanism.

ii. What would be the major products in the same reaction if started with the following carbinol?

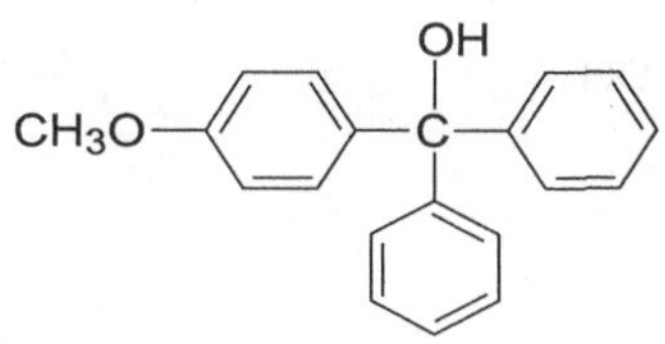

Only One Option Correct Type

Direction (Q. Nos. 2-9) *This section contains 8 multiple choice questions. Each question has four choices (a), (b), (c) and (d), out of which ONLY ONE is correct.*

2 In which of the following reactions phenol will not be produced as a major product?

a. 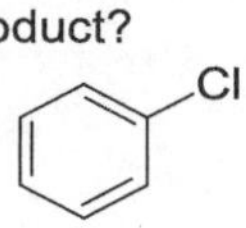 + Mg $\xrightarrow{\text{Ether}}$ $\xrightarrow{O_2}$ $\xrightarrow{HCl}$

b. 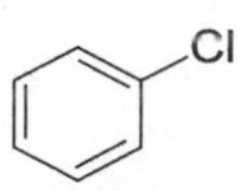 + KOH (*aq*) $\xrightarrow{50°C}$

c. 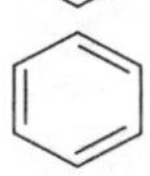 + $CH_3CH{=}CH_2$ $\xrightarrow{H^+}$ $\xrightarrow[H_2SO_4]{H_2O_2}$

d. 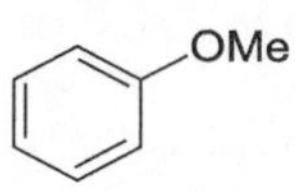 $\xrightarrow[H_2O]{HCl}$

Working Space

3 All of the following may give *meta* nitrophenol except

a. 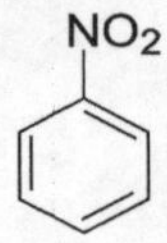 $\xrightarrow[AlCl_3]{Cl_2}$ $\xrightarrow[600°C]{NaOH(aq)}$ $\xrightarrow{HCl}$

b. 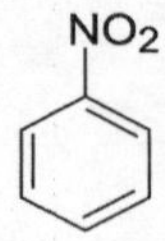 $\xrightarrow[(ii)\ NH_4HS]{(i)\ HNO_3/H_2SO_4}$ $\xrightarrow[(ii)\ H_2O(v)]{(i)\ HNO_2,\ 10°C}$

c. 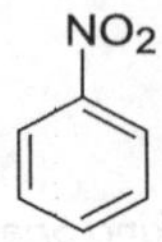 $\xrightarrow[H_2SO_4]{SO_3}$ $\xrightarrow[Fusion]{NaOH(s)}$ $\xrightarrow{HCl}$

d. 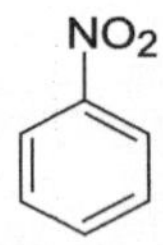 $\xrightarrow[Fusion]{NaOH(s)}$ $\xrightarrow{HCl}$

4 Select the best combination of reagents which will bring the following conversion.

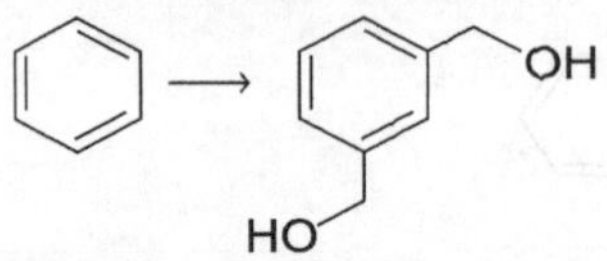

a. $Br_2/FeBr_3; HNO_3/H_2SO_4; NaOH; H_2/Pt$
b. $CH_3COCl/AlCl_3; Br_2/FeBr_3; NaBH_4/CH_3OH$
c. $CH_3Cl/AlCl_3; KMnO_4; CH_3Cl/AlCl_3; NBS/h\nu; LiAlH_4/H_2O; NaOH$
d. $CH_3Cl/AlCl_3; HNO_3/H_2SO_4; NBS/h\nu/NaOH$

5 Which of the following is the best preparation of *meta* cresol, starting from toluene ?

a. $\xrightarrow[H_2SO_4]{HNO_3}$ $\xrightarrow[HCl]{Zn}$ $\xrightarrow[0°C]{HNO_2}$ $\xrightarrow{H_2O(v)}$

b. $\xrightarrow[FeBr_3]{Br_2}$ $\xrightarrow[Heat]{KOH}$ $\xrightarrow{HCl}$

c. $\xrightarrow[H_2SO_4]{HNO_3}$ $\xrightarrow[(ii)\ (CH_3CO)_2O]{(i)\ Zn/HCl}$ $\xrightarrow[H_2SO_4]{HNO_3}$ $\xrightarrow[H_2O]{H^+}$ $\xrightarrow[0°C]{HNO_2}$ $\xrightarrow{H_3PO_2}$ $\xrightarrow[HCl]{Zn}$ $\xrightarrow[0°C]{HNO_2}$ $\xrightarrow{H_2O(v)}$

d. $\xrightarrow[H_2SO_4]{HNO_3}$ $\xrightarrow[(ii)\ (CH_3CO)_2O]{(i)\ Zn/HCl}$ $\xrightarrow[H_2SO_4]{HNO_3}$ $\xrightarrow[HCl]{Zn}$ $\xrightarrow[0°C]{HNO_2}$ $\xrightarrow{H_2O(v)}$

6 Identify the product in the following reaction.

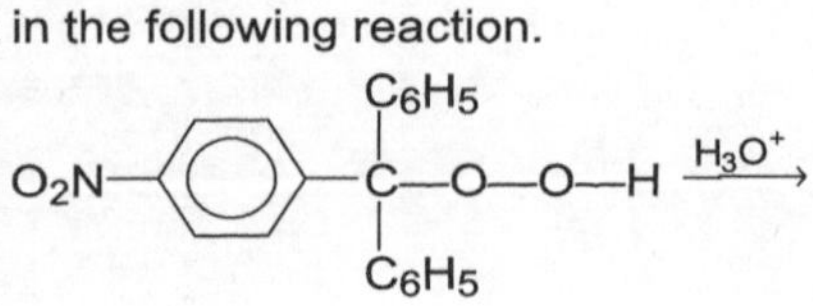

a. Benzophenone + phenol
b. Nitrobenzene + benzophenone
c. *p*-nitrobenzophenone + phenol
d. *p*-nitrobenzophenone + benzene

7 Which of the following sets of reagents used in the order shown, would enable the preparation of *p*-iodophenol from *p*-iodonitrobenzene?

a. 1. Fe, HCl; 2. NaOH; 3. $NaNO_2$, H_2SO_4; 4. H_3PO_2
b. 1. Fe, HCl; 2. NaOH; 3. $NaNO_2$, H_2SO_4; 4. H_2O, heat
c. 1. Fe, HCl; 2. NaOH; 3. $NaNO_2$, H_2SO_4; 4. ethanol
d. 1. NaOH, heat; 2. HCl

8 Which of the following will best give resorcinol (1, 3-dihydroxy benzene)?

a. 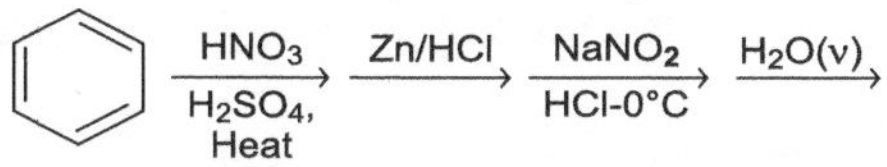

b. 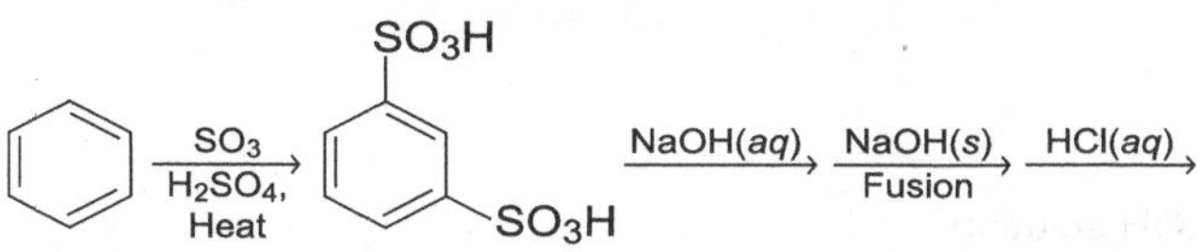

c. 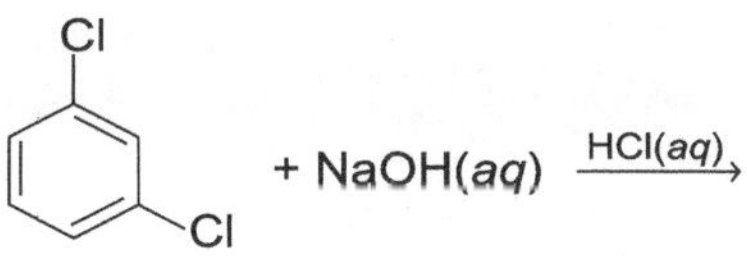

d. 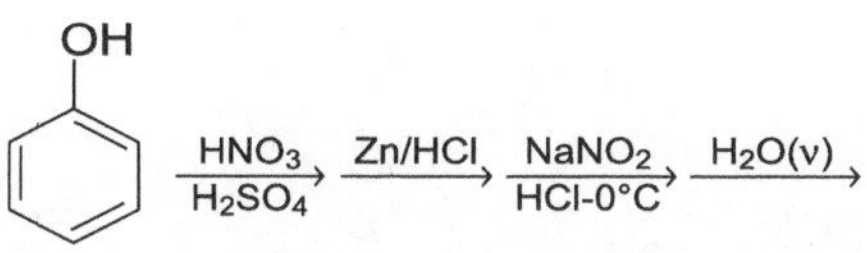

9 The end product *C* of this reaction sequence is

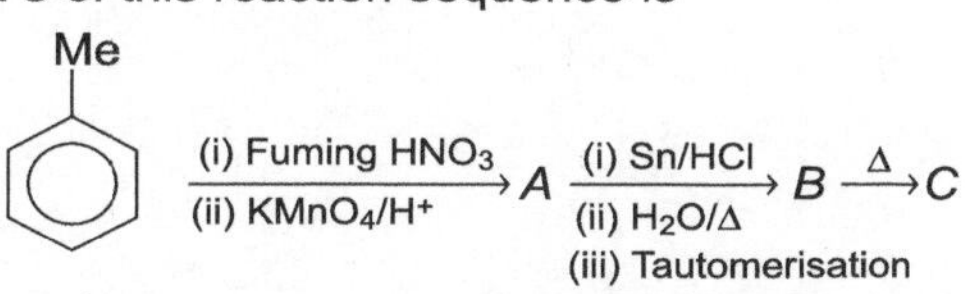

a. resorcinol
b. phloroglucinol
c. catacol
d. fluoroscene

One or More than One Options Correct Type

Direction (Q. Nos. 10-15) *This section contains 6 multiple choice questions. Each question has four choices (a), (b), (c) and (d), out of which ONE or MORE THAN ONE are correct.*

10 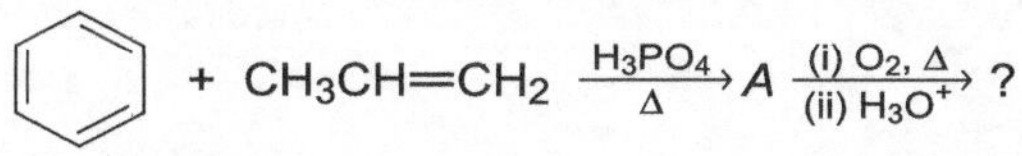

The product formed during the reaction is/are

a. acetone
b. an aromatic alcohol
c. phenol
d. a three membered ketone

11 The reaction that gives phenol is/are

a. 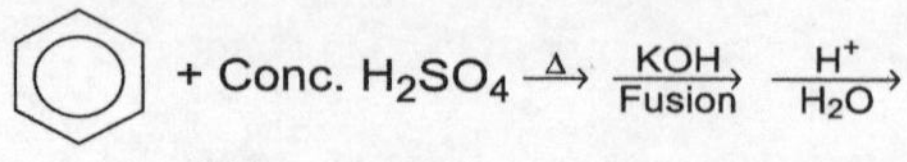

b. 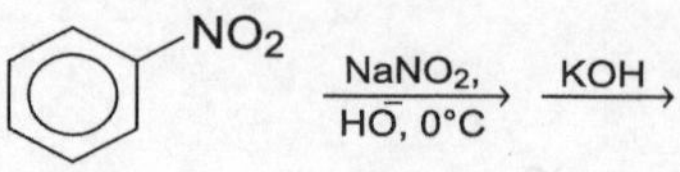

c. 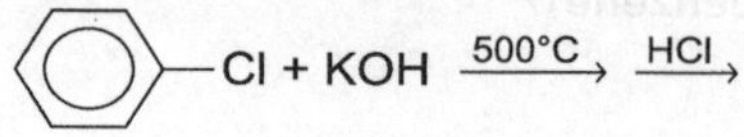

d. 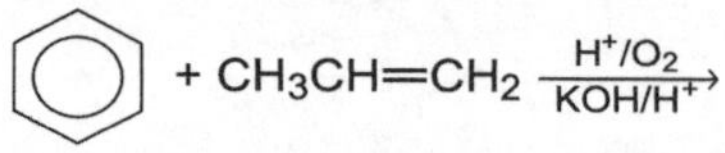

12 Phenol can be prepared by

a. hydrolysis of chlorobenzene with *aq*. KOH solution
b. heating sodium salicylate with (NaOH + CaO)
c. reacting cumene hydroperoxide with dil. H_2SO_4
d. heating benzenediazonium chloride with dil. H_2SO_4

13 Which are the routes that can prepare quinol?

a. 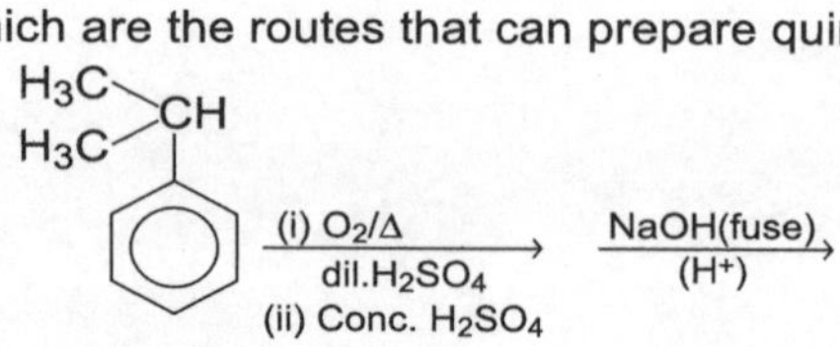

b. 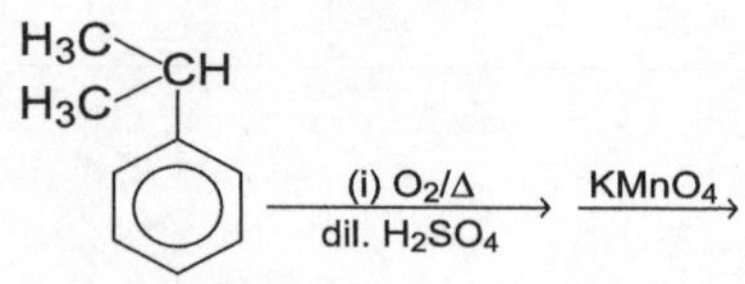

c. 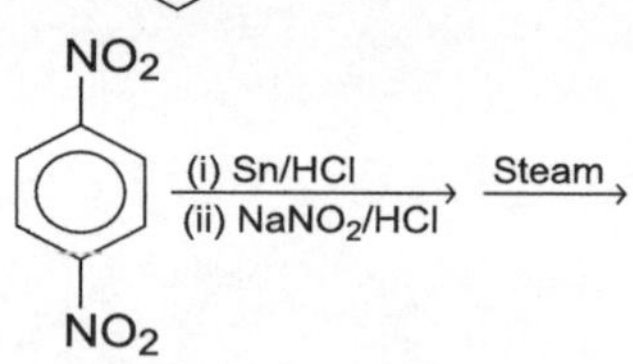

d. 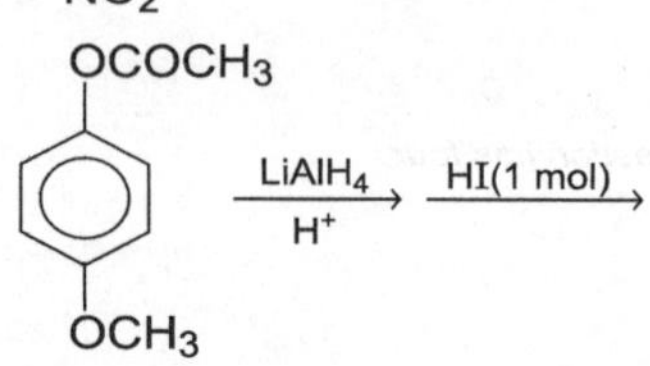

14 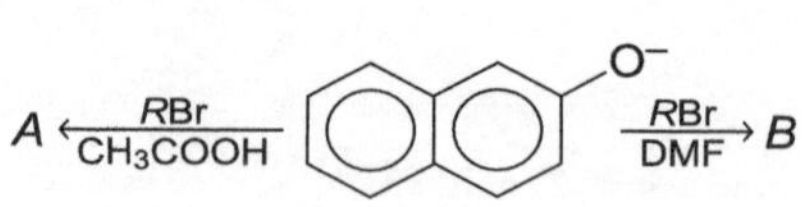

β- or 2-naphthoxide ion

The compounds given are

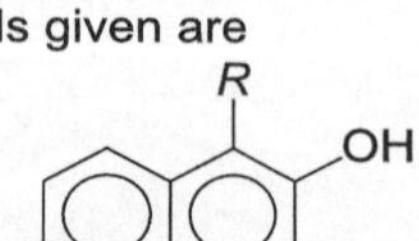

I

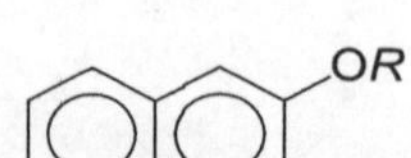

II

Which of the following statement(s) is/are correct?

a. The compound *A* is I
b. The compound *A* is II
c. The compound *B* is I
d. The compound *B* is II

Working Space

15 Which of the following statement(s) is/are correct about the following reaction?

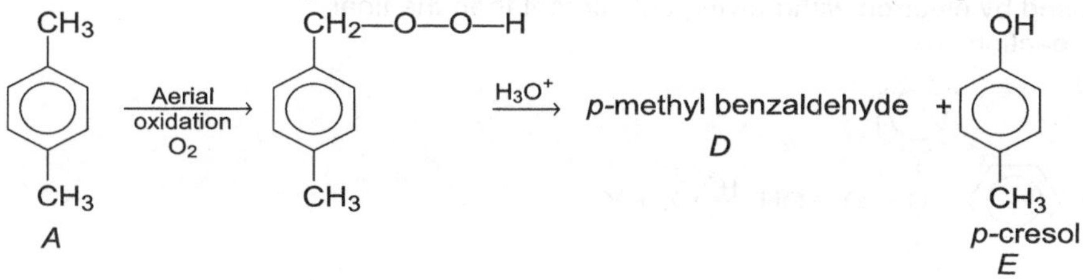

a. Compound *D* is in major amount (68%) and *E* is in minor (32%)
b. Compound *D* is in minor amount (32%) and *E* is in major (68%)
c. The aldehyde (*D*) arises from migration of H and cresol (*E*) comes from migration of *p*-tolyl
d. H is a better migrator

Comprehension Type

Direction (Q. Nos. 16-18) *This section contains a paragraph, describing theory, experiments, data, etc. Three questions related to the paragraph have been given. Each question has only one correct answer among the four given options (a), (b), (c) and (d).*

Passage

Observe the following reaction mechanism of cumene hydroperoxide hydrolysis.

Step I

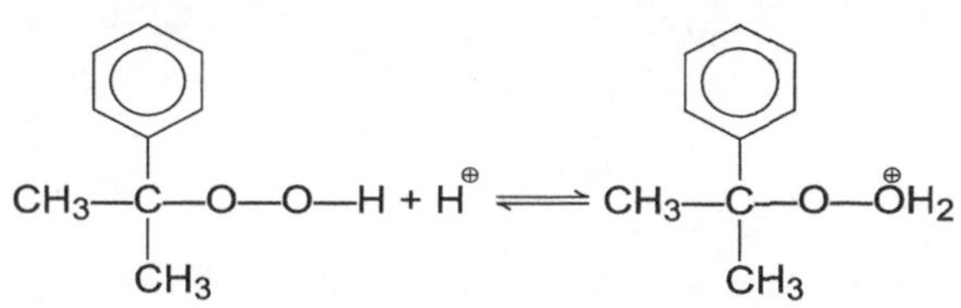

Step II

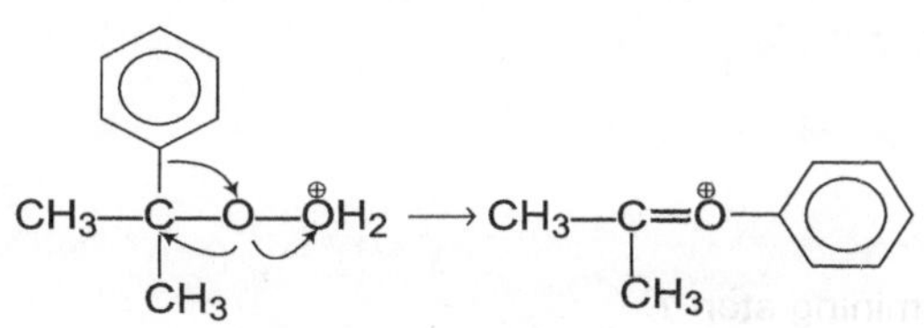

Step III

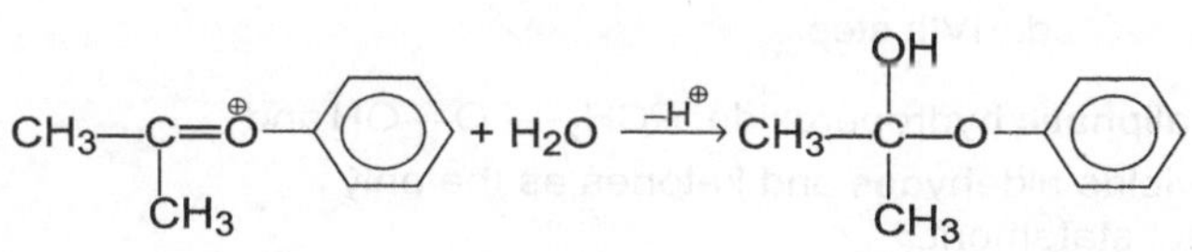

Step IV

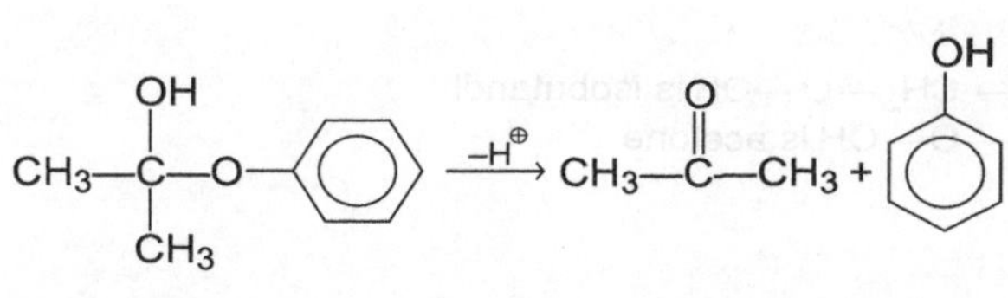

Working Space

16 If the relative migratory aptitude of an aryl group is raised by electron releasing substituents and decreased by electron withdrawing substituent then the final products $(X + Y)$ of the reaction are

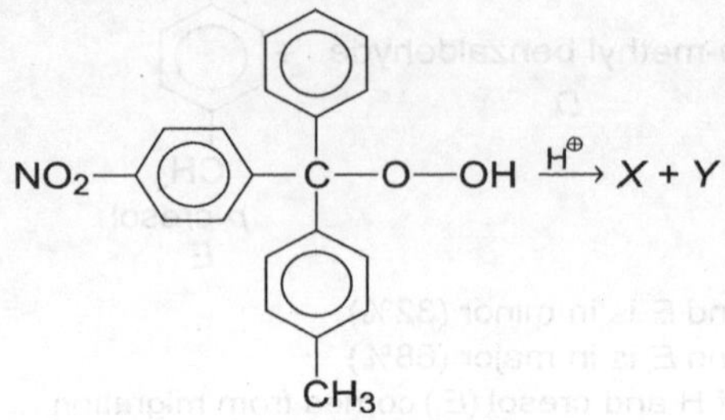

a. 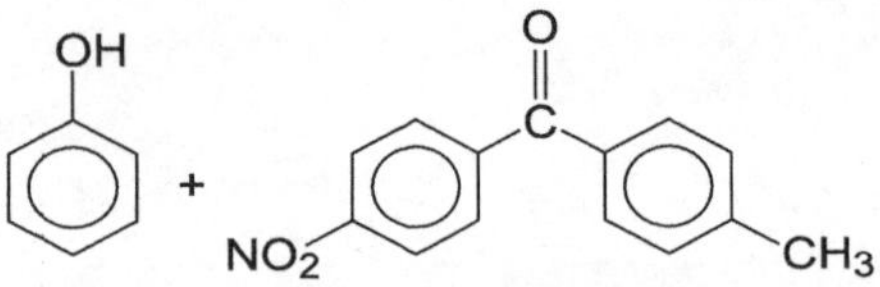

b. 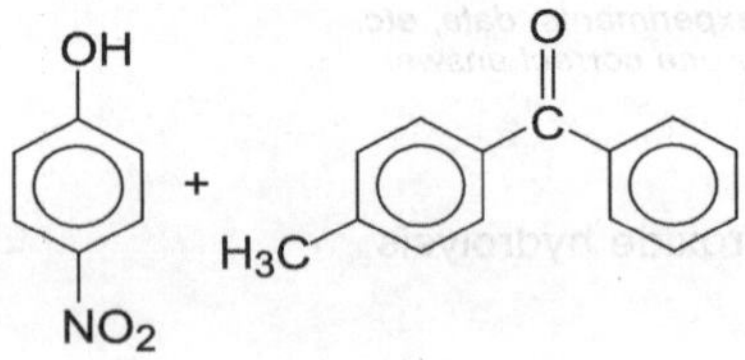

c. 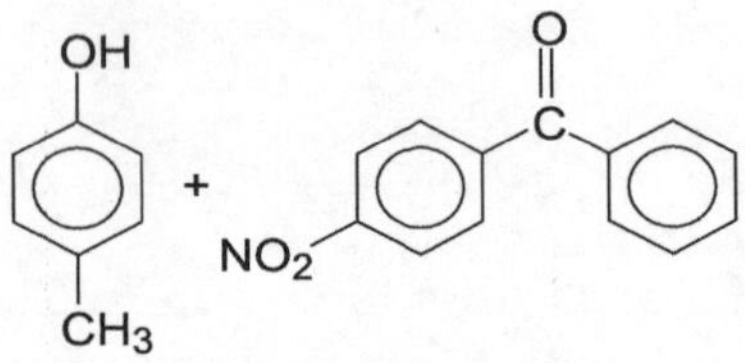

d. 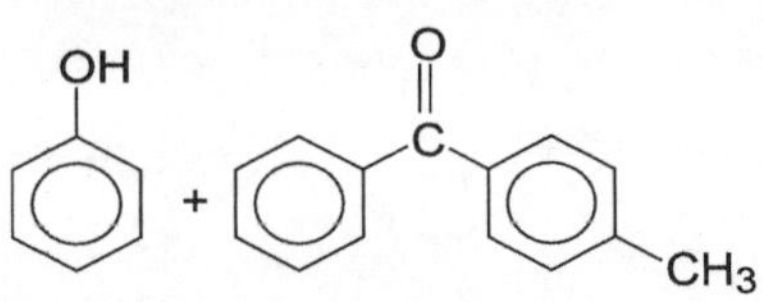

17 Which step of this reaction is rate determining step ?

a. Ist step
b. IInd step
c. IIIrd step
d. IVth step

18 In a similar reaction, treatment of aliphatic hydroperoxide, $RCH_2—O—OH$ and $R_2CHO—OH$, with aqueous acid yields aldehydes and ketones as the only organic products. Select the correct statement.

a. $-H$ migrates slower than $-R$
b. $-R$ migrates faster than $-H$
c. the hydrolysis product of $(CH_3)_2CH—CH_2—O—OH$ is isobutanol
d. the hydrolysis product of $(CH_3)_2CH—O—OH$ is acetone

Working Space

Matching List Type

Direction *(Q. No. 19) Choices for the correct combination of elements from Column I and Column II are given as options (a), (b), (c) and (d), out of which one is correct.*

19 Match the Column I with Column II and mark the correct options from the codes given below.

	Column I (Conversion)		Column II (Reagent in correct sequence)
i.	Benzene ⟶ Resorcinol	p.	$\xrightarrow{\text{Conc. } H_2SO_4} \xrightarrow[H_3O^+]{\text{NaOH}}$
ii.	Benzoic acid ⟶ *m*-hydroxy benzoic acid	q.	$\xrightarrow[H_3O^+ \; CHCl_3]{\text{NaOH}} \xrightarrow[H_3O^+]{\text{NaOH}/H_2O_2}$
iii.	Phenol ⟶ *o*-dihydroxy benzene	r.	$\xrightarrow[CH_3Br]{\text{NaOH}} \xrightarrow{\text{Conc. } cH_2SO_4} \xrightarrow{H_3O^+}$
iv.	Chlorobenzene ⟶ *p*-dihydroxy benzene	s.	$\xrightarrow[H_2SO_4]{HNO_3} \xrightarrow[H_2SO_4]{HNO_3} \xrightarrow[\text{(ii) } HNO_2 \text{ (iii) } H_2O]{\text{(i) Sn/HCl}}$

Codes

	i	ii	iii	iv
a.	p,s	p	q	p
b.	p	p	r	q
c.	p	q	s	r
d.	s	p,s	q,s	p

One Integer Value Correct Type

Direction *(Q. No. 20) This section contains 1 question. When worked out will result in an integer from 0 to 9 (both inclusive).*

20 How many of the following reagents are used to bring about following conversion?

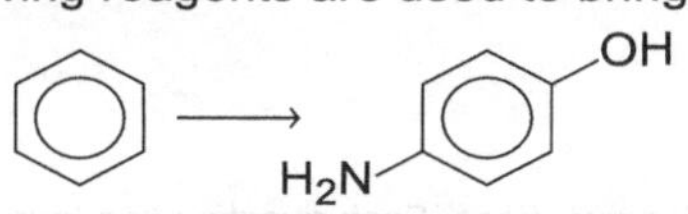

I. $Br_2/FeBr_3$

II. $CH_3Cl/AlCl_3$

III. $KMnO_4$

IV. HNO_3/H_2SO_4

V. NBS/$h\nu$/NaOH

VI. NaOH

VII. Sn/HCl

VIII. NH_4HS

IX. $NaBH_4$

DPP-5 Physical Properties, Acidity and Electrophilic Substitution Reactions of Phenols

Subjective Question

Working Space

Direction (Q. No.1) *This question is subjective in nature, need to be solved completely on notebook.*

1 a. Order the following in the increasing order of acidic strength:

i. p-O_2N—C_6H_4OH (I), m-O_2N—C_6H_4OH (II), *ortho*-nitrophenol (III)

ii. carbonic acid (I), H_2SO_4 (II), phenol (III), H_2O (IV)

iii. *m*-bromophenol (I), *m*-cresol (II), *m*-nitrophenol (III), phenol (IV)

iv. *p*-chlorophenol (I), 2, 4-dichlorophenol (II), 2, 4, 6-trichlorophenol (III)

b. Explain the following.

i. Which has higher boiling point? Phenol or benzenethiol.

ii. Which has higher melting point? Hydroquinol or catechol.

iii. Although *o*-fluorophenol has greater dipole moment than *o*-chlorophenol, boiling point of later is much higher than former.

iv. Between *meta* and *para* isomers of nitrophenols, later has much higher boiling point and decomposes before boiling at 1.0 atm.

v. The order of the solubility of nitrophenols is $o < m < p$

Only One Option Correct Type

Direction (Q. Nos. 2-4) *This section contains 3 multiple choice questions. Each question has four choices (a), (b), (c) and (d), out of which ONLY ONE is correct.*

2 pK_a value is greater for

a. 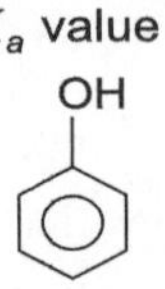

b. 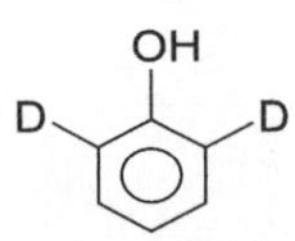

c. 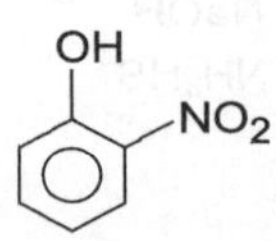

d. picric acid

3 The product (Y) of the following sequence of the reactions would be

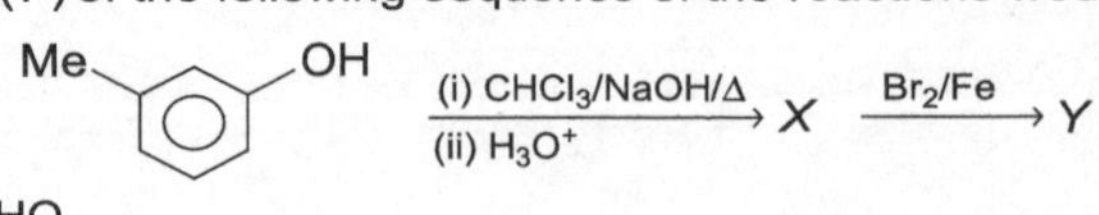

a. 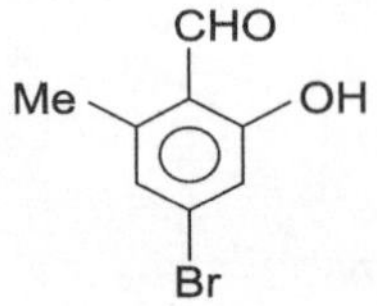

b. 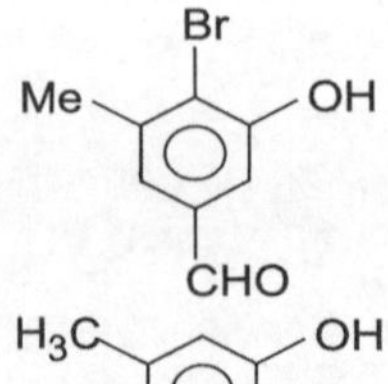

c. 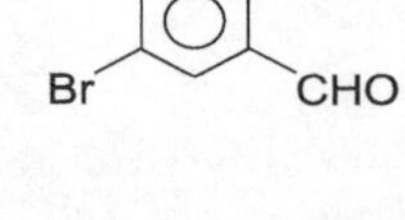

d. 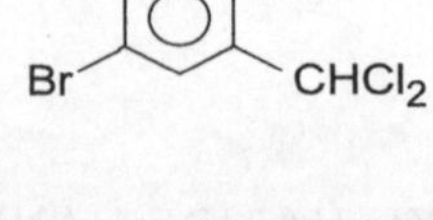

Working Space

4 Observe the following reactions,

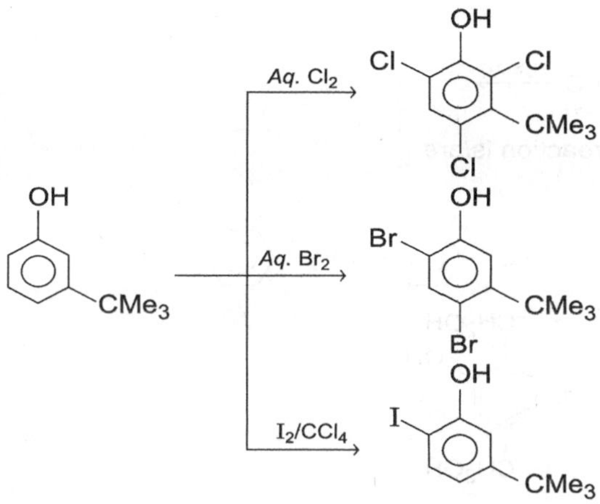

The main factor for such observation with different halogens can be

a. polar factor

b. steric factor

c. ionisation of X_2

d. ionisation of phenol

One or More than One Options Correct Type

Direction *(Q. Nos. 5-10) This section contains 6 multiple choice questions. Each question has four choices (a), (b), (c) and (d), out of which ONE or MORE THAN ONE are correct.*

5 Select the electrophilic substitution reactions

a. 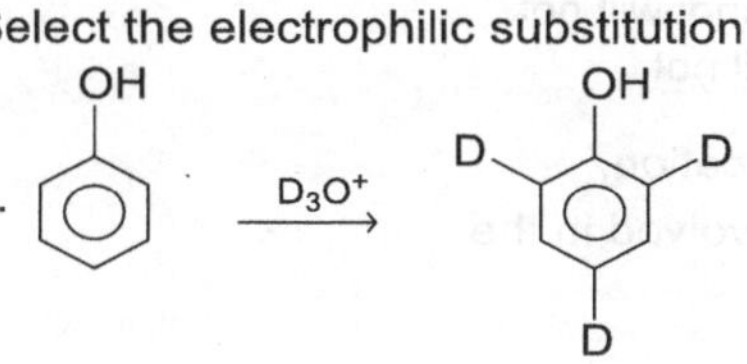

b. 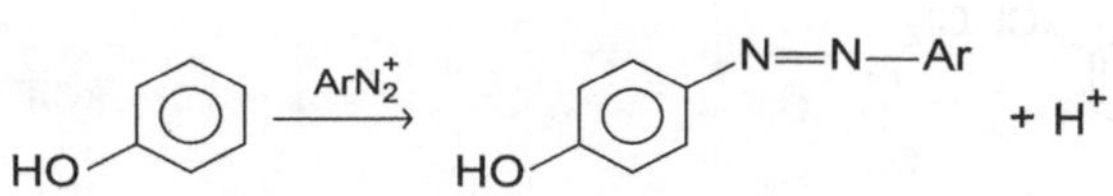

c. 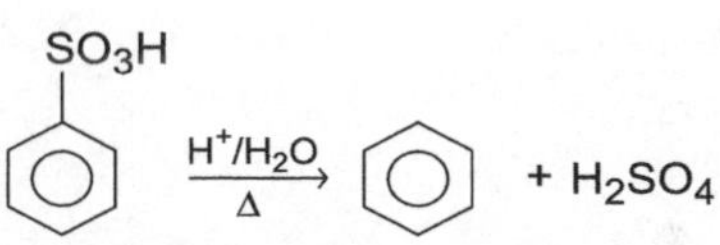

d. 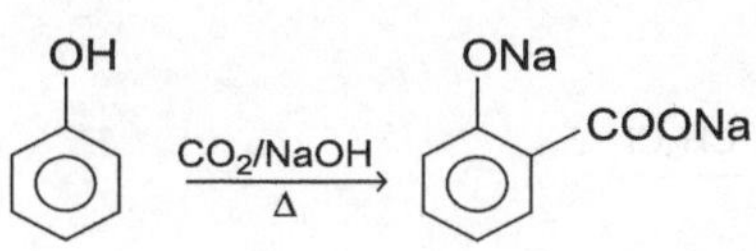

Working Space

6 Consider the following reaction,

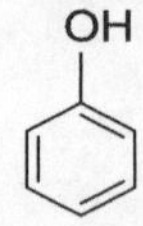 $+ CH_2O \xrightarrow{H_2SO_4}$
Excess

The expected product(s) of the above reaction is/are

a. 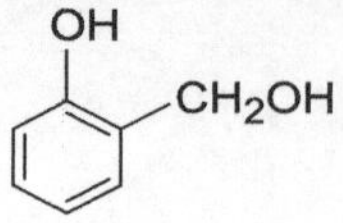

b. 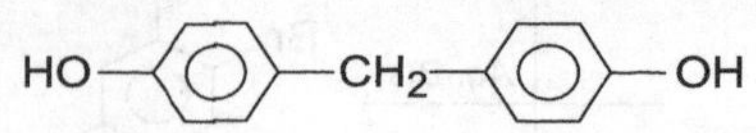

c. 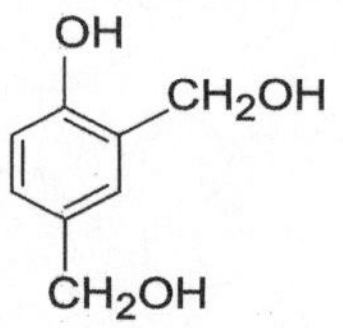

d. 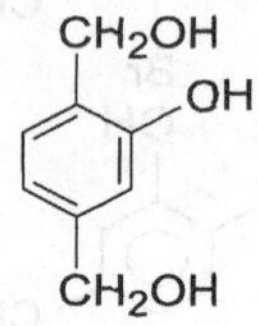

7 Which of the following acid-base reactions is/are feasible?

a. $PhONa + CH_3CH_2OH \longrightarrow PhOH + CH_3CH_2ONa$

b. $PhOH + NaOH \longrightarrow PhONa + H_2O$

c. $PhONa + aq\,HCl \longrightarrow PhOH + NaCl$

d. $PhONa + H_2O + CO_2 \longrightarrow$ 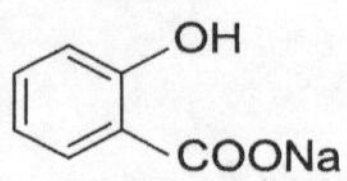

8 Which of the following statement is/are correct?

a. 4-chlorophenol will dissolve in NaOH but 4-chloro-1-methyl benzene will not

b. 4-methyl benzoic acid will dissolve in *aq* $NaHCO_3$ but 4-methyl phenol will not

c. 2, 4, 6-trinitrophenol will dissolve in *aq.* $NaHCO_3$ but 4-methyl phenol will not

d. 4-ethyl phenol will dissolve in *aq.* NaOH but ethyl phenyl ether will not

9 When phenol is reacted with $CHCl_3$ and NaOH followed by acidification, salicylaldehyde is obtained. Which of the following species are involved in the above mentioned reaction as intermediates?

a. 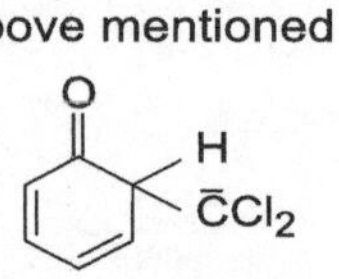

b. 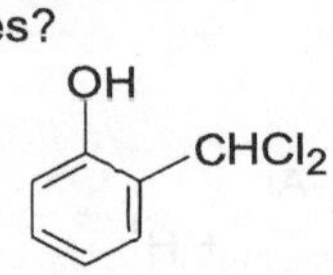

c. 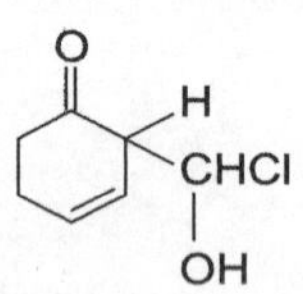

d. 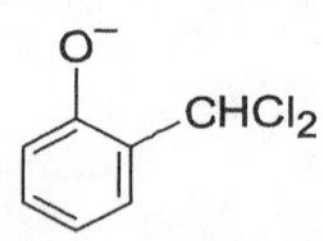

10 Which of the following reaction will give a visible change?

a. 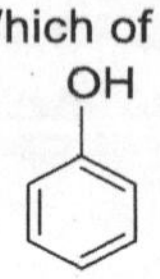 $+ Br_2\text{-}H_2O \longrightarrow$

b. 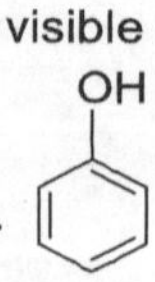 $+ NaOH\,(aq.) \xrightarrow{CH_3Cl}$

c. 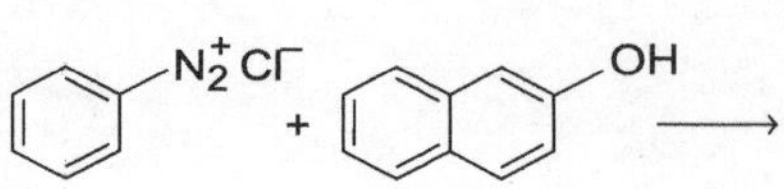

d. 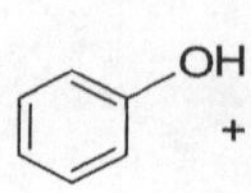 $+ \text{conc. } H_2SO_4 \xrightarrow[(aq.)]{NaNO_2}$

Working Space

Statement Type

Direction *(Q. Nos. 11 and 12) This section is based on Statement I and Statement II. Select the correct answer from the codes given below.*

a. Both Statement I and Statement II are correct and Statement II is the correct explanation of Statement I
b. Both Statement I and Statement II are correct but Statement II is not the correct explanation of Statement I
c. Statement I is correct but Statement II is incorrect
d. Statement II is correct but Statement I is incorrect

11 **Statement I** Phenol gives very poor yield in Friedel-Crafts acylation using $CH_3COCl/AlCl_3$.

Statement II Phenol reacts with $AlCl_3$ forming $C_6H_5OAlCl_2$.

12 **Statement I** $C_6H_5—O—\overset{14}{C}H_2—CH=CH_2$ on heating undergo rearrangement to give

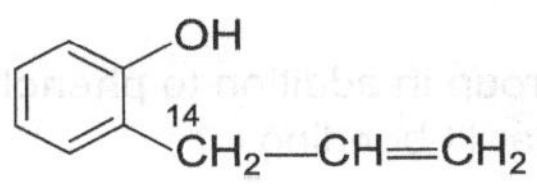

Statement II Allyl aryl ether on heating undergo intramolecular rearrangement to give mainly *ortho* allyl phenol.

Matching List Type

Direction *(Q. No. 13) Choices for the correct combination of elements from Column I and Column II are given as options (a), (b), (c) and (d), out of which one is correct.*

13 Match the Column I with Column II and mark the correct option from the codes given below.

	Column I		Column II (Property of product)
i.	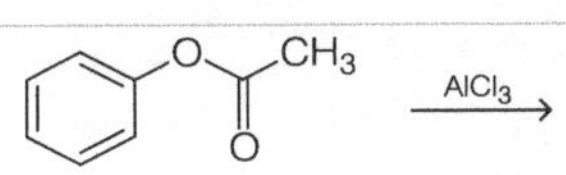	p.	Shows intramolecular H-bonding
ii.	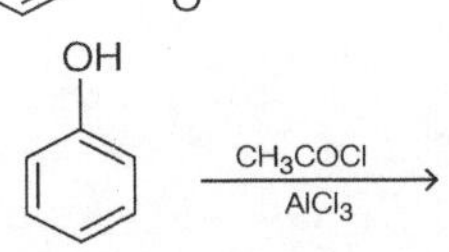	q.	Soluble in water
iii.	Phenol + $CCl_4 \xrightarrow{NaOH} \xrightarrow[H_2O]{H^+}$	r.	o-hydroxybenzoic acid
iv.	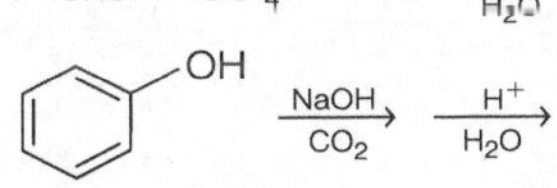	s.	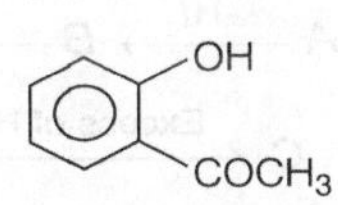

Codes

	i	ii	iii	iv
a.	p,q,s	p,q,s	p,q,r	p,q,r
b.	p,q,s	p,q	q,r	p,r
c.	p,r	p,q,s	r,s	q,r
d.	p,q,s	q,s	r,s	p,q,s

Comprehension Type

Working Space

Direction *(Q. Nos. 14-19) This section contains 3 paragraphs, each describing theory, experiments, data, etc. Six questions related to the paragraphs have been given. Each question has only one correct answer among the four given options (a), (b), (c) and (d).*

Passage I

An organic compound *A* gives positive Libermann's reaction and forms compounds *B* and *C* on treatment with trihalogen derivative of methane in alkaline medium followed by hydrolysis. *C* is steam volatile and gives no colouration with Schiff's reagent, which is given by compound *B*. *C*, when reduced with lithium aluminium hydride forms *D* with formula $C_7H_8O_2$. *D* when oxidised forms *E*, which on reaction with anhydride in the presence of corresponding acid forms a pain reliever *F*.

14 *B* and *C* are isomeric but *C* does not give colour with Schiff's reagent. It is because of the fact that

a. *B* contains an aldehyde group and *C* contains keto group in addition to phenolic group
b. *B* contains keto group and *C* contains aldehyde group in addition to phenolic group
c. Carbonyl reactivity in *C* is marked by intramolecular H-bonding
d. Carbonyl group is lost in hydrolysis step

15 Compound *D*, *E* and *F* respectively are

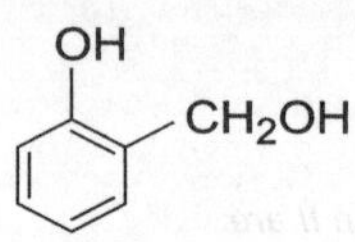

I.

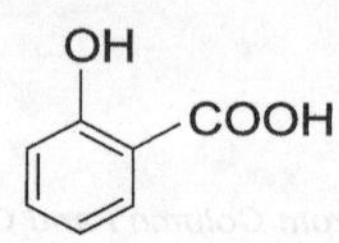

II.

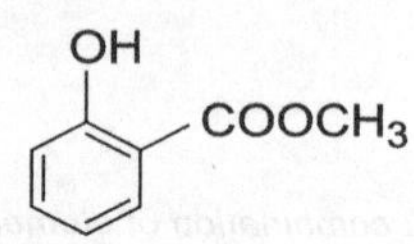

III.

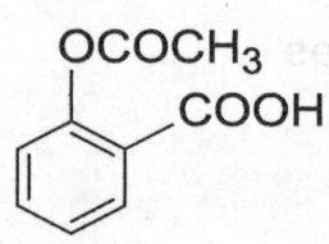

IV.

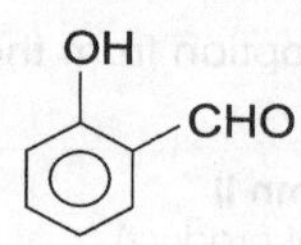

V.

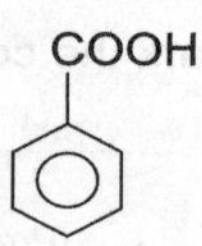

VI.

The correct answer is

a. I, II and III
b. I, II and IV
c. V, II and IV
d. II, VI and III

Passage II

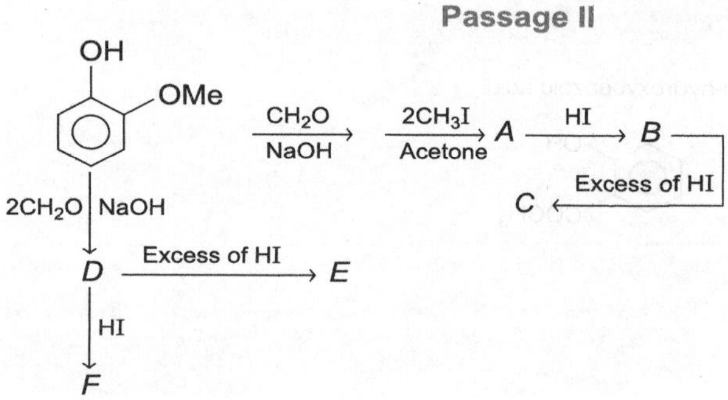

16 Here compound *B* is

a. 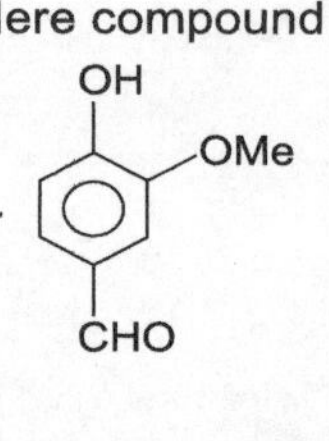

b. 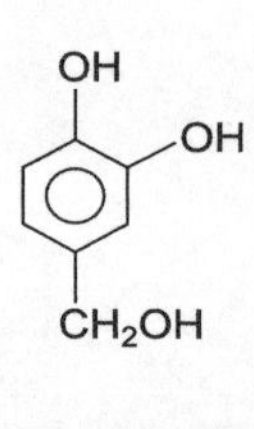

c. 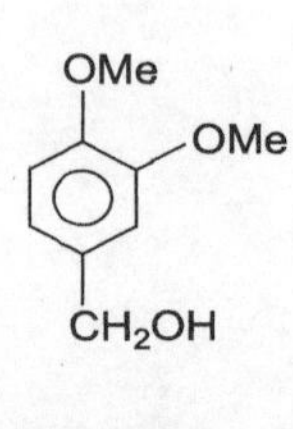

d. 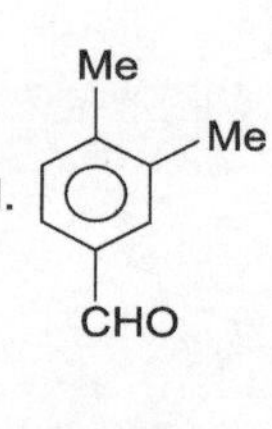

17 Compound E is

a. 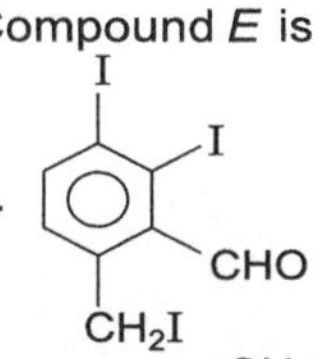

b. 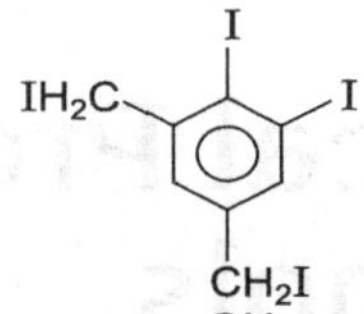

c. 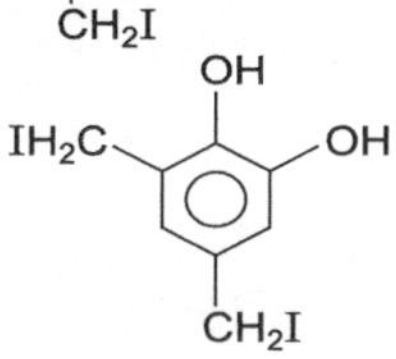

d. 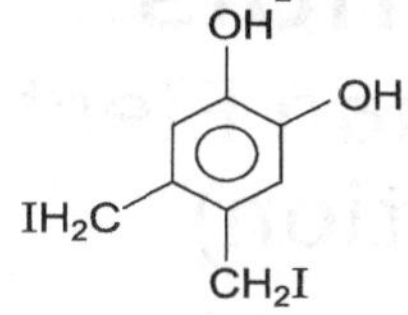

Passage III

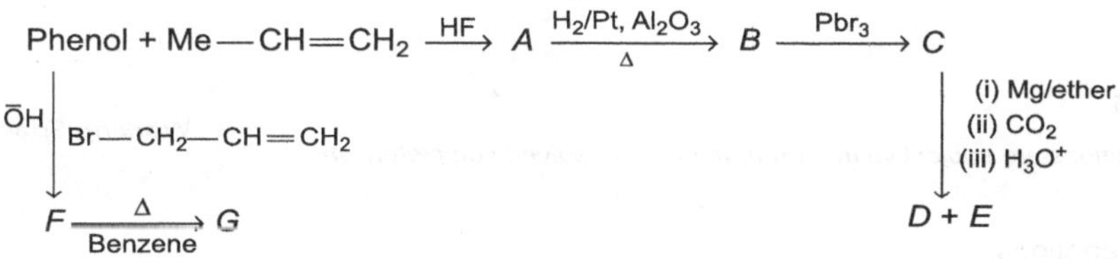

18 The name of the reaction and the intermediate species involved in the formation of A are

a. Friedel-Crafts alkylation, carbocation
b. Friedel-Crafts acylation, carbocation
c. Friedel-Crafts alkylation, radical anion
d. Friedel-Crafts acylation, radical anion

19 The compound D and E are geometrical isomers. The stable form is

a. 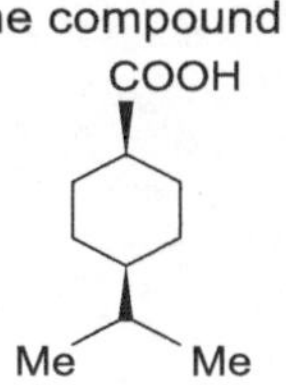

b. 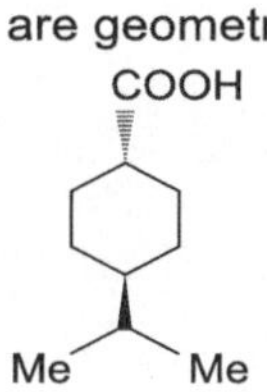

c. 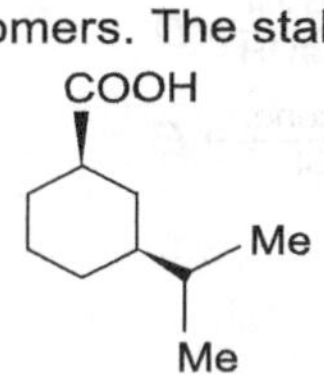

d. 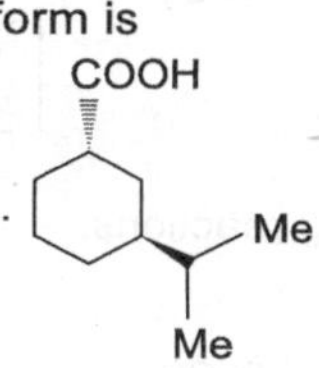

One Integer Value Correct Type

Direction (Q. No. 20) This section contains 1 question. When worked out will result in an integer from 0 to 9 (both inclusive).

20 Predict the value of x, y and z and fill the following row.

	x	y	z
Value			

i. Resveretrol, an antioxidant present in red wine, has the following structure

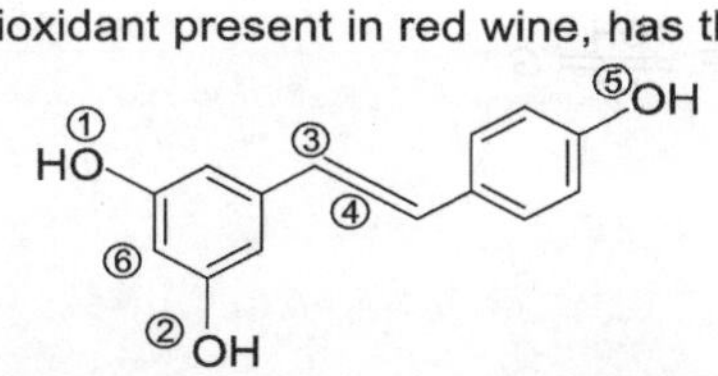

when treated with only 1 mole of sodium the reaction takes place at position x.

ii. y phenols are less acidic than p-nitrophenol among the given o-nitrophenol, m-nitrophenol, phenol, dinitrophenol, picric acid, p-cresol

iii. The ΔT_f (depression in freezing point) of 0.2 m solution of Ph_3C—OH in cold conc. H_2SO_4 is found to be $\frac{zxK}{10}$ [K_f = x K kgmol^{1}]. What is the value of z?

DPP-6 Chemical Properties of Phenols (Other than Electrophilic Substitution)

Subjective Questions

Working Space

Direction *(Q. Nos. 1-3) These questions are subjective in nature, need to be solved completely on notebook.*

1 i. **Complete the following reactions.**

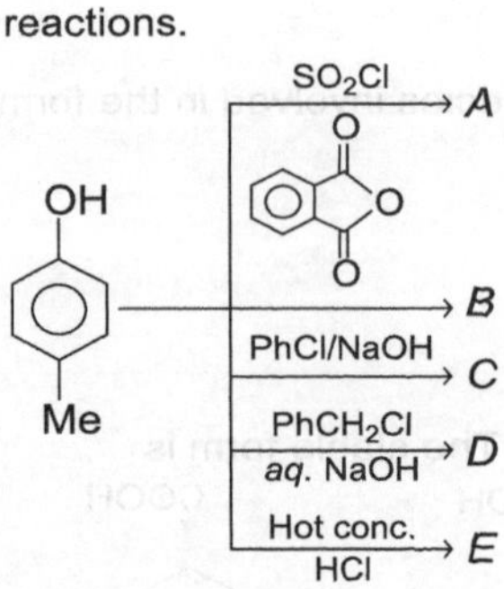

ii. **Complete the following reactions.**

a. 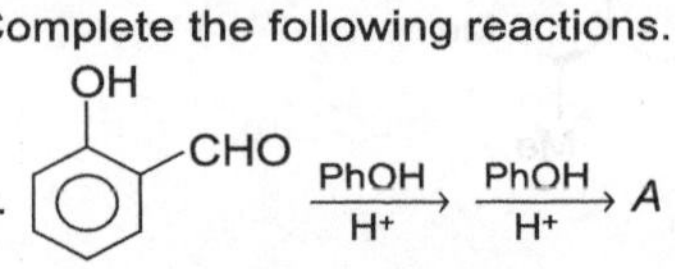

b. $PhOH \xrightarrow[\text{20 atm, 200°C}]{H_2/Ni} \xrightarrow{KMnO_4/H^+} B \xrightarrow[\Delta \text{ (ii) } P_2O_5 \text{ (iii) } H_2/Ni]{\text{(i) } NH_3} C$

c. $PhOH \xrightarrow[\text{(ii) } H_2/Ni,\ \Delta/\text{pressure}]{\text{(i) } CH_2O,\ H^+} D$

d. 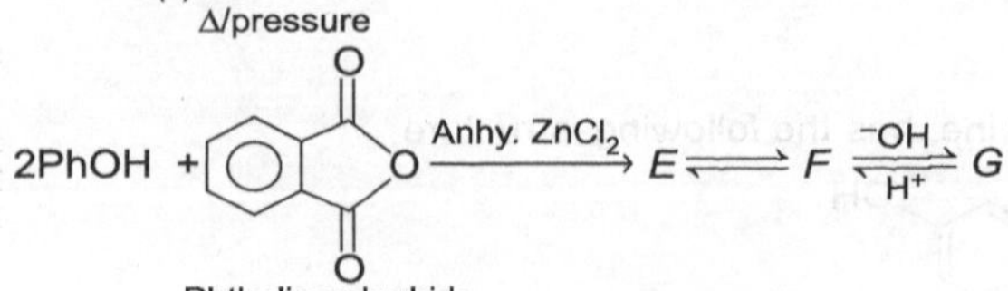

Phthalic anhydride

Working Space

2 i. How will you convert C_6H_5OH into phenyl-2, 4-dinitrophenyl ether?

ii. What product do you expect in the following reactions?

a. 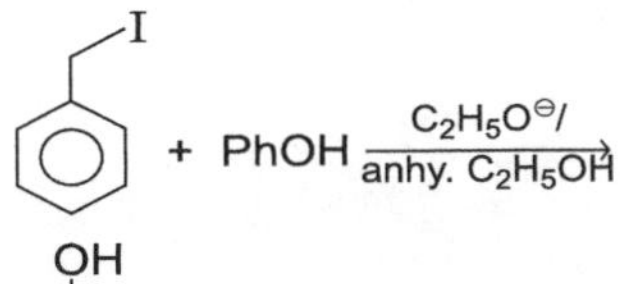

b. 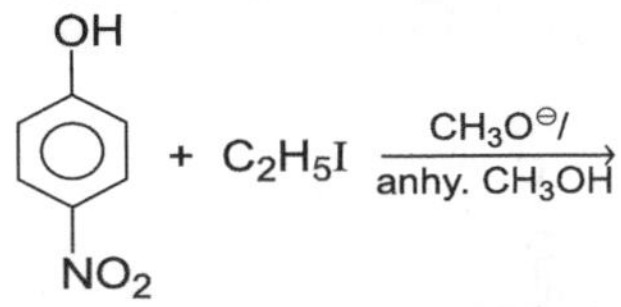

c. 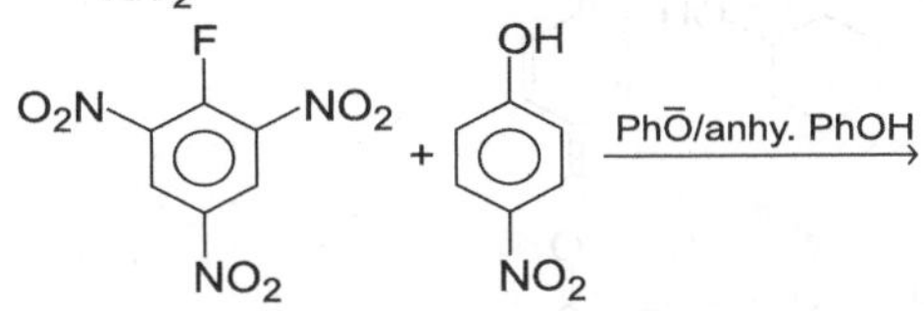

3 i. How will you separate the following mixtures by using a single chemical method?

a. Phenol and *p*-xylene.

b. *p*-ethylphenol and *p*-methylanisole.

c. *m*-dinitrobenzene, *m*-nitroaniline, *m*-nitrobenzoic acid, *m*-nitrophenol.

ii. Identify the lettered compounds in the following synthetic road maps.

a. 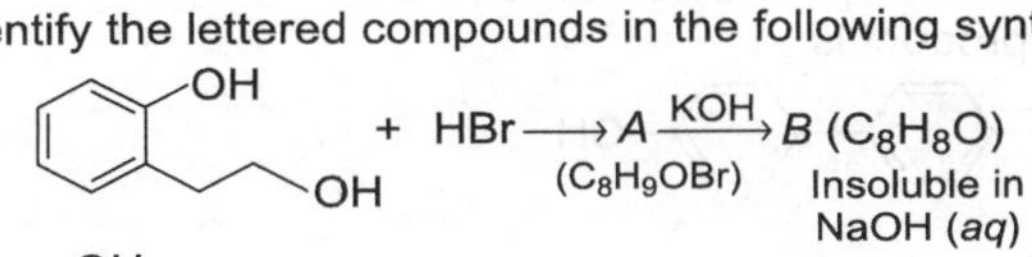

b. 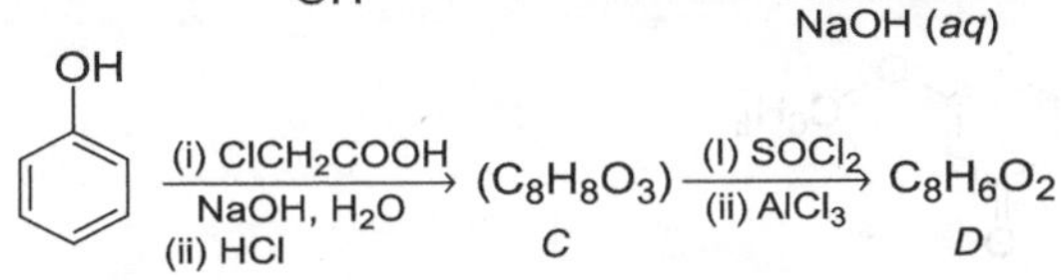

c. 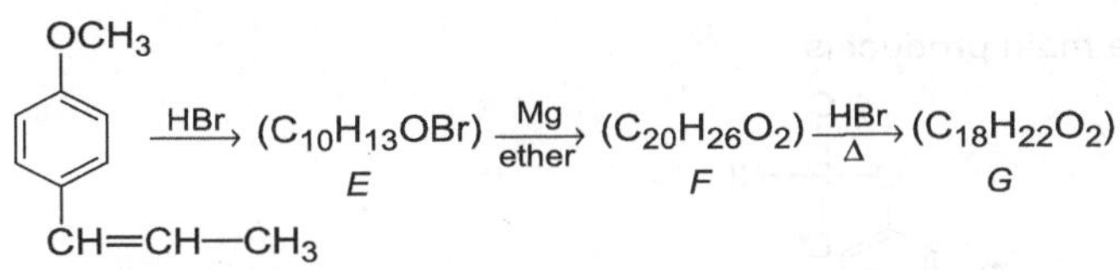

Only One Option Correct Type

Direction (Q. Nos. 4-11) *This section contains 8 multiple choice questions. Each question has four choices (a), (b), (c) and (d), out of which ONLY ONE is correct.*

4 Identify the reagent(s) needed to carry out the following reaction.

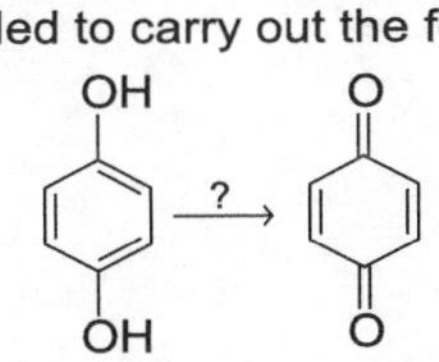

a. Na

b. $LiAlH_4$

c. $Na_2Cr_2O_7$, H_2SO_4/H_2O

d. NaOH

Working Space

5 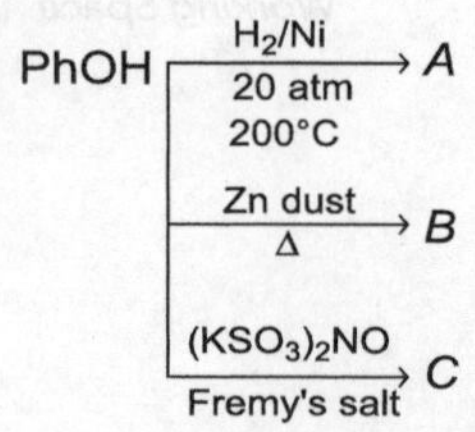

The compounds *A*, *B* and *C* respectively are

	A	B	C

a. 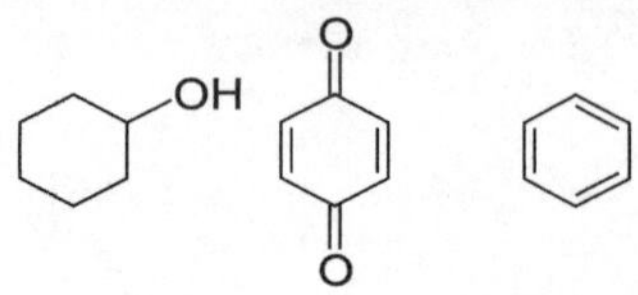

b. 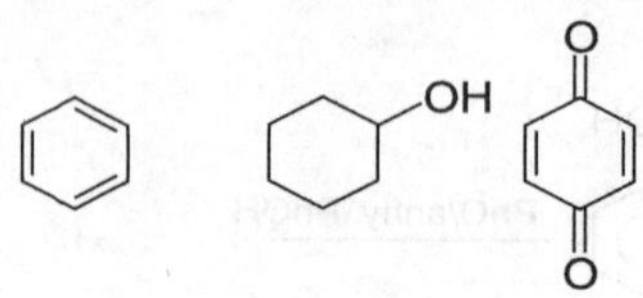

c. 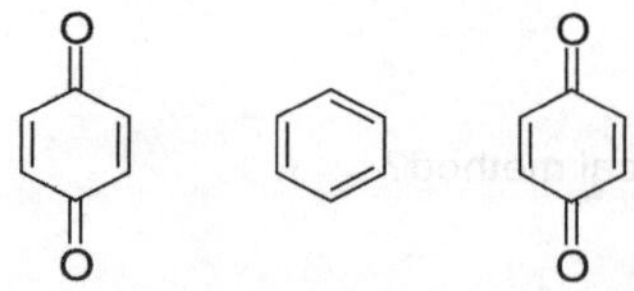

d. 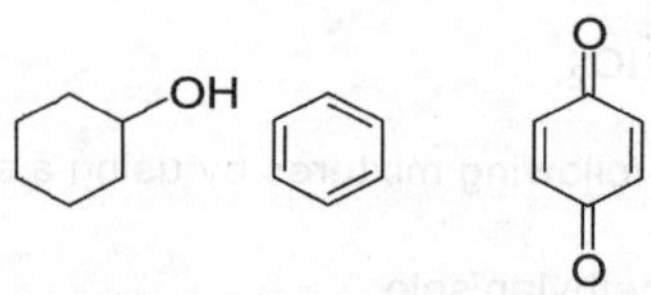

6 Phenol reacts with acetone in the presence of conc. sulphuric acid to form a $C_{15}H_{16}O_2$ product. The structural formula of the product is

a. 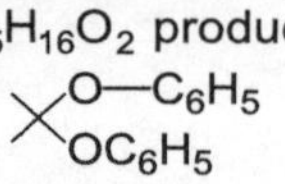

b. 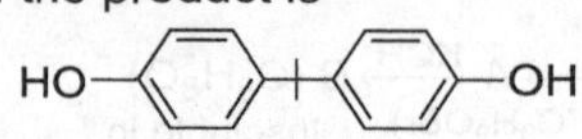

c. 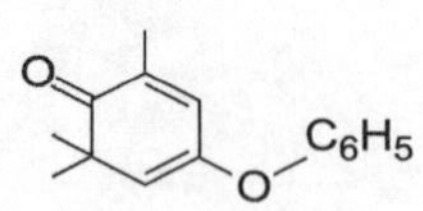

d. 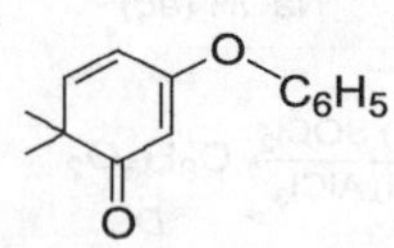

7 When phenol reacts with PCl_5, the main product is

a. 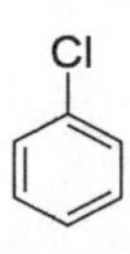

b. 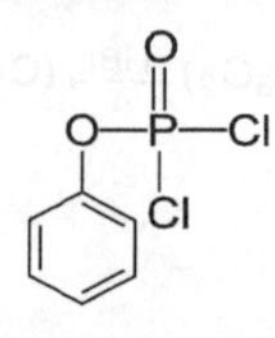

c. 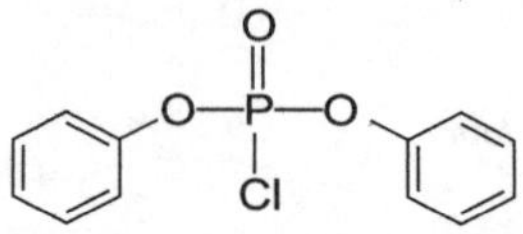

d. 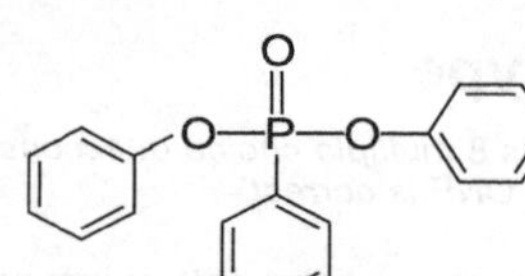

8 Phenol $\xrightarrow[\text{oxidation}]{\text{Aerial}}$ Coloured product.

The colour of the product is due to the formation of

a. 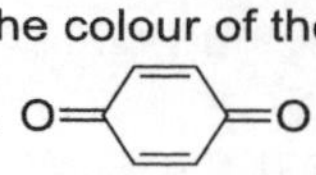

b. 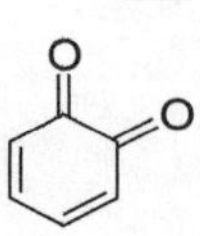

c. 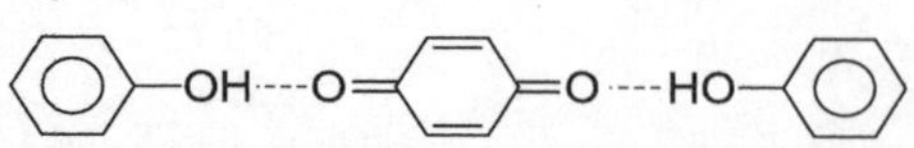

d. All of the above

Working Space

9 Give the decreasing order of reactivity of the following compounds with HBr.

I. $C_6H_5CH_2OH$ II. 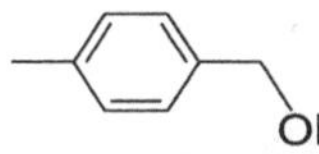 III. 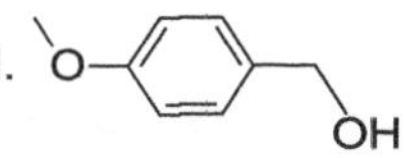 IV. 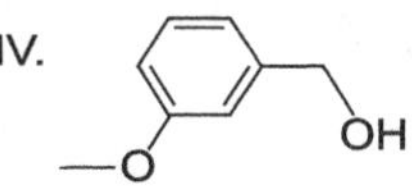

a. III > IV > II > I
b. III > II > IV > I
c. III > II > I > IV
d. II > III > IV > I

10 In the following reaction, *A* is

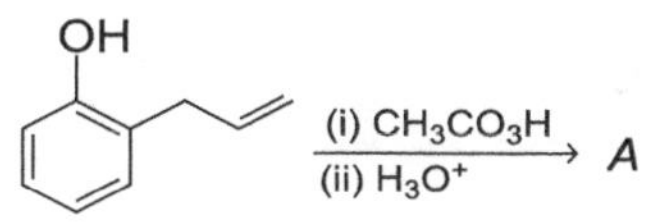

a. 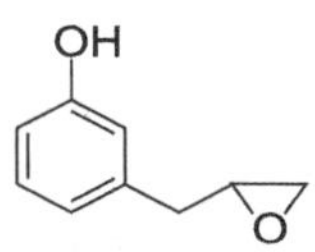
b. 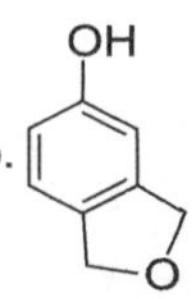
c. 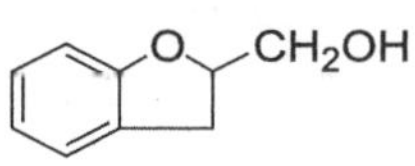
d. 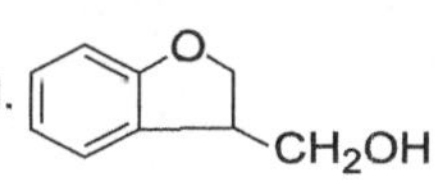

11 Phenols are generally remain unaffected by the action of $NaBH_4/H_3O^+$ like 1, 3 and 1, 4-benzenediols, however 1, 3, 5-benzenetriol (phloroglucinol) gives a high yield of *X* in the reaction. The compound *X* is

a. 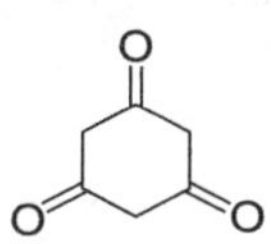
b. 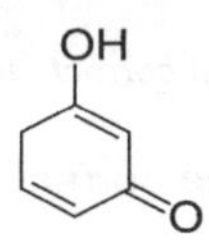
c. 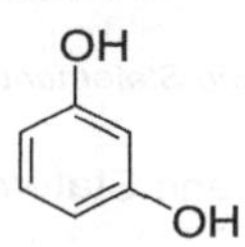
d. Both (a) and (b)

One or More than One Options Correct Type

Direction (Q. Nos. 12-14) *This section contains 3 multiple choice questions. Each question has four choices (a), (b), (c) and (d), out of which ONE or MORE THAN ONE are correct.*

12 What is/are true about phenol?

a. It is stronger acid than water
b. It forms salt with $NaHCO_3$
c. Passing CO_2 through aqueous solution of sodium phenoxide gives phenol
d. On exposure to air, it turns pink coloured

13 The products of the following reaction cannot be

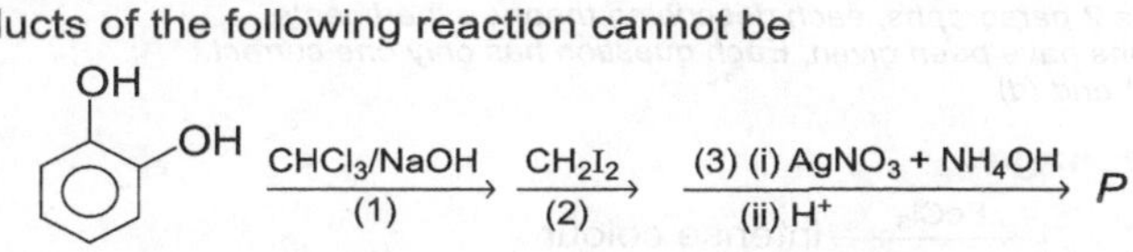

a. 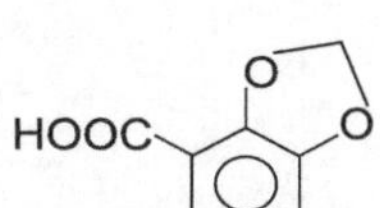
b. 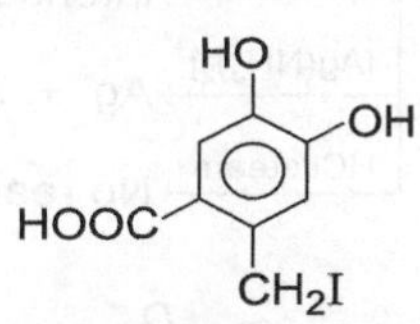
c. 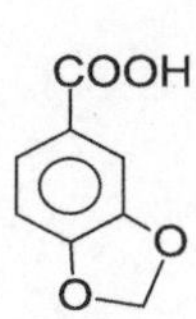
d. 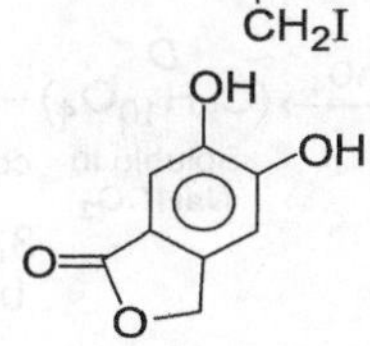

14 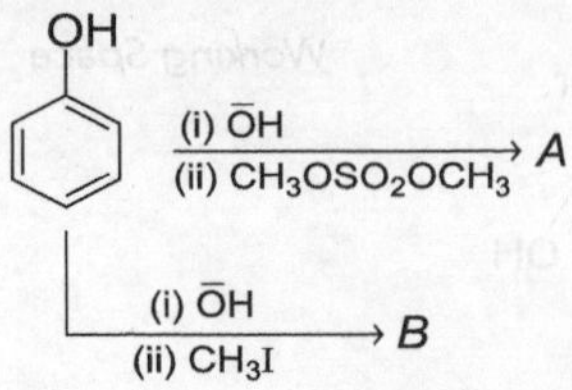

The true statement about *A* and *B* is/are

a. $A = Ph—O—\overset{O}{\underset{O}{\overset{\|}{\underset{\|}{S}}}}—OCH_3$; $B = Ph—O—\overset{O}{\underset{O}{\overset{\|}{\underset{\|}{S}}}}—CH_3$

b. $A = Ph—O—\overset{O}{\underset{O}{\overset{\|}{\underset{\|}{S}}}}—OCH_3$; $B = Ph—O—\overset{O}{\underset{O}{\overset{\|}{\underset{\|}{S}}}}—OCH_3$

c. $A = Ph—OCH_3$; $B = Ph—OCH_3$

d. $A = \bar{O}—\overset{O}{\underset{O}{\overset{\|}{\underset{\|}{S}}}}—OCH_3$; $B = I^-$

Statement Type

Direction *(Q. No. 15) This section is based on Statement I and Statement II. Select the correct answer from the codes given below.*

a. Both Statement I and Statement II are correct and Statement II is the correct explanation of Statement I
b. Both Statement I and Statement II are correct but Statement II is not the correct explanation of Statement I
c. Statement I is correct but Statement II is incorrect
d. Statement II is correct but Statement I is incorrect

15 **Statement I** A mixture of phenol and benzoic acid can be separated by extracting its etheral solution with *aq.* $NaHCO_3$ solution.

Statement II Phenol is a weaker base than carbonic acid.

Comprehension Type

Direction *(Q. Nos. 16-20) This section contains 2 paragraphs, each describing theory, experiments, data, etc. Five questions related to the paragraphs have been given. Each question has only one correct answer among the four given options (a), (b), (c) and (d).*

Passage I

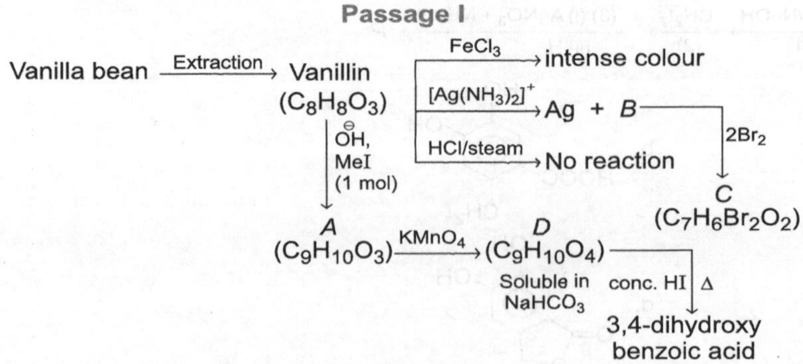

Working Space

16 The functional groups present in vanillin are

a. —OH, —OMe, —COOH
b. —CHO, Me, —OH
c. —OH, —CHO, —OMe
d. —CHO, —CH_2OH, —OH

17 The structure of product *C* is

a. 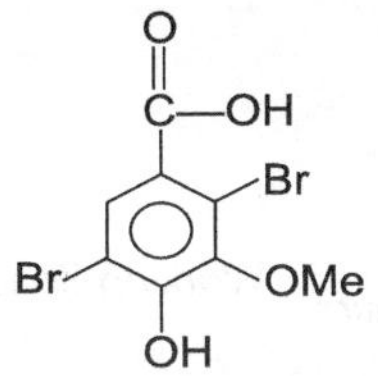
b. 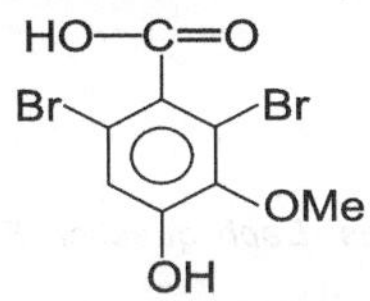
c. 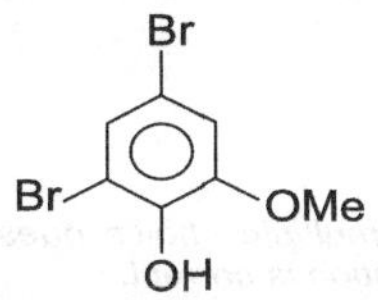
d. 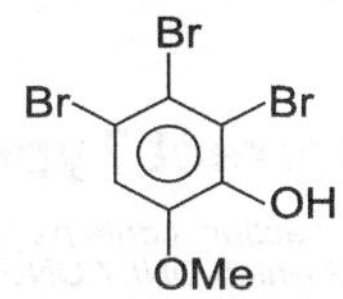

Passage II

Consider the following reactions to answer the next three questions.

$A \xrightarrow{Na}$ No effervescence

$A \xrightarrow{H^+/H_2O} B$

$B \xrightarrow{FeCl_3}$ Violet colouration

$B \xrightarrow{I_2/NaOH}$ Yellow precipitate

$B \xrightarrow{NaCN(aq)} \xrightarrow[H_2O]{H_3O^+} C \xrightarrow[Heat]{H^+} D\ (C_{11}H_{10}O_3)$

$D \xrightarrow{SOCl_2} \xrightarrow{AlCl_3}$ 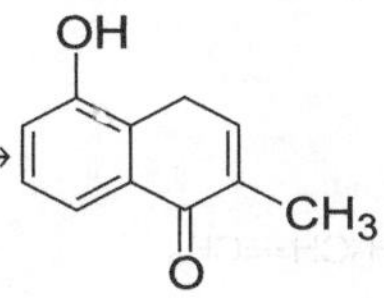

18 The most likely structure of *A* is

a. 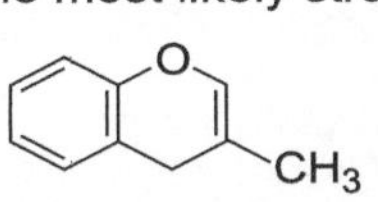
b. 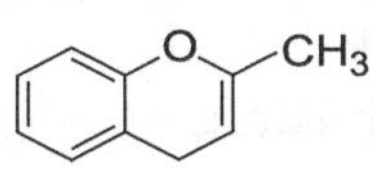
c. 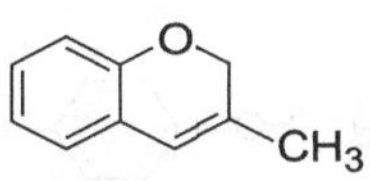
d. 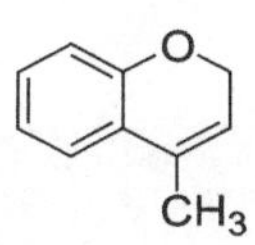

19 Which of the following is incorrect regarding *C* and *D*?

a. *C* is a racemic mixture
b. *D* has two stereomers
c. Both *C* and *D* decolourise Br_2-CCl_4 solution
d. *D* can also be obtained as one of the product in the following reaction

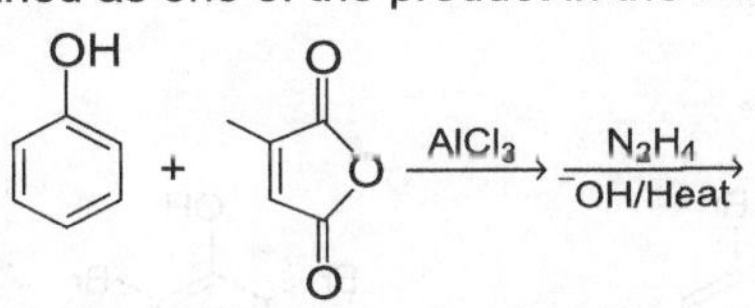

20 The incorrect statement about *B* is

a. When oxidised with $KMnO_4$, it gives salicylic acid
b. On treatment with NaOH/I_2 followed by hydrolysis, gives, *E* which on heating with H_2SO_4 gives

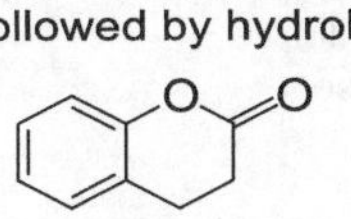

c. *E* formed in part (b) on reaction with $SOCl_2$ followed by $AlCl_3$ gives

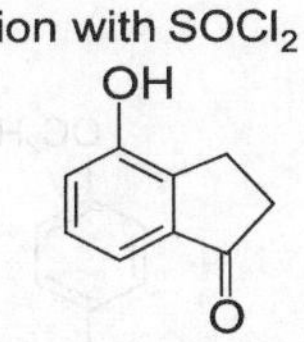

d. Treatment with $NaBH_4$ result in the formation of a single enantiomer in pure form

Revisal Problems for JEE Main

Only One Option Correct Type

Working Space

***Direction** (Q. Nos. 1-28) This section contains 28 multiple choice questions. Each question has four choices (a), (b), (c) and (d), out of which ONLY ONE option is correct.*

1 Which one is the missing reagent of Kolbe Schmidt reaction?

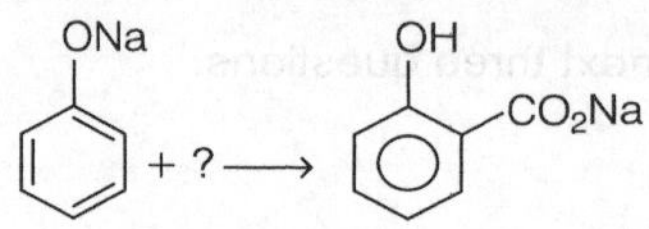

a. HCO_2Et b. $(EtO)_2C{=}O$ c. CO_2 d. HCO_2Na

2 What is the product obtained by heating the following allylic ether of phenol?

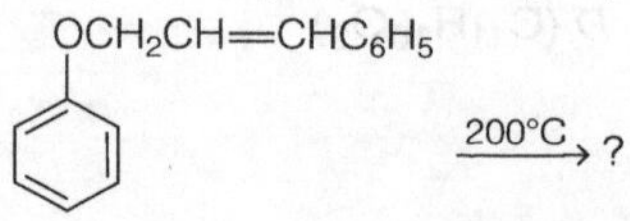

a. 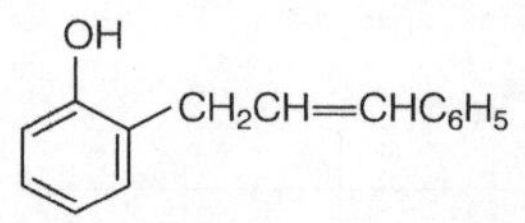

b. 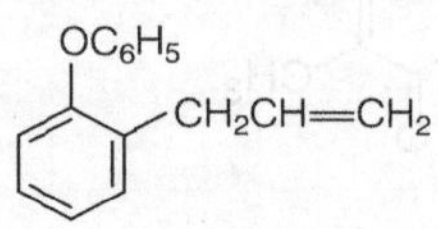

c. 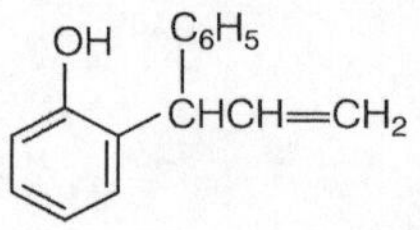

d. 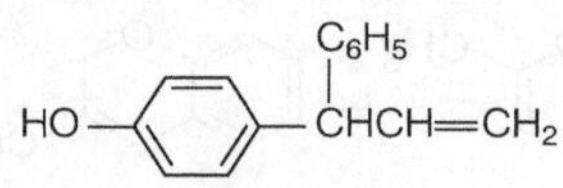

3 Which one of the reagents readily react with bromobenzene?

a. $NaOCH_2CH_3$ at 25°C
b. $NaNH_2/NH_3$ at −33°C
c. NaCN/DMSO at 25°C
d. $(CH_3)_2NH$ at 0°C

4 Indicate the correct product of the following reaction.

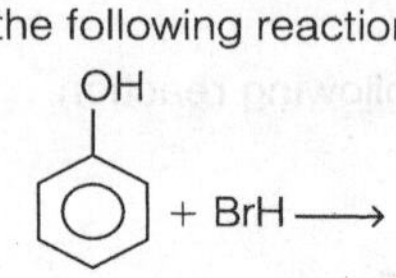

I. 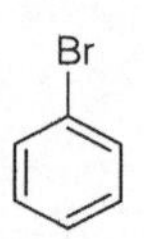

II. 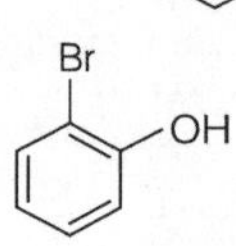

III. 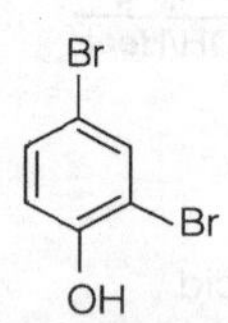

IV. 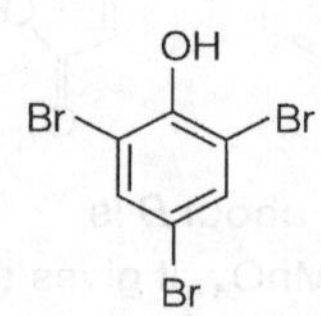

a. Only I
b. Both II and III
c. Both III and IV
d. None of these

5 Give the major product of the following reaction.

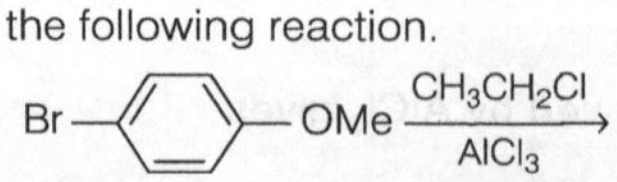

a. 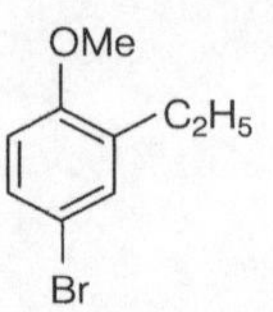

b. 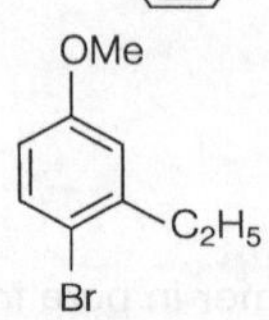

c. 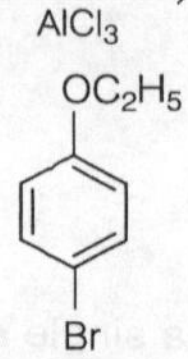

d. 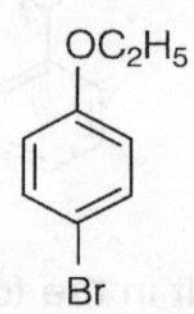

6 The product of the following reaction is

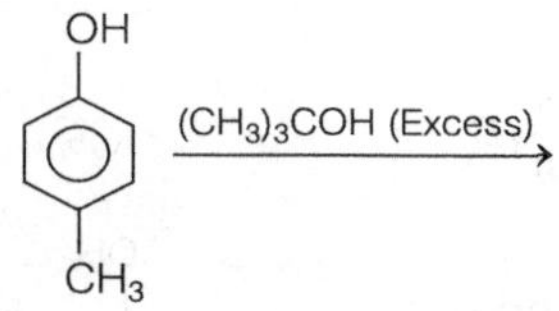

a. 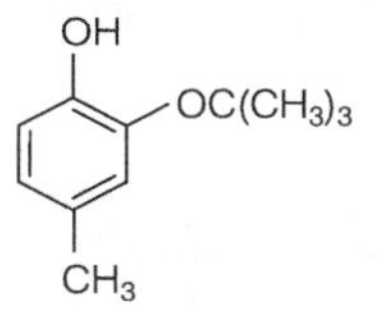

b. 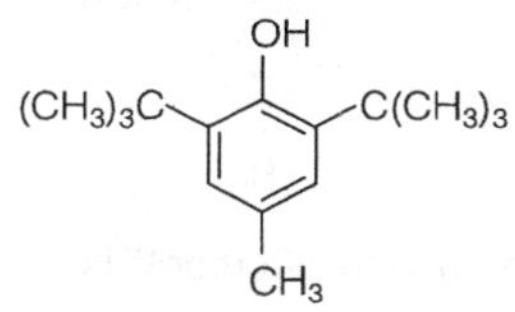

c. 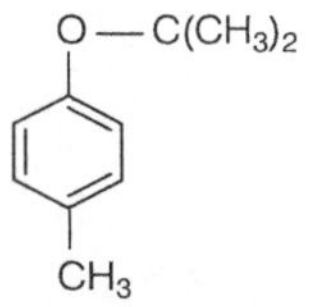

d. 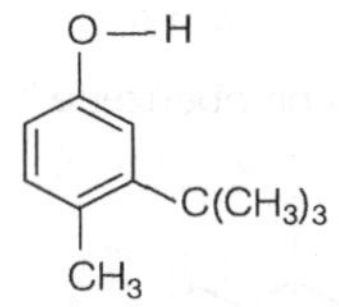

7 Consider the following reaction,

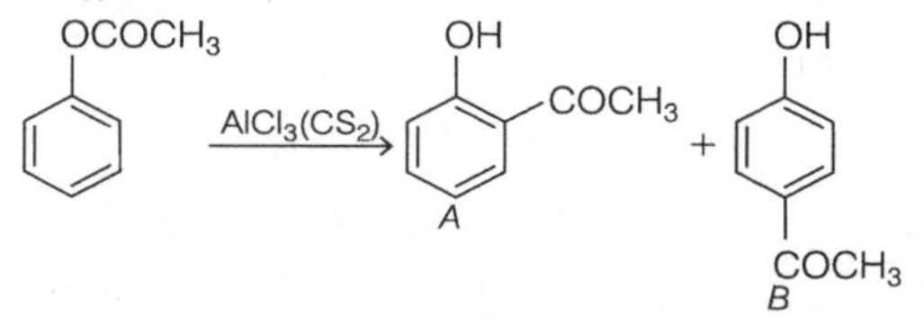

Choose the correct statement.

a. *A* is more volatile than *B*
b. *B* is more volatile than *A*
c. *B* is formed more rapidly at a higher temperature
d. *A* is formed in higher yield at a low temperature

8 What is the product of the following reaction?

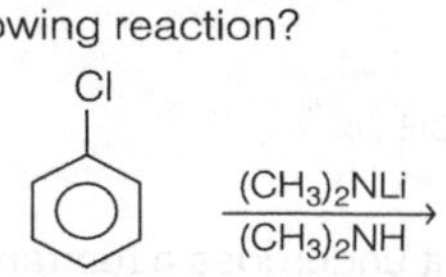

a. N, N-dimethylaniline
b. Phenyl lithium
c. *p*-chloro-N, N-dimethylaniline
d. *m*-chloro-N, N-dimethylaniline

9 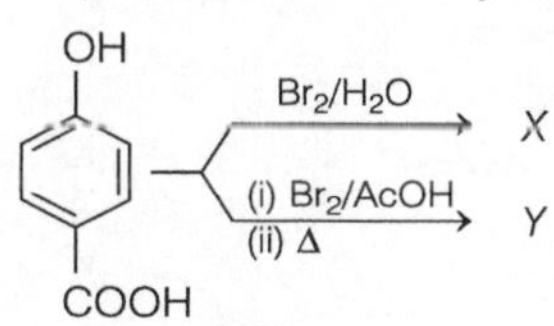 ; The products are

a. 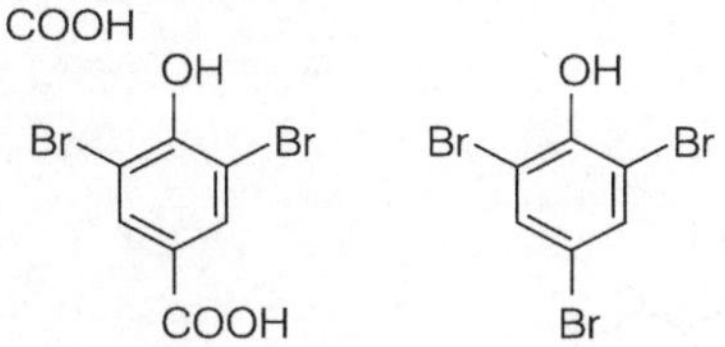
X Y

b. 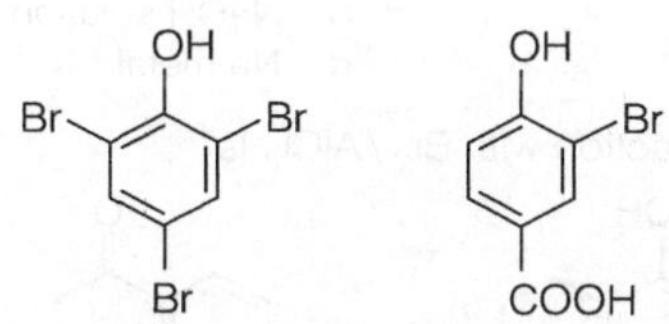
X Y

c. 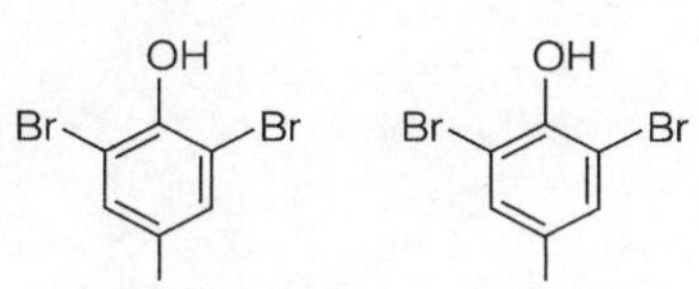
X Y

d. 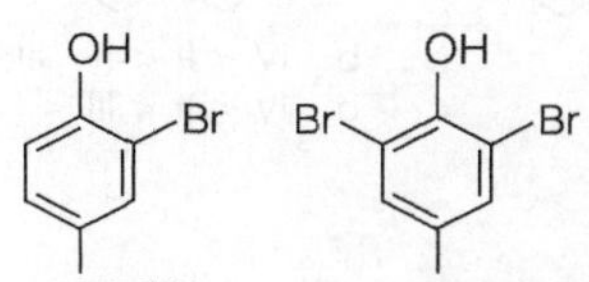
X Y

10 Identify the end product *B* of the following sequence of reactions.

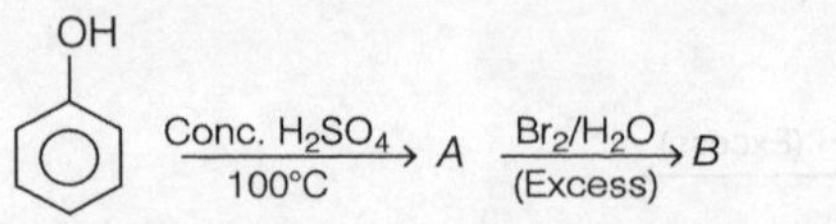

a. 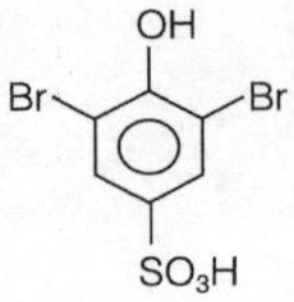

b. 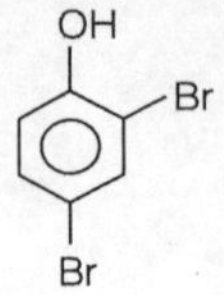

c. 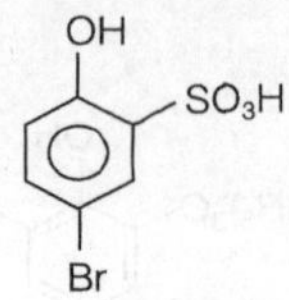

d. 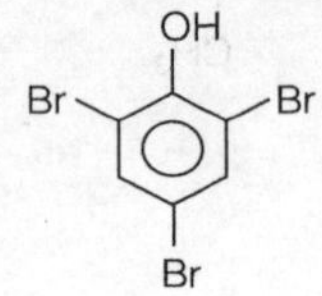

11 The order of increasing pK_a values of *o*, *m*, and *p*-cresols is

a. $o < p < m$ b. $m < p < o$ c. $m < o < p$ d. $p < o < m$

12 Which of the following reaction will best convert nitrobenzene into 2,4,6-trichlorobromobenzene?

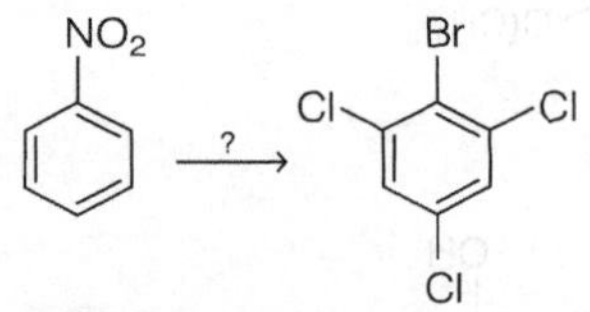

a. $SnCl_2$/HCl, $Cl_2(aq)$, $NaNO_2$/HCl-0°C, CuBr
b. $Cl_2/AlCl_3$, $SnCl_2$/HCl, $NaNO_2$/HCl-0°C, CuBr
c. $SnCl_2$/HCl, Br_2-H_2O, $NaNO_2$/HCl-0°C, CuCl
d. $SnCl_2$/HCl, $NaNO_2$/HCl-0°C; H_2O(v), $Cl_2(aq)$, HBr

13 Select the best combination of reagents which will bring the following conversion.

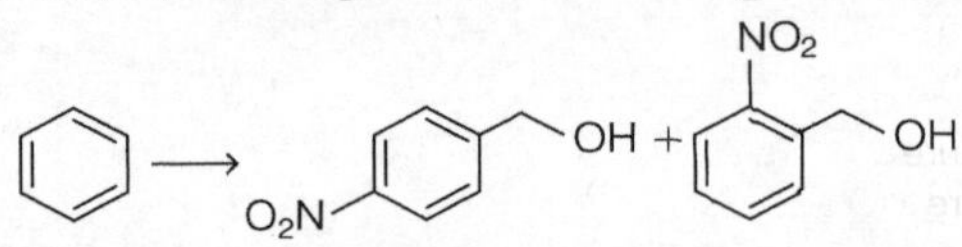

a. $Br_2/FeBr_3$, HNO_3/H_2SO_4, NaOH, H_2/Pt
b. $CH_3COCl/AlCl_3$, $Br_2/FeBr_3$, $NaBH_4/CH_3OH$
c. $CH_3Cl/AlCl_3$, $KMnO_4$, $CH_3Cl/AlCl_3$, NBS/$h\nu$, $LiAlH_4/H_2O$, NaOH
d. $CH_3Cl/AlCl_3$, HNO_3/H_2SO_4, NBS/$h\nu$/NaOH

14 When the following compound is heated with a reagent *X*, it undergoes a rearrangement reaction due to phenyl shift. The reagent *X* can be

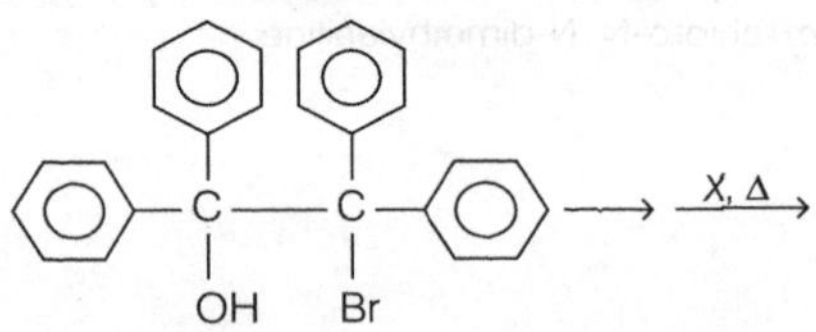

a. NaI/Acetone b. NaOH solution
c. dil. $AgNO_3$ solution d. Na metal

15 Increasing order of rate of reaction with $Br_2/AlCl_3$ is

I. 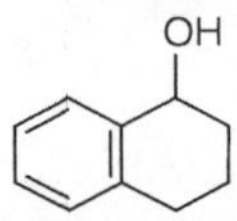

II. 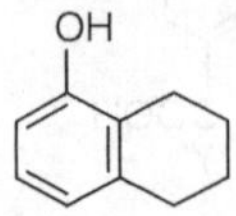

III. 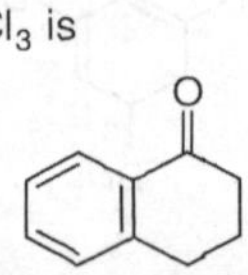

IV. 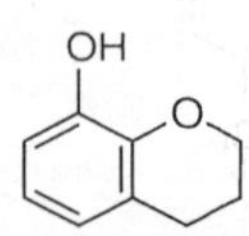

a. III < I < II < IV b. IV < II < I < III
c. II < IV < III < I d. IV < II < III < I

16 Rank the following in the order of decreasing acidic strength (most → least acidic).

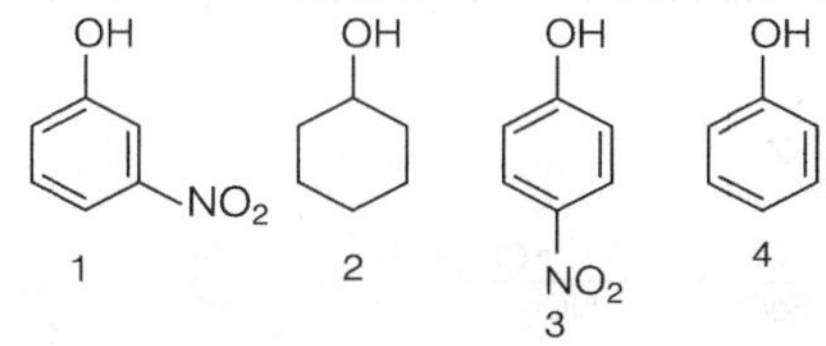

a. 2 > 4 > 1 > 3
b. 3 > 1 > 2 > 4
c. 1 > 3 > 4 > 2
d. 3 > 1 > 4 > 2

17 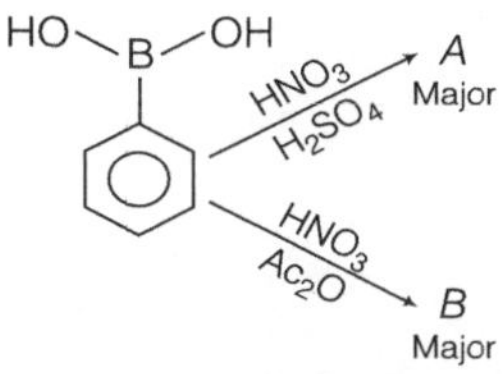

Correct statement about *A* and *B* is

a. *A* is *para* derivative, *B* is *ortho* derivative
b. *A* is *meta* derivative, *B* is *ortho* derivative
c. *A* is *meta* derivative, *B* is *meta* derivative
d. *A* is *ortho* derivative, *B* is *ortho* derivative

18 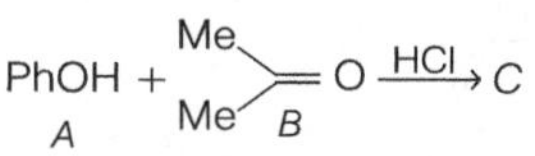

The compound *C* is

a. 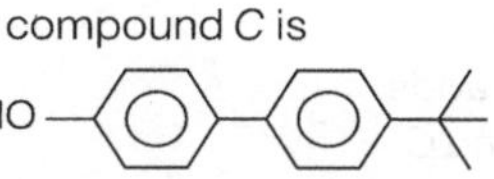
b. 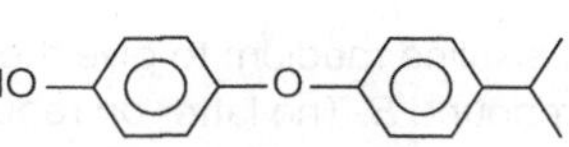
c. 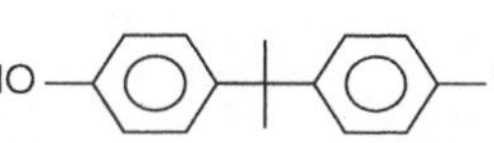
d. 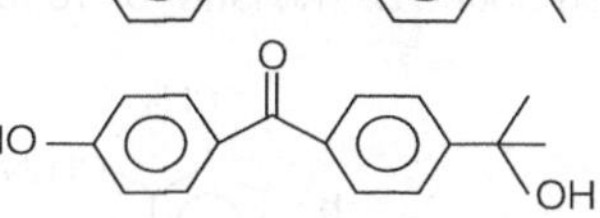

19 A compound *X* (C_7H_8O) is insoluble in water, dilute HCl and aqueous $NaHCO_3$ but dissolves in dilute NaOH. When *X* is treated with bromine water, it is converted into a compound of formula $C_7H_5OBr_3$. Compound *X* is

a. 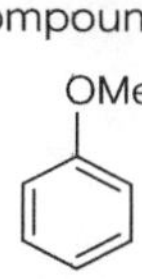
b. 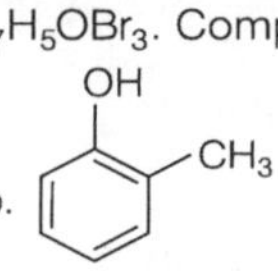
c. 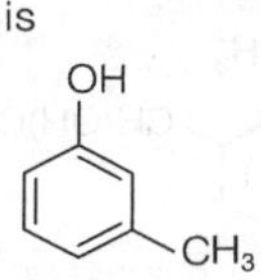
d. 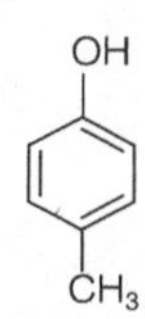

20 Which has maximum pK_a value in phenols?

a. 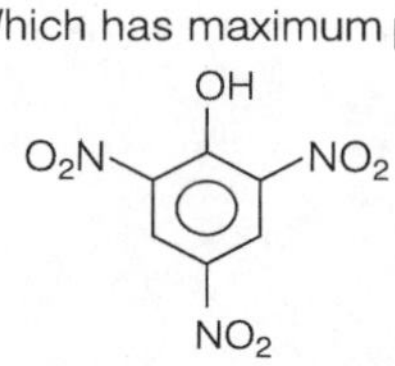
b. 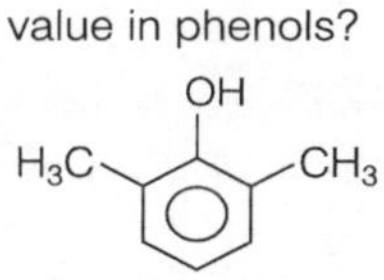
c. 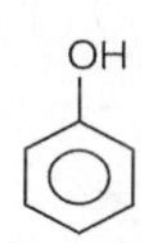
d. 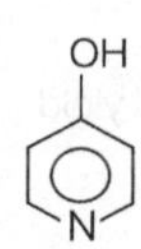

21 What is the major product from the following series of reactions?

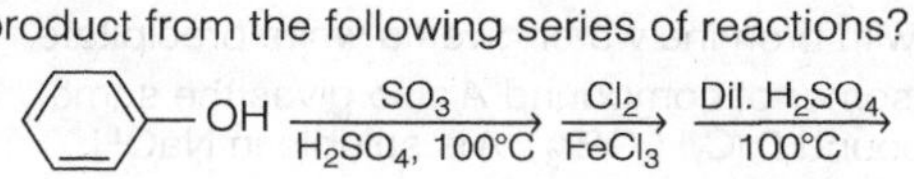

a. 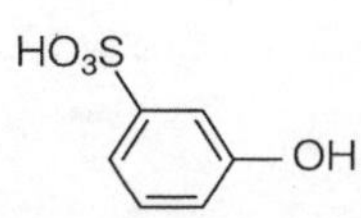
b. 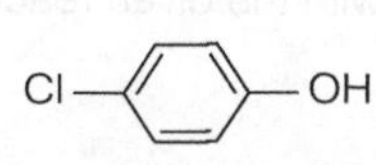
c. 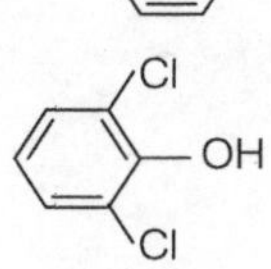
d. 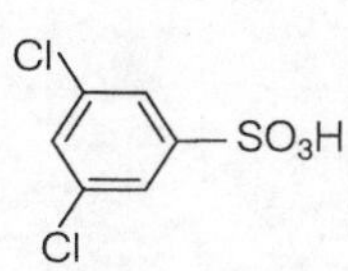

Working Space

Working Space

22 Choose the best reaction sequence that could be used to perform the following transformation.

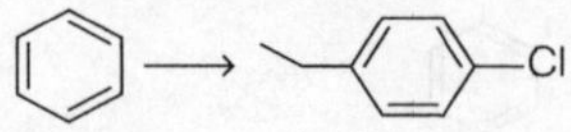

a. 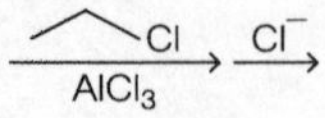

b. 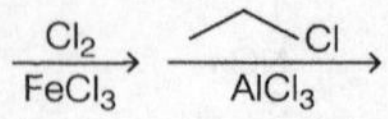

c. 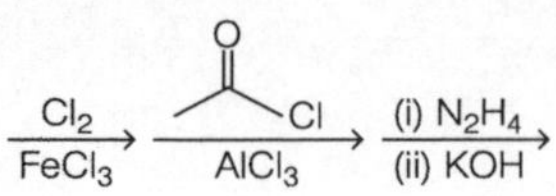

d. 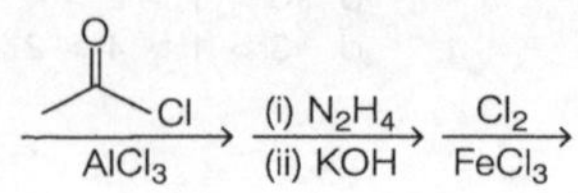

23 The major product of the following reaction sequence will be

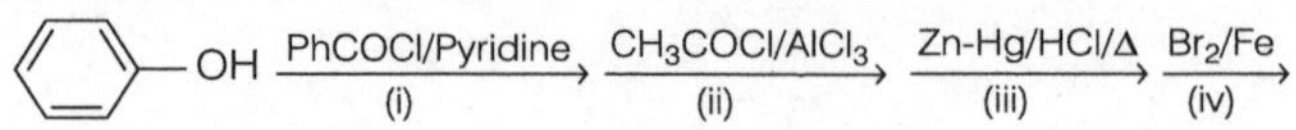

a. 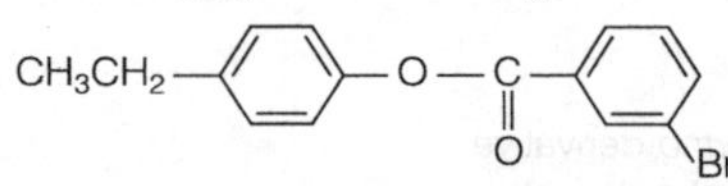

b. 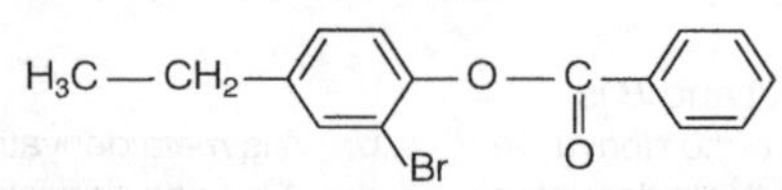

c. 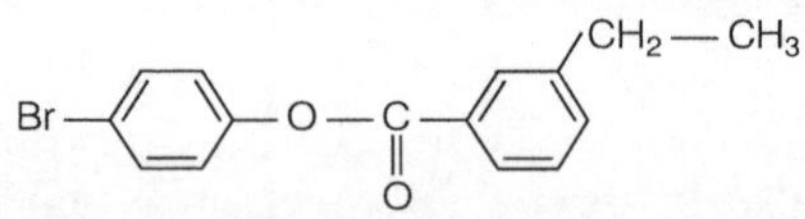

d. 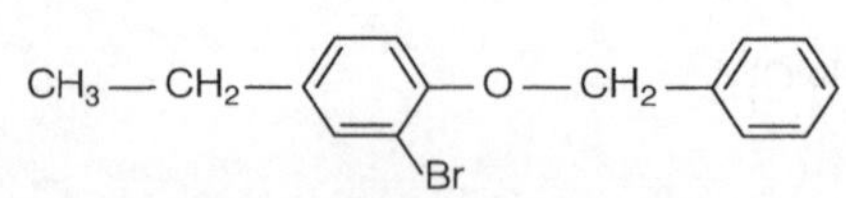

24 *p*-cresol reacts with chloroform in alkaline medium to give the compound *A* which adds hydrogen cyanide to form the compound *B*. The latter on reduction gives compound *C* which is

a. 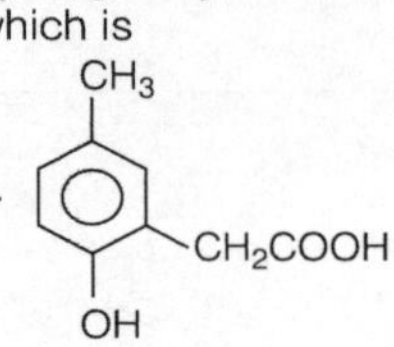

b. 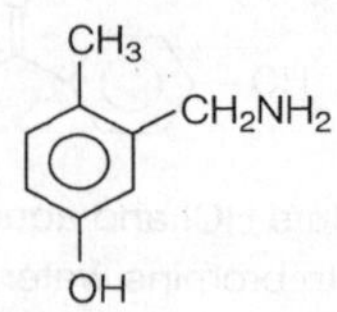

c. 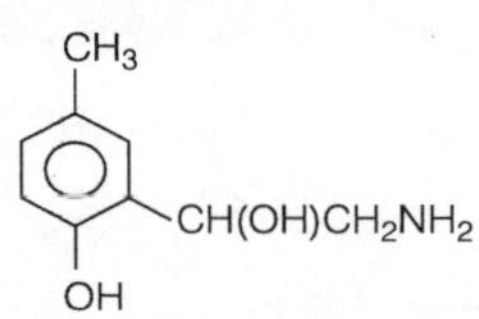

d. 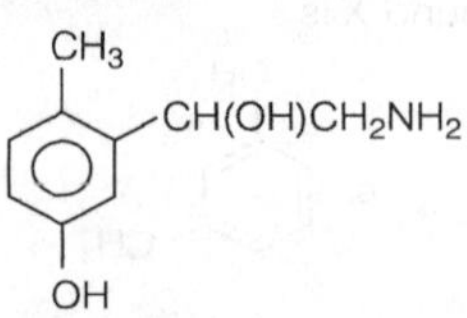

25 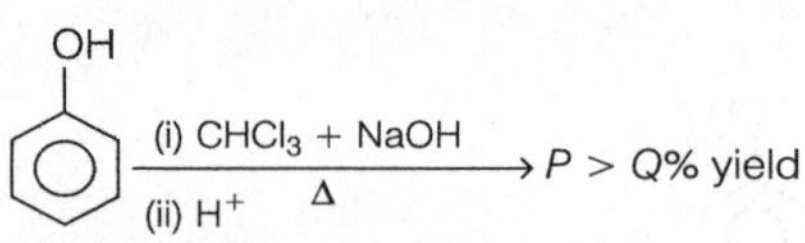

Select the incorrect option.

a. Boiling point ($P < Q$)

b. Melting point ($Q > P$)

c. Water solubility ($P < Q$)

d. Acidic strength ($Q < P$)

26 An aromatic compound *A*, C_7H_8O on reaction with bromine water gives a white precipitate of compound *B* ($C_7H_5OBr_3$). Compound *C*, an isomer of compound *A* also gives the same reaction and gives a white precipitate of compound *D* ($C_7H_5OBr_3$). *A* is soluble in NaOH while *C* does not. The true statement related with the given reaction is

a. *B* and *D* are same compounds

b. *B* is 2, 4, 6-tribromo-3-methyl phenol

c. *C* gives violet colouration with $FeCl_3$

d. All of the above

Working Space

27 Arrange the following in the increasing order of reactivity in aromatic nucleophilic substitution reaction.

I. 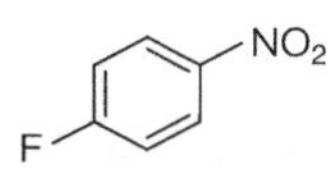

II. 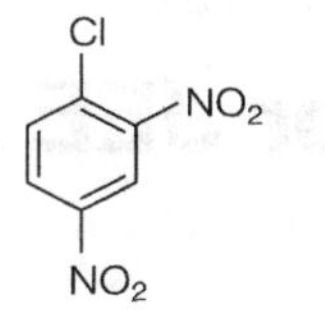

III. 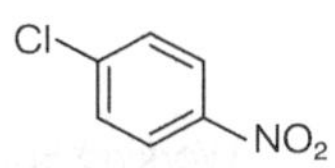

IV. 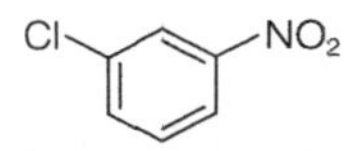

a. I < IV < III < II
b. IV < III < I < II
c. I < IV < II < III
d. IV < I < II < III

28 Identify the product in the following reaction.

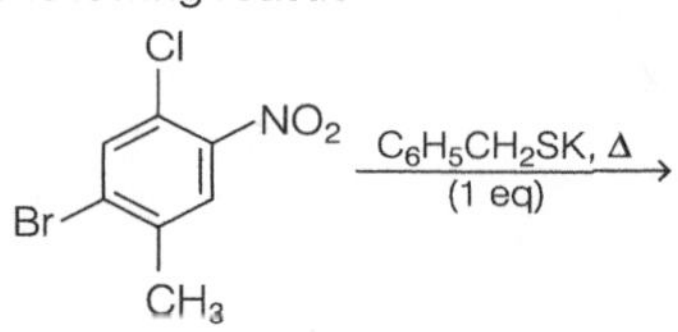

a. 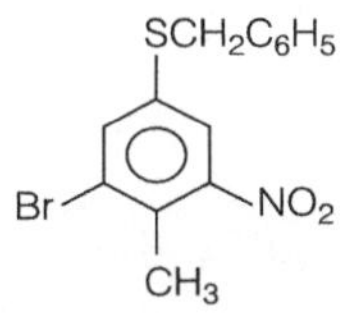

b. 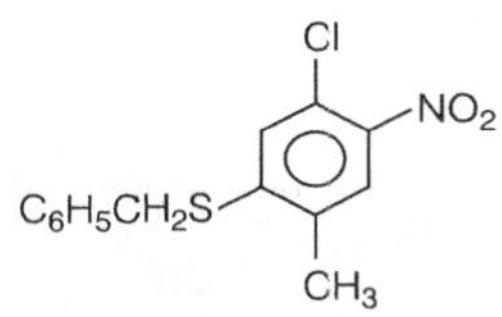

c. 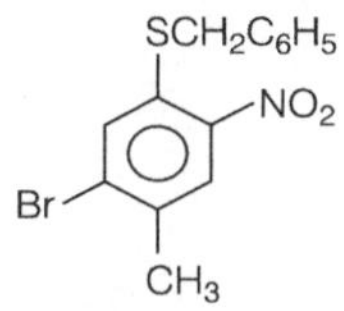

d. 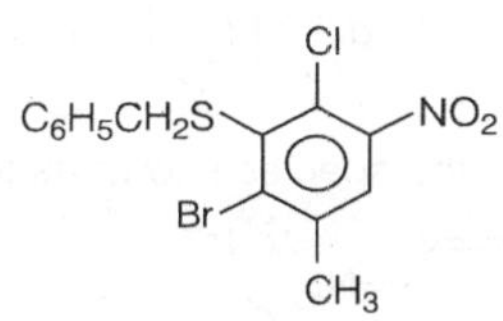

Statement Type

Direction (Q. Nos. 29 and 30) ***This section is based on Statement I and Statement II. Select the correct answers from the codes given below.***

a. If Statement I and Statement II both are correct and Statement II is the correct explanation of Statement II
b. If Statement I and Statement II both are correct and Statement II is not the correct explanation of Statement I
c. If Statement I is correct but Statement II is incorrect
d. If Statement I is incorrect but Statement II is correct

29 **Statement I** Phenol is trichlorinated in aqueous chlorine even in the absence of a Lewis acid catalyst but anisole require Lewis acid catalyst for chlorination under the similar condition.

Statement II In aqueous solution, phenoxide ion is formed from phenol that undergo very rapid aromatic electrophilic substitution reaction.

30 **Statement I** *p*-nitrochlorobenzene is more reactive than *p*-nitrofluorobenzene in aromatic nucleophilic substitution reaction.

Statement II Fluorine forms stronger bond than chlorine with carbon of benzene ring.

SET - 2

Revisal Problems
for JEE Advanced

Only One Option Correct Type

Direction (Q. Nos. 1-11) *This section contains 11 multiple choice questions. Each question has four choices (a), (b), (c) and (d), out of which ONLY ONE is correct.*

Working Space

1 Identify the *X*, *Y*, *Z* in the following reaction

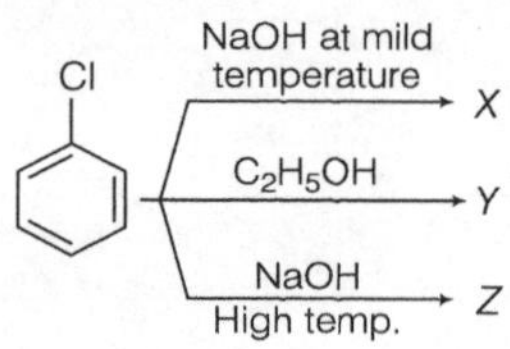

	X	Y	Z
a.	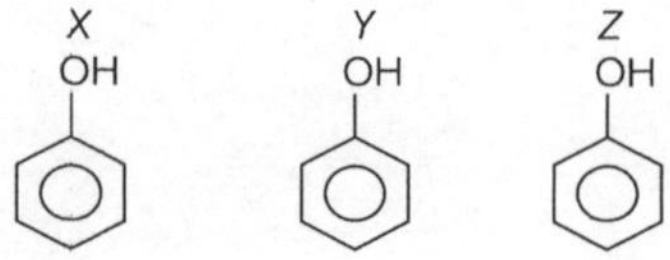		
b.	no reaction	no reaction	no reaction
c.	no reaction	no reaction	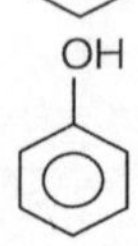
d.	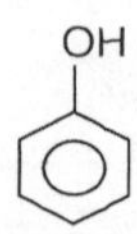	no reaction	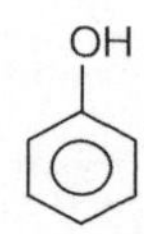

2 What combination of reactants will give the species shown as a reactive intermediate?

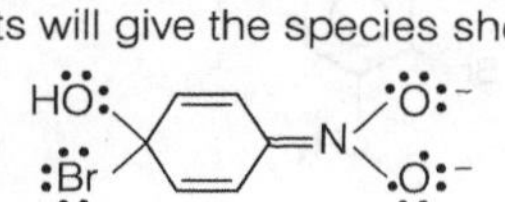

a. 1-bromo-4-nitrobenzene and NaOH
b. 4-nitrophenol and HBr
c. 4-nitrophenol, Br_2 and $FeBr_3$
d. Bromobenzene and $HONO_2$

3 Give the major product from the following reaction sequence.

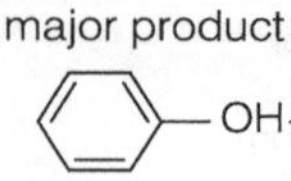 $\xrightarrow[NaOH]{CH_3I}$ $\xrightarrow[H_2SO_4]{HNO_3}$ $\xrightarrow{Zn(Hg),\ HCl}$ $\xrightarrow{CH_3COCl}$ Major product

a. 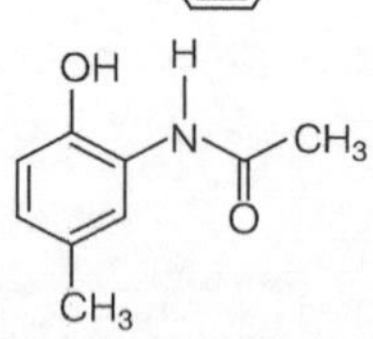
b. 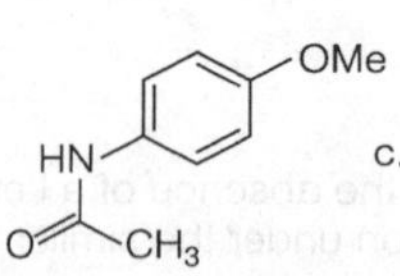
c. 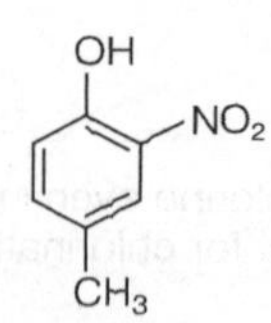
d. 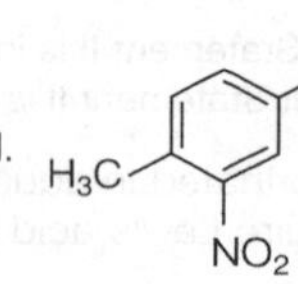

4 Choose the statement that is correct concerning the following reaction.

$$\text{Phenol} + CH_3I/NaOH \longrightarrow \text{Major product}$$

a. This reaction results in methylation of oxygen in phenol. It is utilized frequently because the hydroxide group is often highly activated. This reaction can be reversed with HI
b. This reaction results in methylation of oxygen in phenol. The resulting product is activated approximately the same amount as in phenol.The reaction can be reversed with HI
c. This reaction results in methylation of oxygen in phenol. The resulting product is more highly activated than phenol. This reaction can be reversed with HI
d. This reaction results in an S_N2 reaction between methyl iodide and NaOH. The resulting iodide ion substitutes at *ortho* and *para* positions of phenol. Treatment of this product with Mg will give Grignard reagent

5 Which products would actually form in the reaction shown below?

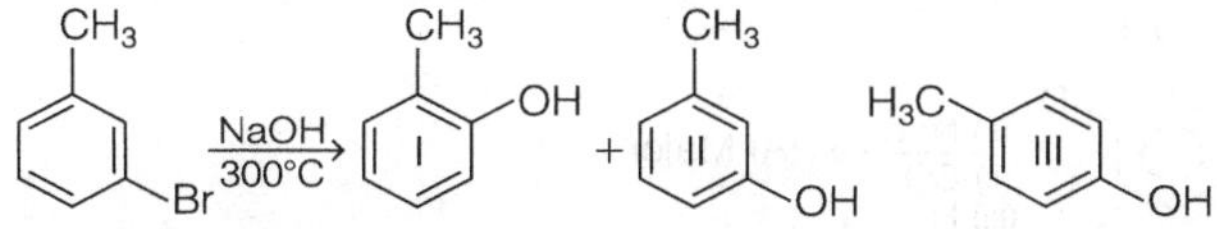

a. Only I b. Only II c. Only III d. Both I and II

6 The decreasing order of ArS_N reaction with $C_2H_5O^-$/EtOH is

I. 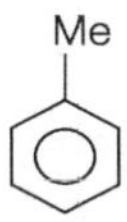

II. 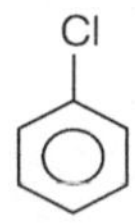

III. 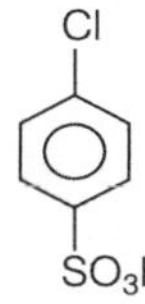

IV. 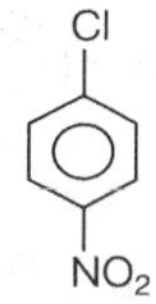

a. I > II > III > IV b. IV > III > II > I
c. III > IV > II > I d. IV > III > I > II

7 The principal product(s) formed when 1 mole of methyl magnesium iodide reacts with 1 mole of *p*-hydroxy acetophenone is/are

I. Methane

II. 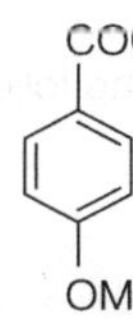

III. 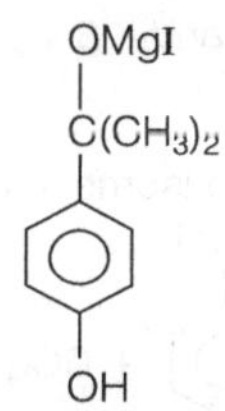

IV. 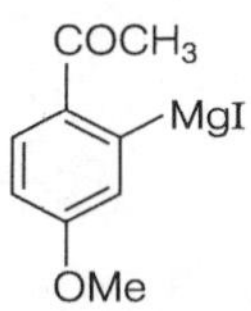

V. 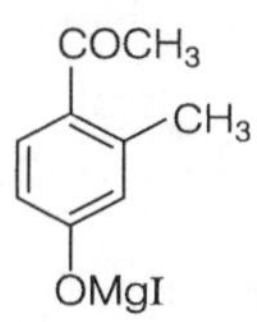

a. Both I and II b. Both I and III
c. I, IV and V d. Only I

8 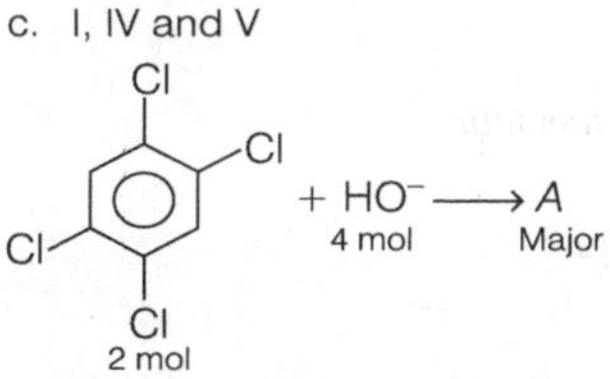

Product *A* of the given reaction is

a. 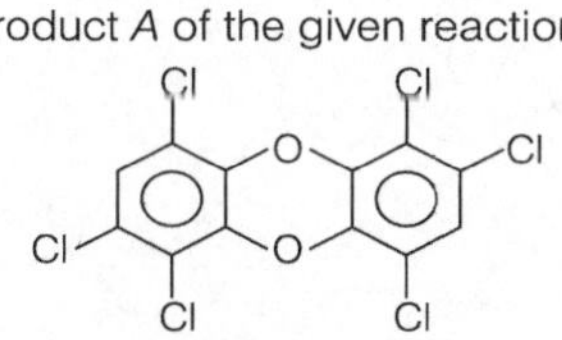

b. 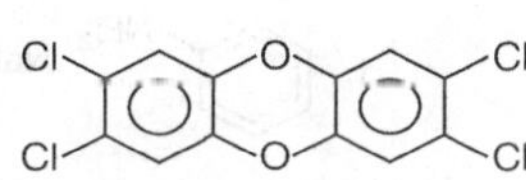

c. 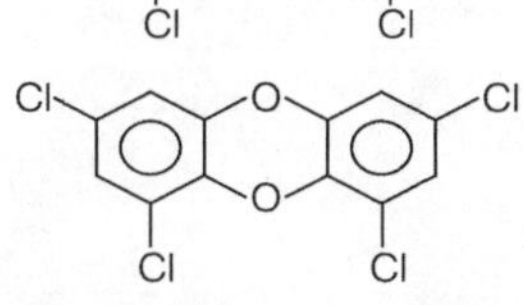

d. 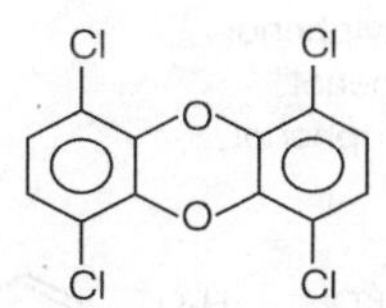

9 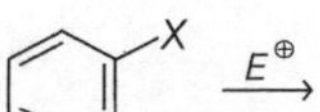

$\xrightarrow{E^{\oplus}}$ *ortho* and *para* products. X = F, Cl, Br and I

If X is fluorine, then which of the following is the most appropriate *ortho* and *para* $\left(\frac{f_o}{f_p}\right)$ ratio?

a. $\frac{f_o}{f_p} = 0.07$ b. $\frac{f_o}{f_p} = 0.22$ c. $\frac{f_o}{f_p} = 0.30$ d. $\frac{f_o}{f_p} = 0.32$

10 Complete the following reaction.

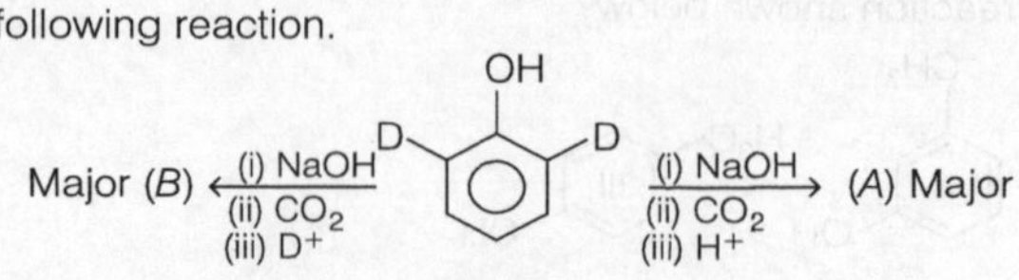

a. *A* and *B* are the same product

b. $B =$ 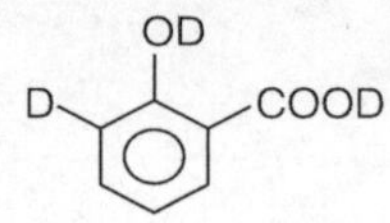 and $A =$ 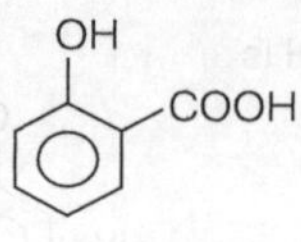

c. $A =$ 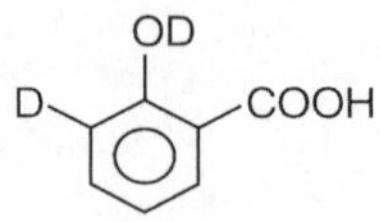 and $B =$ 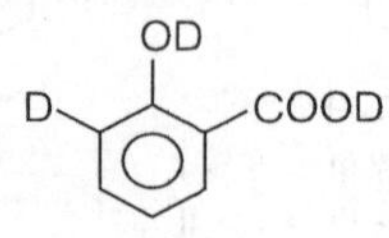

d. $A =$ 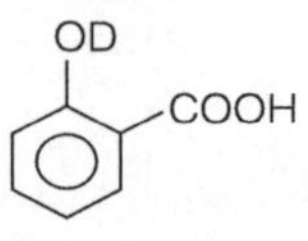 and $B =$ 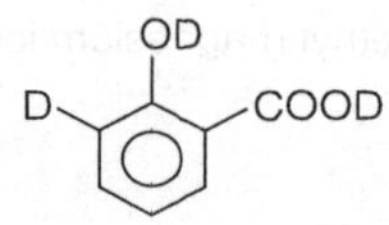

11 Compare the properties of two isomeric products *x* and *y* formed in the following reaction.

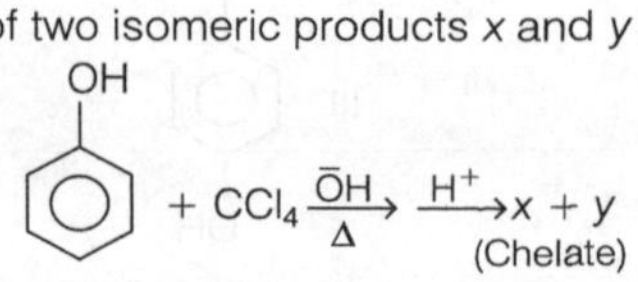

	Acidic strength	H_2O solubility	Volatility
a.	$y > x$	$y > x$	$x > y$
b.	$x > y$	$x > y$	$y > x$
c.	$y > x$	$x > y$	$y > x$
d.	$x > y$	$y > x$	$x > y$

One or More than One Options Correct Type

***Direction** (Q. Nos. 12-21) This section contains 10 multiple choice questions. Each question has four choices (a), (b), (c) and (d), out of which ONE or MORE THAN ONE are correct.*

12 Which of the following reaction will give bromobenzene as a major organic product?

a. 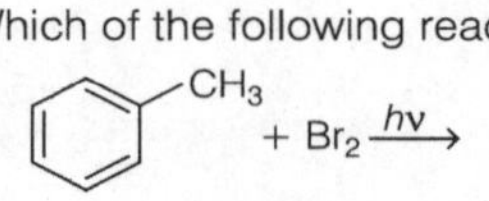

b. 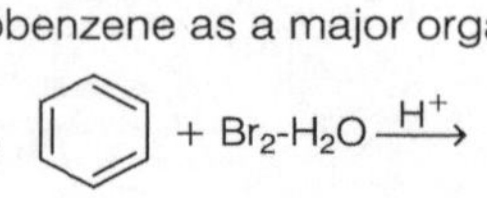

c. 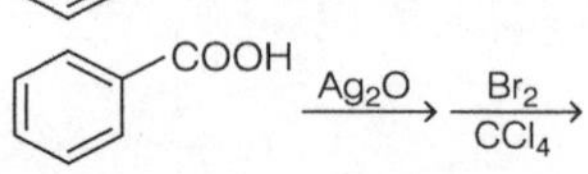

d. 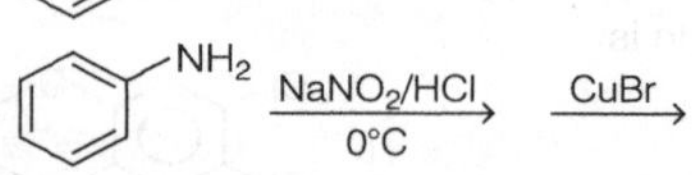

13 Which of the following statements is/are correct?

a. Phenyl vinyl ether will decolourise Br_2/CCl_4 solution but ethyl phenyl ether will not

b. 4-fluorophenol is a stronger acid than 4-methylphenol

c. 3-nitrophenol is a stronger acid than 4-nitrophenol

d. 4-chlorophenol is a stronger acid than 4-fluorophenol

14 In the following reaction,

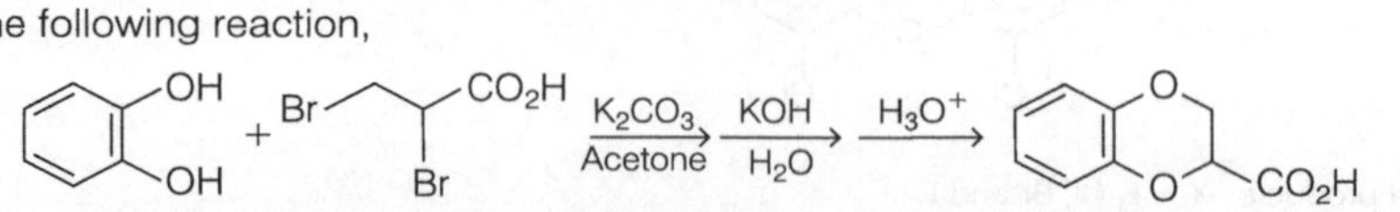

a. one oxyanion is formed at a time and displaces a bromide ion in an S_N2 reaction

b. both bromide react by S_N2 mechanism because one is primary and the other, though secondary, is next to a carbonyl group

c. primary bromide is displaced by S_N2 and the secondary bromide (next to CO_2H) is displaced by S_N1 mechanism

d. both bromide react by S_N1 mechanism

Working Space

15 Consider the following S_NAr reaction, 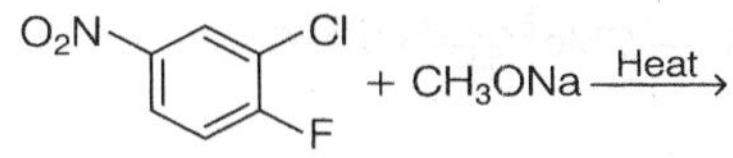

The correct statement is (are)

a. Cl will be substituted because it is a better leaving group

b. F will be substituted because nitro from *para* position facilitates this

c. After substitution of one halogen, substitution of the second halogen would become further more difficult

d. First chlorine and then fluorine would be substituted subsequently

16 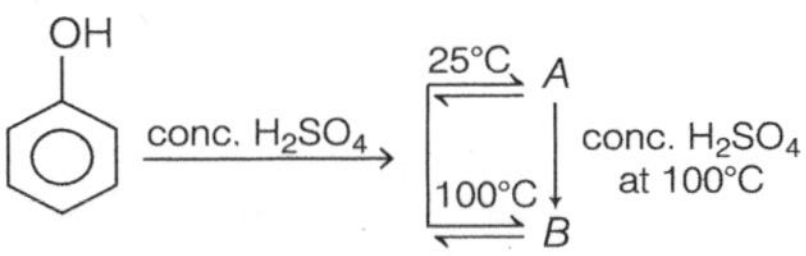

The compounds given are

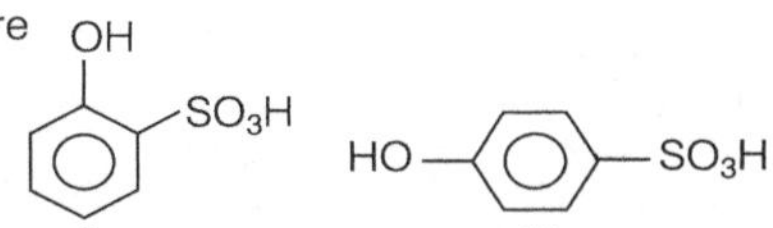

(I) (II)

Which of the following statements is/are correct?

a. Compounds (*A*) and (*B*) are (I) and (II) respectively

b. Compound (I) is the major product at lower temperature when the reaction is under equilibrium control

c. Compound (II) is the major product at high temperature, when the reaction is under rate control

d. Compound (II) is more stable

17 Consider the following two compounds undergoing aromatic nucleophilic substitution reaction.

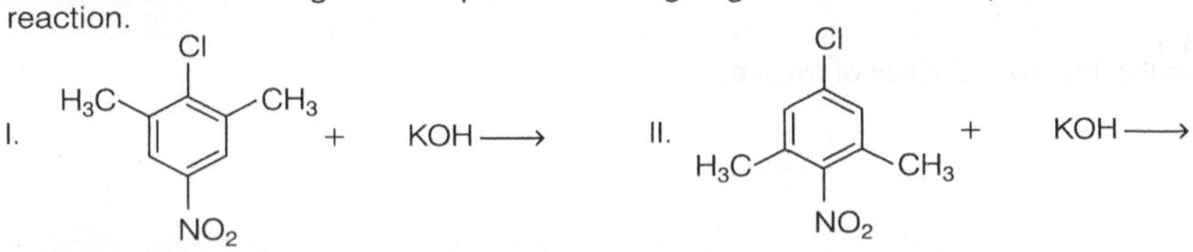

The correct statement(s) is/are

a. α-C to Cl in I is more electrophilic than the same in II

b. II occurs faster than I

c. reaction II occurs at slower rate than I due to steric inhibition to resonance of —NO_2

d. reaction I occurs at slower rate due to greater steric hindrance by two methyl groups at adjacent positions

18 Which of the following statement(s) is/are correct?

a. The nitration of phenol is faster than phenyl acetate

b. The reaction of $PhO^{\ominus}$ is faster than 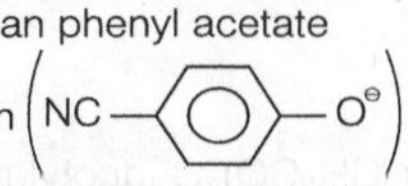 with $PhCH_2Cl$

c. Base catalysed hydrolysis of *p*-nitrophenyl acetate is faster than phenyl acetate

d. Acid catalysed esterification of PhOH is faster than *p*-nitrophenol with MeCOOH

19 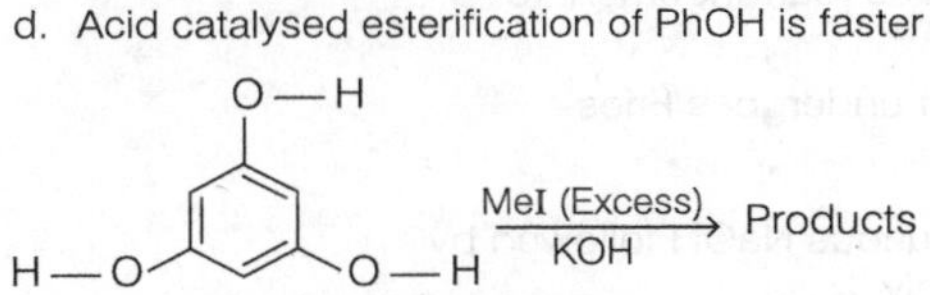

Possible products are

a. 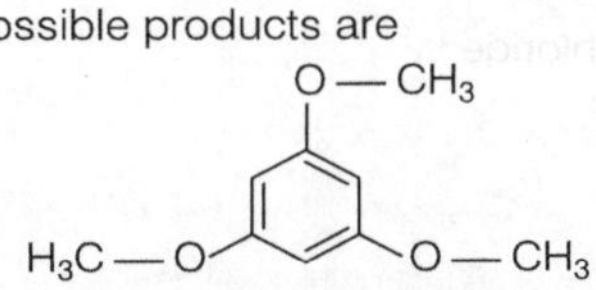

b. 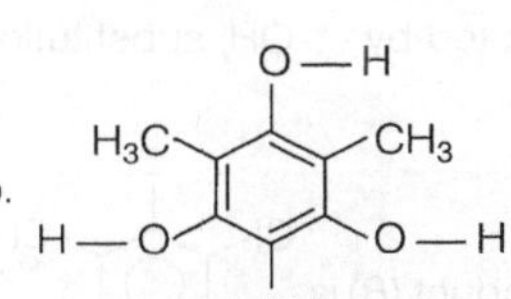

c. 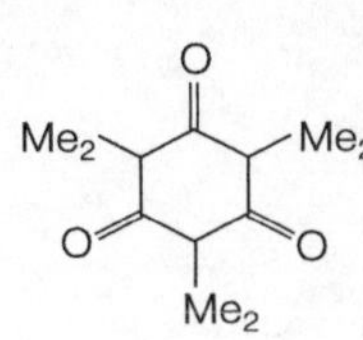

d. 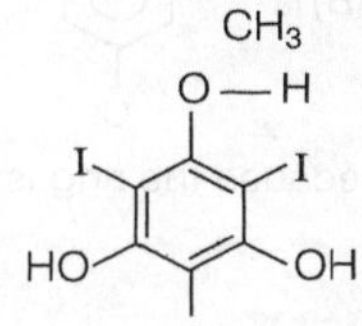

20 I. 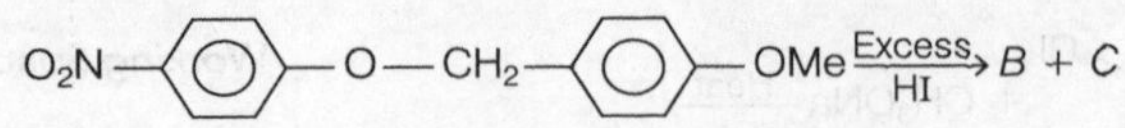

II. 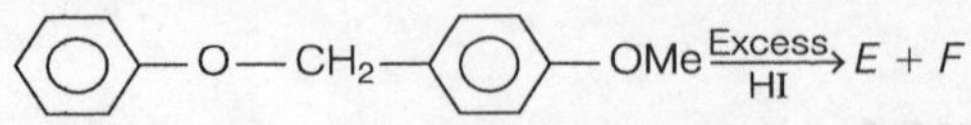

Which of the following statements is/are correct about the above reactions?

a. The compounds B and C respectively are

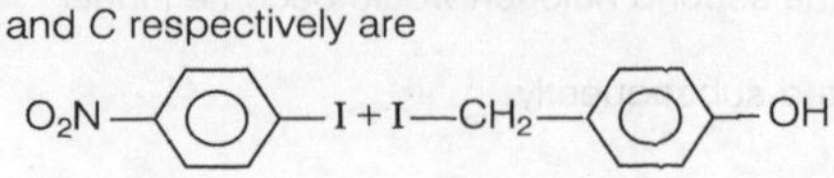

b. The compounds E and F respectively are

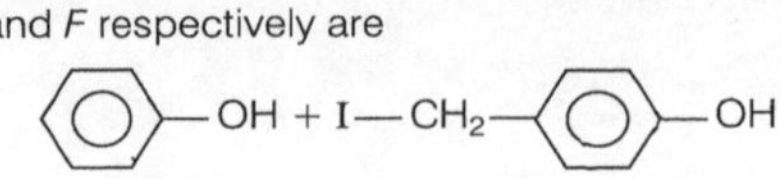

c. The compounds B and C respectively are

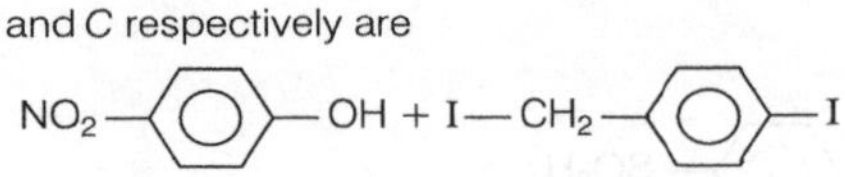

d. The compounds E and F respectively are

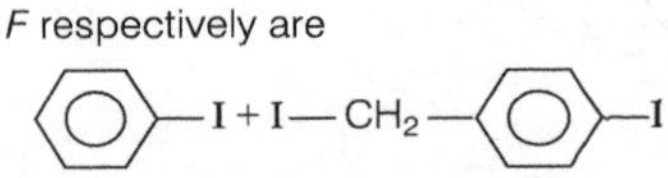

21 p-methoxy phenol + $\mathrm{Me_2C{=}CH_2} \xrightarrow{H^+} A$

p-methyl phenol + $\mathrm{Me_2C{=}CH_2} \xrightarrow{H^+} B$

Select the correct statement(s).

a. A and B are used to preserve the food by the action of oxygen
b. A is BHT and B is BHA
c. A is BHA and B is BHT
d. Both A and B are BHT

Statement Type

***Direction** (Q. Nos. 22-24) This section is based on Statement I and Statement II. Select the correct answer from the codes given below.*

a. Both Statement I and Statement II are correct and Statement II is the correct explanation of Statement I
b. Both Statement I and Statement II are correct but Statement II is not the correct explanation of Statement I
c. Statement I is correct but Statement II is incorrect
d. Statement II is correct but Statement I is incorrect

22 **Statement I** Phenol in aqueous alkali, when treated with $(CH_3CO)_2O$,acetylated at oxygen to give $C_6H_5OCOCH_3$ which on further reaction with $AlCl_3$ undergoes rearrangement to form *ortho* and *para* acylated phenol.

Statement II Phenolic ester on treatment with Lewis acid catalyst undergoes Fries rearrangement.

23 **Statement I** When 4-nitro-1, 2-dichlorobenzene is boiled with aqueous NaOH followed by hydrolysis of product, 2-chloro-4-nitrophenol were produced mainly.

Statement II Once a chloride is substituted by —OH, substitution of second chloride becomes very difficult.

24 **Statement I** 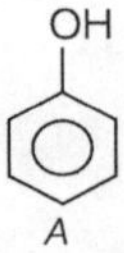 $\xrightarrow{3\,Cl_2/Fe}$ B. Here product (B) is 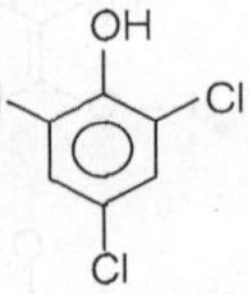

Statement II Phenol cannot be chlorinated because the ring is susceptible to oxidation by chlorine.

Comprehension Type

Working Space

Direction (Q. Nos. 25-29) *This section contains 2 paragraphs, each describing theory, experiments, data, etc. Five questions related to the paragraphs have been given. Each question has only one correct answer among the four given options (a), (b), (c) and (d).*

Passage I

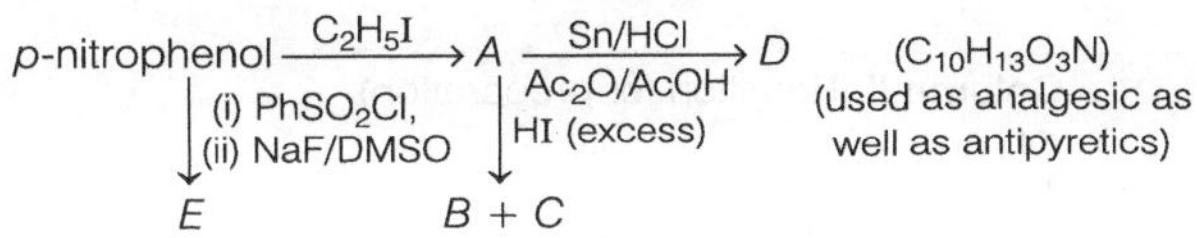

25 Compound *D* is

a. 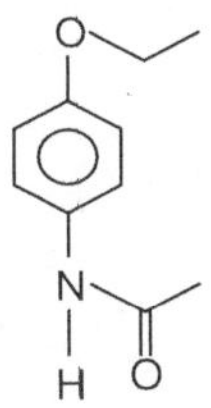

b. 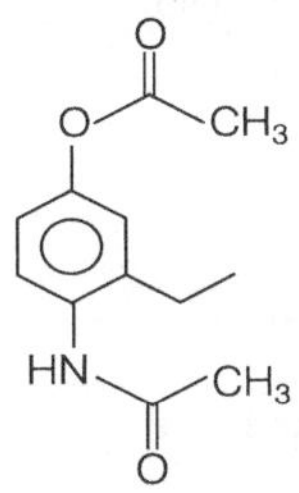

c. 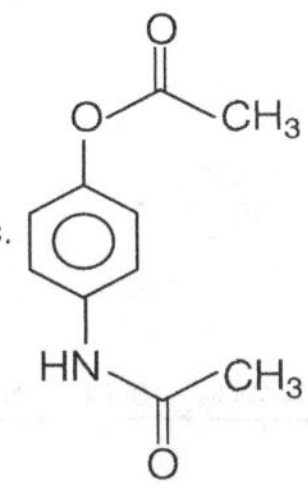

d. 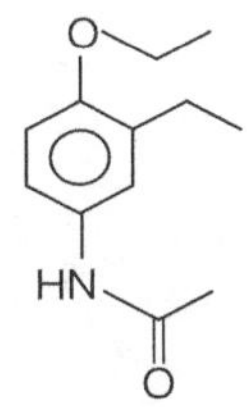

26 The compounds *B* and *C* respectively are

a. 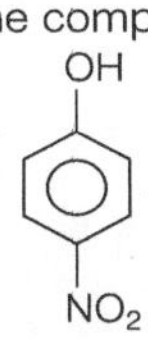 + C_2H_5I

b. 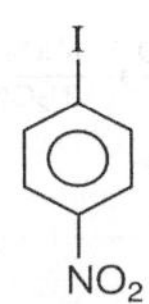 + ethanol

c.  + ethyl iodide

d. 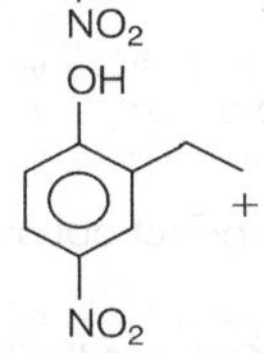 + ethyl iodide

27 The compound *E* has the following structure

a. 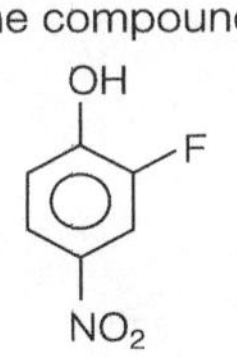

b. 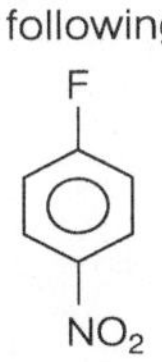

c. 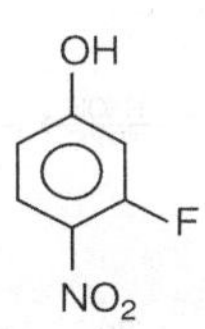

d. 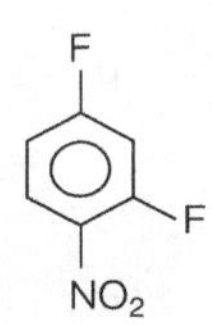

Passage II

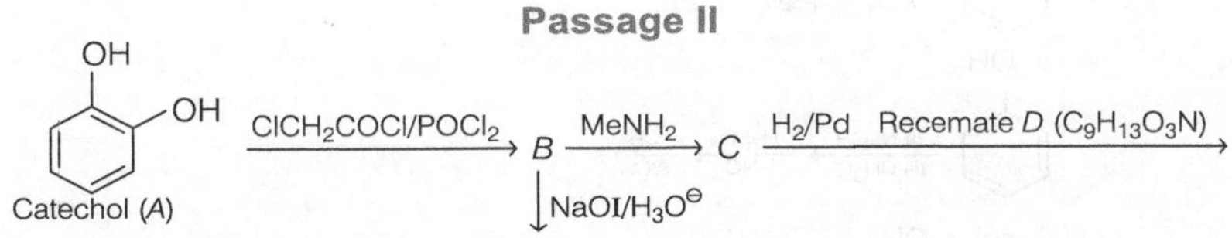

28 The incorrect statement about the compound '*B*' is

a. it is obtained from *A* by acylation of activated ring at *para* position to one of the —OH
b. it is obtained from *A* by esterification of one of the —OH groups
c. it is obtained from *A* by the alkylation in the activated ring at *para* position to one of the —OH groups
d. it is obtained from *A* by esterification of both the —OH groups

29 The end product of the above reaction sequence has the structure

a. 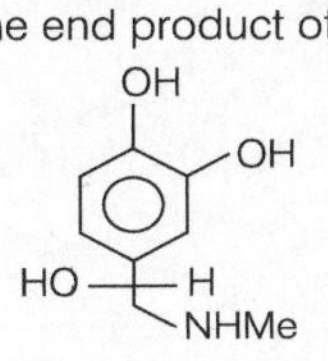

b. 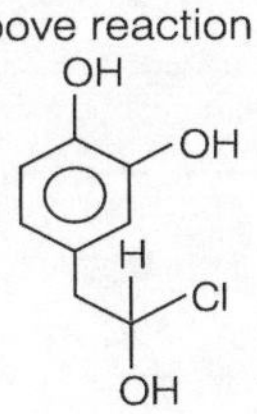

c. Both (a) and (b) d. None of these

Matching List Type

Working Space

***Direction** (Q. Nos. 30-32) Choices for the correct combination of elements from Column I and Column II are given as options (a), (b), (c) and (d), out of which one is correct.*

30 Match the Column I with Column II and mark the correct option from the codes given below.

	Column I (Product)		**Column II** (Reaction for preparation)
i.	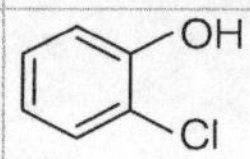	p.	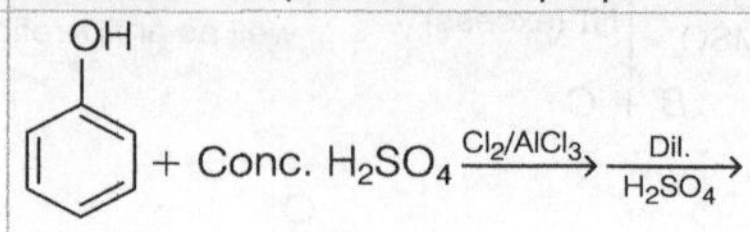
ii.	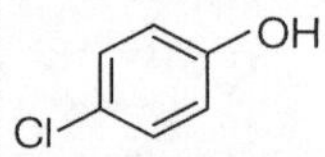	q.	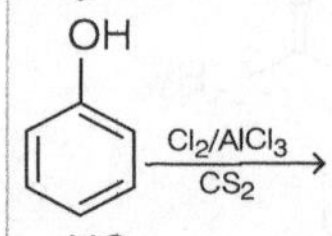
iii.	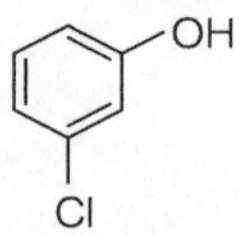	r.	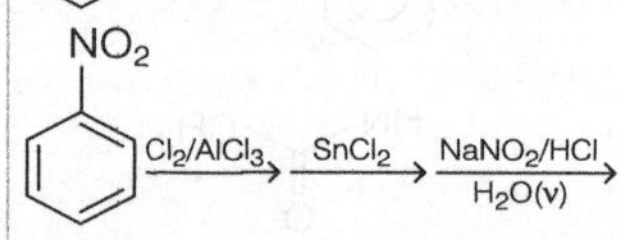
iv.	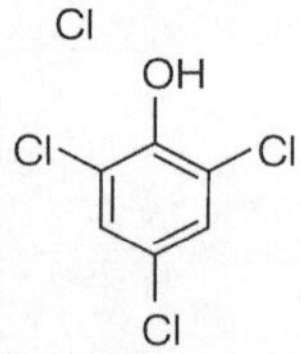	s.	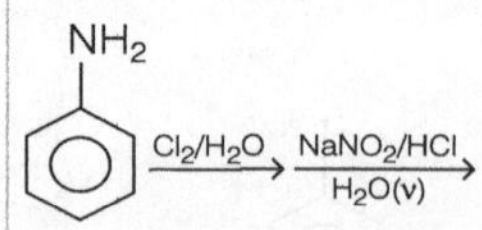

Codes

	i	ii	iii	iv		i	ii	iii	iv
a.	p	r	s	q	b.	p,q	q	r	s
c.	r	p	q	s	d.	s	p	q	r

31 Match the Column I with Column II and mark the correct option from the codes given below.

	Column I (Product)		**Column II** (Reactions)
i.	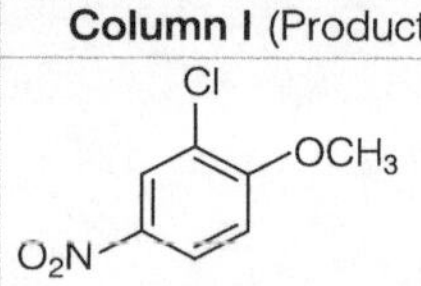	p.	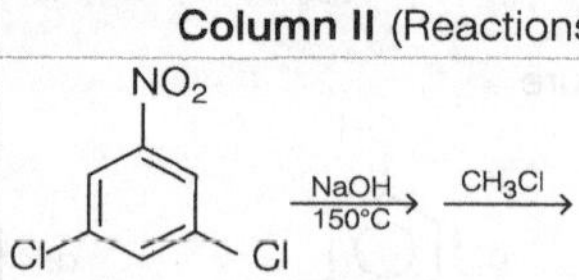
ii.	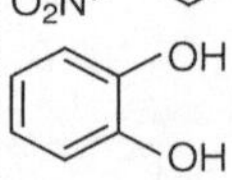	q.	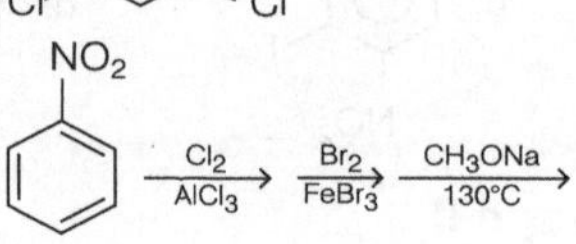
iii.	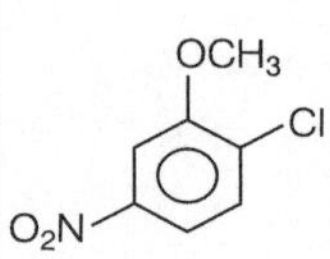	r.	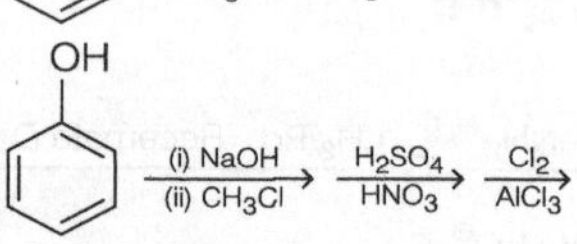
iv.	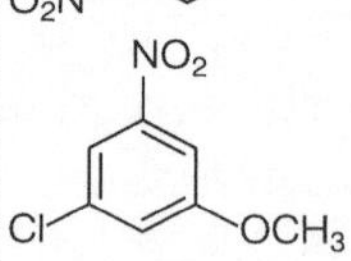	s.	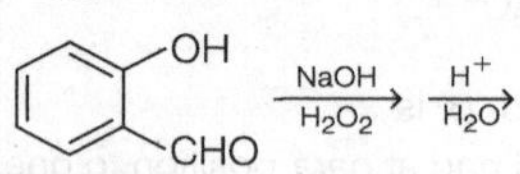
		t.	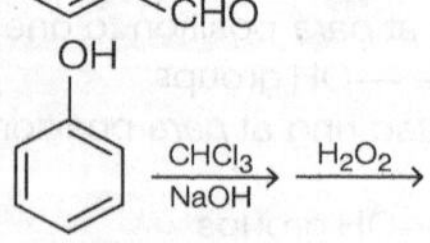
		u.	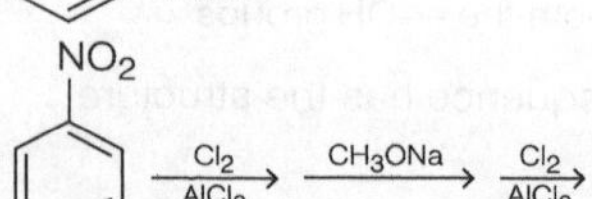

Codes

	i	ii	iii	iv		i	ii	iii	iv
a.	q,r	s,t	u	p	b.	p	q	r	s,t
c.	q	r	p	s,u	d.	r	q	s	p,t

32 Match the Column I with Column II and mark the correct option from the codes given below.

	Column I (Reactions)		Column II (Intermediate involved)
i.	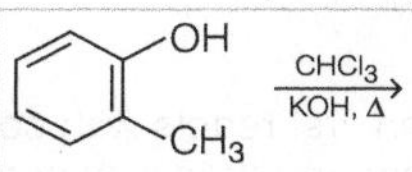	p.	Carbocation
ii.	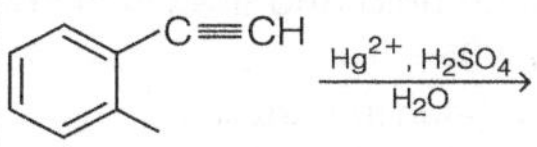	q.	Carbanion
iii.	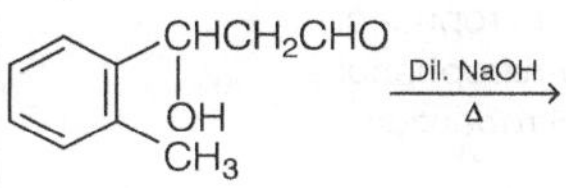	r.	Carbene
iv.	$(CH_3)_3C{-}CH_2OH \xrightarrow{\text{Conc.} H_2SO_4}$	s.	Enol

Codes

	i	ii	iii	iv
a.	q,r	p,s	q,r	p,s
b.	s	q	p	r
c.	q,r	p,s	q	p
d.	p,s	q,r	r,p	s

One Integer Value Correct Type

Direction *(Q. Nos. 33-35) This section contains 3 questions. When worked out will result in an integer from 0 to 9 (both inclusive).*

33 How many of the following show chelation?

(i) *o*-cresol (ii) Oil of wintergreen (iii) *o*-iodophenol (iv) *o*-fluorophenol (v) *o*-cyanophenol (vi) Salol

34 How many products are formed during this reaction?

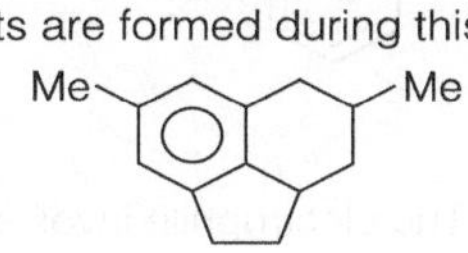

$\xrightarrow{NBS + h\nu}$ Products

35 Acetone + Phenol (Excess) $\xrightarrow{HCl} A\,(C_{15}H_{16}O_2)$

$A \xrightarrow{Na}$ Effervescence

$\xrightarrow[\text{(ii) } CH_3\,Cl]{\text{(i) NaOH}} B\,(C_{17}H_{20}O_2) \xrightarrow{NO}$ No reaction

If *A* is treated with bromine water, how many bromine atoms would be incorporated into *A*?

JEE Main & AIEEE Archive

(Compilation of Last 13 Years' Questions)

1 Arrange the following compounds in the order of decreasing acidity : **(JEE Main 2013)**

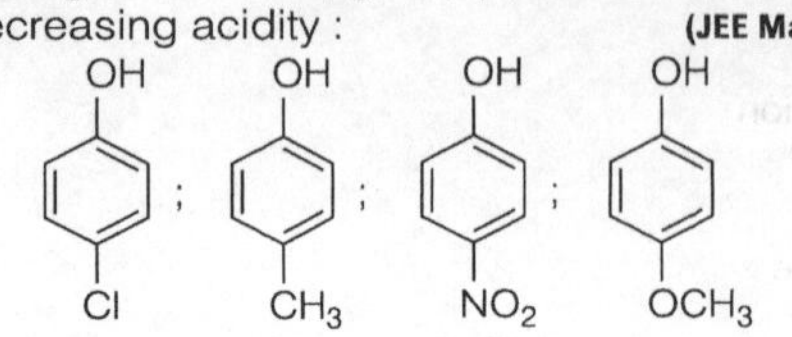

I II III IV

a. II > IV > I > III
b. I > II > III > IV
c. III > I > II > IV
d. IV > III > I > II

2 Aspirin is known as **(AIEEE 2012)**

a. acetyl salicylic acid
b. phenyl salicylate
c. acetyl salicylate
d. methyl salicylic acid

3 *Ortho*-nitrophenol is less soluble in water than *p*- and *m*-nitrophenols because

a. *o*-nitrophenol is more steam volatile than those of *m*- and *p*-isomers
b. *o*-nitrophenol shows intramolecular H-bonding
c. *o*-nitrophenol shows intermolecular H-bonding
d. melting point of *o*-nitrophenol is lower than those of *m*- and *p*-isomers **(AIEEE 2012)**

4 Phenol is heated with a solution of mixture of KBr and $KBrO_3$. The major product obtained in the above reaction is **(AIEEE 2011)**

a. 2-bromophenol
b. 3-bromophenol
c. 4-bromophenol
d. 2, 4, 6-tribromophenol

5 Which of the following reagents may be used to distinguish between phenol and benzoic acid? **(AIEEE 2011)**

a. Aqueous NaOH b. Tollen's reagent
c. Molisch reagent d. Neutral $FeCl_3$

6 The correct order of acidic strength of the following compounds is **(AIEEE 2011)**

I. Phenol II. *p*-cresol
III. *m*-nitrophenol IV. *p*-nitrophenol

a. III > II > I > IV b. IV > III > I > II
c. II > IV > I > III d. I > II > IV > III

7 The major product obtained on interaction of phenol with sodium hydroxide and carbon dioxide is **(AIEEE 2009)**

a. benzoic acid b. salicylaldehyde
c. salicylic acid d. phthalic acid

8 Phenol, when first reacts with concentrated sulphuric acid and then with concentrated nitric acid gives **(AIEEE 2008)**

a. 2, 4, 6-trinitrobenzene
b. *o*-nitrophenol
c. *p*-nitrophenol
d. nitrobenzene

9 Phenyl magnesium bromide reacts with methanol to give **(AIEEE 2006)**

a. a mixture of anisole and Mg(OH)Br
b. a mixture of benzene and Mg(OMe)Br
c. a mixture of toluene and Mg(OH)Br
d. a mixture of phenol and Mg(Me)Br

10 The structure of the compound that gives a tribromo derivative on treatment with bromine water is **(AIEEE 2006)**

a. 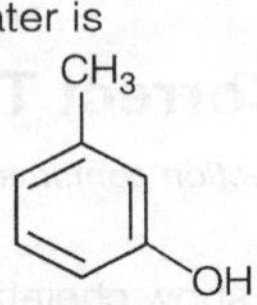

b. 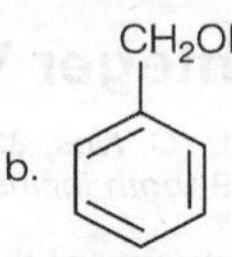

c. 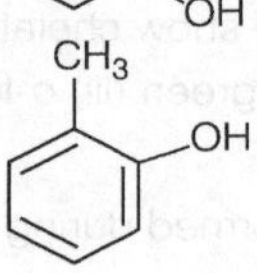

d. 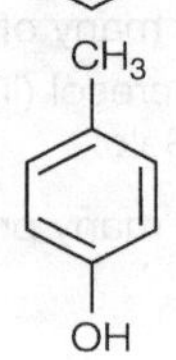

11 The electrophile involved in the above reaction is **(AIEEE 2006)**

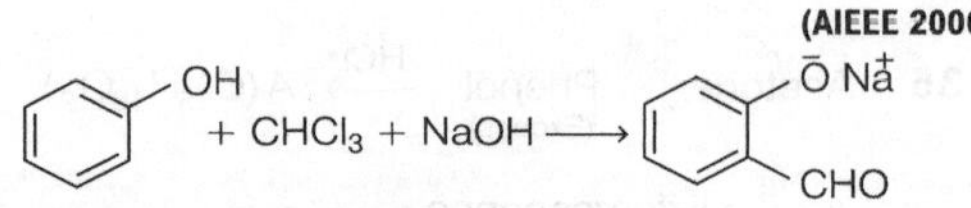

a. dichloromethyl cation ($\overset{\oplus}{C}HCl_2$)
b. dichloro carbene ($:CCl_2$)
c. trichloromethyl anion ($\overset{\ominus}{C}Cl_3$)
d. formyl cation ($\overset{\oplus}{C}HO$)

12 Picric acid is **(AIEEE 2002)**

a. 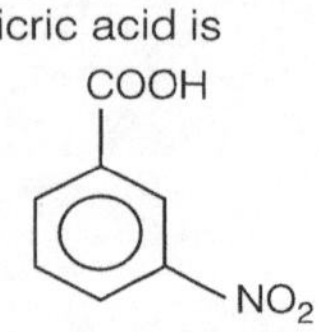

b. 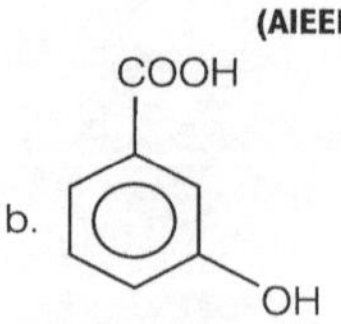

c. 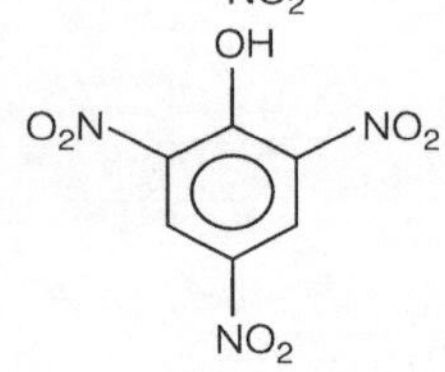

d. 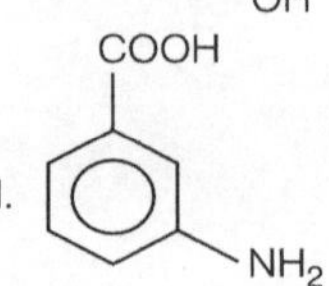

JEE Advanced & IIT JEE Archive

(Compilation of Last 10 Years' Questions)

1 For the identification of β-naphthol using dye test, it is necessary to use

(2014 Adv. Only One Option Correct Type)

a. dichloromethane solution of β-naphthol
b. acidic solution of β-naphthol
c. neutral solution of β -naphthol
d. alkaline solution of β-naphthol

2 In the following reaction, the product(s) formed is (are)

(2013 Adv. One or More than One Options Correct Type)

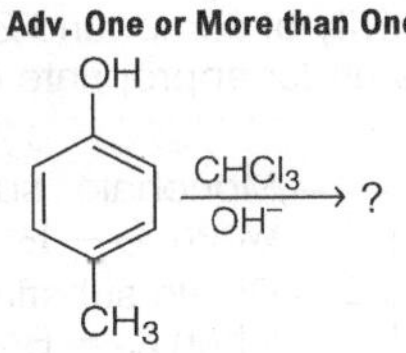

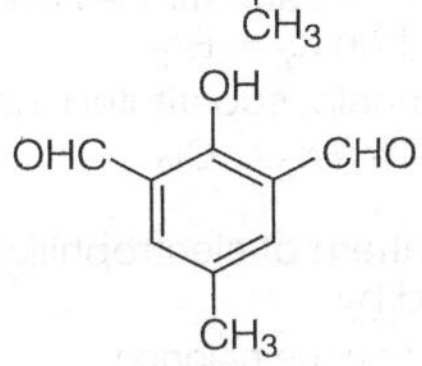

P

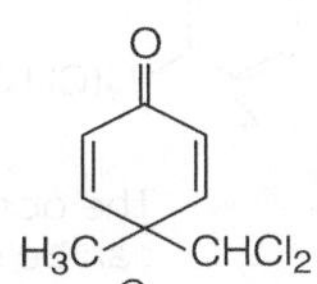

Q

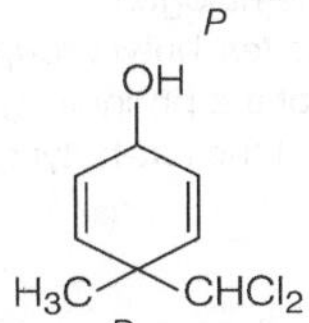

R

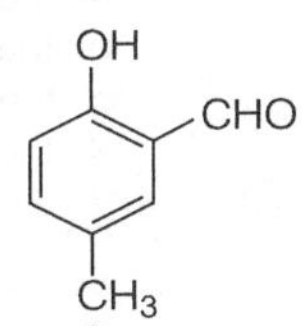

S

a. *P* (major)   b. *Q* (minor)
c. *R* (minor)   d. *S* (major)

3 The major product(s) of the following reaction is (are) **(2013 Adv. One or More than One Options Correct Type)**

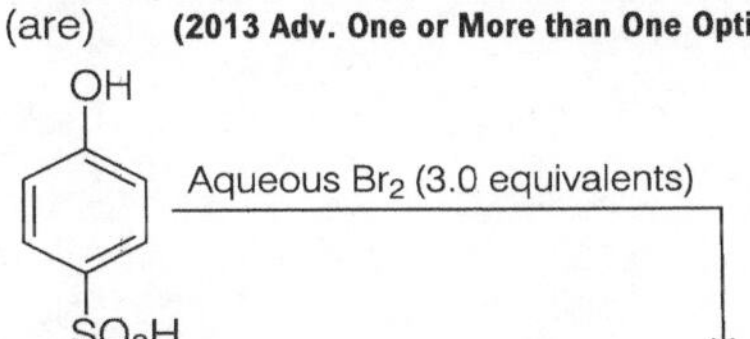

P

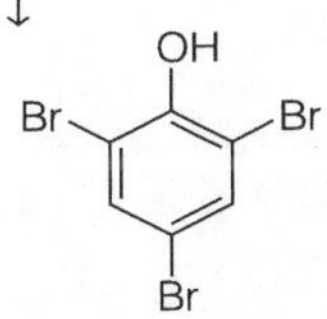

Q

R

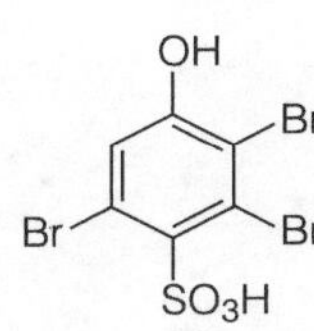

S

a. *P*   b. *Q*   c. *R*   d. *S*

4 In the reaction, 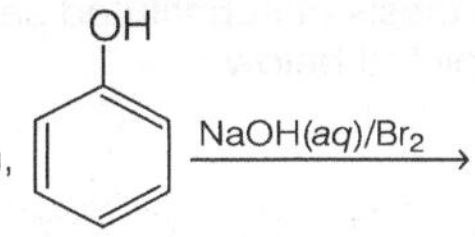

(2010, One or More than One Options Correct Type)

the intermediate(s) is (are)

a. 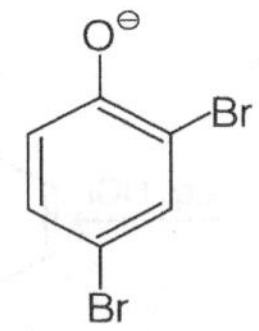

b. 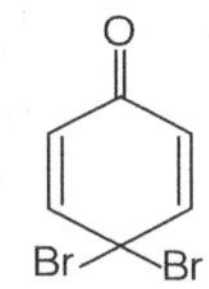

c. 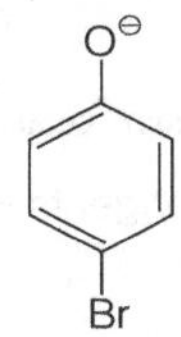

d. 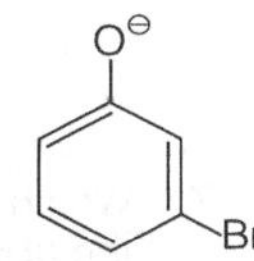

5 The major product of the following reaction is

(2008,Only One Option Correct Type)

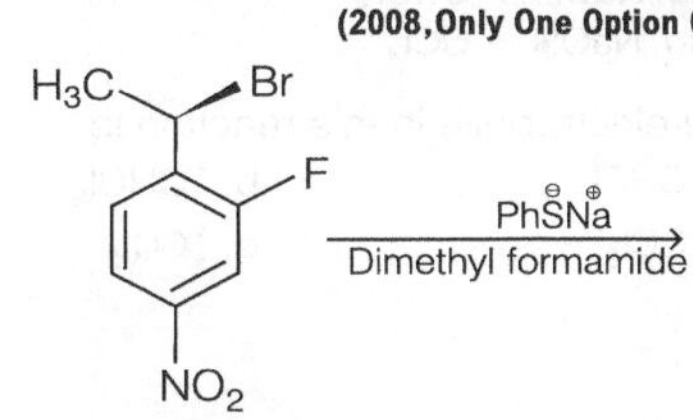

a. 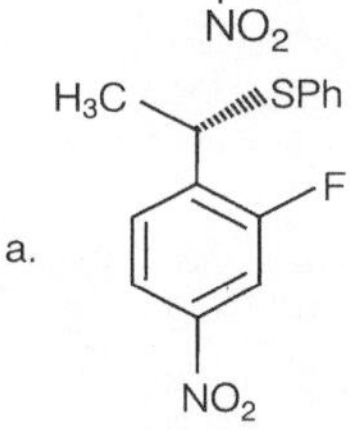

b. 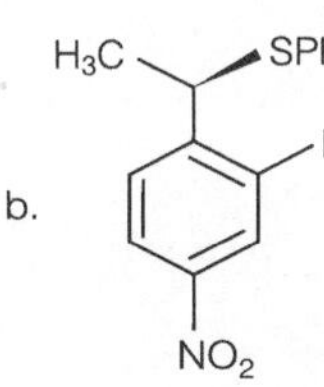

c. 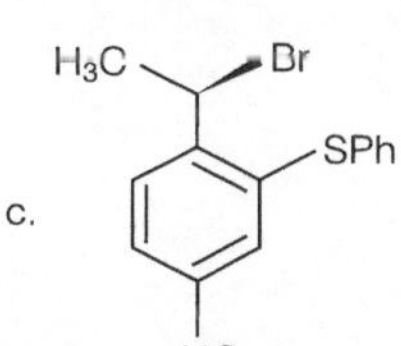

d. 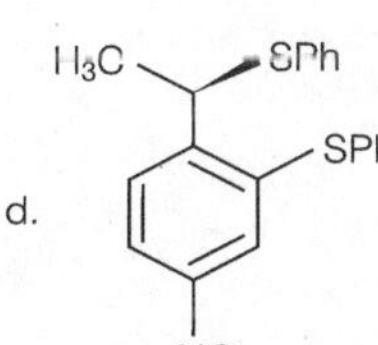

6 **Statement I** Bromobenzene, upon reaction with Br_2 / Fe gives 1,4-dibromobenzene as the major product.

Statement II In bromobenzene, the inductive effect of the bromo group is more dominant than the mesomeric effect in directing the incoming electrophile. **(2008, Statement Type)**

Passage 1

Reimer-Tiemann reaction introduces an aldehyde group, on to the aromatic ring of phenol, *ortho* to the hydroxyl group. This reaction involves electrophilic aromatic substitution. This is a general method for the synthesis of substituted salicylaldehydes as depicted below.

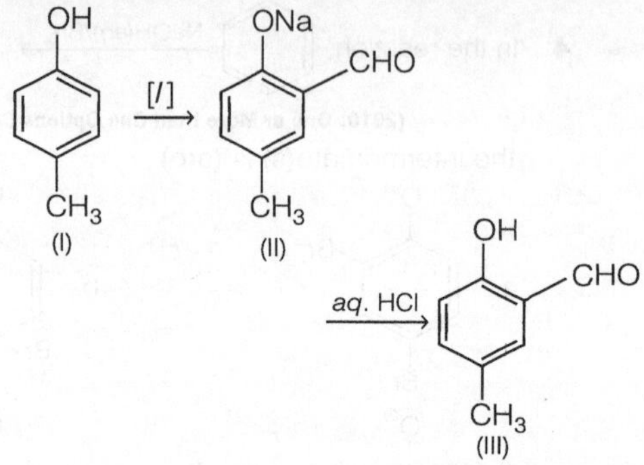

(2007, Comprehension Type)

7 Which one of the following reagents is used in the above reaction?

a. *aq.* NaOH + CH_3Cl
b. *aq.* NaOH + CH_2Cl_2
c. *aq.* NaOH + $CHCl_3$
d. *aq.* NaOH + CCl_4

8 The electrophile in this reaction is

a. $:CHCl$ b. $^{+}CHCl_2$
c. $:CCl_2$ d. $:CCl_3$

9 The structure of the intermediate *I* is

a. 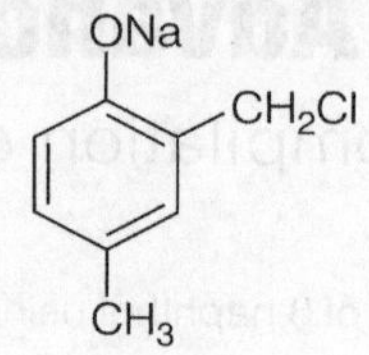
b. 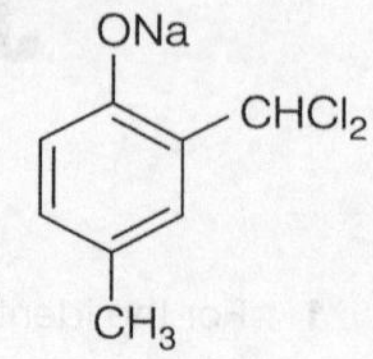
c. 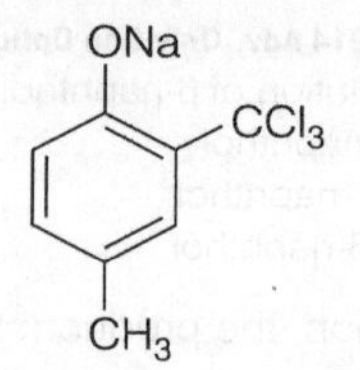
d. 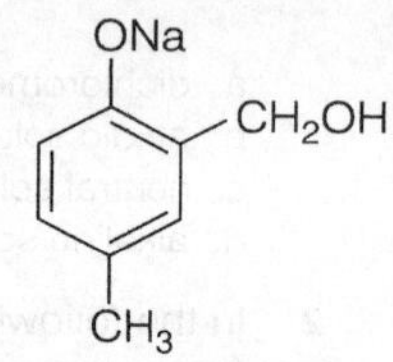

10 The reactivity of compound *Z* with different halogens under appropriate conditions is given below

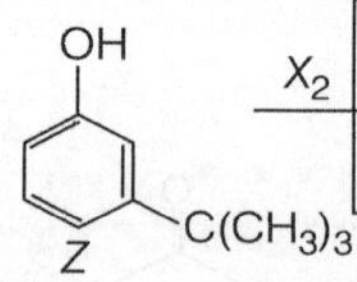

X_2 →
- Monohalo substituted derivative when $X_2 = I_2$
- Dihalo substituted derivative when $X_2 = Br_2$
- Trihalo substituted derivative when $X_2 = Cl_2$

The observed pattern of electrophilic substitution can be explained by

a. the steric effect on the halogen
b. the steric effect of the *tert*-butyl group
c. the electronic effect of the phenolic group
d. the electronic effect of the *tert*-butyl group

2. Aromatic Aldehydes, Ketones and Acids

DPP

DAILY PRACTICE PROBLEMS

DPP-1 Methods of Preparation of Aromatic Carbonyl Compounds

Subjective Questions

Working Space

Direction *(Q. No. 1) This question is subjective in nature, need to be solved completely on notebook.*

1 i. Identify the missing species.

a. 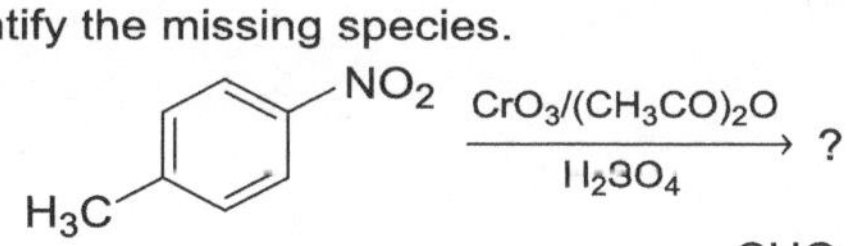

b. 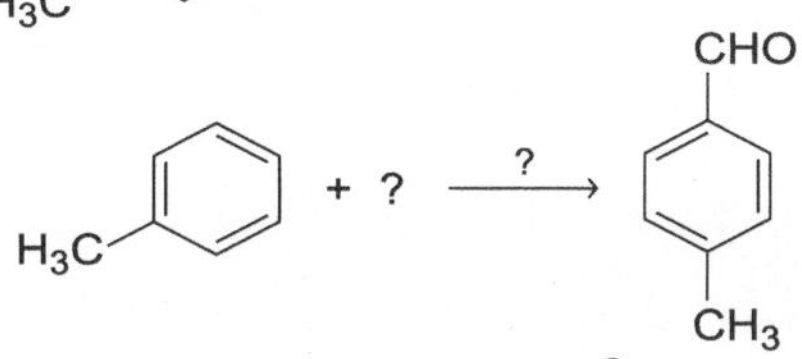

c. 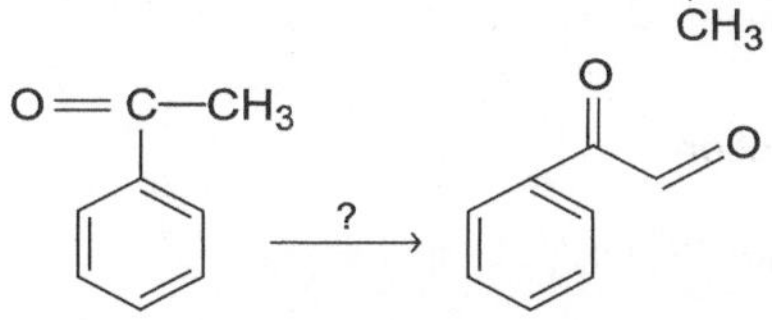

d. 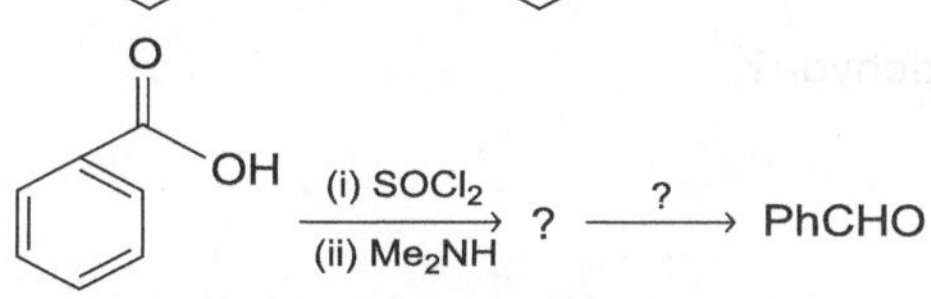

ii. Complete the following.

a. 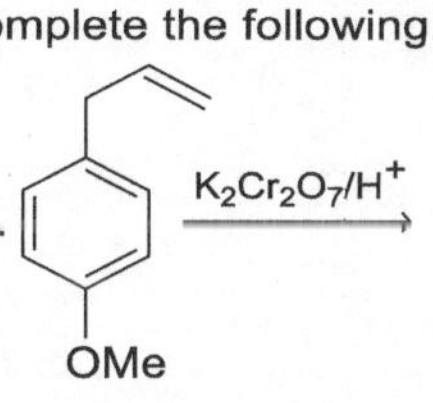

b. 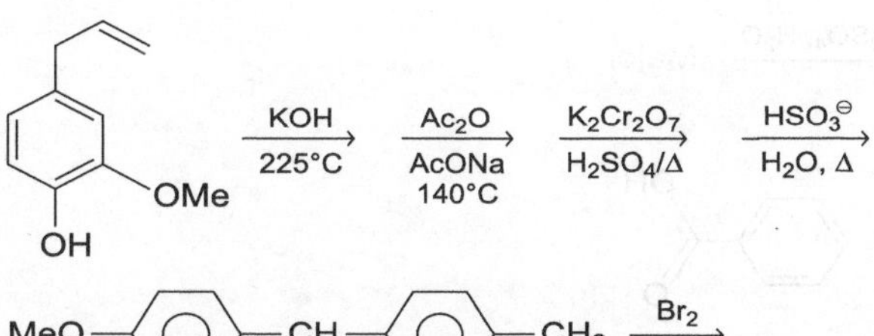

c. 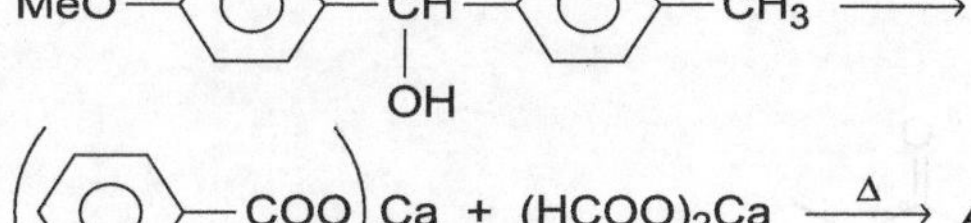

d. 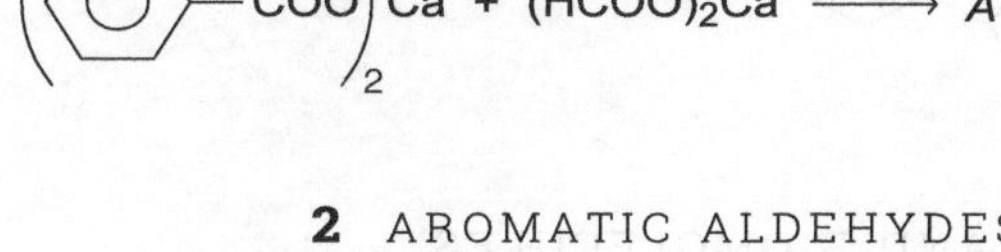

Working Space

iii. Starting from benzene, how will you synthesise the following compound *A*?

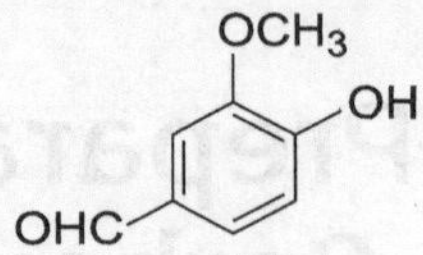

A

Only One Option Correct Type

Direction *(Q. Nos. 2-12) This section contains 11 multiple choice questions. Each question has four choices (a), (b), (c) and (d), out of which ONLY ONE is correct.*

2 The product *A* in the following reaction sequence is

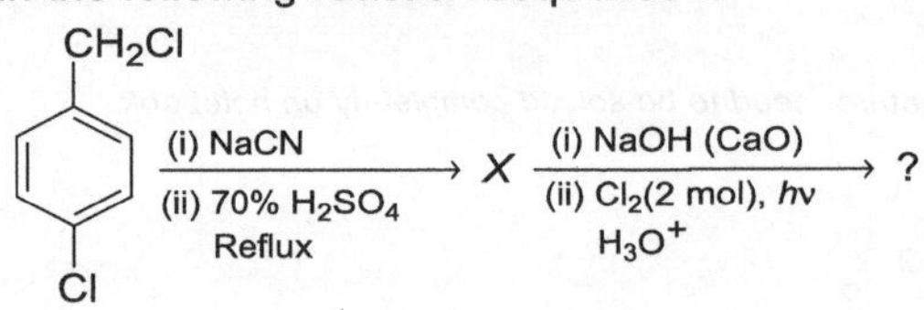

a. 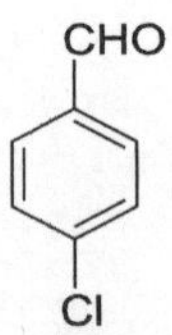

b. 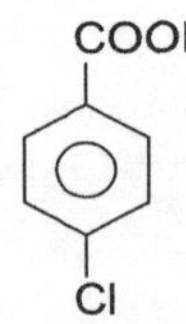

c. 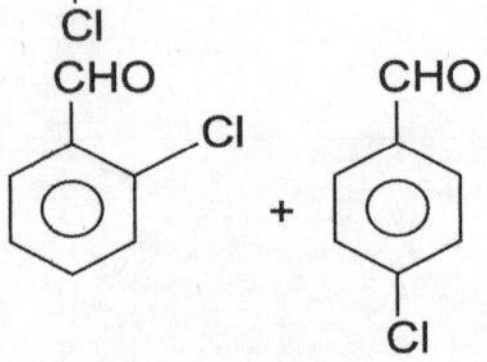

d. *X* gives no reaction

3 Which of the following reaction would yield benzaldehyde?

I. $C_6H_5CH_2Cl \xrightarrow[H_2O,\ \Delta]{\overline{O}H}$

II. $C_6H_5CH(OCH_3)_2 \xrightarrow{H_3O^+}$

III. $C_6H_5COOH \xrightarrow[(ii)\ H_2O]{(i)\ LiAlH_4,\text{ether}}$

a. I and II
b. Only II
c. II and III
d. All of these

4 Give the major product of the following reaction

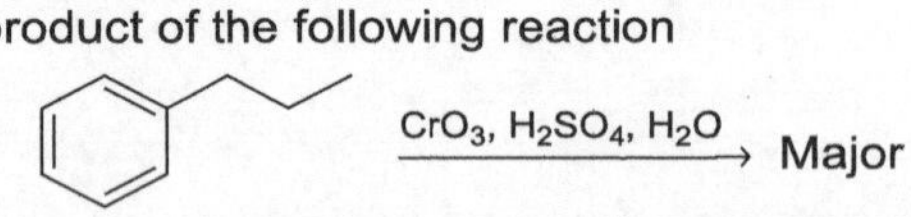

a. 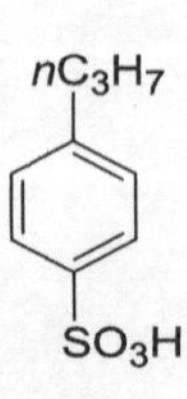

b. 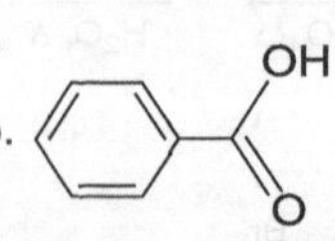

c. 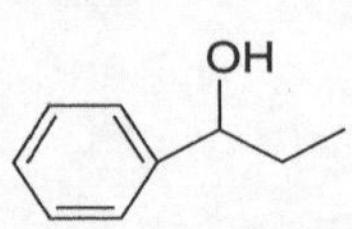

d. 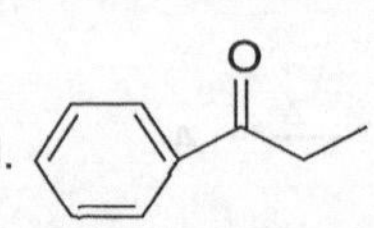

Working Space

5 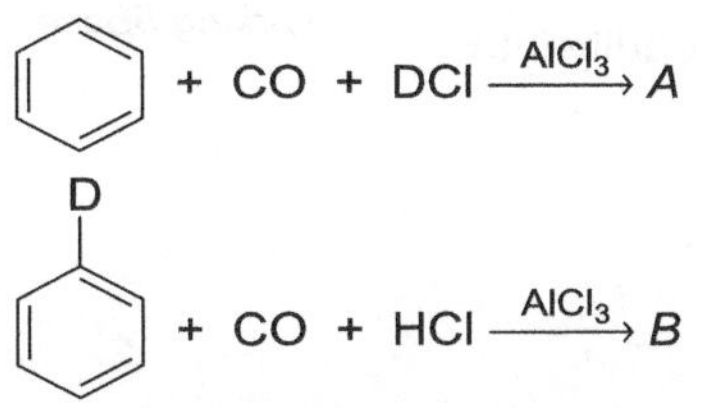

The true statement about *A* and *B* is

a. *A* = 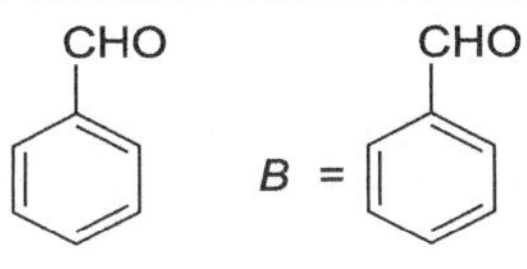

b. *A* = 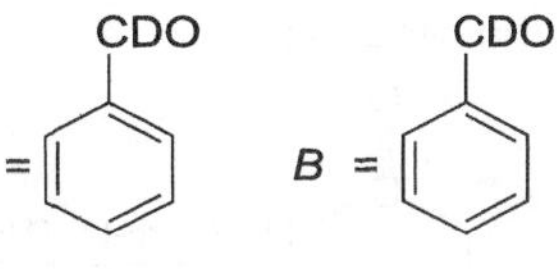

c. *A* = 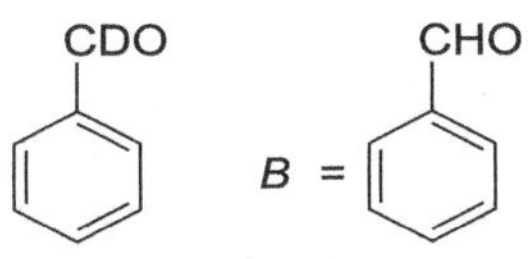

d. *A* = 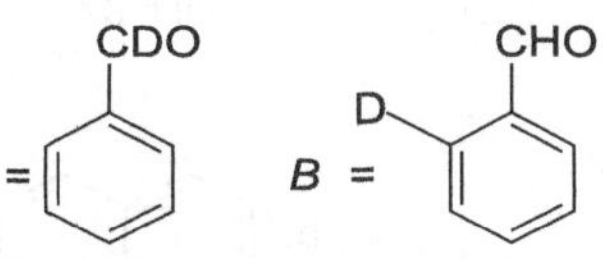

6 Which of the following is not a synthesis of benzophenone $C_6H_5COC_6H_5$?

a. $C_6H_6 + C_6H_5COCl \xrightarrow{AlCl_3}$

b. $(C_6H_5)_2CHOH \xrightarrow[\text{Acetone}]{H_2CrO_4}$

c. $(C_6H_5)_2C{=}CH_2 \xrightarrow[\text{(ii) Zn, HOAc}]{\text{(i) } O_3}$

d. $C_6H_5COOH + 2C_6H_5Li \xrightarrow[\text{(ii) } H_2O]{\text{(i) Ether}}$

7 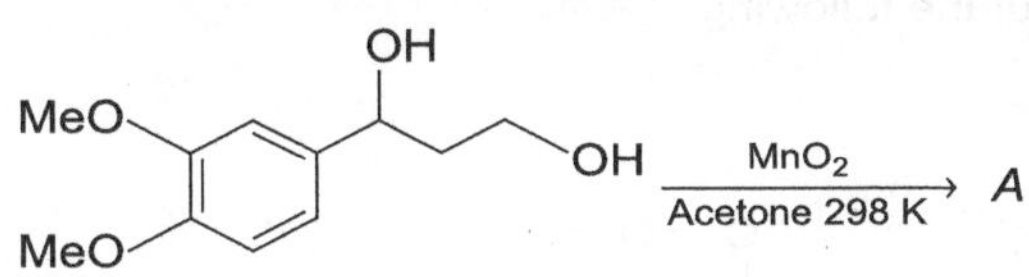

A is

a. 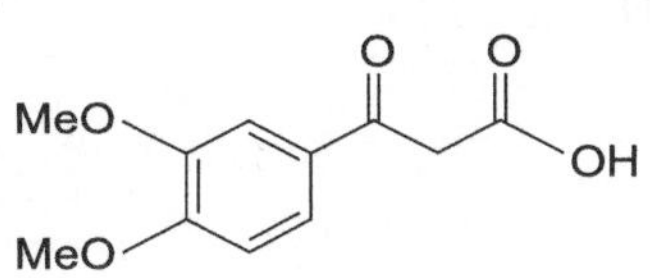

b. 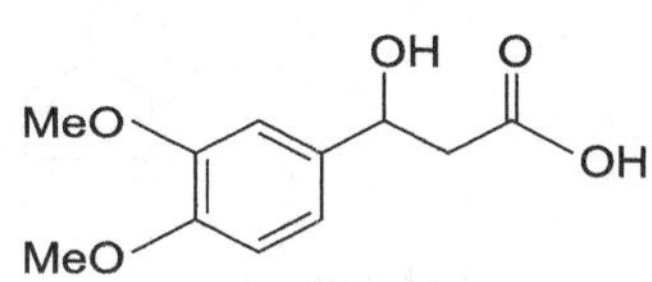

c. 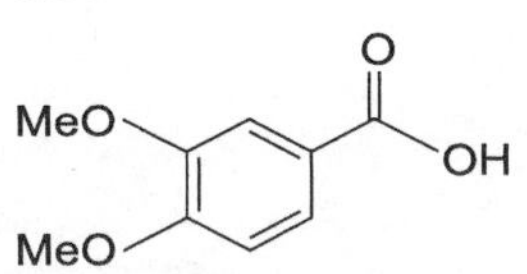

d. 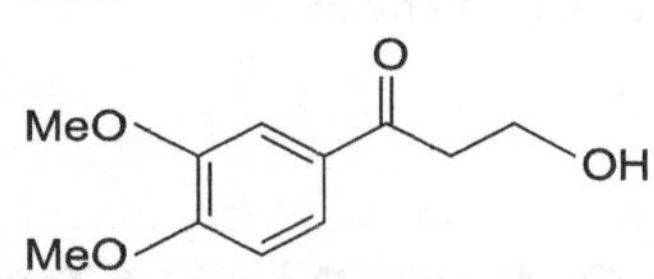

8 Product obtained when benzoyl acetic acid is heated, is

a. 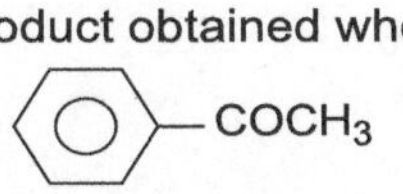

b. 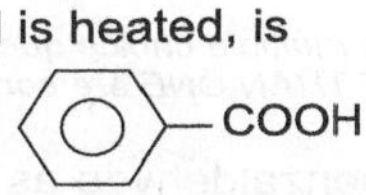

b. 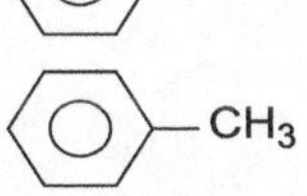

d. 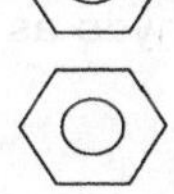

9 What two different combination of starting materials and reagents can be used to prepare benzaldehyde?

I. Benzyl alcohol	A. $LiAlH[OC(CH_3)_3]_3$
II. Benzoyl chloride	B. PCC in CH_2Cl_2
III. Styrene	C. B_2H_6 then H_2O_2 and NaOH
IV. Phenyl acetylene	D. $HgSO_4$ in aqueous H_2SO_4

a. I, A and II, B
b. III, C and IV, D
c. I, B and II, A
d. IV, C and II, A

Working Space

10 Although CrO_3 is a good oxidising agent and oxidises —CHO group readily into —COOH but toluene when treated with $CrO_3/(CH_3CO)_2O, H^+$, gives benzaldehyde, not benzoic acid. This is because

a. of the formation of benzylidene acetate which prevent further oxidation
b. of the deactivation of ring after the insertion of —CHO group
c. presence of $(CH_3CO)_2O$ retard the activity of CrO_3
d. Both (a) and (c)

11 What is the major product of the reaction?

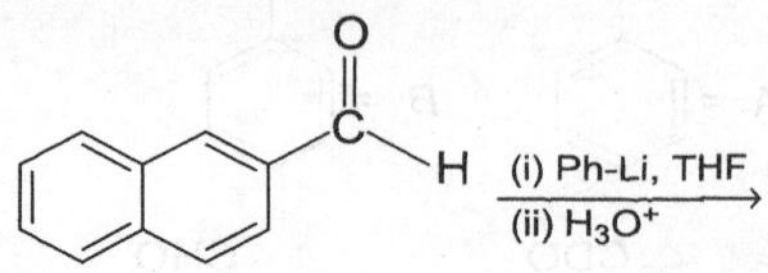

a. 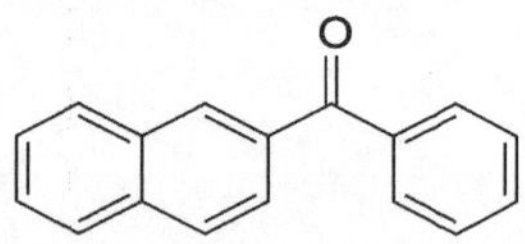

b. 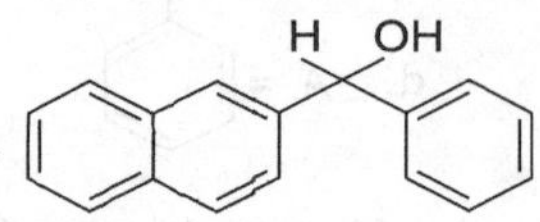

c. 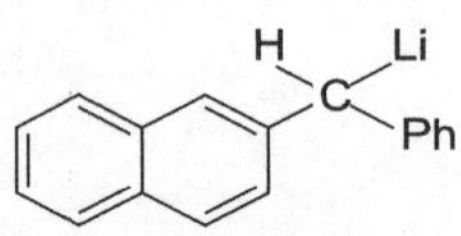

d. None of these

12 Which of the following reagents can be used to carry out the following transformation?

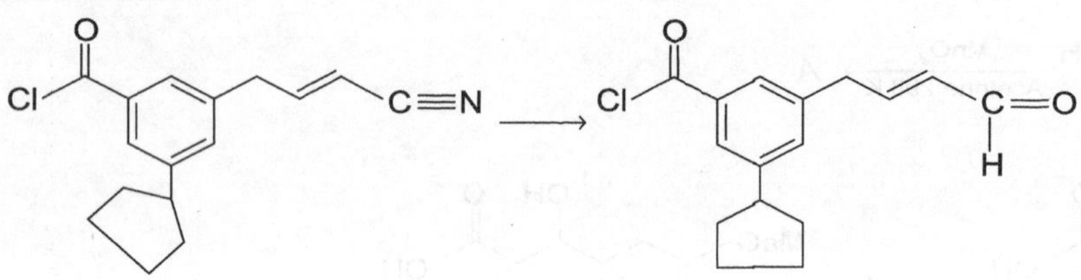

a. DIBAL-H
b. LBAH (lithium tri-*t*-butoxy aluminium hydride)
c. $SnCl_2/HCl$
d. All of the above

One or More than One Options Correct Type

Direction *(Q. Nos. 13 and 14) This section contains 2 multiple choice questions. Each question has four choices (a), (b), (c) and (d), out of which ONE or MORE THAN ONE are correct.*

13 Which of the following reaction will give benzaldehyde as one of the product?

a. 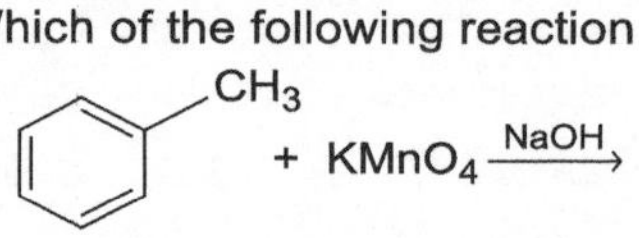

b. 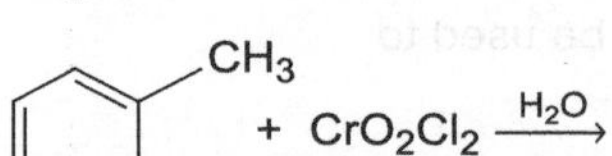

c. $(C_6H_5CO)_2O \xrightarrow{\Delta}$

d. 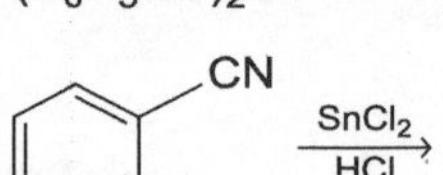

14 $C_6H_5C{\equiv}H \xrightarrow[\text{(ii) } H_2O_2,\ \overline{O}H,\ H_2O]{\text{(i) } B_2H_6} A$

What set of reagent(s) is required to convert *A* into benzaldehyde?

a. PCl_5, NaOH

b. $KMnO_4/OH^-$; $H^+/SOCl_2$; $Pd/BaSO_4$

c. $[Ag(NH_3)_2]^+$; NaOH (CaO); $Cl_2/h\nu$, H_2O

d. $K_2Cr_2O_7/H_2SO_4$; $Pd/BaSO_4$

Working Space

Comprehension Type

Direction *(Q. Nos. 15 and 16) This section contains a paragraph, describing theory, experiments, data, etc. Two questions related to the paragraph have been given. Each question has only one correct answer among the four given options (a), (b), (c) and (d).*

Passage

Consider the following reactions to answer the following two questions.

$$A(C_{12}H_{16}) \xrightarrow[\text{Careful ozonoylsis}]{O_3/Zn\text{-}H_2O} CH_2O + B$$

$$A \xrightarrow[\text{(ii) } H_3O^+]{\text{(i) } KMnO_4/\text{Heat}} \text{Benzoic acid}$$

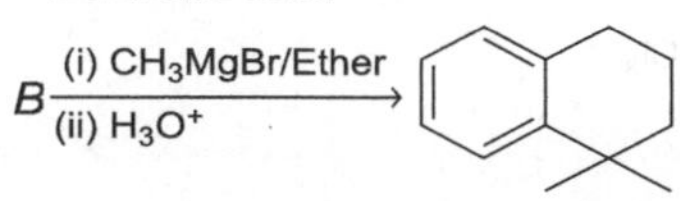

15 The structure of *B* is

a. 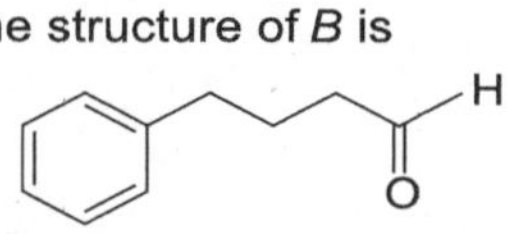

b. 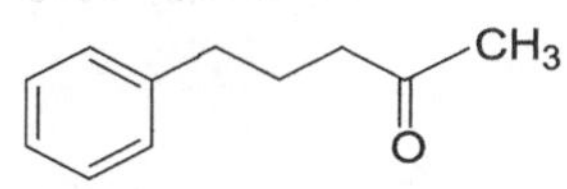

c. 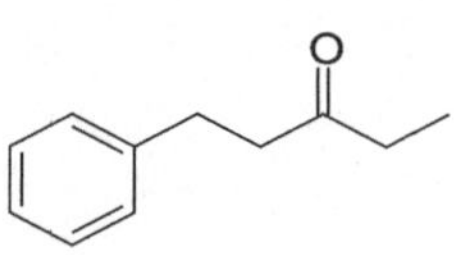

d. 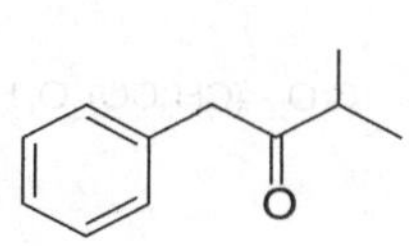

16 What can be said correctly regarding the final step of reaction?

a. The last step is nucleophilic addition of Grignard reagent

b. The last step is an intramolecular Friedel-Craft's alkylation reaction

c. The final step involves nucleophilic attack at aromatic ring

d. None of the above is correct

Statement Type

Direction *(Q. Nos. 17 and 18) This section is based on Statement I and Statement II. Select the correct answer from the codes given below.*

a. Both Statement I and Statement II are correct and Statement II is the correct explanation of Statement I

b. Both Statement I and Statement II are correct but Statement II is not the correct explanation of Statement I

c. Statement I is correct but Statement II is incorrect

d. Statement II is correct but Statement I is incorrect

17 **Statement I** Nitrobenzene is also a solvent for Gattermann's Koch synthesis of benzaldehyde from benzene.

Statement II Nitrobenzaldehyde is also formed as a minor product during Gattermann Koch synthesis if solvent is nitrobenzene.

Working Space

18 **Statement I** When toluene is treated with manganese dioxide in the presence of sulphuric acid, it results in the formation of benzaldehyde.

Statement II The above reaction becomes faster if benzyl alcohol is taken instead of toluene.

Matching List Type

Direction (Q. No. 19) *Choices for the correct combination of elements from Column I and Column II are given as options (a), (b), (c) and (d), out of which one is correct.*

19 Match the column I and Column II and choose the correct option using the codes given below.

	Column I (Substrate in which —CHO group has to be introduced)		Column II (Reagent which introduced a —CHO group)
i.	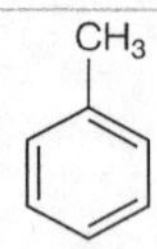	p.	Trichloromethane/$\overline{O}H/\Delta$
ii.		q.	$CO/HCl/AlCl_3$
iii.	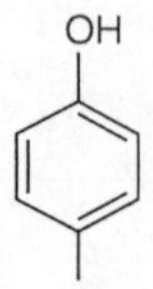	r.	$Cl_2/\Delta, H_2O, 100°C$
iv.	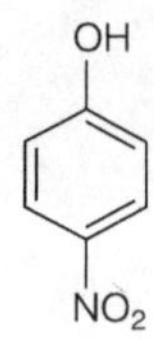	s.	$CrO_3, (CH_3CO)_2O, H_3O^+$
		t.	$HCN, HCl/H_2O, AlCl_3$

Codes

	i	ii	iii	iv
a.	r,s	r,t	p,r,s	s
b.	p,q	s	t	r
c.	r,s	q,t	p,r,s	p
d.	s	t	r	p,q,s

One Integer Value Correct Type

Direction (Q. No. 20) *This section contains 1 question. When worked out will result in an integer from 0 to 9 (both inclusive).*

20 How many products are formed by heating barium benzoate along with barium formate?

DPP-2 Properties of Aromatic Aldehydes

Subjective Questions

Working Space

Direction (Q. Nos. 1-3) *These questions are subjective in nature, need to be solved completely on notebook.*

1 i. What product is expected by the reduction of PhCHO with

a. Tollen's reagent b. Zn/HCl

c. LAH d. Wolff Kishner reduction

e. Two moles of PhCHO with Zn/HCl f. Clemmensen's reduction

ii. Complete the following reactions

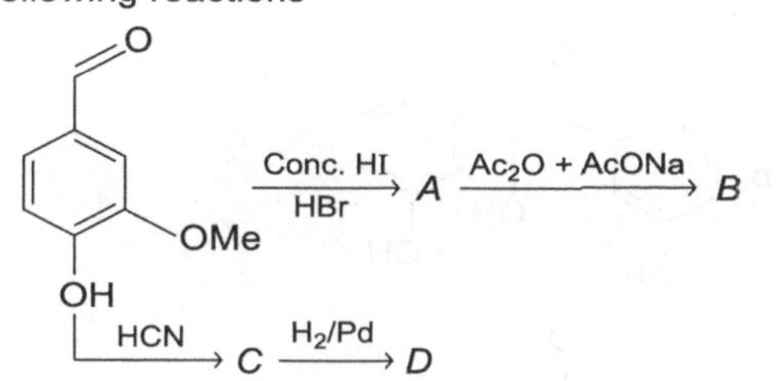

iii. Convert salicylaldehyde into

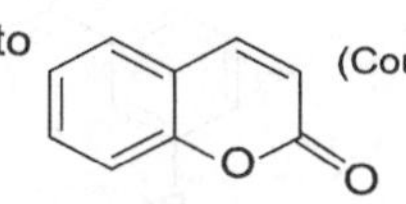

(Coumarin)

2 Complete the following reactions.

i.

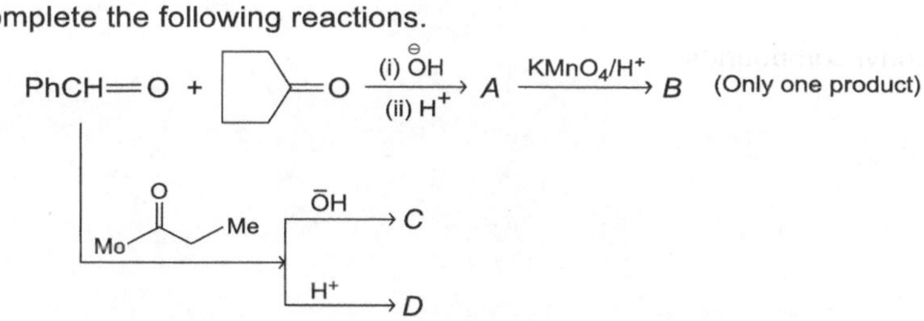

ii.

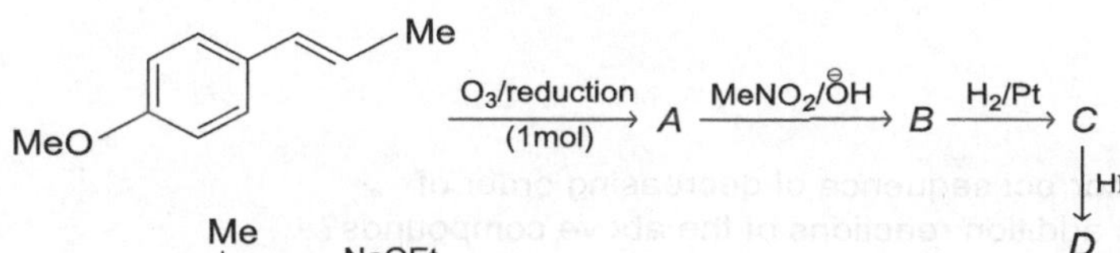

iii.

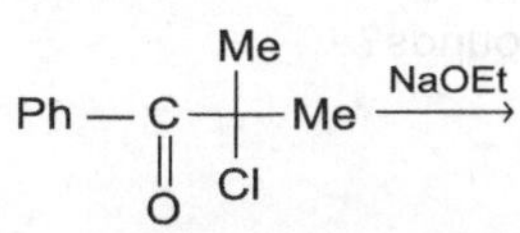

3 An organic compound *A* contains 79.25%C, 5.66% H and 15.09% O. Its vapour density is 53 and it reduces Tollen's reagent. Action of NaOH on *A* produces *B* and *C*. Controlled oxidation of *B* gives *A*, which on further oxidation produces *C*. *C* on distillation with PCl_5 gives *D*. Reaction of *D* with *aq.* ammonia forms *E*, which gives *F* with NaOBr. *F* gives *G* when treated with nitrous acid and hypophosphorus acid. Identify the compounds *A* to *G*.

Only One Option Correct Type

Working Space

Direction *(Q. Nos. 4-9) This section contains 6 multiple choice questions. Each question has four choices (a), (b), (c) and (d), out of which ONLY ONE is correct.*

4 What new compound will be formed when gaseous HCl is added to a solution of benzaldehyde in CH_3OH?

a. $C_6H_5CH(OCH_3)_2$
b. $C_6H_5CHCl(OCH_3)$
c. $C_6H_5COOCH_3$
d. C_6H_5COCl

5 In the reaction given below, the product would be

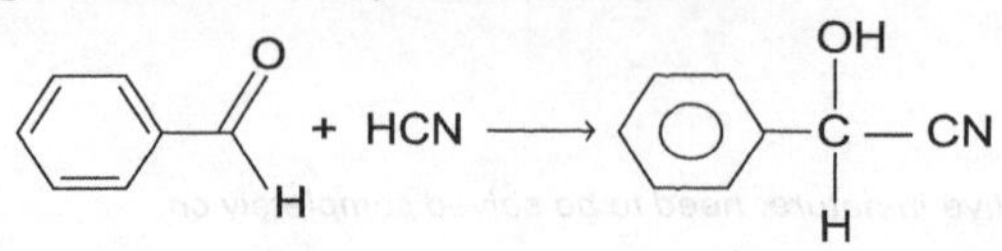

a. a racemate
b. optically active
c. a *meso* compound
d. a mixture of the diastereomers

6 When *m*-bromobenzaldehyde is treated with 50% alkali, the products obtained is (are)

a. 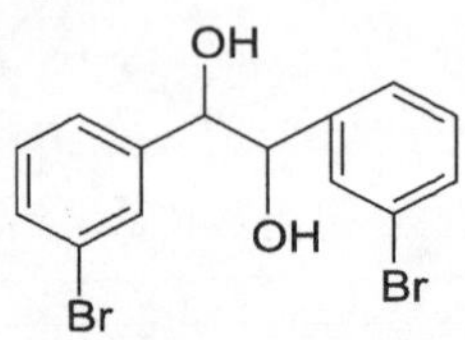

b. 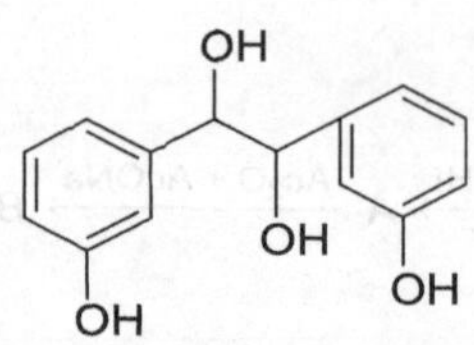

c. 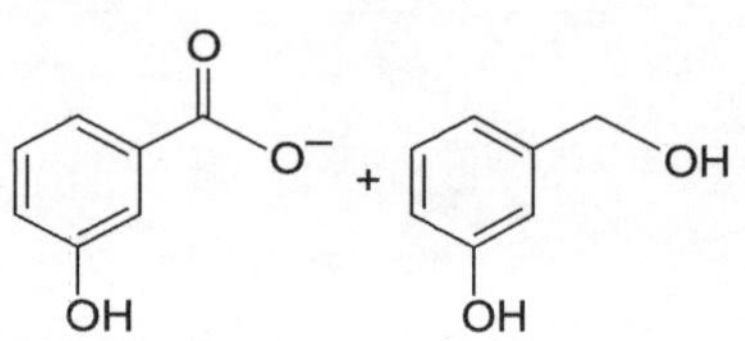

d. 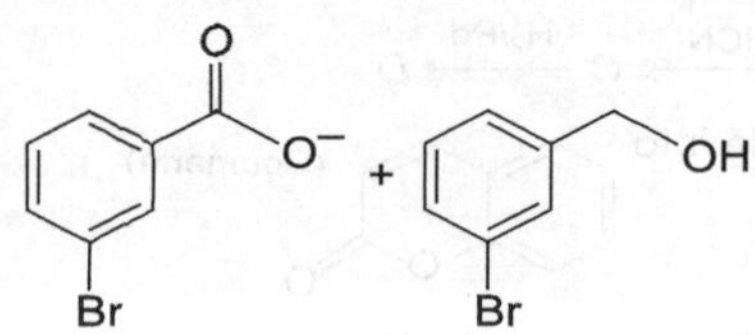

7 Consider the following four carbonyl compounds

I. 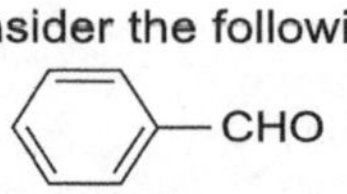

II. 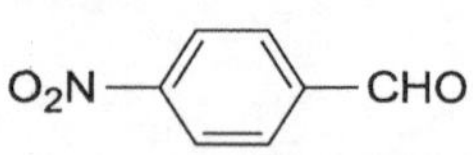

III. 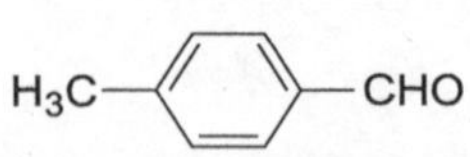

IV. $NO_2—CH_2—CHO$

Which one of the following is the correct sequence of decreasing order of reactivity towards the nucleophilic addition reactions of the above compounds?

a. III > I > II > IV
b. IV > II > I > III
c. II > IV > I > III
d. IV > III > II > I

8 One mole of HCHO and 2 moles of PhCHO react with conc. NaOH.What are the products quantitatively?

a. 1 mole of HCOONa, 1 mole of $PhCH_2OH$ and 1 mole of PhCOONa
b. 1 mole of HCOONa, 1 mole of $PhCH_2OH$ and 0.5 mole of PhCOONa
c. 1 mole of HCOONa, 1.5 mole of $PhCH_2OH$ and 0.5 mole of PhCOONa
d. 1 mole of HCOONa, 2 moles of $PhCH_2OH$ and 2 moles of PhCOONa

Working Space

9 The product expected from the following reaction is/are **[IIT JEE 2003]**

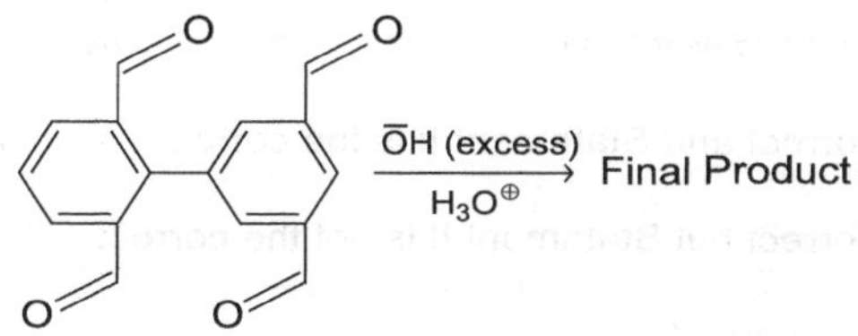

a. 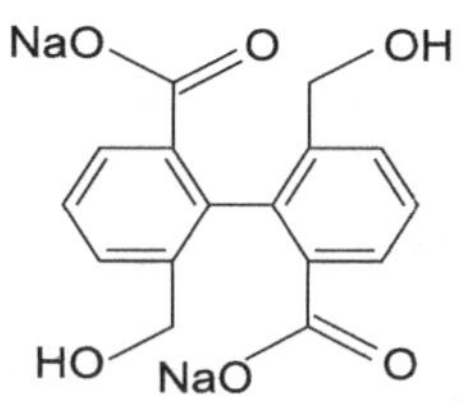

b. 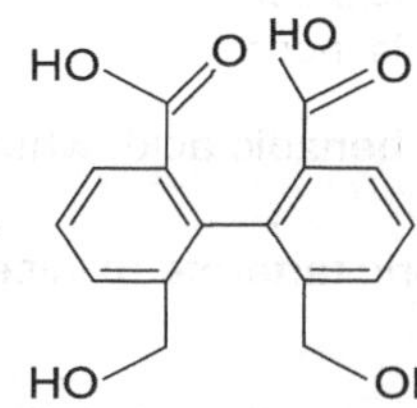

c. 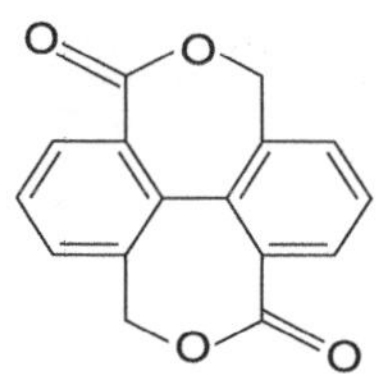

d. 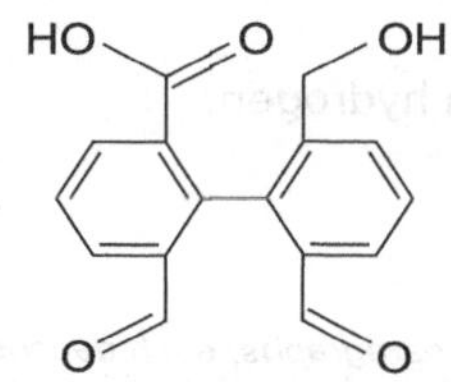

One or More than One Options Correct Type

Direction *(Q. Nos. 10 and 11) This section contains 2 multiple choice questions. Each question has four choices (a), (b), (c) and (d), out of which ONE or MORE THAN ONE are correct.*

10 Which is/are not true regarding benzaldehyde?

a. It reduces Tollen's reagent
b. It reduces Fehling's solution
c. It forms disulphite with $NaHSO_3$
d. It forms imine ($C_6H_5CH{=}NH$), as stable product, with ammonia

11 In which of the following reactions, reactants and products are correctly matched?

a. 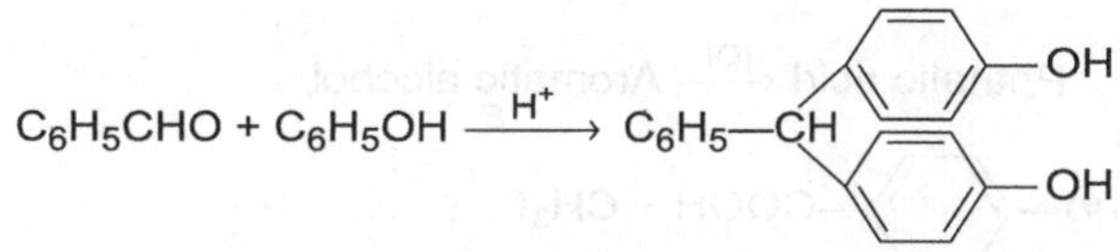

b. $C_6H_5-\underset{\underset{O}{\|}}{C}-CH_3 + Zn \xrightarrow{HCl} C_6H_5-CH_2CH_3$

c. 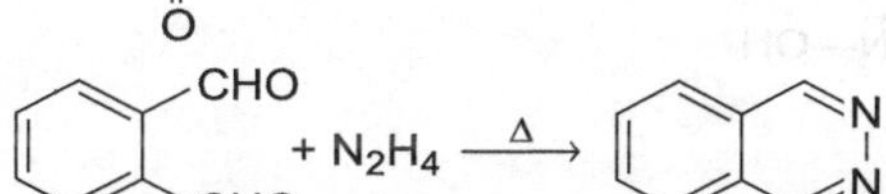

d. 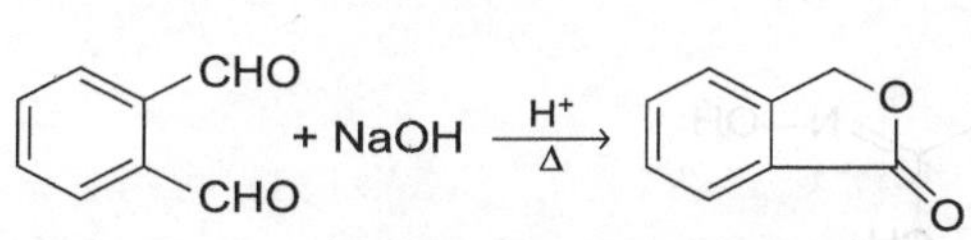

Statement Type

Direction (Q. Nos. 12 and 13) *This section is based on Statement I and Statement II. Select the correct answer from the codes given below.*

a. Both Statement I and Statement II are correct and Statement II is the correct explanation of Statement I
b. Both Statement I and Statement II are correct but Statement II is not the correct explanation of Statement I
c. Statement I is correct but Statement II is incorrect
d. Statement II is correct but Statement I is incorrect

12 **Statement I** Benzaldehyde is oxidised to benzoic acid, when treated with ammoniacal solution of cupric tartarate.

Statement II Ammoniacal solution of cupric tartarate oxidises aldehydic group to carboxylic group.

13 **Statement I** Benzaldehyde when treated with ethanal in alkaline solution yields cinnamaldehyde.

Statement II Benzaldehyde lacks alpha hydrogen.

Comprehension Type

Direction (Q. Nos. 14-18) *This section contains 2 paragraphs, each describing theory, experiments, data, etc. Five questions related to the paragraphs have been given. Each question has only one correct answer among the four given options (a), (b), (c) and (d).*

Passage I

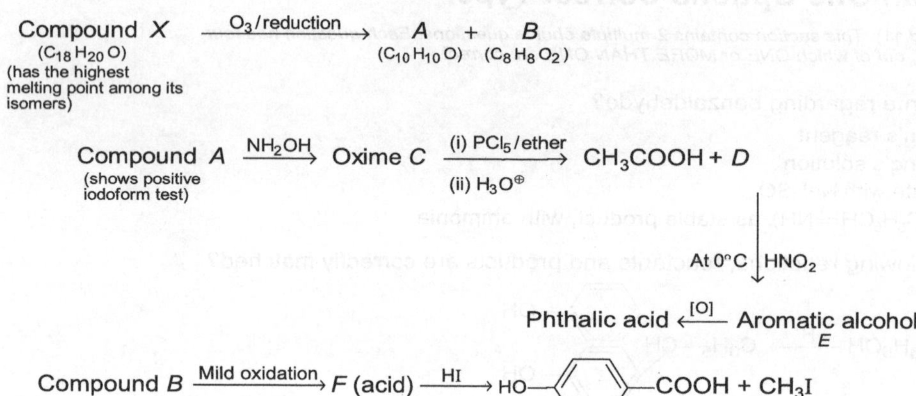

14 Compound C is

a. 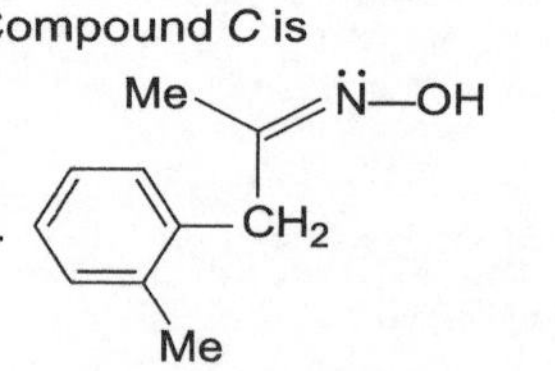

b. 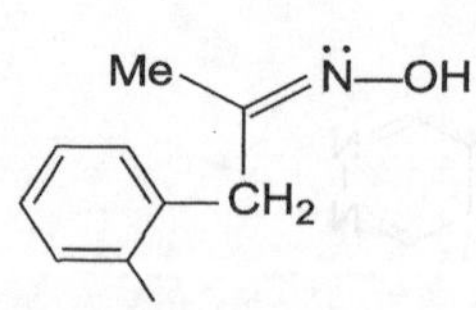

c. 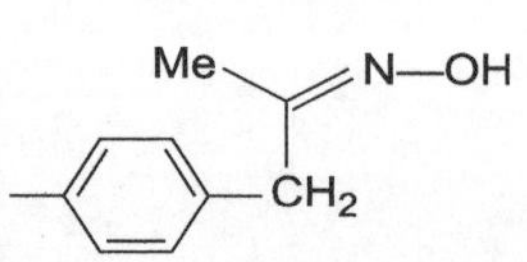

d. 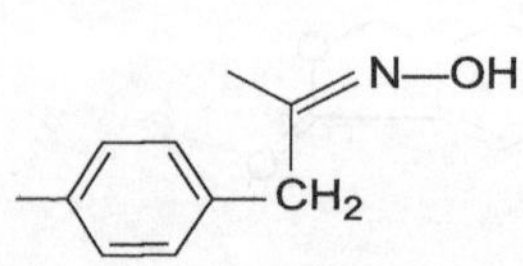

15 Compound *X* is

a 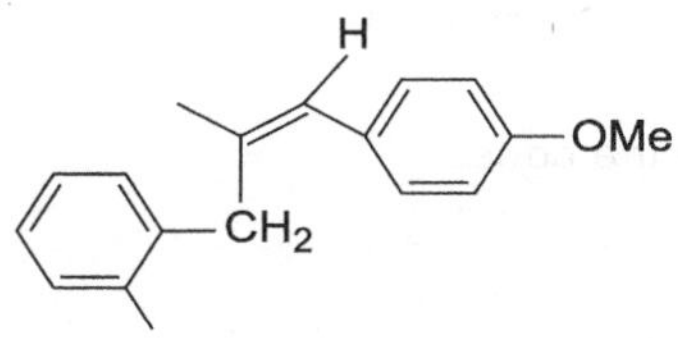

b 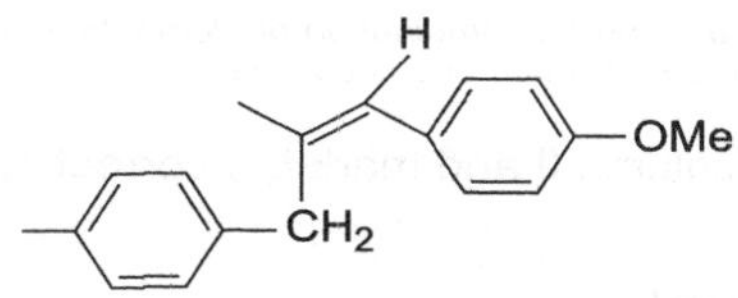

c 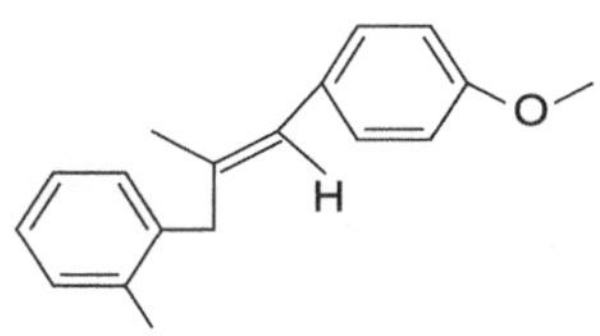

d 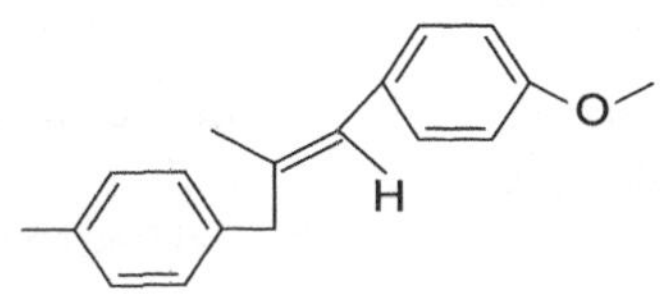

Passage II

Cannizaro reaction is a disproportionation reaction, in which one mole of the aldehyde is oxidised and the other is reduced.

Consider the following Cannizaro reaction:

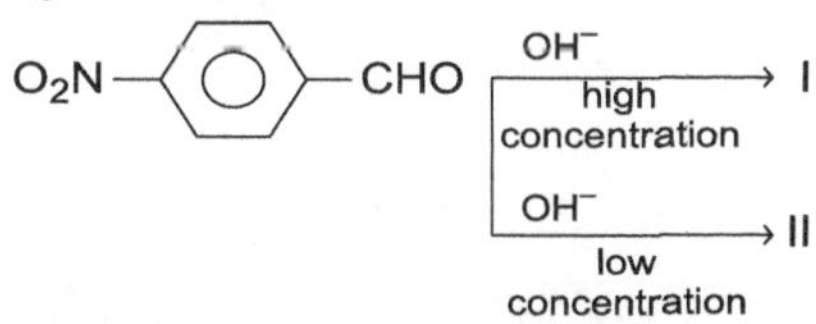

16 The false statement about the given reaction is

a. I and II are different substances

b. I is $O_2N-C_6H_4-COO^-$ and $O_2N-C_6H_4-CH_2O^-$ but second is not

c. II is $O_2N-C_6H_4-COO^-$ and $O_2N-C_6H_4-CH_2OH$ but first is not

d. II is $O_2N-C_6H_4-COO^-$ and $O_2N-C_6H_4-CH_2O^-$ but

I is $O_2N-C_6H_4-COO^-$ and $O_2N-C_6H_4-CH_2OH$

17 In case of (I) the order and molecularity of the reaction are

a. 4th order, bimolecular
b. 3rd order, bimolecular
c. 4th order, tetramolecular
d. 3rd order, trimolecular

18 For the formation of II, the rate law is

a. $R = k\,[\text{Ald}]^2\,[\text{OH}^-]^2$
b. $R = k\,[\text{Ald}]^2\,[\text{OH}^-]$
c. $R = k\,[\text{Ald}]\,[\text{OH}^-]$
d. $R = k\,[\text{Ald}]\,[\text{OH}^-]^2$

Working Space

Matching List Type

Direction *(Q. No. 19) Choices for the correct combination of elements from Column I and Column II are given as options (a), (b), (c) and (d), out of which one is correct.*

19 **Match the column I with column II and mark the correct option from the codes given below.**

	Column I (Reactant of cross Cannizaro reaction)		Column II (Product)
i.	X = *o*-methoxy benzaldehyde Y = *p*-methoxy benzaldehyde	p.	X is oxidised while Y is reduced
ii.	X = *p*-(*i*Pr) benzaldehyde Y = *p*-(*t* – bu)-benzaldehyde	q.	X is reduced and Y is oxidised
iii.	X = *p* (*i*Pr) benzaldehyde Y = *p* – Et benzaldehyde	r.	Y is more reactive
iv.	X = *p*-Me benzaldehyde Y = *p*-Et benzaldehyde	s.	X is more reactive

Codes

	i	ii	i	iv
a.	p,s	q,r	p,s	q,r
b.	p,q	q,r	r	s
c.	q	p,s	r,s	s
d.	p,s	p,r	p,s	q,r

One Integer Value Correct Type

Direction *(Q. No. 20) This section contains a question. When worked out will result in an integer from 0 to 9 (both inclusive).*

20 **How many products are formed during the following reaction?**

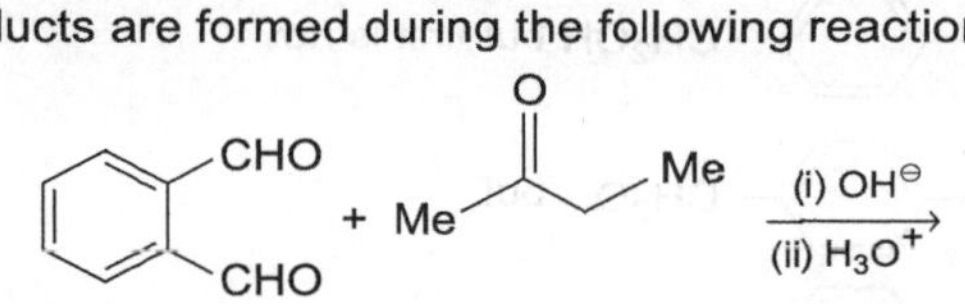

DPP-3 Properties of Aromatic Ketones

Subjective Questions

Working Space

Direction (*Q. Nos.* 1-5) *These questions are subjective in nature, need to be solved completely on notebook.*

1 i. Bring about the following transformation.

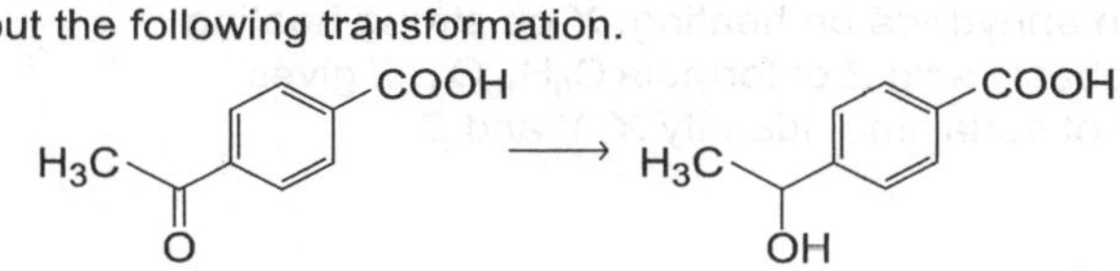

ii. Identify the missing species.

a. $Ph-\overset{O}{\overset{\|}{C}}-CH_3 \xrightarrow{?} Ph-CO-CHO$

b. 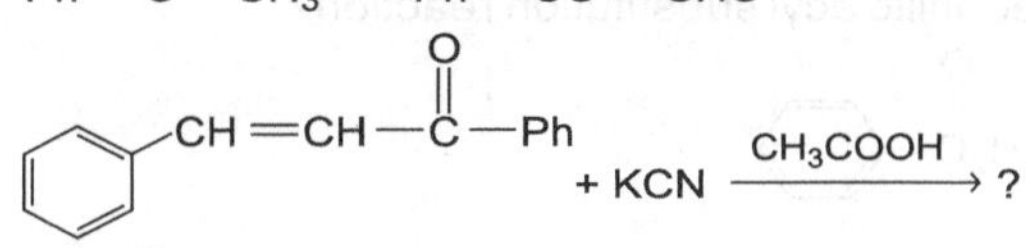

c. $Ph-\overset{O}{\overset{\|}{C}}-Ph \xrightarrow[NaOH]{Zn} ?$

d. $Ph-\underset{CN}{\underset{|}{CH}}-\underset{O}{\underset{\|}{C}}-CH_3 \xrightarrow[Heat]{aq.\ H_2SO_4} ?$

e. $C_6H_5-\underset{O}{\underset{\|}{C}}-CH_2-CH_2-COOH \xrightarrow{?} Ph-CH_2CH_2CH_2COOH$

f. $C_6H_5-\overset{O}{\overset{\|}{C}}-CH_3 \xrightarrow{HNO_2} A \xrightarrow{(CH_3CO)_2O} B \xrightarrow{H^+} C$

2 Identify *A* to *C* and name the stereoisomer of *B*.

$$\underset{\text{(with empirical formula } C_7H_6O)}{\text{Aromatic ketone}} \xrightarrow{\text{Hydroxyl amine}} \underset{(C_{14}H_{13}NO_2)}{B} \xrightarrow{H^+} C \xrightarrow{H_3O^+} \text{Aniline} + p\text{-methoxy benzoic acid}$$

3 Complete the following.

i. 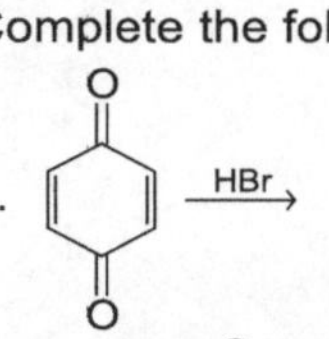

ii. $PhCOCH_3 \xrightarrow{SeO_2}$

iii. 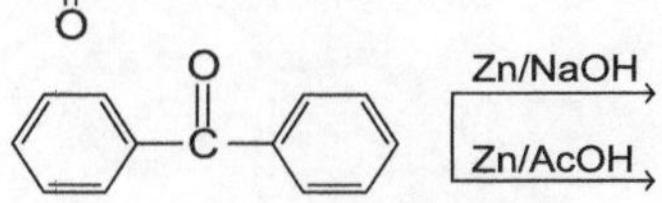

iv. $p\text{-nitrotoluene} \xrightarrow[\text{(ii) } C_2H_5OH\ +\ HCl]{\text{(i) } K_2Cr_2O_7/H^+} \xrightarrow[C_2H_5ONa]{CH_3CO_2C_2H_5} \xrightarrow[\text{(ii) } H_3O^+]{\text{(i) KOH}}$

Working Space

4 i. Suggest a chemical test for distinguishing

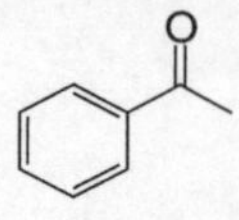 and 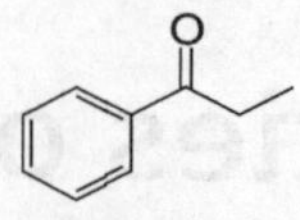

(I) (II)

ii. Consider the following two compounds.

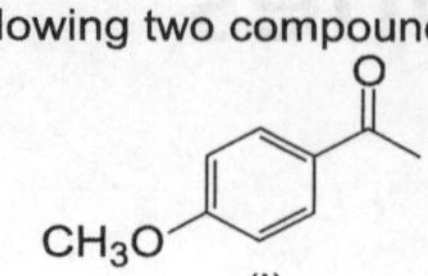 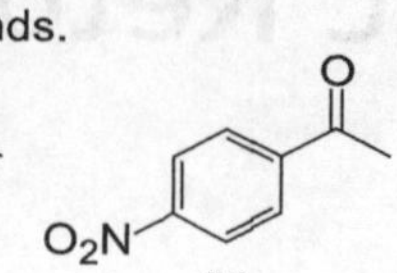

(I) (II)

Carbonyl group of which is protonated more rapidly in acid solution and why?

5 An aromatic ketone *X* with molecular formula $C_{10}H_{12}O_2$ gives a dibasic acid *Y* on vigorous oxidation. *Y* readily forms an anhydride on heating. *X* on strong heating with sodium hypobromite gives monobasic acid *Z* of formula $C_9H_{10}O_3$. *Z* gives 3-methyl anisole, when heated with hot soda lime. Identify *X*, *Y* and *Z*.

Only One Option Correct Type

Direction *(Q. Nos. 6-16) This section contains 11 multiple choice questions. Each question has four choices (a), (b), (c) and (d), out of which ONLY ONE is correct.*

6 Which of these would be the most reactive in nucleophilic acyl substitution reaction?

a. 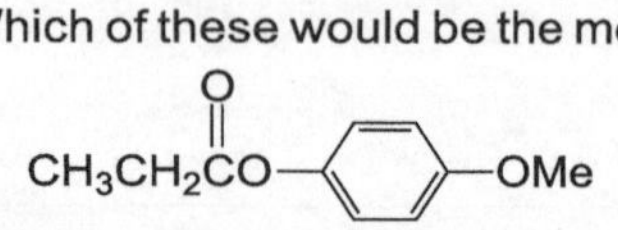

b. 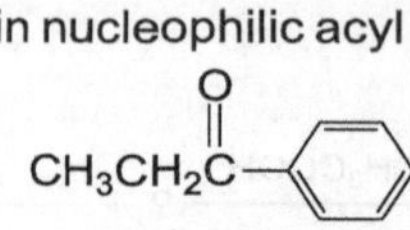

c. 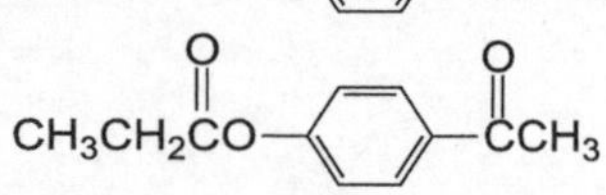

d. All these are equally reactive

7 Acetophenone when reacted with a base, C_2H_5ONa, yields a stable compound which has the structure

a. 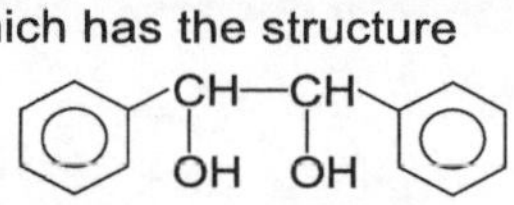

b. 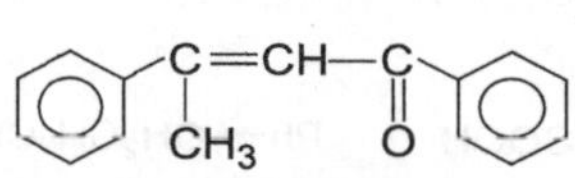

c. 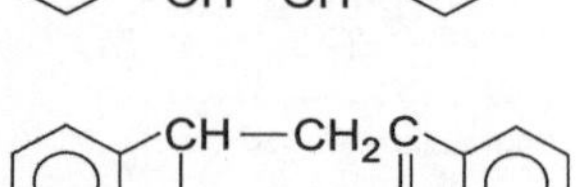

d. 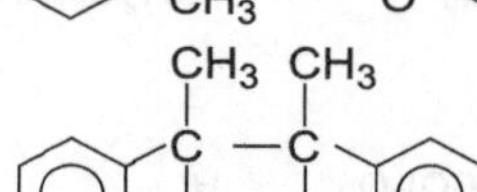

8 In the presence of NaOH, *A* changes to

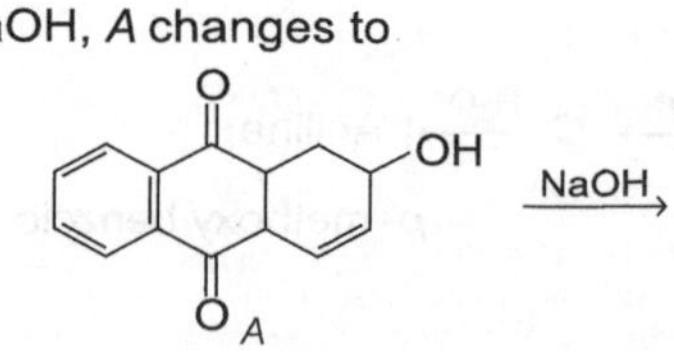

a. 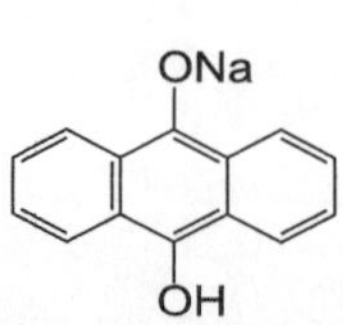

b. 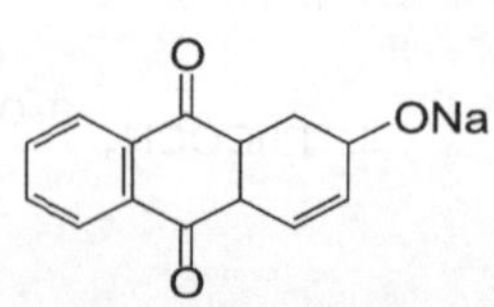

c. 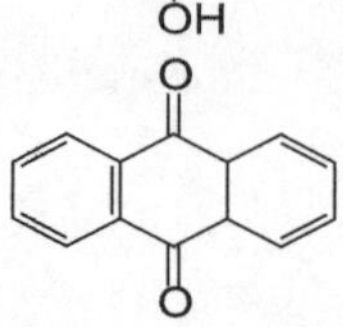

d. 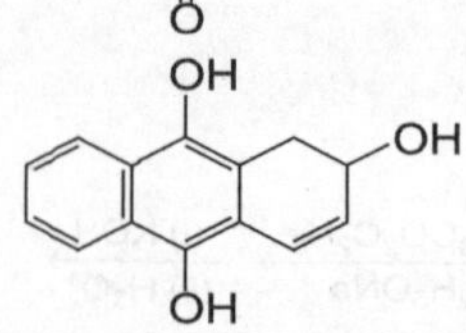

Working Space

9 Which reactant gives the product shown on heating with aluminium chloride?

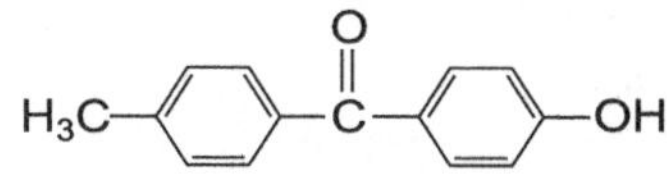

a. 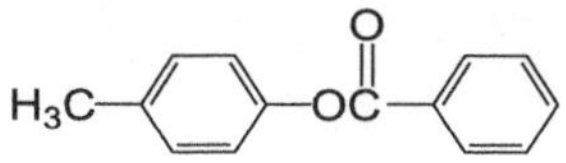

b. 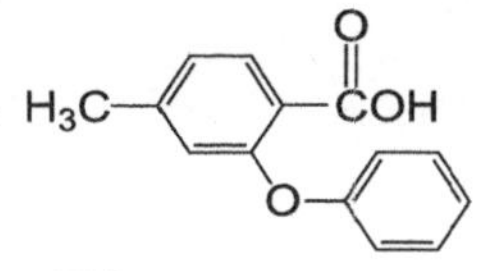

c. 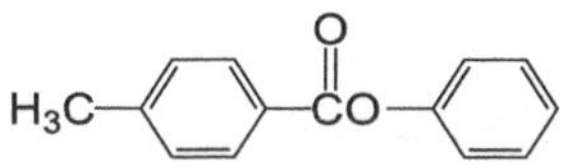

d. 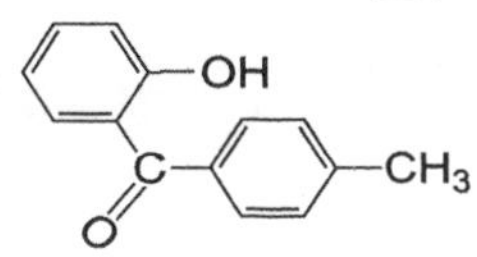

10 The end product of the following reaction is

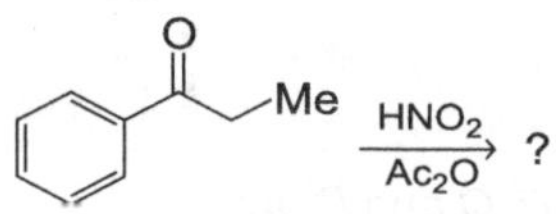

a. 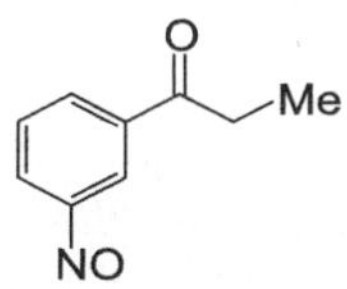

b. 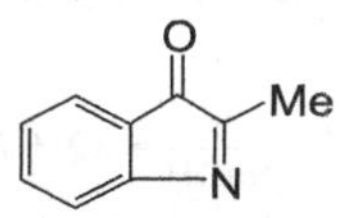

c. 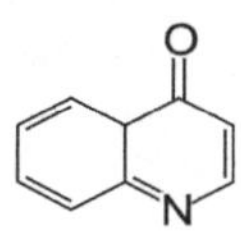

d. 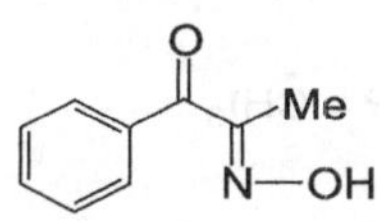

11 Which of the following protecting groups is best used as a protector in the following synthesis?

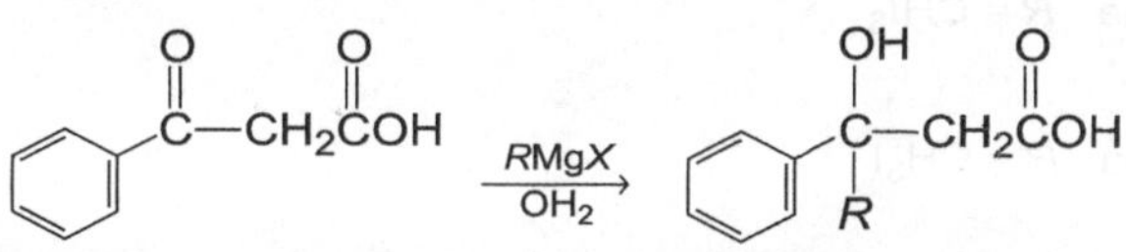

a. $HSCH_2CH_2SH$
b. $HOCH_2CH_2CH_2OH$
c. CH_3CH_2OH
d. CH_3COCl

12 What is the major product of the following reaction sequence?

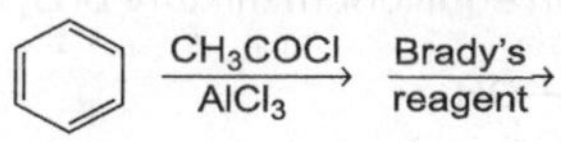

a. 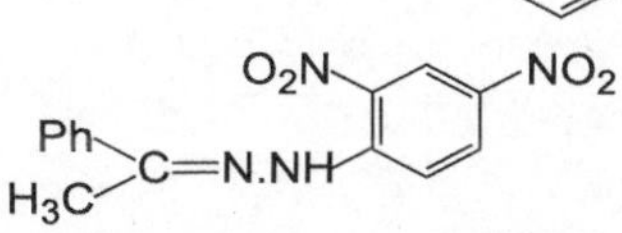

b. 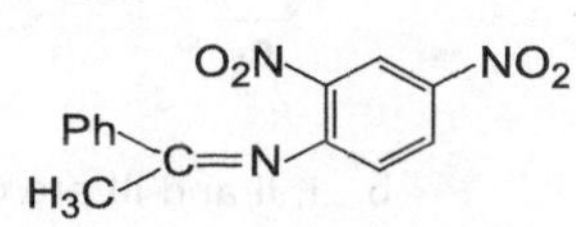

c. 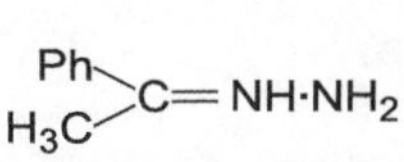

d. 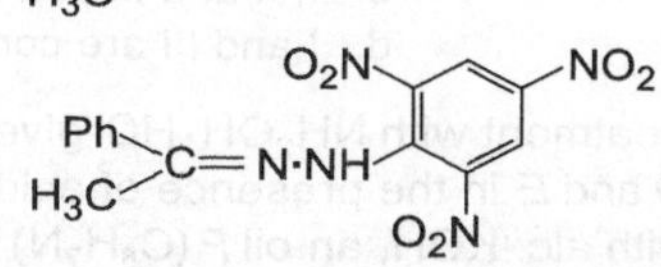

Working Space

13 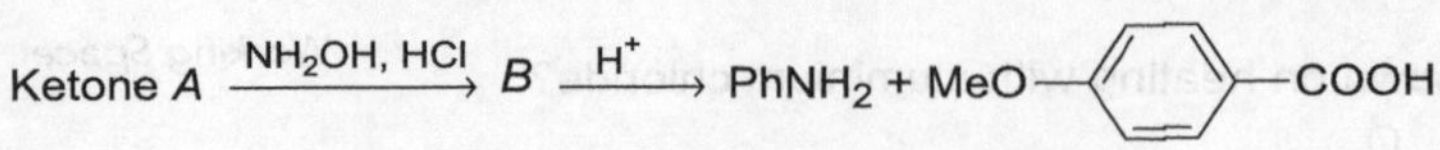

The ketone A and the compound B respectively are

a. 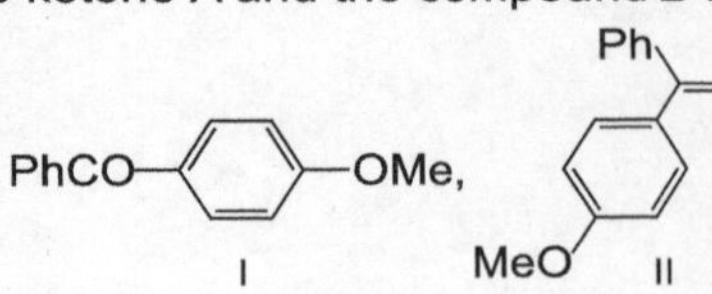

b. I, 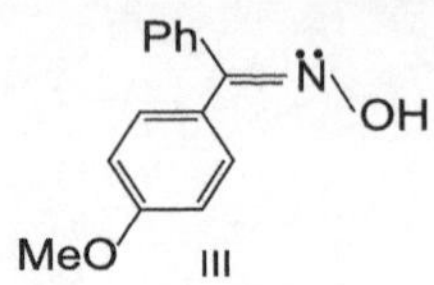

c. 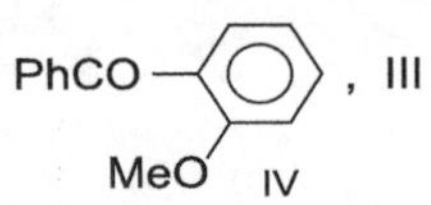 , III

d. III and IV

14 In the following sequence, the correct structure of P, Q and R are

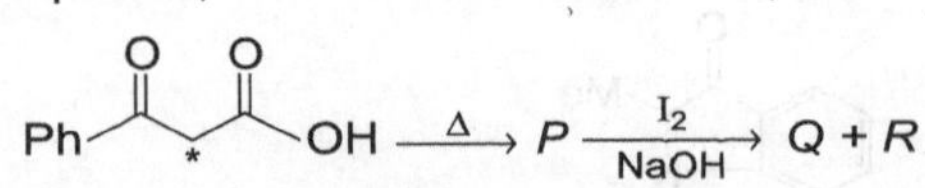

(* = ^{13}C labelled carbon)

a. 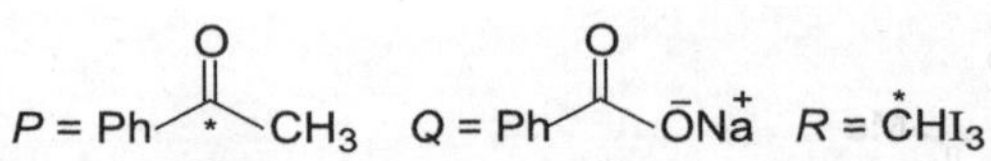

b. 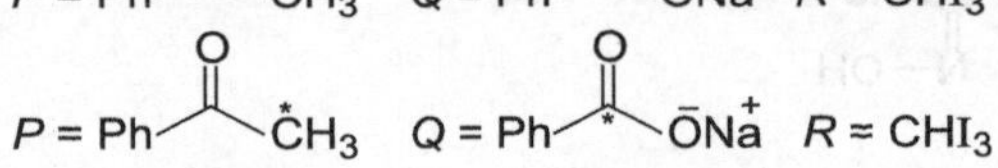

c. 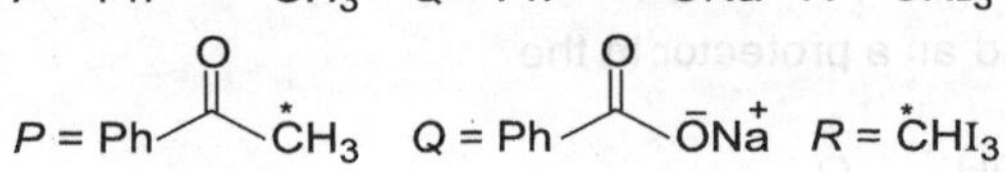

d. 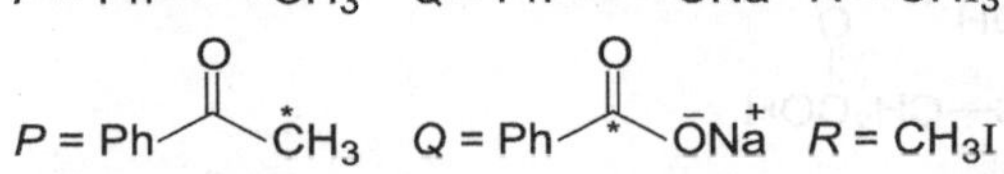

15 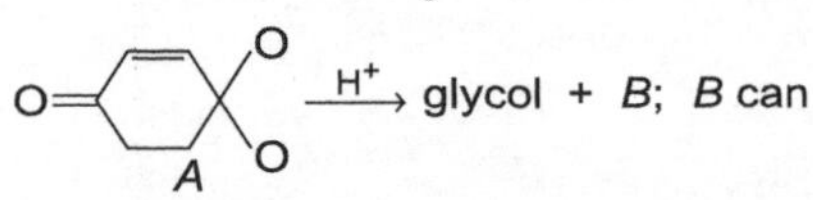

I. undergo tautomerisation

II. form oximes

III. give colour with $FeCl_3$

IV. B is an equilibrium mixture of B_1 and B_2

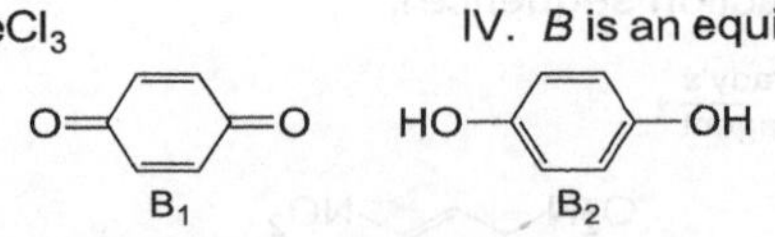

Select the correct statements.

a. All are correct

b. I, II and III are correct

c. II and III are correct

d. I and III are correct

16 A compound $A(C_8H_8O)$ on treatment with $NH_2OH \cdot HCl$ gives B and C, which rearrange themself to give D and E in the presence of acid. B, C, D and E are isomers. When D is boiled with alc. KOH, an oil $F(C_6H_7N)$ separates out. F reacts rapidly with CH_3COCl to give back D. E when boiled with alkali followed by the acidification gives a white solid $G(C_7H_6O_2)$. A is

a. $C_6H_5COCH_3$

b. $C_6H_5CH_2CHO$

c. 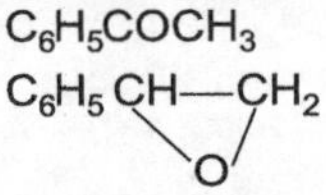

d. $C_6H_5CH_2COOH$

Working Space

Comprehension Type

Direction (Q. Nos. 17-19) *This section contains a paragraph, describing theory, experiments, data, etc. Three questions related to the paragraph have been given. Each question has only one correct answer among the four given options (a), (b), (c) and (d).*

Passage

A carbonyl compound *P*, which gives positive iodoform test, undergoes reaction with MeMgBr followed by dehydration to give an olefin *Q*. Ozonolysis, of *Q* leads to a dicarbonyl compound *R*, which undergoes intramolecular aldol reaction to give predominantly *S*. **[IIT JEE 2009]**

$$P \xrightarrow[\text{(ii) } H^+,\ H_2O \quad \text{(iii) } H_2SO_4,\ \Delta]{\text{(i) MeMgBr}} Q \xrightarrow[\text{(ii) Zn, } H_2O]{\text{(i) } O_3} R \xrightarrow[\text{(ii) } \Delta]{\text{(i) } OH^-} S$$

17 The structure of the carbonyl compound *P* is

a. 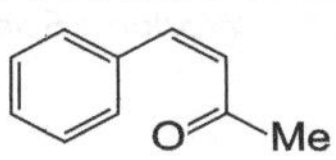

b. 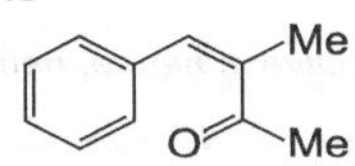

c. 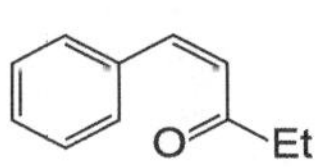

d. 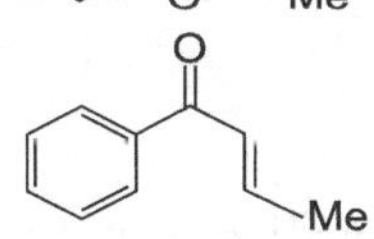

18 The structures of the products *Q* and *R*, respectively are

a. 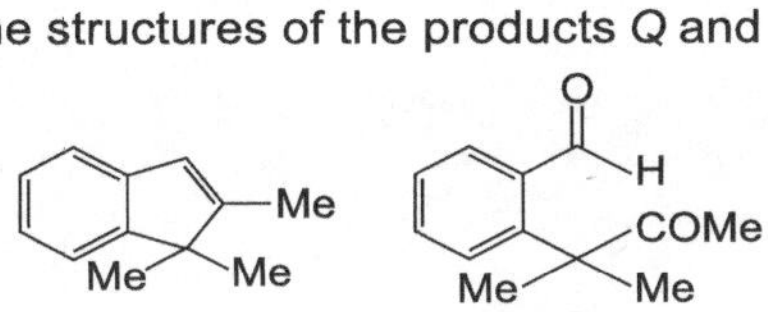

b. 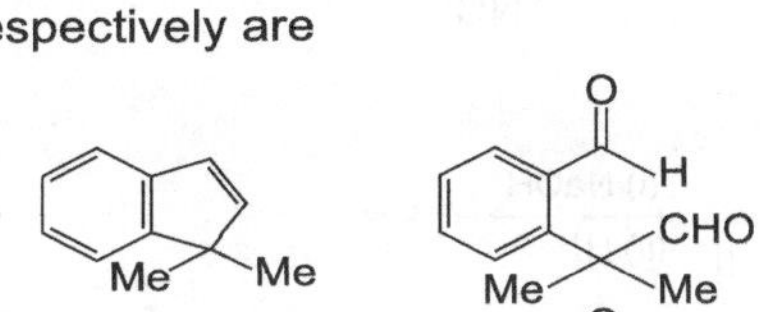

c. 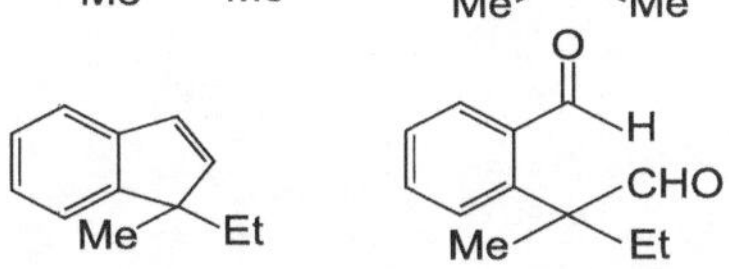

d. 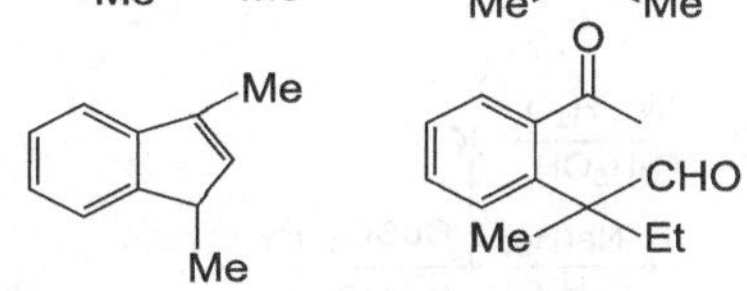

19 The structure of the product *S* is

a. 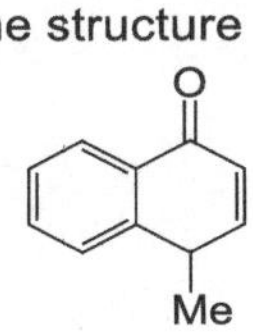

b. 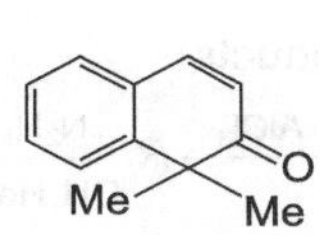

c. 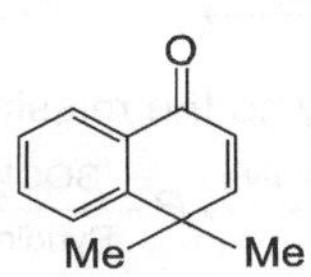

d. 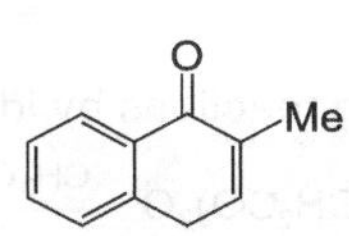

DPP-4 Combined Properties of Aromatic Aldehydes and Ketones

Subjective Questions

Working Space

Direction *(Q. Nos. 1 and 2) These questions are subjective in nature, need to be solved completely on notebook.*

1 Complete the following reactions.

i. 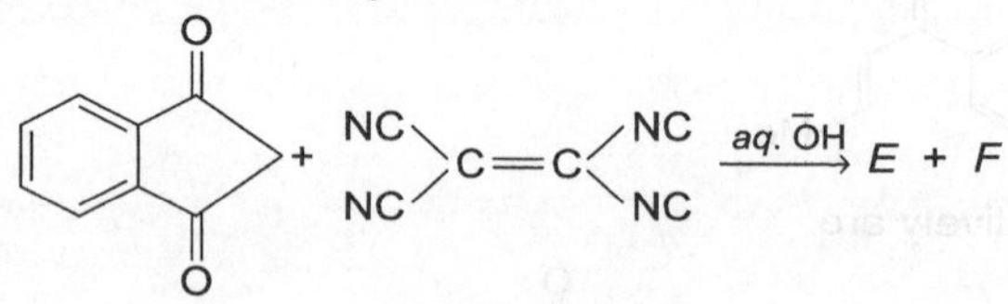

ii. 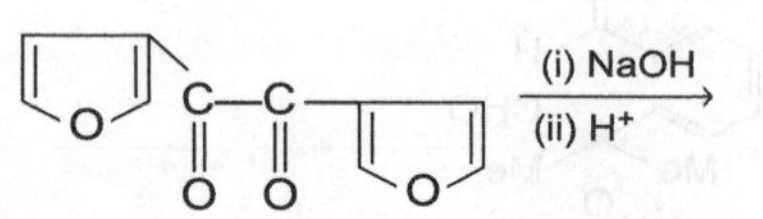

iii. 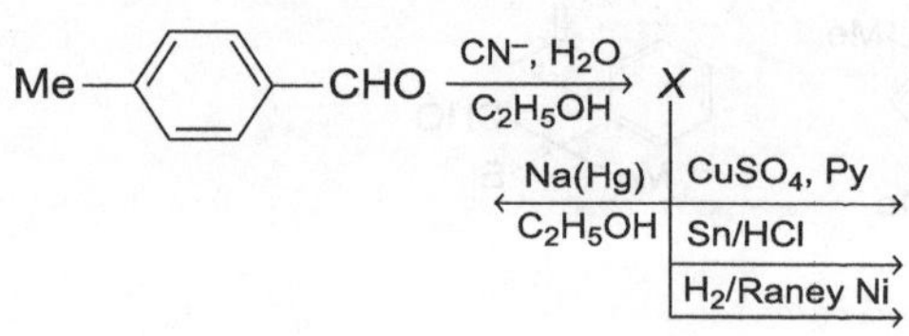

2 Complete the following reactions by identifying the missing products

i. $C_6H_5—CHO + (CH_3CH_2CO)_2O \xrightarrow[\text{Heat}]{CH_3CH_2COONa} P \xrightarrow[\text{Pyridine}]{SOCl_2} Q \xrightarrow{AlCl_3} R \xrightarrow[OH^- \text{Heat}]{N_2H_4} S$

ii. 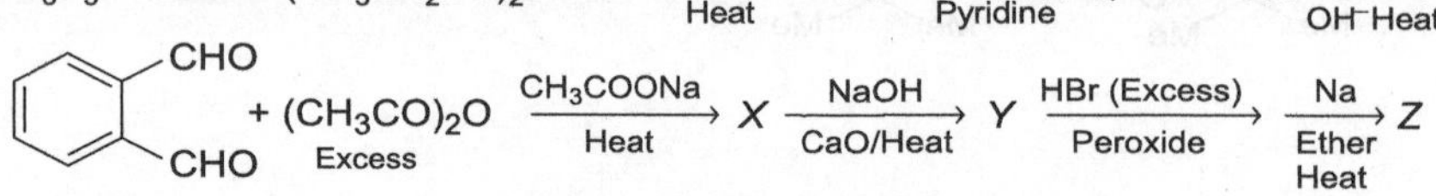

iii. $PhCHO + A \longrightarrow PhCH(OH)CH_2COOEt \xrightarrow{\text{Heat}} B$

iv. $PhCHO \xrightarrow{NH_2NH_2}$

Only One Option Correct Type

Working Space

Direction *(Q. Nos. 3-9) This section contains 7 multiple choice questions. Each question has four choices (a), (b), (c) and (d), out of which ONLY ONE is correct.*

3 Which synthetic method correctly does the following reactions?

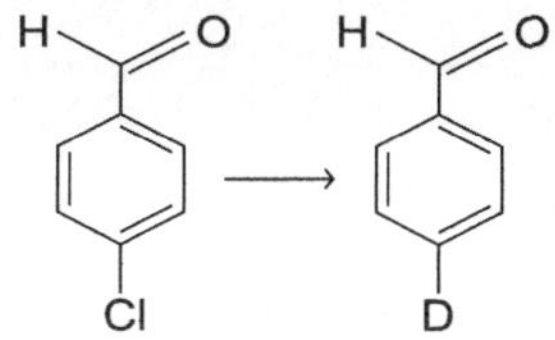

a. (i) Mg, diethylether, (ii) D_2O
b. (i) $LiAlD_4$, diethylether, (ii) D_2O
c. (i) $HOCH_2CH_2OH$, H^+, (ii) Mg, $(CH_3)_2O$, (iii) D_2O, H_2O, H^+
d. (i) $HOCH_2CH_2OH$, H^+, (ii) DCl, (iii) H_2O, H^+

4 In the following reaction,

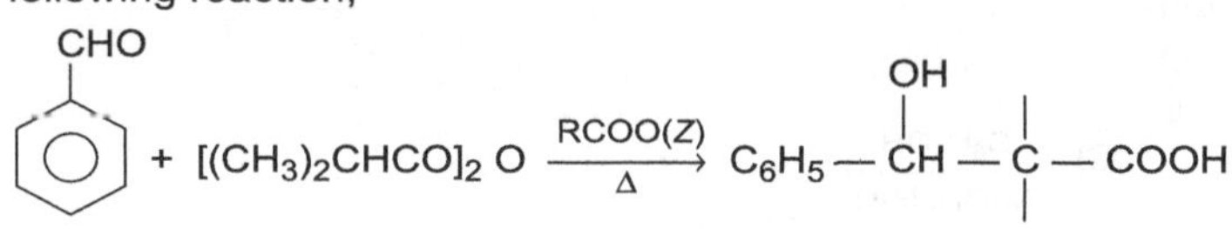

the yield is highest if *Z* is
a. sodium
b. potassium
c. caesium
d. same in all the cases

5 What product is expected from the following reaction.

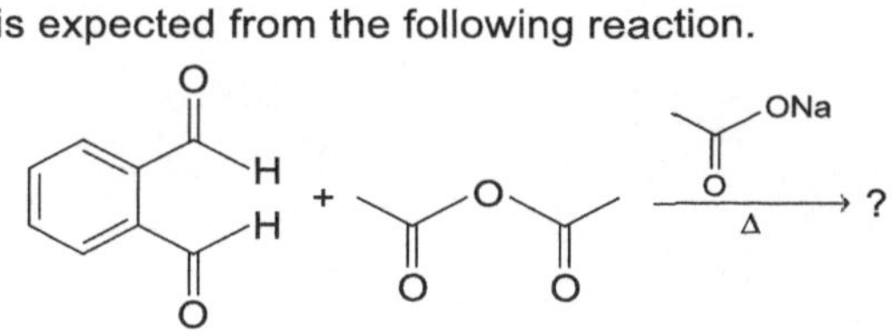

a. 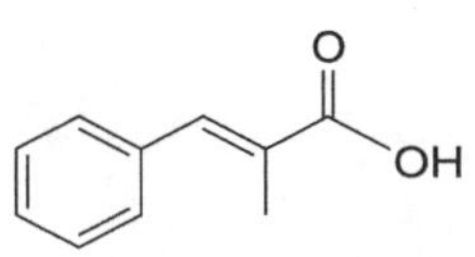
b. 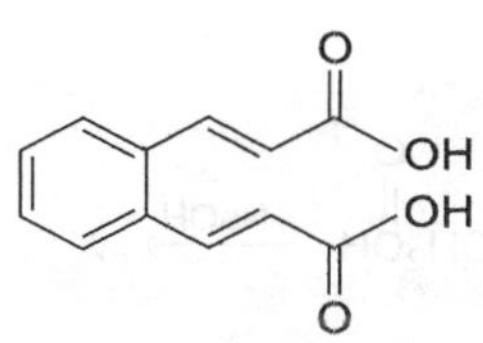
c. 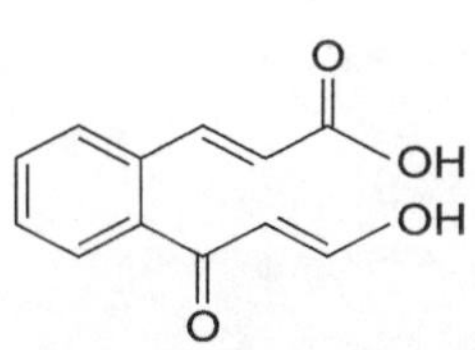
d. 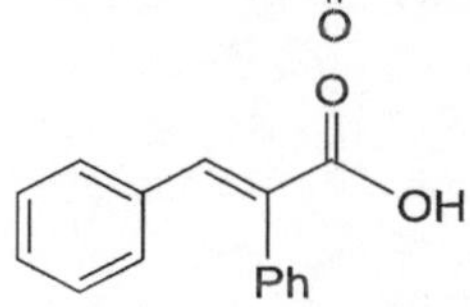

6 When benzaldehyde is refluxed with potassium cyanide in ethanolic solution, the product isolatable on cooling will be
a. PhCH(OH)CN
b. PhCH(OH)CH(OH)Ph
c. PhCOCH(OH)Ph
d. $PhCONH_2$

Working Space

7 The reactivity order of following aromatic aldehydes with $(CH_3CO)_2O$ in the presence of CH_3COONa /heat is

I. 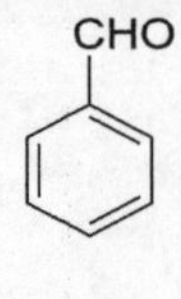 II. 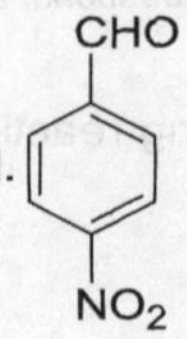

III. 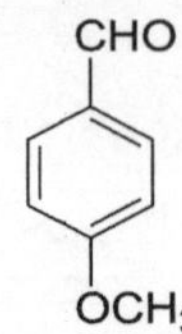 IV. 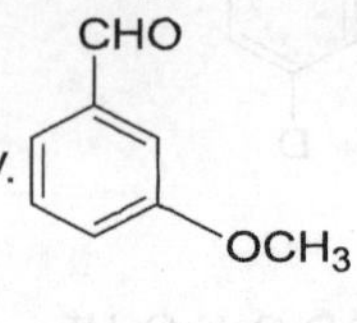

a. II > IV > I > III
b. II > I > IV > III
c. II > III > IV > I
d. III > IV > I > II

8 What is the product of the reaction?

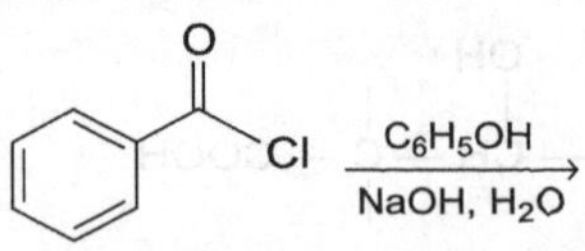

a. 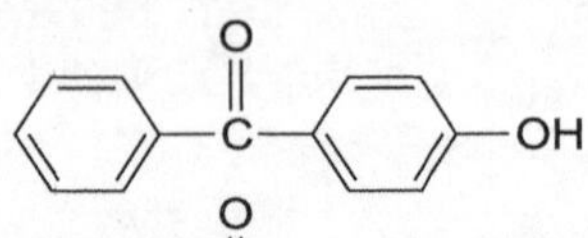

b. 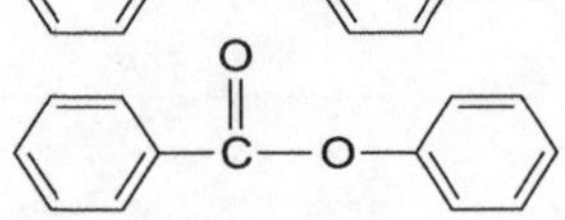

c. 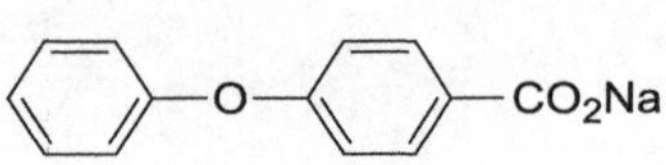

d. 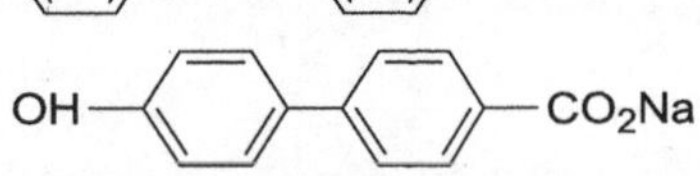

9 In the following reaction,

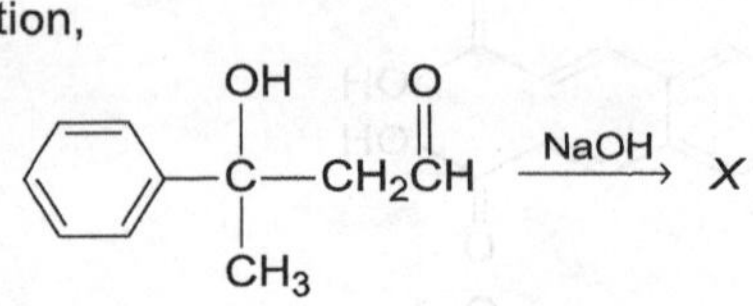

Product X is

a. 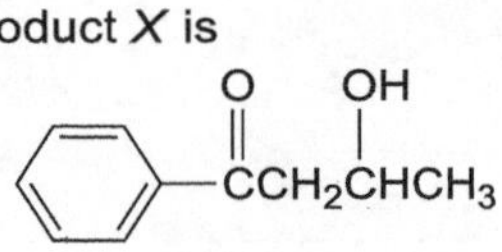

b. 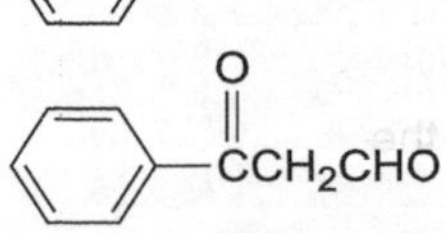

c. 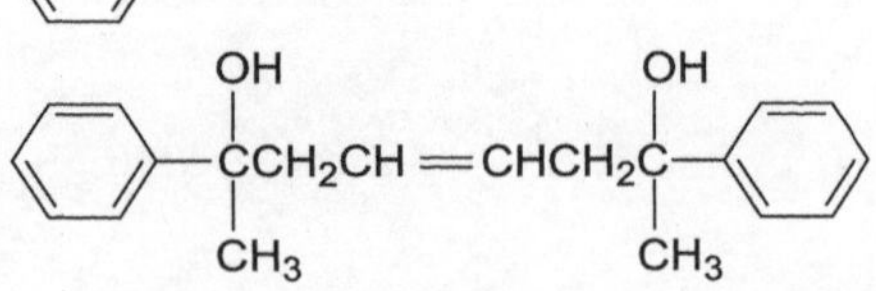

d. No reaction

Working Space

One or More than One Options Correct Type

Direction (Q. Nos. 10-12) This section contains 3 multiple choice questions. Each question has four choices (a), (b), (c) and (d), out of which ONE or MORE THAN ONE are correct.

10 Which of the following reactions do not give product in good yield?

a. $C_6H_5CHO + NH_3 \longrightarrow C_6H_5CH{=}N\underset{|}{\overset{C_6H_5}{C}}H{-}N{=}CHC_6H_5$

b. $C_6H_5CHO + Cl_2 \longrightarrow$ Benzoylchloride

c. $C_6H_5CHO + RMgX \xrightarrow{H_3O^+}$ 3° alcohol

d. C_6H_5CHO + conc. $HNO_3 \xrightarrow{\text{conc. } H_2SO_4}$ 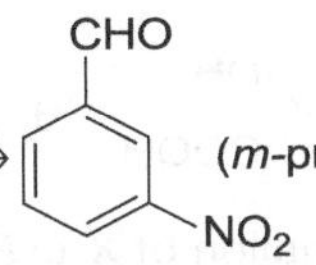 (*m*-product)

11 Choose the correct statement regarding the following reaction

$$C_6H_5{-}CHO + C_2H_5OH + KCN \longrightarrow C_6H_5{-}\underset{OH}{\underset{|}{C}}H{-}\overset{O}{\overset{\|}{C}}{-}C_6H_5 \text{ (Benzoin)}$$

a. Reaction proceeds by initial nucleophilic attack of CN^- on benzaldehyde
b. A carbanion intermediate is involved in this reaction
c. A pure enantiomer of benzoin is formed
d. If reaction is carried out in methanol rather than benzoin, cyanohydrin is formed as major product.

12 Which of the following will undergo faster Perkin condensation than benzaldehyde?

a. 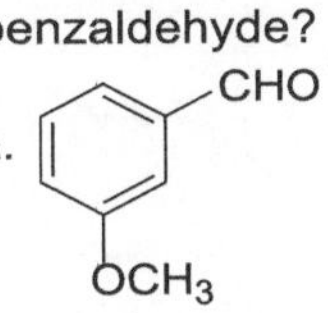

b. 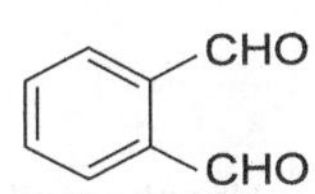

c. 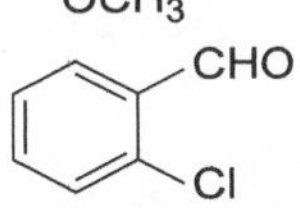

d. 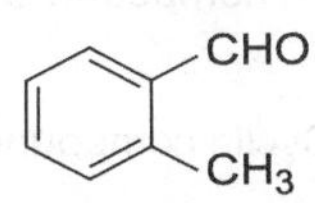

Statement Type

Direction (Q. No. 13) This section is based on Statement I and Statement II. Select the correct answer from the codes given below.

a. Both Statement I and Statement II are correct and Statement II is the correct explanation of Statement I
b. Both Statement I and Statement II are correct but Statement II is not the correct explanation of Statement I
c. Statement I is correct but Statement II is incorrect
d. Statement II is correct but Statement I is incorrect

13 **Statement I** 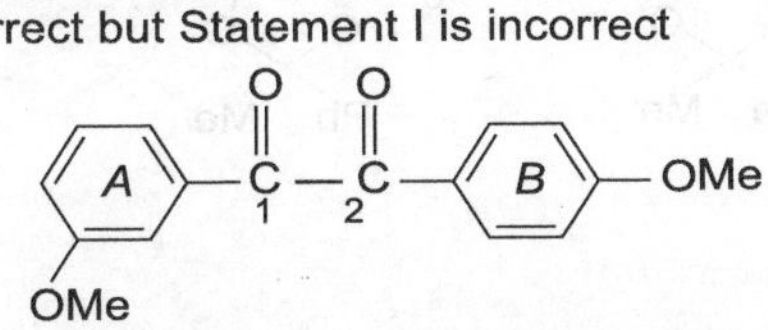

Ring *B* will migrate if this compound is treated with $\overline{O}H$.

Statement II  group attached with ring *A* is more reactive.

Comprehension Type

Working Space

***Direction** (Q. Nos. 14-18) This section contains 2 paragraphs, each describing theory, experiments, data, etc. Five questions related to the paragraphs have been given. Each question has only one correct answer among the four given options (a), (b), (c) and (d).*

Passage I

$$\text{Benzene} \xrightarrow[\text{Hydrochloric acid }(AlCl_3)]{\text{Carbon monoxide}} X \xrightarrow[\text{KCN}]{aq.\text{ Alc}} Y \xrightarrow{\text{PCC}} Z \xrightarrow{\bar{O}H/H_3O^+} P$$

14 The structure of compound P is

a. 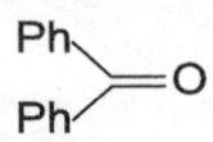 b. 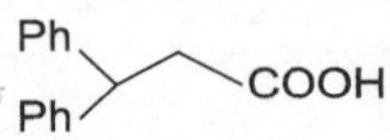 c. 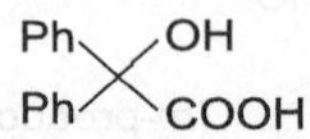 d. 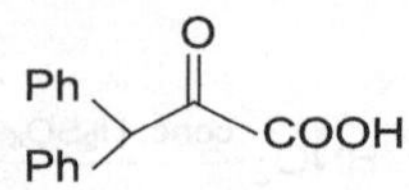

15 Match the name of reactions involved in the formation of X to P

	Column I (Product)		Column II (Name reaction)
i.	X	p.	Oxidation
ii.	Y	q.	Gattermann Koch
iii.	Z	r.	Benzil-Benzilic acid rearrangement
iv.	P	s.	Benzoin condensation

Codes

	i	ii	iii	iv		i	ii	iii	iv
a.	p	q	r	s	b.	q	s	r	p
c.	q	s	p	r	d.	s	q	p	r

Passage II

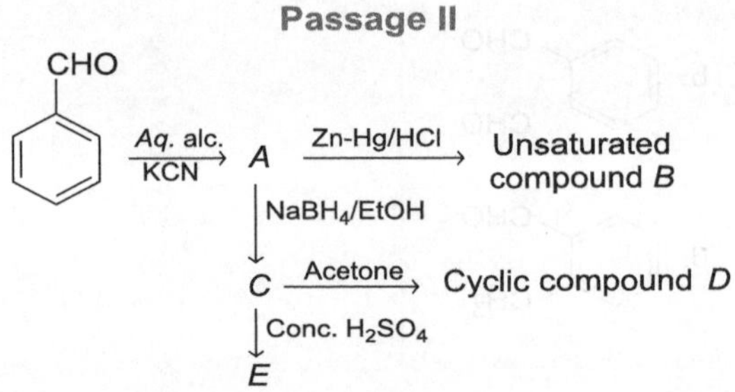

16 Compound B is

a. 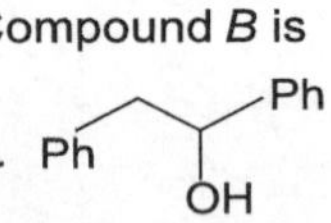 b. 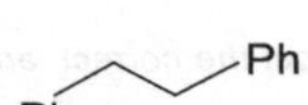 c. 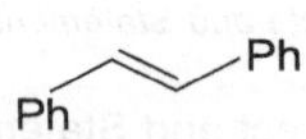 d. 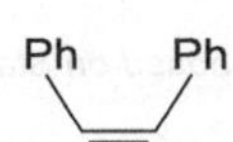

17 Compound D is

a. 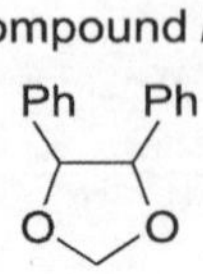 b. 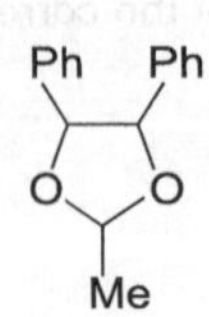 c. 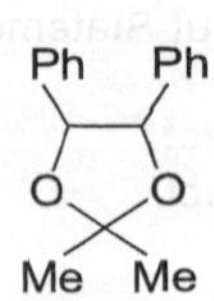 d. 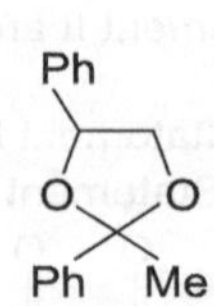

18 The compound E is

a. 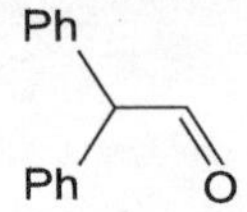 b. 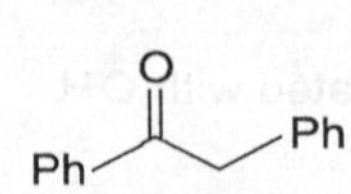

c. Both (a) and (b)

d. None of these

Matching List Type

Direction *(Q. Nos. 19 and 20) Choices for the correct combination of elements from Column I and Column II are given as options (a), (b), (c) and (d), out of which one is correct.*

19 Match the Column I and Column II and mark the correct option from the codes:
In the benzoin condensation reaction

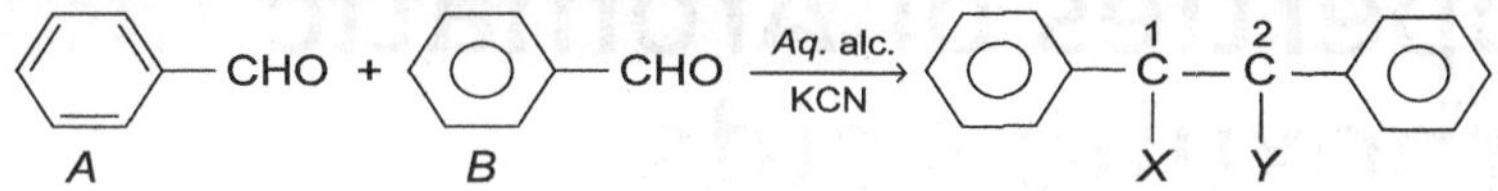

	Column I		Column II
i.	If *B* has Me group at *meta* position	p.	*X* = (═ O) and *Y* = (—OH)
ii.	If *A* has Me group at *ortho* position and *B* has Me at *para*	q.	*X* = (═ O) and *Y* = (═ O)
iii.	If *A* has OMe at *meta* position and *B* has —OMe at *para*.	r.	*X* = (—OH) and *Y* = (═ O)
iv.	If *A* has Me at *para* and *B* has *Et* at *para*	s.	*B* becomes more active

Codes

	i	ii	iii	iv		i	ii	iii	iv
a.	q,r	p	p	p,r	b.	p	r,s	p	r,s
c.	q	p,s	r	r,s	d.	p	p,q	r,s	s

20 Match the Column I and Column II and mark the correct option from the codes given below.

	Column I (Reaction)		Column II (Steps involved)
i.	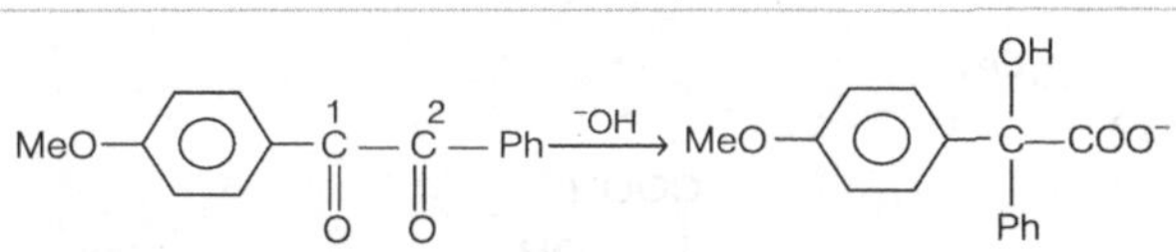	p.	Attack of $\overline{O}H$ at C-1
ii.	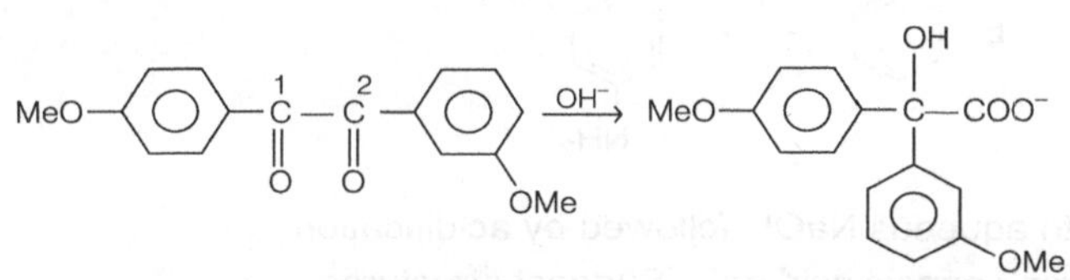	q.	Attack of $\overline{O}H$ at C-2
iii.	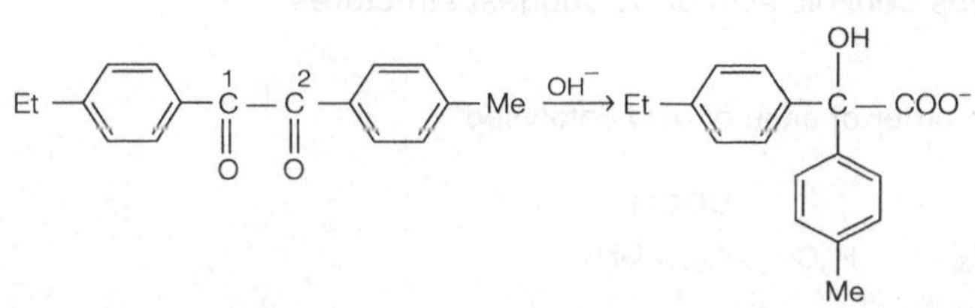	r.	*m*-Me O-C_6H_4 is migrating
iv.	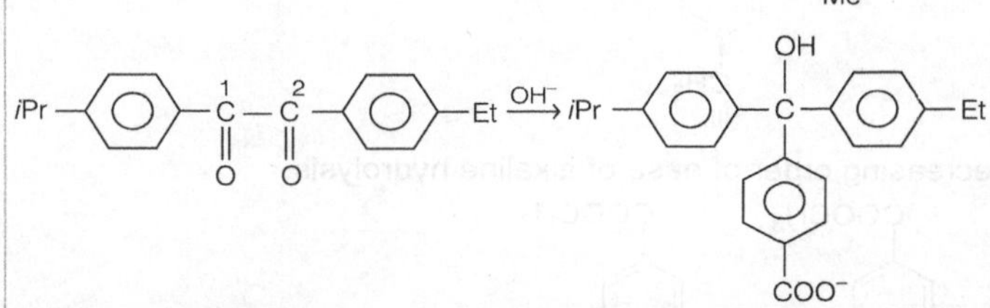	s.	*p*-Et C_6H_4 is migrating

Codes

	i	ii	iii	iv		i	ii	iii	iv
a.	q	q,r	p,s	p	b.	p	q	p,q	r,s
c.	r	r,s	p	p,q	d.	q,s	p,q	r,s	q,s

DPP-5 Preparation and Properties of Aromatic Carboxylic Acids

Subjective Questions

Working Space

Direction *(Q. Nos. 1-3) These questions are subjective in nature, need to be solved completely on notebook.*

1 i. Complete the following reactions.

a. $Ph_2C(OH)COOH \xrightarrow{I_2/P}$

b. 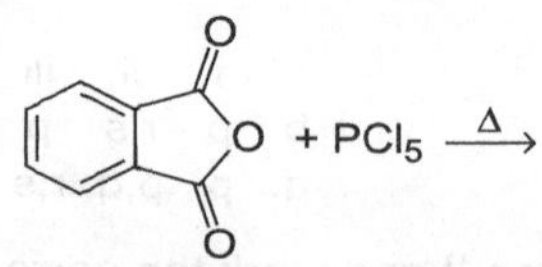

c. 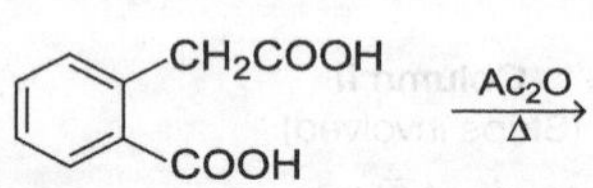

d. 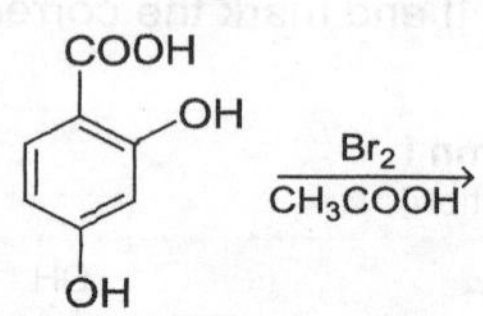

ii. How to carry out the following conversions?

a. 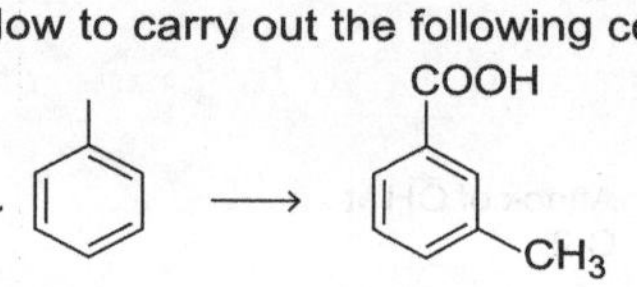

b. 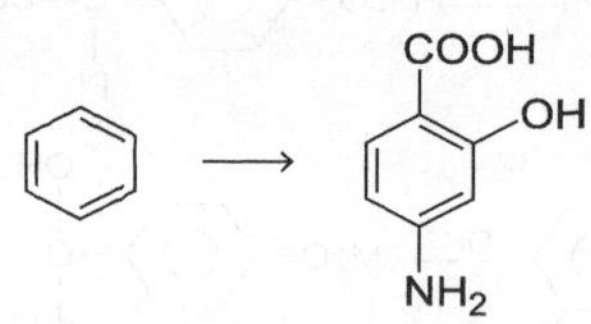

2 Compound *A* ($C_8H_6O_2$) on treatment with aqueous NaOH followed by acidification gives *B*, $C_8H_8O_3$, which on oxidation gives benzoic acid only. Suggest structures for *A* and *B*.

3 i. Arrange the following in the increasing order of ease of acid catalysed esterification.

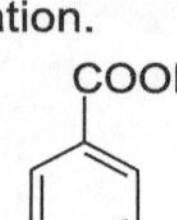

I.

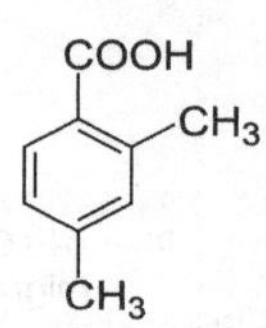

II.

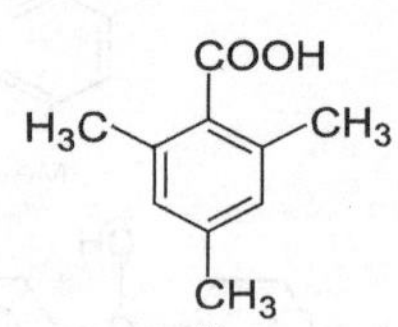

III.

ii. Arrange the following esters in the decreasing order of ease of alkaline hydrolysis.

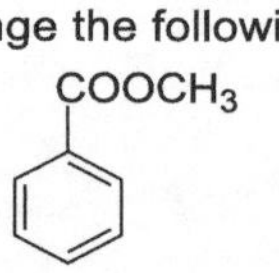

I.

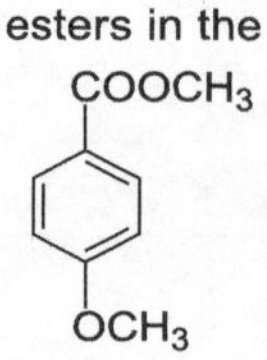

II.

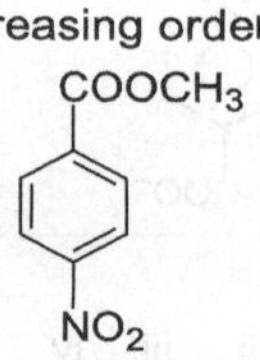

III.

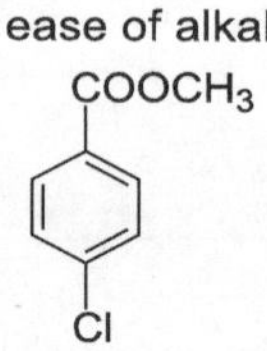

IV.

Working Space

Only One Option Correct Type

Direction (Q. Nos. 4-12) *This section contains 9 multiple choice questions. Each question has four choices (a), (b), (c) and (d), out of which ONLY ONE is correct.*

4 Which will not produce benzoic acid?

a. $C_6H_5CH_2OH + KMnO_4/\bar{O}H, H_2O, \Delta$, then H_3O^+

b. $C_6H_5CH_3 + KMnO_4/\bar{O}H/H_2O, \Delta$, then H_3O^+

c. $C_6H_6 + CO_2$; high pressure

d. $C_6H_5COCl + {}^-OH/H_2O$; then H_3O^+

5 Which compound would be subjected to a haloform reaction to produce *m*-chlorobenzoic acid?

I. 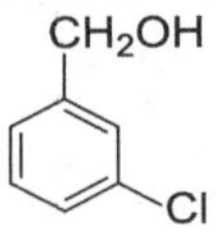

II. 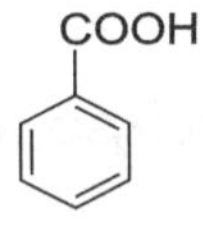

III. 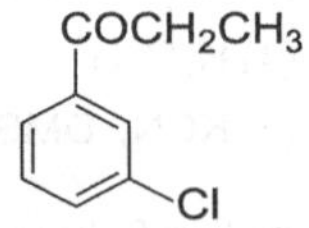

IV. 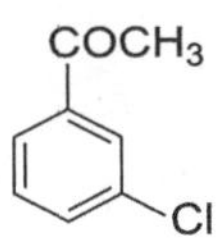

V. 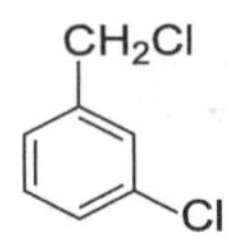

a. I
b. II, V
c. III
d. IV

6 Which of the following compounds would be oxidised to benzoic acid with acidic $Na_2Cr_2O_7$ and heat?

a. 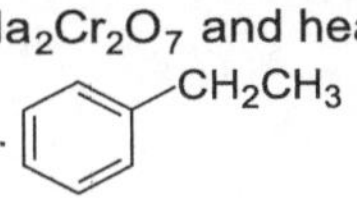

b. 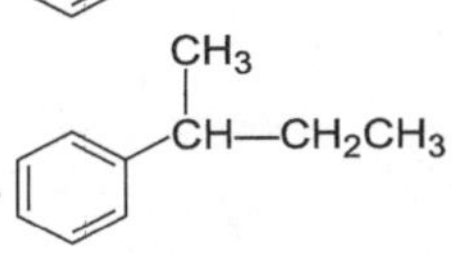

c. 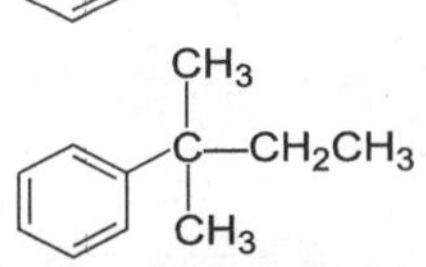

d. None of the above would be oxidised

7 Saponification of an ester *A* followed by neutralisation gives a compound *B*, which gives violet colouration with $FeCl_3$. The ester *A* is

a. ethyl salicylate
b. ethylacetate
c. ethyl benzoate
d. diethyl phthalate

8 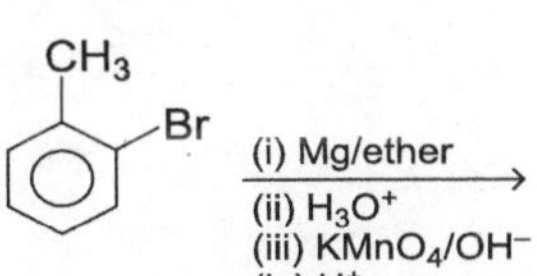

A; *A* is

a. 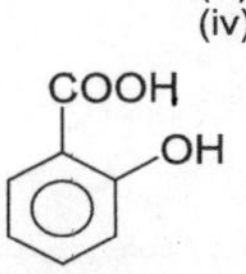

b. 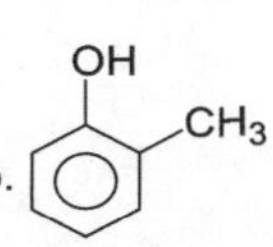

c. 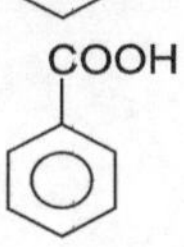

d. 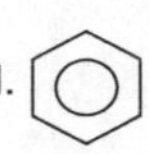

Working Space

9 Choose one of the following. Indicate synthetic route which gives the best yield for the following transformation.

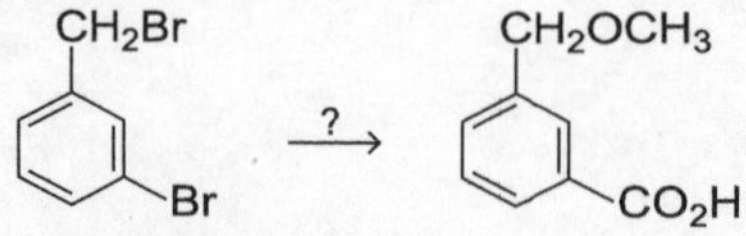

a. (1) Mg/diethyl ether (2) CO_2 (3) H_2O, H^+ (4) $CH_3O^-Na^+$

b. (1) NaOH (2) Mg/diethyl ether (3) CO_2 (4) H_2O, H^+ (5) CH_3I

c. (1) $CH_3O^-Na^+$ (2) Mg/diethyl ether (3) CO_2 (4) H_2O, H^+

d. (1) $CH_3O^-Na^+$ (2) KCN, DMSO (3) H_2O, H_2SO_4, heat

10 What is the end product of the following reaction?

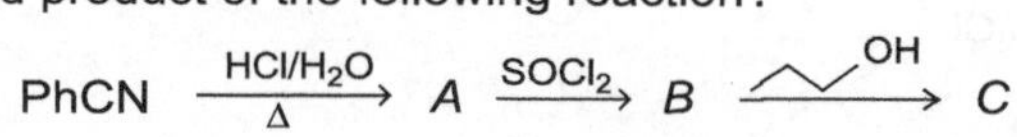

a. 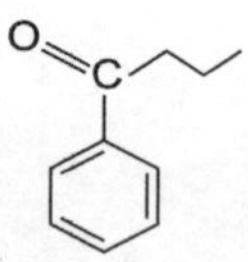 b. 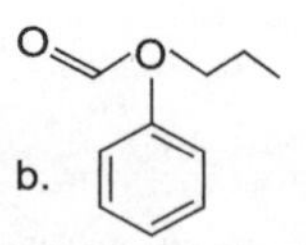 c. 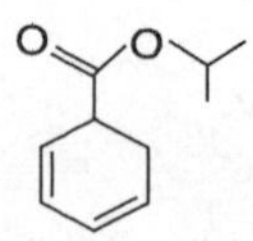 d. 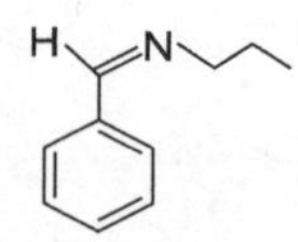

11 Which of the following is the best method for the preparation of following compound *A*?

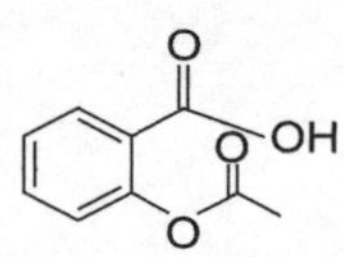

a. 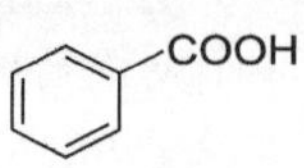 + acetic acid $\xrightarrow{H^+}$

b. 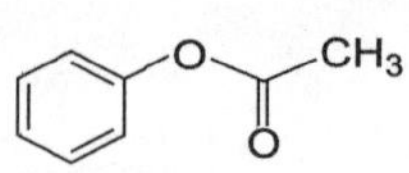 + formic acid $\xrightarrow{H^+}$

c. 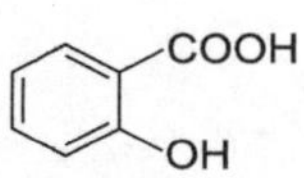 + acetic anhydride $\xrightarrow{H^+}$

d. 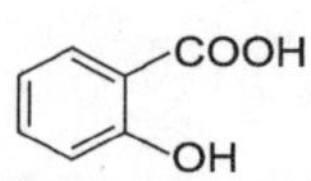 + acetamide $\xrightarrow{H^+}$

12 What is the major product of the reaction?

2 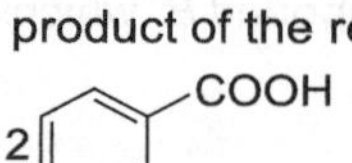 + $HOCH_2CH_2OH \xrightarrow{H^+}$?

a. 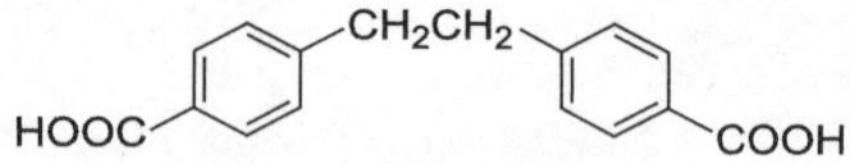

b. 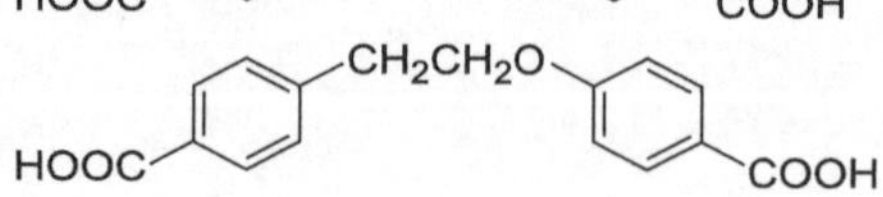

c. 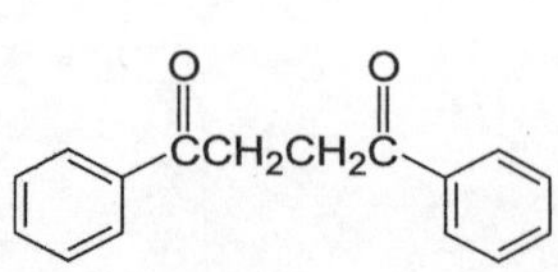

d. 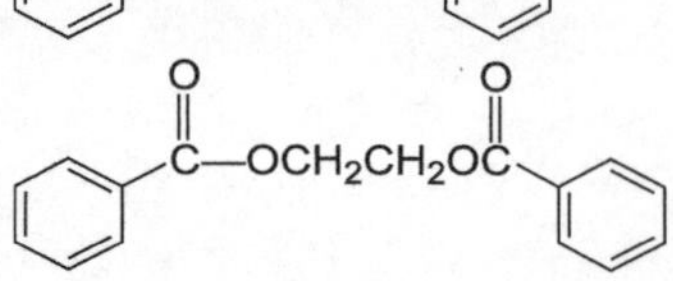

Working Space

One or More than One Options Correct Type

Direction (Q. Nos. 13-15) This section contains 3 multiple choice questions. Each question has four choices (a), (b), (c) and (d), out of which ONE or MORE THAN ONE are correct.

13 Which of following reaction will produce benzoic acid as one of the product?

a. 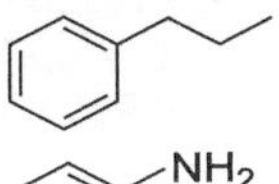 $+ KMnO_4 \xrightarrow[\Delta]{NaOH} \xrightarrow{H_3O^+}$

b. 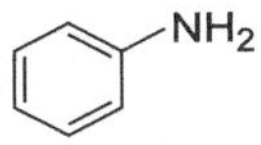 $\xrightarrow[HCl/0°C]{NaNO_2} \xrightarrow{CuCN} \xrightarrow[H_2O]{H^+}$

c. $C_6H_5MgBr + HCOOH \xrightarrow{H_2O}$

d. $C_6H_5CHO + NaOH \xrightarrow[H_2O]{H^+}$

14 Select the correct statement regarding benzoic acid.

a. It is a weaker acid than ethanoic acid.
b. Its sodium salt in aqueous solution does not yield biphenyl in Kolbe's electrolytic method.
c. Its solubility is lower but boiling point is higher than ethanoic acid.
d. It does not form salt with *aq.* $NaHCO_3$

15 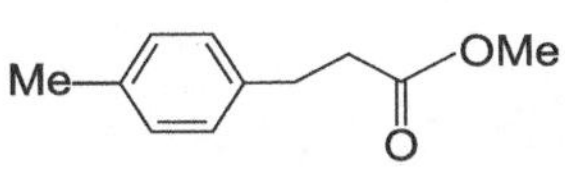 $\xrightarrow{NaOMe}$ *A* + MeOH $\xrightarrow{H_3O^+}$ *B* + MeOH; *B* $\xrightarrow[150° - 200°C]{\Delta}$ *C*

The true statement(s) about the above reaction is/are

a. Compound *B* is β-keto acid.
b. *C* is 1, 5-bis (*p*-tolyl) hexane-3-one
c. *A* gives effervescence of CO_2 when treated with $NaHCO_3$
d. *C* does not form iodoform when treated with NaOI

Statement Type

Direction (Q. Nos. 16 and 17) This section is based on Statement I and Statement II. Select the correct answer from the codes given below.

a. Both Statement I and Statement II are correct and Statement II is the correct explanation of Statement I
b. Both Statement I and Statement II are correct but Statement II is not the correct explanation of Statement I
c. Statement I is correct but Statement II is incorrect
d. Statement II is correct but Statement I is incorrect

16 **Statement I** Decarboxylation takes place when *o*-hydroxy benzoic acid is nitrated forming picric acid.

Statement II —OH is strongly activating group.

17 **Statement I** The following reaction is feasible.

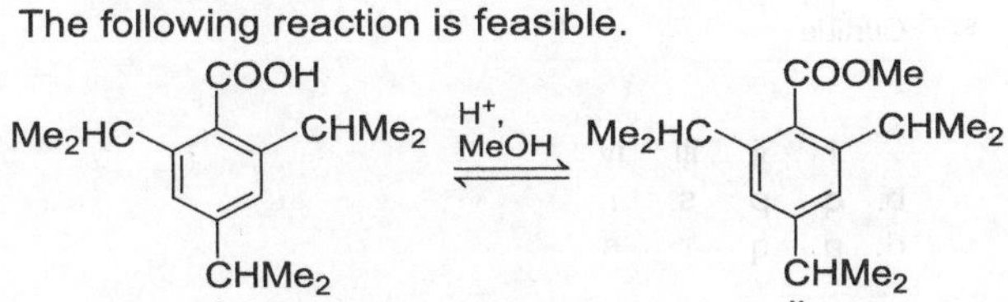

I. II.

Statement II Acylium ion is resonance stabilised.

Comprehension Type

Working Space

Direction *(Q. Nos. 18-20) This section contains a paragraph, each describing theory, experiments, data, etc. Three questions related to the paragraph have been given. Each question has only one correct answer among the four given options (a), (b), (c) and (d).*

Passage

Four rearrangement reaction are given below.

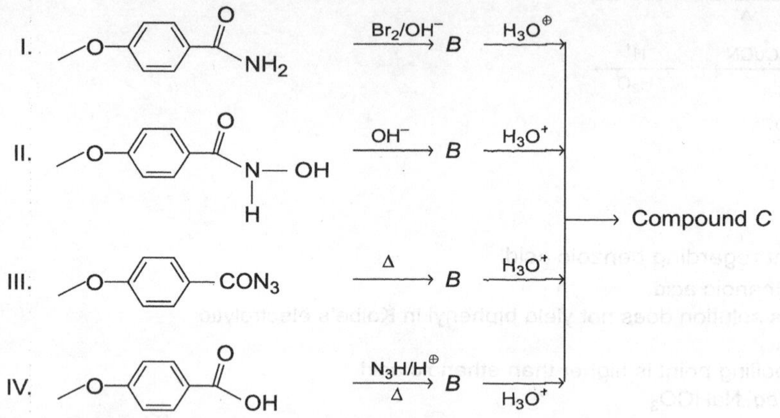

18 Which statement(s) is/are wrong about the above reaction?

a. All the reactions proceed *via* the formation of acyl nitrene

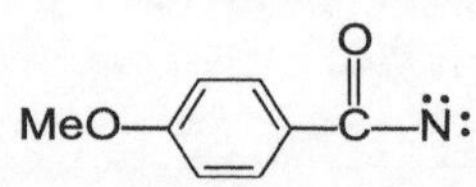

as intermediate species

b. All the reactions proceed *via* the formation of alkyl isocyanate as the intermediate compound

c. In all the reactions the product *C* is *p*-methoxy aniline

d. In all the reactions, the product *C* is aniline

19 Which statement(s) is/are wrong about the reaction?

a. EDG (electron donating group) at *o*/*p* position of a migrating aryl group accelerate the reaction

b. Slow and rate determining step is the formation of isocyanate from acyl nitrene.

c. The overall reaction involves the elimination of CO as CO_2

d. Using optically active reactant, there is a retention of configuration in the product.

20 Match the Column I and Column II choose the correct option using the codes given below :

	Column I (Reaction)		Column II (Name of reaction)
i.	I	p.	Lossen
ii.	II	q.	Hofmann bromamide
iii.	III	r.	Claisen-Schmidt rearrangement
iv.	IV	s.	Curtius

Codes

	i	ii	iii	iv		i	ii	iii	iv
a.	p	q	r	s	b.	q	p	s	r
c.	q	p	r	s	d.	p	q	r	s

DPP-6 Acidity of Carboxylic Acids

Only One Option Correct Type

Working Space

Direction (Q. Nos. 1-8) *This section contains 8 multiple choice questions. Each question has four choices (a), (b), (c) and (d), out of which ONLY ONE is correct.*

1 Among the following compounds, the one having highest acidity constant K_a is

a. 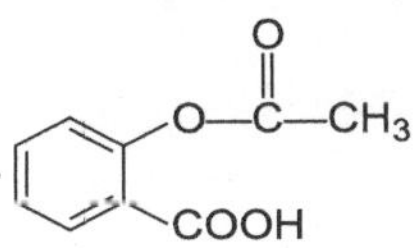

b. 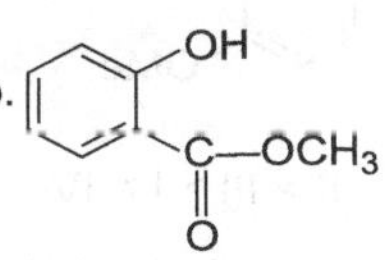

c. 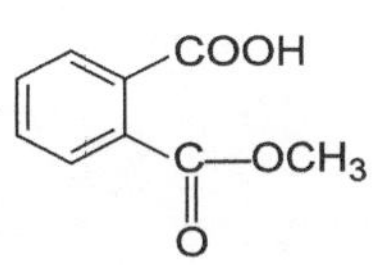

d. 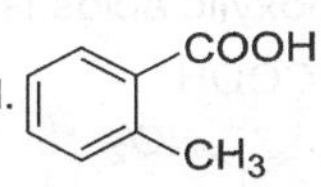

2 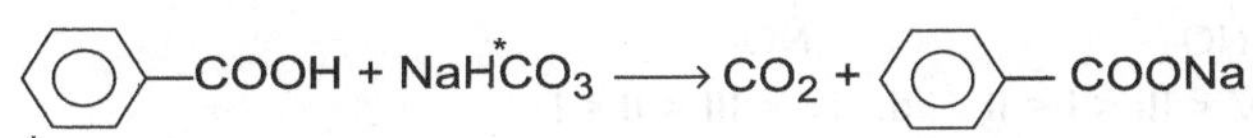

$\overset{*}{C}$ is with the product

a. CO_2

b. 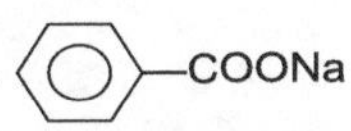

c. Both (a) and (b)

d. None of these

3 Rank the following in decreasing order of their acidic strength and select the correct option.

I. 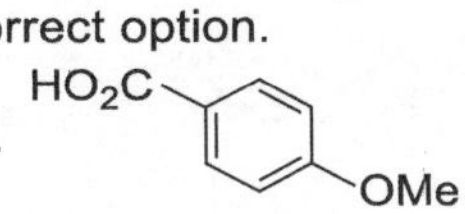

II. 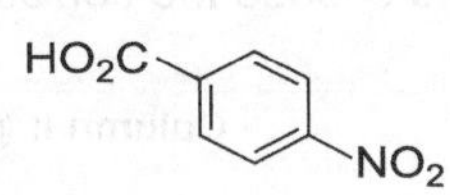

III. 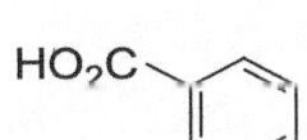

IV. 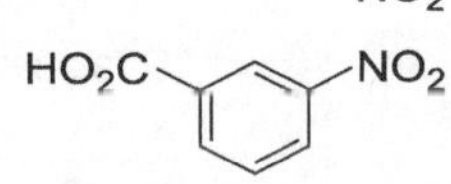

a. III > IV > II > I

b. II > IV > III > I

c. IV > II > I > III

d. IV > III > II > I

4 The order of K_b value of the following compound is

I. 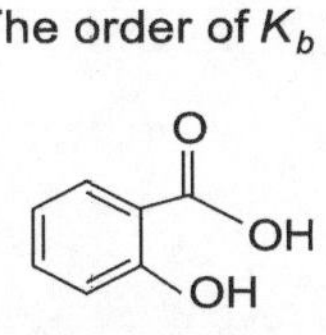

II. 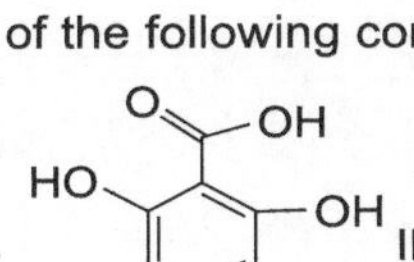

III. 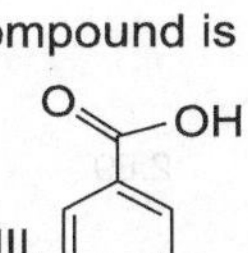

IV. 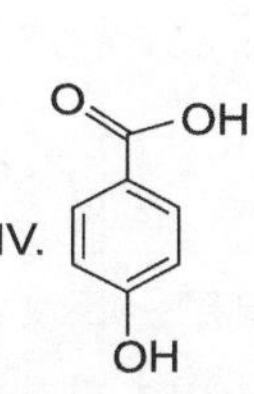

a. II > I > IV > III

b. IV > III > I > II

c. IV > I > III > II

d. III > IV > I > II

Working Space

5 Arrange the following in the decreasing order of acidity.

I. 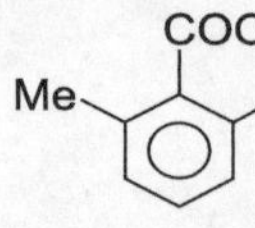 II. 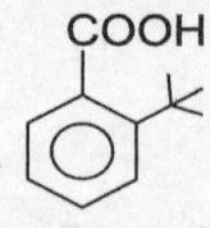 III. 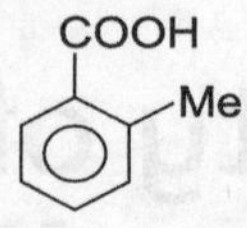

a. I > II > III b. II > I > III c. III > I > II d. II > III > I

6 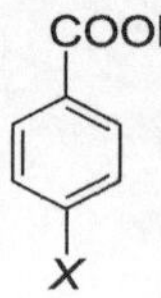 If X is substituted by different halogen atoms, the order of acidity is represented as

a. F > Cl > Br > I b. Cl > F > Br > I c. I > Br > Cl > F d. Cl > Br > F > I

7 Arrange the following in the decreasing order of their acidic strength.

I. 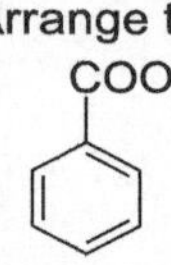 II. 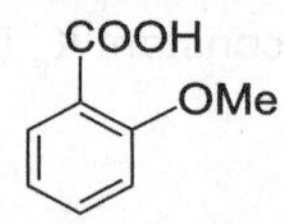 III. 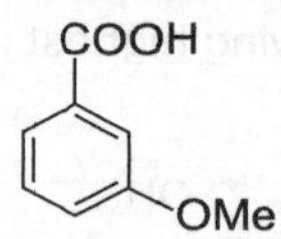 IV. 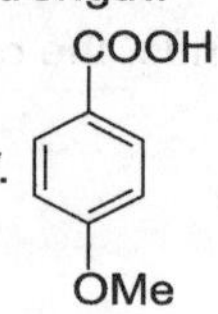

a. II > III > I ≈ IV b. II ≈ III > I > IV c. II > III > I > IV d. I > II > III > IV

8 The increasing order of pK_a of the given carboxylic acids is

I. 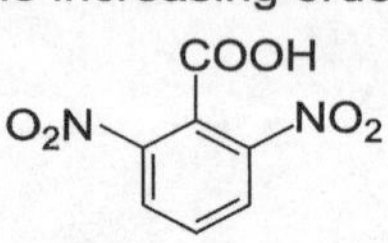 II. 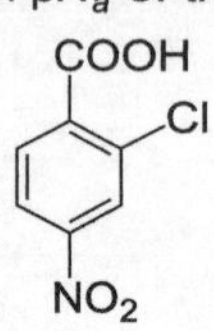 III. 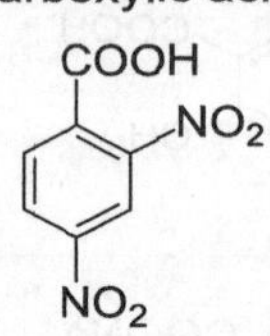 IV. 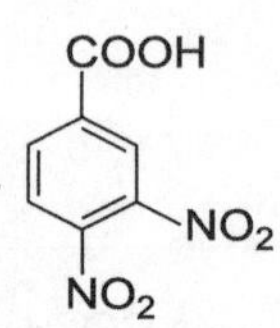

a. III < I < IV < II b. I < III < II < IV c. IV < III < I < II d. IV < III < II < I

Matching List Type

Direction *(Q. No. 9) Choices for the correct combination of elements from Column I and Column II are given as options (a), (b), (c) and (d), out of which one is correct.*

9 Match the Column I and Column II and choose the correct answer using the codes given below :

	Column I (Acid)		Column II (pK_a)
i.	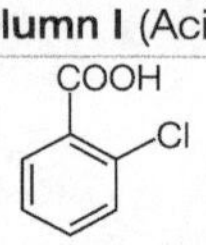	p.	3.98
ii.	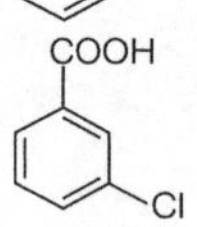	q.	4.17
iii.	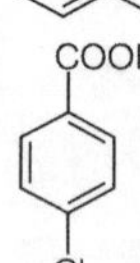	r.	2.89
iv.	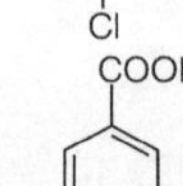	s.	3.82

Codes

	i	ii	iii	iv		i	ii	iii	iv
a.	s	r	q	p	b.	p	q	r	s
c.	r	s	p	q	d.	s	r	p	q

Statement Type

Working Space

Direction *(Q. Nos. 10-14) This section is based on Statement I and Statement II. Select the correct answer from the codes given below.*

a. Both Statement I and Statement II are correct and Statement II is the correct explanation of Statement I
b. Both Statement I and Statement II are correct but Statement II is not the correct explanation of Statement I
c. Statement I is correct but Statement II is incorrect
d. Statement II is correct but Statement I is incorrect

10 **Statement I** *o*-nitrobenzoic acid is a stronger acid than 3, 5-dinitrobenzoic acid.

Statement II NO_2 is a $-I$ and $-R$ showing group.

11 **Statement I** Benzoic acid is stronger than acetic acid.

Statement II In benzoic acid, C-atom in benzene ring is sp^2-hybridised and in acetic acid, C-atom in methyl group is sp^3-hybridised.

12 **Statement I** *o*-hydroxybenzoic acid (salicylic acid) is far weaker than the corresponding *m*-and *p*-isomers.

Statement II The carboxylate ion of salicylic acid is stabilised by intramolecular hydrogen bonding.

13 **Statement I** Benzoic acid is a stronger acid than formic acid.

Statement II Ph group is an electron attracting group and presence of such a group results in increased acidity.

14 **Statement I** Phthalic acid is a stronger acid than either isophthalic or terephthalic acid.

Statement II Phthalate monoanion is weaker than isophthalate and terephthalate monoanions.

One or More than One Options Correct Type

Direction *(Q. Nos. 15 and 16) This section contains 2 multiple choice questions. Each question has four choices (a), (b), (c) and (d), out of which ONE or MORE THAN ONE are correct.*

15 Which of the following statement(s) is/are true?

a. *m*-amino benzoic acid is a stronger acid than its *o* and *p*-counter parts.
b. 2, 6-dihydroxy benzoic acid is weaker than salicylic acid but stronger than *m*-hydroxy benzoic acid.
c. Carboxylic acids are stronger than $\left[R-\overset{\overset{\displaystyle O}{\|}}{C}-O-OH \right]$
d. The order of acidity of some compounds is

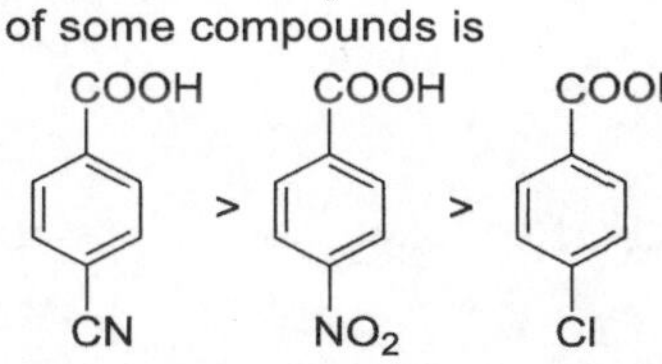

16 Which of the following order of acidity is/are not true?

a. *o*-toluic acid > benzoic acid > *meta*-toluic acid > *para* toluic acid
b. *o*-nitrobenzoic acid > *p*-nitrobenzoic acid > *m*-nitrobenzoic acid > benzoic acid
c. *m*-methoxy benzoic acid > benzoic acid > *o*-methoxy benzoic acid > *p*-methoxy benzoic acid
d. *p*-nitrobenzoic acid > 2, 6-dimethyl-4-nitrobenzoic acid > 3, 5-dimethyl-4-nitro benzoic acid > *p*-carboxylic benzene sulphonic acid

One Integer Value Correct Type

Direction *(Q. Nos. 17-20) This section contains 4 questions. When worked out will result in an integer from 0 to 9 (both inclusive).*

17 How many of the following reacts with NaOH faster than benzoic acid?

I. Picric acid
II. 2, 4-dinitrobenzoic acid
III. *m*-methoxybenzoic acid
IV. *o*-methoxybenzoic acid
V. *p*-methoxybenzoic acid
VI. *o*-aminobenzoic acid
VII. *p*-cyanobenzoic acid
VIII. *o*-toluic acid

18 If the following compound *A* is treated with 4 moles of sodium metal, the amount of hydrogen released is

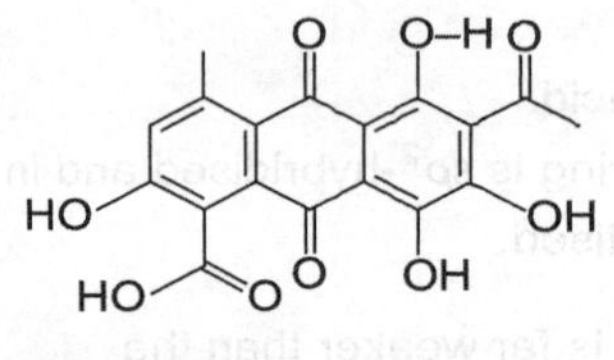

19 The Na will go at which position?

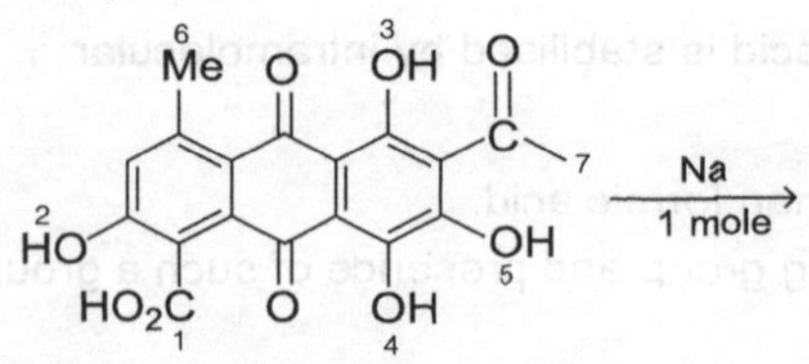

20 How many moles of sodium bicarbonate is required to react with *Y* completely?

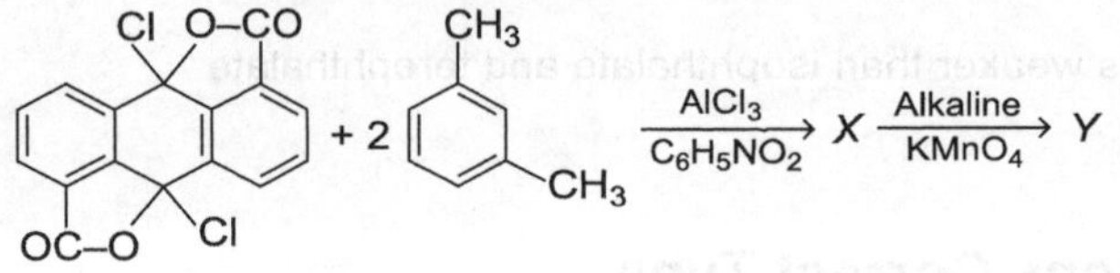

Revisal Problems for JEE Main

SET - 1

Only One Option Correct Type

Working Space

Direction *(Q. Nos. 1-25) This section contains 25 multiple choice questions. Each question has four choices (a), (b), (c) and (d), out of which ONLY ONE is correct.*

1 By using which reagent the following transformation is carried out?

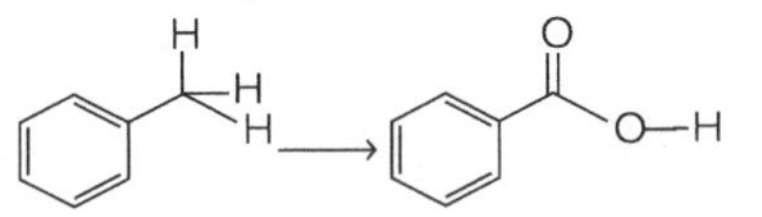

a. HNO_3/H_2SO_4 b. $Cl_2, FeCl_3$

c. SO_2/H_2SO_4 d. $KMnO_4, \bar{O}H/\Delta; H_3O^+$

2 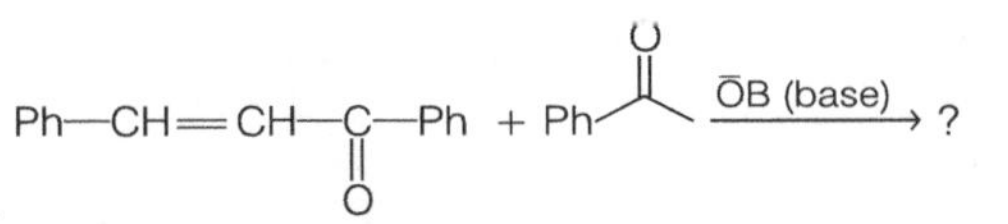

a. a 1,3-diketone b. a 1,5-diketone

c. a 1,2-diketone d. None of these

3 Which aromatic compound behave most like benzoic acid when it undergoes reaction with conc. HNO_3 and H_2SO_4?

a. Phenol b. Phenyl methyl ketone

c. Benzene d. Chlorobenzene

4 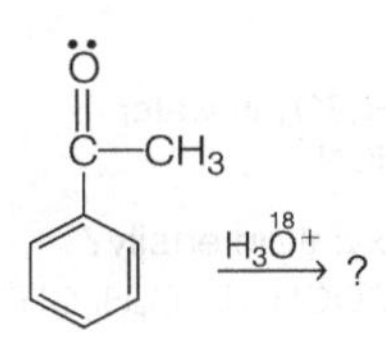

Product is

a. 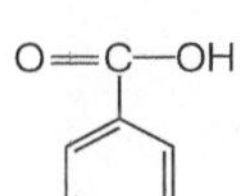 b. 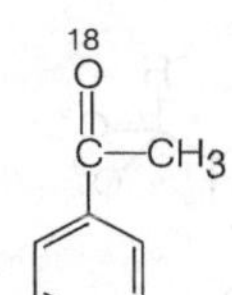

c. 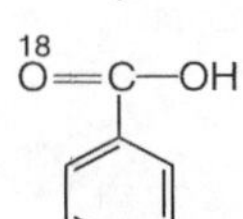 d. No reaction takes place at all

5 The $NaOCH_3$ catalysed reaction between $CH_3CH_2CO_2CH_3$ and benzaldehyde gives

a. 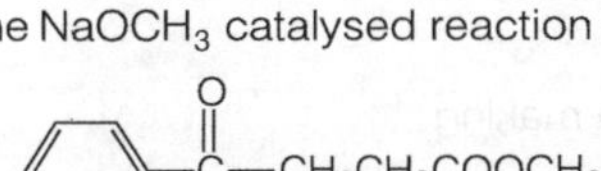 b. 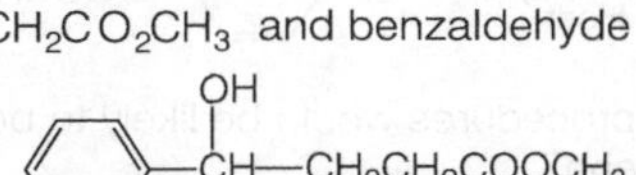

c. 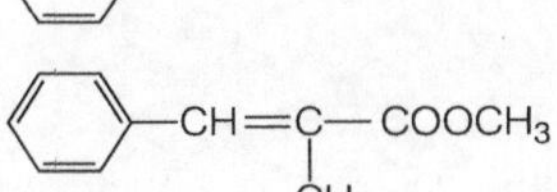 d. 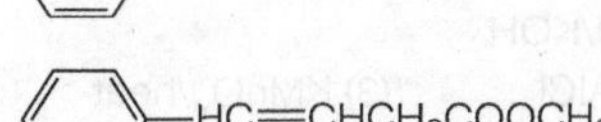

Working Space

6 Which of the following will have the highest K_{eq} value for cyanohydrin formation?

a. 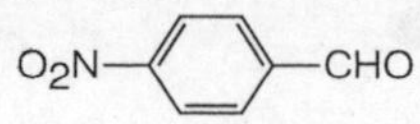
b. 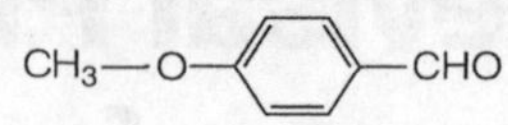
c. 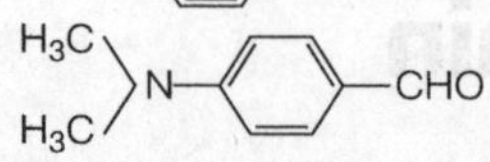
d. 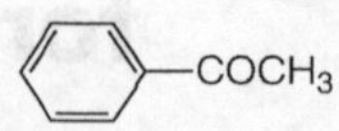

7 Which compound exists to greatest extent as its hydrate when dissolved in aqueous solution?

a. 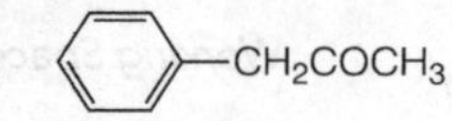
b. 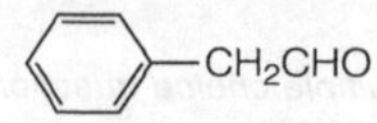
c. $C_6H_5COCH(CH_3)_2$
d. Benzophenone

8 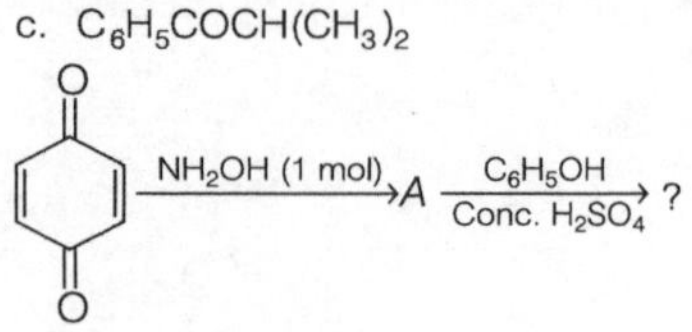

Product is

a. Parathymol b. Indophenol c. Anthranilic acid d. None of these

9 The best procedure for the following transformation is

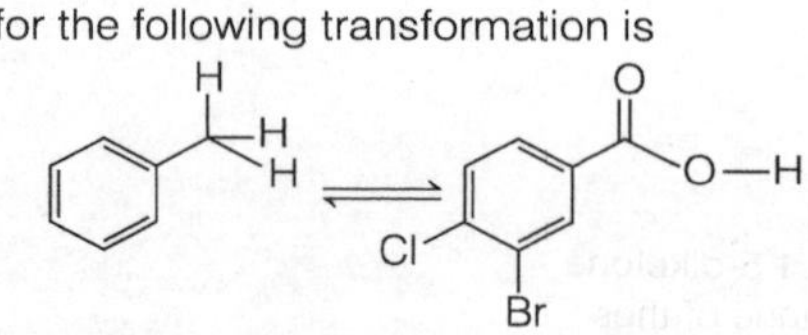

a. $Cl_2/FeCl_3$; $Br_2/FeCl_3$; $KMnO_4/FeCl_3$; H_3O^+
b. $Cl_2/FeCl_3$; $KMnO_4/\bar{O}H$; H_3O^+; $Br_2/FeCl_3$
c. $KMnO_4/\bar{O}H$; H_3O^+; $Br_2/FeCl_3$; Cl_2; $FeCl_3$
d. $KMnO_4/\bar{O}H$; H_3O^+; $Cl_2/FeCl_3$; $Br_2/FeCl_3$

10 Which is the best method for transforming benzoic acid to benzonitrile?

a. Treatment with $SOCl_2$; then with NH_3; then with $SOCl_2$ again.
b. Treatment with Br_2 + $FeBr_3$; then with CuCN
c. Treatment with CH_3OH + trace HCl; then with NH_3; then with H_2SO_4 in water.
d. Treatment with $SOCl_2$; then with benzene + $AlCl_3$; then with HCN

11 Which of the following carboxylic acids undergoes decarboxylation easily?

a. $C_6H_5COCH_2COOH$ b. $C_6H_5COCOOH$ c. $C_6H_5CH(OH)—COOH$ d. $C_6H_5CH(NH_2)—COOH$

12 Among the following the best H^- ion donor is

a. 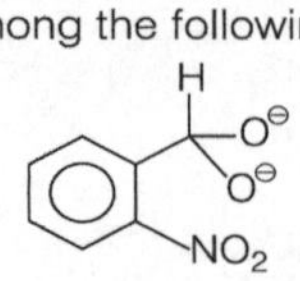
b. 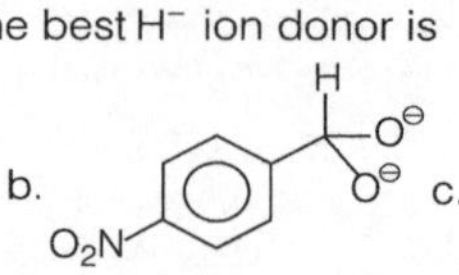
c. 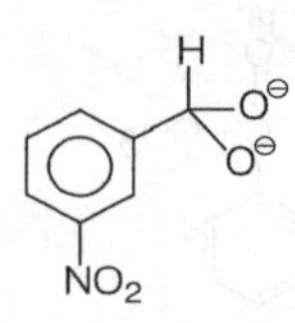
d. 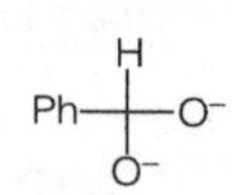

13 Which of the following reactions will not yield benzoic acid as the major product?

a. 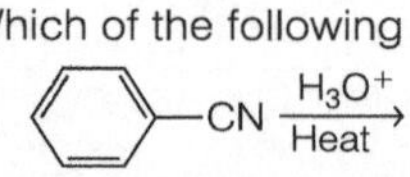
b. 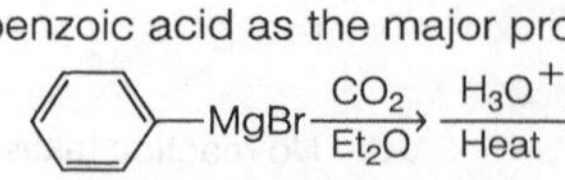
c. 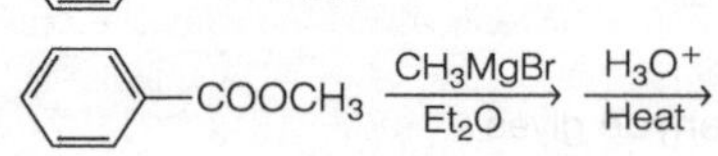
d. 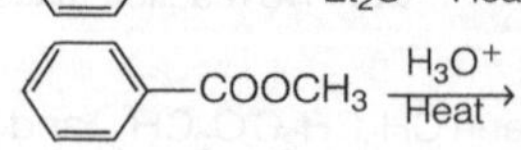

14 Which of the following synthetic procedures would be likely to be successful in making *p*-bromobenzoic acid from benzene?

I. (1) Br_2/Fe, (2) HCOOH/KOH
II. (1) Br_2/Fe, (2) $CH_3Br/AlCl_3$, (3) $KMnO_4$/heat
III. (1) HCOOH/KOH, (2) Br_2/Fe,
IV. (1) $CH_3Br/AlCl_3$, (2) Br_2/Fe, (3) $KMnO_4$/heat
V. (1) $CH_3Br/AlCl_3$, (2) $KMnO_4$/heat, (3) Br_2/Fe

a. I and II b. I and III c. II and IV d. III and IV

15 By which reaction benzaldehyde is directly converted into benzyl benzoate?

a. Tischenko reaction
b. Knoevenagel reaction
c. Wittig reaction
d. None of these

16 The best way for the preparation of *p*-nitrobenzoic acid is

a. 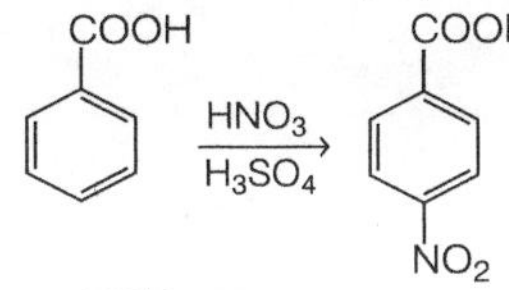 (+ *ortho* isomer)

b. 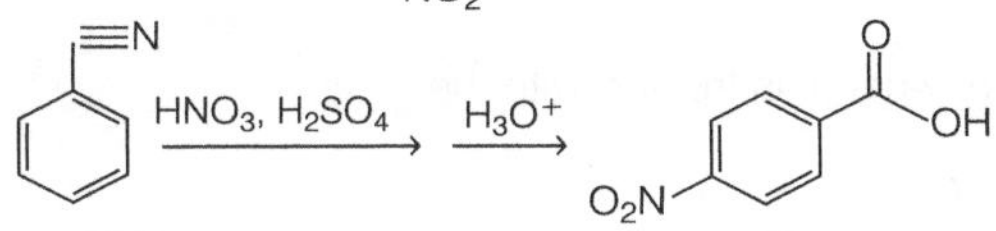

c. 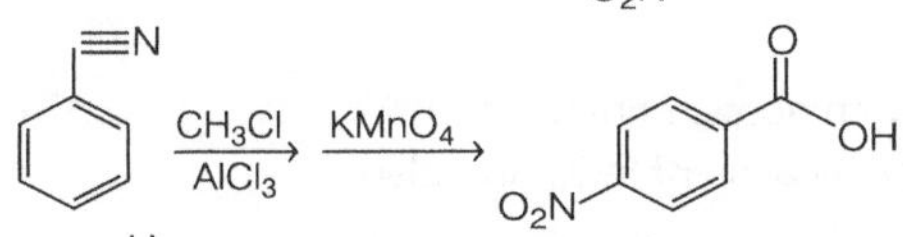 (+ *ortho* isomer)

d. 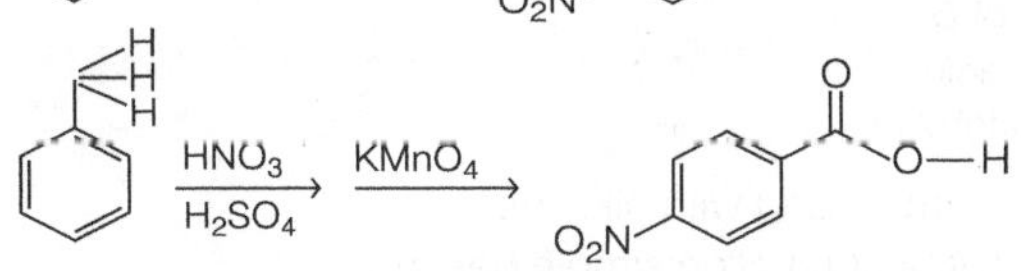

17 Select the best synthesis of *p*-*tert*-butyl acetophenone.

I. 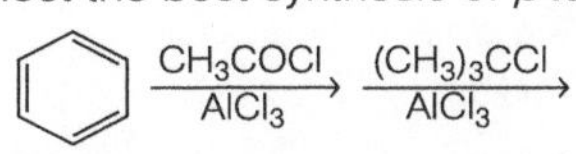

II. 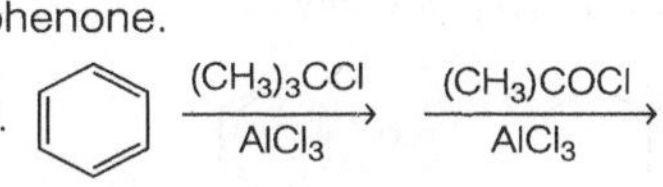

III. 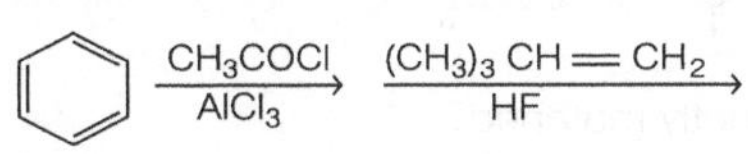

a. Only I
b. Only II
c. Only III
d. More than one of these

18 Which of the statements is true about the internal Cannizaro reaction given below?

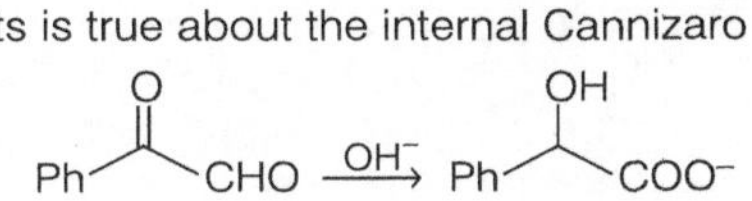

a. $\bar{O}H$ attacks at C-1 of —CHO group
b. $\bar{O}H$ attacks at C-2 of (C=O) group
c. H^+ transfer takes place from C-1 to C-2
d. H^+ exchanges from acid to alcohol ion

19 What would be the best way to prepare the ketone shown below?

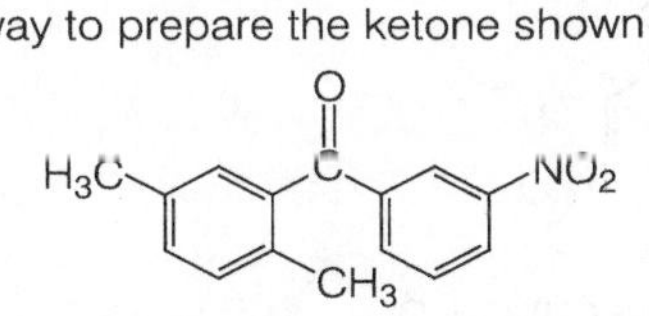

a. 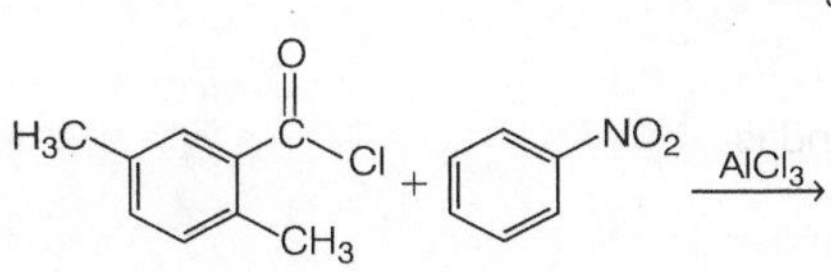

b. 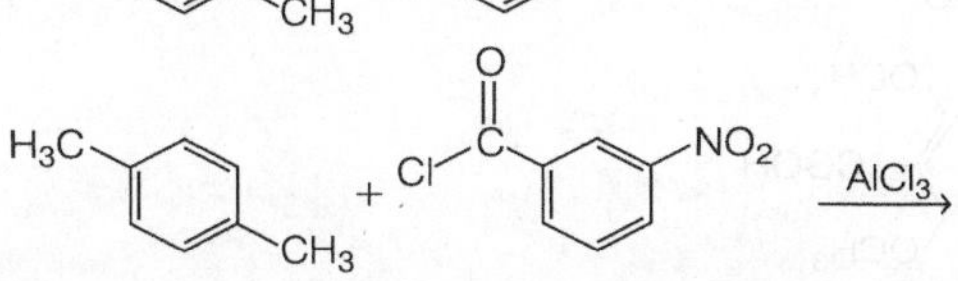

c. 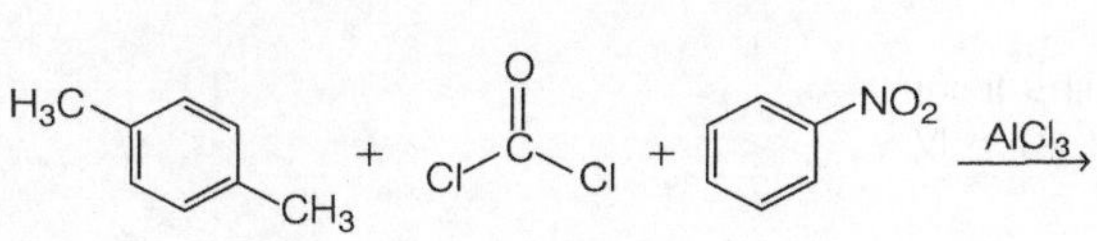

d. All of the above

Working Space

Working Space

20 $PhCOCH_3 \xrightarrow[NaNH_2]{ClCH_2CO_2Et/} A \xrightarrow[\substack{H_3O^+ \\ Warmed}]{OH^-} B.$

Here *B* is

a. 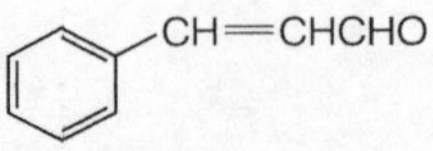

b. 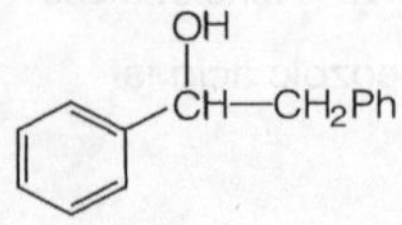

c. 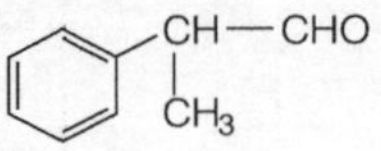

d. 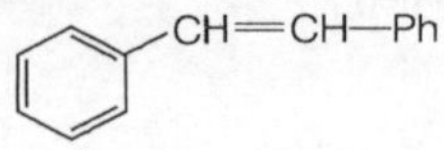

21 Which of the following compounds yields *n*-propyl benzene, when it is treated with the amalgam in HCl?

a. $PhCH_2CH_2COOH$
b. $PhCH_2CH{=}CH_2$
c. $PhCH_2—CH(OH)—CH_3$
d. $PhCOCH_2CH_3$

22 A natural compound, $A\,(C_7H_6O)$ forms an oxime. When treated with acetic anhydride and sodium salt of corresponding acid *A* gives *B* which decolourises alkaline $KMnO_4$ and also gives effervescence with $NaHCO_3$. Predict the structures of *B*.

a. Crotonic acid
b. Cinnamic acid
c. β, γ-unsaturated acid
d. Cinnamaldehyde

23 An aromatic hydrocarbon *A* has molecular formula C_8H_{10}. *A* on boiling with alkaline permanganate solution followed by acid hydrolysis gives $B\,(C_8H_6O_4)$. *B* on simple heating undergoes dehydration to give $C(C_8H_4O_3)$. Identify *A*.

a. 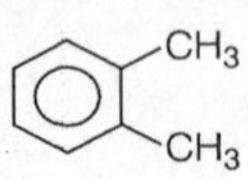
b. 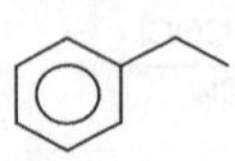
c. 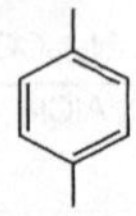
d. 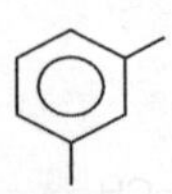

24 In which of the following reactions, reactants and products are not correctly matched?

a. 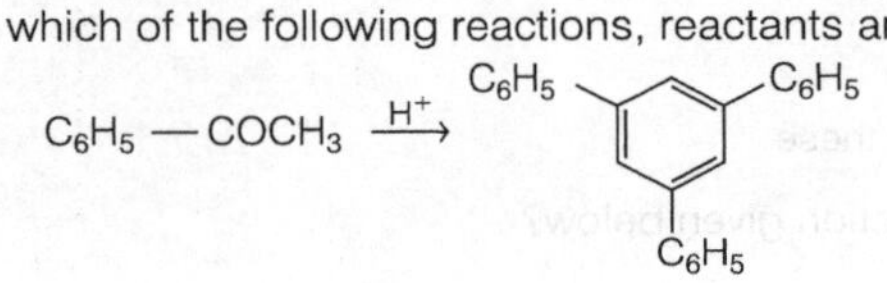

b. 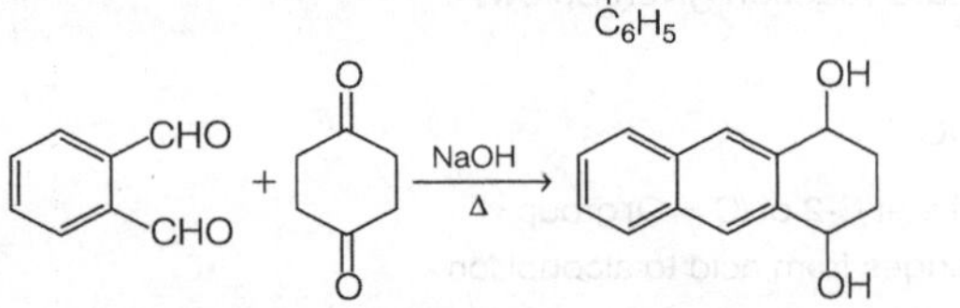

c. 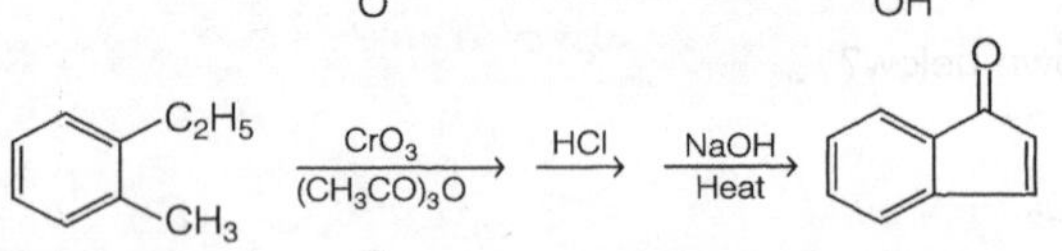

d. $C_6H_5CHO + CH_3—\overset{O}{\overset{\|}{C}}—CH_3 \xrightarrow[C_2H_5OH]{C_2H_5ONa} (C_6H_5—CH{=}CH)_2C{=}O$

25 The correct order of ease of esterification of the following compound is

I. 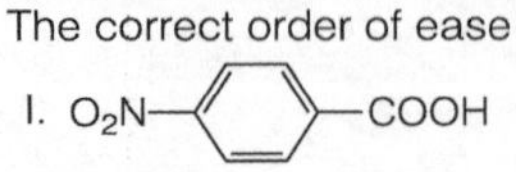

II. 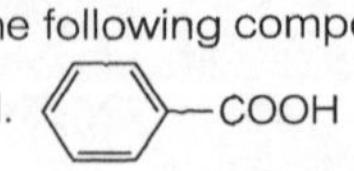

III. 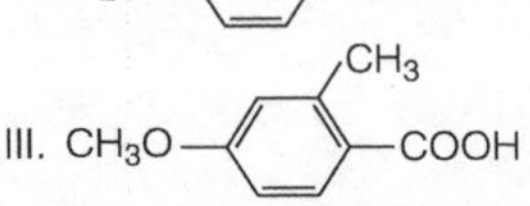

IV. 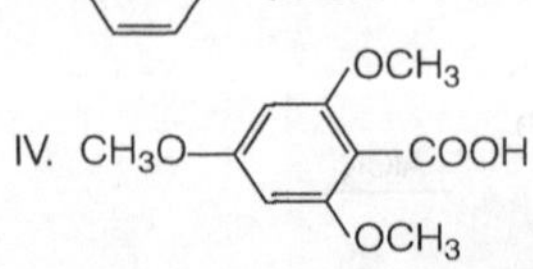

Codes

a. I > II > III > IV
b. IV > III > II > I
c. II > I > IV > III
d. I > III > II > IV

Working Space

Matching List Type

***Direction** (Q. Nos. 26 and 27) Choices for the correct combination of elements from Column I and Column II are given as options (a), (b), (c) and (d), out of which one is correct.*

26 Match the following Columns and choose the correct answer using the codes given below

	Column I (Reaction)		Column II (Intermediate)
i.	Claisen condensation	p.	Free radical
ii.	Allylic bromination	q.	Carbon anion
iii.	Fries rearrangement	r.	Alkyl migration
iv.	Wolff rearrangement	s.	Acyl migration

Codes :

	i	ii	ii	iv		i	iii	iii	iv
a.	p	q	r	s	b.	q	p	s	r
c.	p	q	s	s	d.	q	p	r	s

27 Match the Column I with Column II and select the correct answer using the codes given below the Columns

	Column I		Column II
i.	Aldol condensation	p.	Hydride ion transfer
ii.	Cannizaro reaction	q.	Organo zinc compound
iii.	Reformatsky reaction	r.	CN^- as catalyst
iv.	Benzoin condensation	s.	Enolate ion

Codes

	i	ii	iii	iv		i	ii	iii	iv
a.	s	p	q	r	b.	q	r	s	p
c.	s	r	q	p	d.	q	p	s	r

Statement Type

***Direction** (Q. Nos. 28 and 29) This section is based on Statement I and Statement II. Select the correct answer from the codes given below.*

a. Both Statement I and Statement II are correct and Statement II is the correct explanation of Statement I
b. Both Statement I and Statement II are correct but Statement II is not the correct explanation of Statement I
c. Statement I is correct but Statement II is incorrect
d. Statement II is correct but Statement I is incorrect

28 **Statement I** Sodium 2–OH–*cis* cinnamate gives coumarin on esterification but sodium 2–OH–5–nitro–*cis*–cinnamic acied does not.

Statement II —NO_2 group deactivates the ring towards electrophilic attack.

29 **Statement I** Oxime of benzaldehyde does not give normal Beckmann rearrangnement.

Statement II Oxime of benzaldehyde contains acidic hydrogen at carbonyl carbon.

Revisal Problems for JEE Advanced

Only One Option Correct Type

Working Space

***Direction** (Q. Nos. 1-8) This section contains 8 multiple choice questions. Each question has four choices (a), (b), (c) and (d), out of which ONLY ONE option is correct.*

1 Phthalic acid $\xrightarrow{\Delta} A \xrightarrow{NH_3} B \xrightarrow[\text{(ii) } Br_2/KOH]{\text{(i) NaOH, } H_2O} C \xrightarrow{HCl} D.$

The end product *D*, of the above reaction sequence is

a. *o*-nitrobenzoic acid
b. salicylic acid
c. anthranilic acid
d. crotonic acid

2 *p*-hydroxy benzaldehyde does not undergo Cannizaro reaction because

a. —OH is a deactivating group
b. —OH is a highly activating group
c. $+R$ effect of —OH group neutralises the positive charge of CHO
d. $-I$ effect of —OH group neutralises, the positive charge of —CHO

3 $2HCHO + PhCHO \xrightarrow{\overline{O}H} xPhCH_2OH + yCH_3OH + zHCOONa$

Values of x, y and z respectively are

a. 1, 2, 1
b. 1, 1.5, 0.5
c. 1, 0.5, 1.5
d. 1.5, 1, 1

4 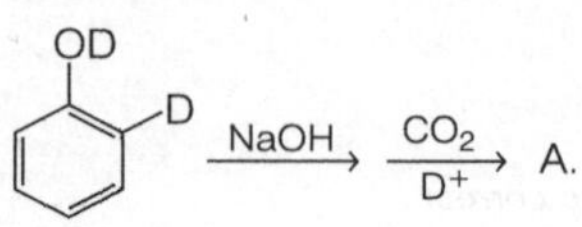

Here, *A* is

a. 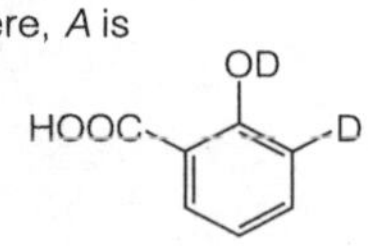

b. 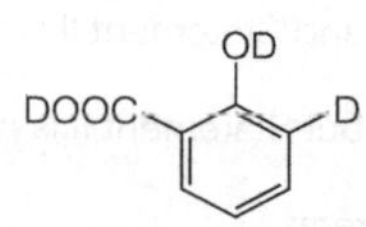

c. 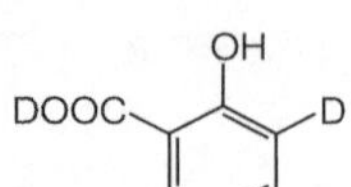

d. Reaction not possible

5 The mechanism of ester formation in acidic medium is as follows :

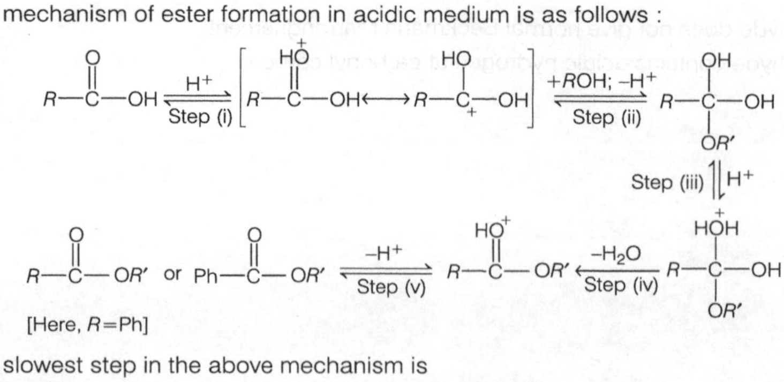

The slowest step in the above mechanism is

a. Step (i)
b. Step (ii)
c. Step (iii)
d. Step (iv)

6 Which of the following can be the product(s) of the following reaction?

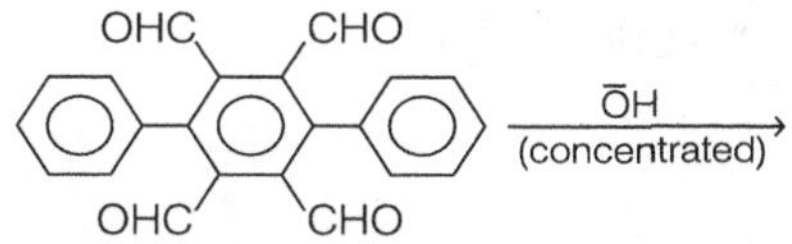

I. 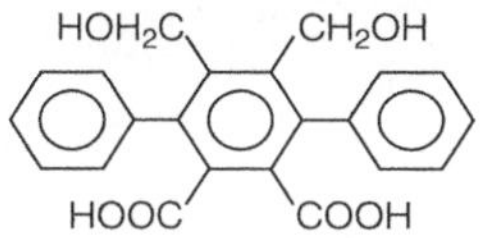

II. 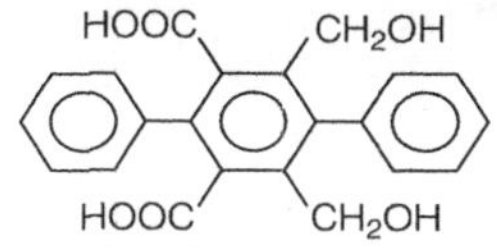

III. 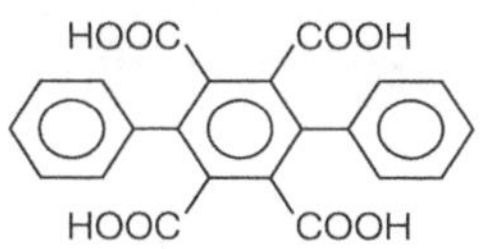

IV. 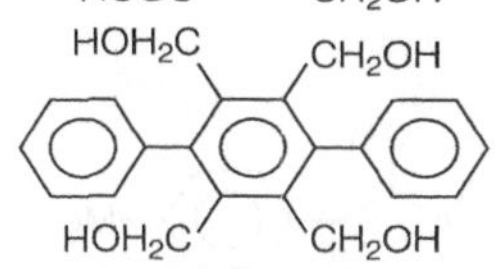

V. 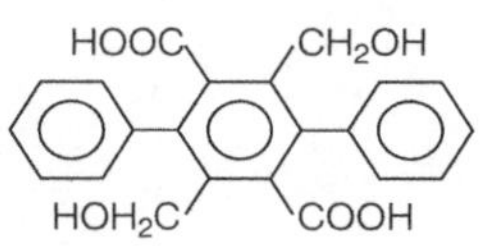

a. I, II and IV
b. III and IV
c. II and V
d. I and V

7 Which of the following reactions can be used to prepare acetophenone ?

I. Benzene and aluminium chloride and propionyl chloride.
II. Benzoyl chloride and diethyl lithium cuprate.
III. Benzene and aluminium chloride and acetic anhydride.
IV. Diphenyl lithium cuprate and acetyl chloride.

a. I and II   b. III and IV   c. I and III   d. II and IV

8 $PhCOOH + PhCH_2OH \overset{H^+}{\rightleftharpoons} PhCOOCH_2Ph + H_2O$

The true statements about this reaction are

I. This reaction is called Fischer esterification reaction.
II. Removal of water shifts the equilibrium towards right hand side.
III. The reaction of PhCOCl and $PhCH_2OH$ to give ester ($PhCOOCH_2Ph$) is irreversible and more feasible than esterification of *R*COOH.
IV. If $PhCH_2OH$ is in excess, the equilibrium is shifted towards right.

a. I and II   b. I, II and III   c. I, II and IV   d. All of these

One or More than One Options Correct Type

Direction *(Q. Nos. 9-15) This section contains 7 multiple choice questions. Each question has four choices (a), (b), (c) and (d), out of which ONE or MORE THAN ONE are correct.*

9 Choose the correct match of reactants and products from the following list.

a. 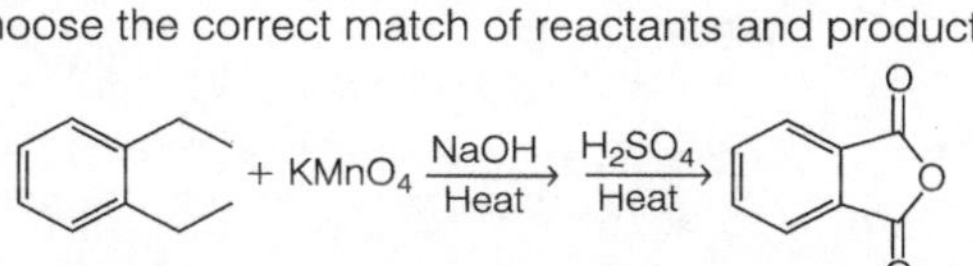

b. 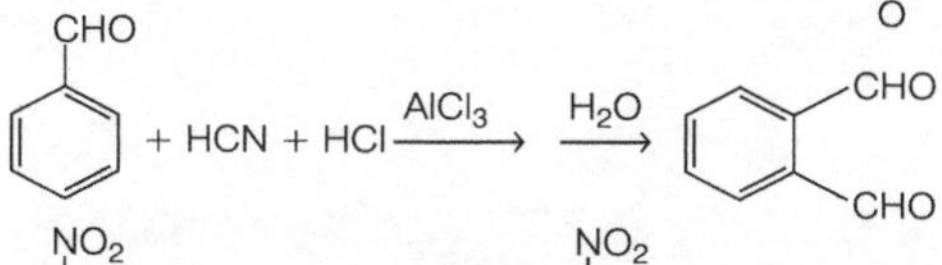

c. 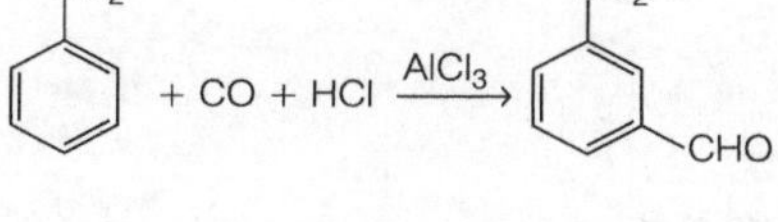

d. 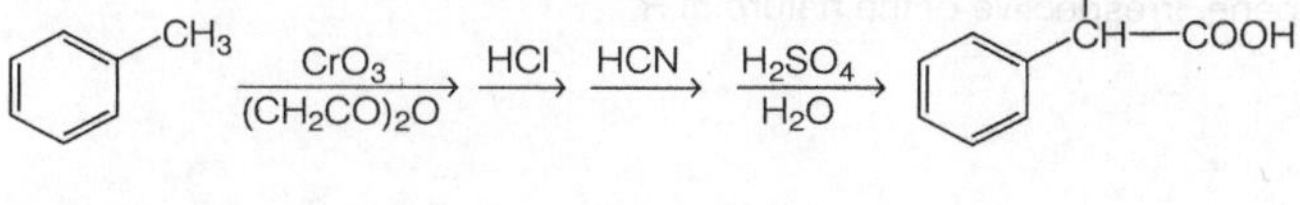

Working Space

10 Which of the following statements are true about the benzoin reaction?

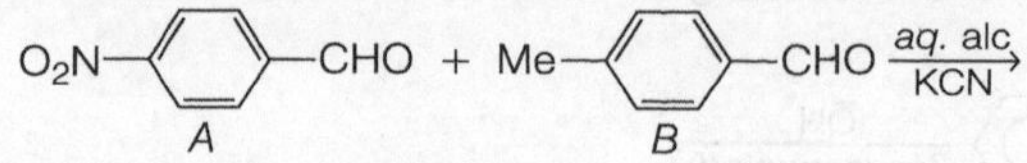

a. $\bar{C}N$ attacks at the C=O group of *A*
b. CN^- attacks at the (C=O) group of *B*
c. The major product formed is

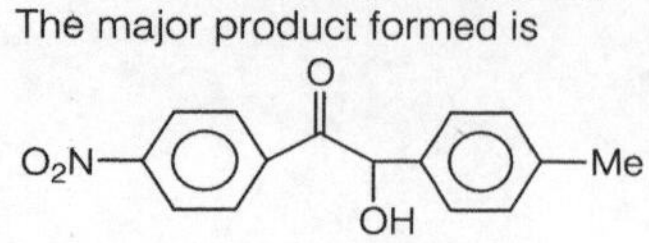

d. The major product formed is

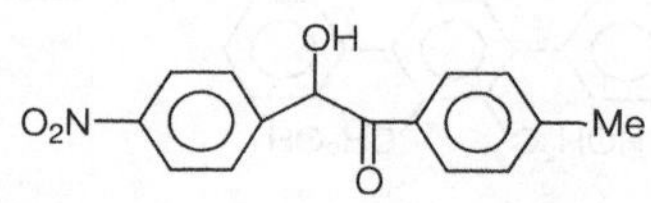

11 Which of the following reactions are Perkin reactions?

a. 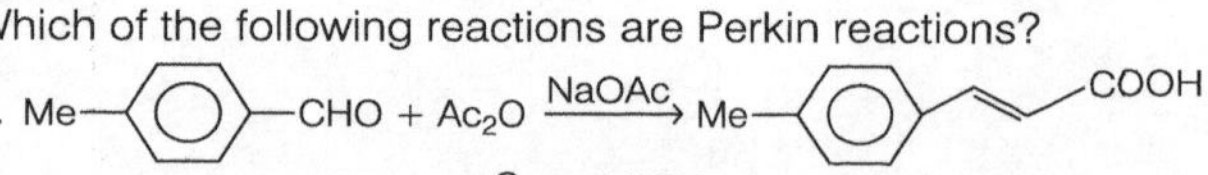

b. 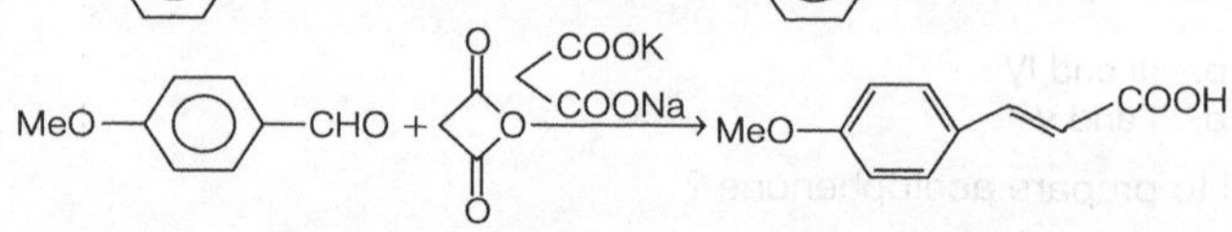

c. 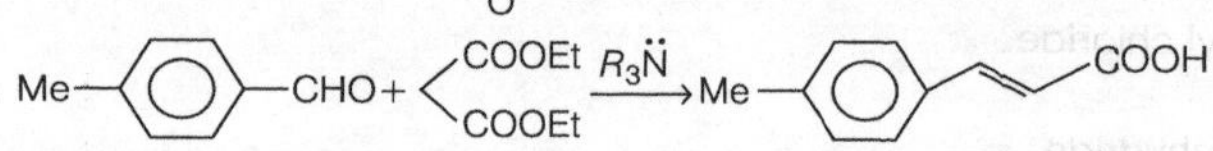

d. 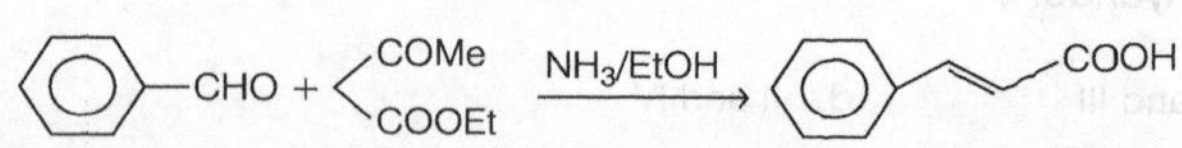

12 Which of the following statement(s) is/are correct about the following reaction?

$$\mathrm{Ph{-}\overset{O}{\overset{\|}{C}}{-}OH + R'OH \underset{}{\overset{H^+}{\rightleftharpoons}} PhCOOR' + H_2O}$$

a. Rate = k[PhCOOH] [R'OH]H^+. First order in acid, alcohol and H^+; overall a third order reaction
b. Rate = k[PhCOOH] [R'OH]., First order in acid and alcohol; overall second order reaction
c. Slow and rate determining step is the addition of alcohol to the protonated acid
d. Slow and rate determining step is the addition of acid to the protonated alcohol

13 In which of the following reactions, reactants and products are not correctly matched ?

a. $C_6H_5{-}CHO + CH_3{-}CH_2{-}CHO \xrightarrow{NaOH} C_6H_5{-}CH{=}CH{-}CH_2{-}CHO$

b. $C_6H_5CHO + (CH_3CH_2CO)_2O \xrightarrow[\Delta]{CH_3CH_2COONa} C_6H_5{-}CH{=}\underset{CH_3}{\underset{|}{C}}{-}COOH$

c. $C_6H_5CHO + HCN \longrightarrow C_6H_5{-}\underset{OH}{\underset{|}{CH}}{-}\overset{O}{\overset{\|}{C}}{-}C_6H_5$

d. $C_6H_5{-}CHO + Zn \longrightarrow C_6H_5{-}\underset{OH}{\underset{|}{CH}}{-}\underset{OH}{\underset{|}{CH}}{-}C_6H_5$

14 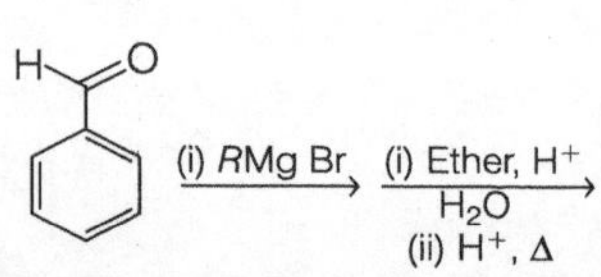

a. If $R = CH_3$, product is 1-phenyl propene
b. If $R = C_2H_5$, product is 1-phenyl propene
c. If $R = CH_3$ or C_2H_5, product is 2-phenyl propene, irrespective of the nature of R
d. If $R = CH_3$, product is 2-phenyl propene

15 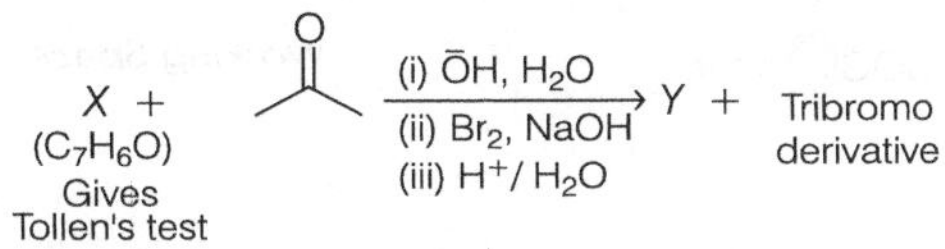

The true statements about the above reaction is/are

a. *Y* decolourises Br_2 water and 1% alkaline $KMnO_4$

b. Here, reaction of *X* and 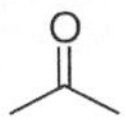 is called Claisen condensation

c. *Y* reacts with Na, NaOH and $NaHCO_3$ all

d. *Y* gives positive haloform reaction

Statement Type

***Direction** (Q. No. 16-18) This section is based on Statement I and Statement II. Select the correct answer from the codes given below.*

a. Both Statement I and Statement II are correct and Statement II is the correct explanation of Statement I

b. Both Statement I and Statement II are correct but Statement II is not the correct explanation of Statement I

c. Statement I is correct but Statement II is incorrect

d. Statement II is correct but Statement I is incorrect

16 **Statement I** Benzyl alcohol on treatment with acidic dichromate solution gives benzoic acid.

Statement II Dichromate is a strong oxidising agent, oxidises the intermediate aldehyde further into acid.

17 **Statement I** 2,6 – $Me_2C_6H_3COMe$ forms iodoform when treated with I_2/NaOH

Statement II Compounds containing —COMe group forms haloform when treated with X_2/NaoH.

18 **Statement I** (E) – oxime of $Ph_2CHCOCH_3$ does not undergo normal Beckmann rearrangement in acidic medium (with conc. H_2SO_4).

Statement II (E) – oxime forms diphenylmethyl cation in acidic medium.

Comprehension Type

***Direction** (Q. Nos. 19-27) This section contains 3 paragraphs, each describing theory, experiments, data, etc. Nine questions related to the paragraphs have been given. Each question has only one correct answer among the four given options (a), (b), (c) and (d).*

Passage I

Three isomeric tetra carboxylic acids I, II and III are given.

I. 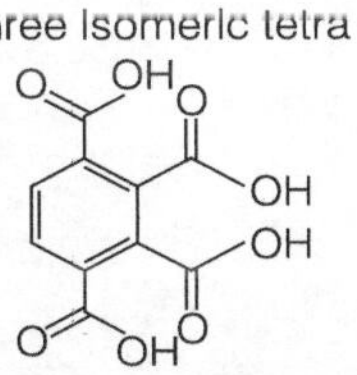

II. 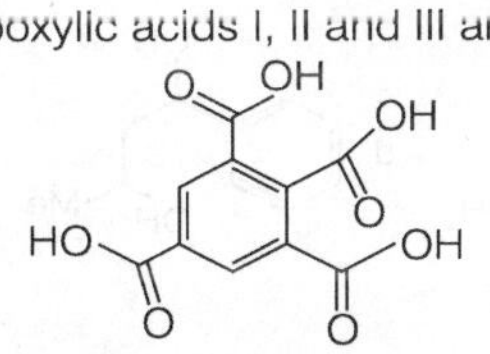

III. 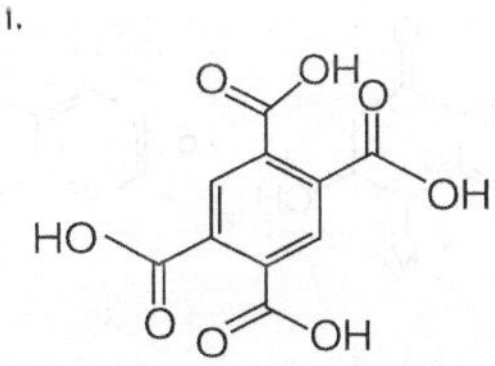

They can be distinguished by anhydride formation by treating with one or two equivalent of $SOCl_2$.

19 Which tetracarboxylic acid on reaction with one equivalent of $SOCl_2$ gives two isomeric monoanhydride?

a. Only I b. Only II
c. Only III d. All of these

20 Which tetracarboxylic acid on reaction with one equivalent of $SOCl_2$ gives only one monohydride?

a. Only I b. Only II c. Only III d. Both II and III

Working Space

Working Space

21 Which tetracarboxylic acid on reaction with second equivalent of $SOCl_2$ to the monohydrides of I, II and III gives a dianhydride ?

a. Only I
b. Only II
c. Only III
d. Both I and III

Passage II

Following few questions are based on the following reaction road map.

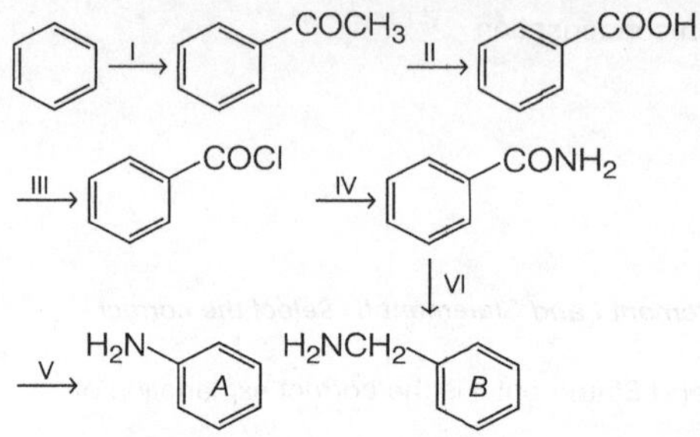

22 Which combination of reagents could be used to bring about step I (the use of water, dilute acid or alkali can be assumed at any stage) ?

a. CH_3COOH/Zn
b. $CH_3COCl/AlCl_3$
c. $CH_3CH_2Br/AlCl_3$ then $KMnO_4$
d. CH_3CHO/OH^-

23 Which reaction type would be involved in step II ?

a. Mannich reaction
b. Clemmensen reaction
c. Iodoform reaction
d. Friedel-Craft's reaction

24 On treatment with an excess of nitrous acid at 0° C, followed by heating of the resulting solution an organic product is obtained. The product is soluble in dilute aqueous caustic soda solution but not in aqueous sodium bicarbonate solution, this behaviour be shown by

a. *A* and *B*
b. *A* only
c. *B* only
d. Neither *A* nor *B*

Passage III

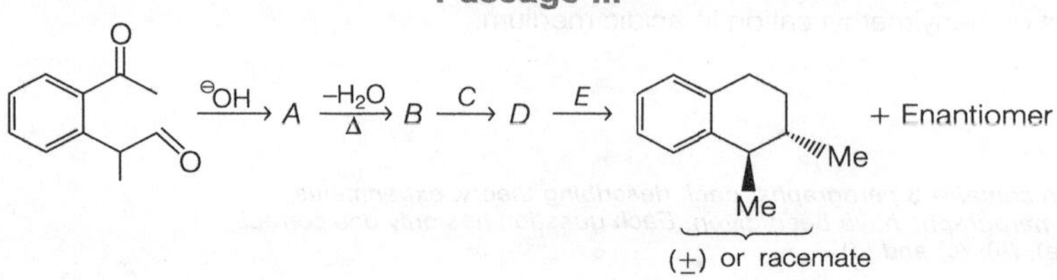

25 Compound *A* is

a. 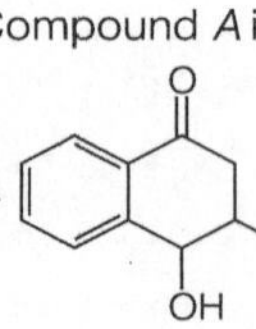
b. 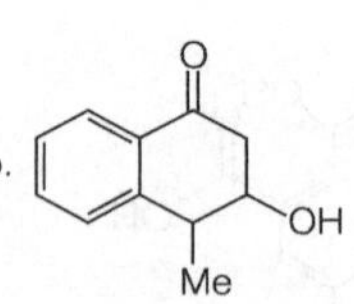
c. 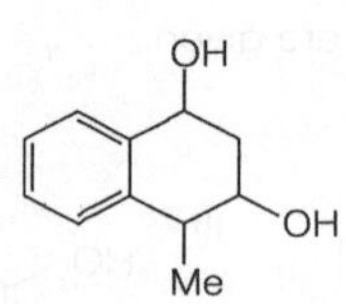
d. 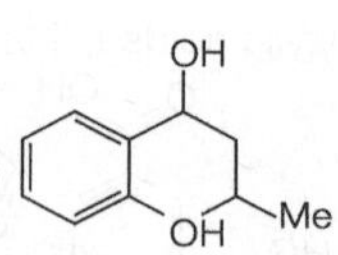

26 Compound *D* is

a. 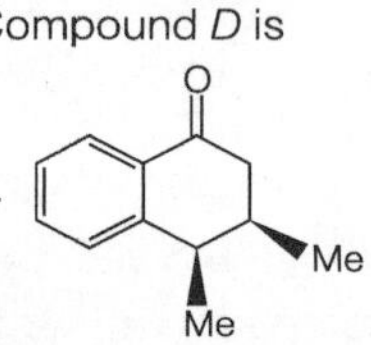
b. 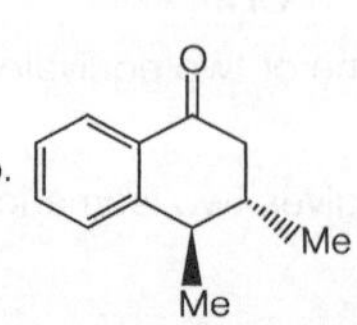
c. 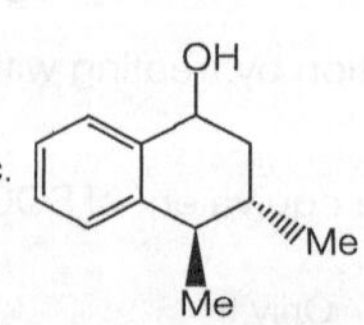
d. 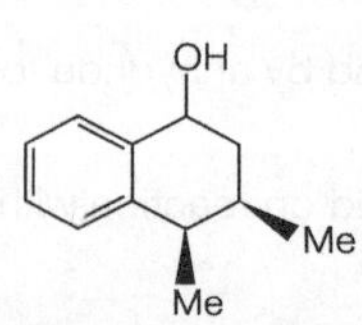

27 Reagent *E* is

a. Zn-Hg /HCl
b. NH_2-NH_2/$\overset{\ominus}{O}H$
c. HI/P
d. All of these

Working Space

Matching List Type

Direction *(Q. Nos. 28-30) Choices for the correct combination of elements from Column I and Column II are given as options (a), (b), (c) and (d), out of which one is correct.*

28 Match the Column I with Column II and mark the correct option from the codes given below.

	Column I (Pair to be tested)		Column II (Reagent/Test)
i.	Acetone and benzaldehyde	p.	Fehling's solution
ii.	Acetaldehyde and benzaldehyde	q.	Tollen's reagent
iii.	Formaldehyde and acetaldehyde	r.	Iodoform test
iv.	Acetophenone and benzaldehyde	s.	Baeyer's reagent

Codes

	i	ii	iii	iv		i	ii	iii	iv
a.	q,r,s	p,r	r	q,r,s	b.	q,r	p,r	r	q,r
c.	s,p	q,r	p	s	d.	p	q	r	s

29 Match the Column I with Column II and mark the correct option from the codes given below.

	Column I (Reaction)		Column II (Product)
i.	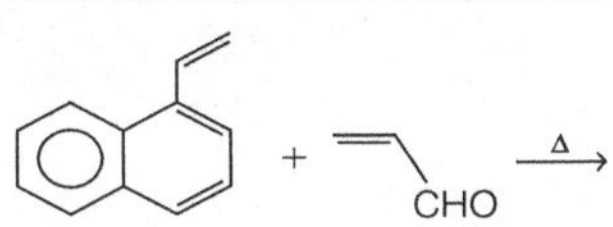	p.	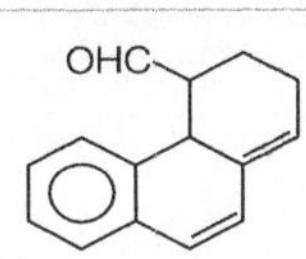
ii.	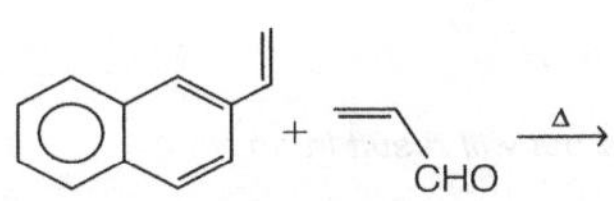	q.	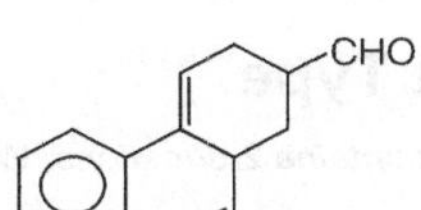
iii.	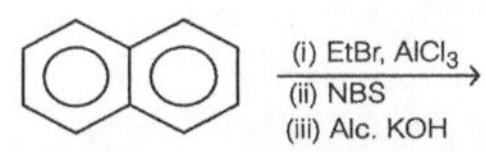	r.	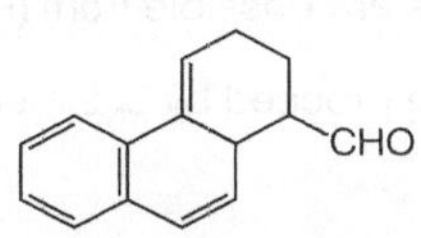
iv.	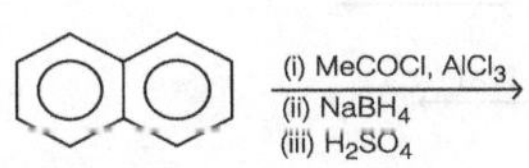	s.	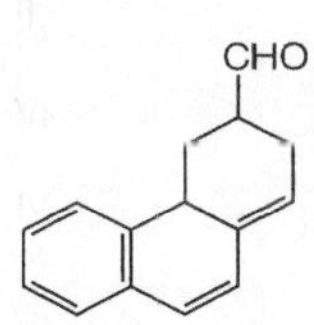
		t	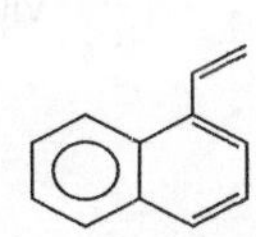
		u	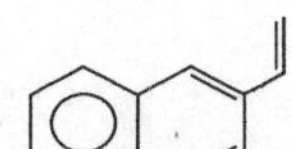

Codes

	i	ii	iii	iv		i	ii	iii	iv
a.	p	q	s,r	t,u	b.	q,r	p,s	t	u
c.	s,p	q	r	s,t	d.	p,t	s	q	r,u

30 Match the column I with column II and mark the correct option from the codes given below.

Working Space

	Column I (Reaction)		Column II (Steps involved)
i.	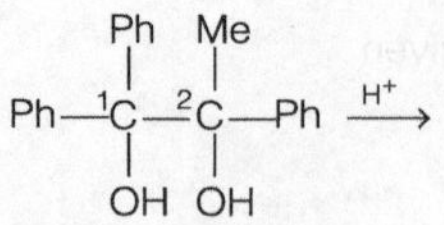	p.	Carbocation is formed at C-1.
ii.	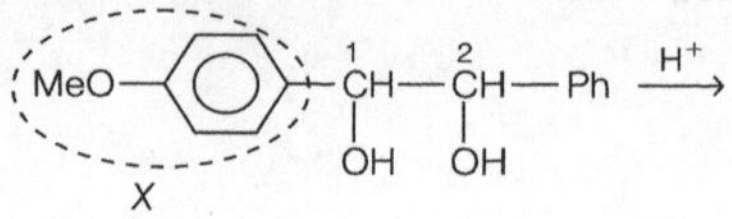	q.	Carbocation is formed at C-2.
iii.	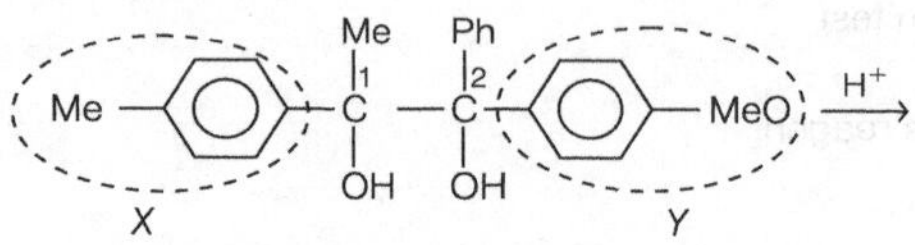	r.	Ph group migrates.
iv.	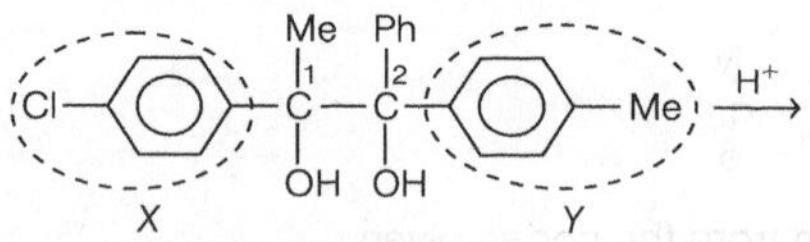	s.	*X* migrates
		t.	*Y* migrates

Codes

	i	ii	iii	iv
a.	p,r	p	q,s	q
b.	t,s	r	p	q
c.	p,r	p,r	q,s	q,s
d.	s,p	q	r	t

One Integer Value Correct Type

Direction *(Q. Nos. 31 and 32) This section contains 2 questions. When worked out will result in an integer from 0 to 9 (both inclusive).*

31 How many monoximes and dioximes are possible from $(PhCO)_2$?

32 How many reactions given below are proceed by S_N2 mechanism?

I. $CH_3{-}\overset{O}{\overset{\|}{C}}{-}OH \xrightarrow{NaOH}$

II. $CH_3{-}\overset{O}{\overset{\|}{C}}{-}Cl \xrightarrow{NaOH}$

III. $CH_3COOC_2H_5 \xrightarrow{H_2O}$

IV. $CH_3CONH_2 \xrightarrow{NaI}$

V. $CH_3CO(I) \xrightarrow{NaNH_2}$

VI. $CH_3COCl + CH_3COONa \longrightarrow$

VII. $CH_3COOH \xrightarrow{C_2H_5ONa}$

VIII. $CH_3{-}\overset{O}{\overset{\|}{C}}{-}OC_2H_5 \xrightarrow{NaNH_2}$

IX. $CH_3{-}\overset{O}{\overset{\|}{C}}{-}OC_2H_5 \xrightarrow{NaOH}$

JEE Main & AIEEE Archive

(Compilation of Last 13 Years' Questions)

1 Sodium phenoxide when heated with CO_2 under pressure at 125° C yields a product which on acetylation produces *C*. **(JEE Main 2014)**

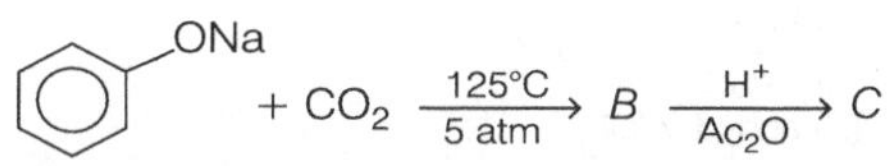

The major product *C* would be

a. 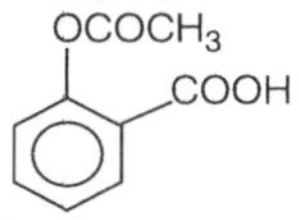

b. 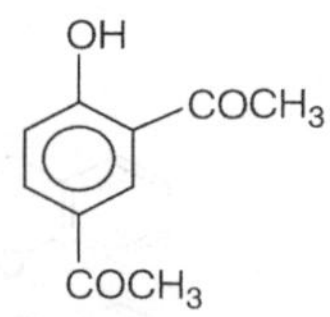

c. 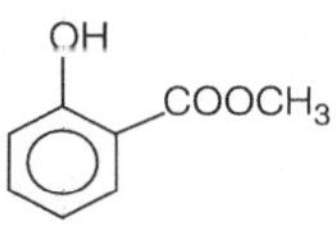

d. 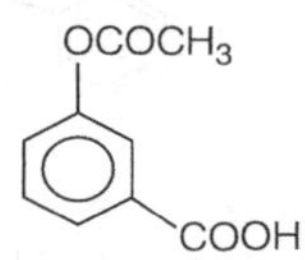

2 *p*-cresol reacts with chloroform in alkaline medium to give the compound *A* which adds hydrogen cyanide to form the compound *B*. The latter, on acidic hydrolysis gives chiral carboxylic acid. The structure of the carboxylic acid is **(AIEEE 2005)**

a. 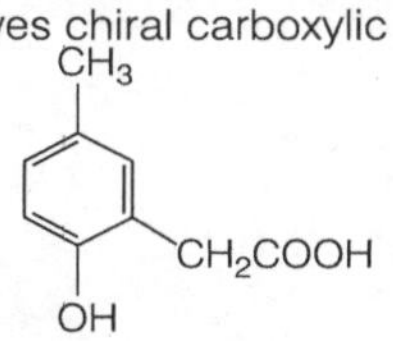

b. 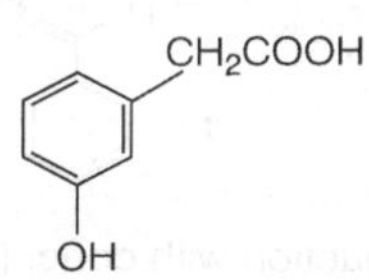

c. 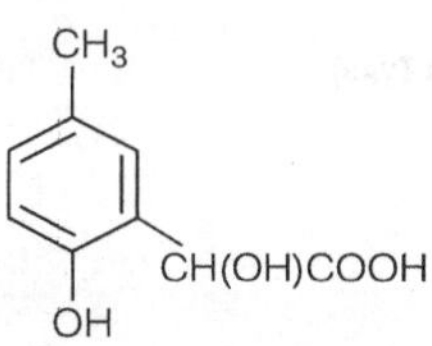

d. 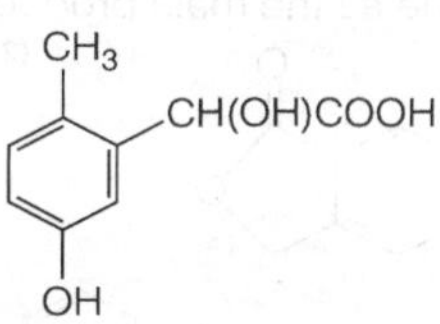

3 Consider the acidity of the carboxylic acids :

i. PhCOOH
ii. *o* - $NO_2C_6H_4COOH$
iii. *p* - $NO_2C_6H_4COOH$
iv. *m*- $NO_2C_6H_4COOH$

Which of the following order is correct? **(AIEEE 2004)**

a. (i) > (ii) > (iii) > (iv)
b. (ii) > (iv) > (iii) > (i)
c. (ii) > (iv) > (i) > (iii)
d. (ii) > (iii) > (iv) > (i)

JEE Archive

JEE Advanced & IIT JEE Archive

(Compilation of Last 10 Years' Questions)

Passage for Q. Nos. (1-2)

In the following reactions sequence, the compound J is an intermediate.

$$I \xrightarrow[CH_3COONa]{(CH_3CO)_2O} J \xrightarrow[\text{(ii) } SOCl_2 \text{ (iii) anhyd. } AlCl_3]{\text{(i) } H_2, Pd/C} K$$

J $(C_9H_8O_2)$ gives effervescence on treatment with $NaHCO_3$ and positive Baeyer's test. **(2012, Comprehension Type)**

1 The compound K, is

a. 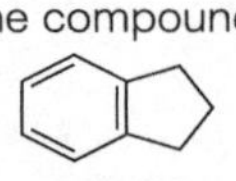

b. 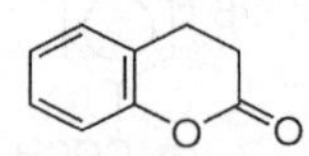

c. 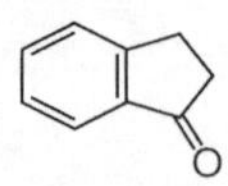

d. 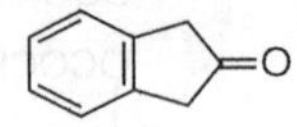

2 The compound I, is

a. 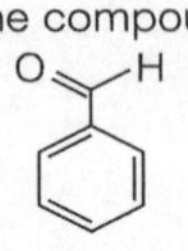

b. 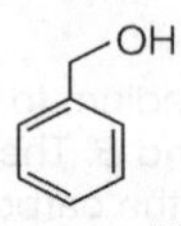

c. 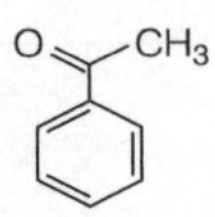

d. 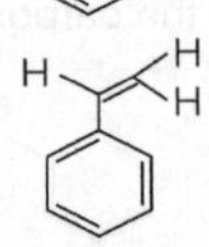

3 Which of the following reactants on reaction with conc. NaOH followed by acidification gives following lactone as the main product? **(2006, Only One Option Correct Type)**

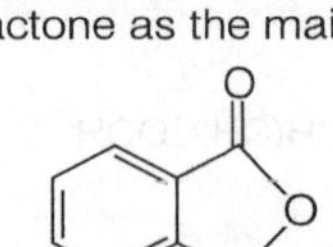

a. 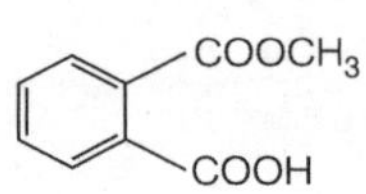

b. 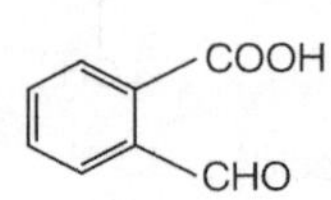

c. 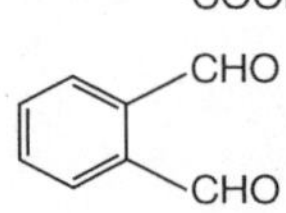

d. 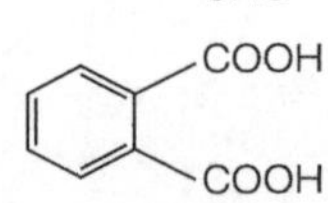

4 4-methyl benzene sulphonic acid reacts with sodium acetate to give **(2005, Only One Option Correct Type)**

a. 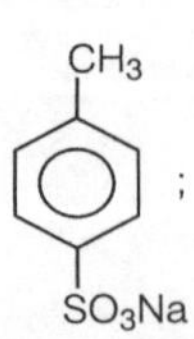

; CH_3CO_2H

b. 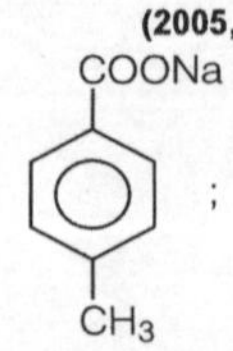

; SO_3

c. 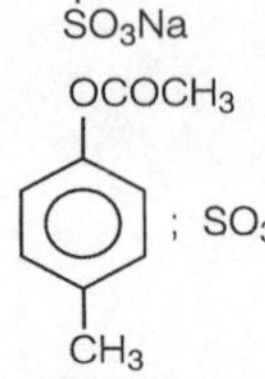

; SO_3

d. 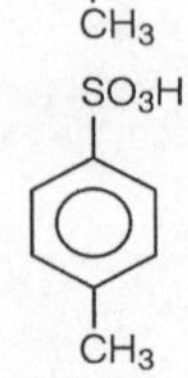

; NaOH

3. Aliphatic Nitrogen Containing Compounds

DPP
DAILY PRACTICE PROBLEMS

DPP-1 Classification, Nomenclature and Isomerism of Amines

Subjective Questions

Working Space

Direction *(Q. Nos. 1-4) These questions are subjective in nature, need to be solved completely on notebook.*

1 Write the types of functional groups in the following compounds.

i. 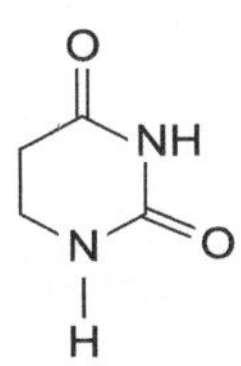

ii. 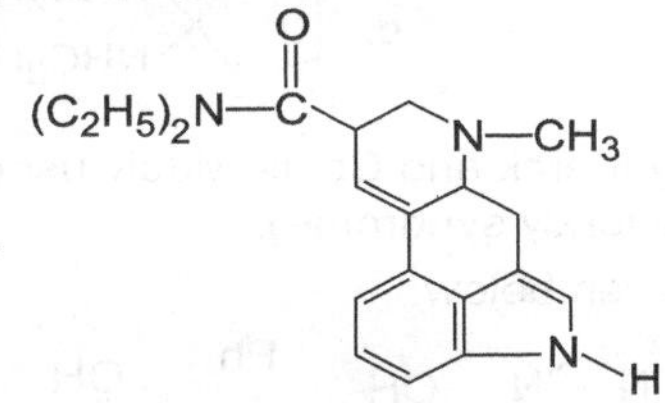

iii. 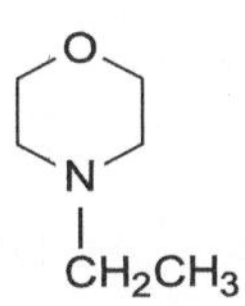

iv. 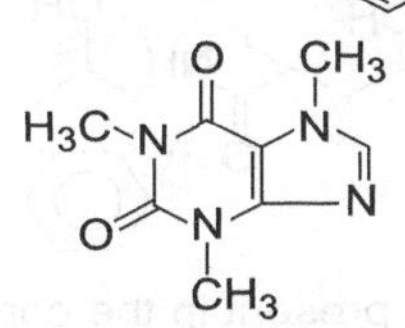

2 Write the common or IUPAC name of the following :

i. $CH_3\overset{+}{N}H_3\ CH_3CO_2^-$

ii. 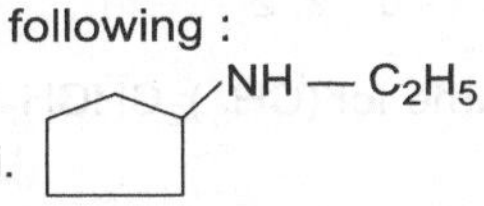

iii. 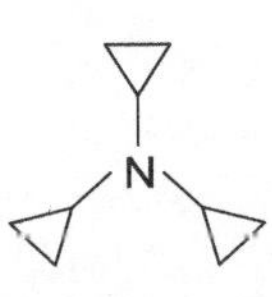

iv. 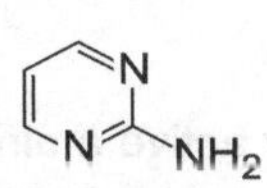

3 Write the structural formulae for each of the following compounds.

i. Benzylmethylamine
ii. Dimethyl ammonium chloride
iii. 2-methyl imidazole
iv. N-ethyl piperidine
v. N-ethyl pyridinium bromide
vi. tetraethyl ammonium hydroxide

4 Draw structures and write IUPAC names of the compounds that meet the following descriptions.

i. A secondary amine with one isopropyl group
ii. A tertiary amine with one phenyl and one ethyl group
iii. A quarternary ammonium salt with four different groups bonded to nitrogen

Only One Option Correct Type

Direction (*Q. Nos. 5-14*) *This section contains 10 multiple choice questions. Each question has four choices (a), (b), (c) and (d), out of which ONLY ONE is correct.*

5 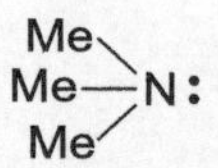 In this molecule C—N—C bond angle is 108°. The hybridisation and shape of the above molecule are

a. sp^2, angular
b. sp^3, tetrahedral
c. sp^3, pyramidal
d. sp^2, pyramidal

6 Which of the following compounds is cyclopentyldiethylamine?

a. 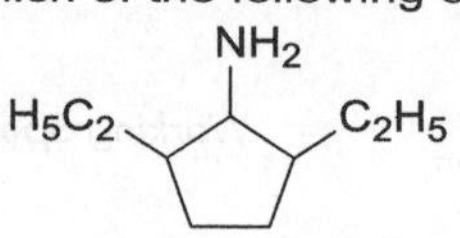
b. 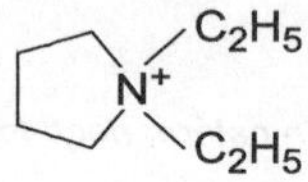
c. 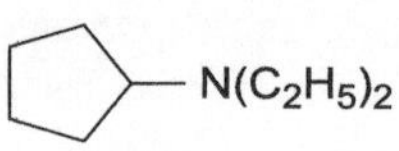
d. 

7 Crixivan, a drug produced by Merck and Co, is widely used in the fight against AIDS (acquired immune deficiency syndrome).
The structure of crixivan is given below:

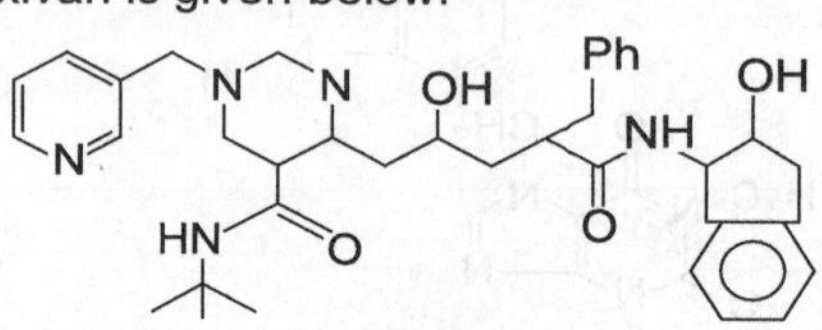

How many 2° and 3° amine groups are present in the compound?

a. 0, 2
b. 2, 0
c. 3, 1
d. 2, 2

8 Which of the following is a proper name for $(CH_3)_2CHCH_2NHCH_2CH_2CH(CH_3)_2$?

a. 2, 7-dimethyl-4-azaoctane
b. 3-methyl-N-(2-methylpropyl)-1-butanamine
c. 2, 7-dimethyl propylbutylamine
d. 3-amino-2,7-dimethyloctane

9 The structural formulae of four physiologically active amines are displayed below. Out of these choline is a nerve transmission agent, benzedrine is an amphetamine, benadryl is an antihistamine and conine is a toxic alkaloid found in hemlock. Which of these would be classified as a tertiary amines?

a. 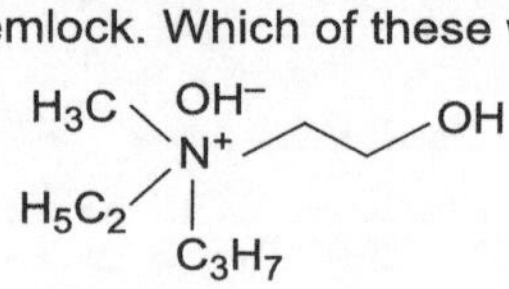
b. 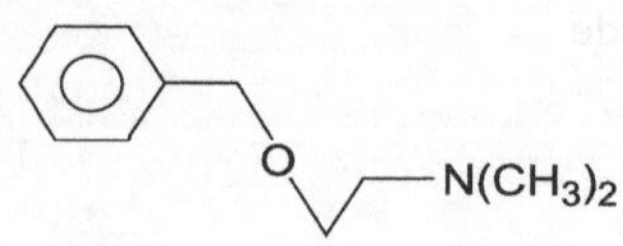
c. 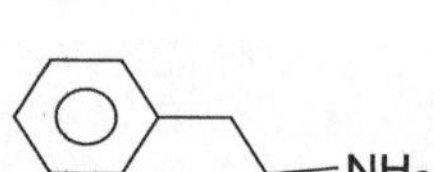
d. 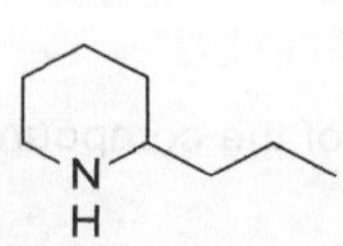

Working Space

Working Space

10 Tetra methylammonium iodide is also known by the name

a. tetramethyl methyl amine iodide
b. N, N, N-trimethyl methylamine iodide
c. trimethylamine methiodide
d. trimethylamine methaneiodide

11 Select the compound being capable of resolution into enantiomers?

a. 2-methylpiperidine
b. 1-methylpiperidine
c. 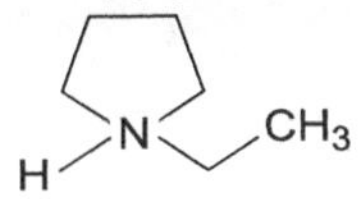
d. 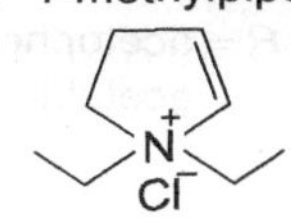

12 Ethyl methylamine cannot be resolved under normal conditions because

a. the favoured configuration is not chiral
b. it isomerises rapidly with the achiral isomer trimethylamine
c. the nitrogen atom rapidly inverts its configuration leading to a racemic mixture
d. the C — N bond is not stable under conditions used for resolution

13 How many 3° amines are possible for the formula $C_4H_{11}N$?

a. 1
b. 2
c. 3
d. 4

14 Which of the following is a set of enamine?

I. 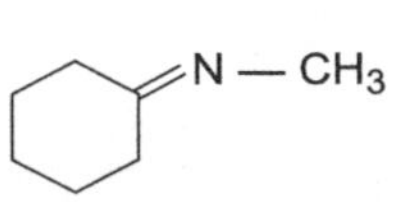

II. 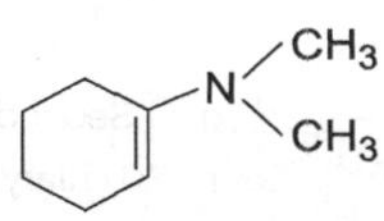

III. 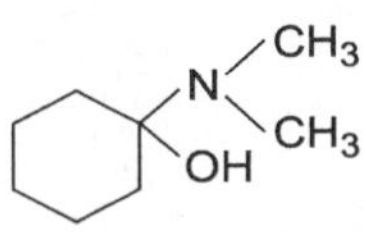

IV. 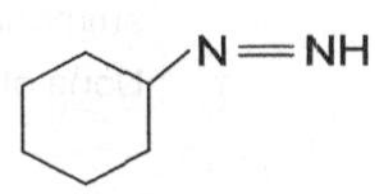

V. 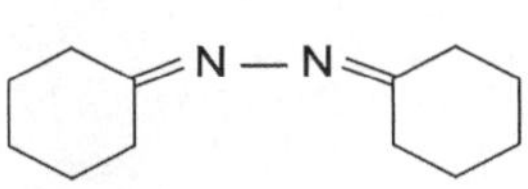

a. I and II
b. II, III and IV
c. Only II
d. II, IV and V

One or More than One Options Correct Type

Direction *(Q. Nos. 15-17) This section contains 3 multiple choice questions. Each question has four choices (a), (b), (c) and (d), out of which ONE or MORE THAN ONE are correct.*

15 Which of the following compound will undergo amine inversion?

a. 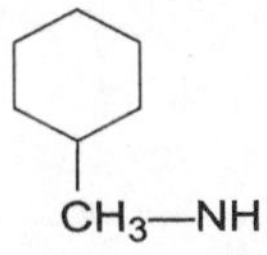
b. $CH_3—NH—C_2H_5$
c. 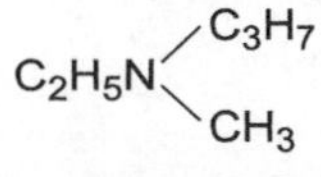
d. 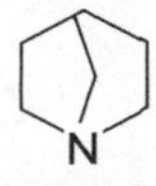

16 Which of the following is not the secondary amine?

a. 2-butanamine
b. N-ethyl-2-pentanamine
c. N-methylpiperidine
d. N, N-dimethylcyclohexylamine

17 Compound containing (=NOH) group are formed when $>C=O$ group reacts with NH_2OH with the elimination of water.

$$\underset{R}{>C=O} + NH_2OH + O \xrightarrow[-H_2O]{H^+/\Delta} >C=NOH$$

In which of the following case(s) product will show geometrical isomerism?

a. If R = acetone
b. If R = acetophenone
c. If R = benzaldehyde
d. If R = acetaldehyde

Matching List Type

Direction *(Q. No. 18) Choices for the correct combination of elements from Column I and Column II are given as options (a), (b), (c) and (d), out of which one is correct.*

18 Match the Column I with Column II and mark the correct option from the codes given below.

	Column I		Column II
i.	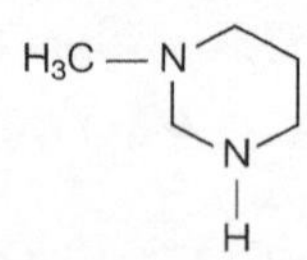	p.	Primary amine
ii.	Morpholine	q.	Secondary amine
iii.	2-ethylpyrolidine	r.	Tertiary amine
iv.	Piperidine-3-carboxylic acid	s.	Quarternary ammonium salt
		t.	None of the above

Codes

	i	ii	iii	iv
a.	q,r	q	q	q
b	q	p	r	q
c.	p,q	r	t	q
d.	p	q,r	q	r,s

One Integer Value Correct Type

Direction *(Q. Nos. 19 and 20) This section contains 2 questions. When worked out will result in an integer from 0 to 9 (both inclusive).*

19 In IUPAC name of x-amino-y, z-dimethylpentane, amino group is present at x position ($x \neq 1$) and $y = z + 1$. The value of x is

20 The number of optically active 1° amines of the formula $C_5H_{13}N$ can be......

DPP-2 General Methods of Preparation of 1° Amines and other than 1° Amines

Subjective Questions

Working Space

Direction (Q. Nos. 1-3) *These questions are subjective in nature, need to be solved completely on notebook.*

1 i. An aqueous acidic solution is often used to carry out the protonolysis step that follows the $LiAlH_4$ reduction $RCONH_2 \xrightarrow[\text{(ii) } H_3O^+]{\text{(i) } LiAlH_4} A$.

What is A and how is it converted into RCH_2NH_2?

ii. Gabriel phthalimide synthesis can use alkyl sulphonate ester as shown :

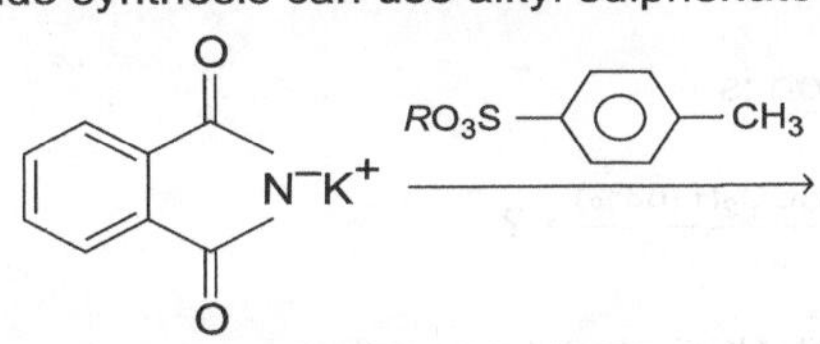

Which act as nucleophile in the above reaction? Also give the type (mechanism involve) in the reaction.

2 What are A to E in the following?

i. 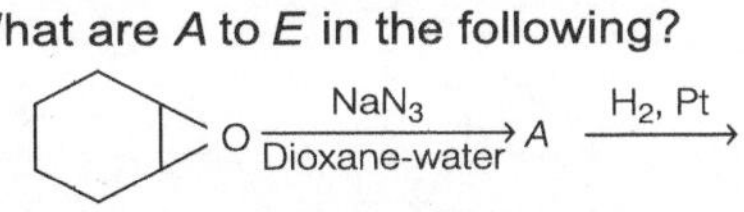

ii. 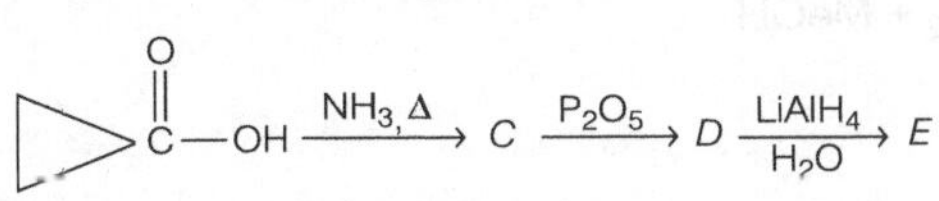

3 Bring about the following transformations

i. 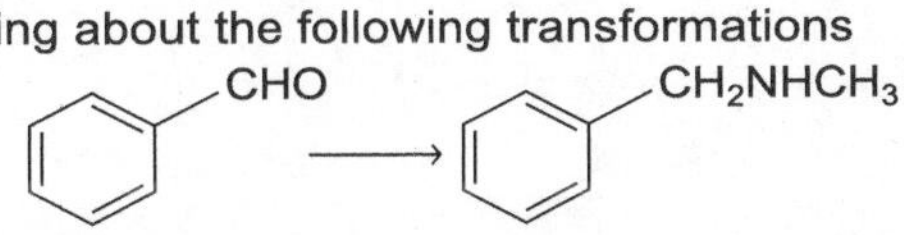

ii. 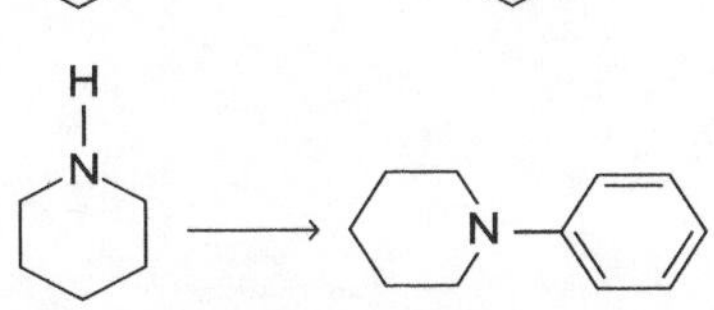

iii. 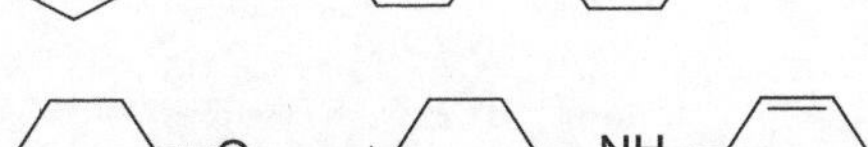

Only One Option Correct Type

Working Space

***Direction** (Q. Nos. 4-12) This section contains 9 multiple choice questions. Each question has four choices (a), (b), (c) and (d), out of which ONLY ONE is correct.*

4 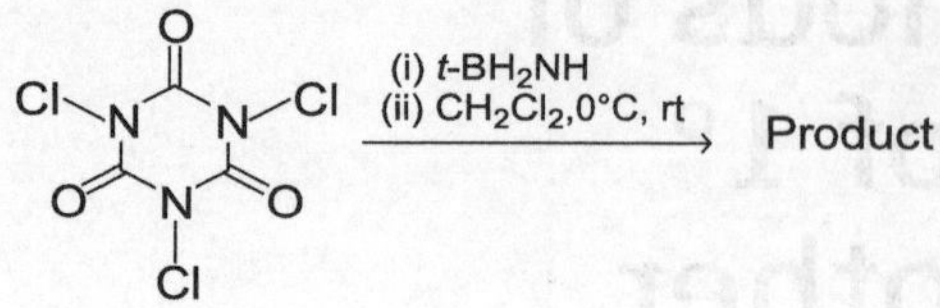

The ratio of number of σ to π bonds in the product is

a. 4 : 1 b. 8 : 1 c. 5 : 3 d. 7.5:1

5 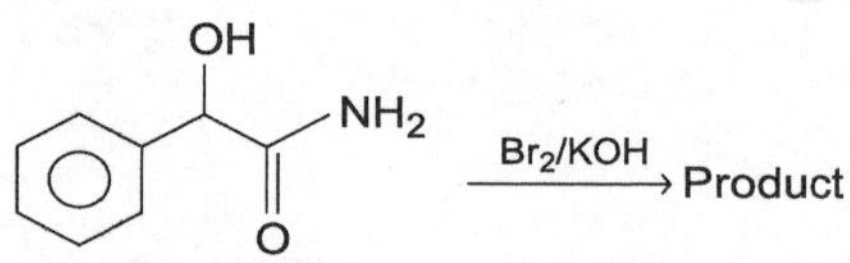

(α-hydroxy amide)

Product of this Hofmann bromamide reaction is

a. Ph—C(=O)—CH_3 b. Ph—CHO

c. Ph—CH(OH)(NH_2) d. Ph—CH_2—NH_2

6 The product expected in the following reaction is

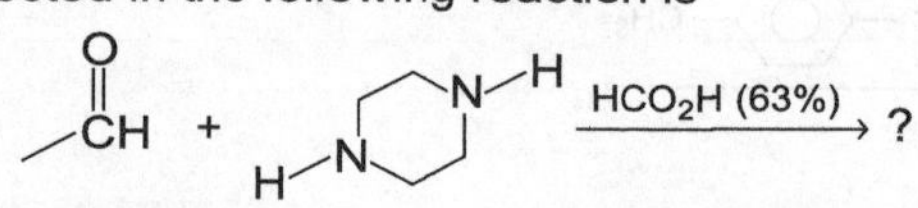

a. N, N′-dibutoxy piperadine b. N, N′-di-*n*butyl piperadine
c. 2 moles of pentanamide d. pentanimine

7 Thermal decomposition of 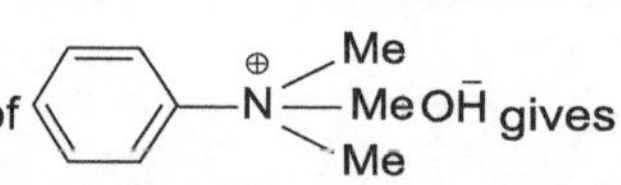 gives

a. 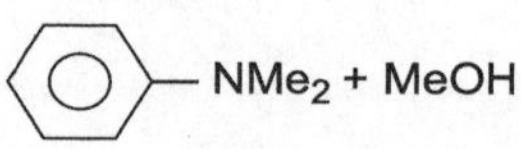 b. 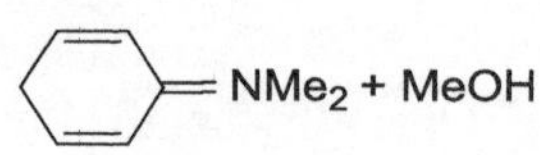

c.  d. No reaction

8 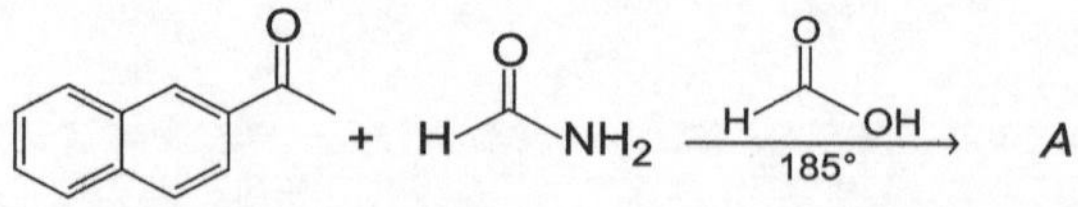

Methyl β-naphthyl ketone

Here, the product *A* is

a. 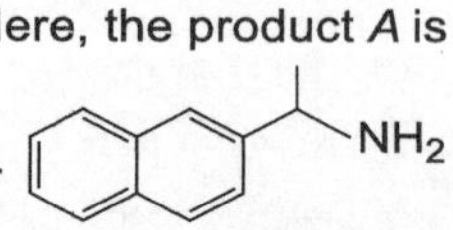 b. 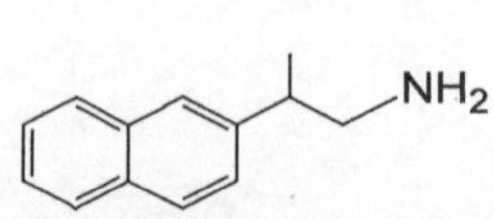

c. 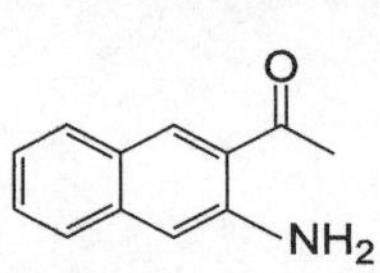 d. Both (a) and (b)

Working Space

9

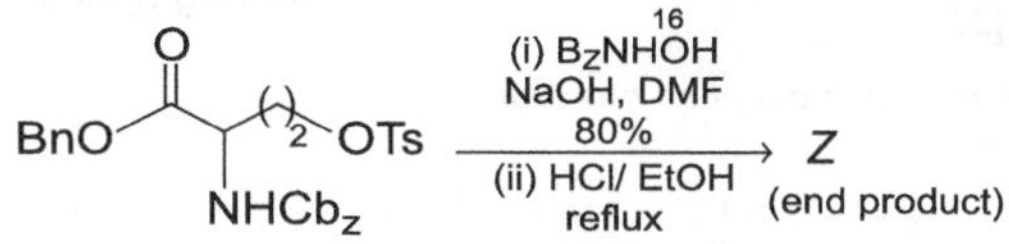

Z is

a. 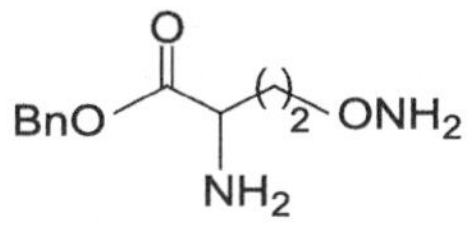

b. 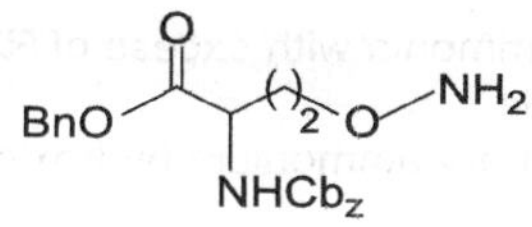

c. 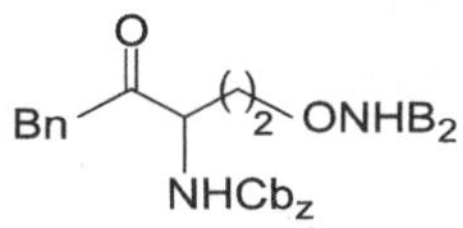

d. 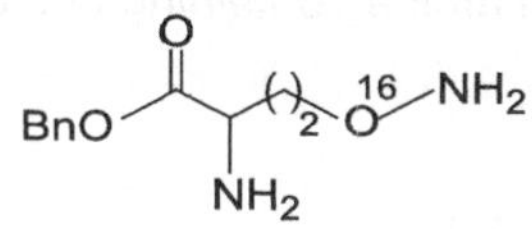

10 Identify the product *D* in the following reaction sequence.

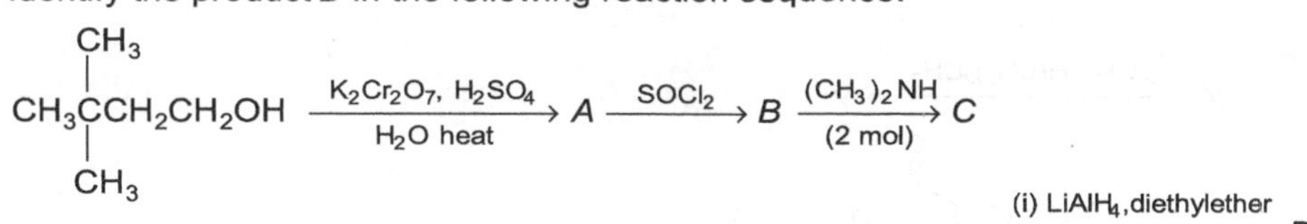

a. 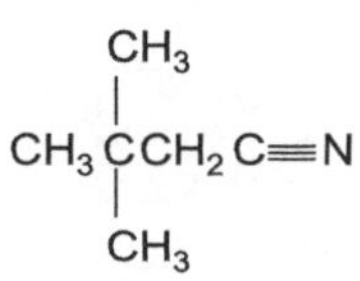

b. 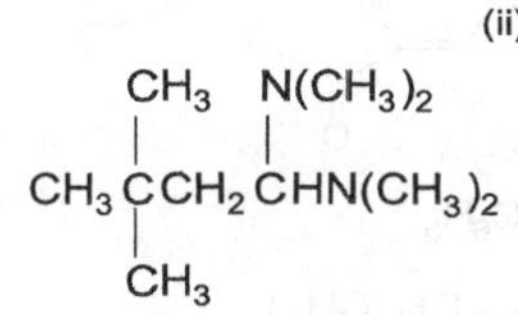

c. 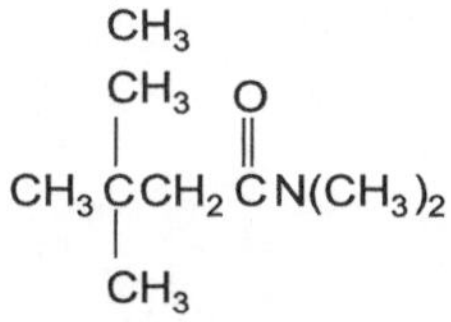

d. 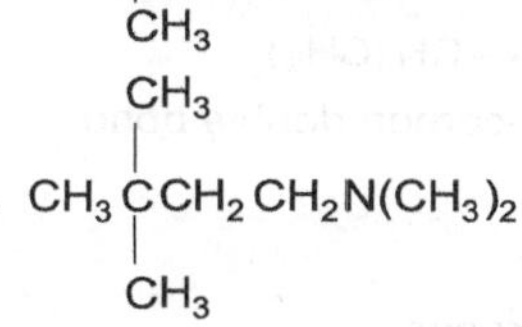

11 The nitrogen atom in each of the following tertiary amines may be removed as trimethyl amine by repeated Hofmann eliminations (exhaustive methylation followed by heating with AgOH). Which of the amines requires the greater number of Hofmann sequences to accomplish this?

a. 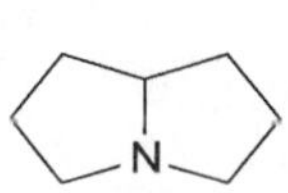

b. 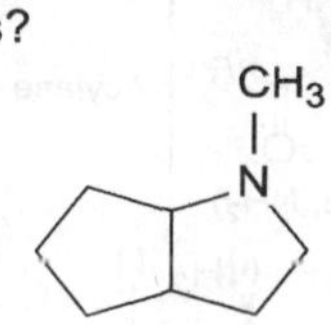

c. 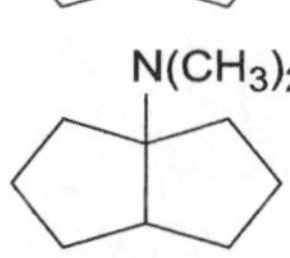

d. 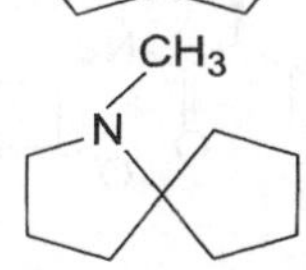

12

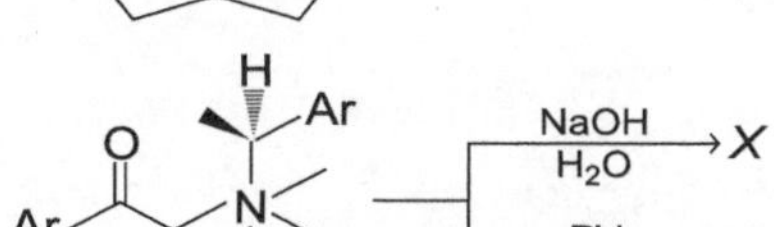

The true statement about *X* and *Y* is/are

a. *X* is a mixture of diastereomeric 3° amines but *Y* is not.
b. *Y* is *a* mixture of enantiomeric 3° amines but *X* is not.
c. *X* and *y* are identical mixtures
d. *X* is a mixture of enantiomeric 3° amines while *Y* is a mixture of diastereomeric 3° amines

One or More than One Options Correct Type

Working Space

Direction *(Q. Nos. 13-17) This section contains 5 multiple choice questions. Each question has four choices (a), (b), (c) and (d), out of which ONE or MORE THAN ONE are correct.*

13 Tertiary amines may be obtained by

a. Gabriel phthalimide synthesis
b. heating an alcoholic solution of ammonia with excess of *RX*
c. the hydrolysis of dialkyl cyanamide
d. thermal decomposition of quarternary ammonium hydroxide

14 In case of which of the following an amine containing one carbon atom less than the parent compound is formed?

a. Hofmann rearrangement
b. Lossen's rearrangement
c. Curtius rearrangement
d. Schmidt rearrangement

15

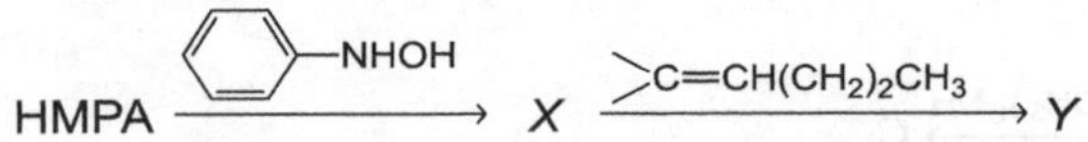

Y is HMPA=

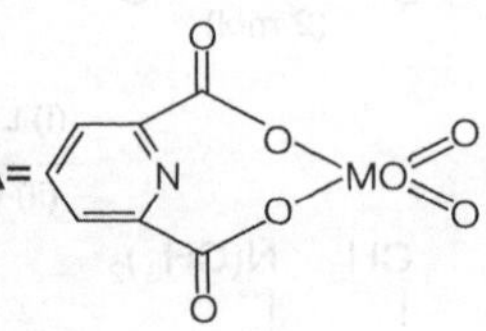

In the above reaction

a. the product is $CH_2=CHCH_2\underset{|}{\overset{NHC_6H_5}{CH}}-CH(CH_3)_2$.
b. there is rearrangement of carbon-carbon double bond.
c. *X* is a good aminating agent.
d. *X* is a good leaving group.

16 Consider the following synthesis process

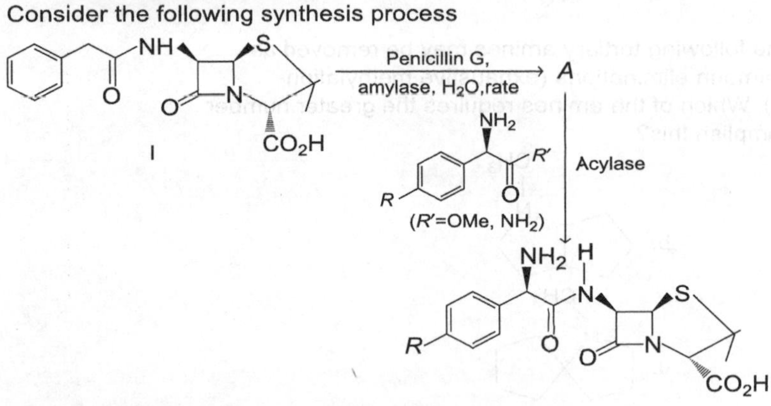

The true statement(s) about the above process is/are

a. In the preparation of *A* from I, protection of group is necessary.
b. In *R* = H, product is ampicillin
c. If *R* = OH, product is amoxicillin
d. Formation of I takes place by enzymatic deacylation

Working Space

17 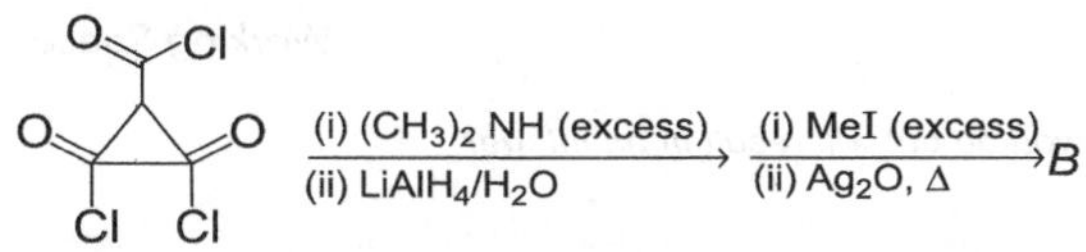

The true statements about *B* is /are

a. It also has a degree of unsaturation
b. It is isomeric with benzene
c. It decolourises the bromine water
d. It undergoes electrophilic substitution reactions like benzene

Statement Type

Direction (*Q. No.* 18) *This section is based on Statement I and Statement II. Select the correct answer from the codes given below.*

a. Both Statement I and Statement II are correct and Statement II is the correct explanation of Statement I
b. Both Statement I and Statement II are correct and Statement II is not the correct explanation of Statement I
c. Statement I is correct but Statement II is incorrect
d. Statement II is correct but Statement I is incorrect

18 **Statement I** $(CH_3)_3CNH_2$ cannot be prepared by the Gabriel phthalimide reaction.

Statement II A tertiary alkyl halide is required in first step of S_N2 reactions.

Comprehension Type

Direction (*Q. Nos.* 19 *and* 20) *This section contains a paragraph, describing theory, experiments, data, etc. Two questions related to the paragraph have been given. Each question has only one correct answer among the four given options (a), (b), (c) and (d).*

Passage

Carboxylic acid derivatives can be converted into primary amines with loss of one carbon atom, by both the Hofmann rearrangement and the Curtius rearrangement. Both proceed through similar mechanism.

19 Consider the following statements,

I. Hofmann rearrangement involves a primary amide.
II. Hofmann rearrangement involve complete retention of configuration during the migration of alkyl group, but Curtius does not.
III. Curtius rearrangement involves an acyl azide.

The true statements is/are

a. I and II
b. II and III
c. I and III
d. I, II and III

20 Following antidepressant drug tranylcypromine *A* is made in the following manner.

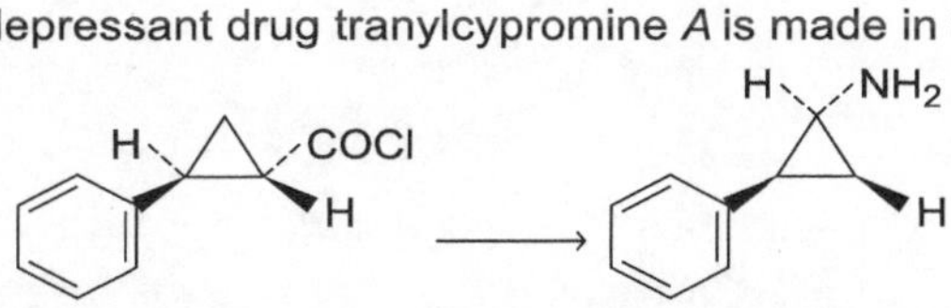

The correct set of reagents/condition for the above conversion is

a. NH_3, KBrO/Δ
b. H_2O, N_3H/Δ
c. NaN_3, Δ, H_2O
d. H_2O, Δ, NH_3

One Integer Value Correct Type

Working Space

Direction *(Q. Nos. 21-23) This section contains 3 questions. When worked out will result in an integer from 0 to 9 (both inclusive).*

21 CH_3COOH by Schmidt reaction forms primary amine. 1.55 g such amine required.........g of CH_3COOH.

22 How many of the following is reduced by lithium aluminium hydride to give 1° amine?

$R—NO_2$, RCN, RN_3 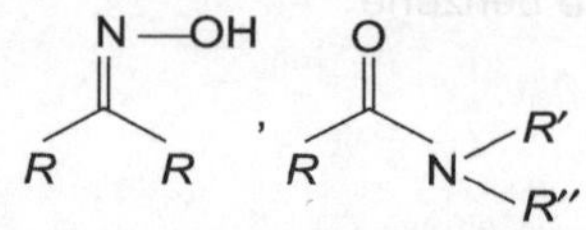 , $R—N \longleftrightarrow C$, 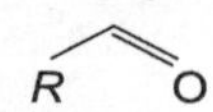

23 *X* and *Y* is the number of times of HEM takes place to remove nitrogen from the compound (HEM = Hofmann Exhaustive Methylation).

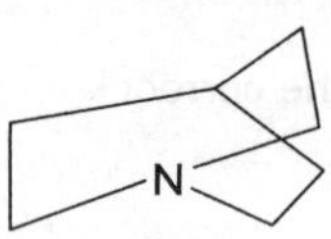 = *X* () ; 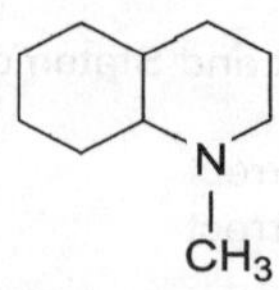 = *Y* ()

Sum of *X* + *Y* = ☐

DPP-3 Physical Properties and Basicity of Amines

Only One Option Correct Type

Direction (Q. Nos. 1-13) *This section contains 13 multiple choice questions. Each question has four choices (a), (b), (c) and (d), out of which ONLY ONE is correct.*

Working Space

1 Among the following amines namely ethylmethyl amine, propyl amine, trimethyl amine, the lowest boiling point is of

a. trimethyl amine
b. ethylmethyl amine
c. propyl amine
d. All of these have same boiling point

2 The compound that will react most readily with NaOH to form methanol is

a. $(CH_3)_4\overset{\oplus}{N}I^-$
b. CH_3OCH_3
c. $(CH)_3S^+I$
d. $(CH_3)_3C—Cl$

3 Strongest base is

a. $(H_2N)_2C{=}NH$
b. $(H_2N)_2C{=}NH_2^{\oplus}$
c. $(H_2N)_2C{=}O$
d. $(H_2N)_2\overset{\oplus}{C}—OH$

4 Which of the following is the strongest acid?

a. 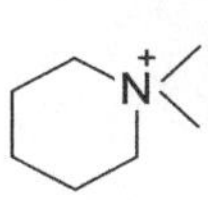
b. 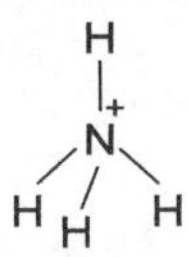
c. 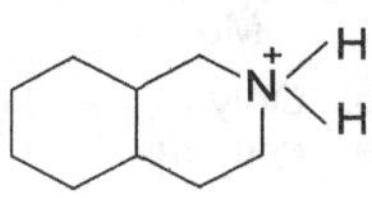
d. 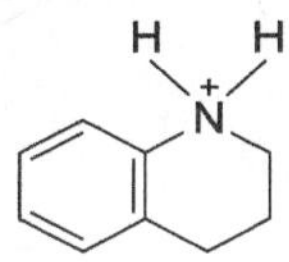

5 Which of the following is the strongest base in aqueous solution?

a. $(CH_3)_3N$
b. $(CH_3)_2NCH_2CH_2OH$
c. $CH_3CH_2CH_2NHOH$
d. $(CH_3)_4NOH$

6 Identify the strongest and weakest bases among the following

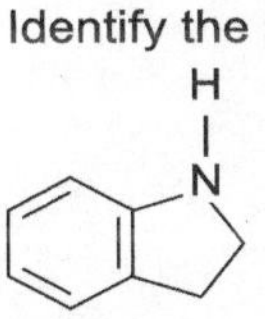
I.

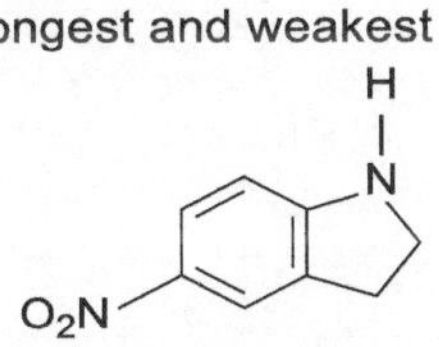
II.

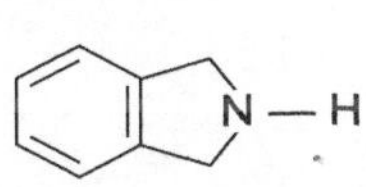
III.

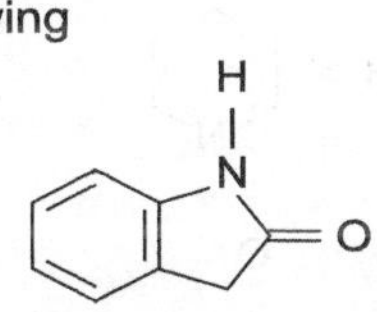
IV.

Strongest and weakest bases respectively are

a. IV and I
b. IV and II
c. I and IV
d. III and IV

7 The decreasing order of basic character at *a*,*b* and *c* in the following compound is

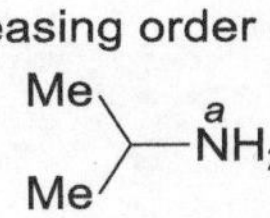
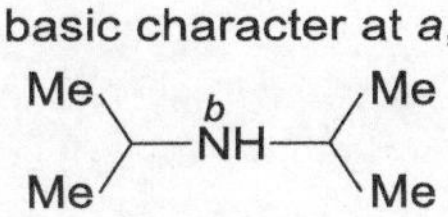
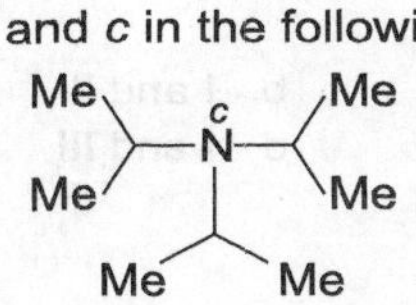

a. $a > b > c$
b. $c > b > a$
c. $b > a > c$
d. $c > a > b$

8 Dipole moment of $CH_3\,CH_2\,CH_3$ (I), $CH_3\,CH_2\,OH$ (II), $CH_3\,CH_2NH_2$ (III) will be in order

a. I < II < III
b. III < II < I
c. I < III < II
d. II < III < I

9 Maximum pK_b value is of

a. 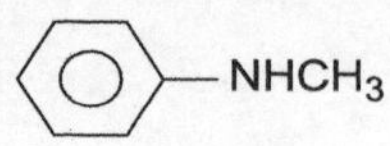
b. $(CH_3CH_2)_2NH$
c. $(CH_3)_2NH$
d. 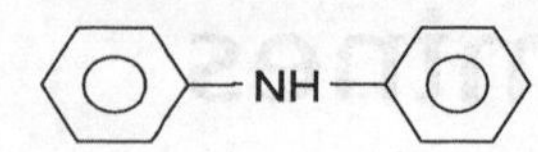

10 Three heterocyclic amines are shown below I = pyridine, II = pyrimidine, III = imidazole. What is the order of increasing basicity? (weaker base < stronger base)

I. 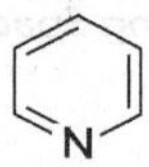
II. 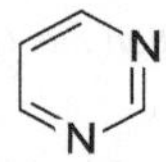
III. 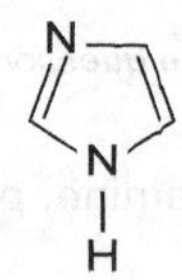

a. II < I < III
b. II < III < I
c. I < II < III
d. III < II < I

11 Which of the following position of the given compound is most activated towards reaction with HCl?

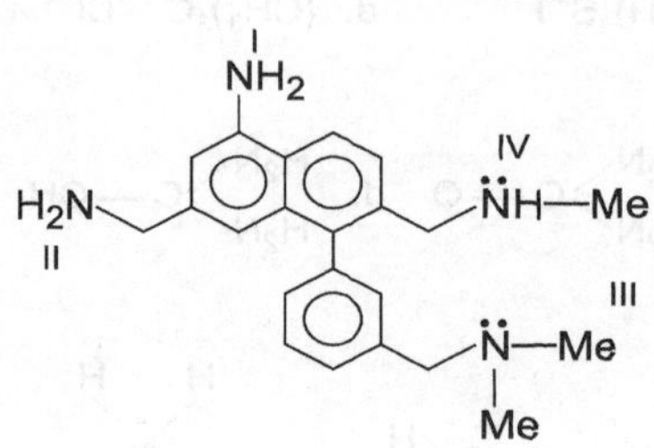

a. Only II
b. Only I
c. Only IV
d. Only III

12 Consider the following comparison of nature of amines

I. 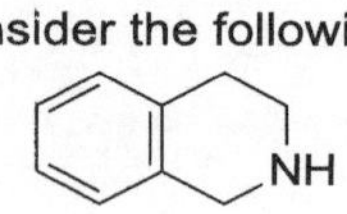 is more stable than 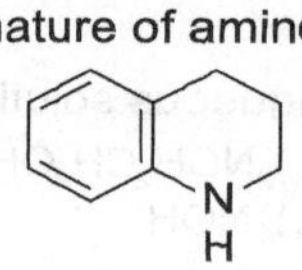

II. 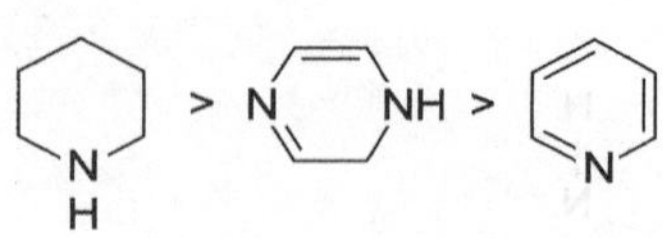 (Basic nature)

III. 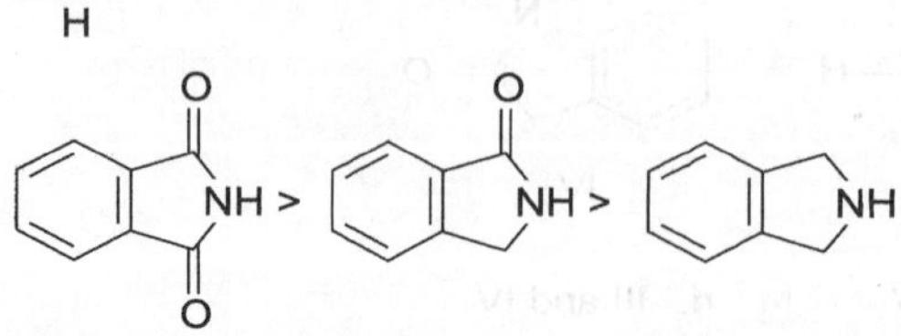 (Basic nature)

Correct comparisons are

a. I, II and III
b. I and II
c. II and III
d. I and III

Working Space

13 The product of this reaction is

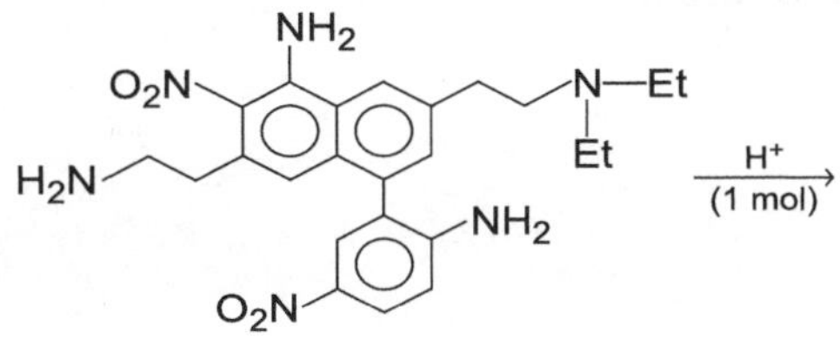

a. 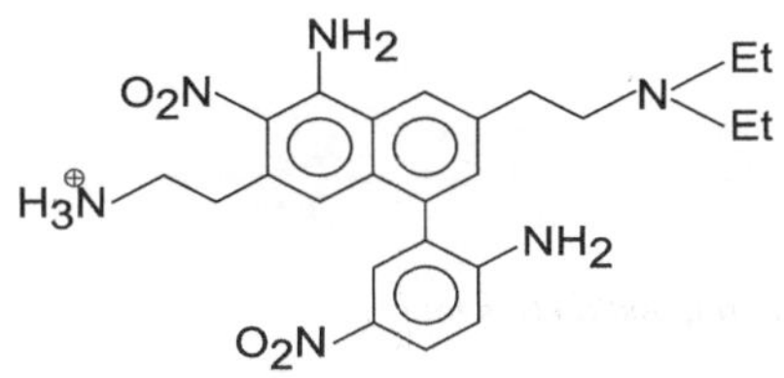

b. 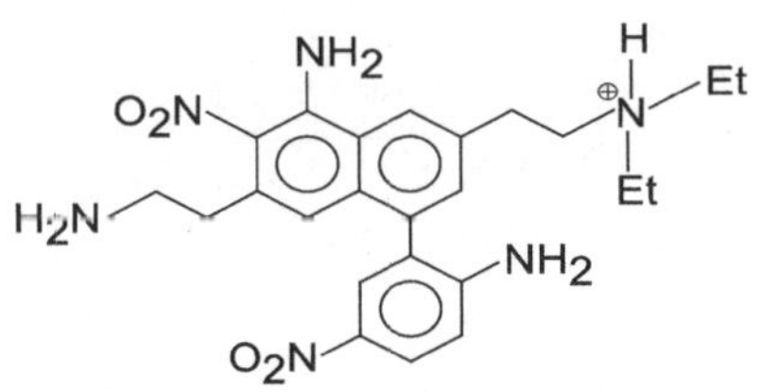

c. 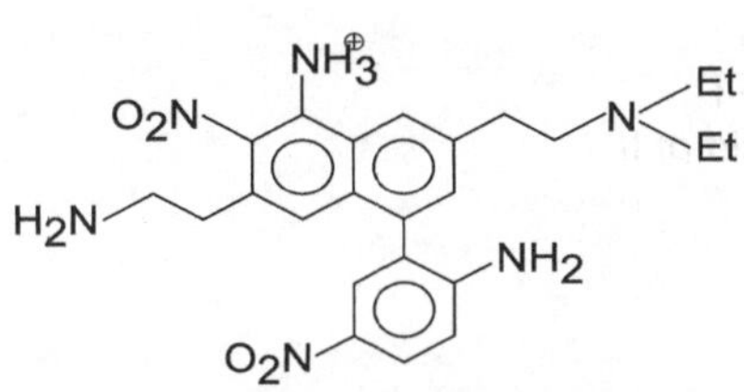

d. 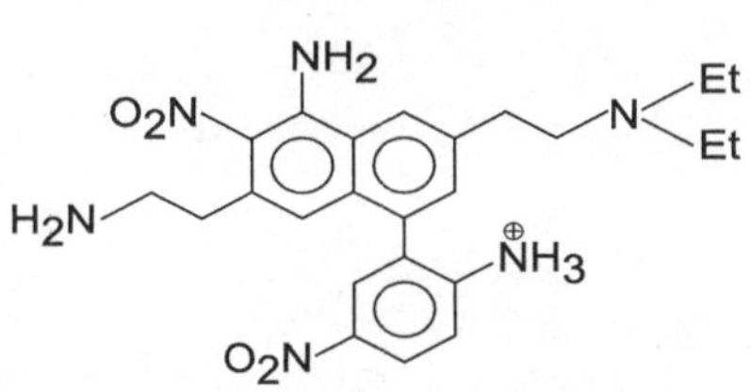

Subjective Questions

Direction *(Q. Nos. 14 and 15) These questions are subjective in nature, need to be solved completely on notebook.*

14 Write the basicity order of atoms, I, II and III?

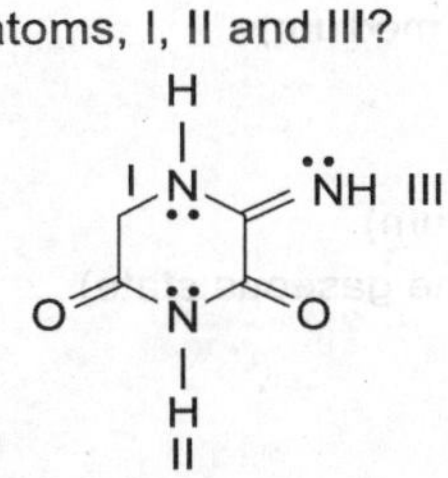

Working Space

15 i. 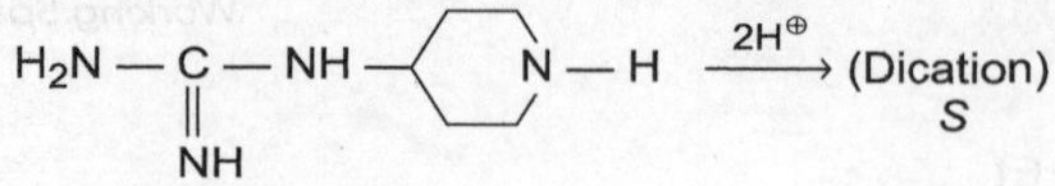

Write the structure of *S*.

ii. 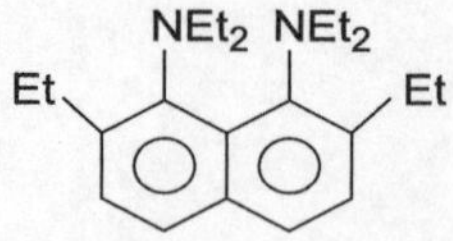

is usually strong base. Explain, why?

One or More than One Options Correct Type

Direction *(Q. Nos. 16-18) This section contains 3 multiple choice questions. Each question has four choices (a), (b), (c) and (d), out of which ONE or MORE THAN ONE are correct.*

16 Which of the following statements is/are correct?

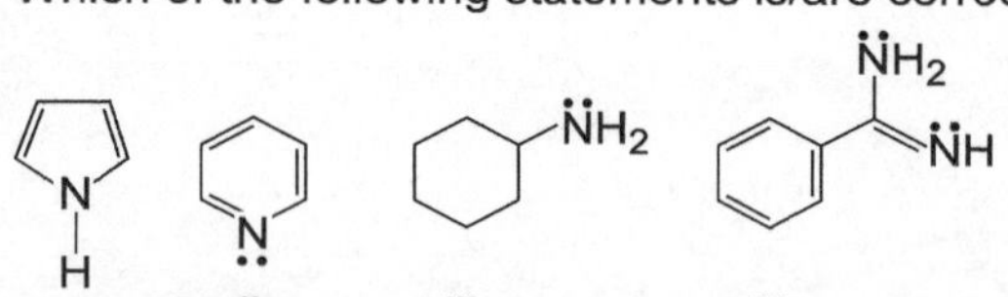

I. II. III. IV.

a. I and II are aromatic and have qual basic strength.
b. I is aromatic and II is antiaromatic but II is stronger base than I
c. The order of basicity of the above compounds is IV > III > II > I
d. The conjugate acid of IV is more stabilised the conjugate acid of II

17 Which of the following basically exist(s) as dipolar ion(s)?

a. 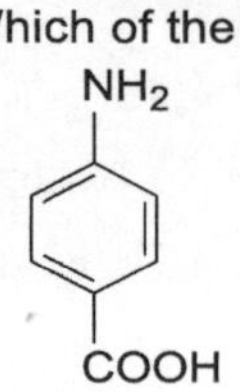

b. 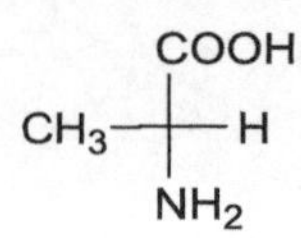

c. 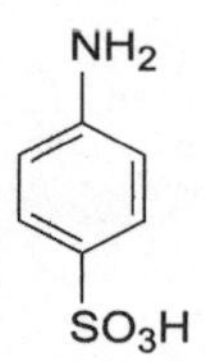

d. 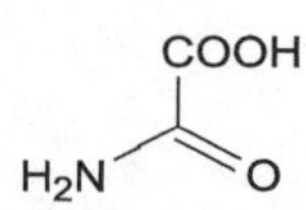

18 Choose the correct comparison(s).

a. $CH_3-C(CH_3)_2-NH_2 < CH_3-NHCH_3$ (basicity in aqueous medium)

b. $CH_3CH_2CH_2NH_2 > (CH_3)_3N$ (basicity in aqueous medium)

c. $CH_3-CH(CH_3)-NH_2 < CH_3-NHCH_2CH_3$ (basicity in the gaseous state)

d. 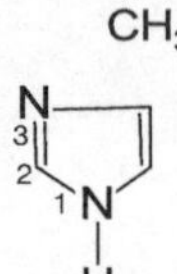 N(1) > N(3) (basicity in aqueous medium)

Working Space

Matching List Type

Direction *(Q. No. 19) Choices for the correct combination of elements from Column I and Column II are given as options (a), (b), (c) and (d), out of which one is correct.*

19 Match the column I with column II and mark the correct options from the codes given below.

	Column I (Amines)		Column II (pK_b values)
i.	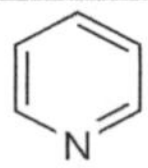	p.	3.3
ii.	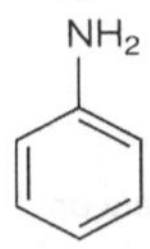	q.	8.8
iii.	$CH_3CH_2NH_2$	r.	3.35
iv.	CH_3NH_2	s.	9.4

Codes

	i	ii	iii	iv
a.	p	q	s	r
b.	q	s	p	r
c.	r	q	s	p
d.	q	p	r	s

One Integer Value Correct Type

Direction *(Q. No. 20) This section contains 1 question. When worked out will result in an integer from 0 to 9 (both inclusive).*

20

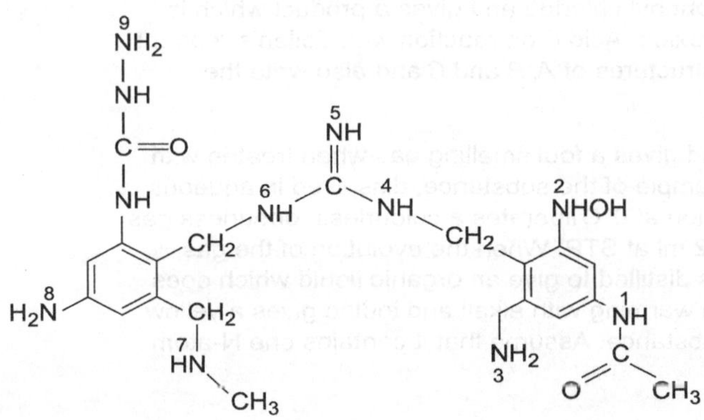

Which N-atom (marked 1 to 9) in the above molecule will accept proton most readily?

DPP-4 Distinction and Separation of 1°, 2° and 3° Amines

Subjective Questions

Working Space

Direction (Q. Nos. 1-3) *These questions are subjective in nature, need to be solved completely on notebook.*

1 Write chemical reactions that could be used in laboratory for distinction between the following pairs of compounds.

i. Allyl amine and propylamine
ii. Tetrapropyl ammonium chloride and tetrapropyl ammonium hydroxide
iii. Tripropylammonium chloride and tetrapropyl ammonium chloride
iv. Triethylamine and diethylamine
v. Cyclohexylamine and piperidine
vi. Pyridine and benzene

2 An organic compound *A*, C_3H_7NO on hydrolysis gives on amine *B* and a carboxylic acid *C*. *B* reacts with benzene sulphonyl chloride and gives a product which is insoluble in aqueous sodium hydroxide. Acid *C* on reaction with Tollen's reagent gives a silver mirror. Identify the structures of *A*, *B* and *C* and also write the reactions involved.

3 A basic volatile nitrogen compound gives a foul smelling gas when treated with $CHCl_3$ and alc. KOH. A 0.295 g sample of the substance, dissolved in aqueous HCl and treated with $NaNO_2$ solution at 0°C liberates a colourless, odourless gas whose volume corresponds to 112 ml at STP. When the evolution of the gas ceased, the aqueous solution was distilled to give an organic liquid which does not contain nitrogen and which on warming with alkali and iodine gives a yellow precipitate. Identify the original substance. Assume that it contains one N-atom per molecule.

Only One Option Correct Type

Direction (Q. Nos. 4-10) *This section contains 7 multiple choice questions. Each question has four choices (a), (b), (c) and (d), out of which ONLY ONE is correct.*

4 Which amine yields N-nitroso amine after treatment with nitrous acid ($NaNO_2$, HCl)?

a. 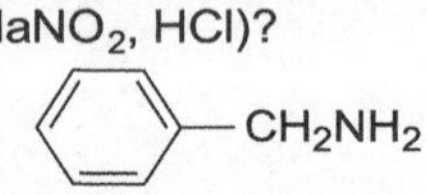

b. 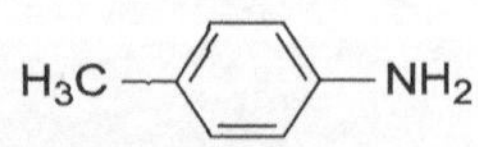

c. 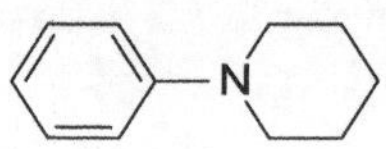

d. 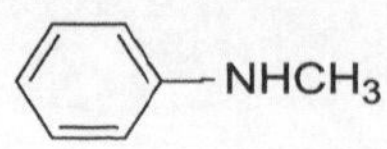

Working Space

5 A compound with formula $C_5H_{13}N$ gives a base soluble derivative in the Hinsberg test(C_6H_5 SO_2Cl in base). Which of the following best satisfy this condition?

a. N, N-dimethylpropylamine
b. Isopropyldimethylamine
c. 2,2-dimethylpropylamine
d. N-methyl-2-methylpropylamine

6 A nitrogen containing compound dissolves in 10% aqueous sulphuric acid. The Hinsberg test ($C_6H_5SO_2Cl$ in base) gives a solid product that is not soluble in 10% aqueous NaOH. Which of the following would best fit in these facts?

a. N, N-dimethylaniline,
b. N-methylbenzamide,
c. N-methylaniline,
d. Benzylamine,

7 The Hinsberg test of a compound with formula $C_5H_{14}N_2$ produces a solid that is insoluble in 10% aqueous NaOH. This solid derivative dissolves in 10% aqueous sulphuric acid. Which of the following would best fit in these facts?

a. $NH_2CH_2CH_2CH_2N(CH_3)_2$
b. $(CH_3)_2NCH_2CH_2NHCH_3$
c. $NH_2CH_2C(CH_3)_2CH_2NH_2$
d. $(CH_3)_2NCH_2N(CH_3)_2$

8 Observe the following reactions, $A + C_6H_5SO_2Cl \xrightarrow{^-OH} \xrightarrow{KOH}$ Clear solution

$\xrightarrow{H^+}$ Substance is soluble in acidic medium

$B + C_6H_5SO_2Cl \xrightarrow{-OH}$ Insoluble substance in H^+/OH^-

$C + C_6H_5SO_2Cl \xrightarrow{-OH}$ Insoluble in base but soluble in $H^+ \xrightarrow{H^+}$ Clear solution.

A, *B* and *C* are respectively

a. 1°, 2° and 3° amines
b. 1°, 3° and 2° amines
c. 1°, aniline and 2° amine
d. 1° amine, 2° amine and aniline

9 Compound *A* has the formula $C_9H_{13}N$ and forms phthalic acid with $KMnO_4/OH^-$. It forms a compound (in liquid state) with $(COOC_2H_5)_2$ which gets decomposed by KOH. The possible structure of compound *A* is

a. 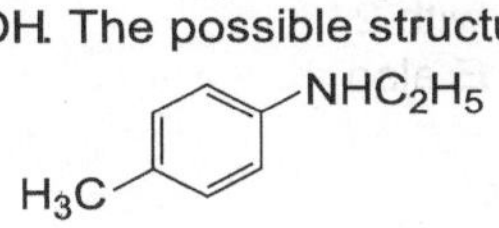

b. 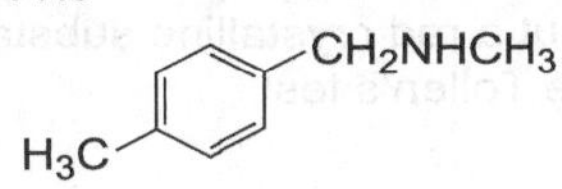

c. 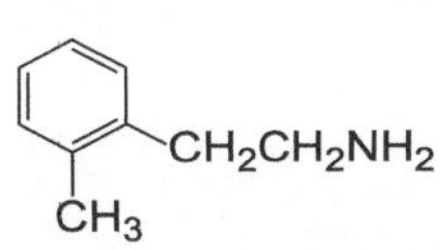

d. 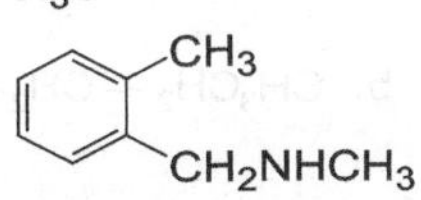

10 Compound *A* releases an inert gas with HNO_2 form isocyanate with $HgCl_2/CS_2$ and a solid with diethyl oxalate. It can be

a. an optically active compound containing three C-atoms
b. a compound formed by the reduction product of reaction of Ag CN (alc) with *R*Cl.
c. a compound which is formed by acylazide in alkaline medium.
d. a compound having —NH_2 group at carbonyl carbon.

One or More than One Options Correct Type

Direction *(Q. Nos. 11-13) This section contains 3 multiple choice questions. Each question has four choices (a), (b), (c) and (d), out of which ONE or MORE THAN ONE are correct.*

11

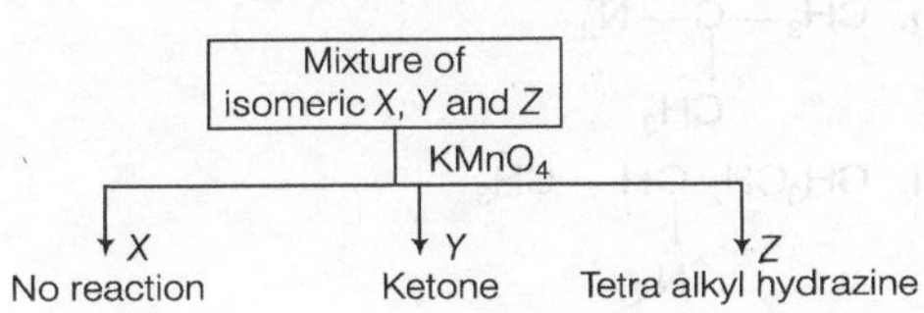

Working Space

The true statement(s) about *X*, *Y* and *Z* is/are

a. *Z* forms an amine oxide when treated with Caro's acid.
b. $X = (CH_3)_3N$, $Y = (CH_3)_2NCH_3$ or
c. *Y* gives smell of mustard oil when treated with $CS_2/HgCl_2$.
d. *X* forms an amine axide when treated with Caro's acid.

12 $\underset{X}{C_4H_{11}N} + HNO_2 \longrightarrow C_4H_{10}O$ (3° alcohol). Hence, *X* will give

a. carbylamine reaction
b. Hofmann mustard oil reaction
c. diazonium salt (as the intermediate) with HNO_2
d. Hofmann bromamide reaction

13 An amine *A* ($C_6H_{15}N$) on treatment with MeI and then KOH gives *B* ($C_8H_{20}N^+OH^-$). This on heating produces *C*, an alkene 'isobutene' and *D* an amine. The possible structure of amine *A* is/are

a. $(CH_3)_2CHCH_2NHCH_2CH_3$
b. $(CH_3)_3CNHCH_2CH_3$
c. $CH_3CH_2CH_2CH_2NHCH_2CH_3$
d. $CH_3(CH_2)_5NH_2$

Comprehension Type

Direction *(Q. Nos. 14-16) This section contains a paragraph, describing theory, experiments, data, etc. Three questions related to the paragraph have been given. Each question has only one correct answer among the four given options (a), (b), (c) and (d).*

Passage

An organic compound *A* is treated with HBr. The product is racemate and one of the optically active isomer *B* is resolved. The compound *B* on treatment with NH_3 gives another compound *C* ($C_4H_{11}N$). Treatment of *C* with HNO_2 gives N_2 and compund *D*.

Compound *D* is oxidised to *E* with acidified $K_2Cr_2O_7$. Compound *E* reacting with 2, 4-dinitrophenyl hydrazine gives out a red crystalline substance. Compound *E* also gives iodoform test but a negative Tollen's test.

14 The compound *C* is

a. $CH_3CH_2—CH(NH_2)—CH_3$
b. $CH_3CH_2—CH_2—NH_2$
c. $CH_3—C(CH_3)_2—NH_2$
d. $CH_3—CH(CH_3)—NH—CH_3$

15 Compound *C* + compound *E* $\xrightarrow{NaBH_3CN}$ Product. The product is

a. 1° amine
b. 2° amine
c. 3° amine
d. quarternary amine

16 Compound *C* $\longrightarrow$ Compound *D*. The intermediate of the reaction is

a. $CH_3CH_2CH(NH_2)—CH_3$
b. $CH_3—C(CH_3)_2—N_2^+$
c. $CH_3—CH_2—CH_2—CH_2^+$
d. $CH_3CH_2CH(^+N_2)—CH_3$

Working Space

Matching List Type

Direction *(Q. Nos. 17 and 18) Choices for the correct combination of elements from Column I and Column II are given as options (a), (b), (c) and (d), out of which one is correct.*

17 Match the Column I with Column II and mark the correct option from the codes given below.

	Column I		Column II
i.	Primary amine	p.	Hinsberg's reagent
ii.	Secondary amine	q.	Diethyl oxalate
iii.	Tertiary amine	r.	Carbylamine reaction
iv.	Ethylamine	s.	Nitrous acid

Codes

	i	ii	iii	iv
a.	p,q,r,s	p,q,s	s	p,q,r,s
b.	p,q,r,s	p,s	q,r	p,q
c.	q,s	p,q	p,r,s	p,q,r
d.	p,q,r	p	q,r,s	q,r,s

18 Match the Column I with Column II and mark the correct option from the codes given below.

	Column I		Column II
i.	$Ph-C(CH_3)_2-NH_2$	p.	Treatment of CS_2, $HgCl_2$ gives out alkyl isothiocyanate
ii.	$CH_3CH_2NH_2$	q.	Treatment of *p*-toluene sulphonyl chloride produces the compound insoluble in alkali
iii.	$CH_3CH_2N(CH_3)CH_3$	r.	Treatment of H_2O_2; heat gives out alkene
iv.	$CH_3CH_2NHCH_3$	s.	Treatment of carbon disulphide produces dithiocarbamic acid

Codes

	i	ii	iii	iv
a.	p,s	r,s	q	p,q
b.	p,s	q	r	q
c.	p,s	p,s	r	q
d.	p	p,s	q	r,s

One Integer Value Correct Type

Direction *(Q. Nos. 19 and 20) This section contains 2 questions. When worked out will result in an integer from 0 to 9 (both inclusive).*

19 Consider the following reaction.

$$CH_3CH_2CH_2NH_2 \xrightarrow{HNO_2} A \text{ (mixture)}$$

The mixture contains how many constituents?

20 The difference in the number of carbon atoms in the oxidation reaction of 2° amine with $KMnO_4$ is an integer multiple of ?

DPP-5 Chemical Properties of Aliphatic Amines

Subjective Questions

Working Space

Direction *(Q. Nos. 1 and 2) These questions are subjective in nature, need to be solved completely on notebook.*

1 Compound *X* containing chlorine, on treatment with strong ammonia gives a solid *Y*, which is free from chlorine. Compound *Y* contains C = 49.13%, H = 9.59% and N = 19.18% and reacts with bromine and NaOH to give a basic compound *Z*, which reacts with HNO_2 to give ethanol. Suggest structures for *X*, *Y* and *Z*.

2 Provide structures of the product in the following sequence of reaction.

i. 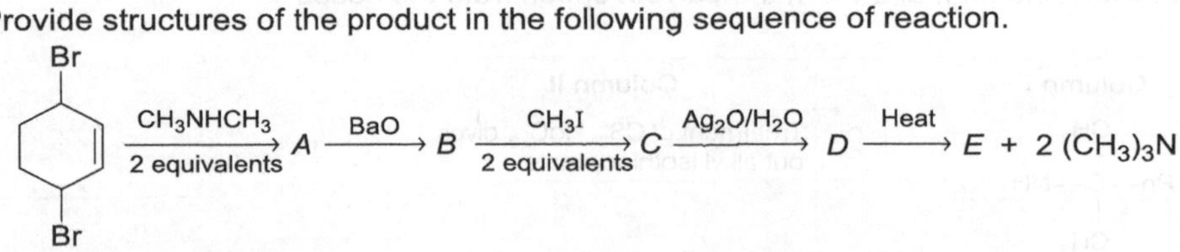

ii. 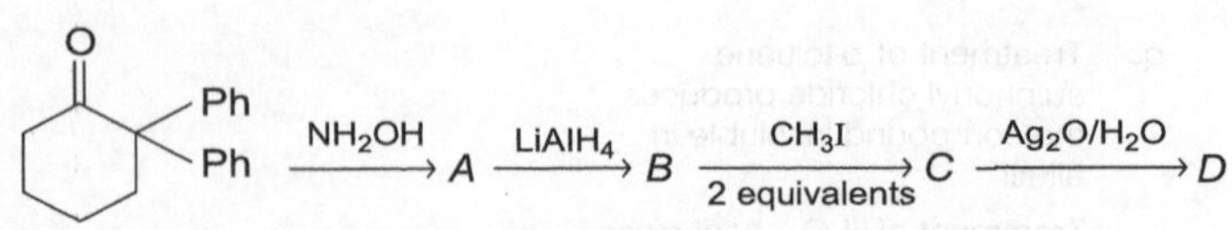

Only One Option Correct Type

Direction *(Q. Nos. 3-13) This section contains 11 multiple choice questions. Each question has four choices (a), (b), (c) and (d), out of which ONLY ONE is correct.*

3 Which of the following compounds will be obtained in the end of the following reaction ?

$$\text{Ethyl amine} \xrightarrow{HNO_2} A \xrightarrow{PCl_5} B \xrightarrow{NH_3} C$$

a. Ethyl cyanide b. Ethyl amine c. Methyl amine d. Acetamide

4 What sequence of reactions would best accomplish the following reaction?

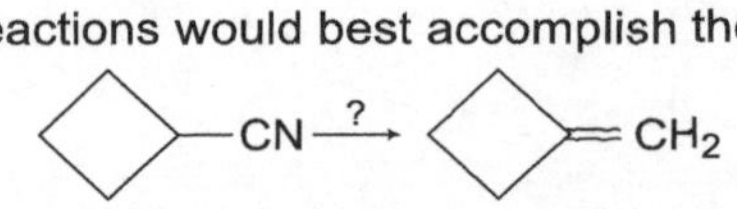

a. (i) $LiAlH_4$ in ether; (ii) $3CH_3I$ followed by heating with AgOH
b. (i) $LiAlH_4$ in ether; (ii) P_2O_5 and heat
c. (i) 20% H_2SO_4 and heat; (ii) P_2O_5 and heat
d. H_2 and Lindlar catalyst

5 In the following reactions, reactants *A*, *B* and *C* are respectively

$$C_2H_5NH_2 + A \longrightarrow C_2H_5N{=}CH{-}C_6H_5 + H_2O$$
$$\text{Urea} + B \longrightarrow H_2NN\,HCO\,NH_2 + NH_3$$
$$C_2H_5NH_2 + C \longrightarrow C_2H_5Cl + H_2O + N_2$$

a. CH_3CHO, $NH_2{-}NH_2$ and PCl_5 b. C_6H_5CHO, NH_2NH_2, $SOCl_2$
c. C_6H_5CHO, $NH_2{-}NH_2$ and NOCl d. CH_3CHO, $NH_2{-}NH_2$ and PCl_3

6 Amines are well known to be stronger bases and nucleophiles than alkenes. Why do enamines, such as 1-dimethylaminocyclopentene, preferentially reacts with electrophiles at a double bond carbon rather than at nitrogen?

a. The nitrogen is sterically hindered by alkyl substituents
b. Nitrogen is more electronegative than carbon
c. The carbocation formed by electrophilic attack at C-2 is stabilised by pi-bonding with the lone pair of electrons on nitrogen
d. Ammonium cations are less stable than carbocations

7

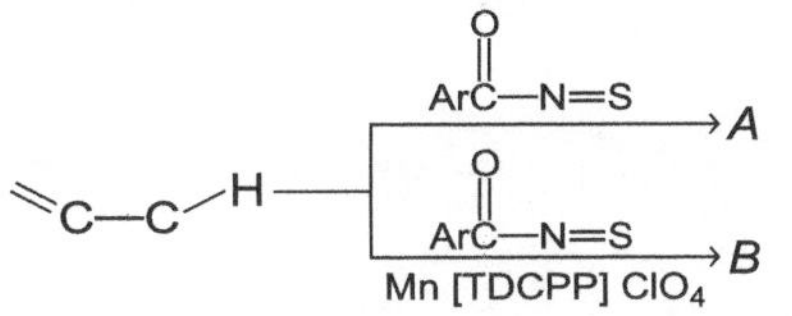

The true statement about $A \leftarrow B$ is

a. *A* and *B* are identical compounds
b. *B* is the deoxygenated product of *A*.
c. *A* is the deoxygenated product of *B*.
d. *A* and *B* are not related in any manner

8 NBS $\xrightarrow[KOH]{Br_2}$ *A*. Product *A* is

a. 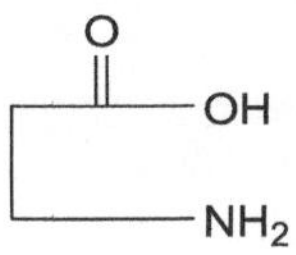

b. 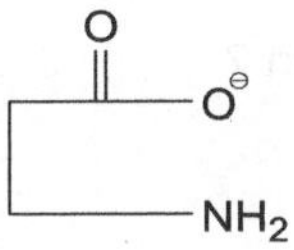

c. 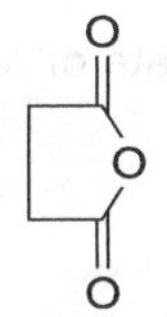

d. 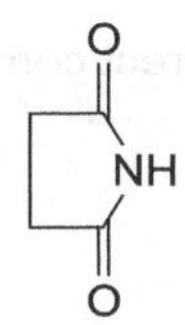

9 N_2 gas is liberated when (HCl + $NaNO_2$) react with the following compounds.

I. $CH_3CH_2NH_2$
II. Urea
III. CH_3CONH_2
IV. $C_6H_5CH_2NH_2$

The answer is

a. I, II, III and IV
b. II, III and IV
c. I, III, and IV
d. I, II, and IV

10 What product is expected from the following reaction.

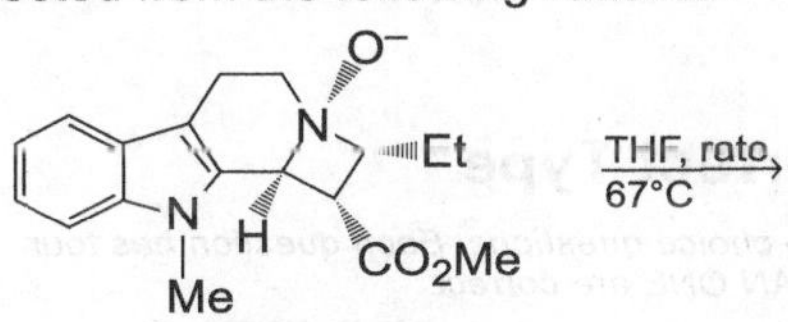

Azetidine-1-oxide

a. 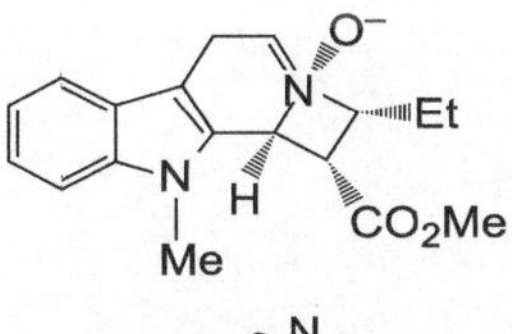

b. 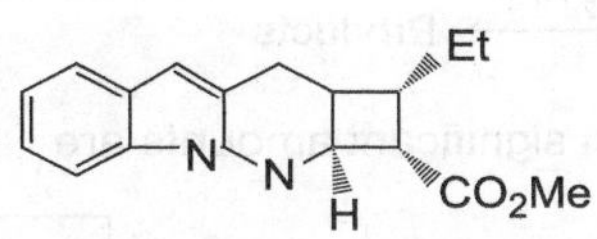

c. 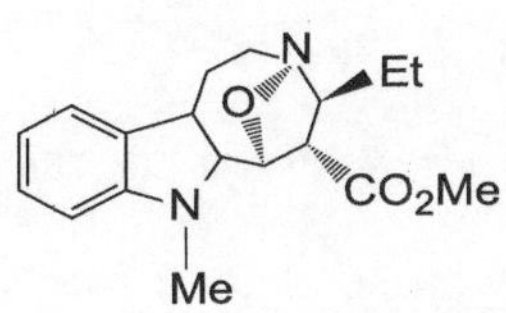

d. 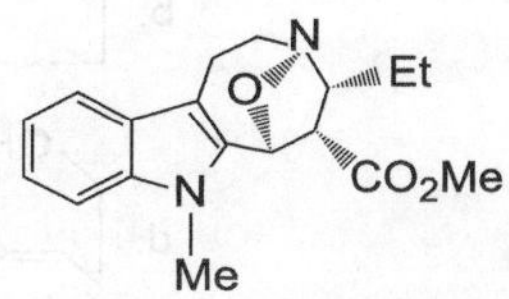

Working Space

11 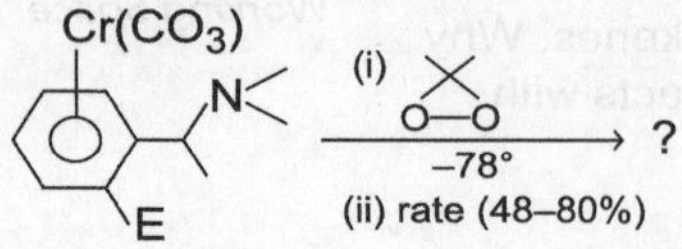

a. 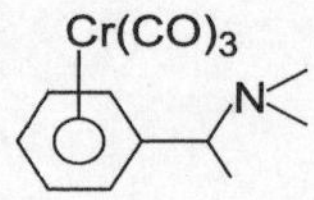

b. 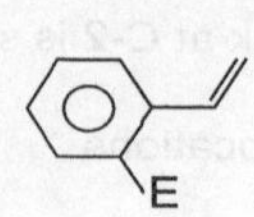

c. 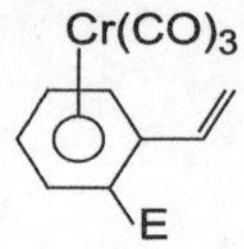

d. All of these

12 Consider the following reactions

Reaction I 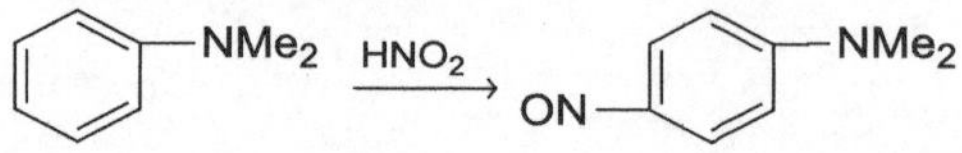

Reaction II 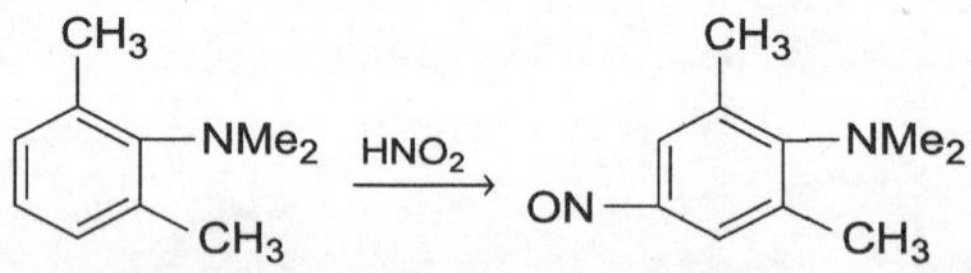

Which of the following is a correct comparison of rate of reaction ?

a. $r_I > r_{II}$
b. $r_I < r_{II}$
c. $r_I = r_{II}$
d. Reactions are not possible

13 A hydrocarbon $A\,(C_4H_8)$ on reaction with HCl gives a compound $B\,(C_4H_9Cl)$ which on reaction with 1 mole of NH_3 gives a compound $C\,(C_4H_{11}N)$. On reacting with $NaNO_2$ and HCl followed by treatment with water, compound C yields an optically active alcohol, D. Ozonolysis of A gives 2 moles of acetaldehyde.

The structure of compound C is

a. butylamine
b. N, N-dimethylethanamine
c. diethylamine
d. 2-aminobutane

One or More than One Options Correct Type

Direction *(Q. Nos. 14-19) This section contains 6 multiple choice questions. Each question has four choices (a), (b), (c) and (d), out of which ONE or MORE THAN ONE are correct.*

14 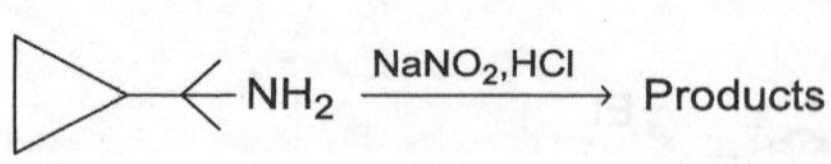

The products found in significant amounts are

a. 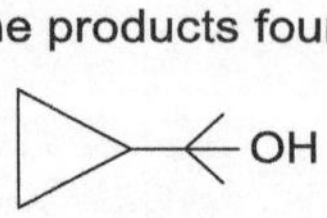

b. 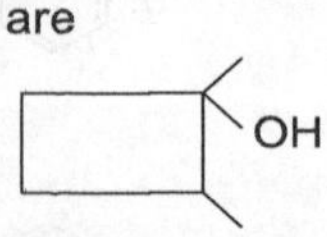

c. 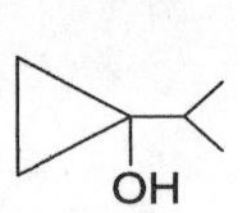

d. 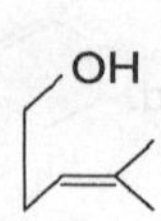

15 Amine with molecular weight wt. 31 A 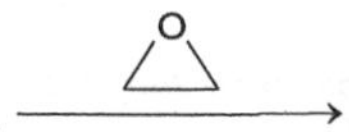 B

Which of the following statements are true for the above reaction ?

a. A is α-amino alcohol
b. A is more reactive than B
c. A can form H-bonding (intramolecular)
d. B is less nucleophilic than A

16 What kind of bonding is/are present in the product of following reaction?

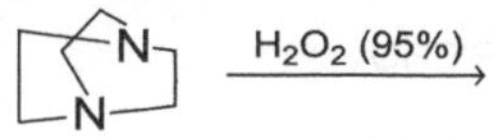

1, 4-diazabicyclo [2, 2, 2] octane

a. covalent bonding
b. hydrogen bonding
c. electrostatic attraction
c. covalent coordinate bonding

17 I. $C_2H_5NH_2 \xrightarrow[\text{(ii) AgNO}_2]{\text{(i) NOCl}} W$

II. $(CH_3)_2CH \cdot NH_2 \xrightarrow[\text{(ii) AgNO}_2]{\text{(i) NOCl}} X$

III. $(CH_3)_3CNH_2 \xrightarrow[\text{(ii) AgNO}_2]{\text{(i) NOCl}} Y$

IV. $CH_3CH(NH_2)C_2H_5 \xrightarrow[\text{(ii) AgNO}_2]{\text{(i) NOCl}} Z$

Which of the product show tautomerism ?

a. W b. X
c. Y d. Z

18 Which of the following reaction represents major product?

a. 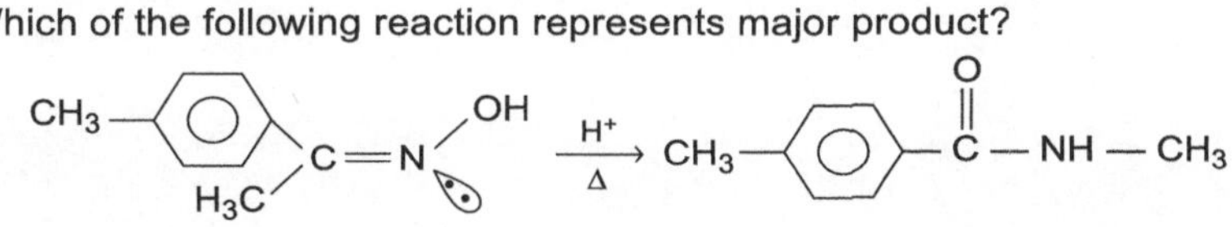

b. 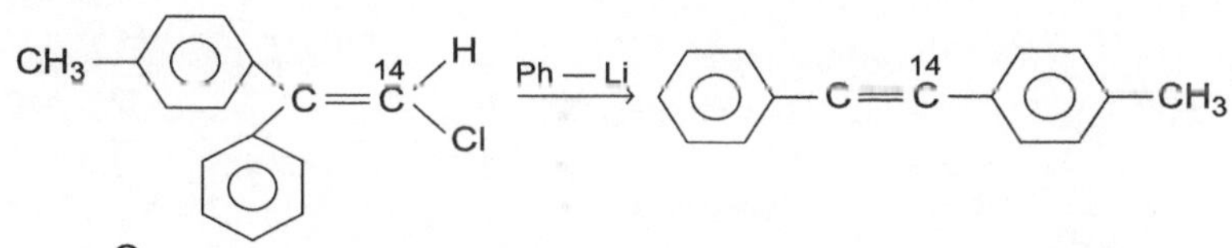

c. $Ph—\overset{O}{\overset{\|}{C}}—NH_2 \xrightarrow{Br_2 + KOH} Ph—NH_2$

d. 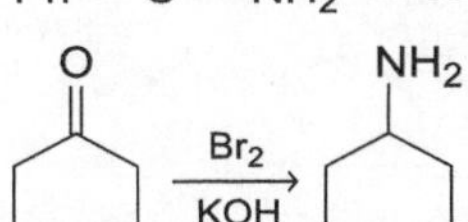

19 Dichlorocarbene is involved as an intermediate in the reaction of

a. phenol + $CHCl_3$ + 4KOH
b. ethylamine + $CHCl_3$ + KOH
c. phenol + CCl_4 + 4KOH
d. $CHCl_3$ + KOH

Matching List Type

Working Space

***Direction** (Q. No. 20) Choices for the correct combination of elements from Column I and Column II are given as options (a), (b), (c) and (d), out of which one is correct.*

20 **Match the Column I with Column II and mark the correct option from the codes given below.**

	Column I		Column II
i.	$RNH_2 \xrightarrow{KMnO_4}$	p.	$R_3N \longrightarrow O$
ii.	$R_2NH \xrightarrow{H_2SO_5}$	q.	R_3CNO_2
iii.	$R_3N \xrightarrow{H_2SO_5}$	r.	$R'CHO$
iv.	$R_3CNH_2 \xrightarrow{H_2SO_5}$	s.	R_2NOH

Codes

	i	ii	iii	iv
a.	r	s	p	q
b.	p	r	s	q
c.	s	p	q	r
d.	q	p	r	s

One Integer Value Correct Type

***Direction** (Q. No. 21) This section contains a question. When worked out will result in an integer from 0 to 9 (both inclusive).*

21 Compound *A* is chiral with one Me as branch and is a liquid with formula $C_5H_{11}O_2N$. Acidification of alkali solution of chiral *A* gives racemic *A*. Reduction of chiral *A* with H_2/Ni gives chiral *B* with formula $C_5H_{13}N$. The number of carbon atom in the main chain of this chiral *B* is

SET - 1

Revisal Problems
for JEE Main

Only One Option Correct Type

Working Space

Direction *(Q. Nos. 1-30) This section contains 30 multiple choice questions. Each question has four choices (a), (b), (c) and (d), out of which ONLY ONE option is correct.*

1 Which is the best leaving group?

a. N_2 b. $OH^{\ominus}$ c. $NH_2^{\ominus}$ d. $CH_3COO^{\ominus}$

2 In order to prepare a 1° amine incorporating an additional CH_2 group from an alkyl halide, what reagent is often used as the nitrogen source?

a. Sodium amide, $NaNH_2$
b. Sodium azide, NaN_3
c. Potassium cyanide, KCN
d. Potassium phthalimide, $C_6H_4(CO)_2NK$

3 What reagent would be best for converting the amide of (*R*)-2-phenylpropanoic acid, into (*R*)-1-phenylethylamine?

a. Excess H_2 and Pt
b. NaOBr in aqueous base
c. $NaBH_4$
d. $LiAlH_4$ in ether

4 In the following sequence of reactions, deduce structures of *B* and *C*:

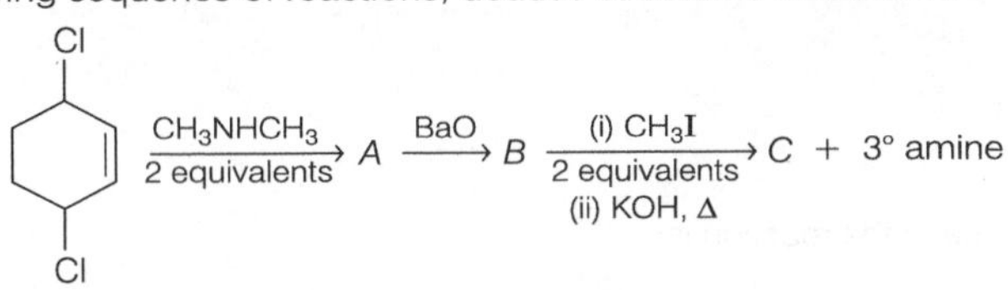

a. 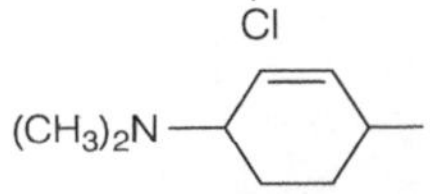 iso-propylamine

b. 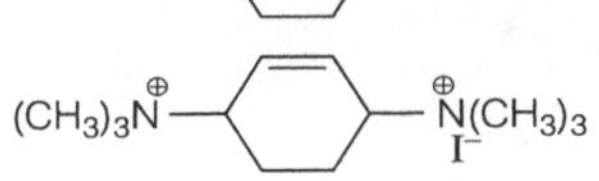 benzene

c. 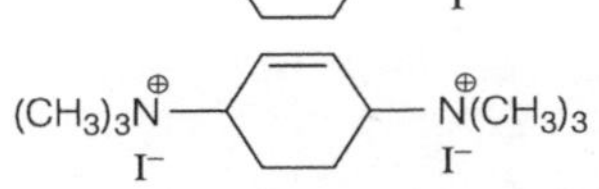 iso-propylamine

d. 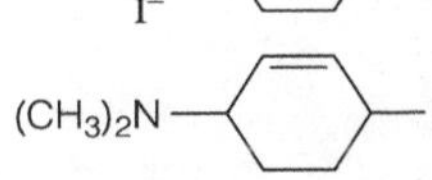 benzene

5 Which of the following compounds when heated with KOH and primary amine gives the carbylamine test?

a. $CHCl_3$ b. CH_3Cl c. CH_3OH d. CH_3CN

6 Hofmann's elimination product of *A* is

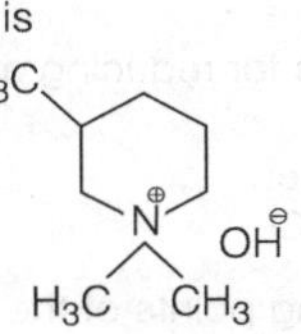

a. 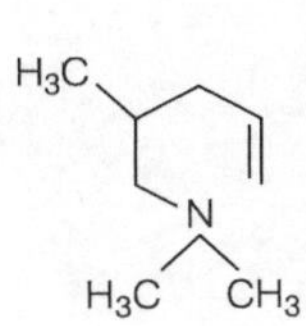

b. 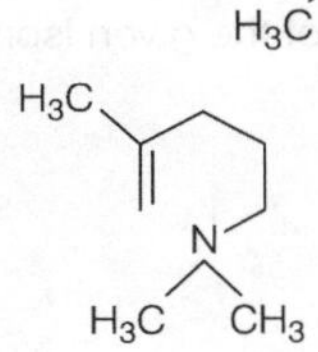

c. 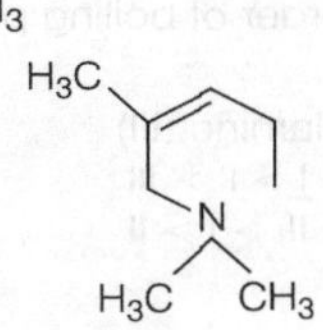

d. 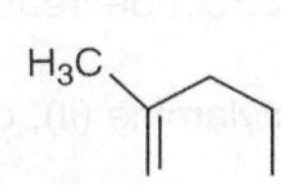

7 $CH_3-C(CH_3)_2-CH_2NH_2 \xrightarrow{HNO_2} A$ (major product), A is

a. $CH_3-C(CH_3)_2-CH_2OH$ b. $CH_3-C(CH_3)(OH)-CH_2CH_3$

c. $CH_3-C(CH_3)_2-CH_2N_2^+$ d. $CH_3-C(CH_3)(H)-CH=CH_2$

8 Which of the following is the end product D of the reaction sequence?

$$CH_3CH_2NH_2 \xrightarrow{3\,CH_3I} B \xrightarrow[H_2O]{Ag_2O} C \xrightarrow{\Delta} D + CH_2{=}CH_2 + H_2O$$

a. $(CH_3)_3N$ b. $(CH_3)_2NH$
c. $(CH_3CH_2)_2NH$ d. $(CH_3CH_2)_3N$

9 Which of the following procedures would be best for preparing isobutyl isopropylamine?

a. $(CH_3)_2CHBr + (CH_3)_2CHCH_2NH_2$
b. (I) $(CH_3)_2CHBr + (CH_3)_2CHCONHNa$ (ii) $LiAlH_4$ in either
c. $(CH_3)_2CHCN + (CH_3)_2CHNH_2$
d. (I) $(CH_3)_2CHNH_2 + (CH_3)_2CHCH{=}O$ (ii) H_2 and Pt.

10 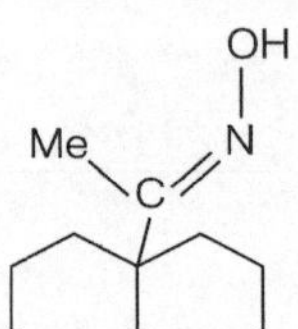 $\xrightarrow{H_2SO_4}$ Product and name of the reaction is

a. 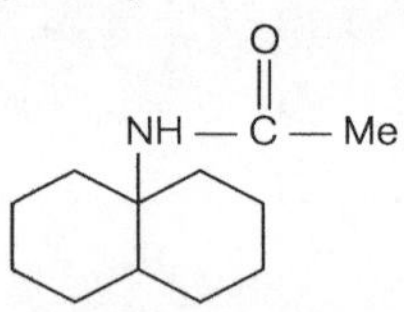
(Hofmann bromamide reaction)

b. 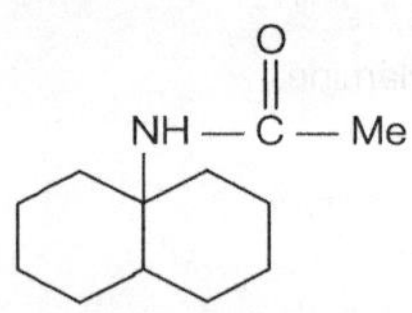
(Beckmann rearrangement)

c. 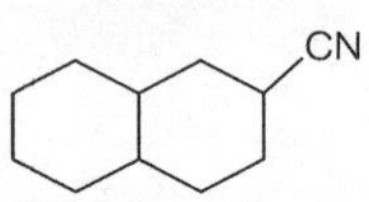
(Curtius reaction)

d. None of these

11 Which of the following compounds is the strongest base?

a. $(CH_3)_2NCH_2CO_2CH_3$ b. $(CH_3)_2NCOCH_3$
c. $(CH_3)_2NC \equiv N$ d. $(CH_3)_2N-N{=}O$

12 Which of the following reagents would not be a good choice for reducing an aryl nitro compound to an amine ?

a. H_2 (excess) and Pt catalyst b. $LiAlH_4$ in ether
c. Fe in 15% HCl d. Zn in 10% HCl

13 Which of the following is a correct decreasing order of boiling points of the given isomeric compounds?

Ethyldimethyl amine (I), *n*-butylamine (II), diethylamine (III)

a. II > III > I b. I > II > III
c. III > II > I d. III > I > II

14 Choose the correct reaction,

a. 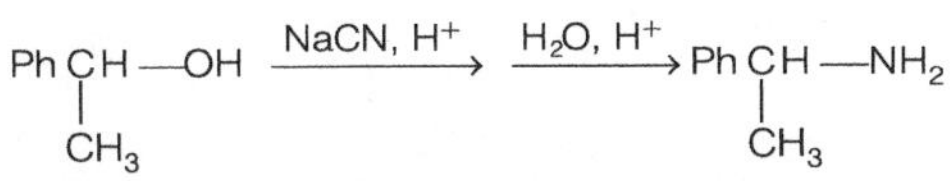

b. 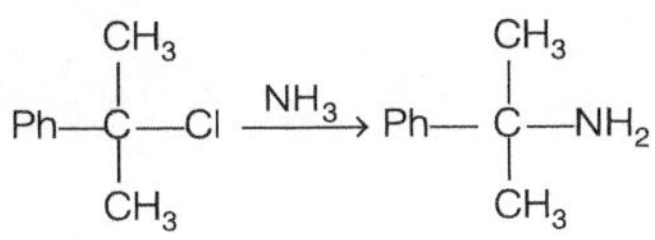

c. 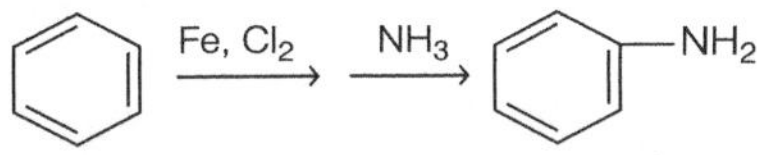

d. 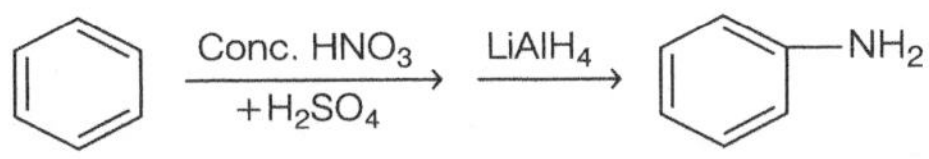

15 What is the chief product from the Hofmann elimination of 4-methyl-2-aminopentane?

a. 4-methyl-1-pentene
b. 4-methyl-2-pentene
c. 2-methyl-1-pentene
d. 2-methyl-2-pentene

16 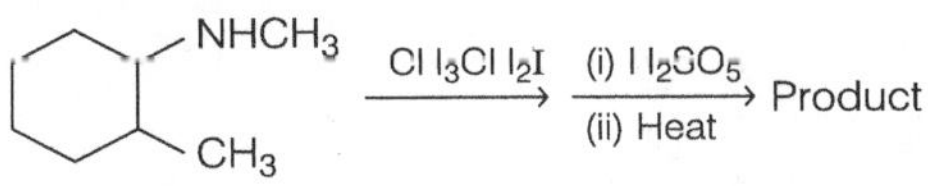

The main product is

a. 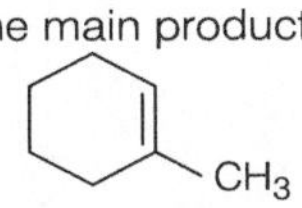

b. 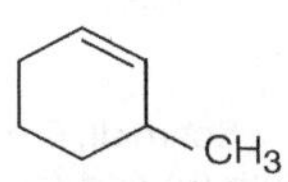

c. 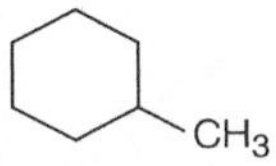

d. $CH_2{=}CH_2$

17 In the reaction,

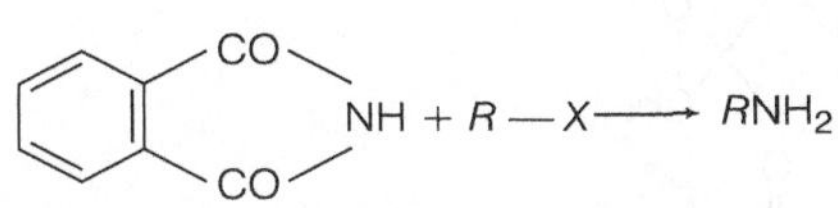

To prepare $CH_3\ CH(CH_3)NH_2$, *R* should be

a. Et
b. Me
c. $(CH_3)_2CH$
d. $CH_3CH_2CH_2$

18 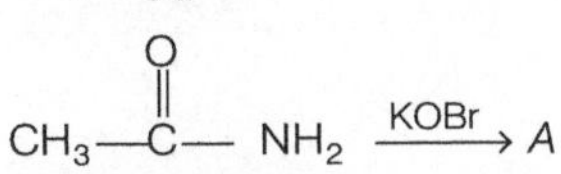

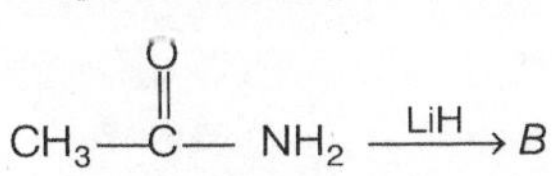

Relation between *A* and *B* is

a. Chain isomer
b. Homologues
c. Functional isomer
d. Identical

19 Which of the following is the strongest Bronsted base?

a. 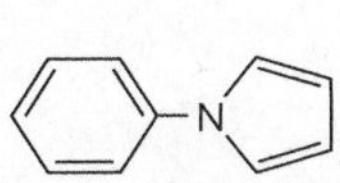

b. 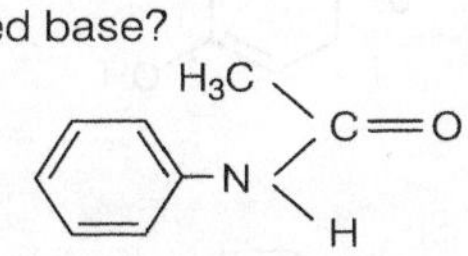

c. 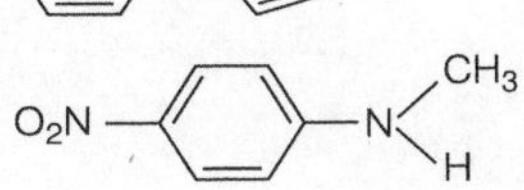

d. 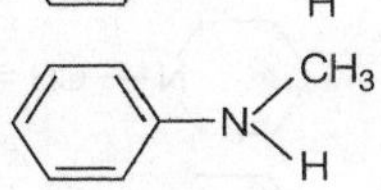

Working Space

Working Space

20 Give the product that would be obtained by treating the following tertiary amine with hydrogen peroxide followed by heat.

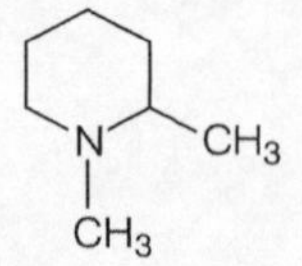

a. 

b. 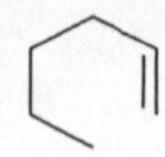

c. 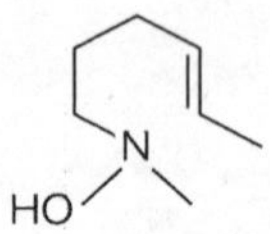

d. 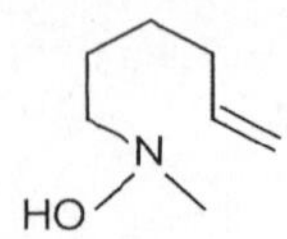

21 Acid $A \xrightarrow[\Delta]{N_3H\ \text{conc.}\ H_2SO_4} B \xrightarrow[\Delta]{CHCl_3 + KOH} C \xrightarrow{H_3O^+} B + HCOOH$

$B \xrightarrow{HNO_2} D$ (various type)

$C \xrightarrow{\text{Reduction}} CH_3-C(CH_3)_2-NHCH_3$

Reaction involve in the formation of *B* and *C* are

a. *B* = carbylamine, *C* = Schmidt
b. *B* = Schmidt, *C* = carbylamine
c. *B* = hydrolysis, *C* = carbylamine
d. Both (b) and (c)

22 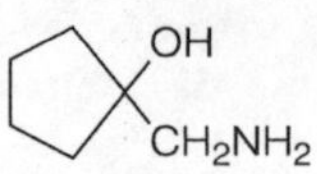

$\xrightarrow{HNO_2}$ *A* (major product), *A* is

a. 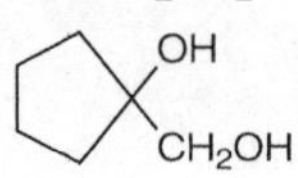

b. 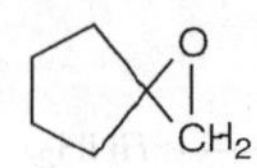

c. 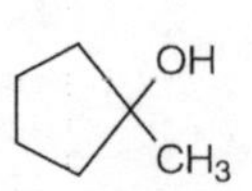

d. 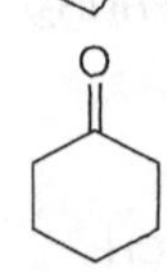

23 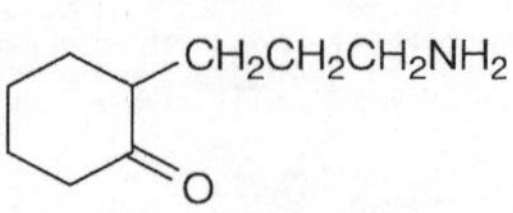

$\xrightarrow[\text{Raney Ni}]{H_2}$ *A*, *A* is

a. 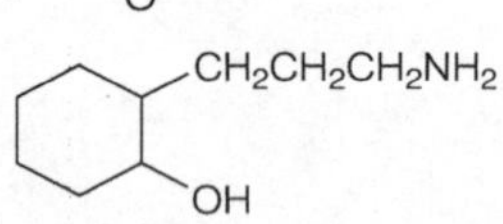

b. 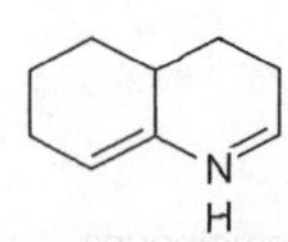

c. 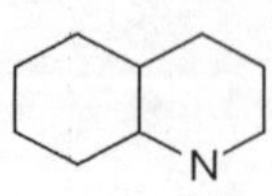

d. 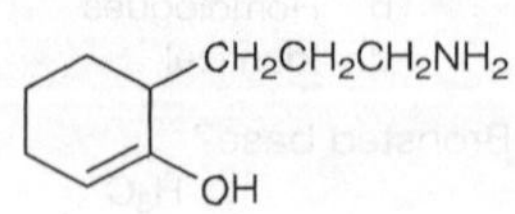

24 Which has maximum pK_a value?

a. 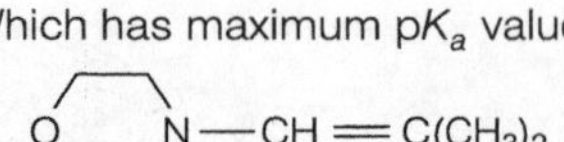

b. 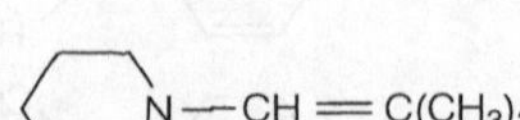

c. 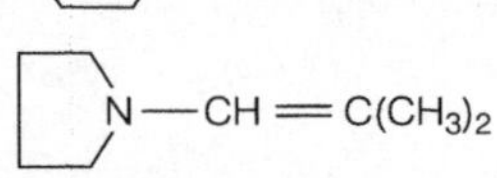

d. 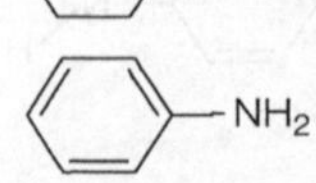

25 What is the chief product from the following reaction?

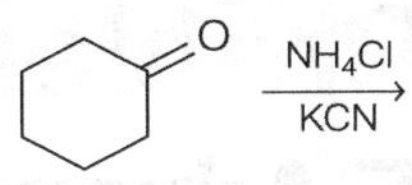

a. 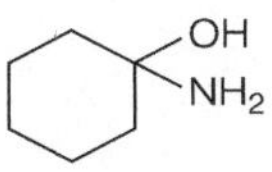 b. 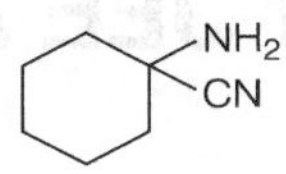

c. 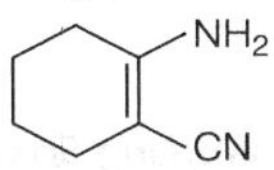 d. 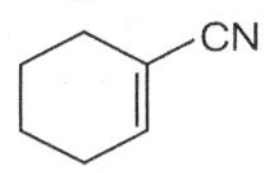

26 The amine, 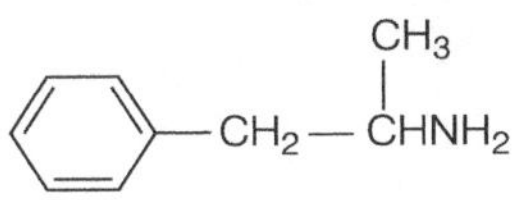 can best be prepared by

a. 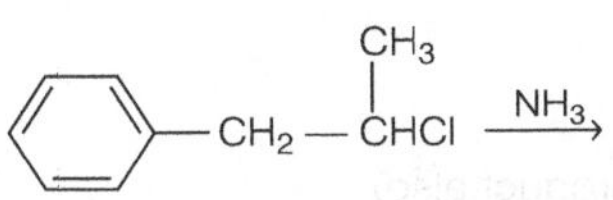 b. 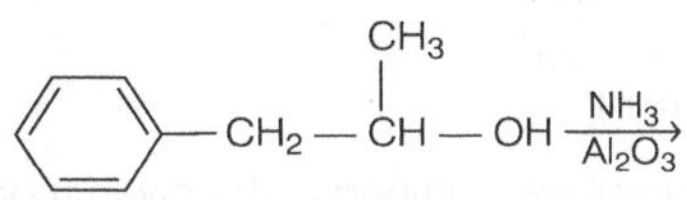

c. 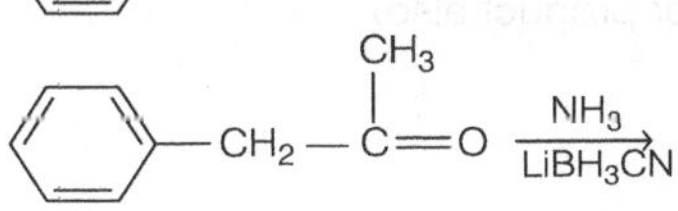 d. 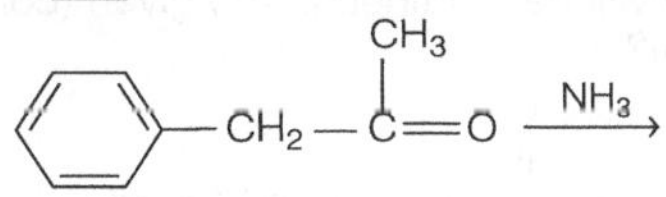

27 For the following compounds, which is the strongest base and which is the strongest acid?

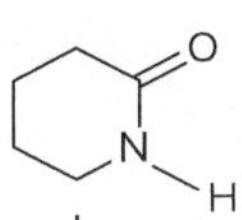
I

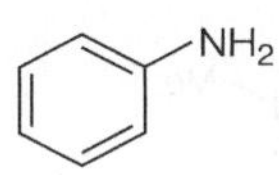
II

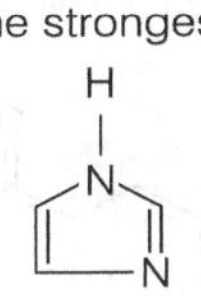
III

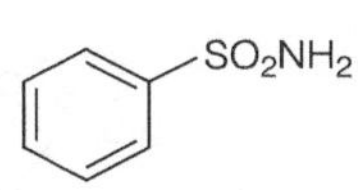
IV

a. II = strongest base, I = strongest acid
b. IV = strongest base, III = strongest acid
c. III = strongest base, IV = strongest acid
d. II = strongest base, III = strongest acid

28 What is the likely product from the following reaction?

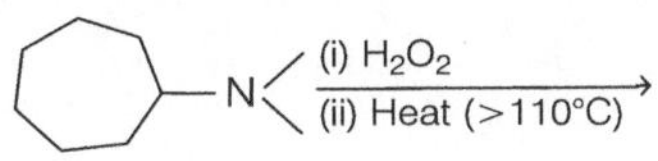

a. 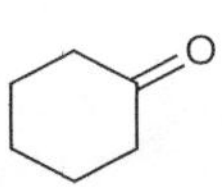 b. 

c. 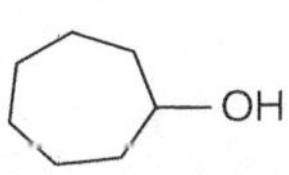 d. 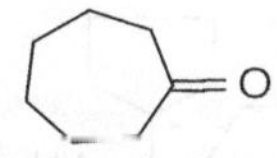

29 Dibutylamine, $(C_4H_9)_2NH$, and anisole, $C_6H_5OCH_3$, have similar boiling points, and are relatively insoluble in water. How might a mixture of these compounds be separated into the pure components?

a. (i) Dissolve mixture in ether, (ii) Extract the anisole into 10% aqueous NaOH
b. (i) Dissolve mixture in ether, (ii) Extract the amine into 10% aqueous HCl
c. (i) Dissolve mixture in ether, (ii) Extract the amine into 10% aquous NaOH
d. (I) Dissolve mixture in ether, (ii) Extract the anisole into 10% aqueous HCl

30 An optically active amine *A* of molecular formula $C_4H_{11}N$ is subjected to Hofmann's exhaustive methylation process followed by hydrolysis. An alkene *B* is produced which upon ozonolysis and subsequent hydrolysis yields formaldehyde and propanal. The amine *A* is

a. $CH_3CH(NH_2)CH_2CH_3$
b. $CH_3—NH—CH(CH_3)—CH_3$
c. $CH_3—N(CH_3)CH_2CH_3$
d. $CH_3CH_2CH_2CH_2NH_2$

Working Space

SET - 2

Revisal Problems
for JEE Advanced

Only One Option Correct Type

Working Space

Direction *(Q. Nos. 1-11) This section contains 11 multiple choice questions. Each question has four choices (a), (b), (c) and (d), out of which ONLY ONE option is correct.*

1 Which of the following order of basicity is correct?

a. $R—C\equiv N: > R'—CH=NR > RNH_2$

b. $R—C\equiv N: < R'—CH=NR < RNH_2$

c. $RNH_2 < R—C\equiv N: < R'—CH=NR$

d. $RCH=NH < R—C\equiv N: < RNH_2$

2 Which of the following product will be obtained in the given (consider minor product also) Beckmann type rearrangement?

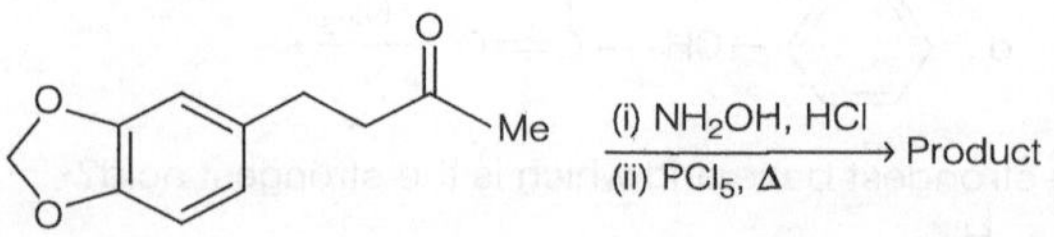

a. 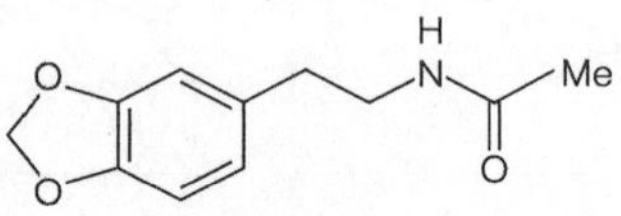

b. 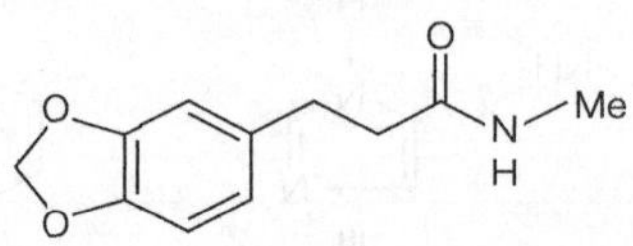

c. 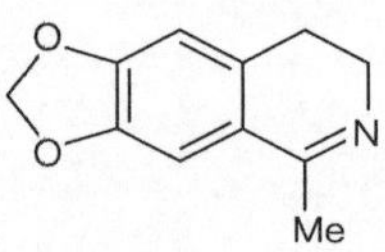

d. All of these

3 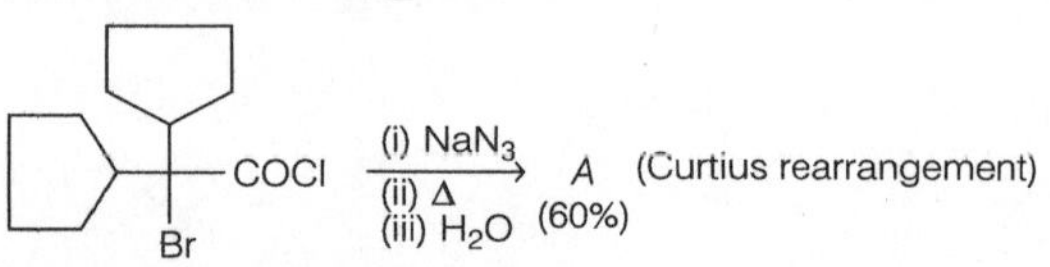

Product *A* is

a. 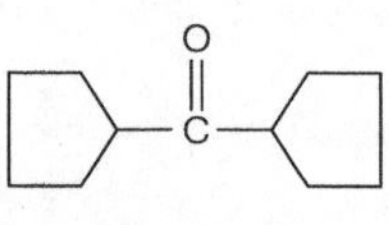

b. 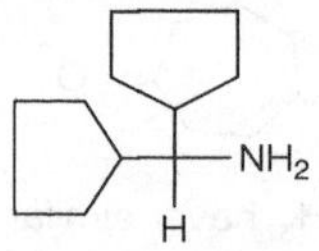

c. 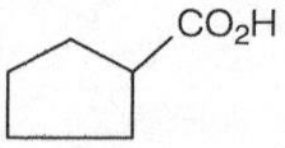

d. 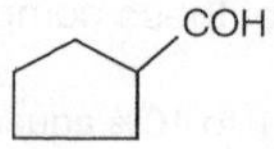

4 In the reaction below

$$CH_3COCH_2Cl + CH_2N_2 \longrightarrow C_4H_7ClO(X) + C_4H_7ClO(Y)$$

X is a ketone while *Y* is an epoxide.

Deduce structure of *X* and *Y*.

a. 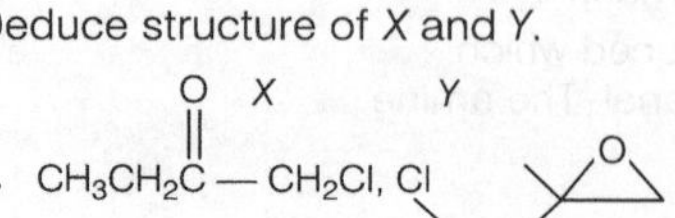

b. 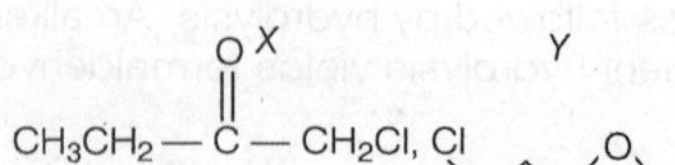

c. 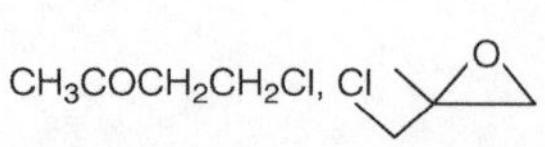

d. 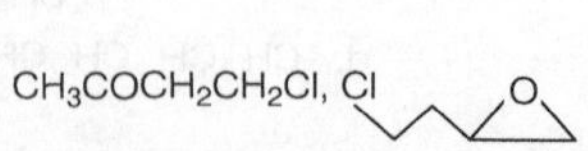

Working Space

5 What product mixture is expected from reaction of (*S*)-2 aminobutane with 2-butanone in the presence of $NaBH_3CN$?

a. A pair of enantiomers
b. A pair of diastereomers, both *meso*
c. A pair of diastereomers, one *meso*, one chiral
d. A pair of diastereomers, both chiral (not enantiomers.)

6 Increasing order of basic character of the following is

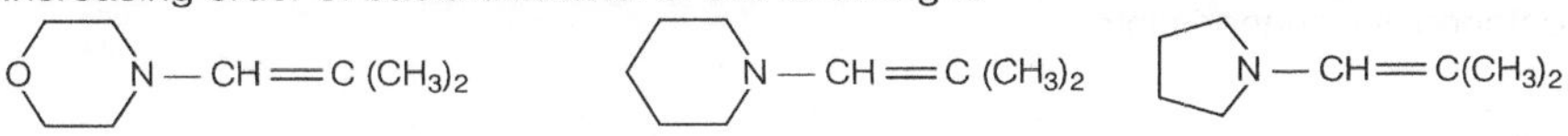

I II III

a. I < II < III b. III < II < I c. I < III < II d. III < I < II

7 Only one of the following amines will lose its nitrogen atom as trimethyl amine by repeated Hofmann elimination reactions (exhaustive methylation followed by heating with AgOH). Identify that amine.

a. 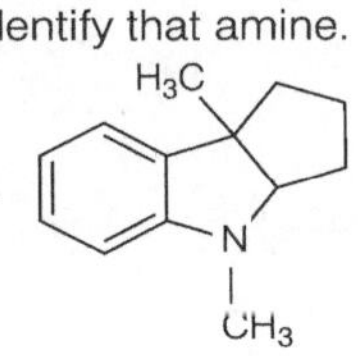

b. 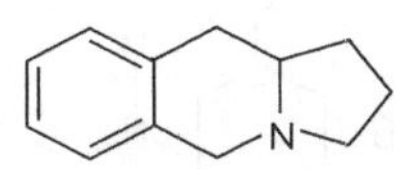

c. 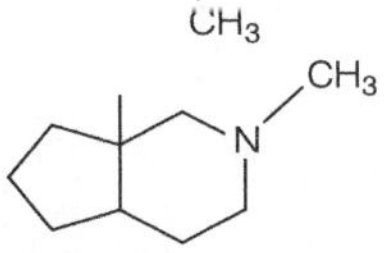

d. 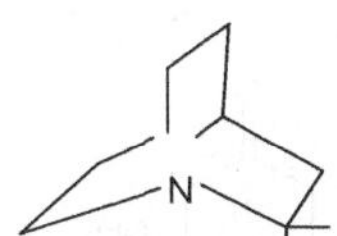

8 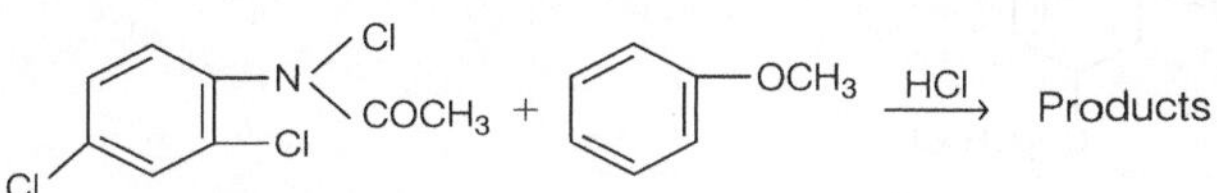

N-chloro-2,4-dichloro acetanilide Anisole

Choose the correct set of possible product from the following.

I. 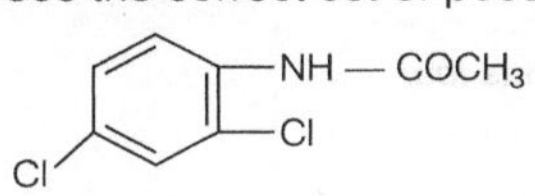

II. 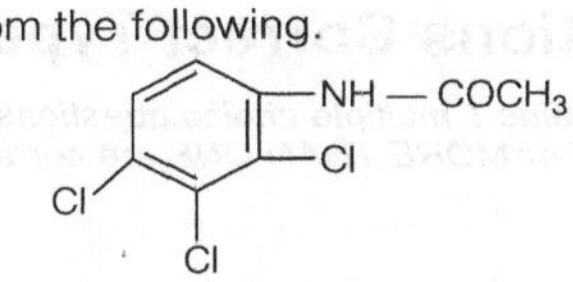

III. 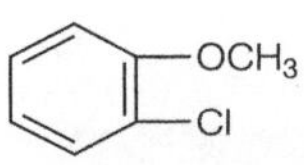

IV. 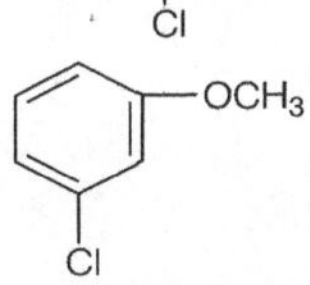

V. 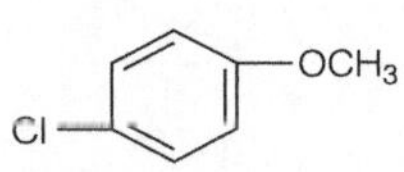

a. I and II b. I, III and V c. II, III and V d. All of these

9 Final product of the following reaction is

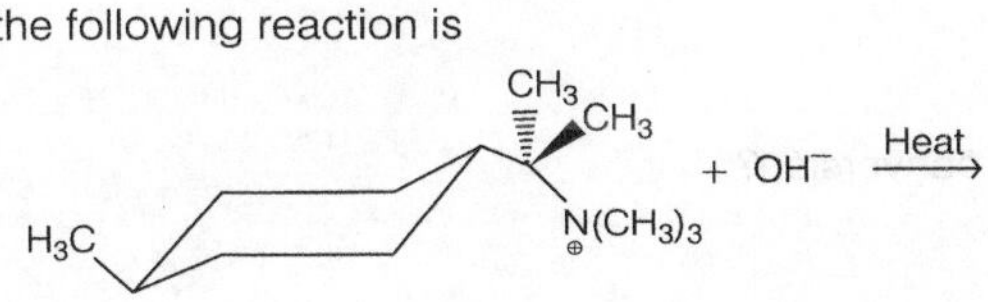

a. 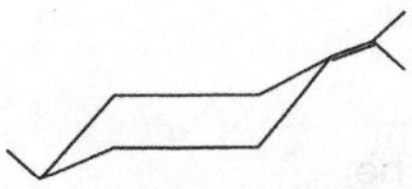

b. 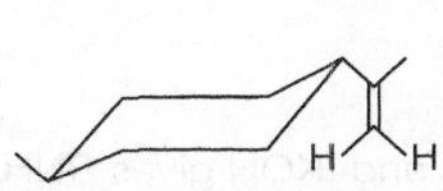

c. 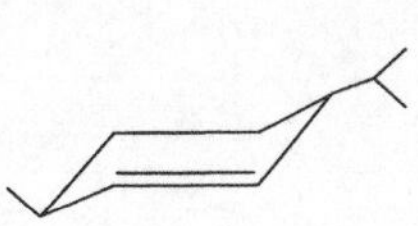

d. 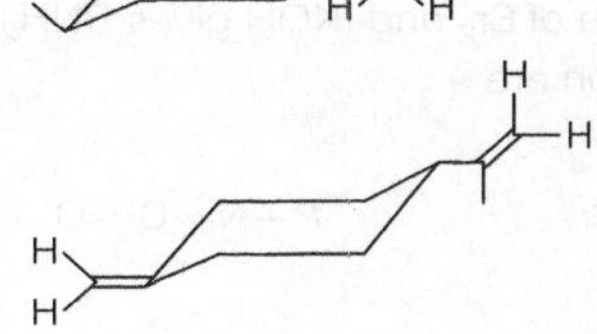

10 Basic reagents are used in many organic reactions. Examples are : I triethylamine, II DBU, III potassium phthalimide and IV LDA. Which of these is the strongest base?

Working Space

a. 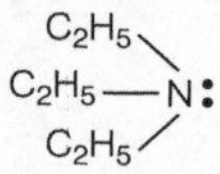 b. 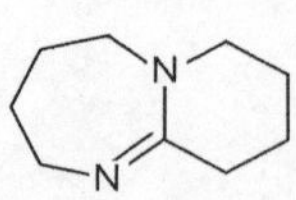 c. 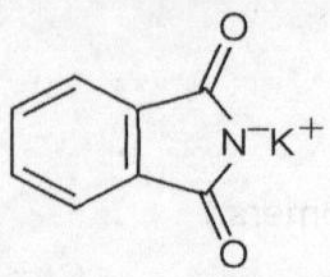 d. 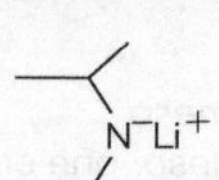

11 Following compound *A* spontaneously changes into

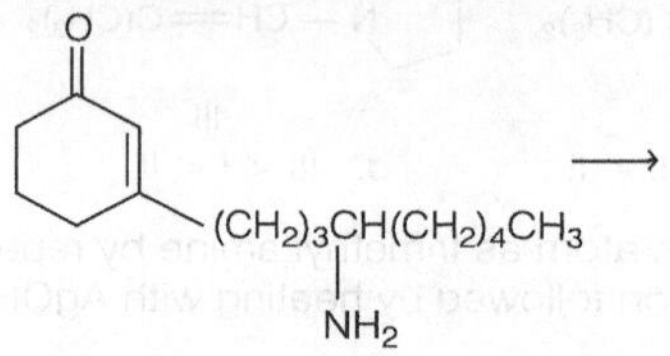

a. 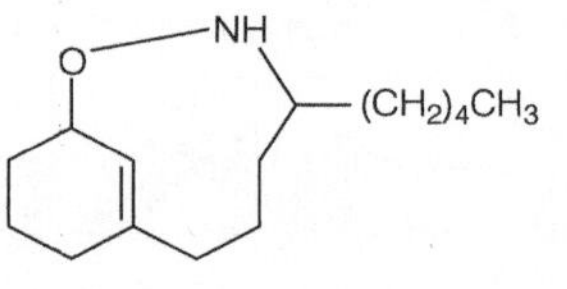

b. 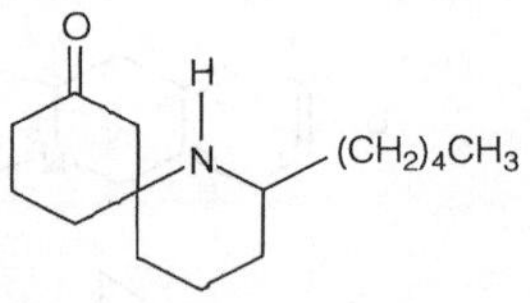

c. 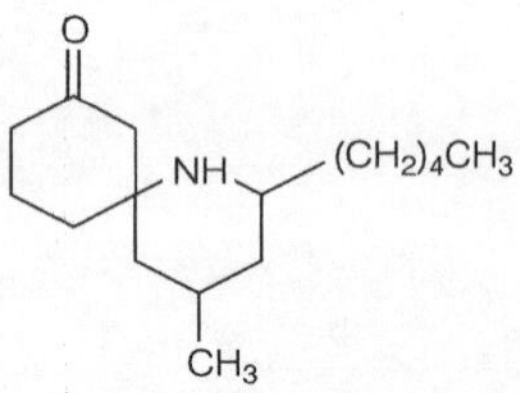

d. 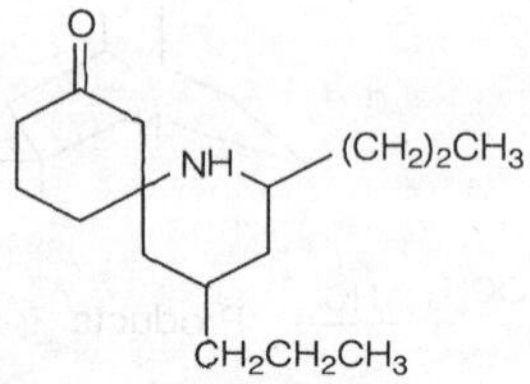

One or More than One Options Correct Type

Direction (Q. Nos. 12-18) *This section contains 7 multiple choice questions. Each question has four choices (a), (b), (c) and (d), out of which ONE or MORE THAN ONE are correct.*

12 $A \xleftarrow[\text{Heat}]{H_3O^+} CH_3\underset{\underset{CH_3}{|}}{C}HNC \xrightarrow[\text{Heat}]{N_2,Ni} B$

Product *A* and *B* can be distinguished by

a. the treatment of $CHCl_3$, ^-OH
b. the action of HNO_2; *A* liberates N_2 gas while *B* does not
c. the action of $CS_2/HgCl_2$ *B* gives odour of mustard oil while *A* does not
d. the treatment of *p*-toluene sulphonyl chloride; *A* gives alkali soluble product

13 $CH_3CH_2\overset{\overset{O}{||}}{C}—OH \longrightarrow CH_3CH_2NH_2$

Which of the following is/are effective for this conversion?

a. NH_3 (excess), heat, Na, C_2H_5OH
b. N_3H cold conc. H_2SO_4
c. NH_3 (excess), heat, NaOH, Br_2
d. $NH_2—NH_2, \overset{+}{H}$, $NaNO_2$, HCl; $\overset{-}{O}H$, H_2O

14 Reaction of $RCONH_2$ with a mixture of Br_2 and KOH gives RNH_2 as the main product. The intermediates inolved in the reaction are

a. $R—\overset{\overset{O}{||}}{C}—NHBr$ b. $R—NHBr$ c. $R—N{=}C{=}O$ d. $R—\overset{\overset{O}{||}}{C}—N^-—Br$

15 Which of the following statement(s) is/are true regarding the following compound?

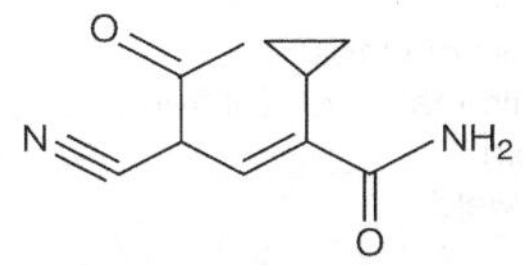

a. Total number of substituents in the given compound is 1
b. Double bond equivalent (DBE) of given compound is 6
c. Total number of carbon in the parent chain is 4
d. Amide is used as a secondary suffix in the IUPAC name of given compound.

16 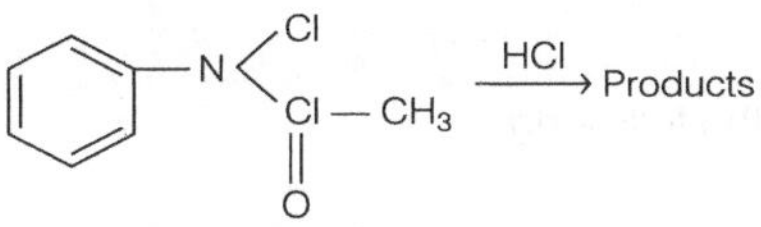

The probable products is/are

a. 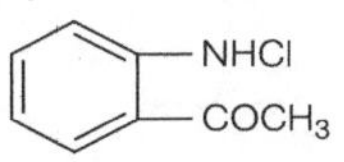
b. 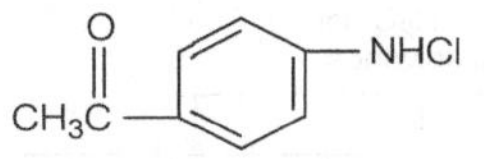
c. 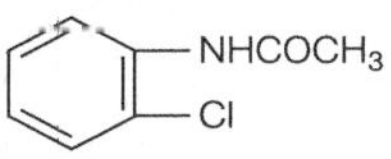
d. 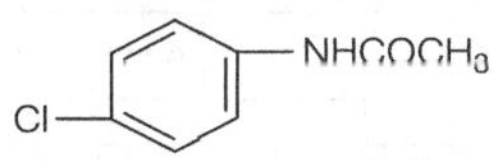

17 Select the correct statement(s).

a. N-methyl piperidine *A* has higher boiling point than 4-methylpiperidine *B*

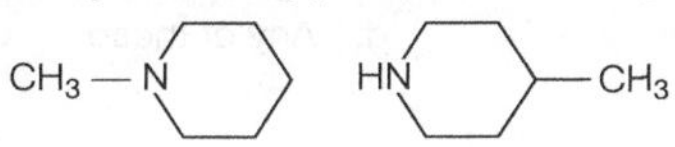

A *B*

b. *B* has higher boiling point than *A*
c. *A* and *B* have same boiling point being isomeric amines
d. There is formation of two diastereoisomeric quarternary ammonium chloride when 4-tert-butyl-N methylpiperidine is treated with benzoyl chloride

18 In the following conversion process $RCH_2CN \xrightarrow{H_2/Ni} RCH_2CH_2NH_2$ the intermediate species involved is/are

a. 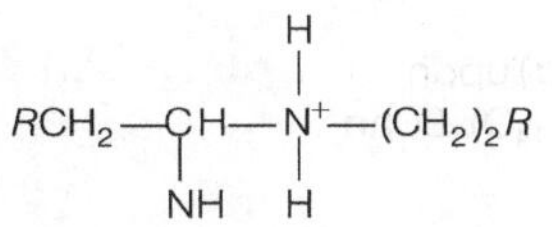
b. 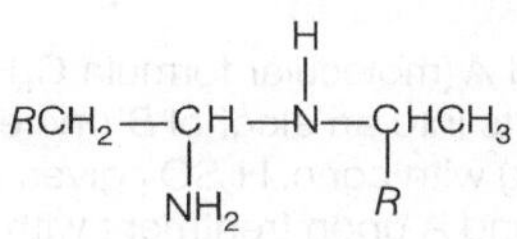
c. $RCH_2—CH = NCH_2CH_2R$
d. $RCH_2—CH = NCH_2CH_2R$

Comprehension Type

Direction (Q. Nos. 19-26) *This section contains 3 paragraphs, each describing theory, experiments, data, etc. Eight questions related to the paragraphs have been given. Each question has only one correct answer among the four given options (a), (b), (c) and (d).*

Passage I

Two synthetic routes were planned to synthesise cyclohexylamine

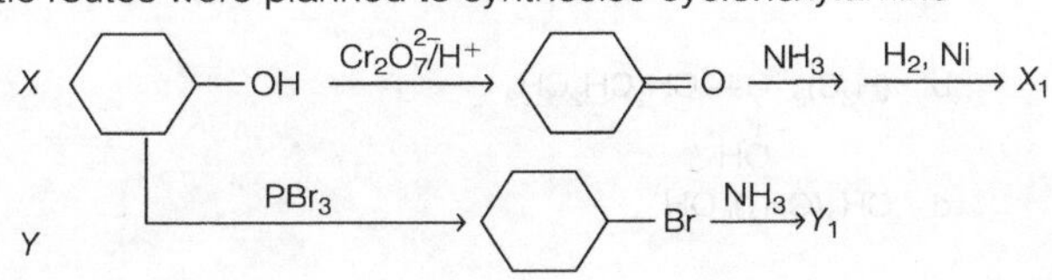

Taking all the aspects of 2° alcohols and 2° alkyl halide, answer the following questions.

Working Space

19 Select the correct statement(s).

a. Plan *Y* is preferred to plan *X* since, it involves lesser number of steps
b. Plan *X* is preferred to plan *Y* since, plan *Y* also involves the elimination of HBr from 2° alkyl halide
c. Plan *X* involves reductive amination which gives better yield
d. Plan *Y* involves S_N reaction and being faster gives better yield

20 Cyclohexylbromide can also be converted into cyclohexylamine in good yield using

a. NaN_3 followed by reduction with $LiAlH_4$ since, N_3^- is a good nucleophile
b. NaCN followed by reduction using Pt / H_2 since, CN^- is a strong nucleophile
c. Both (a) and (b)
d. None of the above

Passage II

It is often possible to take advantage of their basicity to purify amines by the following scheme.

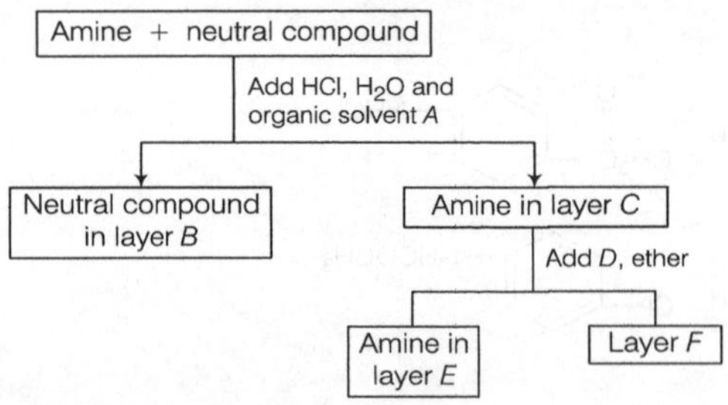

21 Organic solvent *A* can be

a. ketone b. alcohol c. ether d. Any of these

22 Amine is in the form of

a. $RNH_3^+Cl^-$ in aqueous layer *C*
b. $RNH_3^+Cl^-$ in etheral layer *C*
c. RNH_2 in etheral layer *C*
d. RNH_2 in aqueous layer

23 Select the incorrect statement(s).

a. Strong base *D* (say NaOH) is added
b. Amine goes in etheral layer *E*
c. *F* is aqueous layer containing NaCl (if *D* is NaOH)
d. Amine goes in aqueous layer

Passage III

An optically active organic compound *A* (molecular formula $C_6H_{15}N$) gives off $N_2(g)$ upon treatment with HNO_2 and thus converts into an alcohol *B* (molecular formula $C_6H_{14}O$). *B*, an optically inactive alcohol upon heating with conc. H_2SO_4 gives a compound *C* (molecular formula C_6H_{12}). Compound *A* upon treatment with excess CH_3I and subsequently with Ag_2O followed by heating yields an alkene *D* as the single alkenic product. The compound *B* is unoxidisable to carbonyl compound.

24 The compound *A* is

a. $(CH_3)_3C—CH_2—CH_2NH_2$
b. 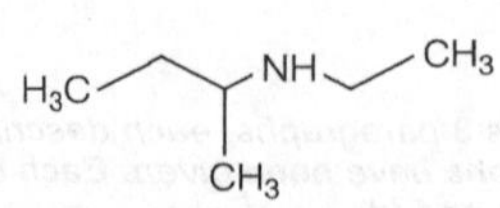

c. $(H_3C)_3 \cdot C—CH(NH_2) \cdot CH_3$
d. $(H_3C)_2 \cdot CH—CH(NH_2)—CH_2 \cdot CH_3$

25 The compound *B* is

a. $(H_3C)_2 \cdot C(OH)—CH(CH_3)_2$
b. $(H_3C)_2 \cdot C(OH)—CH_2CH_2CH_3$
c. $(CH_3)_3C—CH_2—CH_2—OH$
d. $CH_3(CH_2)_5OH$

26 The compound *C* is

a. $((CH_3)_2 = CHCH_2CH_3$
b. $(CH_3)_2C = C(CH_3)_2$
c. $CH_3CH = CH—CH(CH_3)_2$
d. $CH_3CH_2CH = CHCH_2CH_3$

Working Space

Matching List Type

***Direction** (Q. No. 27-29) Choices for the correct combination of elements from Column I and Column II are given as options (a), (b), (c) and (d), out of which one is correct.*

27 Match the Column I with Column II and mark the correct option from the codes given below.

	Column I		Column II
i.	Intramolecular alkyl shift	p.	Curtius degradation
ii.	Synthesis of 1° amine	q.	Hofmann degradation
iii.	Retention of stereochemical configuration	r.	Schmidt
iv.	Decarboxylation takes place	s.	Gabriel

Codes

	i	ii	iii	iv
a.	p,q	p,q,r,s	p,q	r
b.	p	q,s	p,s	q,s
c.	r	s	q	p,s
d.	p,r	q,s	p	s

28 Match the Column I with Column II and mark the correct option from the codes given below.

	Column I (Pair of distinction)		Column II (Reagent/test)
i.	$CH_3CH_2NH_2$, $(CH_3)_2NH$	p.	HNO_2 + β-naphthol
ii.	CH_3NH_2, 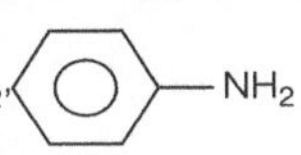	q.	$NaNO_2$ + HCl
iii.	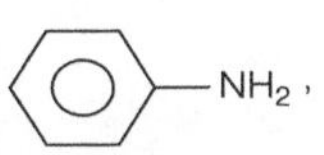, 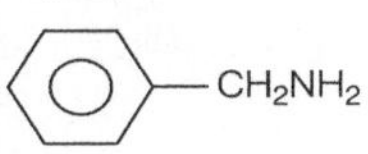	r.	Hofmann mustard oil reaction
iv.	$(CH_3)_3N$, $CH_3CH_2CH_2CH_2NH_2$	s.	Carbylamine

Codes

	i	ii	iii	iv		i	ii	iii	iv
a.	q,r,s	p	p	q,r,s	b.	r	p,q	s	p,q
c.	p,s	r	p,q	q,s	d.	p,s	q	r,s	p,q

29 Match the Column I with Column II and mark the correct option from the codes given below.

	Column I (Amines)		Column II (Characteristics)
i.	$CH_3CH_2CH_2NH_2$	p.	Treatment of $NaNO_2$, HCl, gives N-nitroso compound
ii.	$CH_3CH_2NHCH_3$	q.	Treatment of $NaNO_2$, HCl gives diazonium chloride
iii.	$CH_3—N(CH_3)—CH_3$	r.	Treatment of CH_3I (excess) followed by AgOH; heat gives out alkene
iv.	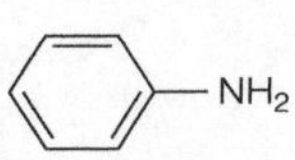	s.	Treatment of HCl, heat gives dealkylation.

Codes

	i	ii	iii	iv		i	ii	iii	iv
a.	p	p,s	p	r,s	b.	p,s	q,s	p	r,s
c.	r,s	p,r,s	s	q	d.	p,s,q	r,s	q,	r

Statement Type

Working Space

Direction *(Q. Nos. 30-32) This section is based on Statement I and Statement II. Select the correct answer from the codes given below.*

a. Both Statement I and Statement II are correct and Statement II is the correct explanation of Statement I
b. Both Statement I and Statement II are correct and Statement II is not the correct explanation of Statement I
c. Statement I is correct but Statement II is incorrect
d. Statement II is correct but Statement I is incorrect

30 **Statement I** $NaBH_3CN$ can be used to reduce imine to 1° amine.

Statement II Protonated imine undergoes reduction in reductive amination reactions.

31 **Statement I** $NaBH_3CN$ is less reactive than $NaBH_4$.

Statement II $NaBH_3CN$ reduces carbonyl compounds to alcohols.

32 **Statement I** Tertiary amine is less basic than secondary amine in aqueous solution.

Statement II Over crowding due to $+I$ effect of three alkyl groups increases strain.

One Integer Value Correct Type

Direction *(Q. Nos. 33 and 34) This section contains 2 questions. When worked out will result in an integer from 0 to 9 (both inclusive).*

33 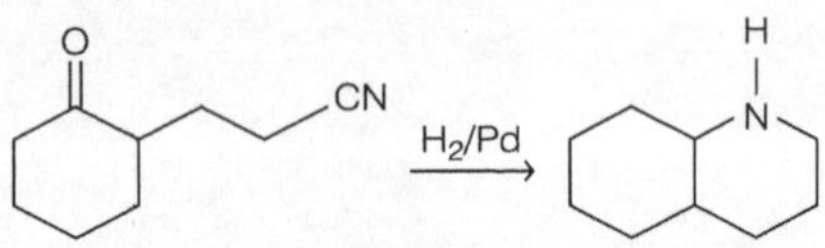

In the above conversion, how many types of amines are involved?

34 Consider the following reaction scheme.

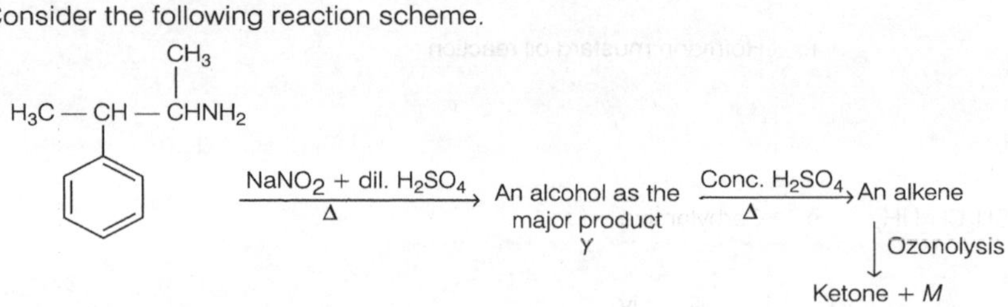

The number of carbon atoms in the ketone is

JEE Main & AIEEE Archive

(Compilation of Last 13 Years' Questions)

1 On heating an aliphatic primary amine with chloroform and ethanolic potassium hydroxide, the organic compound formed is **(JEE Main 2014)**

a. an alkanol
b. an alkanediol
c. an alkyl cyanide
d. an alkyl isocyanide

2 Considering the basic strength of amines in aqueous solution, which one has the smallest pK_b value? **(JEE Main 2014)**

a. $(CH_3)_2NH$ b. CH_3NH_2
c. $(CH_3)_3N$ d. $C_6H_5NH_2$

3 An organic compound *A* on reacting with NH_3 gives *B*. On heating *B* gives *C*. *C* in the presence of KOH reacts with Br_2 to give $CH_3CH_2NH_2$. *A* is **(JEE Main 2013)**

a. $CH_3-\underset{\displaystyle CH_3}{\underset{|}{CH}}-COOH$
b. $CH_3CH_2CH_2CH_2COOH$
c. CH_3CH_2COOH
d. CH_3COOH

4 The gas leaked from a storage tank of the Union Carbide plant in Bhopal gas tragedy was **(JEE Main 2013)**

a. Methyl isocyanate b. Methylamine
c. Ammonia d. Phosgene

5 A compound with molecular mass 180 u is acylated with CH_3COCl, to get a compound with molecular mass 390 u. The number of amino groups present per molecule of the former compound is **(JEE Main 2013)**

a. 6 b. 2
c. 5 d. 4

6 Which one of the following is the strongest base in aqueous solution ? **(AIEEE 2007)**

a. Trimethylamine b. Aniline
c. Dimethylamine d. Methylamine

7 In the chemical reaction,

$$CH_3CH_2NH_2 + CHCl_3 + 3KOH \longrightarrow A + B + 3H_2O,$$

the compounds *A* and *B* are respectively **(AIEEE 2007)**

a. C_2H_5CN and 3KCl
b. $CH_3CH_2CONH_2$ and 3KCl
c. C_2H_5 NC and 3KCl
d. CH_3NC and 3KCl

8 Which one of the following methods is neither meant for the synthesis nor for separation of amines ? **(AIEEE 2005)**

a. Curtius reaction
b. Wurtz reaction
c. Hofmann method
d. Hinsberg method

9 Reaction of cyclohexanone with dimethylamine in the presence of catalytic amount of an acid forms a compound. Water during the reaction is continuously removed. The compound formed is generally known as **(AIEEE 2005)**

a. an amine b. an imine
c. an enamine d. a Schiff 's base

10 The correct order of increasing basic nature for the bases NH_3, CH_3NH_2 and $(CH_3)_2NH$ is **(AIEEE 2003)**

a. $CH_3NH_2 < NH_3 < (CH_3)_2NH$
b. $(CH_3)_2NH < NH_3 < CH_3NH_2$
c. $NH_3 < CH_3NH_2 < (CH_3)_2NH$
d. $CH_3NH_2 < (CH_3)_2NH < NH_3$

JEE Advanced & IIT JEE Archive

(Compilation of Last 10 Years' Questions)

1 Match the Column I with Column II and mark the correct option from the codes given below. **(2009, Matching Type)**

	Column I		Column II
i.	$CH_3CH_2CH_2CN$	p.	Reduction with Pd - C / H_2
ii.	$CH_3CH_2OCOCH_3$	q.	Reduction with $SnCl_2$ / HCl
iii.	$CH_3CH=CHCH_2OH$	r.	Development of foul smell on treatment with chloroform and alcoholic KOH.
iv.	$CH_3CH_2CH_2CH_2NH_2$	s.	Reduction with diisobutylaluminium hydride (DIBAL-H)
		t.	Alkaline hydrolysis

Codes

	i	ii	iii	iv
a.	p,q,s,t	p,s,t	p	r
b.	q,s	p,r	p,q,s	s,t
c.	r	q,s	p,q	r,s,t
d.	p,q	p,q	q,s	s,t

2 $CH_3NH_2 + CHCl_3 + KOH \rightarrow$ Nitrogen containing compound + KCl + H_2O. Nitrogen containing compound is **(2006, Only One Option Correct Type)**

a. CH_3CN b. CH_3NHCH_3

c. $CH_3-\overset{-}{N}\equiv\overset{+}{C}$ d. $CH_3-\overset{+}{N}\equiv\overset{-}{C}$

JEE Archive

4. Aromatic Nitro Compounds, Anilines and Diazonium Salts

DPP

DAILY PRACTICE PROBLEMS

DPP-1 Preparation and Properties of Aromatic Nitro Compounds

Subjective Questions

Working Space

Direction (Q. Nos. 1-4) *These questions are subjective in nature, need to be solved completely on notebook.*

1 Predict the major product (products) that would be obtained when each of the following compounds is nitrated by single nitration.

i. N phonyl acetamide
ii. methyl benzoate
iii. *n*-propyl phenyl ether
iv. 1-methyl naphthalene

2 Complete the following reactions

i. $PhNO_2 \xrightarrow{Zn + aq.\ NH_4Cl} (I)$

ii. $p\text{-Me-}C_6H_4NO_2 \xrightarrow[NaOH]{Aq.\ As_2O_3} (II)$

iii. $p\text{-Me-}C_6H_4NO_2 \xrightarrow{LAH} (III)$

iv. $PhNO_2 \xrightarrow{Zn + Alc.\ NaOH} (IV)$

v. $p\text{-Me-}C_6H_4NO_2 \xrightarrow[\text{Excess or Zn/NaOH}]{H_2NNH_2/\text{Raney Ni}} (V)$

3 Write the reaction mechanism that can account for the formation of each of the following products.

i. 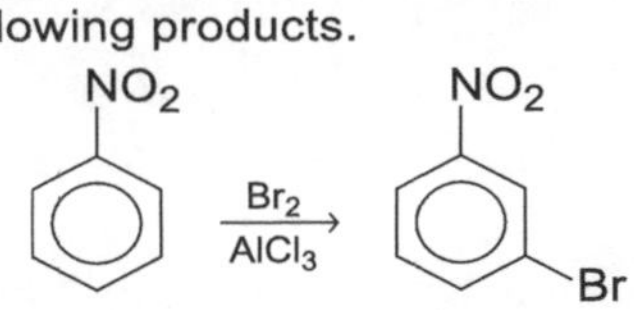

ii. 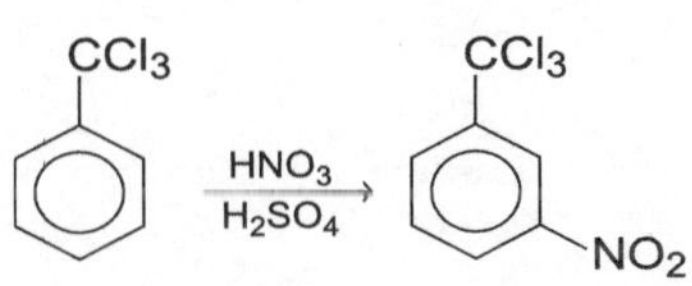

iii. 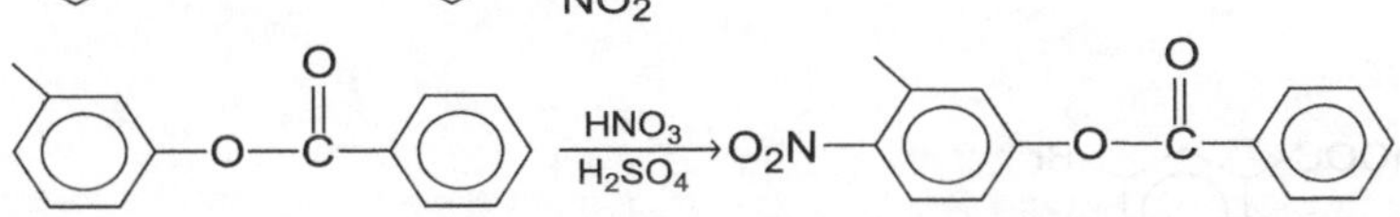

(By single nitration)

4 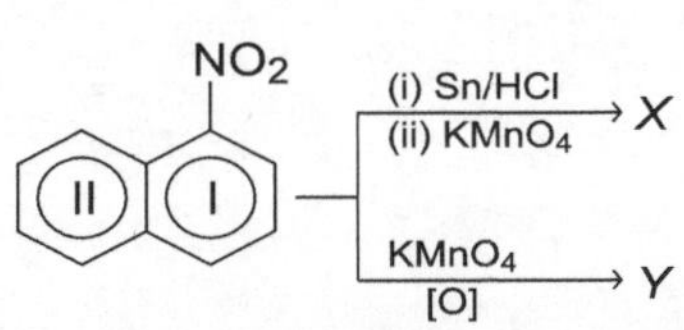

What are *X* and *Y*?

Working Space

Only One Option Correct Type

Direction (Q. Nos. 5-10) *This section contains 6 multiple choice questions. Each question has four choices (a), (b), (c) and (d), out of which ONLY ONE option is correct.*

5 The rate of nitration of toluene in nitromethane with excess of HNO_3 is constant and the reaction is zero order with respect to toluene. If the rate is temperature dependent, which of the following can be expected as a rate determining step ?

a. $HNO_3 + HNO_3 \rightleftharpoons H_2NO_3^+ + NO_3^-$

b. $2HNO_3 \rightleftharpoons NO_3^- + H_2O^+—NO_2$

c. $H_2O^+—NO_2 \rightleftharpoons H_2O + \overset{+}{N}O_2$

d. 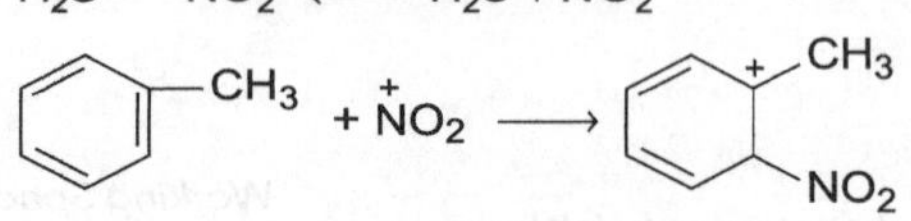

6 Product *C* of the given reaction is

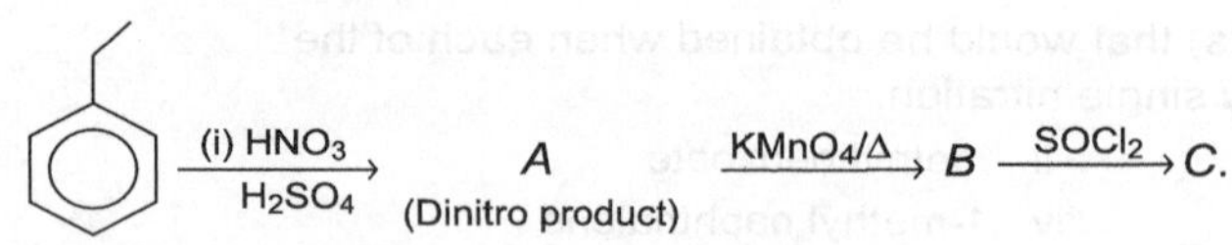

a. 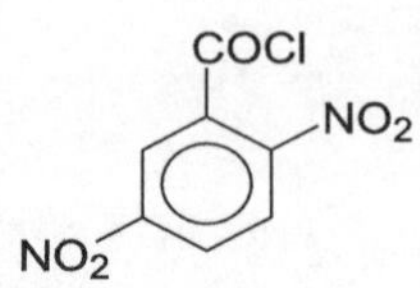
b. 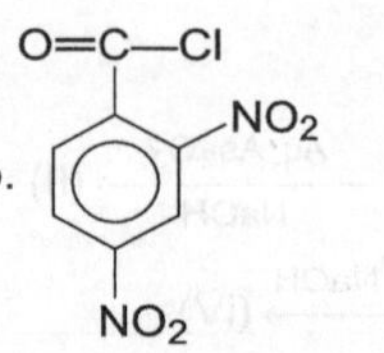
c. 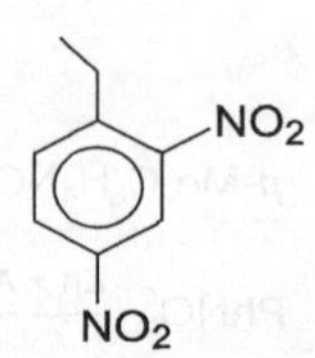
d. 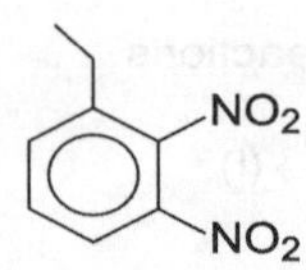

7 The products *R* and *S* are respectively

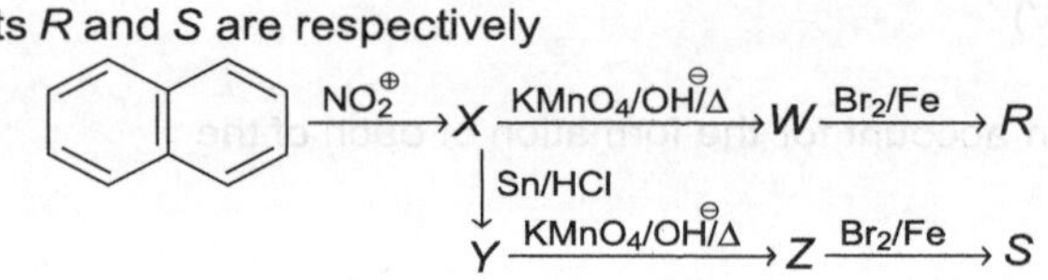

a. 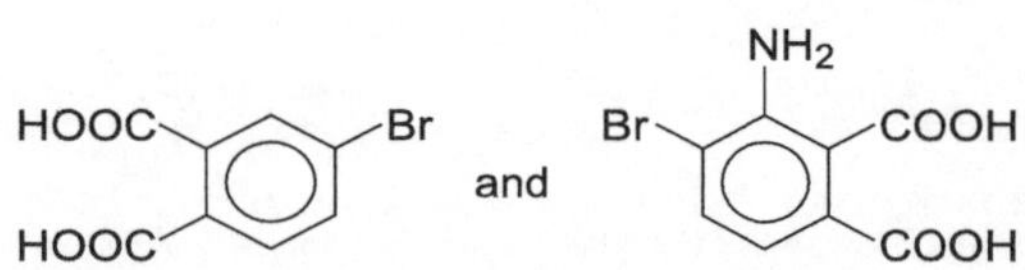

b. 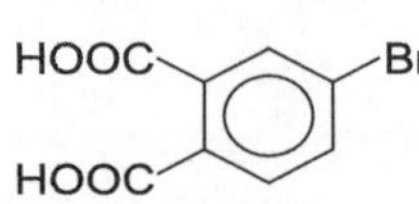 : *R* and *S* are same products

c. 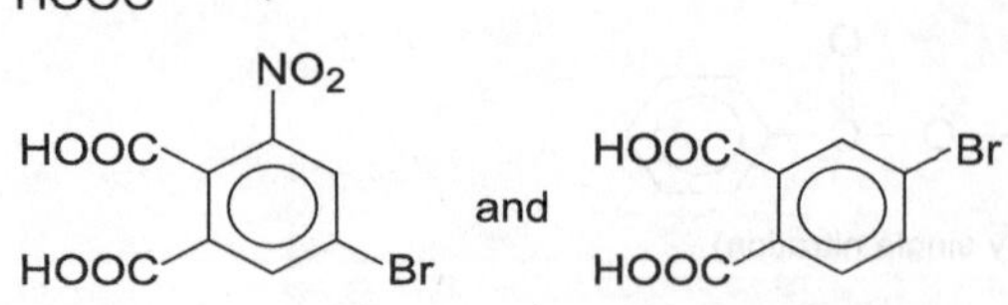

d. 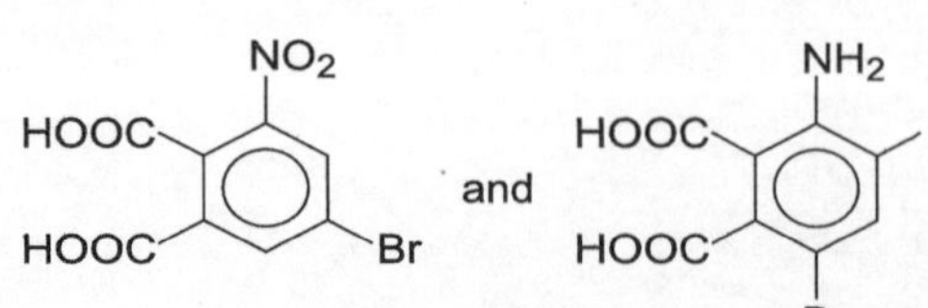

Working Space

8 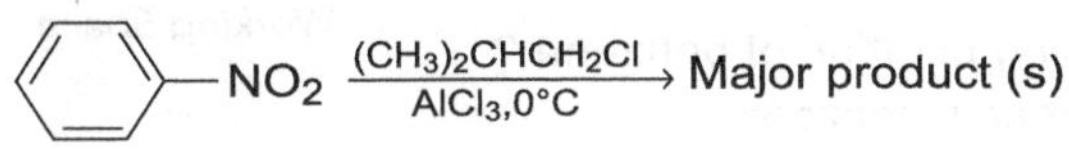

Major product (s) is/are

I. 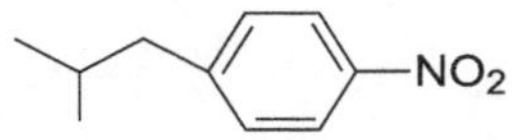 II. 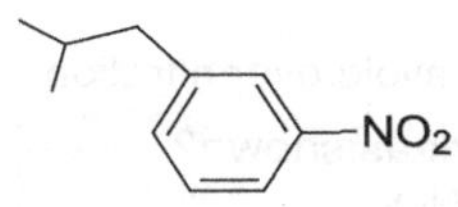 III. 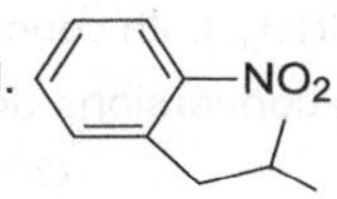

IV. 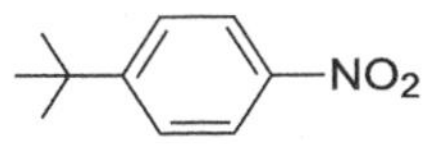 V. 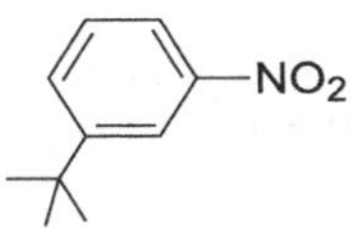 VI. 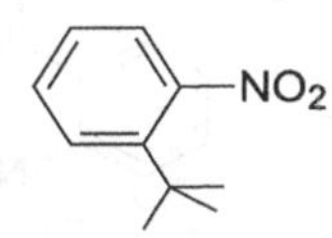

a. Only V
b. II, IV and VI
c. I, III and V
d. No reaction takes place

9 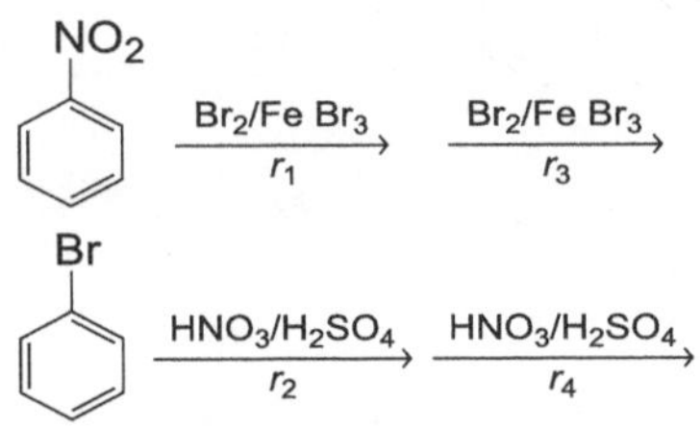

a. $r_1 > r_2$ but $r_3 > r_4$
b. $r_2 > r_1$ but $r_3 > r_4$
c. $r_1 > r_2$ but $r_4 > r_3$
d. $r_2 > r_1$ but $r_4 > r_3$

10 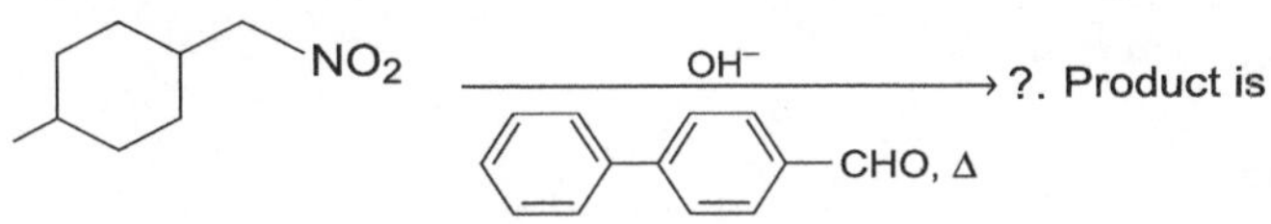

a. 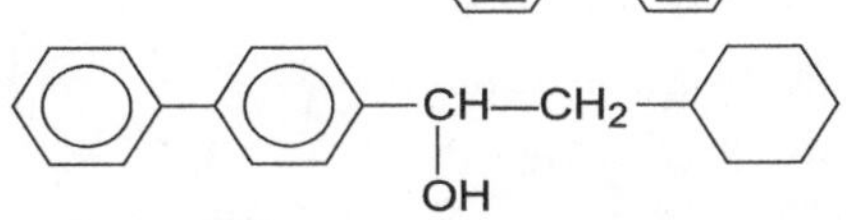

b. 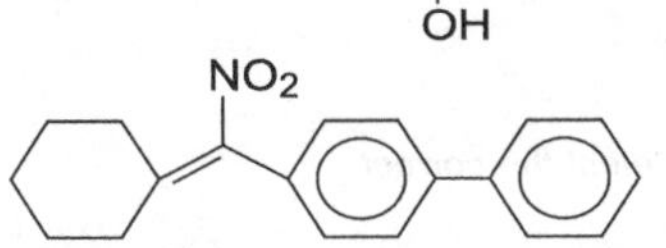

c. 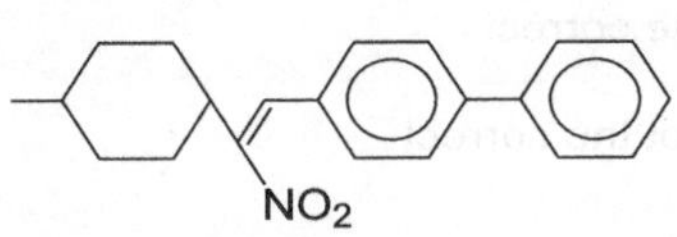

d. All of the above

One or More than One Options Correct Type

Direction *(Q. Nos. 11-14) This section contains 4 multiple choice questions. Each question has four choices (a), (b), (c) and (d), out of which ONE or MORE THAN ONE are correct.*

11 Which of the following reaction(s) is/are correct?

a. $PhNO_2 \xrightarrow{LAH} PhNH_2$

b. $PhNO_2 \xrightarrow[+ NaOH]{Na_2SnO_2} Ph—N{=}N—Ph$

c. $PhNO_2 \xrightarrow{Na_3AsO_3 + NaOH} Ph—\overset{+}{N}(O^{\ominus}){=}N—Ph$

d. $PhNO_2 \xrightarrow{Al\text{-}Hg/H_2O} Ph—NH—NH—Ph$

Working Space

12 Which of the following statement(s) is/are true regarding nitration of benzene ?

a. With HNO_3/H_2SO_4 (concentrated), reaction is zero order in benzene.

b. Concentrated nitric acid alone cannot bring about nitration of benzene.

c. It can be achieved using NO_2BF_4 salt.

d. Use of H_2SO_4 with HNO_3, both concentrated, avoid over nitration.

13 Which of the following conversions do not work as shown?

a. 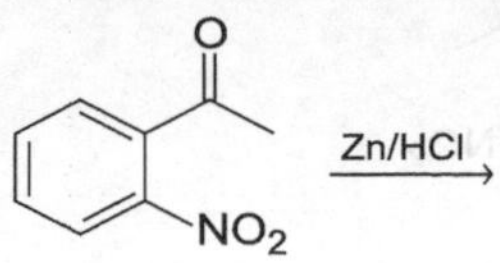

b. 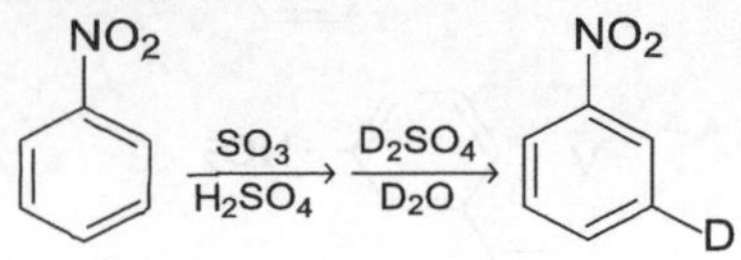

c. 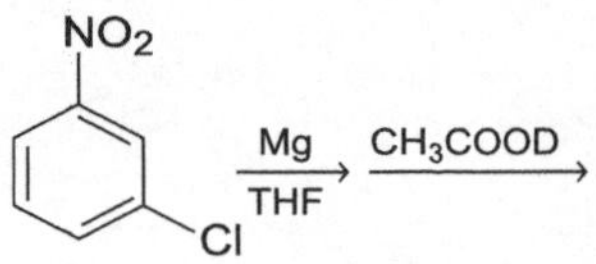

d. 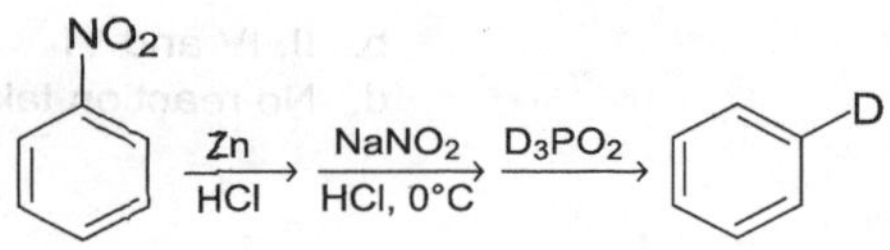

14

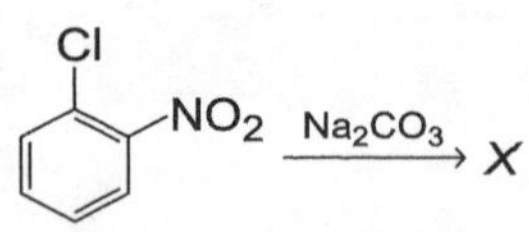

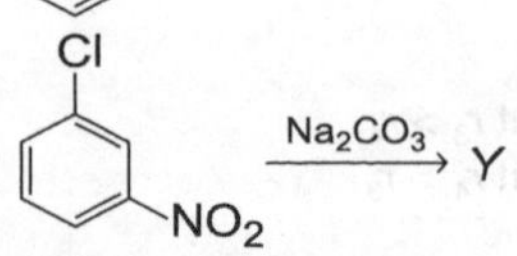

The true statement(s) about *X* and *Y* is/are

a. *X* and *Y* are isomers of each other

b. *Y* is same as that of its original compound

c. *X* is *o*-nitrophenol but *Y* is not

d. *X* is same as that of its original compound

Statement Type

Direction (Q. Nos. 15 and 16) This section is based on Statement I and Statement II. Select the correct answer from the codes given below.

a. Both Statement I and Statement II are correct and Statement II is the correct explanation of Statement I

b. Both Statement I and Statement II are correct and Statement II is not the correct explanation of Statement I

c. Statement I is correct but Statement II is incorrect

d. Statement II is correct but Statement I is incorrect

15 **Statement I** Nitrobenzene on treatment with chloroethane in the presence of $AlCl_3$ gives *meta* nitroethyl benzene.

Statement II Nitro group is *meta* directing in electrophilic aromatic substitution reactions.

16 **Statement I** Direct nitration of benzene to trinitrobenzene is practically impossible.

Statement II Three nitro groups present on benzene ring produces large steric hindrance.

Working Space

Comprehension Type

Direction (Q. Nos. 17 and 18) *This section contains a paragraph, describing theory, experiments, data, etc. Two questions related to the paragraph have been given. Each question has only one correct answer among the four given options (a), (b), (c) and (d).*

Passage

The —NO_2 group in an aromatic ring deactivates the *ortho* and *para* positions for an electrophilic attack.

When —NO_2 group is present at *ortho* or *para* positions of a leaving group (nucleopage), it activates the ring for nucleophilic attack.

The reduction of —NO_2 group by metal in acid causes its reduction to —NH_2 group and then the ring becomes strongly activates for a electrophilic attack. The strong activation of —NH_2 group is moderated by its acylation with CH_3COCl to —NHAc group. Deacylation is carried out by hydrolysis with H_3O^+ or OH^-.

The ring alkylation by using RX/AlX_3 is not possible in the presence of —NO_2 or —NH_2 group but is possible in the presence of —NHAc group.

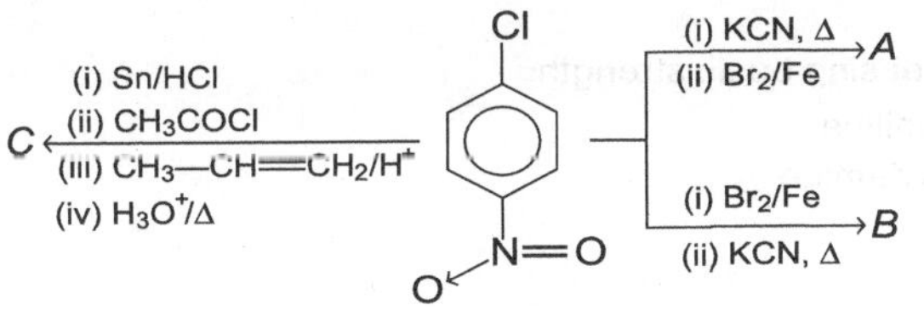

17 *A* and *B* are

a. homologues
b. identical
c. functional isomers
d. different compounds

18 *C* is

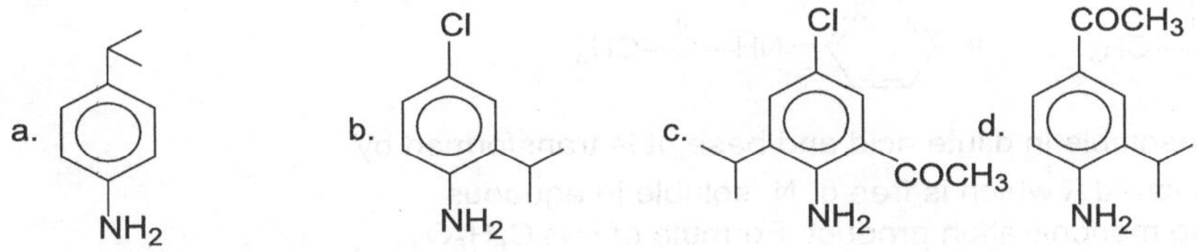

One Integer Value Correct Type

Direction (Q. Nos. 19 and 20) *This section contains 2 questions. When worked out will result in an integer from 0 to 9 (both inclusive).*

19 When 2, 4, 6-trinitrochlorine is warmed with water and then treated with zinc dust, the sum of positions of locants present in the product is

20 Select the correct answer in the form of the number written in parenthesis.

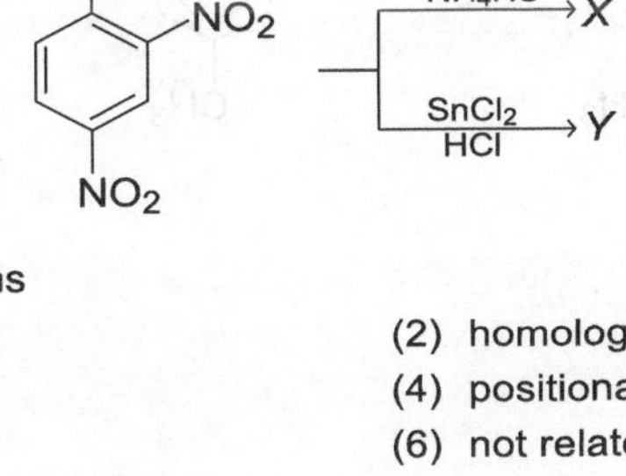

X and *Y* are related as

(1) isomer (functional)
(2) homologue
(3) identical
(4) positional isomers
(5) optical isomers
(6) not related in any manner.

DPP-2 Preparation Methods and Basicity of Anilines

Subjective Questions

Direction (Q. Nos. 1 and 2) *These questions are subjective in nature, need to be solved completely on notebook.*

Working Space

1 a. Arrange the following in order of increasing basic strength.

i. *p*-nitroaniline, aniline, *p*-methyl aniline

ii. diphenylamine, aniline, cyclohexylamine

Also give explanation in each case.

b. Compare the basic strength of

i. I. 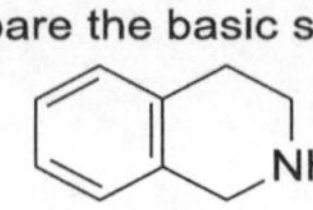 II. 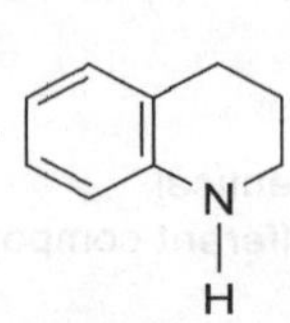

ii. I. 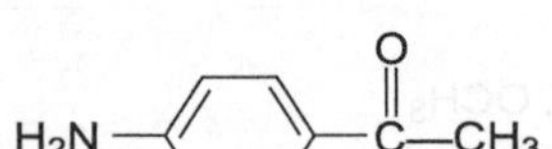 II. 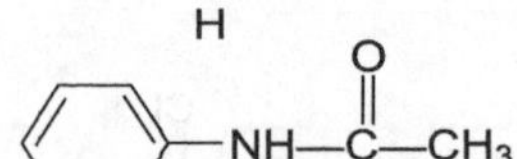

2 Compound *Q*, C_8H_9NO, is insoluble in dilute acid and base. It is transformed by $KMnO_4$ in H_2SO_4 to a compound *R* which is free of N, soluble in aqueous $NaHCO_3$ and gives only one mononitration product. Formula of *R* is $C_8H_6O_4$. Identify *Q* and *R*.

Only One Option Correct Type

Direction (Q. Nos. 3-13) *This section contains 11 multiple choice questions. Each question has four choices (a), (b), (c) and (d), out of which ONLY ONE is correct.*

3 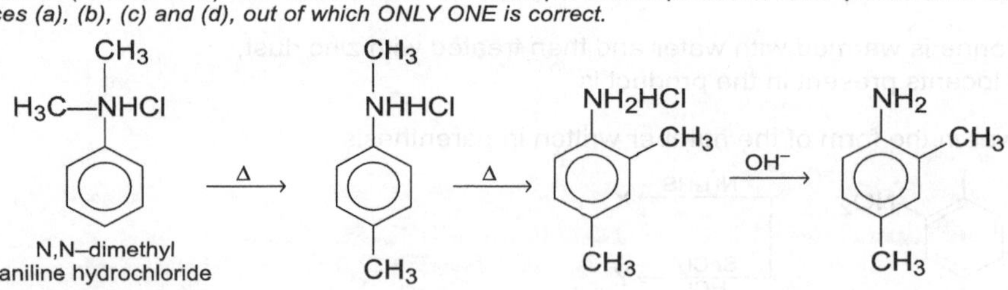

N,N–dimethyl aniline hydrochloride

This is called

a. Hofmann-Martius rearrangement

b. Hofmann-degradation reaction

c. Ulmann reaction

d. Fries migration

4 Product formed in the given reaction is

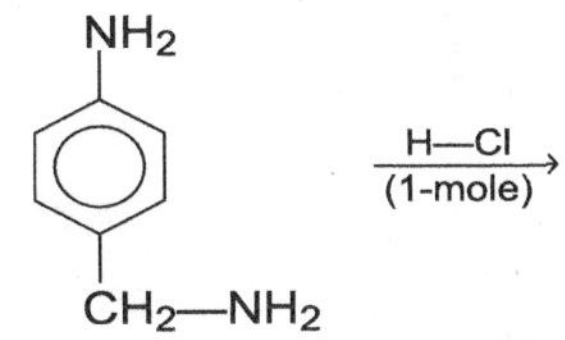

a. 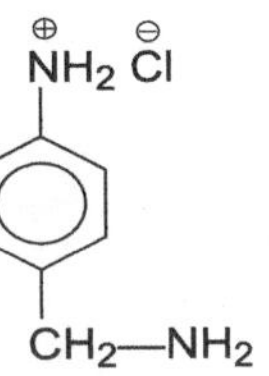
b. 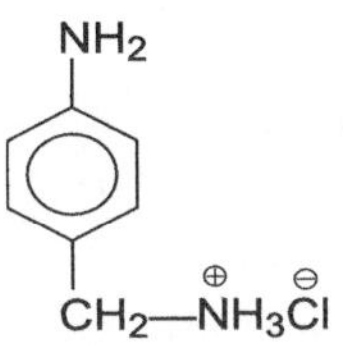
c. 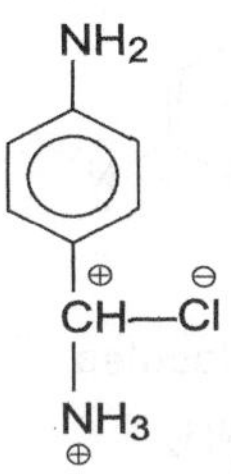
d. 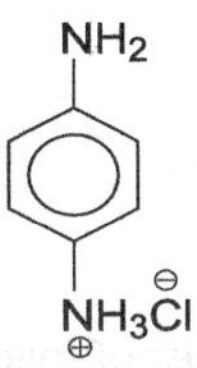

5 Which of the following C_8H_9NO isomers is the weakest base?

a. *o*-aminoacetophenone
b. *m*-aminoacetophenone
c. *p*-aminoacetophenone
d. Acetanilide

6 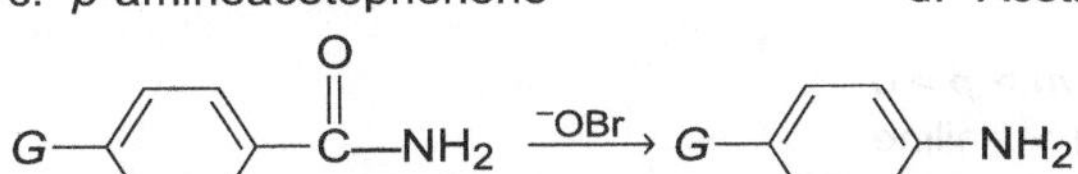

The rate of Hofmann degradation is influenced by the nature of substituent, *G*. Which of the following represent the correct decreasing rate of degradation with various substituents?

a. $—NO_2$ > —Cl > —H > $—CH_3$ > $—OCH_3$
b. $—OCH_3$ > $—CH_3$ > —Cl > —H > $—NO_2$
c. $—CH_3$ > —Cl > —H > $—OCH_3$ > $—NO_2$
d. $—OCH_3$ > $—CH_3$ > —H > —Cl > $—NO_2$

7 How might one best accomplish the following synthesis?

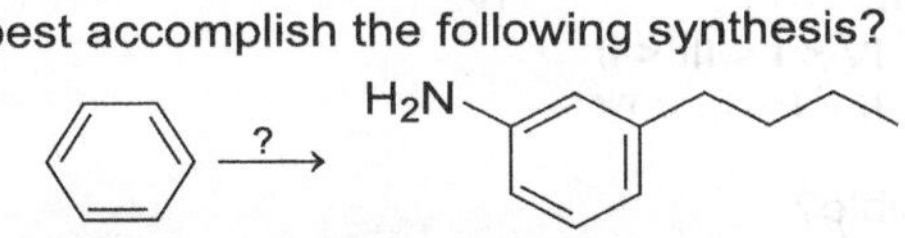

a. (i) $C_4H_9Cl + AlCl_3$ (ii) HNO_3 + heat (iii) $ZnCl_2/HCl$
b. (i) HNO_3 + heat (ii) $C_4H_9Cl + AlCl_3$ (iii) $ZnCl_2/HCl$
c. (i) $C_3H_7COCl + AlCl_3$ (ii) HNO_3 + heat (iii) $ZnCl_2/HCl$
d. (i) HNO_3 + heat (II) $C_3H_7COCl + AlCl_3$ (III) $ZnCl_2/HCl$

8 In which of the following compounds is there the least delocalisation of the lone pair of electrons on the nitrogen atom?

a. 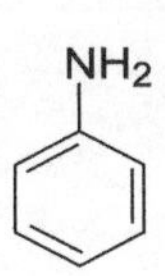
b. 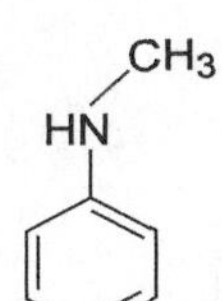
c. 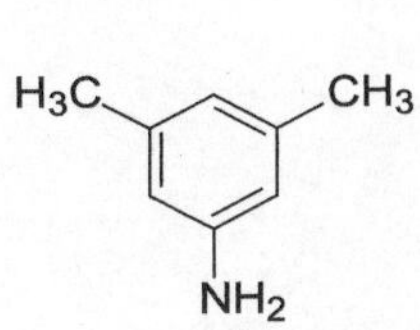
d. 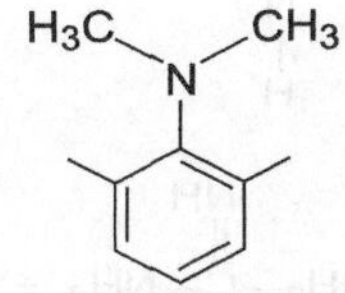

Working Space

Working Space

9 Which of the following will accept H$^+$ ion with a fastest rate?

a. 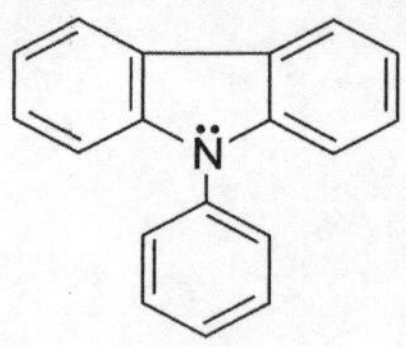

b. 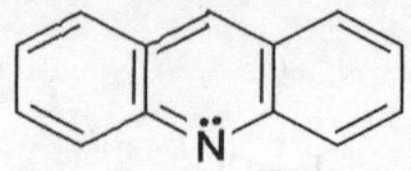

c. 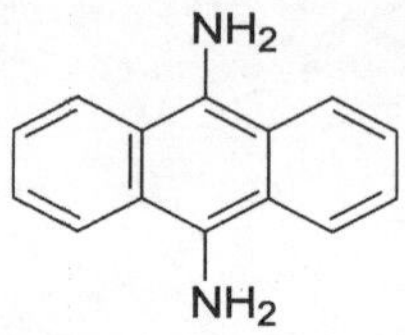

d. 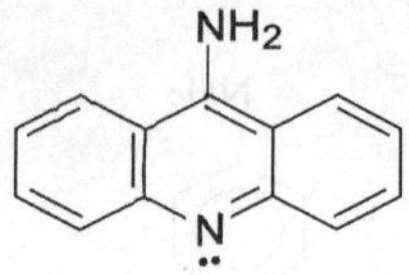

10 Select the basic strength order of following molecules.

(o) 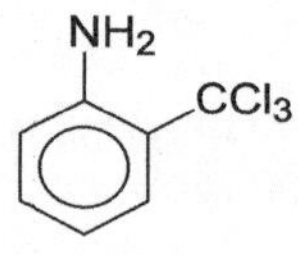 (m) 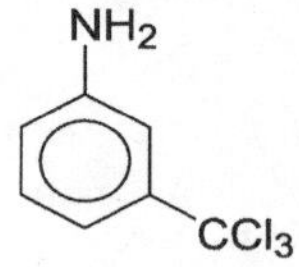 (p) 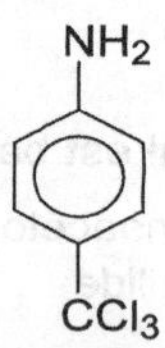 (Aniline) 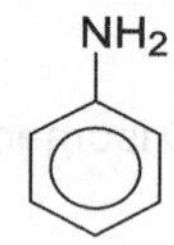

a. Aniline > *p* > *m* > *o*
b. Aniline > *m* > *p* > *o*
c. *p* > *o* > Aniline > *m*
d. *p* > *o* > *m* > Aniline

11 Rank the following compounds in order of increasing basicity (weakest → strongest)

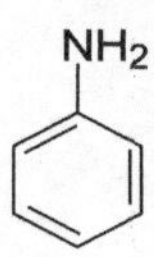
I.

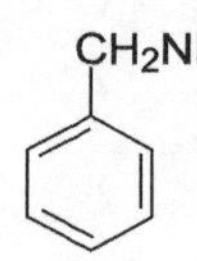
II.

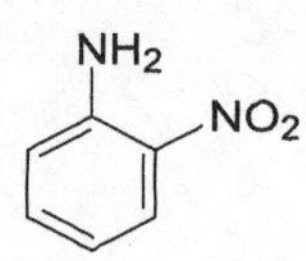
III.

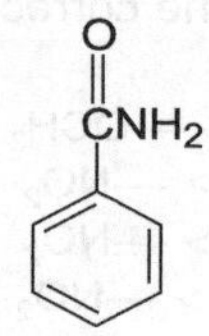
IV.

a. IV < II < I < III
b. IV < I < III < II
c. IV < III < I < II
d. II < I < III < IV

12 Which of the following reactions is/are feasible?

a. 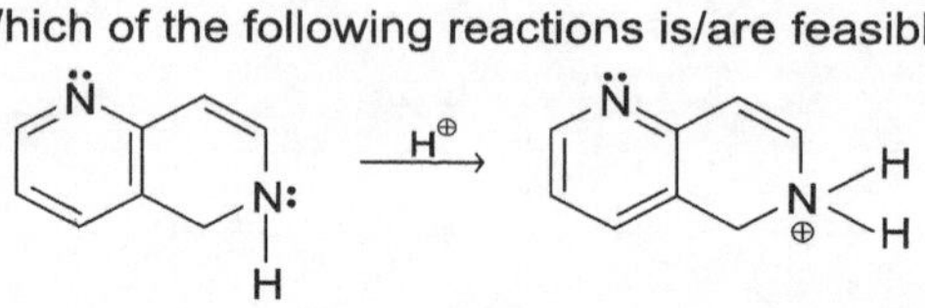

b. 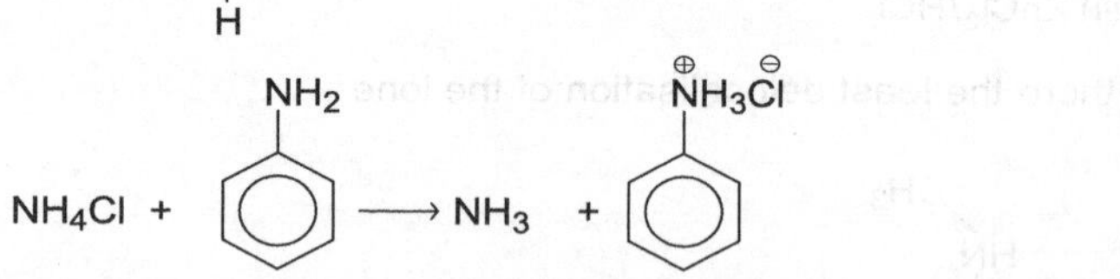

c. 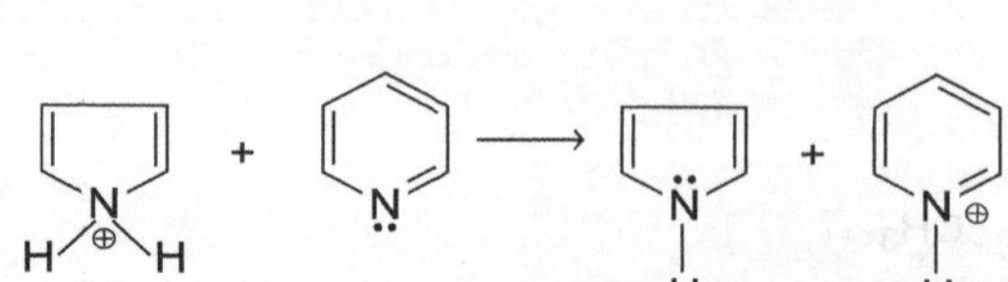

d. $NH_2-\overset{\overset{\oplus}{NH_2}}{\overset{\|}{C}}-NH_2 + CH_3-\overset{CH_2}{\overset{\|}{C}}-NH_2 \longrightarrow NH_2-\overset{NH}{\overset{\|}{C}}-NH_2 + CH_3-\overset{CH_2}{\overset{\|}{C}}-\overset{\oplus}{N}H_3$

Working Space

13 A white solid *A* of the molecular formula $C_{12}H_{10}OClN$, on reaction with PCl_5 in ether at 273 K gives a compound *B* which may be hydrolysed to equimolar quantities of *p*-chloroaniline and benzoic acid. What structure would you assign to *A* ?

a. 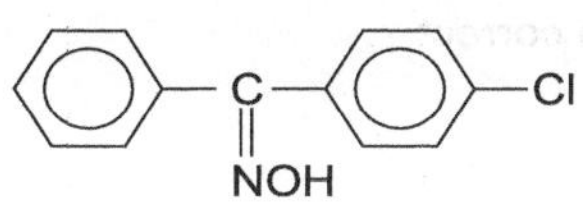

b. 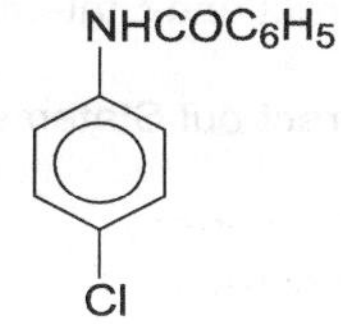

c. 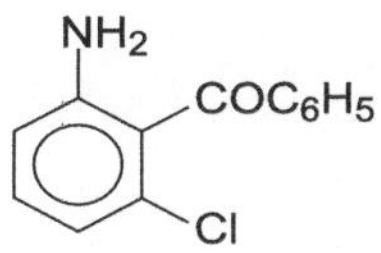

d. All of these

One or More than One Options Correct Type

Direction *(Q. Nos. 14 and 15) This section contains 2 multiple choice questions. Each question has four choices (a), (b), (c) and (d), out of which ONE or MORE THAN ONE are correct.*

14 Which of the following reactions can be used for preparation of aniline?

a. 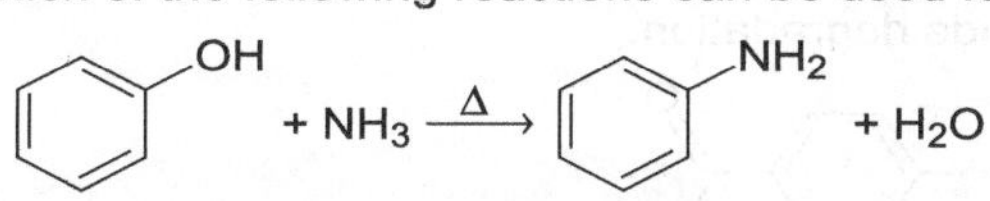

b. 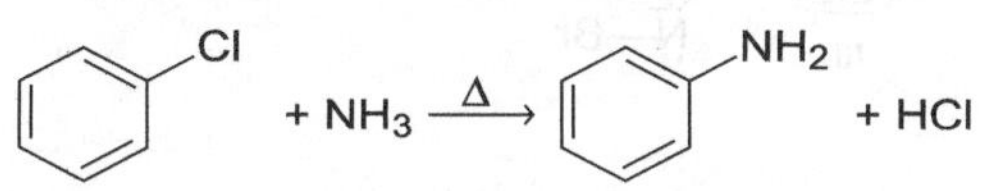

c. 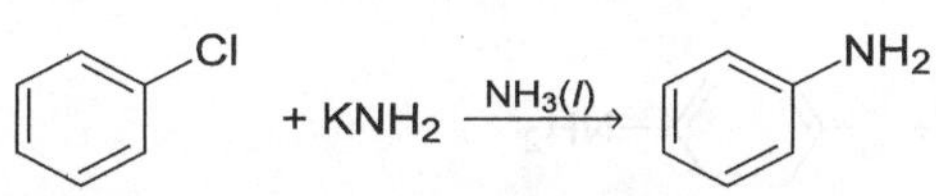

d. 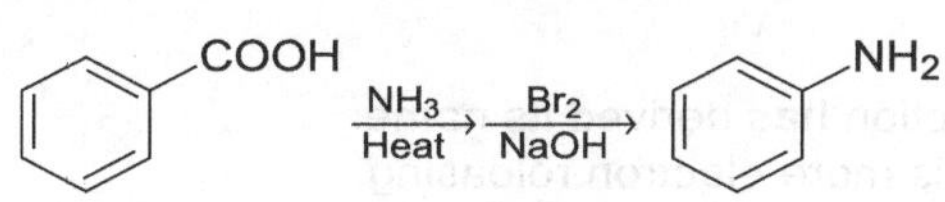

15 Select the correct statement(s).

a. Basic order of different amines shown is

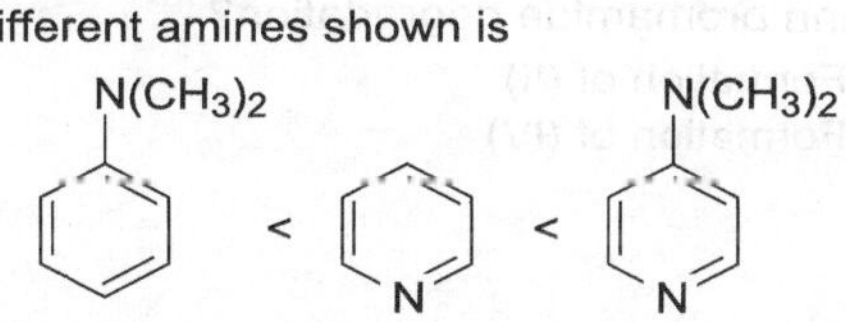

b. Unshared electron pair on nitrogen that is not a part of the π-electron cloud of the aromatic system in central ring is available for protonation

X : 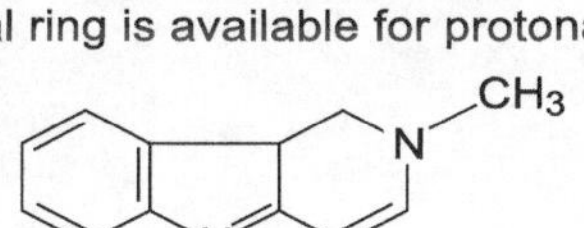

c. Each nitrogen of *X* is protonated

d. N-atom with alkyl group is protonated

Working Space

Statement Type

***Direction** (Q. No. 16) This section is based on Statement I and Statement II. Select the correct answer from the codes given below.*

a. Both Statement I and Statement II are correct and Statement II is the correct explanation of Statement I
b. Both Statement I and Statement II are correct but Statement II is not the correct explanation of Statement I
c. Statement I is correct but Strontement II is incorrect
d. Statement II is correct but Statement I is incorrect

16 **Statement I** N, N-diethylaniline is a stronger base than aniline.

Statement II Ethyl group is electron releasing by $+I$ effect.

Comprehension Type

***Direction** (Q. Nos. 17 and 18) This section contains a paragraph, describing theory, experiments, data, etc. Two questions related to the paragraph have been given. Each question has only one correct answer among the four given options (a), (b), (c) and (d).*

Passage

The conversion of an amide to an amine with one carbon atom less by the action of alkaline hydrohalite is known as Hofmann bromamide degradation.

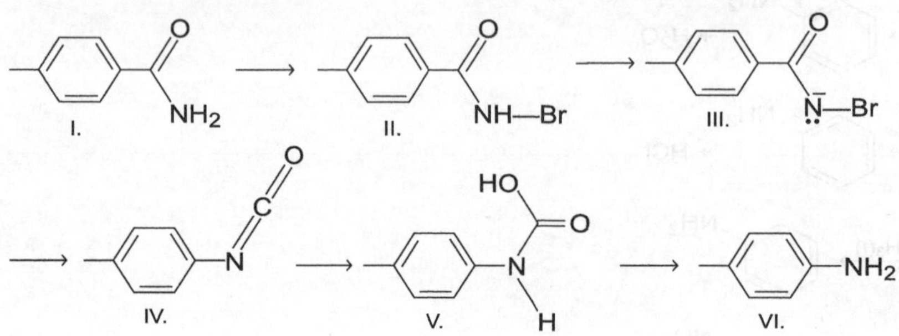

In this reaction RCONHBr is formed from which the reaction has derived its name. Hofmann reaction is accelerated if the migrating group is more electron releasing. Hofmann degradation reaction is an intramolecular reaction.

17 Which is the rate determining step in Hofmann bromamide degradation?

a. Formation of (I)
b. Formation of (II)
c. Formation of (III)
d. Formation of (IV)

18 I. 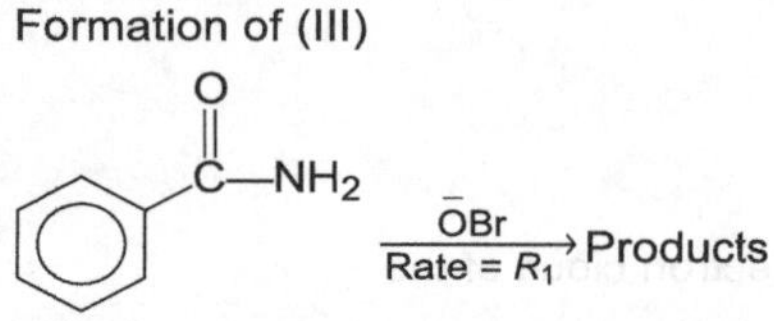

II. 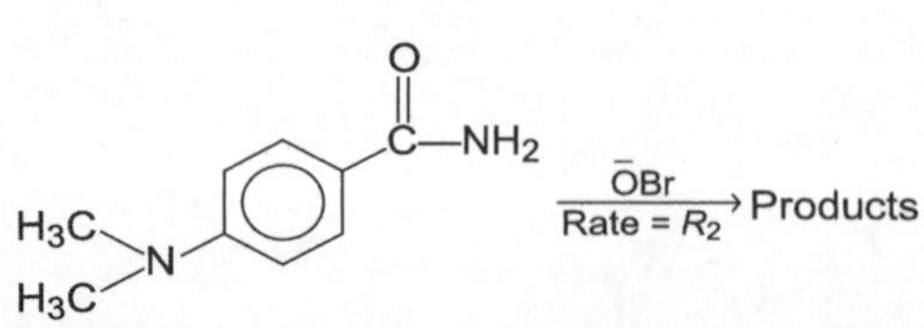

What is the exact relation between R_1, R_2?

a. $R_1 = R_2$
b. $R_1 < R_2$
c. $R_1 > R_2$
d. None of these

Working Space

One Integer Value Correct Type

Direction (Q. Nos. 19 and 20) This section contains 2 questions. When worked out will result in an integer from 0 to 9 (both inclusive).

19 How many of the following are the products of the reaction given below?

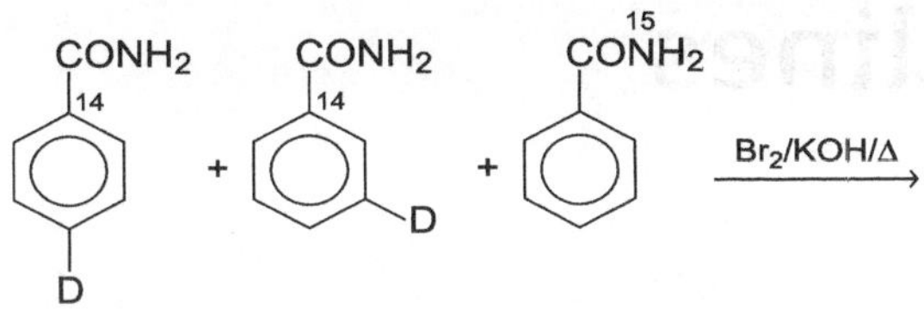

I. 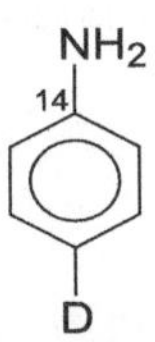

II. 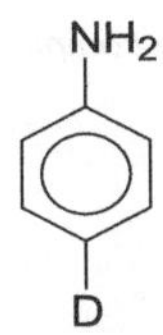

III. 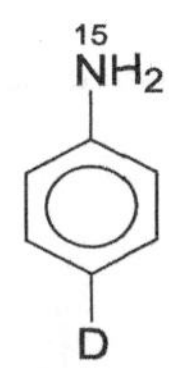

IV. 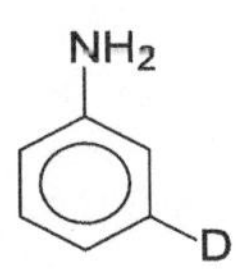

V. 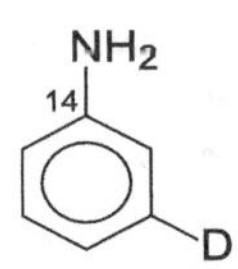

VI. 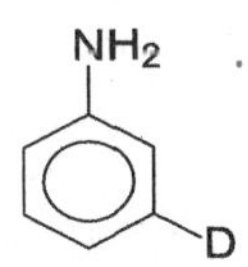

VII. 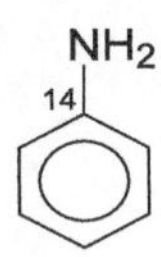

VIII. 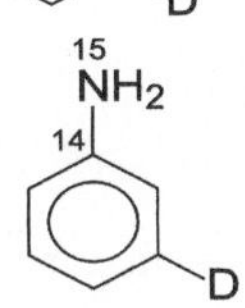

IX. 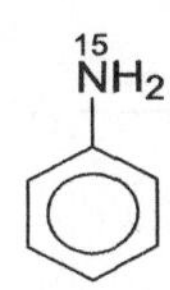

20 How many of them cannot be prepared by Gabriel method from their corresponding halides or tosylates?

I. *p*-toluidine II. *m*-nitroaniline III. Vinylamine
IV. *p*-nitroaniline V. *o*-nitroaniline VI. allylamine

DPP-3 Chemical Properties of Anilines

Subjective Questions

Working Space

Direction (*Q. Nos. 1 and 2*) *These questions are subjective in nature, need to be solved completely on notebook.*

1 Complete the following reactions:

i. 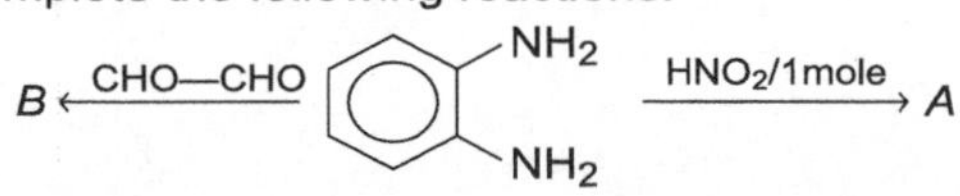

ii. 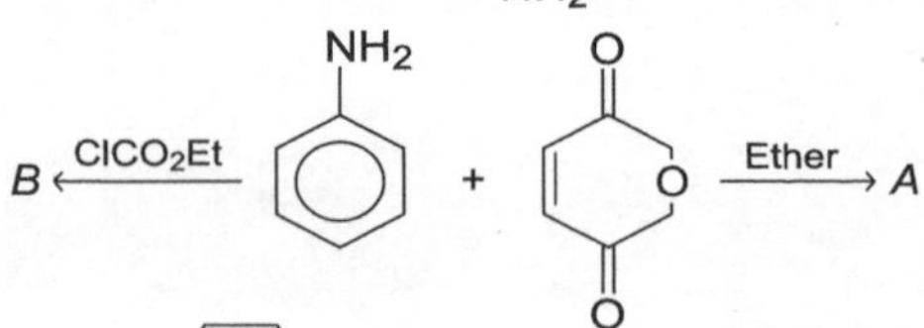

iii. 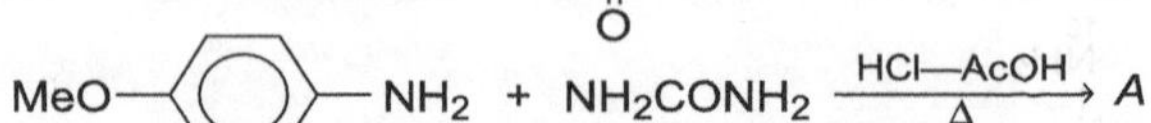

iv. 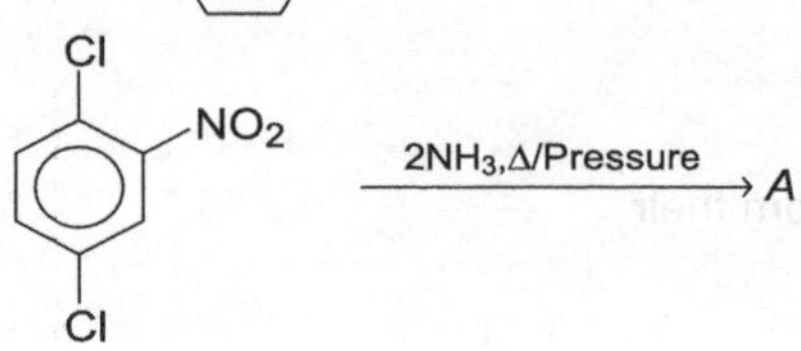

v. 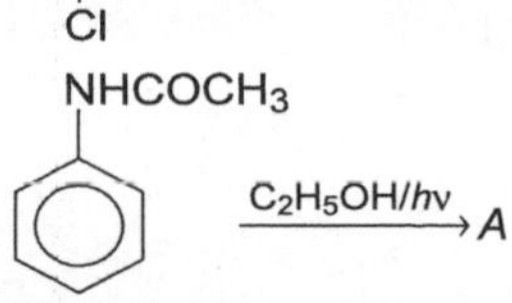

2 The aqueous solution of a nitrogen and chlorine containing organic compound *A* is acidic to litmus. *A* on treatment with aqueous NaOH gives a compound *B* containing nitrogen but not chlorine. Compound *B* on treatment with $C_6H_5SO_2Cl$ in the presence of NaOH gives an insoluble product *C* $C_{13}H_{13}NO_2S$. Give the structures of *A* and *B*. Also give equations for reactions involved.

Only One Option Correct Type

Direction (*Q. Nos. 3-12*) *This section contains 10 multiple choice questions. Each question has four choices (a), (b), (c) and (d), out of which ONLY ONE is correct.*

3 Which of the following represents a resonance form of the intermediate formed during bromination of aniline?

a. 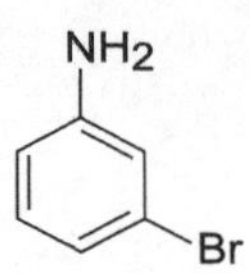

b. 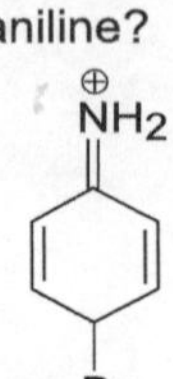

c. 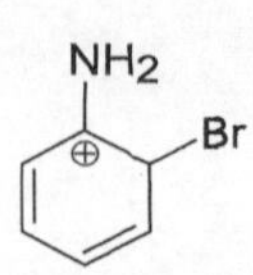

d. 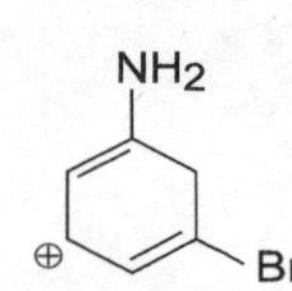

Working Space

4 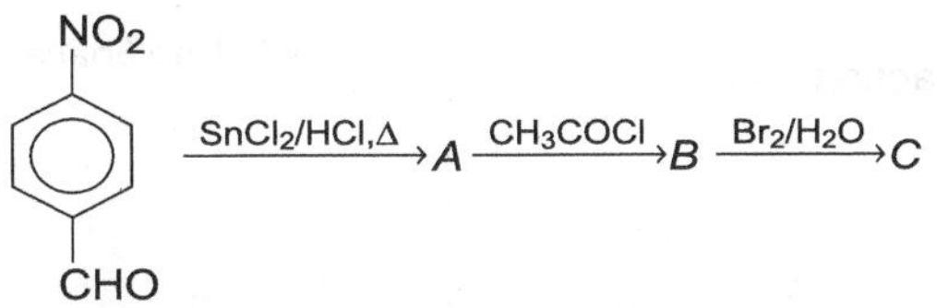

Product *C* is

a. 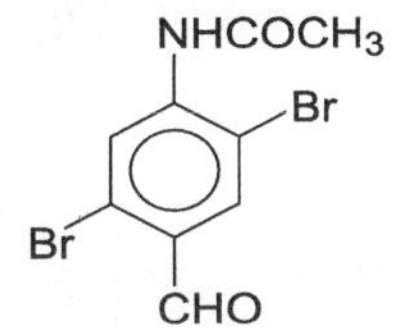

b. 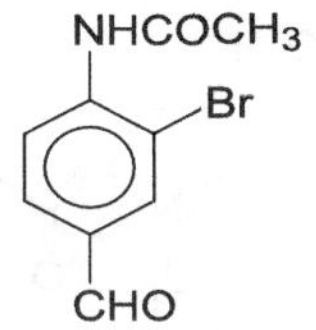

c. Both (a) and (b)

d. None of these

5 Sodium salt of anthranilic acid $\xrightarrow[\Delta]{NaNO_2/HCl}$?

Product is

a. biphenyl

b. 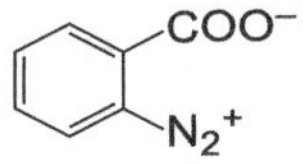

c. biphenylene

d. All of these

6 A student tried to synthesise 4-nitro-N, N-dimethylaniline from N, N-dimethylaniline *via* electrophilic aromatic substitution using nitric acid (HNO_3) and sulphuric acid (H_2SO_4). To his surprise, the major product he obtained was 3-nitro-N, N-dimethylaniline. Why the designed reaction failed to provide the desired product, 4-nitro-N, N-dimethylaniline?

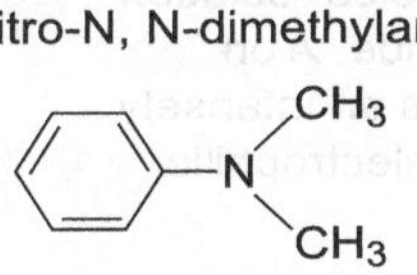

N,N–dimethylaniline

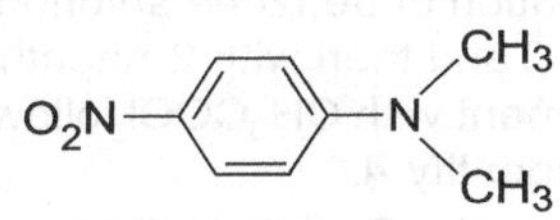

4-nitro-N,N–dimethylaniline

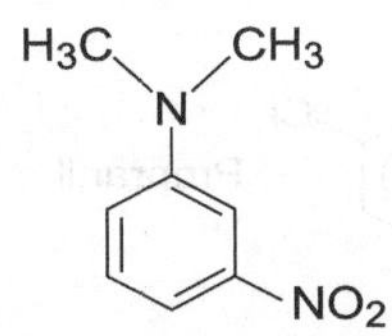

3-nitro-N, N-dimethylaniline

a. The amine group is *meta* directing.
b. The amine group is strong electron withdrawing group by $-I$ effect.
c. Protonation at amine group makes it *meta* directing.
d. Steric inhibition to resonance is responsible for the above result.

7 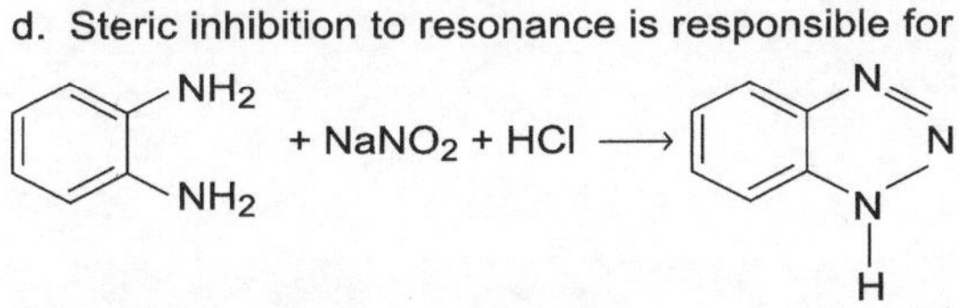

This reaction is an example of

a. intermolecular C—N coupling

b. intramolecular C—N coupling

c. intermolecular N—N coupling

d. intramolecular N—N coupling

Working Space

8 Identify the major product P in the following two step reaction.

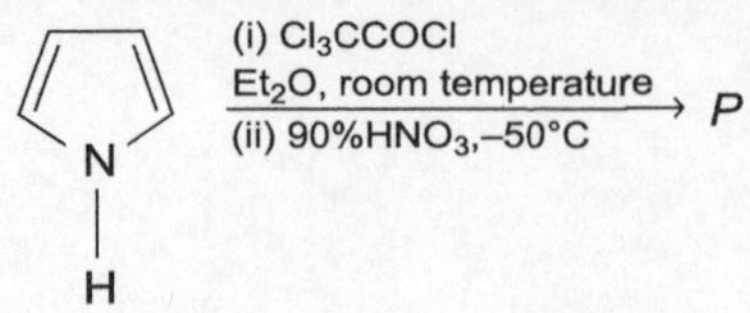

a. 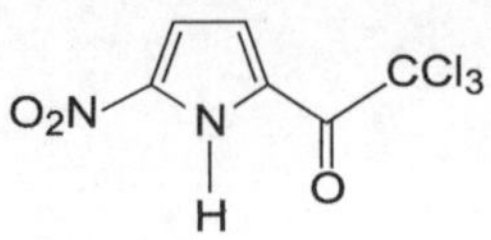

b. 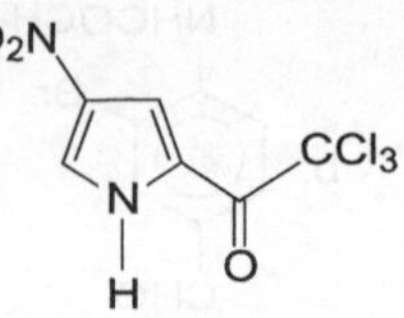

c. 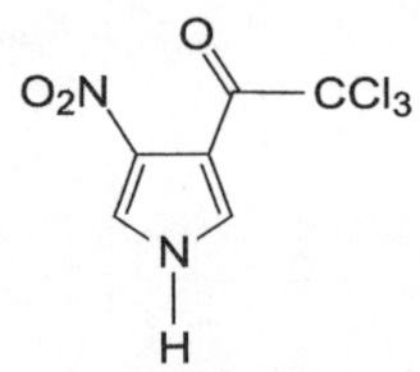

d. 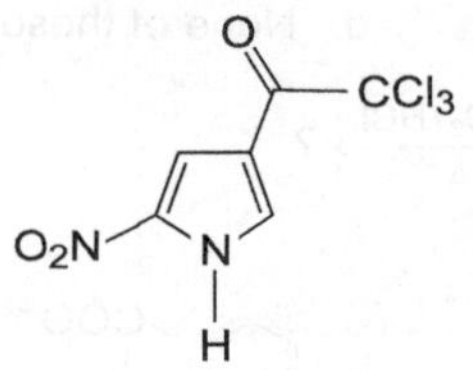

9 *p*-chloro aniline and anilinium hydrochloride can be distinguished by

a. Sandmeyer's reaction
b. Libermann's nitroso reaction
c. $AgNO_3$
d. Carbylamine test

10 An organic compound A has molecular formula C_7H_9N and it forms a clear solution when dissolved in aqueous KOH solution of benzene sulphonyl chloride. A on treatment with $NaNO_2$ and HCl at 0°C and then with 2-naphthol forms an intensely coloured compound. Also A on treatment with CH_3COCl followed by electrophilic substitution yields a single product. Identify A.

a. N-methylaniline
b. 2-methylaniline
c. 3-methylaniline
d. 4-methylaniline

11 The herbicide propranil, shown below is synthesised commercially from chlorobenzene. Which sequence of reactions would be most effective for carrying out this synthesis?

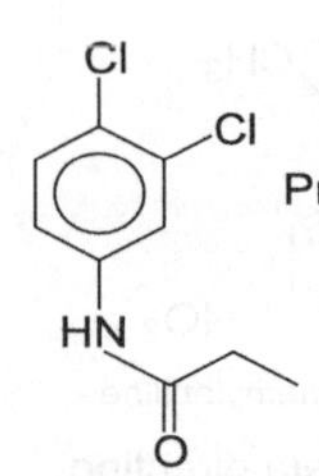

Propranil

a. Chlorination, nitration, reduction, acylation
b. Chlorination, reduction, nitration, acylation
c. Nitration, chlorination, reduction, acylation
d. Nitration, reduction, chlorination, acylation

12 Consider the *p*-amino phenol

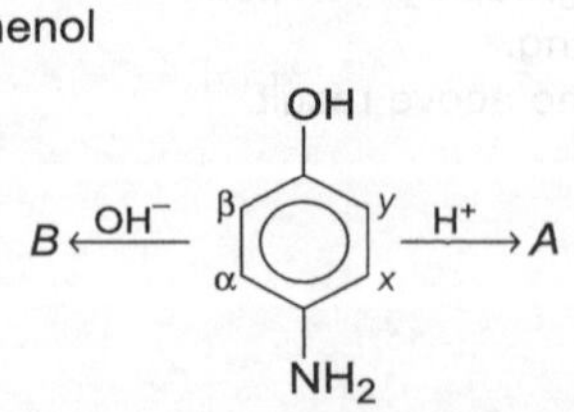

Which positions are activated in acidic and basic media for coupling?

a. x in A and β in B
b. x in A and α in B
c. y in A and α in B
d. y in A and β in B

Working Space

One or More than One Options Correct Type

Direction *(Q. Nos. 13-16) This section contains 4 multiple choice questions. Each question has four choices (a), (b), (c) and (d), out of which ONE or MORE THAN ONE are correct.*

13 Which amongst the following is/are correctly marked for most favoured electrophilic substitution?

a. 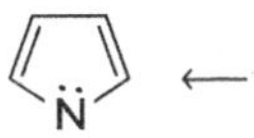

b. 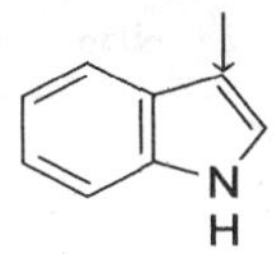

c. 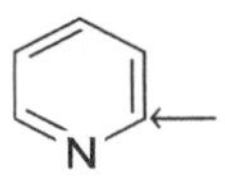

d. 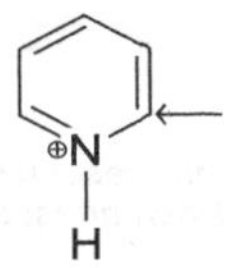

14 What is true about the following reactions?

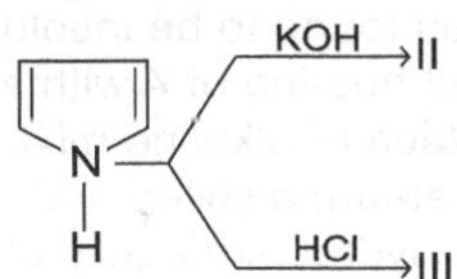

I

a. I is aromatic
b. II is aromatic
c. III is anti-aromatic
d. Out of I, II and III only one compound is non-aromatic.

15 The correct explanation for the following observation is

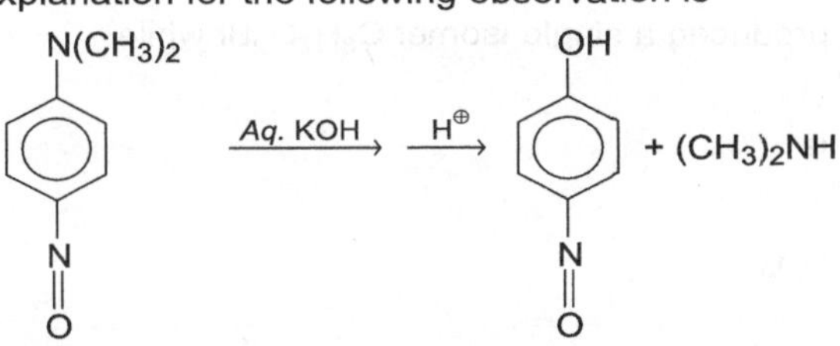

a. Intermediate is

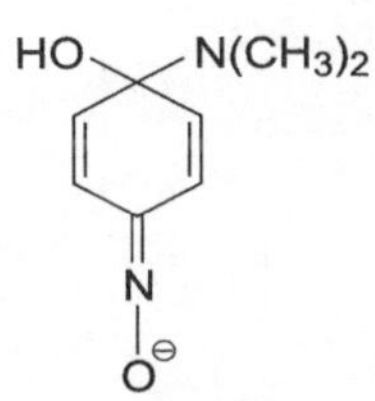

and S_N2 aromatic is favoured by $-M$ effect of —N=O group
b. $+M$ effect of —N=O favours S_N2 aromatic
c. Intermediate is

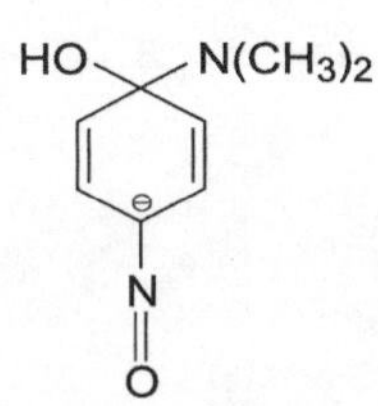

$-I$ effect of —N=O favours S_N2 aromatic
d. strong basic condition favours S_N2 Ar

Working Space

16 $PhNHNHPh \xrightarrow[\text{Strongly acidic}]{H^+}$? Product(s) of this reaction is/are

a. 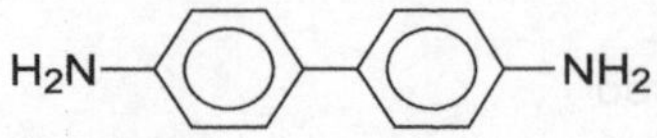
b. 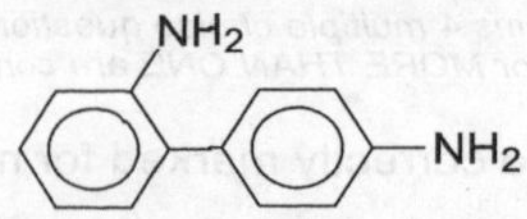
c. 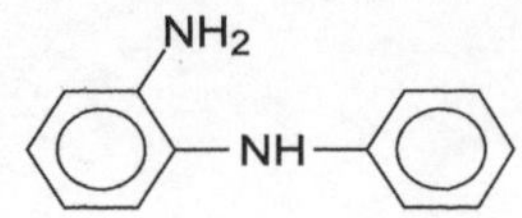
d. 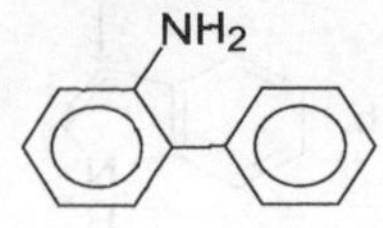

Comprehension Type

***Direction** (Q. Nos. 17 and 18) This section contains a paragraph, describing theory, experiments, data, etc. Two questions related to the paragraphs have been given. Each question has only one correct answer among the four given options (a), (b), (c) and (d).*

Passage

A solid organic compound *P* of formula $C_{15}H_{15}ON$, was found to be insoluble in water, dilute HCl or dil. NaOH in cold. After prolonged heating of *A* with aqueous NaOH, a liquid *R* was found to be floating on the surface of alkaline mixture. *R* did not solidified on cooling to room temperature. It was steam distilled and separated. Also, acidification of alkaline mixture with hydrochloric acid caused precipitation of a white solid *S* ($C_8H_8O_2$). Some additional informations are given below

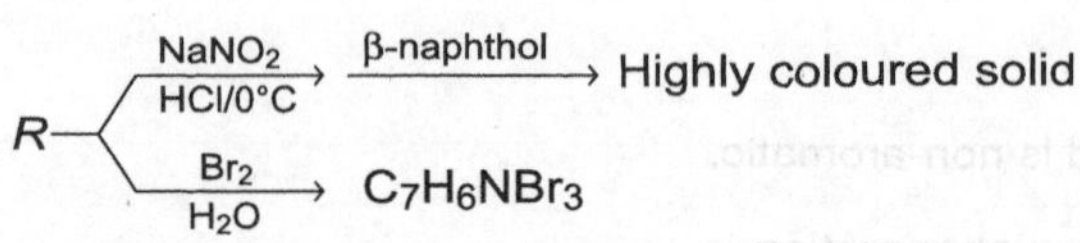

S on treatment with $Br_2/FeBr_3$ in CCl_4 produced a single isomer $C_8H_7O_2Br$ while heating *S* with soda lime gave toluene.

17 The most likely structural formula of *P* is

a. 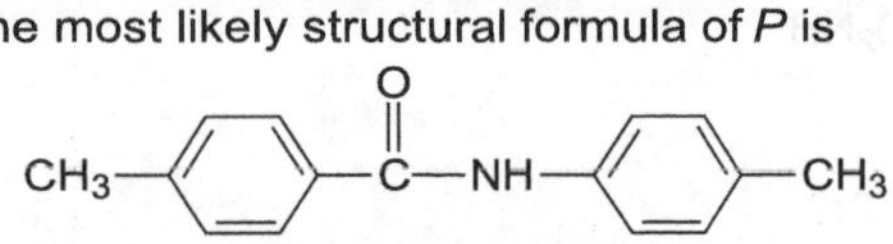
b. 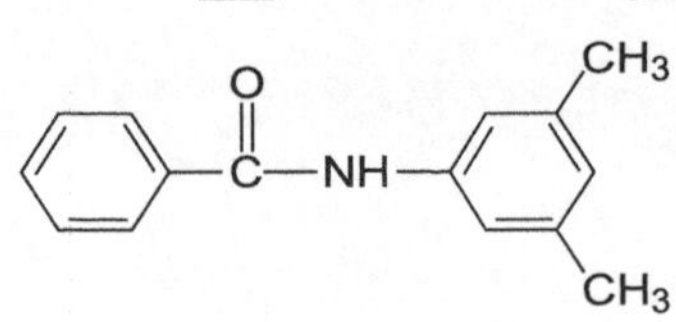
c. 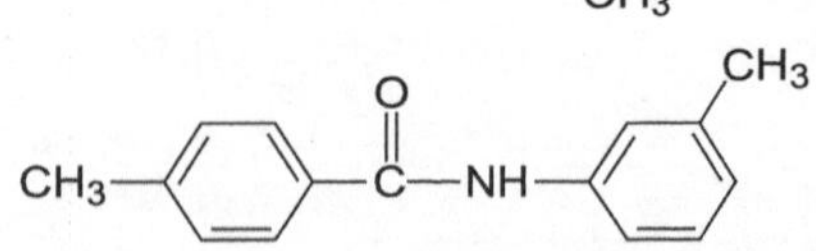
d. 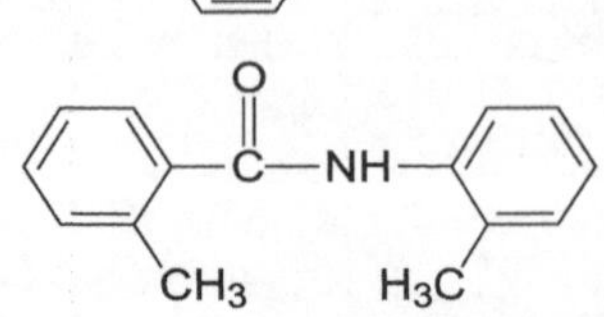

18 If S is treated with $SOCl_2$ followed by addition of benzene and some $AlCl_3$ would produce

a. 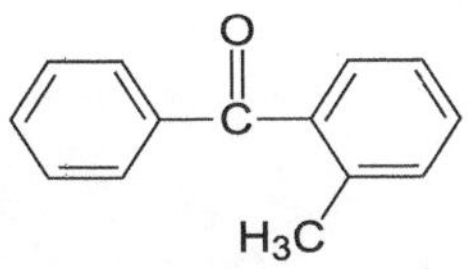

b. 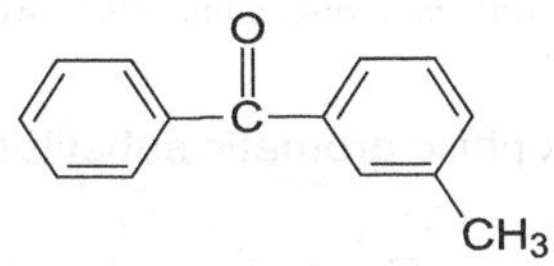

c. 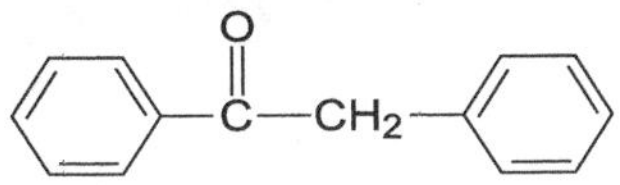

d. 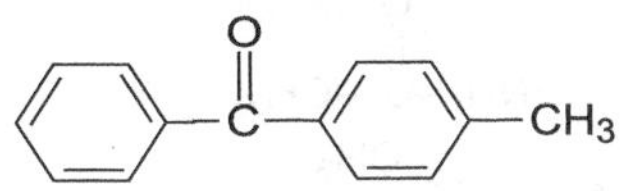

Statement Type

Direction (Q. No. 19) *This section is based on Statement I and Statement II. Select the correct answer from the codes given below.*

a. Both Statement I and Statement II are correct and Statement II is the correct explanation of Statement I
b. Both Statement I and Statement II are correct and Statement II is not the correct explanation of Statement I
c. Statement I is correct but Statement II is incorrect
d. Statement II is correct but Statement I is incorrect

19 **Statement I** Aniline can be estimated by treatment with excess of bromine water solution.

Statement II Aniline reacts quantitatively with bromine water solution forming insoluble 2, 4, 6-tribromoaniline.

Matching List Type

Direction (Q. No. 20) *Choices for the correct combination of elements from Column I and Column II are given as options (a), (b), (c) and (d), out of which one is correct.*

20 Match the Column I with Column II and mark the correct option from the codes given below.

	Column I (Compound)		**Column II** (Test)
i.	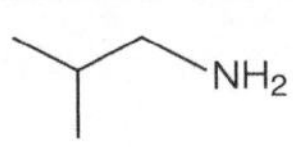	p.	Liebermann's nitroso reaction
ii.		q.	Evolution of N_2 with HNO_2
iii.	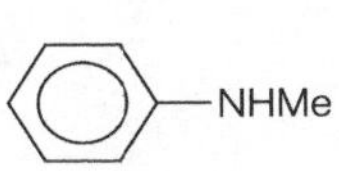	r.	Dye test
iv.	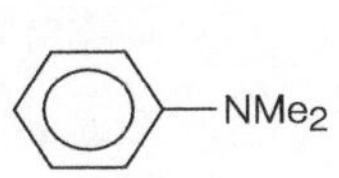	s.	Green colour with HNO_2
		t.	Carbylamine test

Codes

	i	ii	iii	iv		i	ii	iii	iv
a.	q,t	r,t	p	s	b.	p	q,s	q	r,s
c.	q	s	r,t	p	d.	p,s	r	r,t	q,s

Working Space

Working Space

One Integer Value Correct Type

***Direction** (Q. Nos. 21 and 22) This section contains 2 questions. When worked out will result in an integer from 0 to 9 (both inclusive).*

21 Identify the position where electrophilic aromatic substitution (EAS) is most favourable.

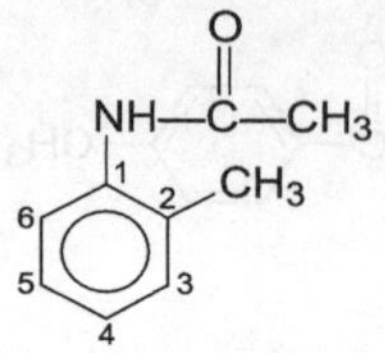

22 Novocaine is synthesised from toluene in the following manner

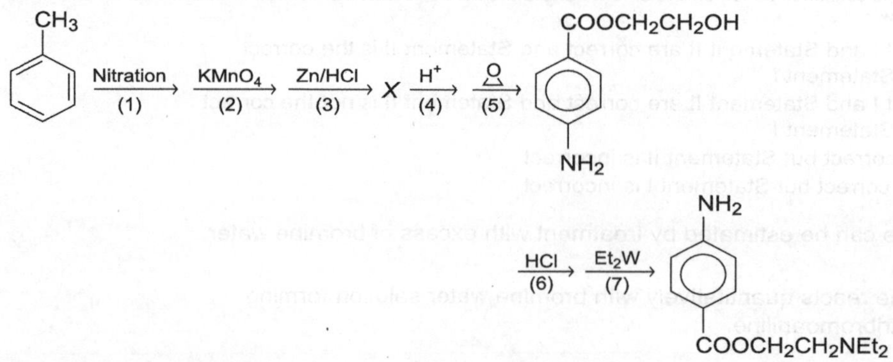

Choose the wrong step in the above synthesis process.

DPP-4 Preparation and Properties of Diazonium Salts

Subjective Questions

Working Space

Direction *(Q. Nos. 1 and 2) These questions are subjective in nature, need to be solved completely on notebook.*

1 i. Give the decreasing order of reactivity of diazonium ion coupling with phenol.

a. I. p-NO_2—$C_6H_4\overset{+}{N_2}$ II. III. IV. V.

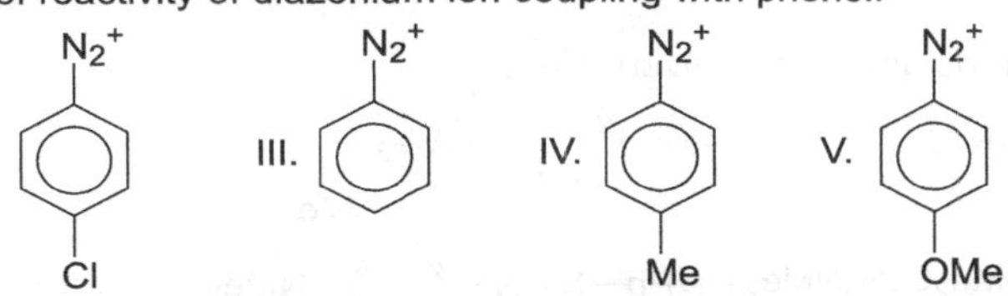

b. I. II. III. IV. V.

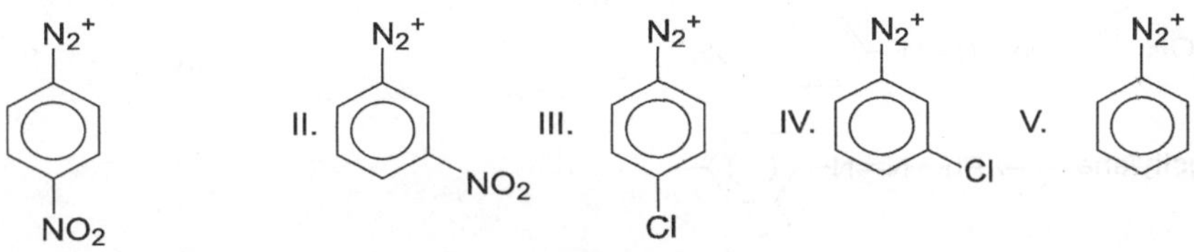

ii. Give the decreasing order of reactivity for the following coupling compounds with PhN_2^+Cl.

a. I. Aniline II. Phenol III. Toluene IV. Chlorobenzene V. Nitrobenzene

b. I. Toluene II. Ethylbenzene III. Cumene IV. *t*-butyl benzene V. Anisole

2 An aromatic compound *A*, $C_7H_7O_2N$ on reduction gives *B*, C_7H_9N. Diazotisation of *B* and subsequent treatment with CuCN/KCN gives *C*, C_8H_7N. Hydrolysis of *C* followed by oxidation forms *D* which has an equivalent mass 83. *D* forms only one mono substitution product. Identify the structures of *A*, *B*, *C* and *D*.

Only One Option Correct Type

Direction *(Q. Nos. 3-9) This section contains 7 multiple choice questions. Each question has four choices (a), (b), (c) and (d), out of which ONLY ONE option is correct.*

3 Which could not be directly prepared from 4-bromobenzenediazonium ion?

a. Bromobenzene

b. 1, 4-dibromobenzene

c. 4-bromofluorobenzene

d. 4-bromoacetophenone

4 Select the final product from this sequence of reactions.

$$C_6H_5CCl_3 \xrightarrow[OH_2]{NaOH} \xrightarrow{H_3O^+} \xrightarrow[H_2SO_4]{HNO_3} \xrightarrow[HCl]{Sn} \xrightarrow{HNO_2} \xrightarrow{C_6H_5OH}$$

a. 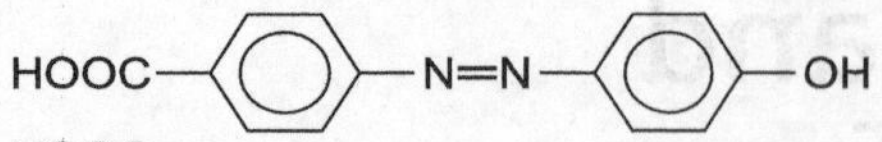

b. 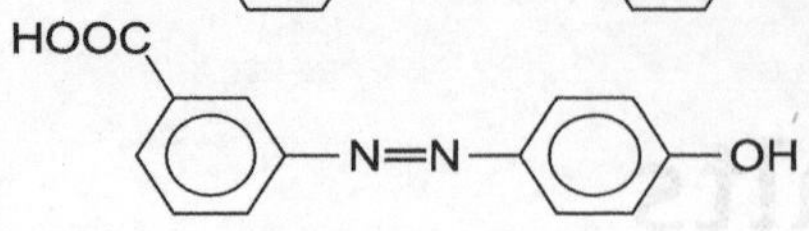

c. 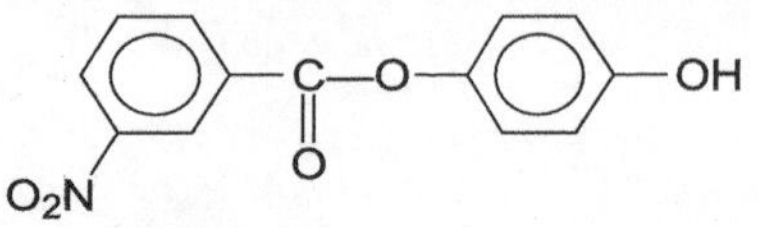

d. 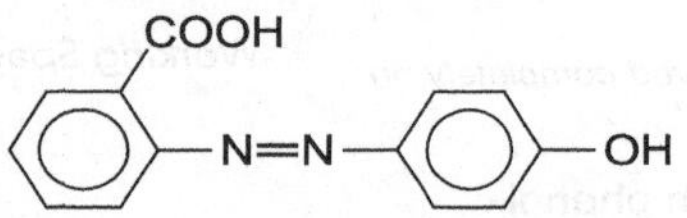

5 The feasible reactions among the given is/are.

a. $PhN_2^+ Cl^- + PhNMe_2 \longrightarrow$ 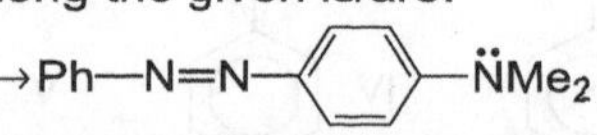

b. $PhN_2^+ Cl^- + 2,6\text{-}Me_2C_6H_3NMe_2 \longrightarrow$ 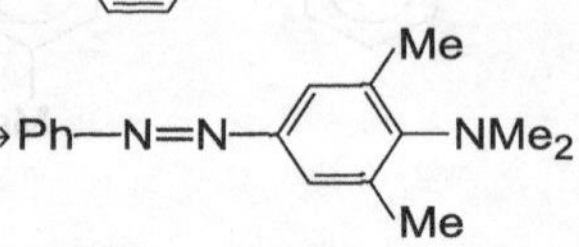

c. $PhN_2^+ Cl^- + PhOH \xrightarrow{HCl}$ 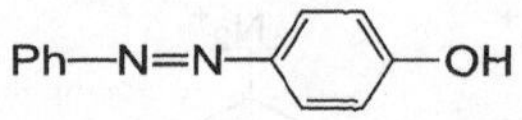

d. $PhN_2^+ Cl^-$ + Mesitylene $\longrightarrow$ 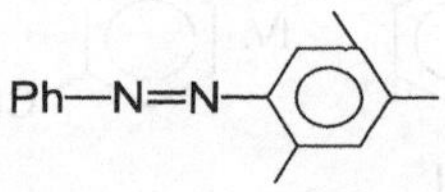

6 Aniline is reacted with Br_2 water and the resulting product is treated with an aqueous solution of $NaNO_2$ in the presence of dil. HCl. The resulting solution is converted into a tetraboronfluorate which is subsequently heated to dry. The final product is

a. 1, 3, 5-tribromobenzene
b. *p*-bromo fluorobenzene
c. *p*-bromo aniline
d. 2, 4, 6-tribromo fluorobenzene

7 Devise a series of reactions to convert benzene into *meta*-chlorobromobenzene. Select reagents and conditions from the following table, listing them in the order of use.

1. sulphuric acid (conc.), heat	2. $Cl_2 + FeCl_3$ and heat
3. $NaNO_2 + H_3O^+$, 0°C	4. H_2, Pd/C catalyst
5. Mg in ether	6. PBr_3
7. H_3PO_2	8. HNO_3 (conc.) + H_2SO_4 (conc.) and heat
9. Cu_2Br_2 + HBr	10. $(CH_3CO)_2O$ + pyridine

a. 1 then 2 then 6
b. 2 then 8 then 4 then 3 then 9
c. 8 then 4 then 10 then 2 then 3 then 9
d. 8 then 2 then 4 then 3 then 9

Working Space

8 Which of the following is regenerated at the end of the reaction?

$$\underset{X}{p\text{-nitrophenol}} \xrightarrow[(1)]{(C_2H_5)_2SO_4/NaOH} \xrightarrow[(2)]{Sn/HCl} Y \xrightarrow[(3)]{NaNO_2/HCl} \xrightarrow[(4)]{C_6H_5OH} Z$$

$$\xrightarrow[(5)]{NaOH/C_2H_5Br} \xrightarrow[(6)]{Sn/HCl} W \xrightarrow[(7)]{NaNO_2/HCl} \xrightarrow[\substack{CuNO_2 \\ (8)}]{NaNO_2} \xrightarrow[(9)]{HI} (?)$$

a. X b. Y c. Z d. W

9 An azo compound X is cleaved at the azo linkage by stannous chloride, $SnCl_2$ to form two amines, 3-bromo-4-amino toluene and 2-methyl-4-aminophenol.

$$A + B \longrightarrow X$$

Which of the following is the suitable combination of A and B?

a. 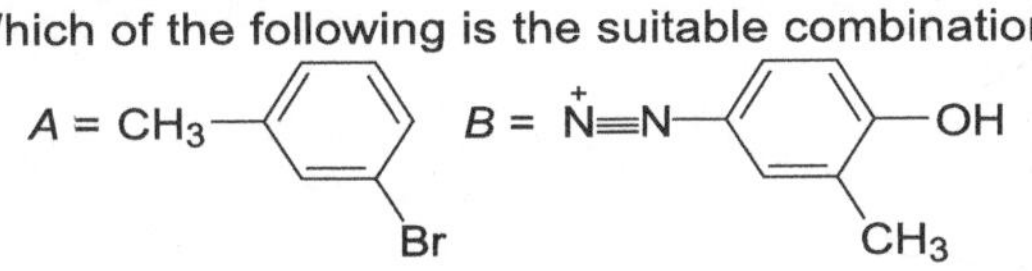

b. 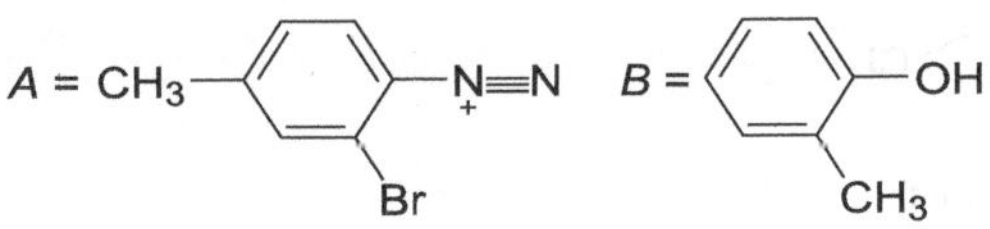

c. 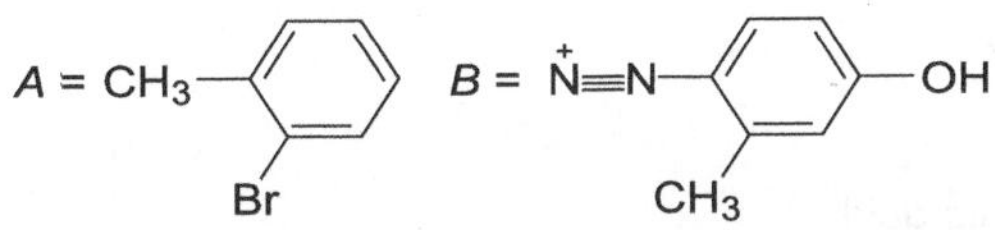

d. 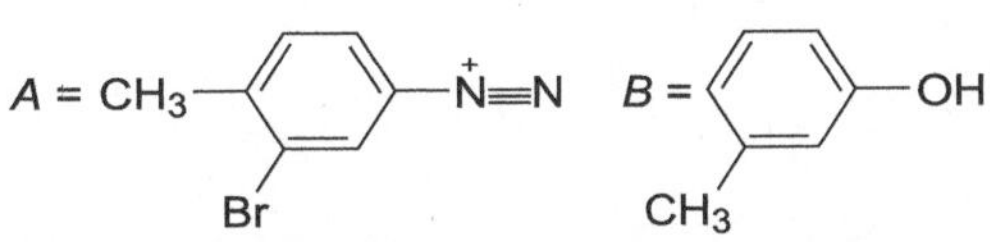

One or More than One Options Correct Type

Direction *(Q. Nos. 10-12) This section contains 3 multiple choice questions. Each question has four choices (a), (b), (c) and (d), out of which ONE or MORE THAN ONE are correct.*

10 In which of the following, reactants and products are correctly matched?

a. 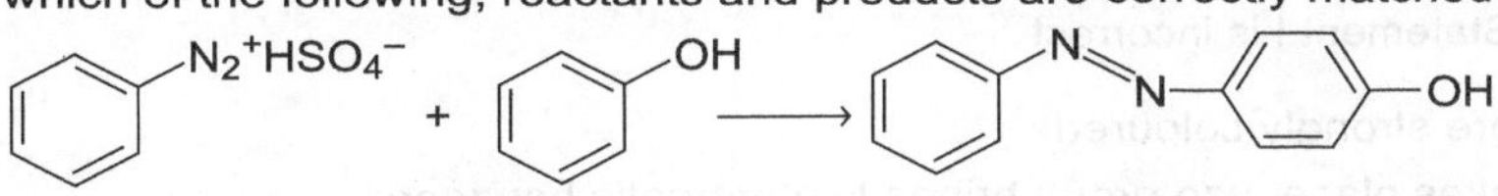

b. 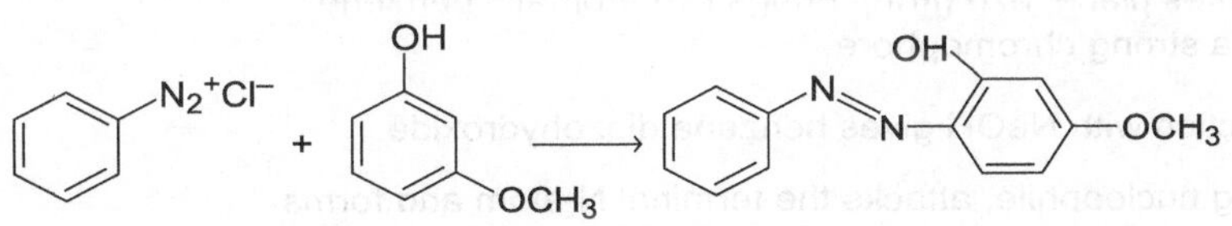

c. 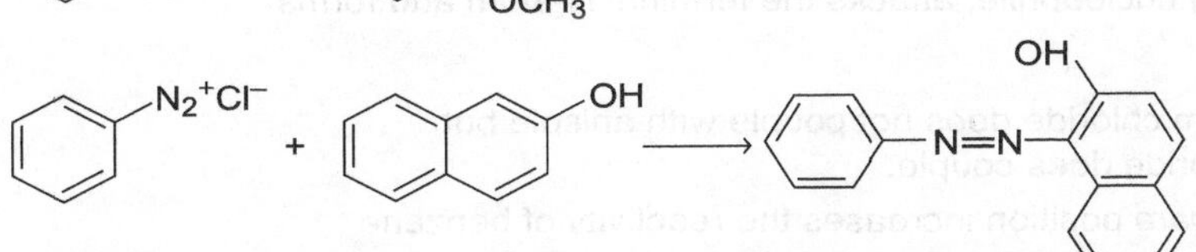

d. 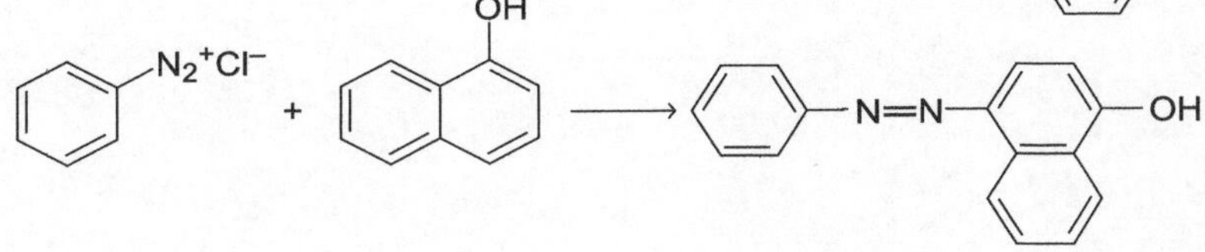

Working Space

11 Which of the following is true about the given reaction sequence?

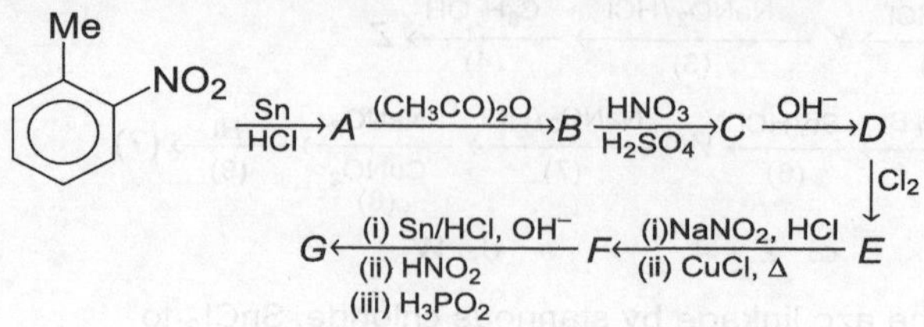

a. $B =$ 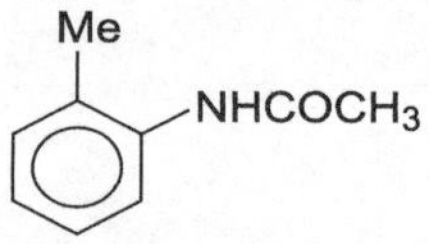

b. $G =$ 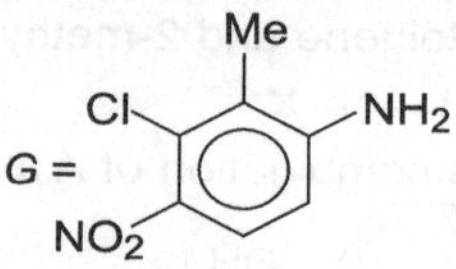

c. $E =$ 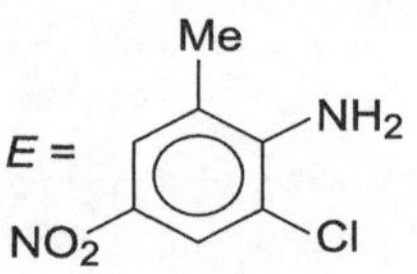

d. $G =$ 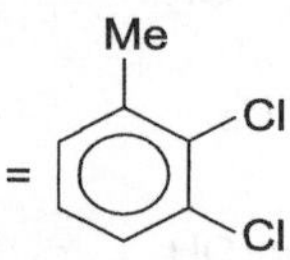

12 The following reaction sequence involve the formation of

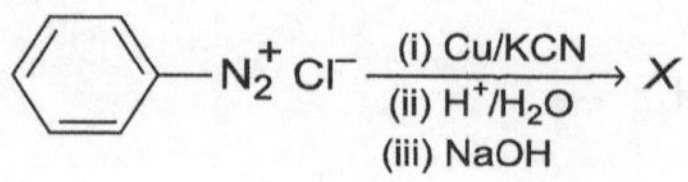

a. a cyanide
b. a carboxylic acid
c. an amide
d. an arene

Statement Type

Direction *(Q. Nos. 13-15) This section is based on Statement I and Statement II. Select the correct answer from the codes given below.*

a. Both Statement I and Statement II are correct and Statement II is the correct explanation of Statement I
b. Both Statement I and Statement II are correct but Statement II is not the correct explanation of Statement I
c. Statement I is correct but Statement II is incorrect
d. Statement II is correct but Statement I is incorrect

13 **Statement I** Azo compounds are strongly coloured.

Statement II When coupling takes place, azo group brings two aromatic benzene rings in conjugation which is a strong chromophore.

14 **Statement I** $Ph^{\oplus}N_2Br^-$ on reaction with NaOH gives benzene diazohydroxide.

Statement II $\overline{O}H$ being a strong nucleophile, attacks the terminal N-atom and forms a covalent bond.

15 **Statement I** Benzene diazonium chloride does not couple with anisole but *p*-nitrobenzenediazonium chloride does couple.

Statement II Nitro group from *para* position increases the reactivity of benzene diazonium ion.

Comprehension Type

Direction (Q. Nos. 16 and 17) *This section contains a paragraph, describing theory, experiments, data, etc. Two questions related to the paragraph have been given. Each question has only one correct answer among the four given options (a), (b), (c) and (d).*

Passage

$$\underset{}{Ar{-}NH_2} \xrightarrow[\text{Diazotisation}]{Aq.\ NaNO_2\ +\ HCl\ (0\text{–}\ 5°C)} \underset{\text{I}}{Ar{-}N_2^+} \xrightarrow[\text{Diazo coupling}]{Ar{-}G} \underset{\text{II}}{Ar{-}N{=}N{-}Ar{-}\ddot{G}}$$

16 Diazocoupling is best carried out by

a. 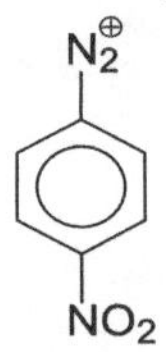

b. 

c. 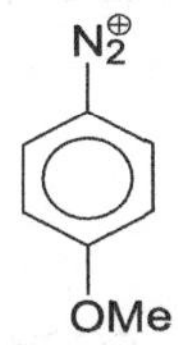

d. 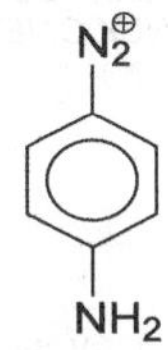

17 Diazo coupling reaction with phenols and aniline require pH conditions respectively.

a. pH < 7, pH > 7
b. pH < 7, pH < 7
c. pH > 7, pH < 7
d. pH > 7, pH > 7

One Integer Value Correct Type

Direction (Q. No. 18) *This section contains 1 question. When worked out will result in an integer from 0 to 9 (both inclusive).*

18

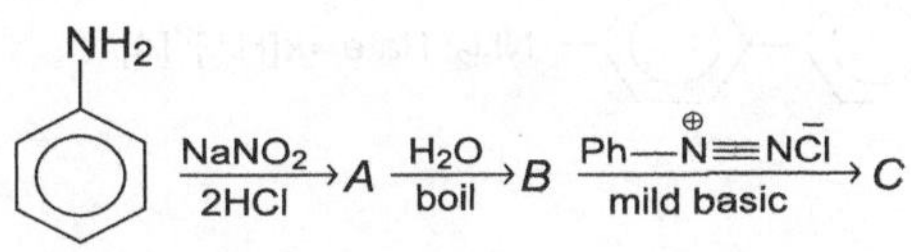

Double bond equivalent of compound *C* is

SET - 1

Revisal Problems for JEE Main

Only One Option Correct Type

Direction (Q. Nos. 1-25) *This section contains 25 multiple choice questions. Each question has four choices (a), (b), (c) and (d), out of which ONLY ONE is correct.*

Working Space

1 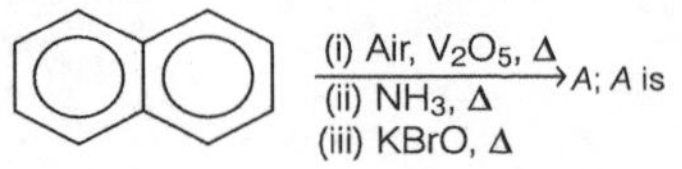

a. 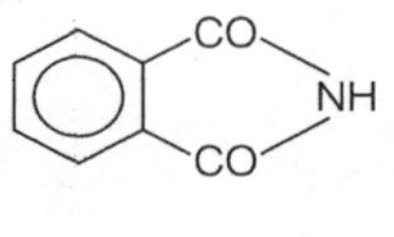 b. 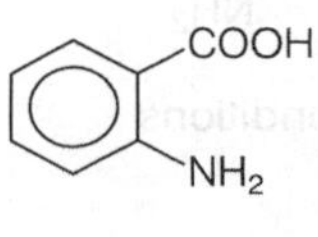 c. 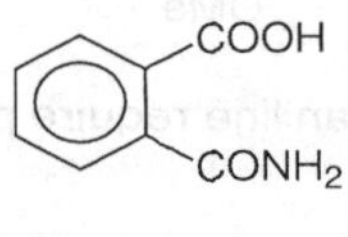 d. 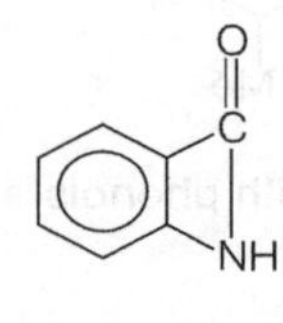

2 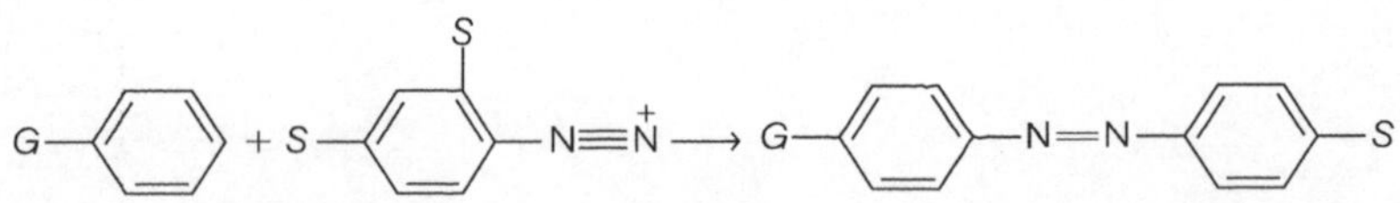

For such kind of diazo coupling reaction, the suitable substituents G and S are respectively

a. $—NH_2$ and $—OCH_3$
b. $—NO_2$ and — CHO
c. $—NH_2$ and $—NHCH_3$
d. $—OCH_3$ and $—NO_2$

3 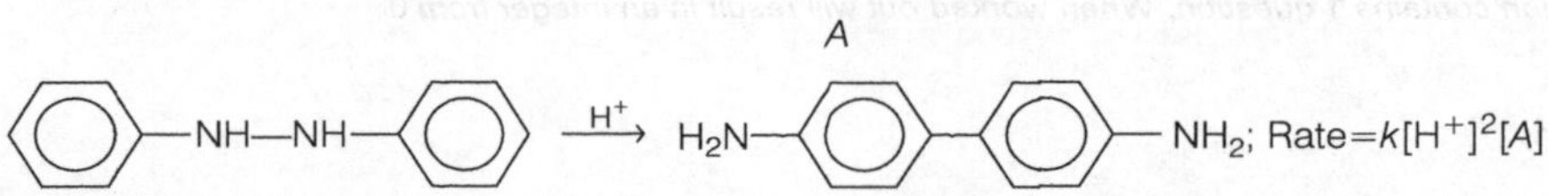

Hence, reactive intermediate is

a. 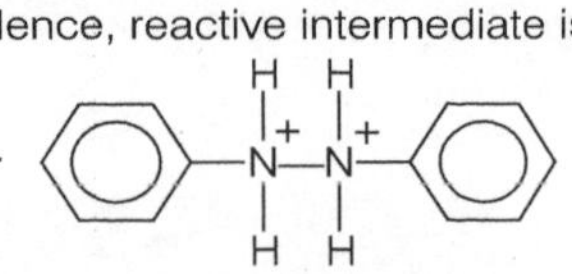

b. 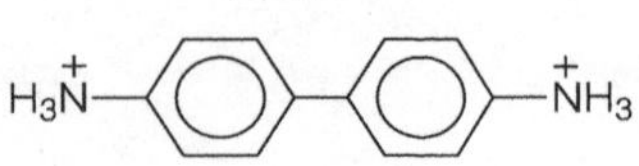

c. Both (a) and (b)
d. None of the above

4 Which compound yields a N-nitroso amine after treatment with nitrous acid?

a. 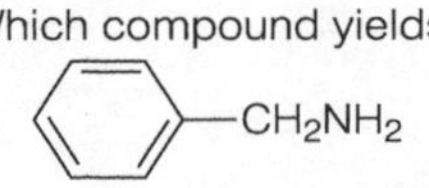 b. 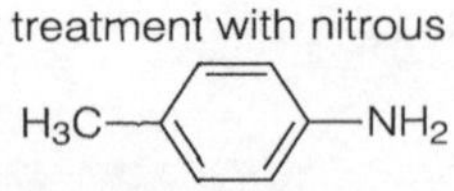

c. 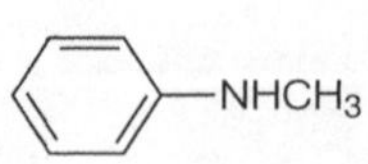 d. 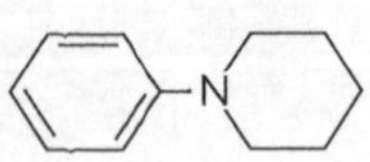

5 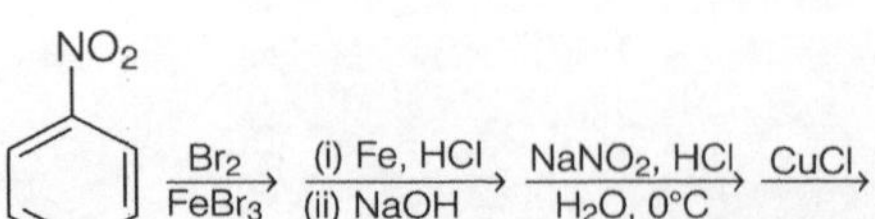

a. *meta*-bromochlorobenzene
b. 1, 3-dibromo-5-chlorobenzene
c. 1, 3, 5-tribromobenzene
d. mixture of *ortho* and *para*-bromochlorobenzene

Working Space

6 A C_8H_{10} hydrocarbon is nitrated by HNO_3 and sulphuric acid. Two and only two $C_8H_9NO_2$ isomers are obtained. Which of the following fits well in this evidence?

a. Ethylbenzene b. *Ortho*-xylene c. *Meta*-xylene d. *Para*-xylene

7 Which of the following when present in water gives precipitate of hydrated ferric oxide with ferric chloride?

a. Methyl amine b. 2, 4-dinitro aniline c. Picric acid d. None of these

8 Which reaction sequence would be best for preparing 3, 5-dibromoaniline from nitrobenzene?

a. (i) $SnCl_2$ / HCl (ii) $2Br_2$ in ether
b. (i) excess Br_2 +$FeBr_3$ and heat (ii) $SnCl_2$ / HCl
c. (i) $SnCl_2$ / HCl (ii) H_2SO_4 and heat (iii) excess HBr
d. (i) H_2SO_4 and heat (ii) excess Br_2 + $FeBr_3$ and heat (iii) $SnCl_2$ / HCl

9 Select the best combination of reagents which will bring the following conversion.

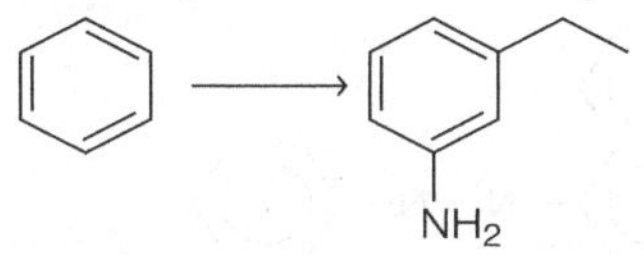

a. Br_2 / $FeBr_3$; HNO_3 / H_2SO_4; NaOH; H_2 / Pt
b. CH_3COCl / $AlCl_3$; HNO_3 / H_2SO_4; Zn / HCl
c. CH_3Cl / $AlCl_3$; $KMnO_4$; CH_3Cl / $AlCl_3$; NBS / *hv*; $LiAlH_4$ / H_2O; NaOH
d. CH_3Cl / $AlCl_3$; HNO_3 / H_2SO_4; NBS / *hv* / NaOH

10 Aniline has high boiling point, high vapour pressure at 100° C and is insoluble in water. Aniline is, therefore separated by

a. steam distillation b. vacuum distillation
c. simple distillation d. alkali treatment

11

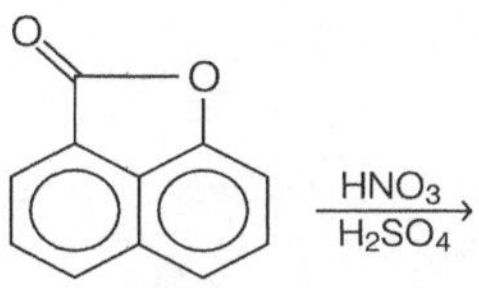

A. Product *A* is

a. 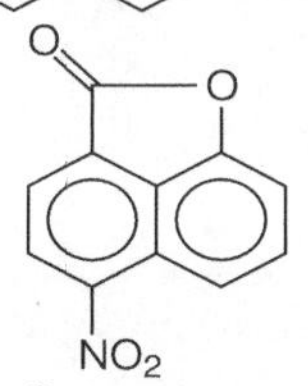

b. 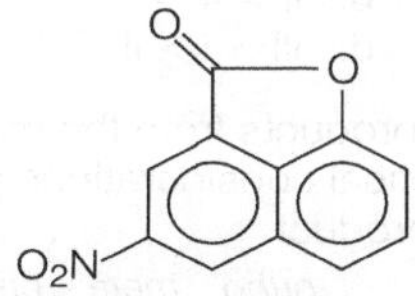

c. 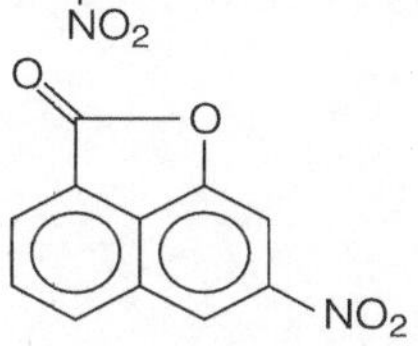

d. 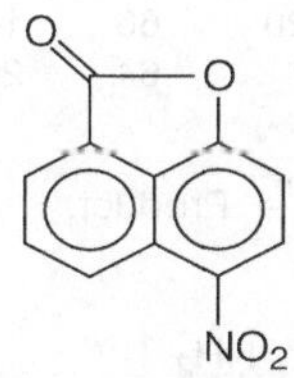

12 Choose the most important resonance structure of the intermediate that is formed, in the mechanism of the reaction, when an electrophile (E^+) attacks dimethylaniline at the *ortho* position.

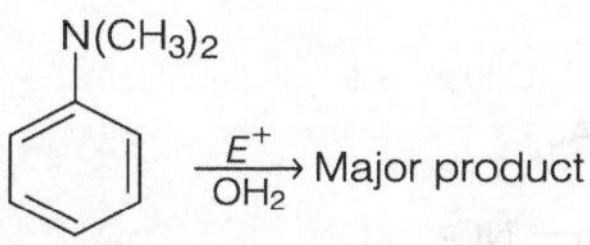

a. 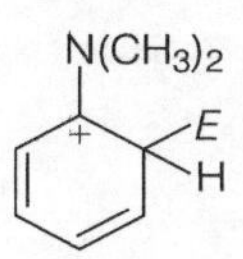

b. 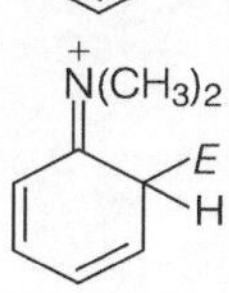

c. 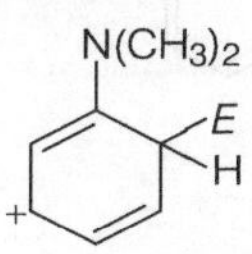

d. 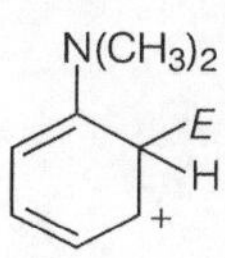

13 Consider the following reactions

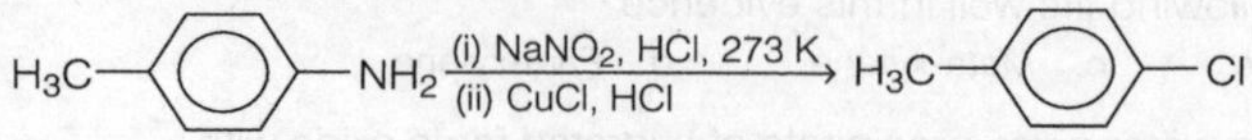

The name of the reaction and the intermediate *via* which it is known to proceed, are respectively

a. Hunsdiecker and benzyne
b. Sandmeyer and a free radical
c. Meerwein and a free radical
d. Sandmeyer and a carbanion

14 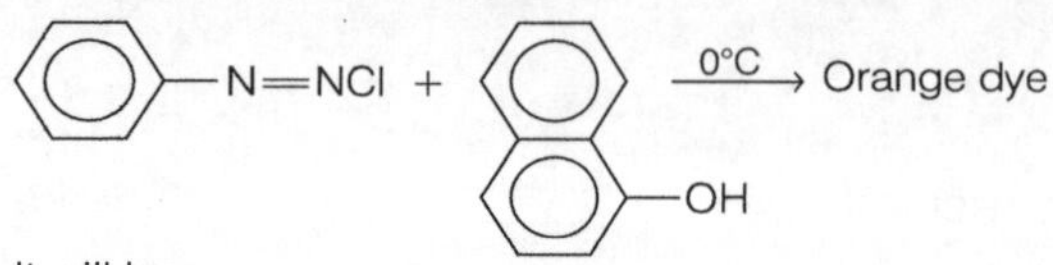

It will be

a. 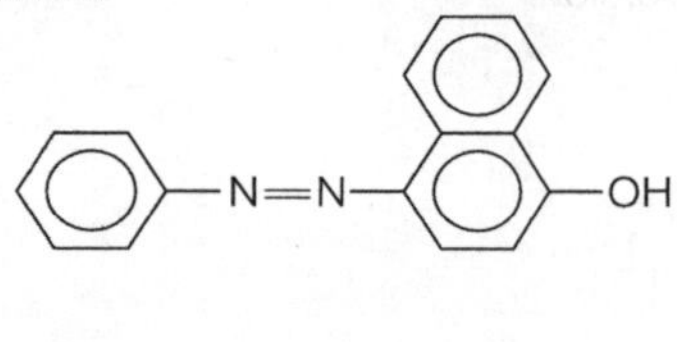

b. 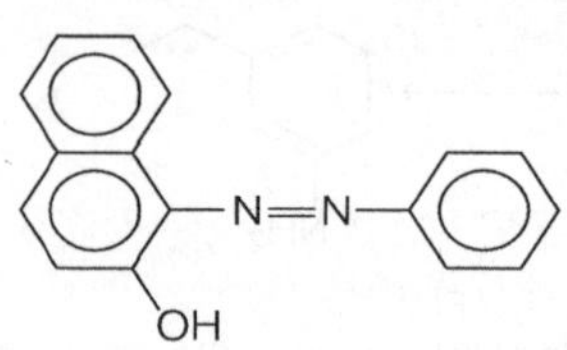

c. 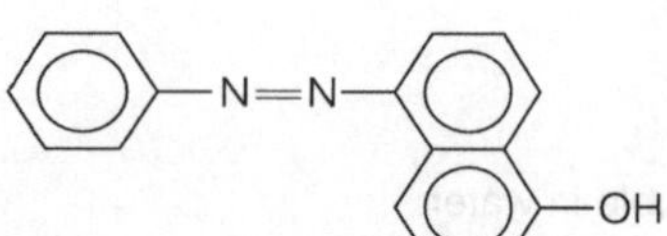

d. None of these

15 Select the basic strength order of following molecule?

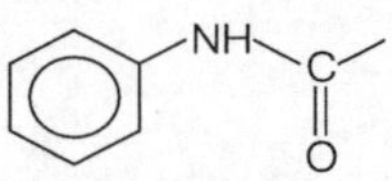

I.

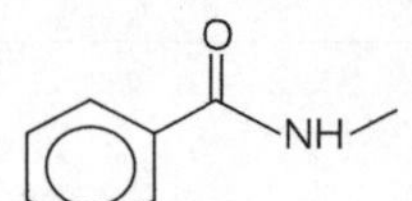

II.

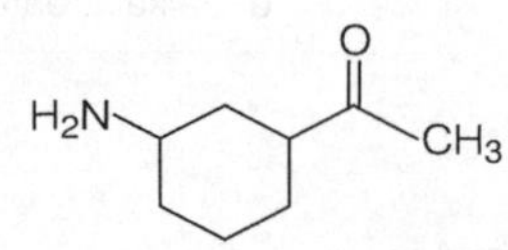

III.

a. III > II > I
b. II > III > I
c. I > III > II
d. III > I > II

16 The following list shows possible yields of products from the mononitration of (trichloromethyl)-benzene. From the theoretical considerations predict the most probable proportions (percentage yield) of nitrated product.

	ortho	*meta*	*para*
a.	25	7	68
b.	20	66	14
c.	46	8	46
d.	7	64	29

17 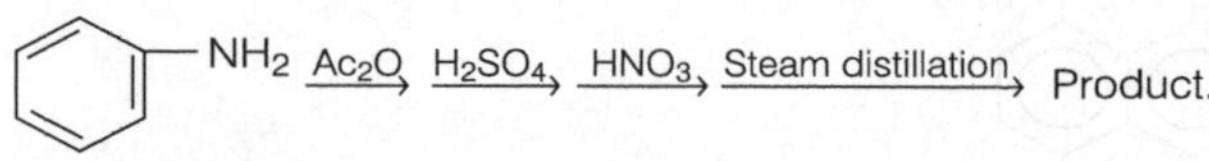

The main product is

a. 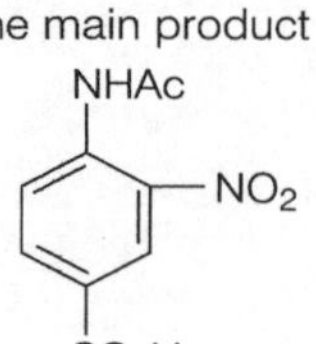

b. 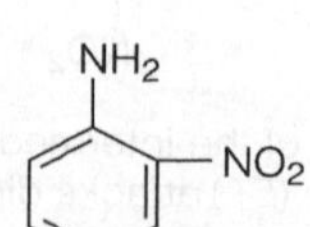

c. 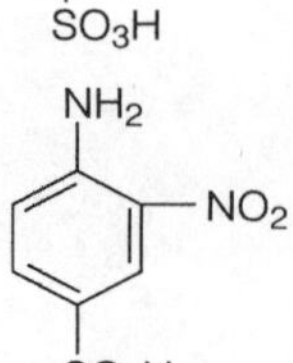

d. 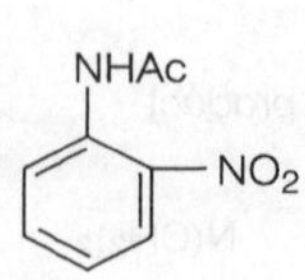

Working Space

18 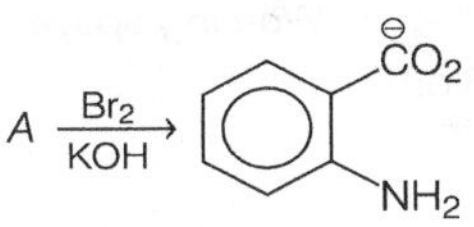

Reactant *A* is

a. 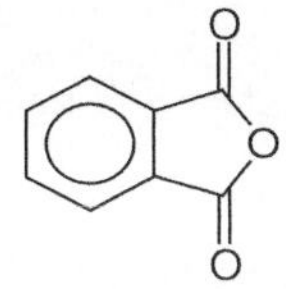

b. 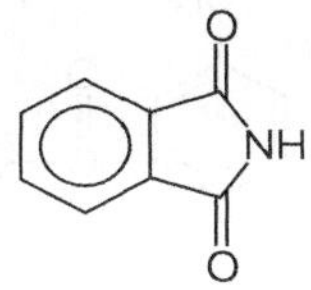

c. 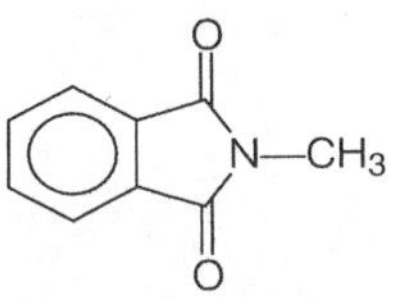

d. 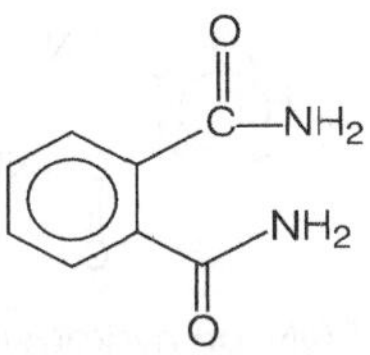

19 The basic character of the substituted anilines (I to IV) is such that

I. 

II. 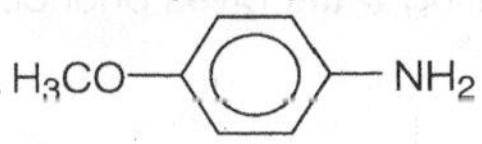

III. 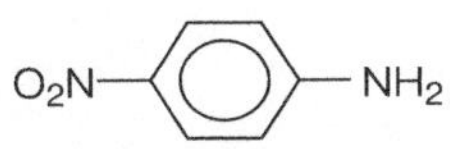

IV. 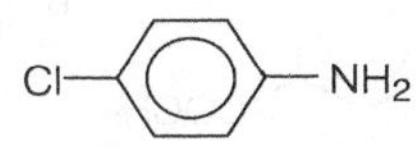

a. III < IV < II < I
b. IV < III < II < I
c. IV < III < I < II
d. III < IV < I < II

20 Aromatic diazonium group can be replaced by various groups/atoms. In this context, match Column I with Column II and select the correct answer using the codes given below the Columns

	Column I (Reagent/reaction type)		Column II (Diazo group is replaced by)
i.	H_3PO_2	p.	Hydrogen
ii.	Sandmeyer	q.	Hydroxyl group
iii.	H_2O/Δ	r.	Aryl group
iv.	Gomberg	s.	Chlorine

Code

	i	ii	iii	iv
a.	s	r	q	p
b.	p	s	q	r
c.	p	q	r	s
d.	q	r	s	p

21 In which of the following reaction, reactants and products are not correctly matched?

a. $C_6H_5N_2^+Cl^- \xrightarrow{H_3PO_3} C_6H_6$

b. $C_6H_5NH_2 \xrightarrow[HBF_4]{NaNO_2} \xrightarrow[0°C]{\Delta} C_6H_5F$

c. $C_6H_5N_2^+Cl^- \xrightarrow[Cu/\Delta]{NaNO_2} C_6H_5NO_2$

d. 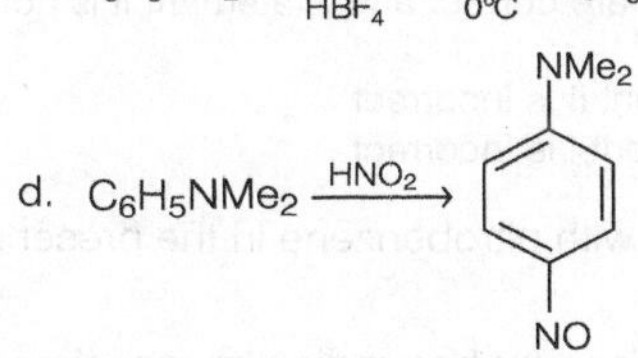

22 pK_b of amines 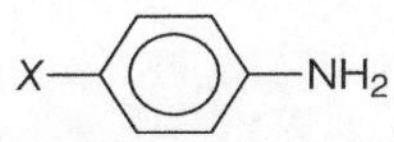

is in order when *X* is

a. $CH_3 < H < CF_3 < NO_2$
b. $H < CH_3 < CF_3 < NO_2$
c. $H < CH_3 < NO_2 < CF_3$
d. $CH_3 < H < NO_2 < CF_3$

Working Space

23 Compound A, $C_{14}H_{10}N_2O$ when heated with dilute sulphuric acid gives ammonium sulphate, compound B, $C_8H_6O_4$ and compound C, C_6H_7N (as its sulphate). Compound B on heating gives D, $C_8H_4O_3$ and on heating with soda lime gives benzene. Compound C with dilute H_2SO_4 / $NaNO_2$ in the cold followed by heating with water gives E, C_6H_6O. Compound E on heating with Zn dust again gives benzene. Compound A is

a. 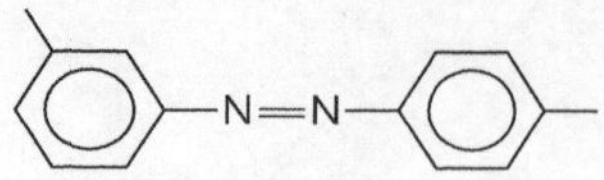

b. 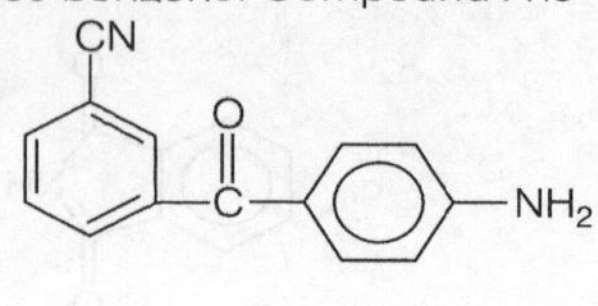

c. 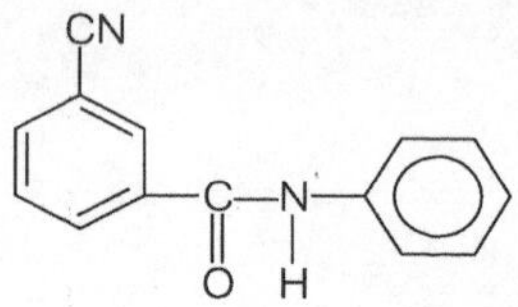

d. 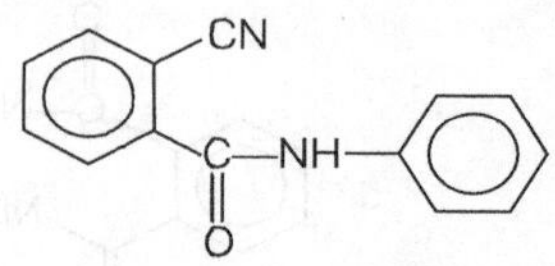

24 A compound A having molecular formula C_8H_9ON, on hydrolysis with NaOH gives two products B and C. The compound C on heating with soda lime gives CH_4. The compound B on diazotisation followed by treatment with boiling water gives phenol. Structure of A is

a. 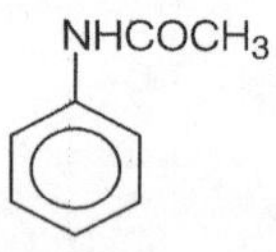

b. 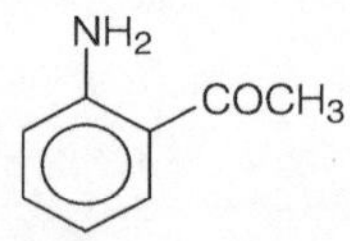

c. 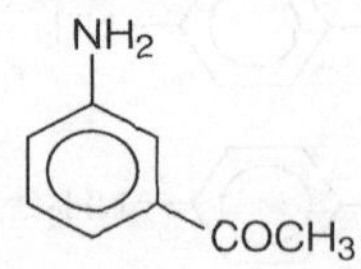

d. 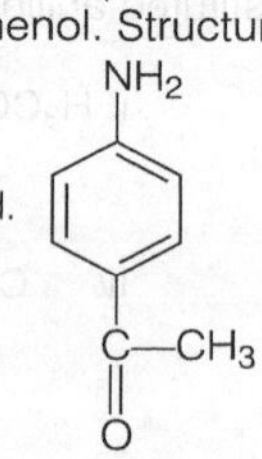

25 Toluene is treated with a mixture of HNO_3 and H_2SO_4. The major product is reduced with $SnCl_2$ in HCl to give A. When A is treated with $NaNO_2$ in HCl at 0° C, followed by reaction with CuCN to give B. B is reduced with $LiAlH_4$ to give C. What is the most probable structure of C?

a. 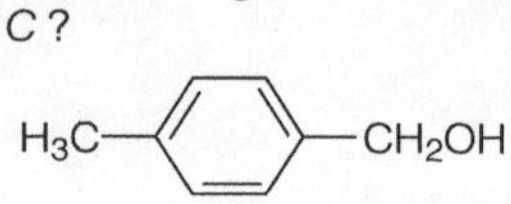

b. 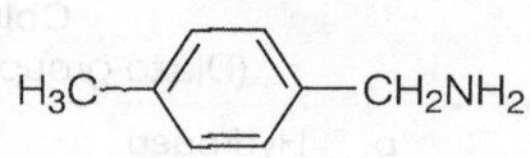

c. 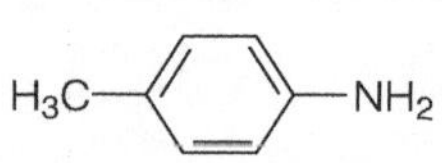

d. 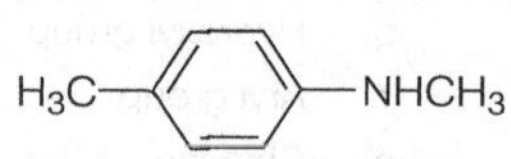

Statement Type

***Direction** (Q. Nos. 26 and 27) This section is based on Statement I and Statement II. Select the correct answer from the codes given below.*

a. Both Statement I and Statement II are correct and Statement II is the correct explanation of Statement I
b. Both Statement I and Statement II are correct and Statement II is not the correct explanation of Statement I
c. Statement I is correct but Statement II is incorrect
d. Statement II is correct but Statement I is incorrect

26 **Statement I** $PhN_2^{\oplus}Br^{\ominus}$ on reaction with nitrobenzene in the presence of alkali gives p-nitrobiphenyl.

Statement II The reaction takes place by free radical mechanism.

27 **Statement I** Coupling of $Ph\overset{\oplus}{N}_2$ with aniline is faster than with phenol.

Statement II Aniline is more electron donating than phenol.

SET - 2

Revisal Problems
for JEE Advanced

Only One Option Correct Type

Working Space

Direction (Q. Nos. 1-13) *This section contains 13 multiple choice questions. Each question has four choices (a), (b), (c) and (d), out of which ONLY ONE is correct.*

1 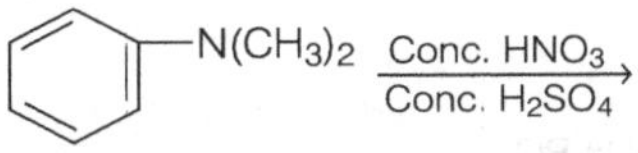 Major product

Major product is

a. 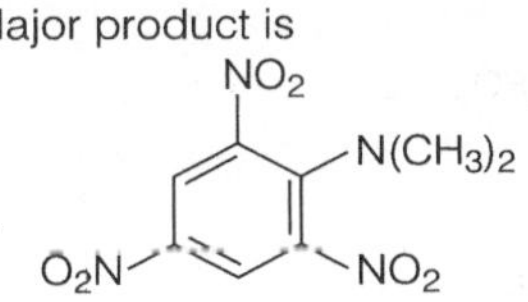

b. 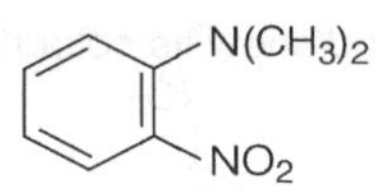

c. 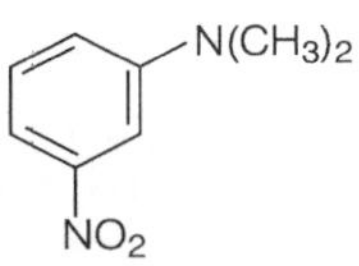

d. 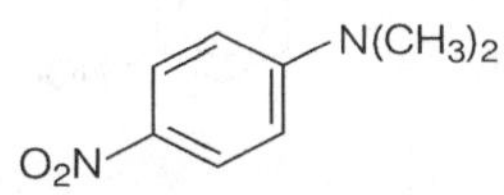

2 In which case, coupling takes place

a. 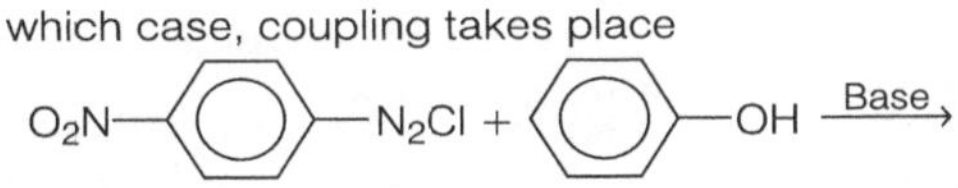

b. 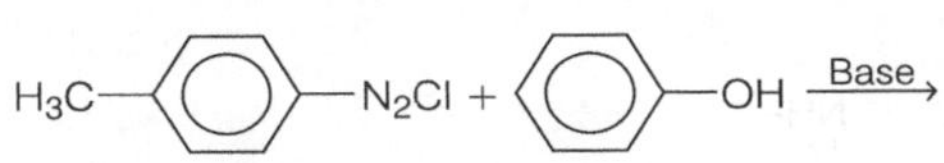

c. Both (a) and (b)
d. All of the above

3 What is the correct order of rate of nitration of the following compounds?

$\underset{A}{C_6H_5CH_3}, \underset{B}{C_6H_6}, \underset{C}{C_6D_6}, \underset{D}{C_6T_6}, \underset{E}{C_6H_5Br_3}, \underset{F}{C_6H_5\overset{+}{N}R_3}, \underset{G}{C_6H_5NMe_2}$

a. $G > A > B > C > D > E > F$
b. $G > B > C > D > A > F$
c. $G > A > B = C = D > E > F$
d. $G > A > B > C = D > E > F$

4 How could one prepare 3, 4, 5-tribromoaniline from *para*-nitroaniline? Note : $HNO_2 = NaNO_2 + 10\%\ H_2SO_4$,

a. (i) $SnCl_2$ / HCl (ii) $2Br_2$ in ether (iii) HNO_2, 0°C (iv) H_3PO_2
b. (i) $SnCl_2$ / HCl (ii) HNO_2, 0°C (iii) excess Cu_2Br_2 + HBr
c. (i) $2Br_2$ in ether (ii) $SnCl_2$ / HCl (iii) HNO_2, 0°C (iv) H_3PO_2
d. (i) $2Br_2$ in ether (ii) HNO_2, 0°C (iii) Cu_2Br_2 (iv) $SnCl_2$ / HCl

5 From the choice below, give the major intermediate which is formed in the bromination of nitrobenzene

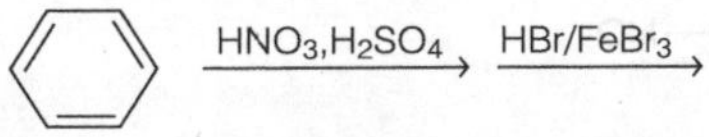 Major substitution product

a. 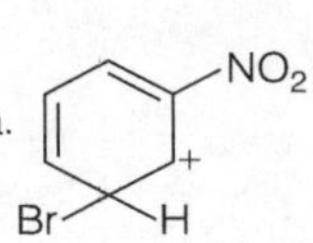

b. 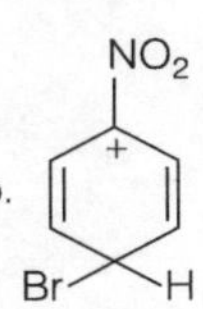

c. 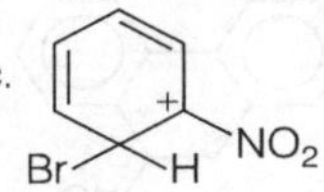

d. 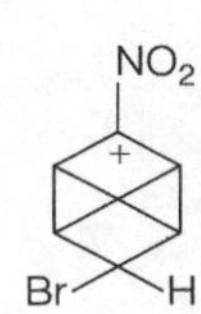

Working Space

6 The basicity constant of

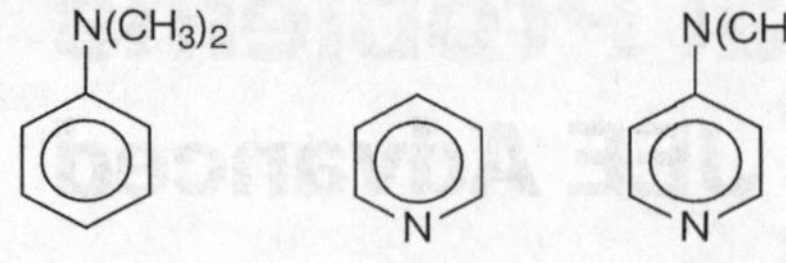

I II III

are in order

a. I < II < III b. III < II < I c. III < I < II d. II < I < III

7 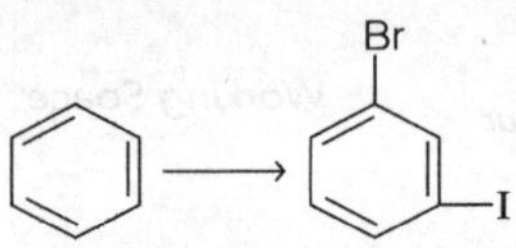

Choose the correct sequence of reactions for the above conversion.

a. Bromination; then iodination in the presence of HNO_3
b. Nitration; bromination; iodination; reduction of NO_2; diazotisation; then treatment of H_3PO_2
c. Nitration; bromination; reduction of NO_2; diazotisation then treatment of KI
d. Nitration; iodination; bromination; reduction of NO_2; diazotisation; then treatment of H_3PO_2

8 Which one is the charge transfer complex and show its colour?

a. 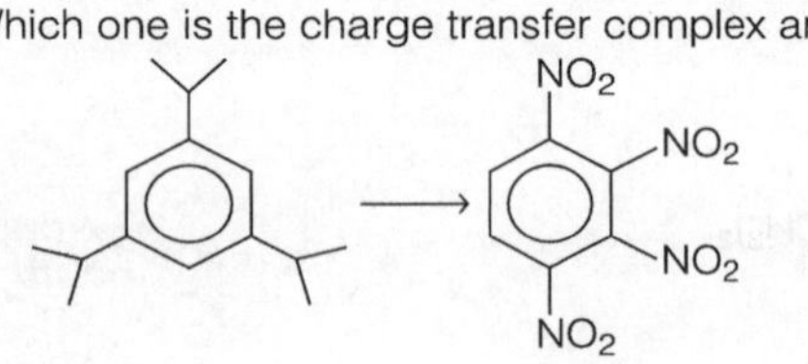

b. 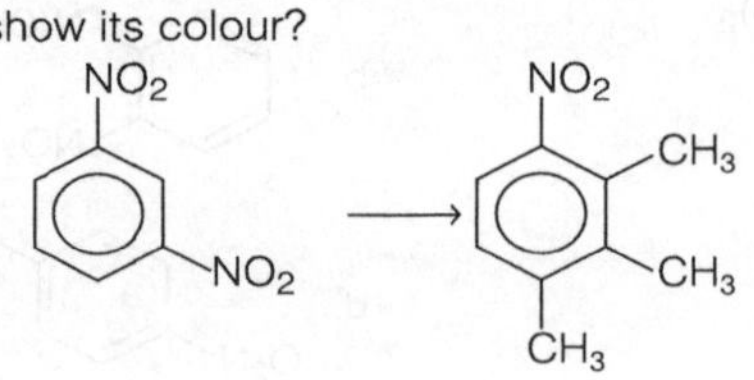

c. 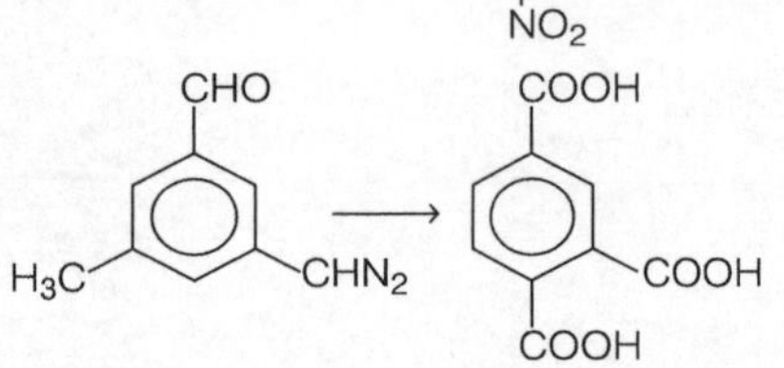

d. None of these

9 What is Q in the given reaction sequence?

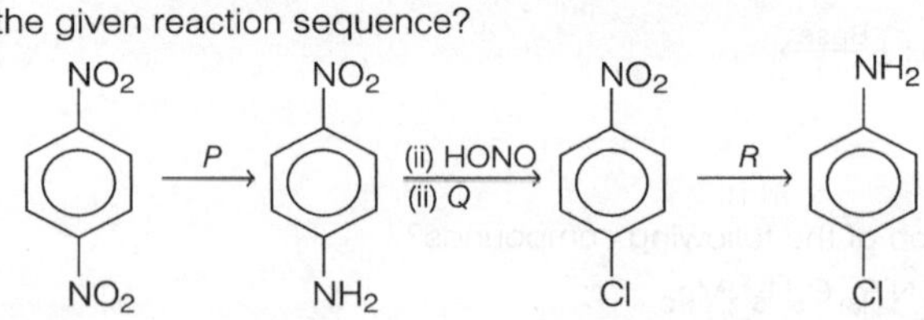

a. Fe / HCl, then $^{-}$OH b. CuCl c. NH_2Cl d. Cl_2 / $h\nu$

10 In the above question 9 what is R?

a. Fe / HCl, then $^{-}$OH b. NH_3, H_2S c. H_3PO_2 d. HONO, then $CuNH_2$

11 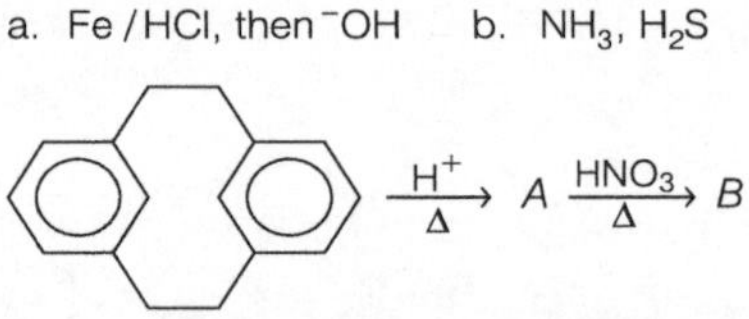

Product B in the above reaction is

a. 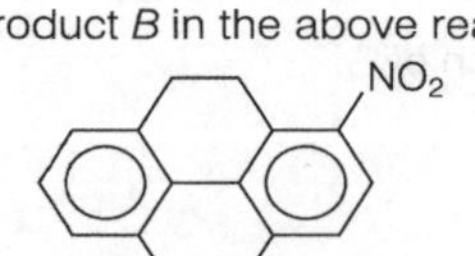

b. 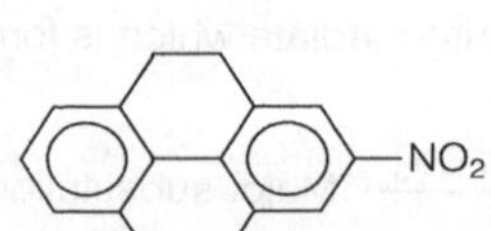

c. 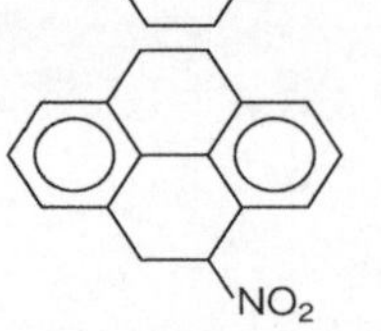

d. 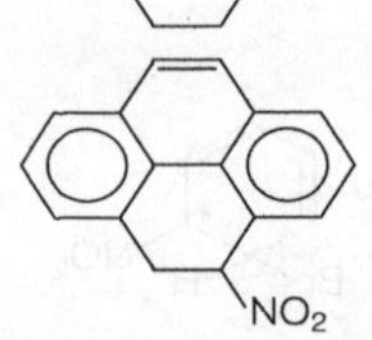

Working Space

12 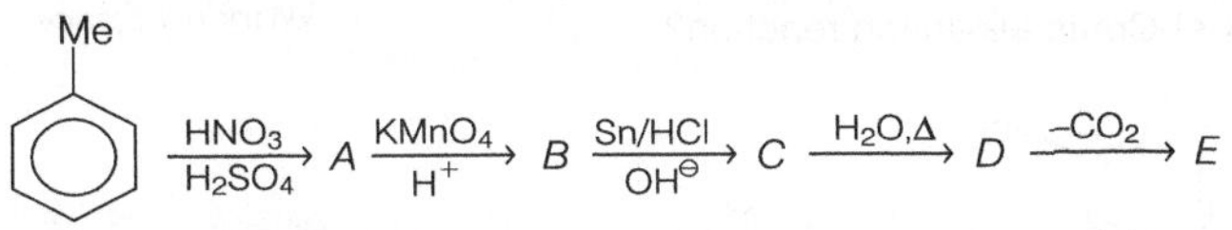

Which compound is not formed in the above reaction sequence?

a. 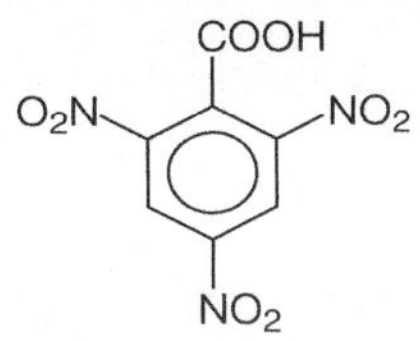

b. 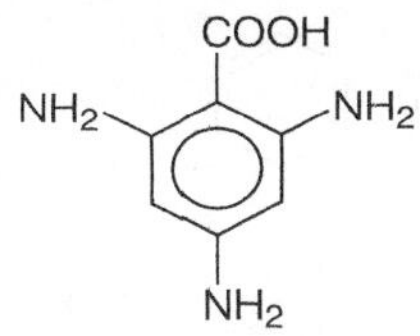

c. 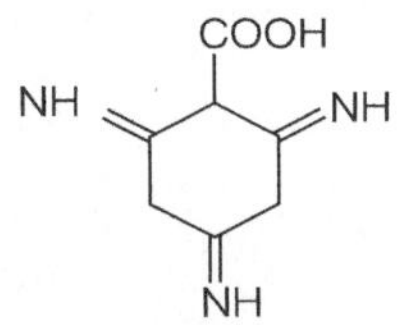

d. 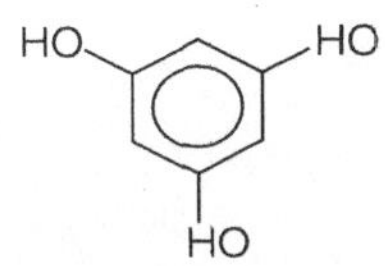

13 Select the decreasing order of relative basic strength of labelled nitrogen atoms?

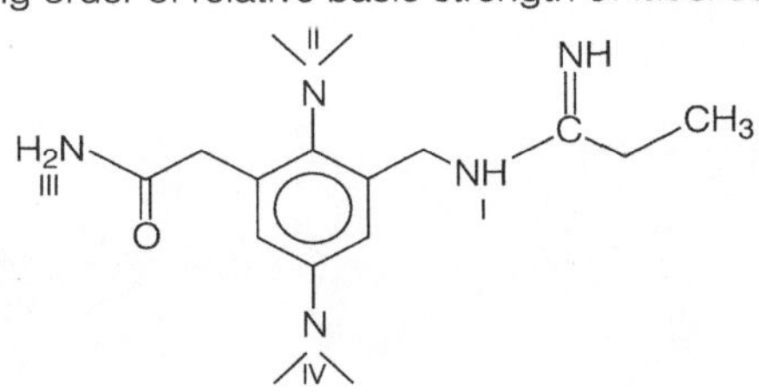

a. I > II > IV > III

b. I > II = IV > III

c. I = III > II = IV

d. II > IV > I > III

One or More than One Options Correct Type

Direction *(Q. Nos. 14-21) This section contains 8 multiple choice questions. Each question has four choices (a), (b), (c) and (d), out of which ONE or MORE THAN ONE are correct.*

14 In which of the following reactant and product are correctly matched?

a. 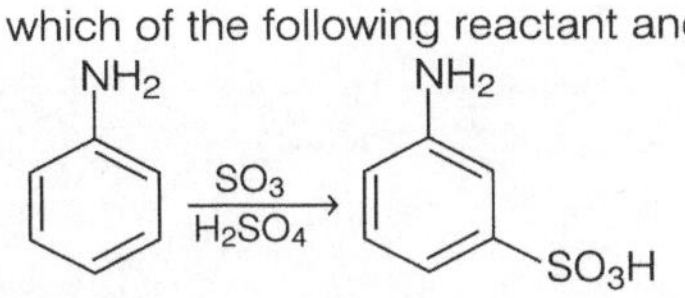

b. 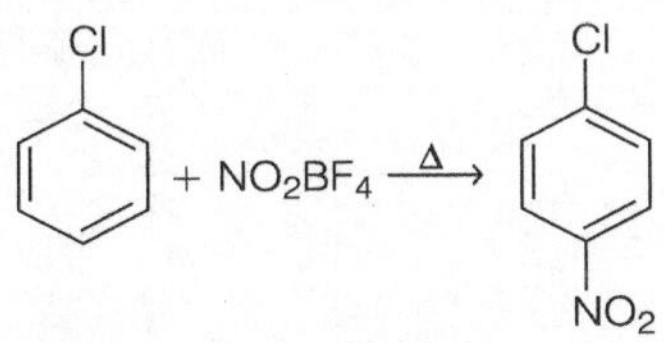

c. 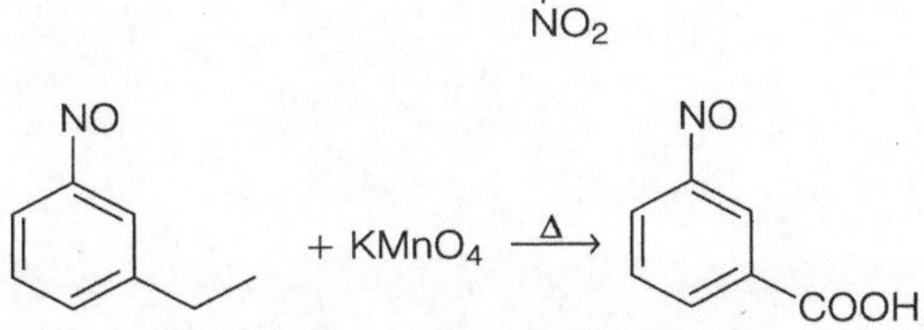

d. 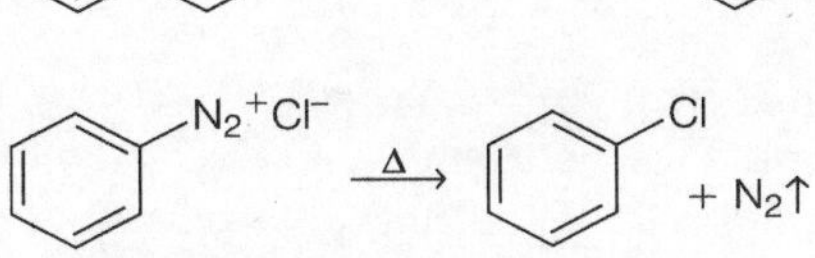

Working Space

15 Which of the following compound(s) fail in Friedel-Crafts alkylation reaction?

a. 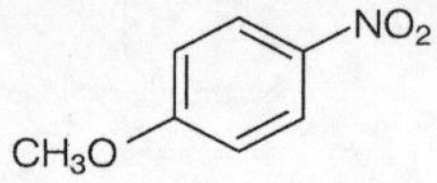

b. 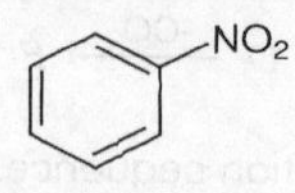

c. 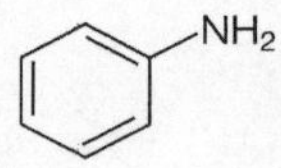

d. 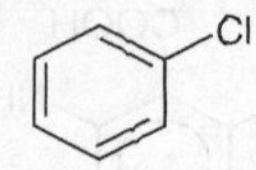

16 Which of the following reactions are wrong?

a. $PhNO_2 \xrightarrow[\text{(pH=5-6)}]{\text{Electrolytic Reduction}}$ p- hydroxyaniline.

b. $PhNO_2 \xrightarrow[\text{(pH=3-4)}]{\text{Electrolytic reduction}}$ Ph NHOH.

c. 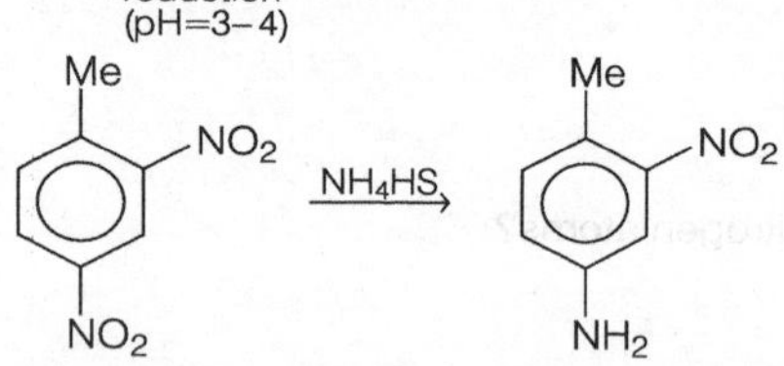

d. 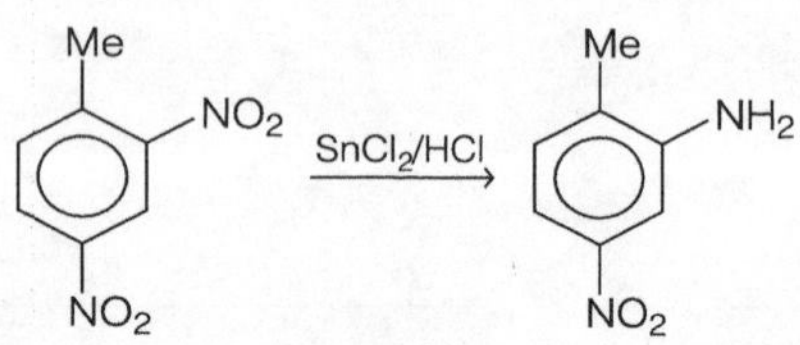

17 In which of the following reactions, a visible change would be observed?

a. $C_6H_5NH_2 + Br_2\text{-}H_2O \longrightarrow$

b. $C_6H_5NH_2 \xrightarrow[HCl/10^{\circ}C]{NaNO_2} \xrightarrow{\beta\text{-naphthol}}$

c. 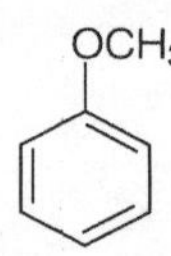 $+ C_6H_5N_2Cl \longrightarrow$

d. 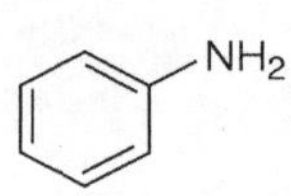 $+ (CH_3CO)_2O \xrightarrow{\Delta}$

18 Oxidation of *A* gives *p*-benzoquinone; *A* can be

a. 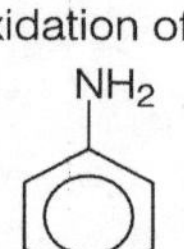

b. 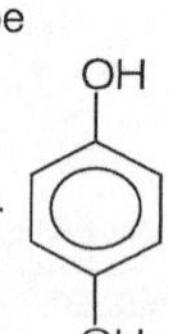

c. 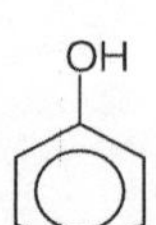

d. 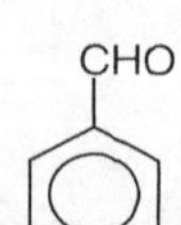

Working Space

19 Which of the following reaction is/are correct?

a. 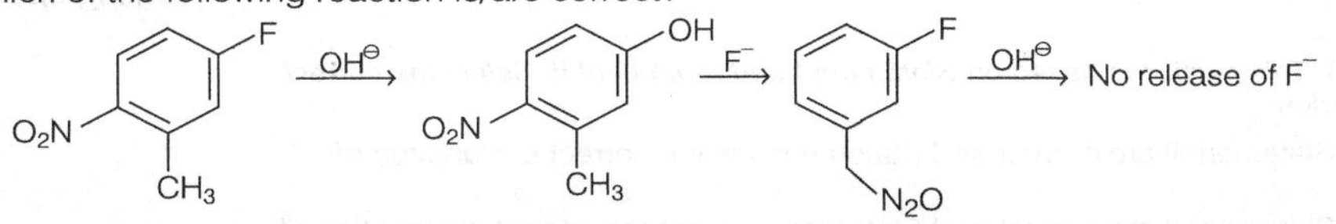

b. 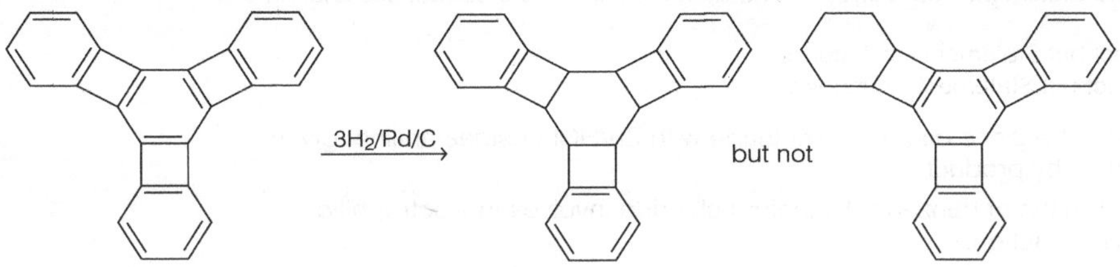

c. 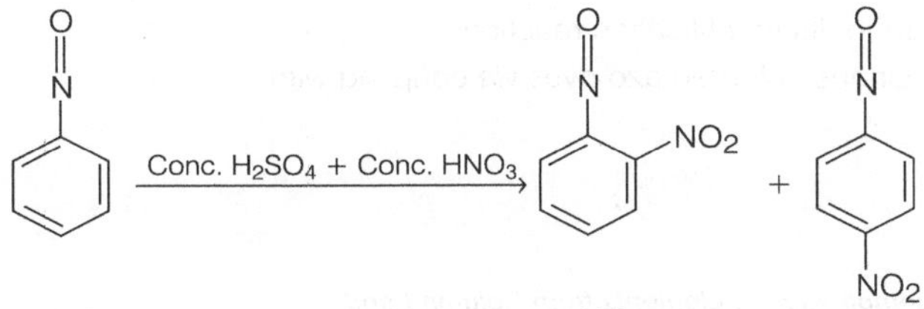

d. 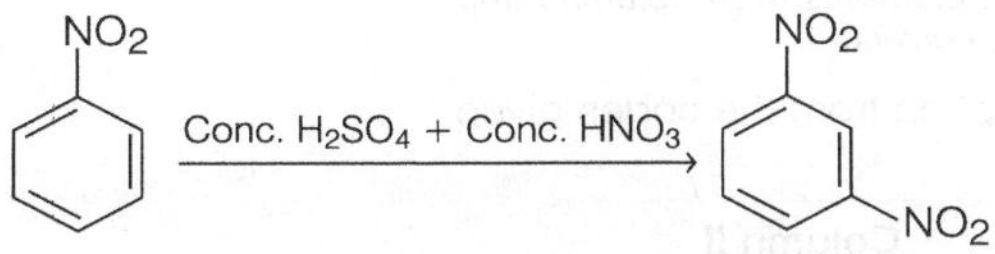

20 Select the correct statement(s) related to following compounds.

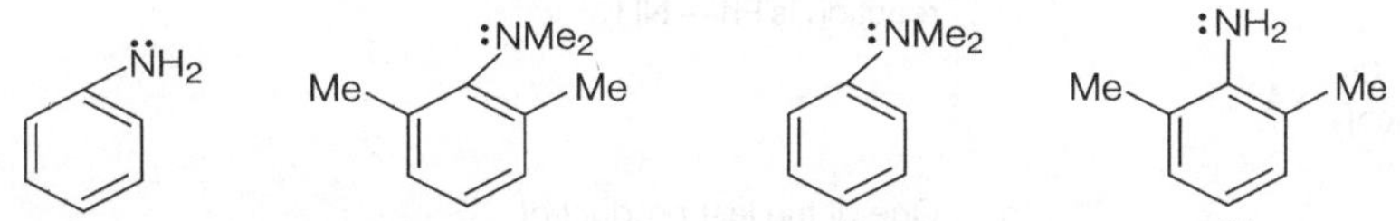

I II III IV

a. II is the most basic compound
b. In II lone pair of nitrogen is less involved in delocalisation.
c. The extent of delocalisation in I and III is same
d. IV is the most basic due to + I effect of two Me groups

21 Which of the following compounds are arranged in decreasing order of their basic strength?

a. 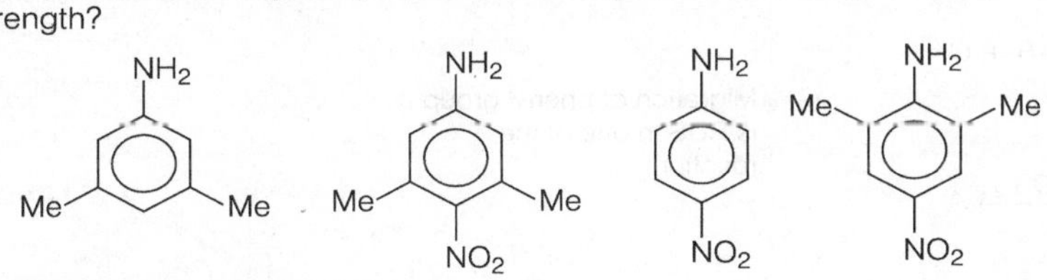

b. $PhCH_2NH_2$, PhNHPh, $PhNH_2$

c. 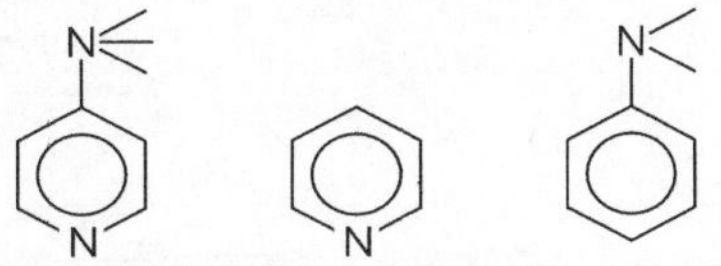

d. $CH_2{=}\ddot{N}{-}CH_3$ $\quad CH_2{=}CH{-}\ddot{N}H_2$ $\quad CH_3{-}C({=}O){-}NH_2$

Working Space

Statement Type

Direction *(Q. Nos. 22 and 23) This section is based on Statement I and Statement II. Select the correct answer from the codes given below.*

a. Both Statement I and Statement II are correct and Statement II is the correct explanation of Statement I
b. Both Statement I and Statement II are correct and Statement II is not the correct explanation of Statement I
c. Statement I is correct but Statement II is incorrect
d. Statement II is correct but Statement I is incorrect

22 **Statement I** Reaction of benzene diazonium chloride with Zn/HCl besides aniline, gives some *p*-chloroaniline as by product.

Statement II Chloronium ion of benzene diazonium chloride involves in electrophilic substitution in a parallel reaction.

23 **Statement I** Diazotisation of aniline can be used for its identification reaction.

Statement II Diazonium salts of aniline forms various coloured azo dyes *via* coupling with suitable reagents.

Matching List Type

Direction *(Q. Nos. 24 and 25) Choices for the correct combination of elements from Column I and Column II are given as options (a), (b), (c) and (d), out of which one is correct.*

24 Match the Column I with Column II and mark the correct option from the codes given below.

	Column I		Column II
i.	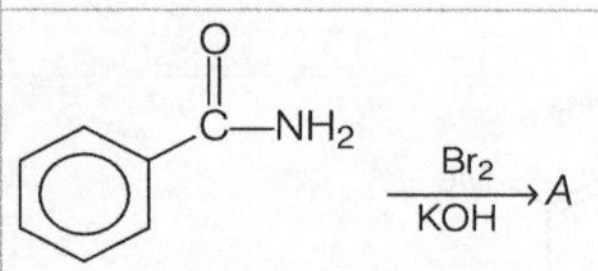	p.	One of the last product of reaction is Ph—NH_2
ii.	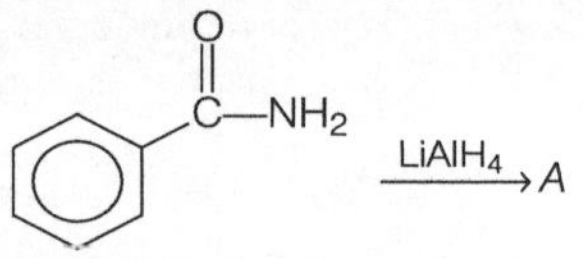	q.	One of the last product of reaction is Ph—CH_2—NH_2
iii.	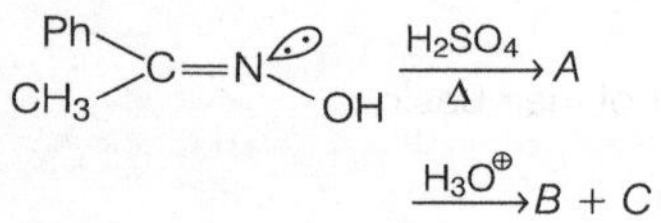	r.	Last product give positive isocyanide test
iv.	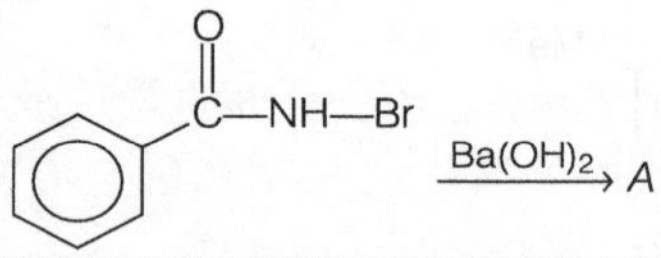	s.	Migration of phenyl group is RDS in one of the reaction

Codes

	i	ii	iii	iv
a.	p,r,s	q,r	p,r,s	p,r,s
b.	p,q	p,q	p,r,s	q,s
c.	p,s	p	q	q,s
d.	q,r	q,s	p,s	q,r

25 Match the Column I with Column II and mark the correct option from the codes given below.

	Column I (Reactants)		Column II (Characteristics of reaction)
i.	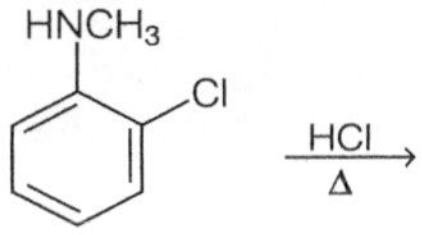	p.	*Para* substituted product
ii.	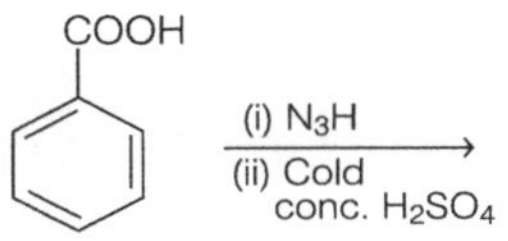	q.	Intramolecular rearrangement
iii.	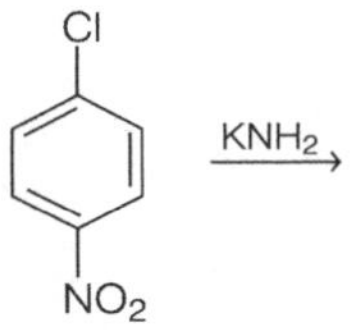	r.	Reaction involves Ph—N $=$ C$=$ O as an intermediate
iv.	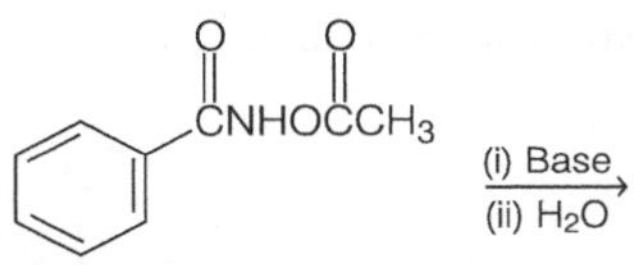	s.	Bimolecular reaction

Codes

	i	ii	iii	iv		i	ii	iii	iv
a.	p,r	r,s	p	q,s	b.	p,s	q,r	p,q	q,r
c.	r,s	p,r	p,r	p,q	d.	p,s	q,s	q,s	r,s

Comprehension Type

***Direction** (Q. Nos. 26-31) This section contains 2 paragraphs, each describing theory, experiments, data, etc. Six questions related to the paragraphs have been given. Each question has only one correct answer among the four given options (a), (b), (c) and (d).*

Passage I

Dr. John directed his graduate students to synthesis *m*-nitrobenzoic acid from nitro benzene.

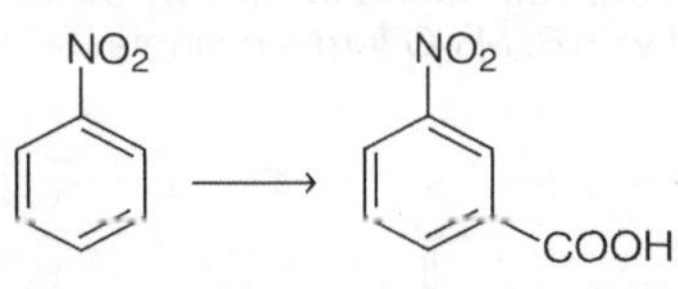

Each student had the idea of orientation but not all the reactions. They put first theoretical concept before their teacher.

Graduate student *A* He is proud of giving two step synthesis

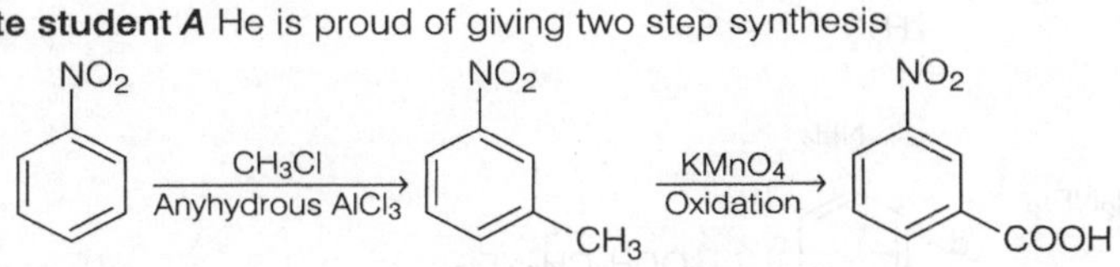

Graduate student *B*

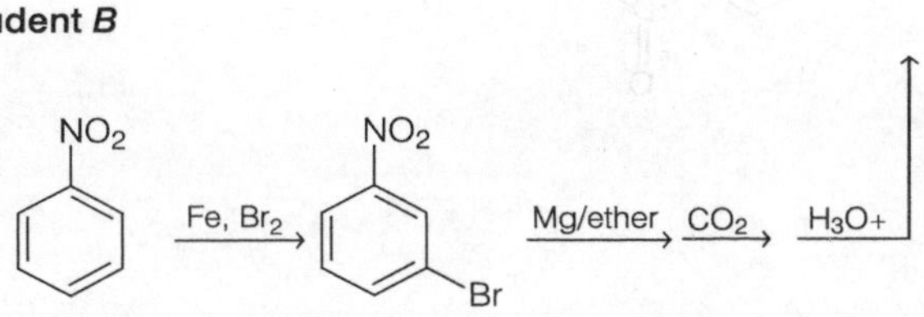

Working Space

Graduate student *C*

Working Space

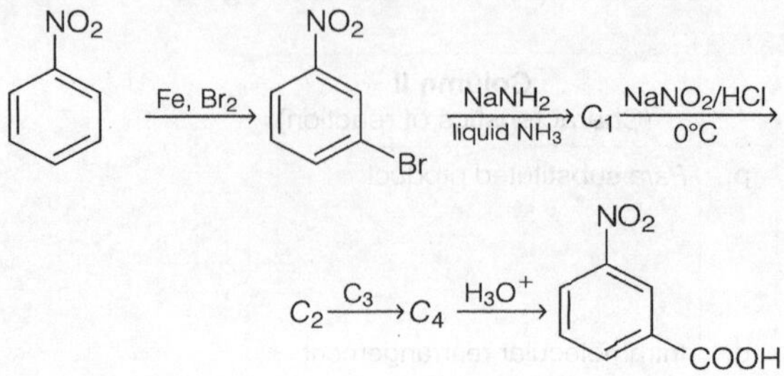

26 Which process is not feasible?

a. By student *A*
b. By student *B*
c. By student *C*
d. By student *B*, *C*

27 C_2 is converted to C_4 by a graduate student *C*, using C_3 which is.........by reaction.

a. NaCN-Gattermann reaction
b. CuCN / HCN-Gattermann reaction
c. CuCN / HCN-Sandmeyer reaction
d. CH_3CN / $AlCl_3$-Friedel-Crafts reaction

28 It happened that during the synthesis, *C* got the original compound from C_2. It was due to

a. heating of C_2
b. hydrolysis of C_2
c. oxidation of C_2
d. reaction of H_3PO_2 and C_2

Passage II

A local anaesthetic *P* has molecular formula $C_{13}H_{20}O_2N_2$. It is insoluble in water, dilute NaOH at room temperature but dissolves in dilute HCl. *P* upon treatment with $NaNO_2$ / HCl / 0° C and then with β-naphthol forms a highly coloured solid.

When *P* is boiled with aqueous NaOH, it dissolves slowly. The alkaline solution is shaken with ether and the layers are separated by extraction. Acidification of aqueous layer causes precipitation of a white solid *Q*, continued acidification causes *Q* to redissolve. Upon isolation, *Q* is found to have molecular formula $C_7H_7O_2N$.

Evaporation of ether layer leaves a liquid *R* of formula $C_6H_{15}ON$. *R* dissolves in water to give a solution that turned colour of moistened litmus paper blue. Treatment of *R* with acetic anhydride gives *S* of formula $C_8H_{17}O_2N$. *S* is insoluble in water and dilute NaOH but dissolves in dilute HCl. *R* is found to be identical with the compound formed by action of diethyl amine on ethylene oxide. Also *Q* on treatment with Br_2-H_2O forms a single monobromo isomer $C_7H_6O_2NBr$

29 What is the most likely structure of *P* ?

a.
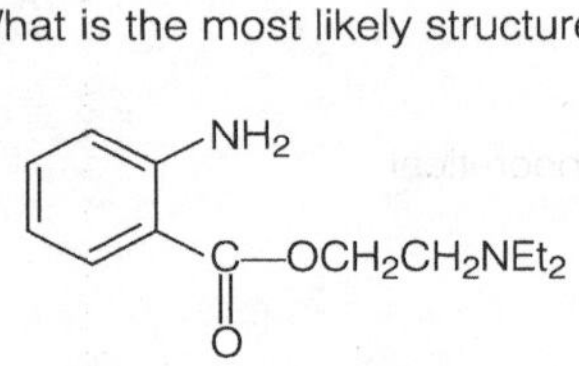

b.
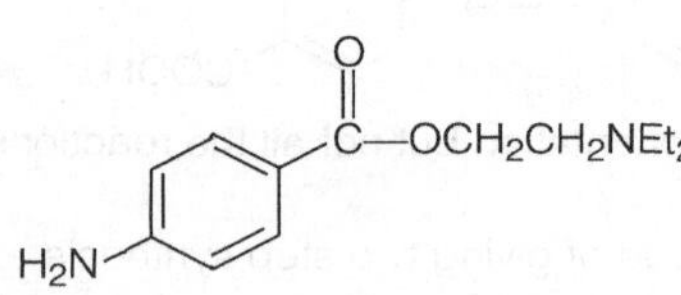

c.
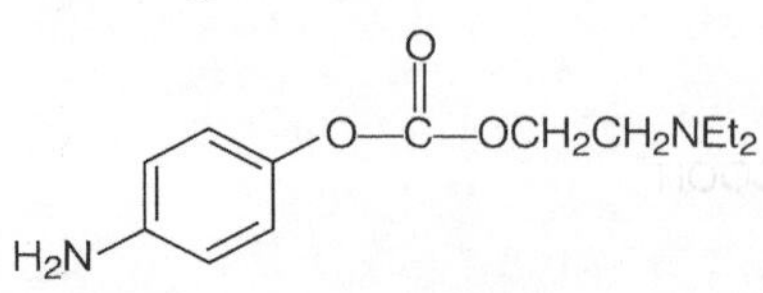

d.
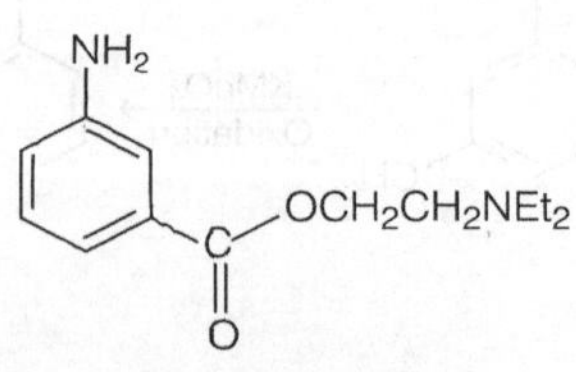

Working Space

30 Compound *Q* is

a. 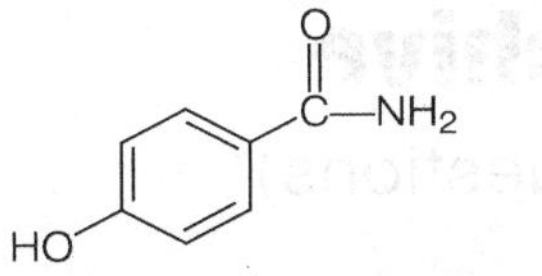

b. 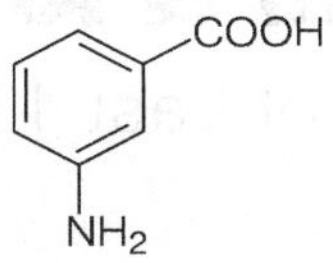

c. 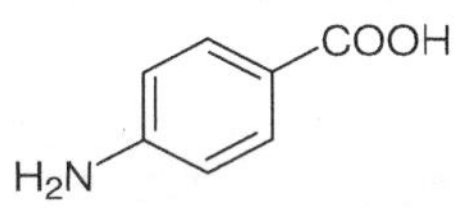

d. 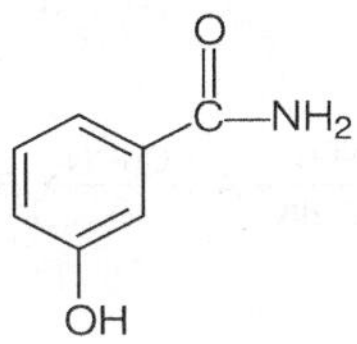

31 Compounds *R* and *S* are respectively

a. $Et_2N-CH_2CH_2OH$ and $Et_2N-CH_2CH_2O-\overset{O}{\overset{\|}{C}}-CH_3$

b. $Et_2N-CH_2CH_2OH$ and $Et_2N-CH_2CH_2COOCH_3$

c. $Et_2N-CH_2OCH_3$ and $Et_2N-CH_2OCH_2-\overset{O}{\overset{\|}{C}}-CH_3$

d. $Et_2N-\underset{OH}{\underset{|}{C}H}-CH_3$ and $Et_2N-\underset{CH_3}{\underset{|}{C}H}-O-\overset{O}{\overset{\|}{C}}-CH_3$

One Integer Value Correct Type

Direction *(Q. Nos. 32 and 33) This section contains 2 questions. When worked out will result in an integer from 0 to 9 (both inclusive).*

32 How many of the following bases are stronger than aniline?

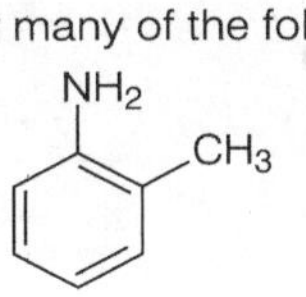 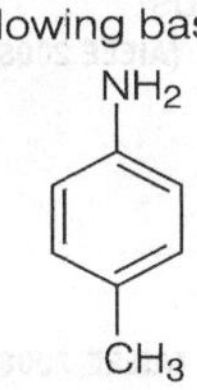 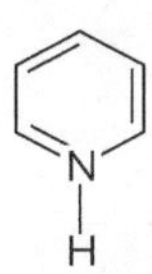 $CH_3-\overset{NH}{\overset{\|}{C}}-NH_2$ CH_3CONH_2 ;

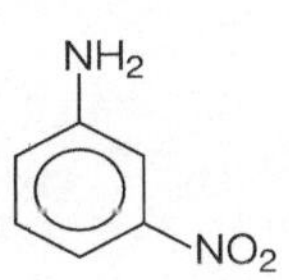 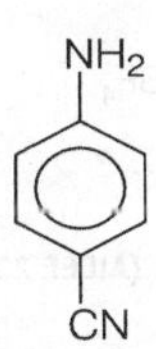 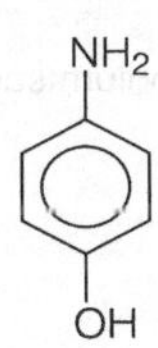

33 How many products are expected when isomeric phenylenediamines are diazotised?

JEE Main & AIEEE Archive

(Compilation of Last 13 Years' Questions)

1 In the chemical reaction,

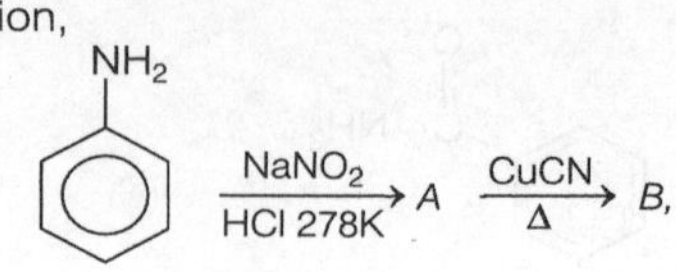

compounds *A* and *B* respectively are **(AIEEE 2011)**

a. fluorobenzene and phenol
b. benzene diazonium chloride and benzonitrile
c. nitrobenzene and chlorobenzene
d. phenol and bromobenzene

2 In the chemical reaction, **(AIEEE 2010)**

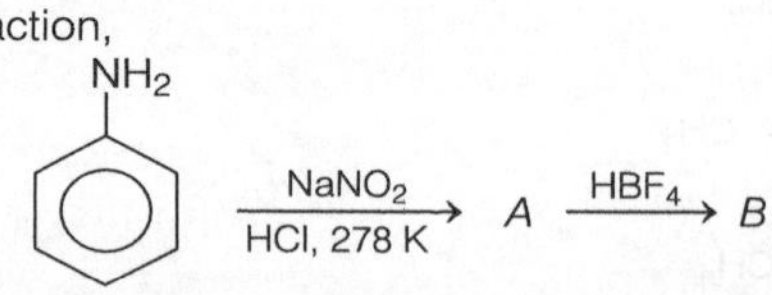

the compounds *A* and *B* respectively are

a. nitrobenzene and fluorobenzene
b. phenol and benzene
c. benzene diazonium chloride and fluorobenzene
d. nitrobenzene and chlorobenzene

3 Toluene is nitrated and the resulting product is reduced with tin and hydrochloric acid. The product so obtained is diazotised and then heated with cuprous bromide. The reaction mixture so formed contains **(AIEEE 2008)**

a. mixture of *o*-and *p*-bromotoluenes
b. mixture of *o*-and *p*-dibromobenzenes
c. mixture of *o*-and *p*-bromoanilines
d. mixture of *o*-and *m*-bromotoluenes

4 Fluorobenzene (C_6H_5F) can be synthesised in the laboratory **(AIEEE 2006)**

a. by heating phenol with HF and KF
b. from aniline by diazotisation followed by heating the diazonium salt with HBF_4
c. by direct fluorination of benzene with F_2 gas
d. by reacting bromobenzene with NaF solution

5 Amongst the following the most basic compound is **(AIEEE 2005)**

a. *p*-nitroaniline
b. acetanilide
c. aniline
d. benzylamine

6 Which of the following is the strongest base? **(AIEEE 2004)**

a. 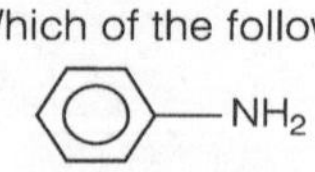

b. 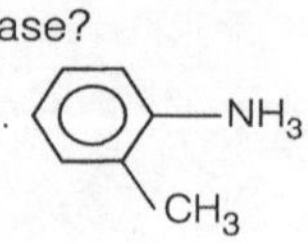

c. 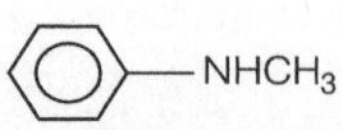

d. 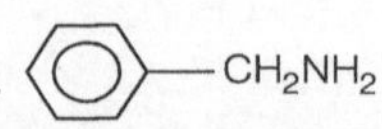

7 On heating benzyl amine with chloroform and ethanolic KOH, product obtained is **(AIEEE 2002)**

a. benzyl alcohol
b. benzaldehyde
c. benzonitrile
d. benzyl isocyanide

JEE Advanced & IIT JEE Archive

(Compilation of Last 10 Years' Questions)

1 In the reaction shown below, the major product(s) formed is/are

(2014 Adv., One or More than One Options Correct Type)

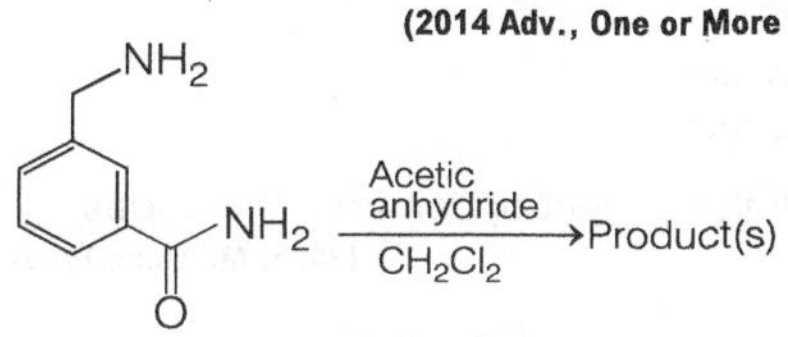

a. 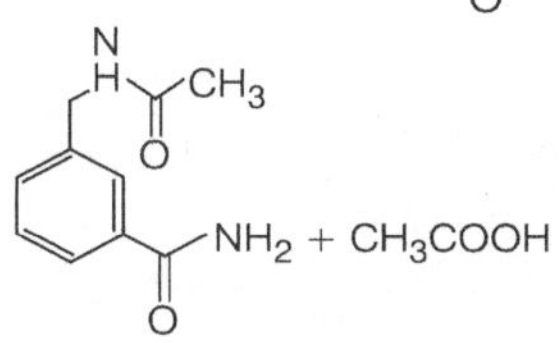

b. 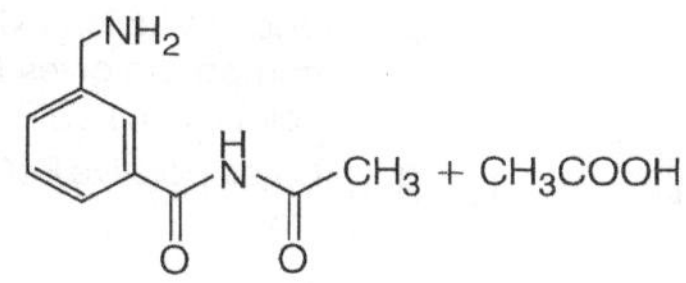

c. 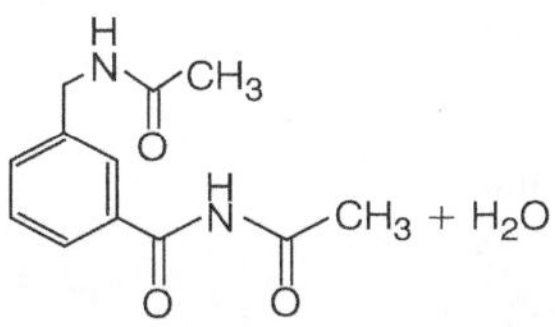

d. 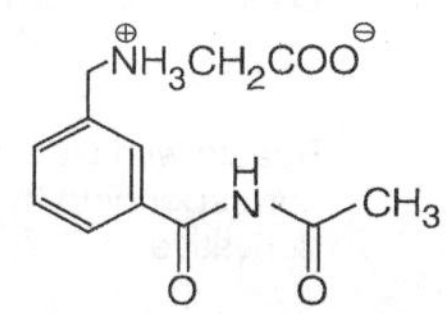

2 Amongst the compounds given, the one that would form a brilliant coloured dye on treatment with $NaNO_2$ in dil. HCl followed by addition to an alkaline solution of β-naphthol is

(2011, Only One Option Correct Type)

a. 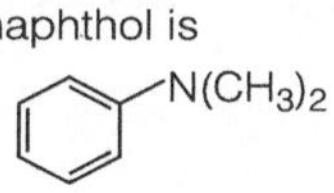

b. 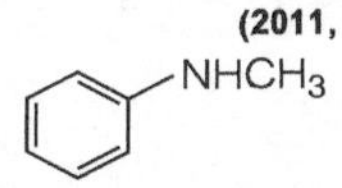

c. 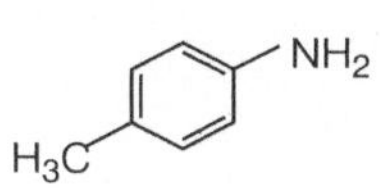

d. 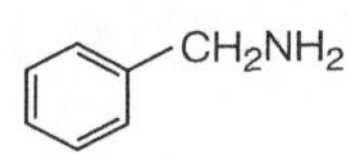

3 The major product of the following reaction is **(2011, Only One Option Correct Type)**

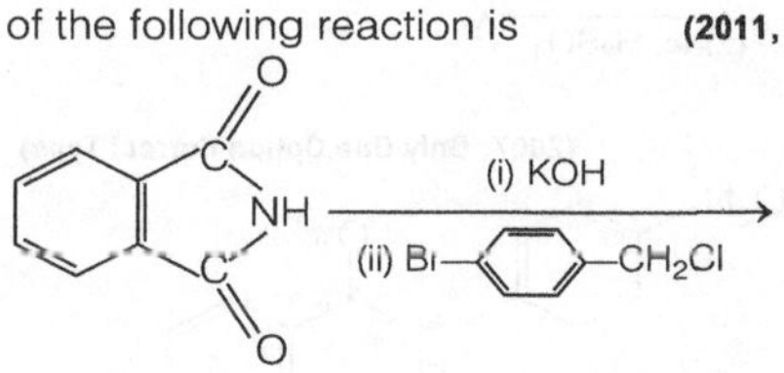

a. 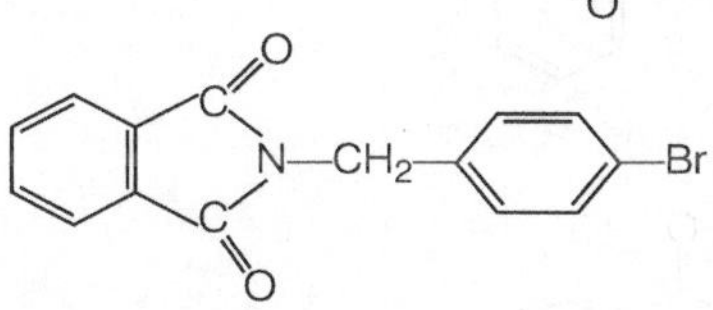

b. 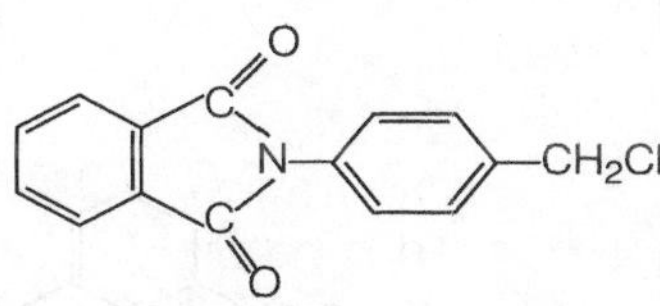

c. 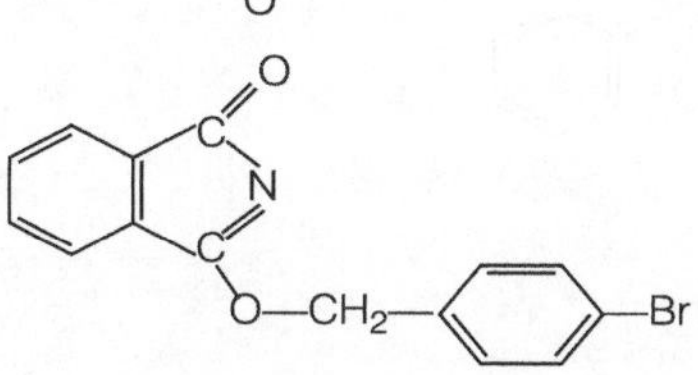

d. 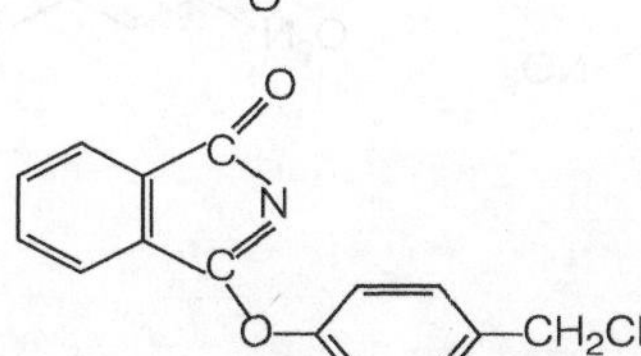

4 **Statement I** Aniline on reaction with $NaNO_2$ / HCl at 0°C followed by coupling with β-naphthol gives a dark blue coloured precipitate.

Statement II The colour of the compound formed in the reaction of aniline with $NaNO_2$ / HCl at 0°C followed by coupling with β-naphthol is due to the extended conjugation. **(2008, Statement Type)**

a. Both Statement I and Statement II are correct and Statement II is the correct explanation of Statement I
b. Both Statement I and Statement II are correct but Statement II is not the correct explanation of Statement I
c. Statement I is correct but Statement II is incorrect
d. Statement II is correct but Statement I is incorrect

5 Match the Column I with Column II and mark the correct option from the codes given below. **(2008, Matching Type)**

	Column I		Column II
i.	$H_2N—\overset{\oplus}{N}H_3\overset{\ominus}{Cl}$	p.	Sodium fusion extract of the compound gives Prussian blue colour with $FeSO_4$
ii.	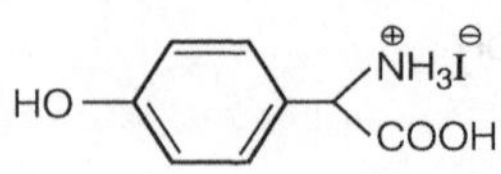	q.	Gives positive $FeCl_3$ test
iii.	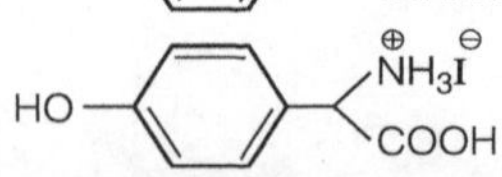	r.	Gives white precipitate with $AgNO_3$
iv.	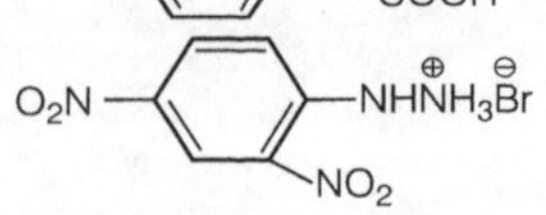	s.	Reacts with aldehydes to form the corresponding hydrazone derivative

Codes

	i	ii	iii	iv
a.	r	p,q	p	q,s
b.	r,s	p,q	p,q,r	p
c.	q,r	r,s	p	p
d.	q,s	q,r,s	p,q	p,q

6 In the following reaction,

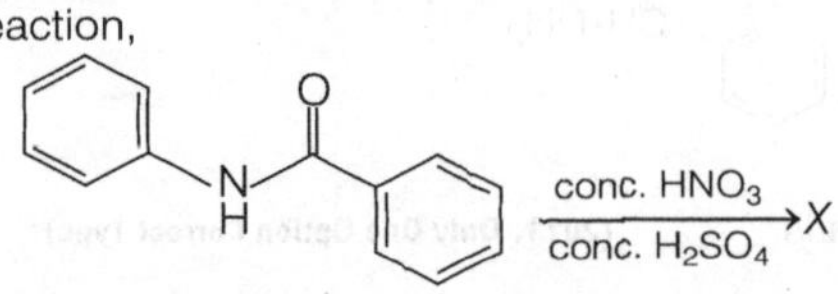

The structure of the major product *X* is **(2007, Only One Option Correct Type)**

a. 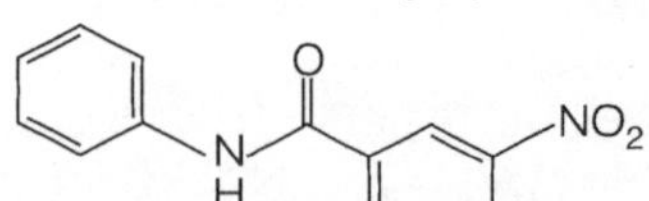

b. 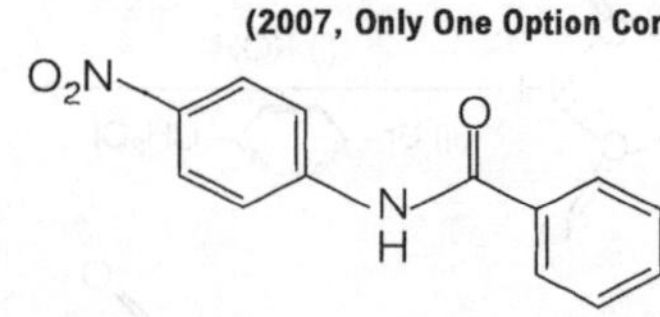

c. 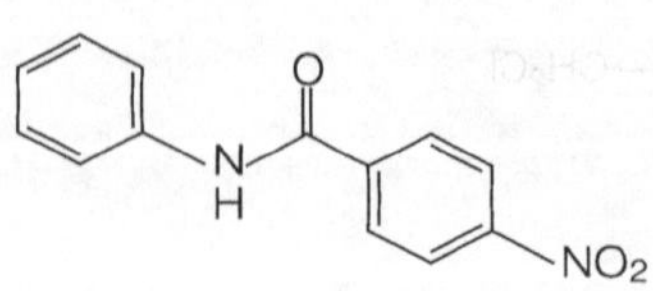

d. 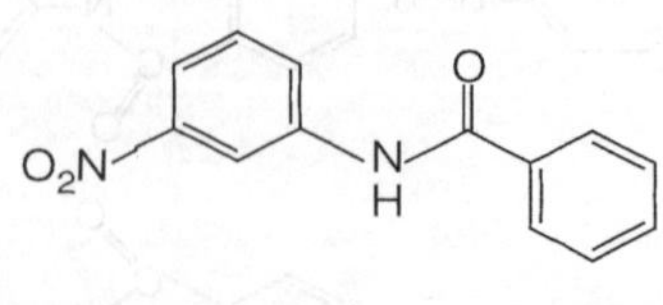

JEE Archive

5. Biomolecules

DPP

DAILY PRACTICE PROBLEMS

DPP-1 Introduction and Classification of Carbohydrates

Subjective Questions

Working Space

***Direction** (Q. Nos. 1-4) These questions are subjective in nature, need to be solved completely on notebook.*

1 i. Complete the following sentences by filling the appropriate words.

a. If two isomers have been classified correctly as anomers, they may also be called ....

b. A glycoside is the carbohydrate form of an ....

c. ...... is a disaccharide but does not undergo mutarotation.

d. In β-D-mannopyranose, only —OH at ...... is in the axial position.

e. In the formation of cyclic hemiacetal from 5-hydroxy hexane-2-one, the position of this chiral centre changes from ...... to ....

ii. Make a table with the Fischer, Haworth and chair conformation (when applicable) for each of the following sugars:

a. β-D-galactose

b. α-D-ribose

c. β-D-fructose

iii. Consider the given molecule.

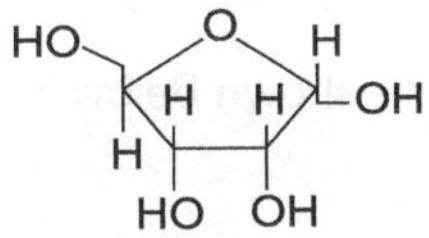

a. Is this sugar a ketose or an aldose?

b. Draw an arrow pointing to the anomeric carbon.

c. Box the OH group attached to C-4.

d. Is this the α or β-anomer?

2 i. Name the following compound using IUPAC rule and indicate the configuration *R* or *S* of each of the chirality centres:

a. D-glucose

b. L-glucose

c. D(–) ribose

ii. Draw the structure of the following compounds.

a. (2*S*), (3*S*), (4*R*), (5*R*)-6-pentahydroxyhexanal

b. 1-(3*S*), (4*R*), (5*R*)-6-pentahydroxy-2-hexanone

Working Space

3 i. Complete the following table for values of p, q, r, s, t, u, v and w (where $p, q \cdots$ = positions of carbon) with the help of hints given below:

Location	*p*	*q*	*r*	*s*	*t*	*u*	*v*	*w*
Value								

Hints

a. D-ribose is C-*p* epimer of D-xylose.
b. L-talose is C-*q* epimer of D-allose.
c. L-allose is C-r epimer of L-gulose
d. D-psicose is C -*s* epimer of D-fructose
e. D-mannose is C -*t* epimer of D-glucose
f. D-allose is C -*u* epimer of D-gulose
g. D-Indose is C -*v* epimer of D-glucose
h. D-galactose is C -*w* epimer of D-glucose

e.g. In (a), D-ribose is a C -3 epimer of D-xylose, so $p = 3$.

ii. Predict the number of stereogenic centre in the following and complete the following table.

A. 2-ketohexose B. An aldohexose C. A ketotriose

A	B	C

4 Write the following sugars in their open chain form. Name them and indicate whether each is a reducing or a non-reducing sugar with explanation.

i. 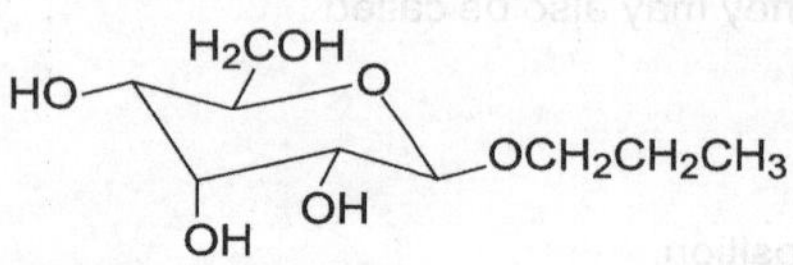

ii. 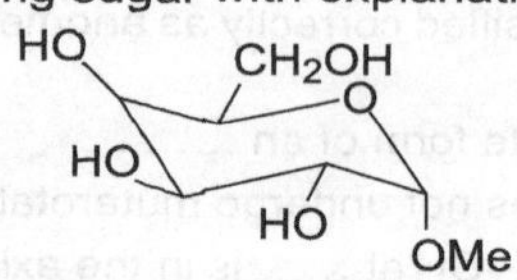

iii. 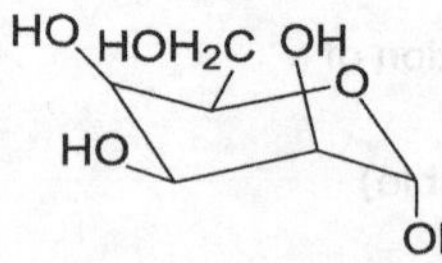

iv. 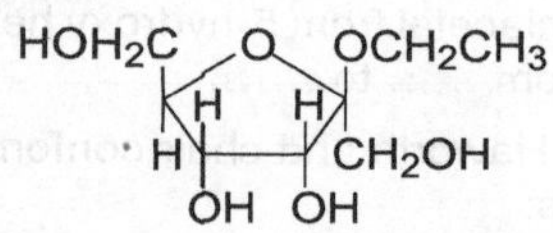

Only One Option Correct Type

Direction *(Q. Nos. 5-9) This section contains 5 multiple choice questions. Each question has four choices (a), (b), (c) and (d), out of which ONLY ONE is correct.*

5 Which set of terms correctly identifies the carbohydrate shown below?

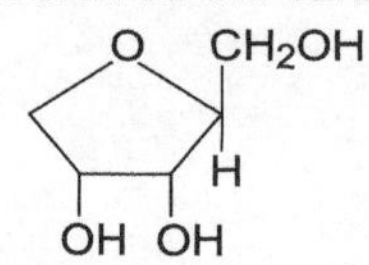

i. Pentose ii. Pentulose iii. Hexulose iv. Hexose
v. Aldose vi. Ketose vii. Pyranose viii. Furanose

a. (ii), (vi) and (viii)
b. (ii), (vi) and (vii)
c. (i), (v) and (viii)
d. A set of term other than these

6 Which of the following structure represents a β-ketohexafuranose?

a. 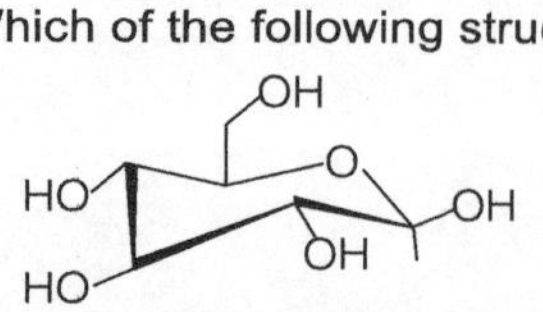

b. 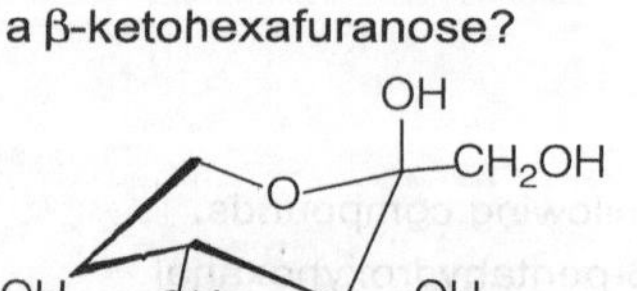

c. 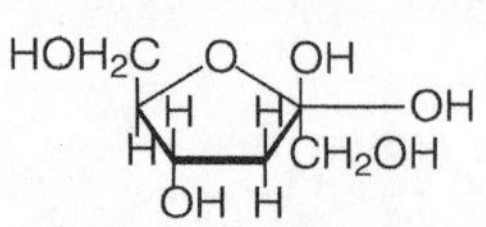

d. 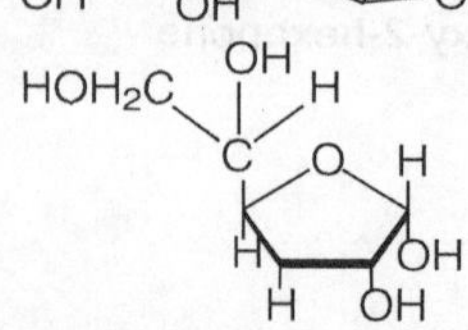

Working Space

7 Which one of the statement concerning the equilibrium shown is true?

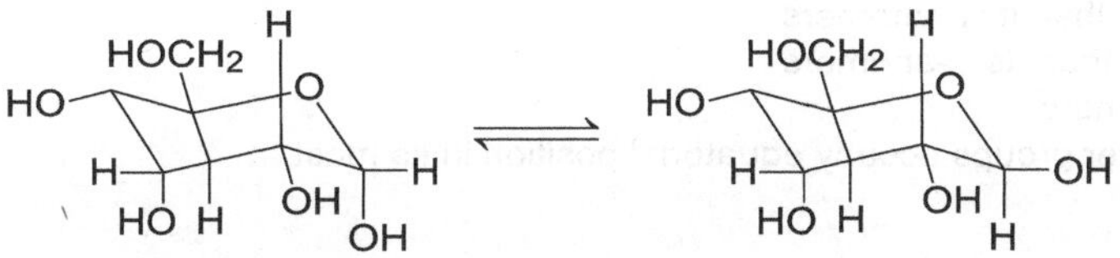

a. The two structures are enantiomers of each other. They have equal but opposite optical rotations and racemise slowly at room temperature

b. The two structures are enantiomers of each other. They racemise too rapidly at room temperature for their optical rotations to be measured

c. The two structures are diastereomers of each other. Their interconversion is called mutarotation

d. The two structures are diastereomers of each other. Their interconversion does not require breaking and making bonds, but involves only a change in conformation

8 Which of the following is the Fischer projection of L-threonine (also known as 2*S*, 3*R*-2-amino-3-hydroxy butanoic acid)?

a. 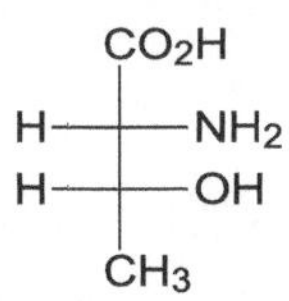 b. 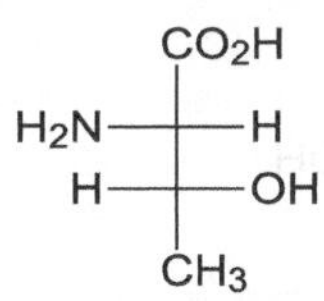 c. 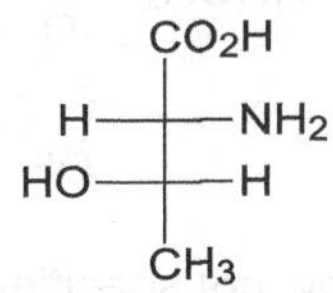 d. 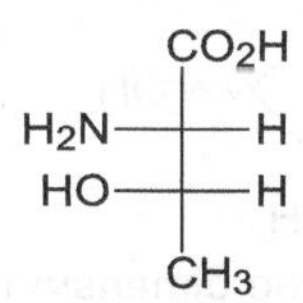

9 Identify the correct set of stereochemical relationships amongst the following monosaccharides I-IV.

I. 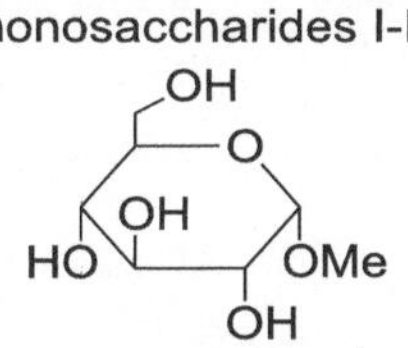 II. 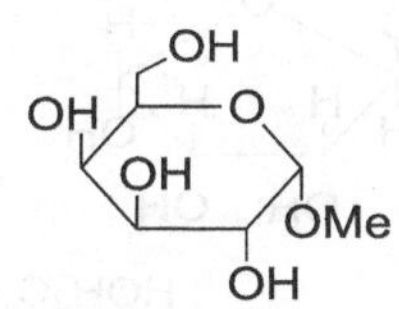

III. 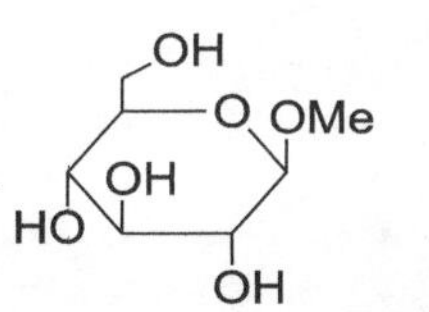 IV. 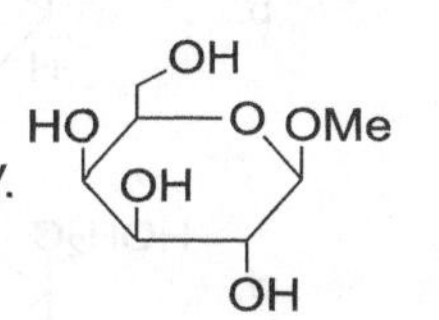

a. I and II are anomers; III and IV are epimers

b. I and III are epimers; II and IV are anomers

c. I and II are epimers; III and IV are anomers

d. I and III are anomers; I and II are epimers

One or More than One Options Correct Type

Direction (Q. Nos. 10 and 11) *This section contains 2 multiple choice questions. Each question has four choices (a), (b), (c) and (d), out of which ONE or MORE THAN ONE are correct.*

10 Which of the following compounds are related as epimers?

a. 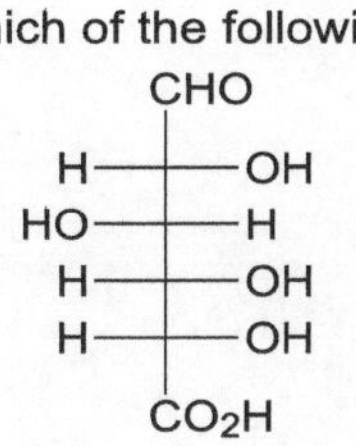 b. 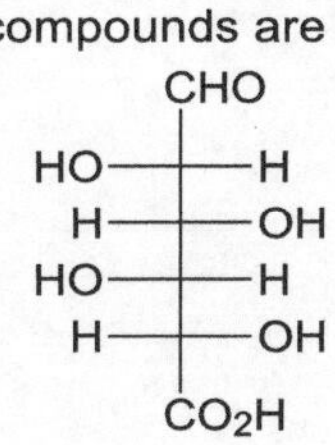 c. 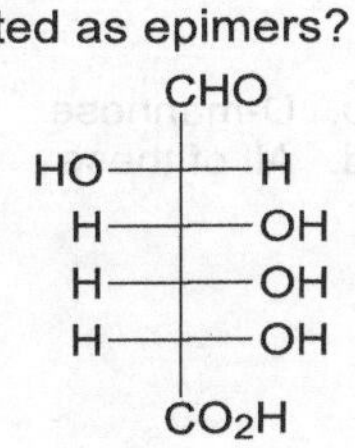 d. 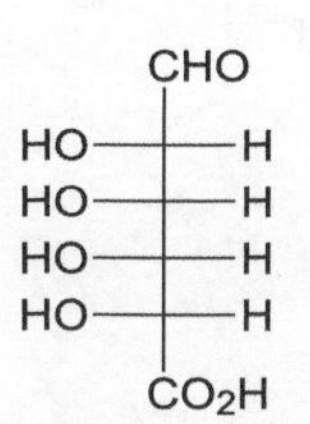

Working Space

11 What is/are true regarding the two anomers (α and β-forms) of D-(+)-glucose?

a. α-D(+) glucose is more stable than its β-anomers
b. β-D(+) glucose is more stable than its α-anomers
c. The two anomers are enantiomers
d. In β-D(+) glucose, all the bigger groups occupy equatorial position in its most stable chair conformation

Comprehension Type

***Direction** (Q. Nos. 12 and 13) This section contains a paragraph, describing theory, experiments, data, etc. Two questions related to the paragraph have been given. Each question has only one correct answer among the four given options (a), (b), (c) and (d).*

Passage

It is convenient at times to represent the cyclic structure of a monosaccharide without specifying whether the configuration of the anomeric carbon atom is α or β. When we do this, we shall use formula such as the following.

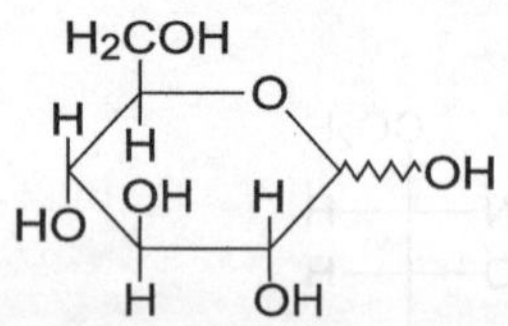

,

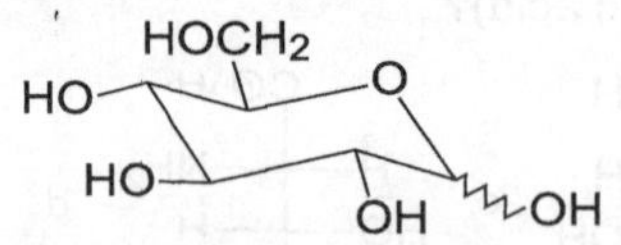

〰 indicates α or β (three dimensional view not specified).

12 Which of the following represent the anomer of the compound shown?

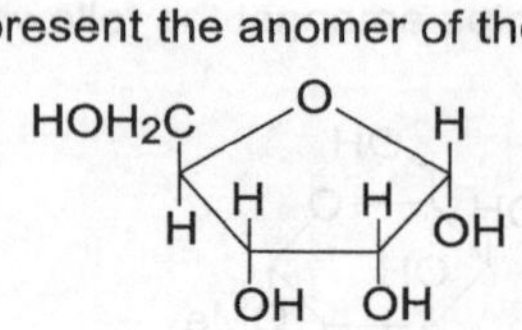

a.

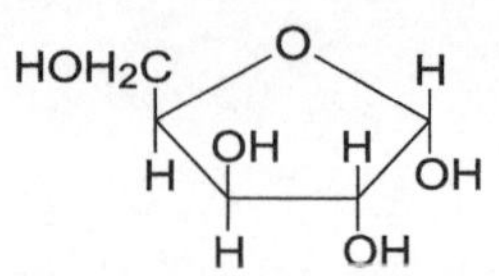

b.

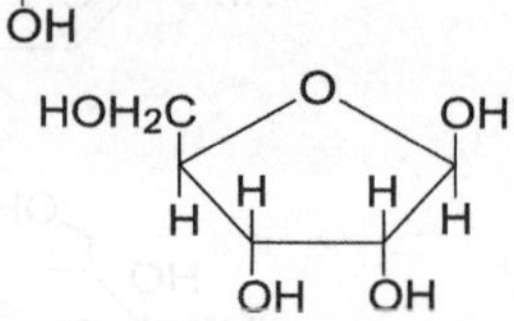

c.

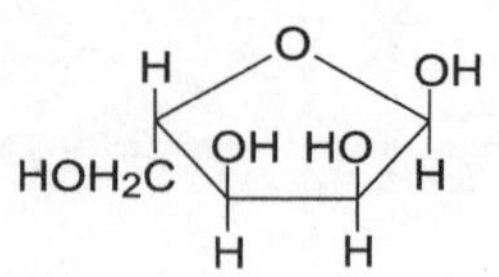

d.

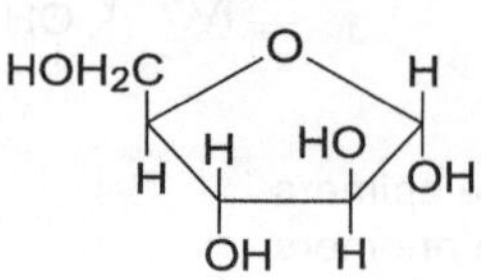

13 Consider the given figure

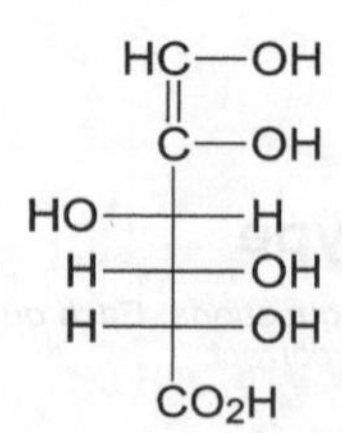

This is the enol form of

a. D-glucose
b. D-mannose
c. D-fructose
d. All of these

Working Space

Statement Type

Direction (Q. Nos. 14 and 15) *This section is based on Statement I and Statement II. Select the correct answer from the codes given below.*

a. Both Statement I and Statement II are correct and Statement II is the correct explanation of Statement I
b. Both Statement I and Statement II are correct but Statement II is not the correct explanation of Statement I
c. Statement I is correct but Statement II is incorrect
d. Statement II is correct but Statement I is incorrect

14 **Statement I** β-D-glucopyranose is the most abundant naturally occurring aldohexose.

Statement II All the ring substituents in the chair conformation are equatorial.

15 **Statement I** D-fructose is used for sweetening cold drinks but not hot ones.

Statement II The sweet form of fructose is fructopyranose. An increase in temperature causes a shift in the pyranose $\rightleftharpoons$ furanose equilibrium towards the less sweet furanose form.

Matching List Type

Direction (Q. No. 16) *Choices for the correct combination of elements from Column I and Column II are given as options (a), (b), (c) and (d), out of which one is correct.*

16 Match the Column I and Column II and mark the correct option from the codes given below.

	Column I (Pair of molecules)		Column II (Characteristics)
i.	Glucose - fructose	p.	Anomers
ii.	Fructose -mannose	q.	Diastereomers
iii.	Glucose - mannose	r.	Functional isomers
iv.	α-D-glucopyranose β-D-glucopyranose	s.	Lobry-De-Bruyn Alberada van Ekenstein transformation

Codes

	i	ii	iii	iv		i	ii	iii	iv
a.	s	p,q	s	r	b.	r,s	r,s	q,s	p,q
c.	p,q	p,s	q,s	r	d.	q	r	r,s	p

One Integer Value Correct Type

Direction (Q. Nos. 17 and 18) *This section contains 2 questions. When worked out will result in an integer from 0 to 9 (both inclusive).*

17 What is the sum of locations of axial hydroxy group in β-D-iodopyranose?

18 Some of the pair of monosaccharides are given below:

I. Allose, atrose
II. Glucose, fructose
III. Glucose, mannose
IV. Mannose, fructose
V. Gulose, indose
VI. Galactose, talose
VII. Galactose, glucose
VIII. Indose, glucose

How many of the above pairs are C-2 epimers?

DPP-2 Preparation and Properties of Monosaccharides

Subjective Questions

Working Space

Direction (Q. Nos. 1-5) *These questions are subjective in nature, need to be solved completely on notebook.*

1 i. Write the equations to show how D(+) glucose could be converted into

a. methyl-β-D-glucoside,
b. methyl-β-2, 3, 4, 6-tetra-O-methyl-D-glucoside,
c. 2, 3, 4, 6-tetra-O-methyl-D-glucose,
d. D-mannose,
e. hexa-O-acetyl-D-glucitol?

ii. Convert D-glucopyranose into 2,3,4-trimethyl glucopyranoside.

2 i. Complete the following reactions.

a. $\text{D-glucose} \xrightarrow[\text{synthesis}]{\text{Kiliani}} \underset{A}{\text{Pair}} \xrightarrow{HNO_3} \underset{B}{\textit{meso}\text{-heptaldaric acid}}$

b. $\text{D-allose} \xrightarrow[\text{synthesis}]{\text{Kiliani}} \underset{C}{\text{Pair}} \xrightarrow{HNO_3} \underset{D}{\textit{meso}\text{-heptaldaric acid}}$

Explain whether the acids *B* and *D* are same or different? Which pair out of *A* and *C* gives *meso*-acids *B* and *D*?

ii. Identify the following.

a. An aldohexose other than D-glucose that is oxidised to D-glucaric acid by HNO_3.
b. An another pair of aldohexose that are oxidised to identical aldaric acid.

3 i. Identify the following sugars.

a. An aldopentose that is not D-arabinose, forms D-arabinitol when it is reduced with $NaBH_4$.
b. A sugar forms the same osazone as D-galactose with phenyl hydrazine, but it is not oxidised by an aqueous solution of Br_2.
c. A sugar that is not D-altrose but forms D-altraric acid when it reacts with nitric acid.

ii. Among the first eight aldopentoses,

a. which are enantiomers?
b. which give identical osazones?

4 When heated to 100°C, D-iodose undergoes a reversible dehydration and exists primarily as 1, 6-anhydro-D-iodopyranose.

i. Draw D-iodose in its pyranose form, showing the more stable chair conformation of the ring.
ii. Which is more stable anomer, α-D-iodopyranose or β-D-iodopyranose?
iii. When heated to 100°C under the same conditions as those used for D-iodose, D-glucose does not undergo dehydration. Explain.

Working Space

5 Complete the following reactions.

i. α-D-glucofuranose $\xrightarrow{\text{Acetone}/H_2SO_4}$ *A* $\xrightarrow[\text{(ii) } H^{\oplus}]{\text{(i) } PhCH_2Cl/\bar{O}H}$ *B*

ii. β-D-glucofuranose $\xrightarrow{\text{Acetone}/H^+}$ *C* $\xrightarrow[\text{(ii) } H^{\oplus}]{\text{(i) } PhCH_2Cl/\bar{O}H}$ *D*

iii. α-D-glucopyranose $\xrightarrow{\text{Acetone}/H^+}$ *E* $\xrightarrow[\text{(ii) } H^{\oplus}]{\text{(i) } PhCH_2Cl/\bar{O}H}$ *F*

iv. β-D-glucopyranose $\xrightarrow{\text{Acetone}/H^+}$ *G*

Only One Option Correct Type

Direction (Q. Nos. 6-11) *This section contains 6 multiple choice questions. Each question has four choices (a), (b), (c) and (d), out of which ONLY ONE is correct.*

6 Observe the following laboratory tests for glucose pentacetate and mention +ve or –ve from the codes given below:

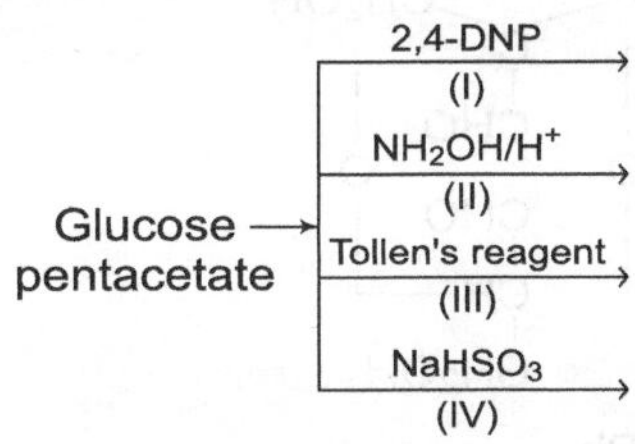

a. + + + +
b. – – – –
c. + – + –
d. + + – –

7 Compound *A*, $C_5H_{10}O_5$, gives a tetraacetate with $(CH_3\overset{O}{\overset{\|}{C}})_2O$ and oxidation of *A* with Br_2-H_2O gives an acid, $C_5H_{10}O_6$. Reduction of *A* with HI gives *iso*-pentane. There are two possible structures for compound *A* which can be distinguished by using

a. HIO_4
b. excess of P, HI
c. phenyl hydrazine
d. Tollen's reagent

8 Which two of the following compounds are reduced to the same chiral alditol by sodium borohydride?

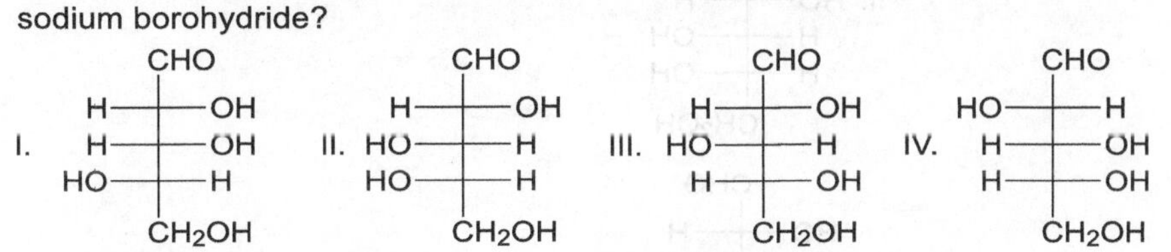

a. I and II
b. II and III
c. III and IV
d. II and IV

9 D-glucose $\xrightarrow[HCl]{CH_3CH_2SH}$ $\xrightarrow[NaOH]{(CH_3)_2SO_4}$ $\xrightarrow[CdCO_3]{HgCl_2}$ *X*. Here the product *X* is

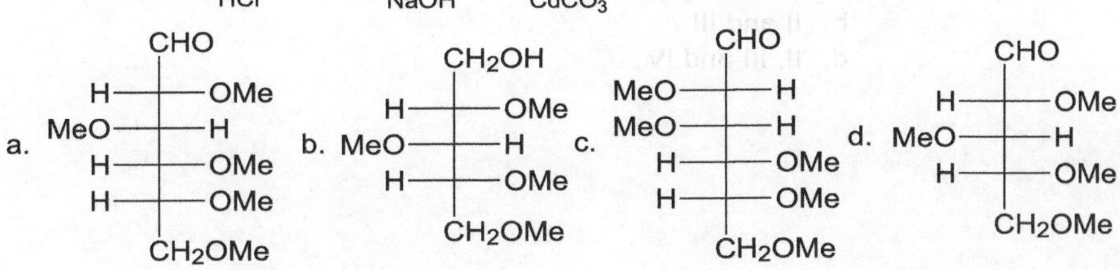

Working Space

10 Mycarose is a rare $C_7H_{14}O_4$ sugar found in some antibiotic natural products. Mycarose gives a positive Tollen's test, does not form an osazone derivative and on exhaustive HIO_4 oxidation gives HCO_2H, CH_3CHO and CH_3COCH_2CHO. Which of the following structures might be mycarose?

a. 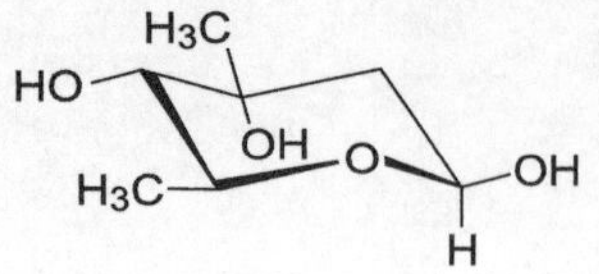

b. 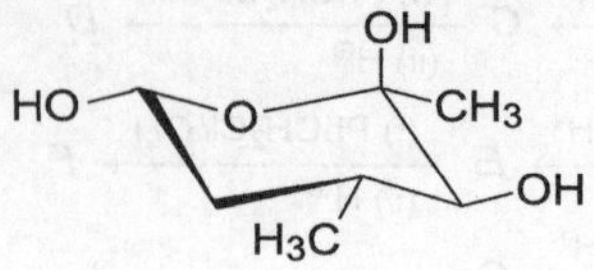

c. 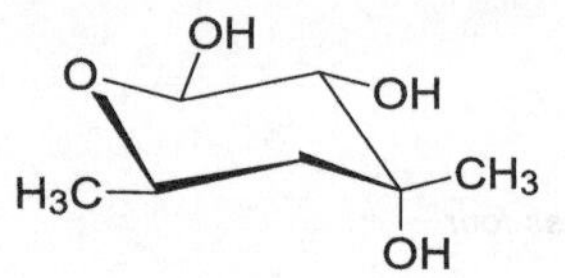

d. 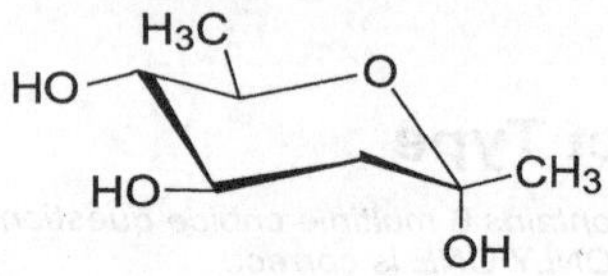

11

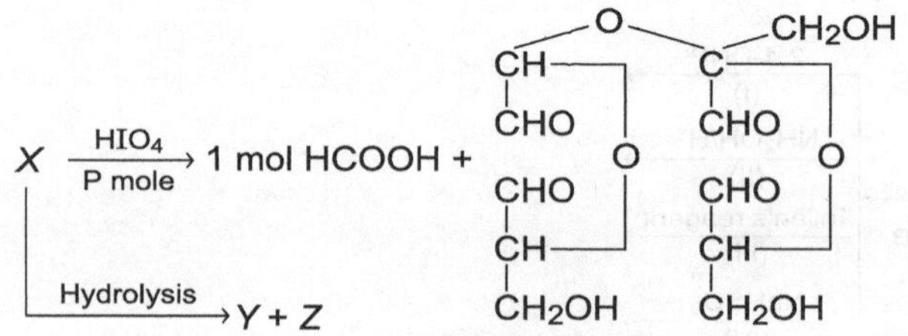

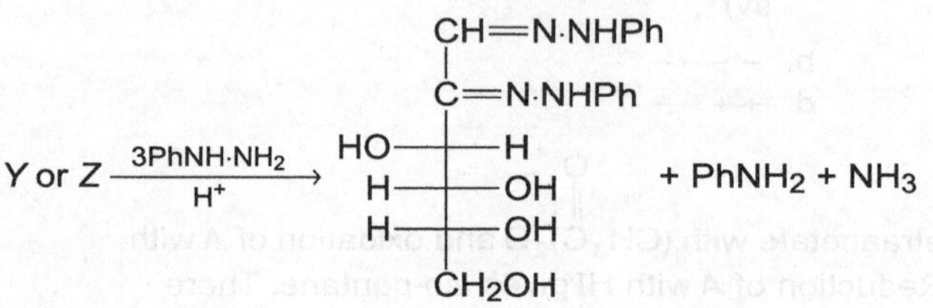

Compounds *Y* and *Z* can be

I. 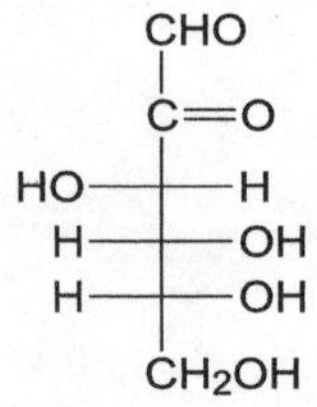

II. 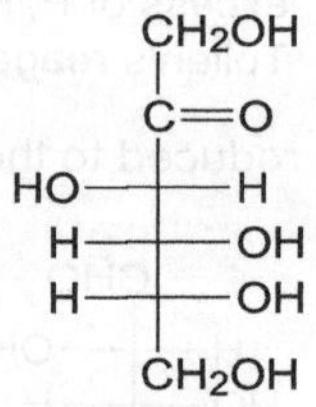

III. 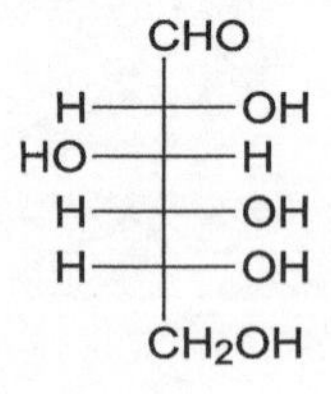

IV. 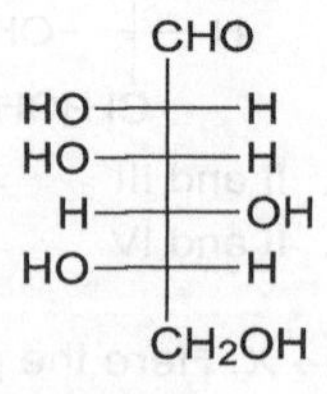

a. Only I
b. II and III
c. I and IV
d. II, III and IV

One or More than One Options Correct Type

Direction (Q. Nos. 12-15) *This section contains 4 multiple choice questions. Each question has four choices (a), (b), (c) and (d), out of which ONE or MORE THAN ONE are correct.*

Working Space

12 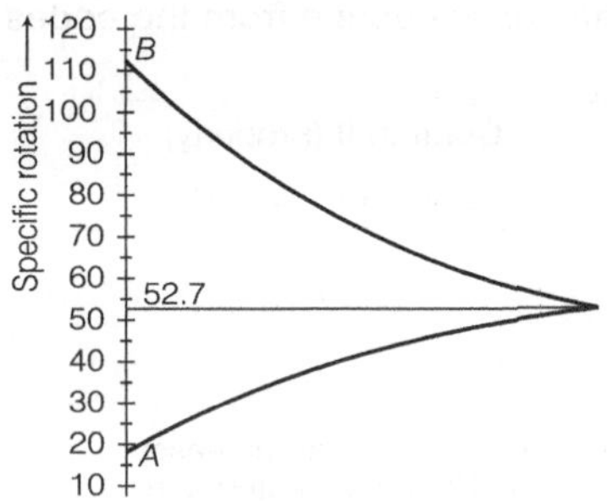

The percentage of A at equilibrium is

a. 36% b. 64%
c. 74% d. 28%

13 D-glucose and D-fructose both form the same osazone. Which statements is/are correct about the above reaction?

a. D-glucose and D-fructose are epimers
b. D-glucose and D-fructose are anomers
c. The configurations of the —OH group at C-3 and C-4 in glucose and fructose are same
d. The configuration of the —OH group at C-4 and C-5 in glucose and fructose are same

14 What is/are true regarding epimers and epimerisation?

a. Epimers are a pair of diastereomers
b. A compound can have more than two epimers
c. Epimers rotate the plane polarised light in opposite directions
d. An aldose on treatment with Br_2-H_2O followed by workup with pyridine in the presence of acid catalyst forms a pair of epimeric aldonic acid

15 α or β-D-glucose on treatment with methanol in the presence of dry HCl gives the same pair of methyl D-glucoside as shown below:

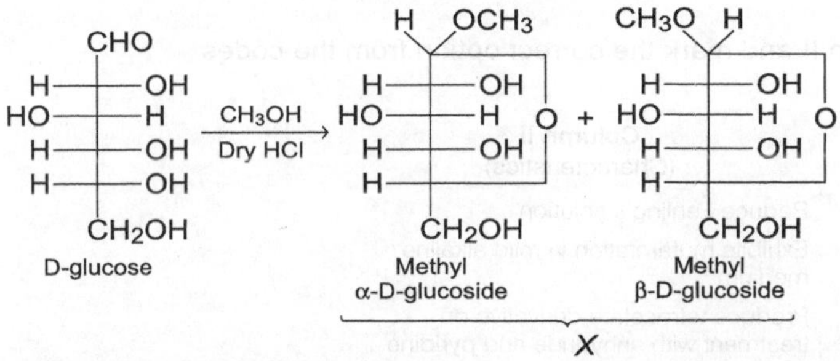

Which of the following statements is/are true regarding the above reaction?

a. X do not reduce Fehling's solution
b. X do not react with HCN or NH_2OH
c. A planar carbocation is involved in the formation of X
d. Behaviour of X as stated in (a) and (b) indicate the presence of free —CHO group

Matching List Type

***Direction** (Q. Nos. 16 and 17) Choices for the correct combination of elements from Column I and Column II are given as options (a), (b), (c) and (d), out of which one is correct.*

16 Match the Column I and Column II and mark the correct option from the codes given below.

	Column I (Compound)		**Column II** (Property)
i.	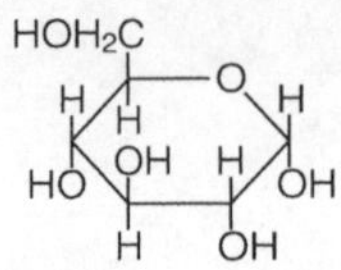	p.	Will form osazone
ii.	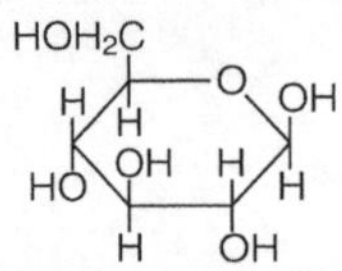	q.	Molecular mass increases by 120 when treated with Ac_2O.
iii.	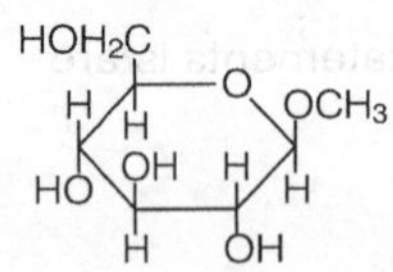	r.	Reduces Tollen's reagent
iv.	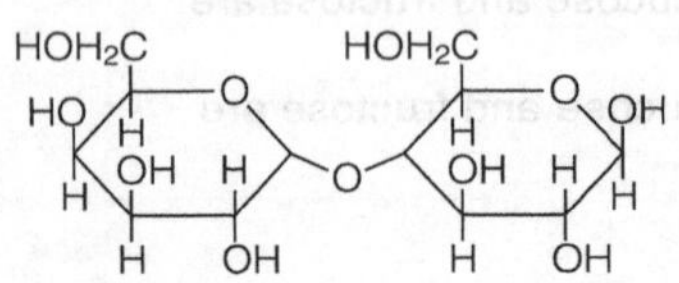	s.	Is α-D-glucopyranose
		t.	Non-reducing sugar

Codes

	i	ii	iii	iv
a.	p,q,r,s	p,q,r	t	p,r
b.	p,q,r,s	p,s	q,t	p,r
c.	p,q,s	p,q,r	p	p,r
d.	p,s	s,t	r	q,t

17 Match the Column I with Column II and mark the correct option from the codes given below.

	Column I (Molecule)		Column II (Characteristics)
i.	Glucose	p.	Reduce Fehling's solution
ii.	Fructose	q.	Exhibits mutarotation in mild alkaline medium
iii.	Mannose	r.	Produce tetracetate derivative on treatment with anhydride and pyridine
iv.	Glucopyranoside	s.	Gets oxidised by Br_2, H_2O

Codes

	i	ii	iii	iv
a.	p,q,s	p,q	p,q,s	r
b.	q,s	q,r	p,s	p,q
c.	p,q	p,r	q,s	q
d.	p	p,s	q,p	r,s

Working Space

Comprehension Type

Direction (Q. Nos. 18-20) This section contains a paragraph, describing theory, experiments, data, etc. Three questions related to the paragraph have been given. Each question has only one or more than one correct answer among the four given options (a), (b), (c) and (d).

Passage

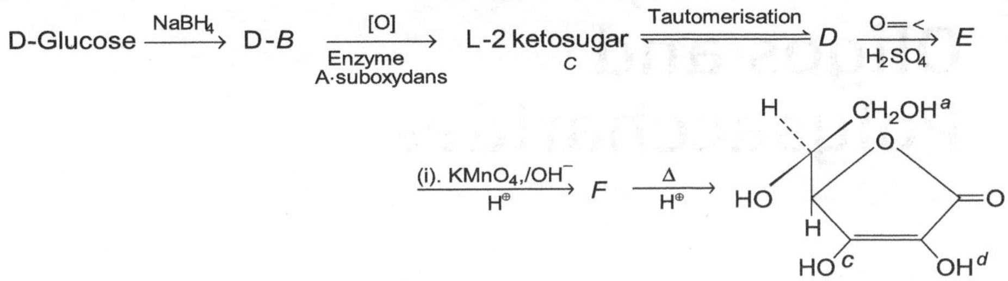

(G) Vitamin C

18 Which of the following is the most acidic hydrogen in case of *G*?

a. H^a b. H^b
c. H^c d. H^d

19 If the lactose formation in compound *G* takes place between C-1 and C-5, then the structure of *G* can be represented as

a. 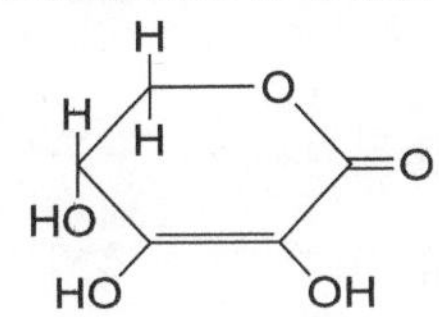

b. 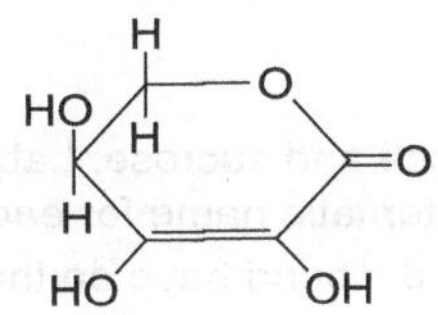

c. 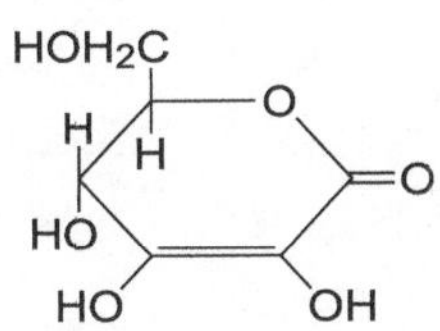

d. 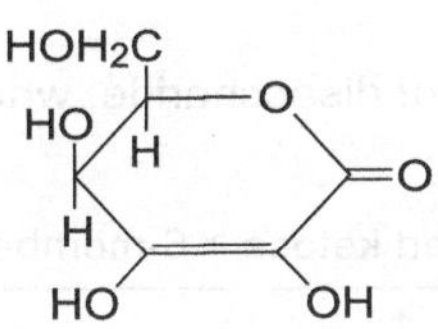

20 Which statement(s) is/are correct about *E*?

a. Both cyclic ketals have five membered ring
b. Both cyclic ketals have six membered ring
c. Cyclic ketals formed between C-2 and C-3 have five membered ring
d. Cyclic ketals formed between C-4 and C-6 have six membered ring

One Integer Value Correct Type

Direction (Q. Nos. 21 and 22) This section contains 2 questions. When worked out will result in an integer from 0 to 9 (both inclusive).

21 How many of the following undergo mutarotation?
Amylose, starch, glucose, maltose, cellulose, fructose, galactose.

22 I. One mole of D-3-deoxy glucose reacts with *x* mole of $PhNHNH_2$ to give osazone.
II. One mole of D-fructose reacts with *y* mole of $PhNHNH_2$ to give osazone.
III. One mole of D-2-deoxy glucose reacts with z moles of $PhNHNH_2$ to give osazone.
The sum of $x + y + z$ is

DPP-3 Some Important Oligos and Polysaccharides

Subjective Questions

Working Space

Direction *(Q. Nos. 1-4) These questions are subjective in nature, need to be solved completely on notebook.*

1 i. Write the types of glycosidic bonds found in each of the following carbohydrates.

a. Glycogen

b. Chitin

c. Amylopectin

d. Cellulose

ii. Draw the structures of maltose, lactose and sucrose. Label the location and type of glycosidic bond and provide a systematic name for each disaccharides.

iii. What impact does the type of glycosidic bond have on the structure of glycogen and cellulose?

2 Give the structures and IUPAC name of disaccharide, which gives the reaction given below:

Disaccharide *A* $\xrightarrow{H_3O^+}$ 5-membered ketose + 6-membered aldose

(Hydrolyse both by maltase or emulsin)

Disaccharide *A* $\xrightarrow{\text{Methylation and hydrolysis}}$

2, 3, 4, 6- tetra-O-methyl-D-glucopyrane (*B*)
+
1, 3, 4, 6-tetra-O-methyl -D- fructofuranose (*C*)

Given, *A* does not reduce Fehling's solution and does not mutarotate.

3 A disaccharide *X* forms silver mirror with Tollen's reagent and is hydrolysed by a β-glycosidase. This disaccharide on treatment with CH_3I (excess)/Ag_2O followed by hydrolysis (in H^+) gives 2, 3, 4 tri-O-methyl mannose and 2, 3, 4, 6-tetra-O-methyl galactose. Identify *X* and also mention the role of Ag_2O.

4 Gentiobiose, a rare disaccharide found in saffron and gentian, is a reducing sugar and forms only D-glucose on hydrolysis with aqueous acid. Reaction of gentiobiose with iodomethane and Ag_2O yields an octamethyl derivative, which on hydrolysis gives 1.0 equivalent of each 2, 3, 4, 6- tetra-O-methyl-D-glucopyranose and 2, 3, 4- tri-O-methyl -D-glucopyranose. If β-glucoside linkage is found in the given disaccharide then predict its structure.

Working Space

Only One Option Correct Type

Direction (Q. Nos. 5-8) This section contains 4 multiple choice questions. Each question has four choices (a), (b), (c) and (d), out of which ONLY ONE is correct.

5 Which of the following molecules fulfills the two requirements stated below?

1. α-(1, 2) - glycosidic linkage
2. Not a reducing sugar

a. 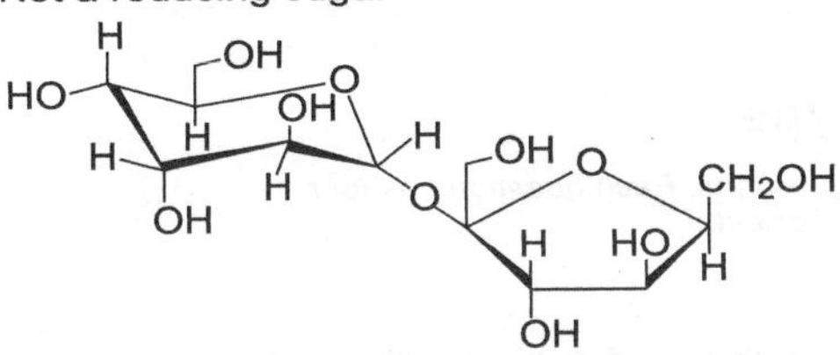

b. 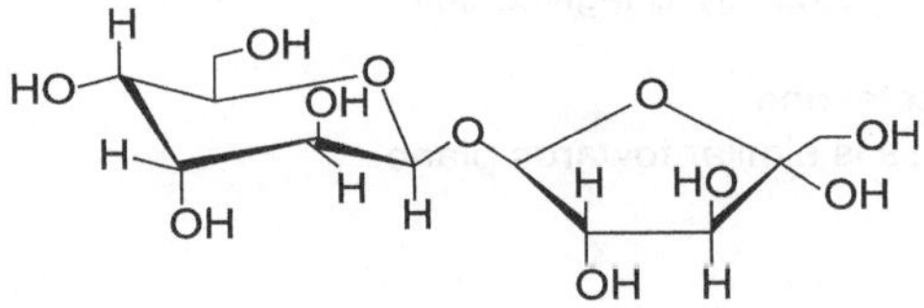

c. 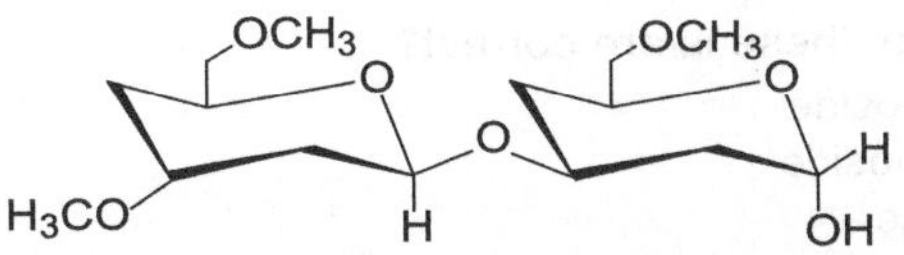

d. 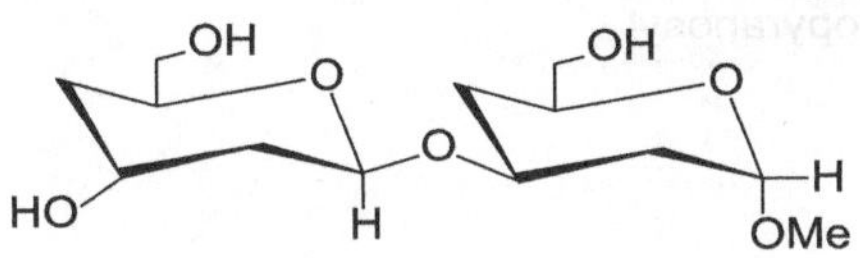

6 How many aldose, ketose, furanose, pyranose and hemiacetal units are present respectively in the following molecule?

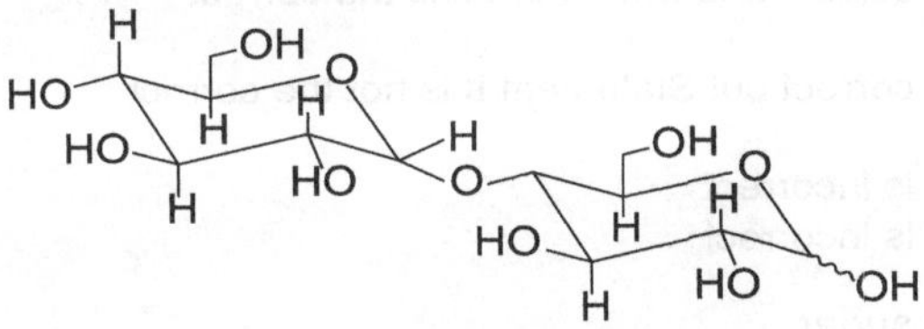

a. 0, 2, 0, 2, 1
b. 1, 1, 1, 1, 2
c. 2, 0, 0, 2, 1
d. 2, 0, 0, 2, 2

7 Select the correct statement about the following disaccharide.

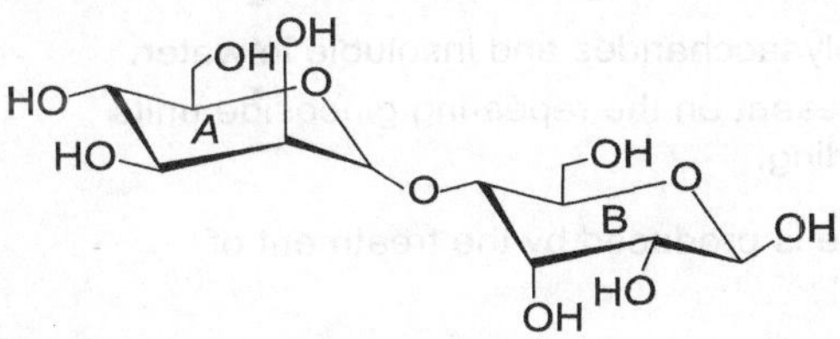

a. Ring *A* is an acetal in the β configuration; ring *B* is a hemiacetal in the α-configuration
b. Ring *A* is a hemiacetal in the α-configuration; ring *B* is an acetal in the β-configuration
c. Ring *A* is an acetal in the α configuration; ring *B* is a hemiacetal in the β-configuration
d. Ring *A* is a hemiacetal in the β configuration; ring *B* is an acetal in the α-configuration

Working Space

8 The best description for the polysaccharides amylose A and cellulose B is

a. $A = 1, 4 - O - \beta$- linked poly D-glucose
$B = 1, 4 - O - \alpha$- linked poly D-glucose

b. $A = 1, 4 - O - \alpha$- linked poly D-glucose
$B = 1, 4 - O - \beta$- linked poly D-glucose

c. $A = B = 1, 4 - O - \beta$- linked poly D-glucose

d. $A = B = 1, 4 - O - \alpha$- linked poly D-glucose

One or More than One Options Correct Type

Direction *(Q. Nos. 9 and 10) This section contains 2 multiple choice questions. Each question has four choices (a), (b), (c) and (d), out of which ONE or MORE THAN ONE are correct.*

9 Which of the following statements is/are true?

a. Sucrose has a pyranose and a furanose ring hooked together by 1, 1′-glycosidic linkage

b. Maltose has one glycosidic linkage and a free hemiacetal end

c. Behaviour of hydrolysis product of maltose and sucrose is similar towards plane polarised light

d. Sucrose is a dexterorotatory sugar

10 Consider the following statements about lactose. Which of these is/are correct?

a. Its IUPAC name is β-D-galactopyranosyl-β-D-glucopyranoside

b. Its IUPAC name is β-D- glucopyranosyl β-D galactopyranoside

c. C_1—β OH of galactose is linked with C_4—OH of β-glucose

d. On methylation with MeOH/HCl, it gives methyl β-D-galactopyranosyl β-D-glucopyranoside

Statement Type

Direction *(Q. Nos. 11-13) This section is based on Statement I and Statement II. Select the correct answer from the codes given below.*

a. Both Statement I and Statement II are correct and Statement II is the correct explanation of Statement I

b. Both Statement I and Statement II are correct but Statement II is not the correct explanation of Statement I

c. Statement I is correct but Statement II is incorrect

d. Statement II is correct but Statement I is incorrect

11 **Statement I** Sucrose is also called invert sugar.

Statement II D(+) sucrose produces one mole of each D(+) glucose and D(−) fructose when subjected to hydrolysis and specific rotation of D(−) fructose is more.

12 **Statement I** Cellulose, starch, etc., are polysaccharides and insoluble in water.

Statement II In cellulose, —OH groups present on the repeating glucoside units are linked through intramolecular H-bonding.

13 **Statement I** 2, 3, 6- tri-O-methyl D-glucose is produced by the treatment of cellulose with CH_3I(excess), Ag_2O and H^+.

Statement II Cellulose is a polymer of D-glucose.

Working Space

Matching List Type

Direction (Q. Nos. 14 and 15) *Choices for the correct combination of elements from Column I and Column II are given as options (a), (b), (c) and (d), out of which one is correct.*

14 Match the Column I with Column II and mark the correct option from the codes given below.

	Column I (Compound)		Column II (Enzyme)
i.	Sucrose	p.	Maltase
ii.	Maltose	q.	Emulsin
iii.	Lactose	r.	Amylase
iv.	Starch	s.	Lactase

Codes

	i	ii	iii	iv
a.	p,q,r,s	p,r	q,s	p,r
b.	p,q,r,s	p,s	q,s	p,r,s
c.	q,s	p,r	p	q,r,s
d.	q,r	p,q,r		p,q,r,s

15 Match the columns and choose the correct answer using the codes given below.

	Column I (Compound)		Column II (Linkage)
i.	Sucrose	p.	$(C_1 - \alpha)$ of glucose $\rightarrow C_4$ of glucose
ii.	Maltose	q.	$(C_1 - \alpha)$ of glucose $\rightarrow C_4$ of a glucose and $(C_1 - \alpha)$ of glucose $\rightarrow C_6$ of glucose in another chain
iii.	Lactose	r.	$(C_1 - \beta)$ of glucose $\rightarrow C_4$ of glucose
iv.	Starch	s.	$(C_1 - \beta)$ of galactose $\rightarrow C_4$ of glucose
		t.	$(C_1 - \alpha)$ of glucose $\rightarrow (C_2 - \beta)$ of fructose

Codes

	i	ii	iii	iv
a.	t	p	s	p,q
b.	t	p	s	q
c.	p	q	r	s
d.	t	p	s	p

One Integer Value Correct Type

Direction (Q. No. 16) *This section contains 1 question. When worked out will result in an integer from 0 to 9 (both inclusive).*

16 Consider an amylose chain of 4000 glucose units. How many cleavage are required to lower the average length to 400 units?

DPP-4 Amino Acids

Subjective Questions

Direction (Q. Nos. 1-5) *These questions are subjective in nature, need to be solved completely on notebook.*

Working Space

1 i. Write the IUPAC names of the following amino acids.

a. Glycine b. Alanine c. Serine d. Cysteine

ii. Draw their structures stereochemically similar to L-glyceraldehyde.

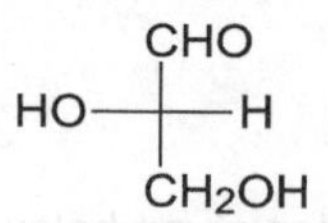

L-glyceraldehyde

(Also assign *R* and *S*-configuration)

iii. Classify them as chiral or achiral.

2 i. Draw the structure of predominant form of

a. valine at pH 11
b. proline at pH 2
c. arginine at pH 7
d. glutamic acid at pH 7
e. a mixture of alanine, lysine and aspartic acid at pH = 6 and at pH = 3.

ii. Draw the form in which each of the following amino acids predominantly exists at physiological pH.

a. Arginine
b. Tyrosine
c. Aspartic acid
d. Glutamine
e. Lysine

iii. What is formed at the end point when glycine is titrated with

a. *aq.* NaOH? b. *aq.* HCl?

3 i. Calculate the *pI* of each of the following amino acids.

a. Asparagine, pK_a [α-COOH = 2.02, α-$\overset{+}{N}H_3$ = 8.84]

b. Arginine, pK_a [α-COOH = 2.17, α-$\overset{+}{N}H_3$ = 9.04, side chain = 12.48]

c. $CH_3-\underset{\underset{\overset{+}{N}H_3}{|}}{CH}-CO_2H$ [pK_a = 2.2, pK_b = 4.4]

d. Serine, pK_a [α-COOH = 2.21, α-$\overset{+}{N}H_3$ = 9.15]

e. $HO\overset{\overset{O}{\|}}{C}-CH_2-CH_2-\underset{\underset{\overset{+}{N}H_3}{|}}{CH}-CO_2H$

pK_a [α-COOH = 2, α-$\overset{+}{N}H_3$ = 9, γ-COOH = 4]

ii. Which amino acid-glycine or methionine has a greater negative charge at pH = 6.201?

Working Space

4 Show the products of these reactions :

i. Proline $\xrightarrow[\text{Excess}]{(CH_3CO)_2O}$

ii. $CH_3CH_2\underset{\displaystyle CH_3}{\underset{|}{C}}HCHO + NH_3 + HCN \xrightarrow{H_2O} \xrightarrow{H_3O^+}$

iii. Isovaleric acid + $Br_2 \xrightarrow{PBr_3} \xrightarrow[\text{excess}]{NH_3}$

iv. $CH_3CH{=}CH_2 \xrightarrow[\text{(ii) } KCN/H_3O^+]{\text{(i) NBS}} \xrightarrow[\text{(ii) } P/Br_2 \text{ (iii) } NH_3/\Delta]{\text{(i) } Pt/H_2}$

v. Lysine $\xrightarrow{\text{Acetic anhydride}}$

5 i. Convert alanine into

a. alanine *iso*-propyl ester

b. N-benzoylalanine

c. *t*-butyloxy carboxylalanine

ii. Write the nature of amino acids (α, β or γ-substituted), which on heating gives

a. $CH_3CH{=}CHCOOH$

b.

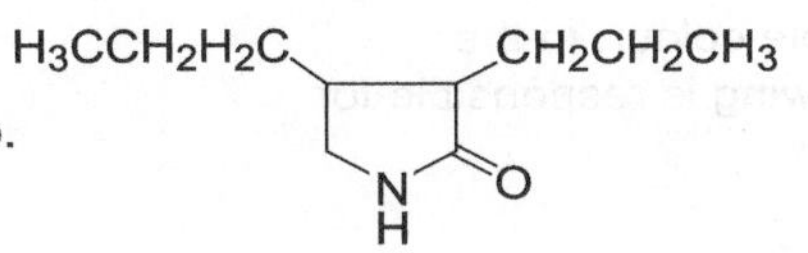

c.

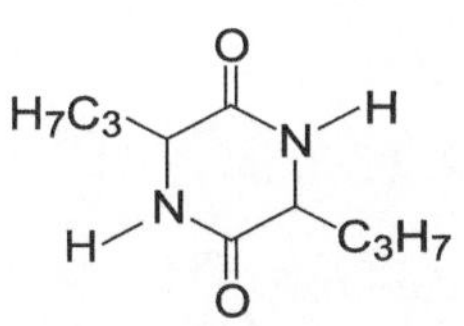

Only One Option Correct Type

Direction *(Q. Nos. 6-12) This section contains 7 multiple choice questions. Each question has four choices (a), (b), (c) and (d), out of which ONLY ONE is correct.*

6 A mixture contains three amino acids A(pI = 3.2), B(pI = 5.7) and C(pI = 9.7). When it is subjected to electrophoresis at pH 7.7, in which direction will each component of the mixture move?

a. A to anode, B and C to cathode

b. A to anode, B no movement, C to cathode

c. A to cathode, B no movement, C to anode

d. A and B to anode and C to cathode

7 Consider the following compounds.

I. Glycine

II. Sulphanilic acid

III. Anthranilic acid

Which can form dipolar (Zwitter) ion?

a. I, II and III   b. I and II

c. II and III   d. I and III

Working Space

8 What is the structure of product formed, when glycine is heated?

a. 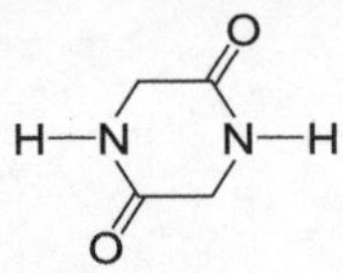

b. 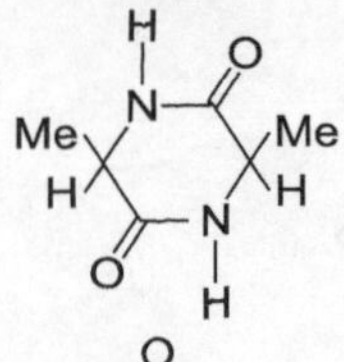

c 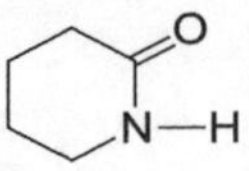

d. 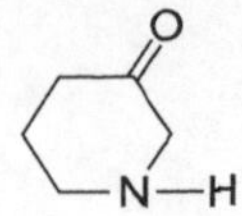

9 The Strecker synthesis of α-amino acids begins with the reaction of an aldehyde with *X*/*Y* followed by the treatment with *Z*. The functional group which is hydrolysed during this reaction is

a. an ester
b. a nitrile
c. an amide
d. an imine derivative

10 All the common amino acids, except the one, react with cold nitrous acid (HNO_2) and evolve nitrogen gas. Which of the following amino acid is that compound?

a. Cysteine
b. Proline
c. Histidine
d. Arginine

11 Ninhydrin reagent reacts with α-amino acids to give a purple colour. In the reaction of ninhydrin with phenylalanine, which of the following is responsible for this colour?

a. 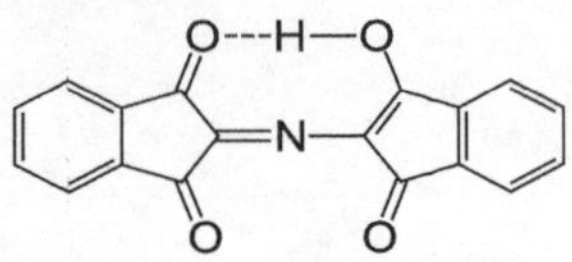

b. 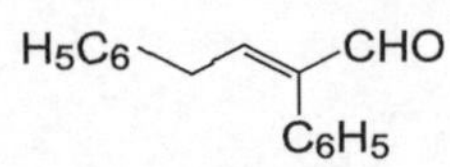

c. 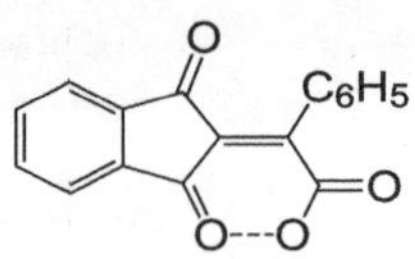

d. 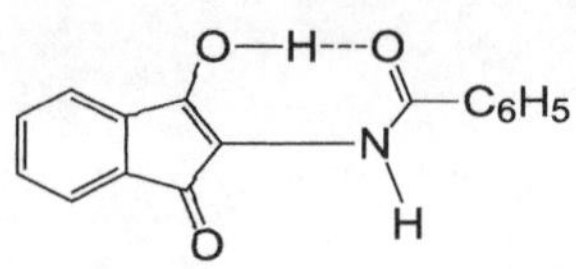

12 Identify *X*, *Y* and *Z* from the information given below:

X = A chiral amino acid which loses its chirality on treatment with lithium aluminium hydride.

Y = The decarboxylation product of α-amino (secondary) acid, $C_5H_9O_2N$.

Z = It is formed when $A(C_3H_4O_3)$ is treated with (i) NH_3/(ii) $NaBH_4$ [*A* forms oxime, gives iodoform test and liberate CO_2 with $NaHCO_3$].

	X	*Y*	*Z*
a.	Alanine	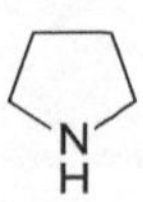	Serine
b.	Serine	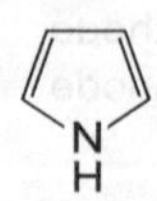	Alanine
c.	Aspartic acid		Serine
d.	Serine	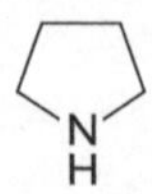	Alanine

Working Space

One or More than One Options Correct Type

Direction (Q. Nos. 13 and 14) *This section contains 2 multiple choice questions. Each question has four choices (a), (b), (c) and (d), out of which ONE or MORE THAN ONE are correct.*

13 In which of the following pairs hydrophobic interactions do not exist?

a. Leucine and alanine
b. Arginine and glutamic acid
c. Glycine and asparagine
d. Glutamic acid and serine

14 In case of which amino acid(s), the isoelectric pH value is very low (i.e. less than 5)?

a. Histidine
b. Aspartic acid
c. Glutamic acid
d. Arginine

Statement Type

Direction (Q. Nos. 15) *This section is based on Statement I and Statement II. Select the correct answer from the codes given below.*

a. Both Statement I and Statement II are correct and Statement II is the correct explanation of Statement I
b. Both Statement I and Statement II are correct but Statement II is not the correct explanation of Statement I
c. Statement I is correct but Statement II is incorrect
d. Statement II is correct but Statement I is incorrect

15 **Statement I** Glycine, as well as other amino acids, are amphoteric.

Statement II The acidic functional group is the ammonium ions ($—\overset{+}{N}H_3$) and the basic functional group is the carboxylate ion ($—CO_2^-$).

Comprehension Type

Direction (Q. Nos. 16 and 17) *This section contains a paragraph, describing theory, experiments, data, etc. Two questions related to the paragraph have been given. Each question has only one correct answer among the four given options (a), (b), (c) and (d).*

Passage

α-amino acids are generally represented by the formula $R—\underset{\underset{NH_2}{|}}{C}H—COOH$. These polymerise and form an amide linkage (peptide linkage) between two monomeric amino acid units. The polymerisation takes place in the following manner.

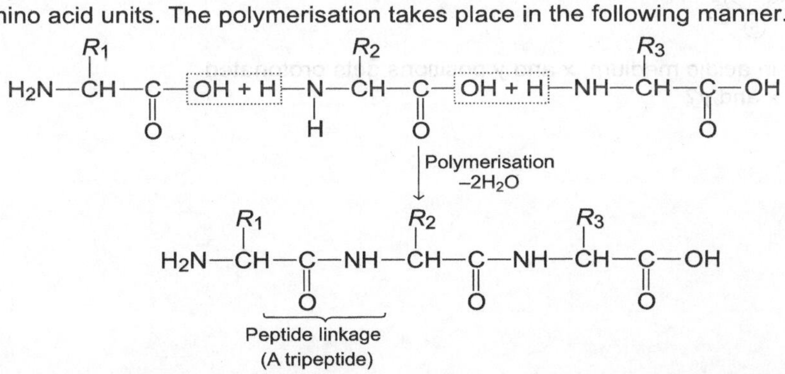

16 In the above trimer, if $R_1 = H$; $R_2 = CH_3$ and $R_3 = CH_2Ph$, then total number of optically active stereoisomers will be

a. 8
b. 6
c. 4
d. 2

Working Space

17 Which statement is incorrect about the given trimer?

a. Liberation of CO_2 with sodium bicarbonate
b. Liberation of N_2 with nitrous acid
c. Formation of yellow ppt with 2, 4-dinitrophenyl hydrazine
d. Rotation of plane polarised light

Matching List Type

Direction (Q. No. 18) *Choices for the correct combination of elements from Column I and Column II are given as options (a), (b), (c) and (d), out of which one is correct.*

18 Match Column I with Column II and mark the correct option from the codes given below.

	Column I (Compound)		Column II (Nature)
i.	Alanine	p.	Chiral
ii.	Cysteine	q.	Neutral
iii.	Aspartic acid	r.	Acidic
iv.	Histidine	s.	Basic

Codes

	i	ii	iii	iv		i	ii	iii	iv
a.	p,q	p,q	p,r	p,s	b.	p,q	p	p,s	p,r
c.	p,q	s	p,r	q	d.	p,s	p	p,q	p,r

One Integer Value Correct Type

Direction (Q. Nos. 19 and 20) *This section contains 2 questions. When worked out will result in an integer from 0 to 9 (both inclusive).*

19 How many of the following amino acids can be synthesised in the human body? Proline, lysine, tyrosine, arginine, valine, histidine, phenylalanine, alanine, tryptophan.

20 Consider the given structure,

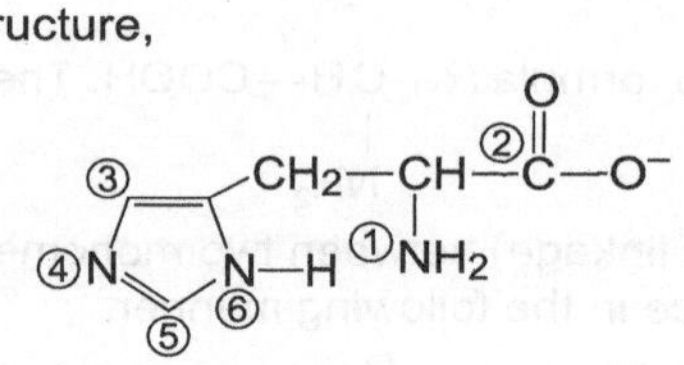

When this molecule is kept in acidic medium, x and y positions gets protonated readily. What is the sum of x and y?

DPP-5 Peptides and Proteins

Subjective Question

Working Space

Direction *(Q. No. 1) This question is subjective in nature, need to be solved completely on notebook.*

1 Consider the following compound.

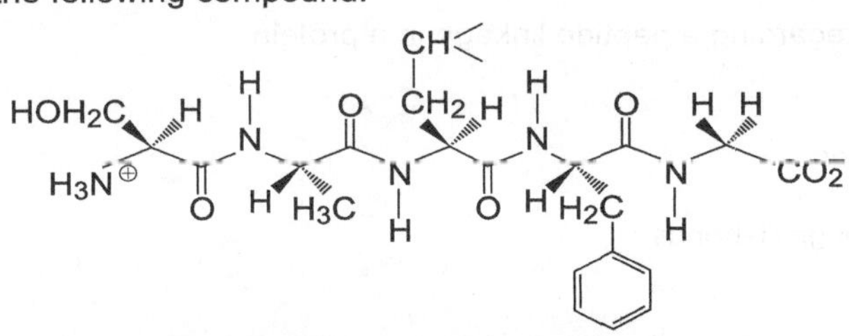

i. What kind of peptide does this structure represent?
ii. How many peptide linkages are present in it?
iii. Write the names of C-terminal and N-terminal amino acids.
iv. Give the sequence of amino acids (using three letters abbreviations).

Only One Option Correct Type

Direction *(Q. Nos. 2-5) This section contains 4 multiple choice questions. Each question has four choices (a), (b), (c) and (d), out of which ONLY ONE option is correct.*

2 Which of the following is the correct structure of the tripeptide gly-pha-ala?

a. 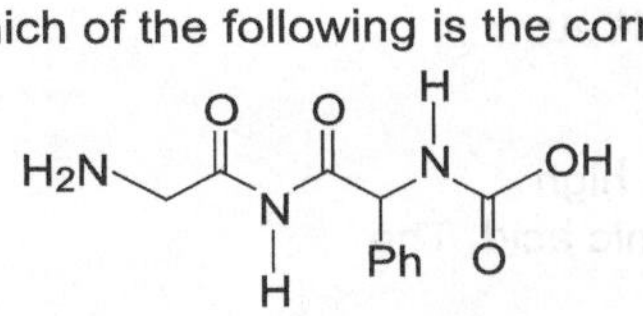

b. 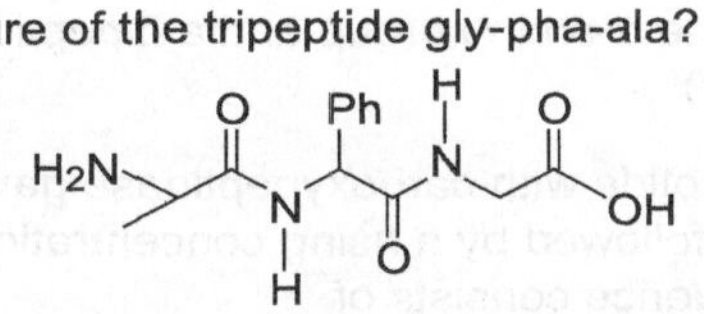

c. 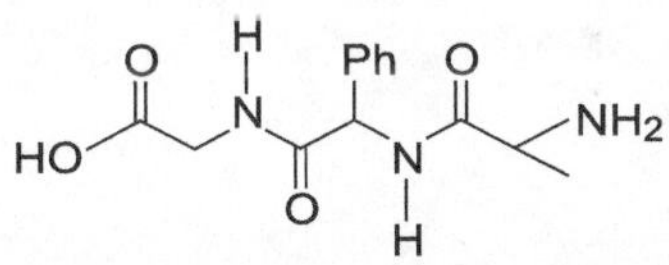

d. 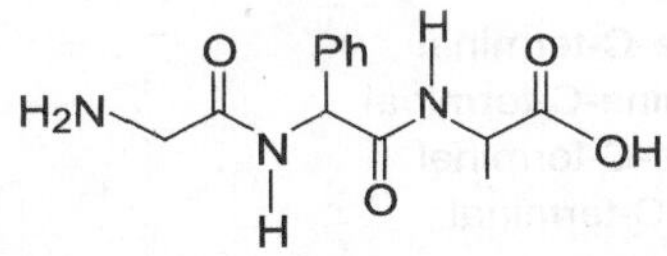

3 Peptides are formed by the joining of amino acids through amide linkages. Which of the following statement is not true in this respect?
a. Amide groups are more resistant to hydrolysis than the similar ester groups
b. $p\pi$ resonance stabilises the amide linkages
c. Amide groups do not contribute in the hydrogen bonding interactions
d. Stable conformations of peptides are restricted to those having planar amide groups

4 Which of the following is not considered a part of the tertiary structure of proteins?
a. The hydrophobic effect
b. The restricted rotation in amide bonds
c. The electrostatic interactions
d. The hydrogen bonds

Working Space

5 The most important contribution to the stability of a protein conformation appears to be the

a. entropy increase from the decrease in ordered water molecules forming a solvent shell around it
b. maximum entropy increase from ionic interactions between the ionised amino acids in a protein
c. sum of free energies of formation of many weak interactions among the hundreds of amino acids in a protein
d. sum of free energies of formation of many weak interactions between its polar amino acids and surrounding water molecules

One or More than One Options Correct Type

Direction *(Q. No. 6) This section contains 1 multiple choice question. It has four choices (a), (b), (c) and (d), out of which ONE or MORE THAN ONE are correct.*

6 Which of the following statements regarding a peptide linkage in a protein molecule is/are correct?

a. It is an amide linkage
b. It has partial double bond character
c. It is hydrophilic in nature
d. It connects protein molecules through H -bonds

Comprehension Type

Direction *(Q. Nos. 7-9) This section contains a paragraph, describing theory, experiments, data, etc. Three questions related to the paragraph have been given. Each question has only one correct answer among the four given options (a), (b), (c) and (d).*

Passage

Complete hydrolysis of a heptapeptide showed that it had the following molecular formula

$$Ala_2, \text{ Glu, Leu, Lys, Phe, Val}$$

Treatment of the heptapeptide with 2, 4-dinitrofluorobenzene followed by incomplete hydrolysis gave, among other products; valine labelled of the α-amino group, lysine labelled at the α-amino group and a dipeptide DNP-ValLeu (DNP = 2 , 4 dinitrophenyl)

7 Hydrolysis of the heptapeptide with carboxypeptidase gave an initial high concentration of alanine, followed by a rising concentration of glutamic acid. The probable amino acid sequence consists of

a. valine-N terminal; alanine-C-terminal
b. valine-N terminal; glutamine-C-terminal
c. alanine-N terminal; valine-C-terminal
d. valine-N terminal; lysine-C-terminal

8 Partial enzymatic hydrolysis of the heptapeptide gave a dipeptide *A* and tripeptide *B*. Complete hydrolysis of *B* gave phenylalanine, glutamic acid and alanine. When *B* was allowed to react with carboxypeptidase, the solution showed an initial high concentration of glutamic acid. Treatment of *B* with 2, 4-dinitrofluorobenzene followed by hydrolysis gave labelled phenylalanine. Thus, the tripeptide *B* is

a. Glu Ala Phe
b. Phe Ala Glu
c. Ala Glu Phe
d. Phe Glu Ala

9 The amino acid sequence of heptapeptide is

a. Val, Leu, Lys, Ala, Phe, Glu, Ala
b. Val, Glu, Lys, Phe, Ala, Leu, Ala
c. Val, Leu, Lys, Phe, Ala, Glu, Ala
d. Ala, Glu, Lys, Ala, Phe, Leu, Val

Working Space

Matching List Type

Direction *(Q. Nos. 10 and 11) Choices for the correct combination of elements from Column I and Column II are given as options (a), (b), (c) and (d), out of which one is correct.*

10 Match the Columns and choose the correct option using the codes given below.

	Column I (Structure of protein)		Column II (Characteristic property)
i.	α-helix structure	p.	The polypeptide chains lie side by side in an open structure having interchain amide H-bonding that holds the chains together.
ii.	β-pleated sheet structure	q.	The polypeptide chain is coiled up into a right handed spiral structure. It is stabilised by intramolecular H-bonding between (C$=$O) of one amino acid residue and the (N—H) of the 4th AA residue in the chain. This is also called 3.6 $_{13}$helix, having 3.6 AA in each turn of the helix and 13-membered ring.
iii.	Parallel pleated sheet structure	r.	This structure has the chains running in opposite directions. The α-C atoms rotate slightly out of the plane of the sheet to minimise repulsions between their bulky (R) groups.
iv.	Antiparallel pleated sheet structure	s.	This structure has chains running in the same direction; all with their N-terminal residue starting at the same end. The (R) groups are at alternate positions below and above the sheet.

Codes

	i	ii	iii	iv
a.	p	q	s	r
b.	q	p	s	r
c.	q	p	r	s
d.	p	q	r	s

11 Match the Column I and Column II and choose the correct option using the codes given below.

	Column I (Peptide reaction)		Column II (Amino acid)
i.	Gly—Ala $\xrightarrow{\text{Carboxy peptidase}}$	p.	Ala
ii.	Ala —Gly $\xrightarrow[\text{amino peptidase}]{\text{Leucine}}$	q.	Gly
iii.	Leu—val—His $\xrightarrow{\text{Hydrazinolysis}}$	r.	Val
iv.	Leu—val —His $\xrightarrow{\text{Leucine amino peptidase enzyme}}$ val $\downarrow$ His + Leu Edmann method	s.	His

Codes

	i	ii	iii	iv
a.	p	q	s	q
b.	p	q	r	s
c.	q	p	r	s
d.	p	q	s	r

Working Space

One Integer Value Correct Type

Direction *(Q. Nos. 12 and 13) This section contains 2 questions. When worked out will result in an integer from 0 to 9 (both inclusive).*

12 How many amino acids are present in the following polypeptide chain?

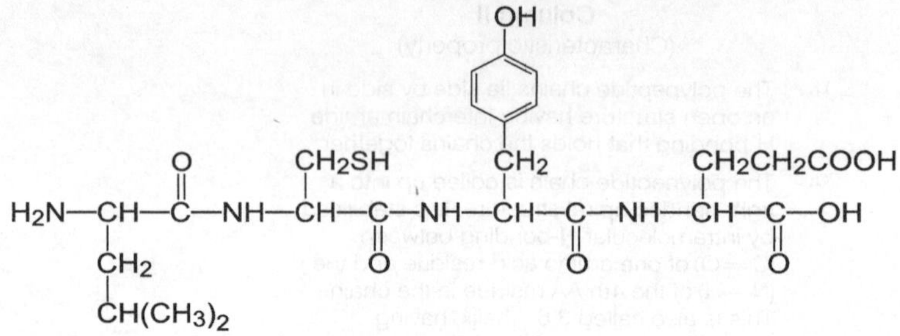

13 A tripeptide is composed equally of L-valine, L-tyrosine and L-alanine (one molecule of each). How many isomeric tripeptides of this kind may exist?

Statement Type

Direction *(Q. Nos. 14-16) This section is based on Statement I and Statement II. Select the correct answer from the codes given below.*

a. Both Statement I and Statement II are correct and Statement II is the correct explanation of Statement I
b. Both Statement I and Statement II are correct but Statement II is not the correct explanation of Statement I
c. Statement I is correct but Statement II is incorrect
d. Statement II is correct but Statement I is incorrect

14 **Statement I** In a protein molecule, the adjacent *R*-groups at α-positions occupy the alternate sides of plane.

Statement II Adjacent *R*-groups in the same plane exert steric repulsion and lowers the stability of conformers.

15 **Statement I** Proteins on prolonged treatment with aqueous alkali undergoes cleavage through peptide linkages.

Statement II Alkali treatment bring about hydrolysis to amides.

16 **Statement I** A peptide bond has about 40% double bond character.

Statement II Steric hindrance causes the *trans*-configuration to be more stable than the *cis*-configuration.

DPP-6 Other Biomolecules (DNA, RNA, Vitamins, Hormones, etc)

Subjective Questions

Working Space

Direction *(Q. Nos. 1 and 2) These questions are subjective in nature, need to be solved completely on notebook.*

1 Consider the following molecule.

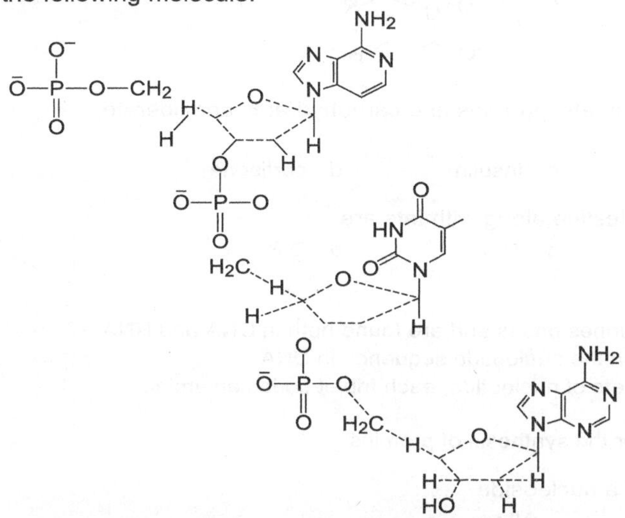

i. Is the above polymer of DNA or RNA? Explain your answer.

ii. Which of the following arrangements is possible in an RNA molecule? The small lines |||| represent hydrogen bonds.

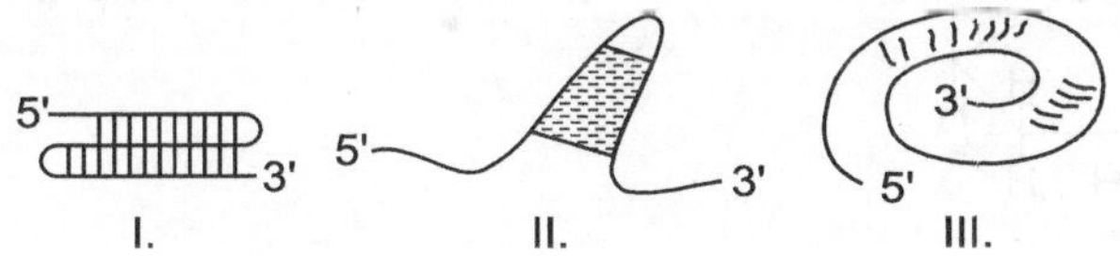

2 i. Identify *A* in the following:

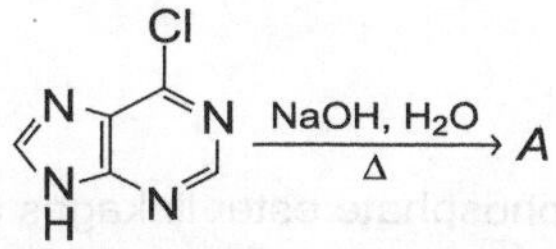

ii. If the sequence of one strand of DNA is ATGCTTCA, what is the sequence of the bases in the complementary strand?

iii. Give the sequence of bases of *m*-RNA molecule synthesised on the DNA strand TATCTACCTGGA.

Working Space

Only One Option Correct Type

Direction *(Q. Nos. 3-14) This section contains 12 multiple choice questions. Each question has four choices (a), (b), (c) and (d), out of which ONLY ONE option is correct.*

3 The false statement about an enzyme is

a. An enzyme is usually a large protein
b. An enzyme is a catalyst for biological reaction
c. An enzyme is a chiral molecule
d. An enzyme changes the equilibrium constant of a reaction

4 Which of the following chemical unit is certainly to be found in enzymes?

a. 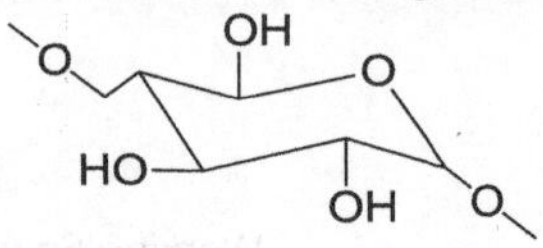

b. 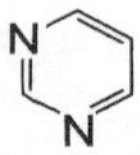

c. 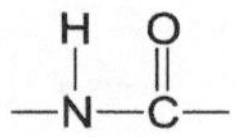

d. 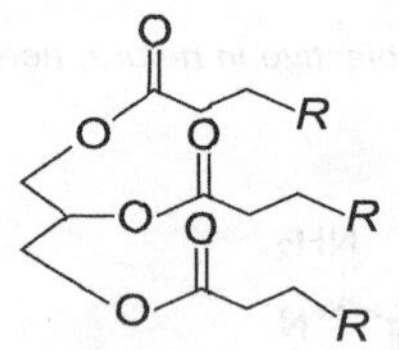

5 *X* controls the process of burning of fats, proteins and carbohydrates and liberate energy in the body *X* is

a. thyroxine b. adrenaline c. insulin d. cortisone

6 The vitamins absorbed from the intestine along with fats are

a. A,D b. A,B c. A,C d. D,B

7 Choose the incorrect statement.

a. Adenine and guanine both are purines bases and are found both in DNA and RNA
b. Genetic information is based upon the nucleotide sequence in DNA
c. The genetic code consists of triplets of nucleotide, each triplet codes an amino acid
d. Transfer RNA carries the code for the synthesis of proteins

8 Consider the following structure of a nucleoside.

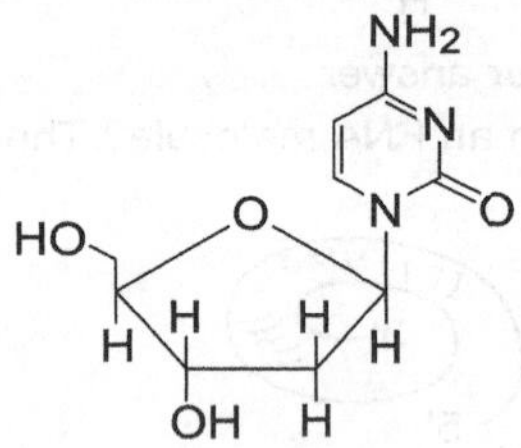

This structure shows

a. an amino acid joined to sucrose
b. β-glycosides of D-deoxyribofuranose
c. β-glycosides of D-ribofuranose
d. α-glycosides of D-ribofuranose

9 In both DNA and RNA, the heterocyclic base and phosphate ester linkages are at

a. C_5 and C_2 respectively of the sugar molecule
b. C_2 and C_5 respectively of the sugar molecule
c. C_1 and C_5 respectively of the sugar molecule
d. C_5 and C_1 respectively of the sugar molecule

Working Space

10 RNA contains uracil, in place of thymine which is a constituent of DNA. Chemically, thymine is

a. 5-methyl uracil
b. 4-methyl uracil
c. 3-methyl uracil
d. 1-methyl uracil

11 Which of the following statements is incorrect?

a. Vitamins are included in diet because they are not synthesised in the human body
b. Most vitamins function as coenzymes
c. A person with diabetes mellitus suffers from hypoglycemia
d. Hypoglycemia can affect the brain due to low blood sugar level

12 In E. coli DNA, the AT/GC ratio is 0.93. If the number of moles of adenine in its DNA sample is 465,000, the number of moles of guanine present is

a. 465,000
b. 500,000
c. 250,000
d. 232,500

13 Consider the following statements about haemoglobin.

I. It has four separate peptide subunits : Two identical α-chains and two identical β-chains each bonded to haeme.
II. In CO poisoning, since much of the haemoglobin is tied up with CO, O_2 transport to the tissues is inhibited.
III. Normal haemoglobin found in red blood cells has a glutamic acid residue.

The correct statements are

a. I and II
b. II and III
c. I and III
d. All of these

14 How many pairs in the gene are needed to code for the enzyme, lyozyme, containing 129 amino acids found in egg white?

a. 3×129 base pairs
b. $(3 \times 129) + (3 \times 2) = 393$ base pairs
c. $(3 \times 129) + (3 \times 3) = 396$ base pairs
d. 4×129 base pairs

One or More than One Options Correct Type

Direction *(Q. Nos. 15 and 16) This section contains 2 multiple choice questions. Each question has four choices (a), (b), (c) and (d), out of which ONE or MORE THAN ONE are correct.*

15 Following structure

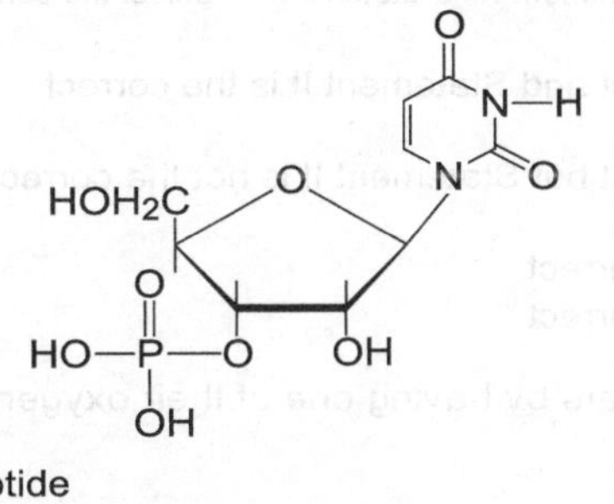

a. represents a nucleotide
b. represents a nucleoside
c. has a base, a sugar and a phosphate group
d. has an acid, a sugar and a phosphate group

16 Which of the following contain disulphide linkages?

a. Oxytocin
b. Vasopressin
c. Insulin
d. Haemoglobin

Working Space

Matching List Type

***Direction** (Q. Nos. 17 and 18) Choices for the correct combination of elements from Column I and Column II are given as options (a), (b), (c) and (d), out of which one is correct.*

17 Match the Column I with Column II and mark the correct option from the codes given below.

	Column I		Column II
i.	RNA	p.	Single stranded α-helix structure
ii.	DNA	q.	Double stranded α-helix structure
iii.	*r*-RNA	r.	Adenine and guanine as purine base
iv.	*t*-RNA	s.	Forms the site of protein synthesis
		t.	Carrier of amino acid

Codes

	i	ii	iii	iv		i	ii	iii	iv
a.	q,t	p,q	s	r,t	b.	p,r	q,r	s	t
c.	p	q	s	s,t	d.	p,s	q,r	s	t

18 Match the Column I and Column II and choose the correct option using the codes given below.

	Column I (Vitamin)		Column II (Deficiency diseases)
i.	Vitamin B_1	p.	Pernicious anaemia
ii.	Vitamin B_{12}	q.	Dermatities, loss of hair and paralysis
iii.	Vitamin H	r.	Blood coagulation
iv.	Vitamin K	s.	Beri-beri

Codes

	i	ii	iii	iv		i	ii	iii	iv
a.	p	s	q	r	b.	s	p	q	r
c.	s	p	r	q	d.	p	s	r	q

Statement Type

***Direction** (Q. Nos. 19 and 20) This section is based on Statement I and Statement II. Select the correct answer from the codes given below.*

a. Both Statement I and Statement II are correct and Statement II is the correct explanation of Statement I

b. Both Statement I and Statement II are correct but Statement II is not the correct explanation of Statement I

c. Statement I is correct but Statement II is incorrect

d. Statement II is correct but Statement I is incorrect

19 **Statement I** Deoxysugars differ from other sugars by having one of their oxygen atoms missing.

Statement II 2-deoxyribose is found in DNA.

20 **Statement I** ATP is the main source of energy of many anaerobic reactions.

Statement II Anaerobic reactions occur in the presence of oxygen.

SET - 1

Revisal Problems for JEE Main

Only One Option Correct Type

Direction *(Q. Nos. 1-27) This section contains 27 multiple choice questions. Each question has four choices (a), (b), (c) and (d), out of which ONLY ONE option is correct.*

Working Space

1 A protein is said to have following groups:

I. $-NH_2$; II. $-\overset{O}{\overset{\|}{C}}OH$; III. $-\overset{O}{\overset{\|}{C}}NH-$

a. I and II
b. I and III
c. II and III
d. I, II and III

2 D-mannose is epimeric with D-glucose at C_2. Which of the following structures represents β-D-mannopyranose?

a. 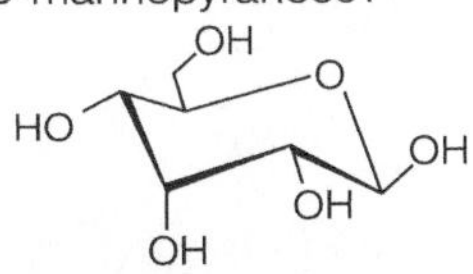
b. 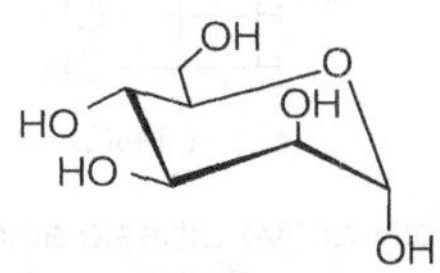
c. 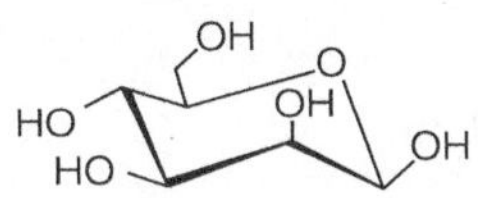
d. 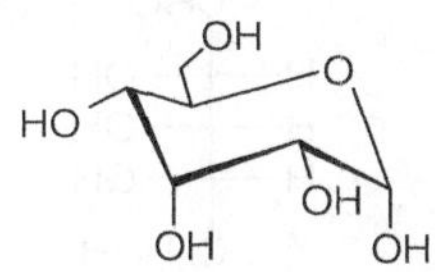

3 Chiral C-atoms in glucose and fructose are

a. 4 in each
b. 3 in each
c. 4 in glucose and 3 in fructose
d. 3 in glucose and 4 in fructose

4 In an α-helix, the *R* groups on the amino acid residues

a. alternate between the outside and the inside of the helix
b. are found on the outside of the helix spiral
c. cause only right-handed helices to form
d. generate the hydrogen bonds that form the helix

5 All of the following are the examples of fibrous proteins except

a. wool
b. fingernails
c. skin
d. insulin

6 The efficiency of an enzyme to catalyse a reaction is due to its capacity to

a. reduce the activation energy of the reaction
b. form strong enzyme substrate complex
c. decrease the bond energies of all the substrate molecules
d. increase the free energy of the catalyst-substrate reaction

7 Which of the following factors has the least influence on the secondary and tertiary structure of proteins?

a. The achiral nature of glycine units
b. Steric hindrance of bulky side chains on the peptide backbone
c. Hydrogen bonding of C=O to N—H groups located near each other in space
d. Conformational restriction imposed by proline units

8 The incorrect statement among the following is

a. Lactose is a disaccharide and is a reducing sugar
b. α-D-glucopyranose and β-D-glucopyranose are epimers
c. Methyl α-D-glucopyranoside has an acetal structure and is non-reducing in nature
d. α-D-glucopyranose has a hemiacetal structure and is reducing in nature

Working Space

9 Which of the following is not a D-sugar?

a. 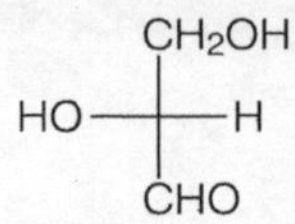

b. 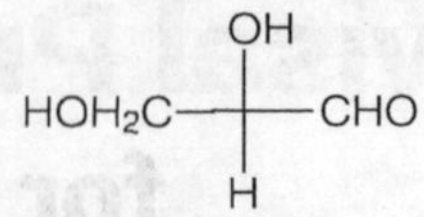

c. 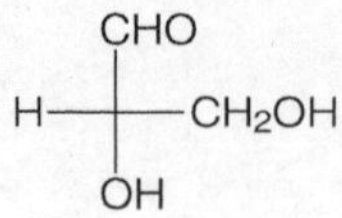

d. 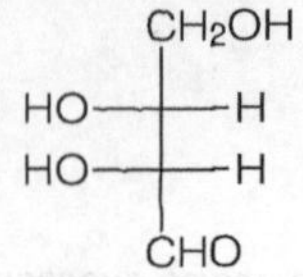

10 Which product will be formed in the following reaction?

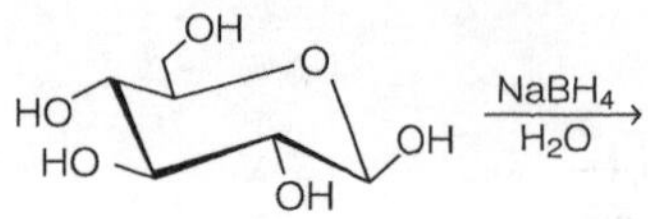

a. 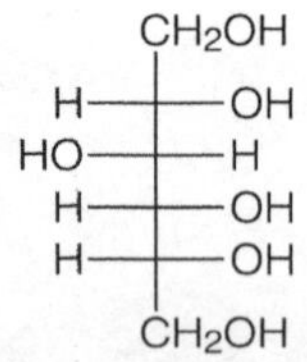

b. 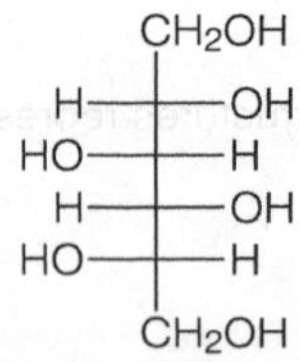

c. 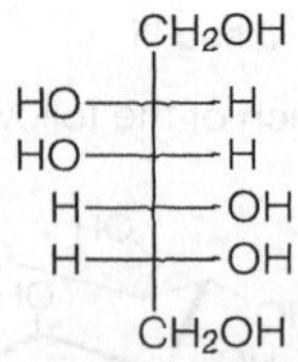

d. 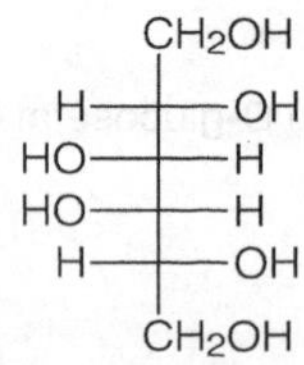

11 Which L-sugar on oxidation gives an optically active dibasic acid?

a. 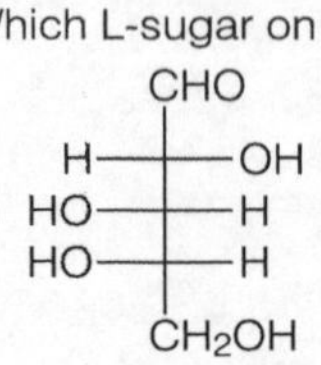

b. 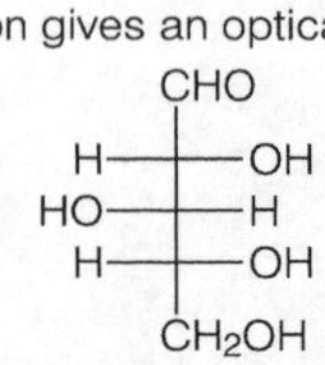

c. 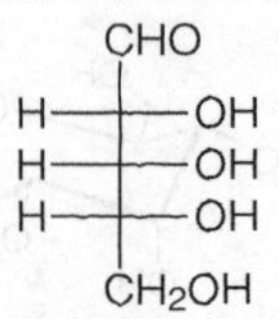

d. 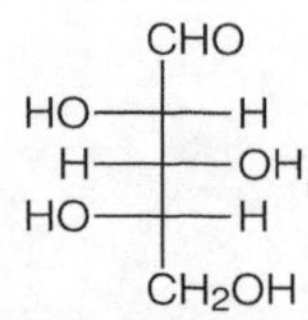

12 Although D-galactose rotates plane polarised light, its oxidation product galactaric acid does not. It is because

a. galactaric acid is a racemic mixture of D and L-isomer
b. galactaric acid is a *meso* compound
c. Both (a) and (b)
d. None of the above

13 Fructose $\xrightarrow[\text{(ii) P/HI}]{\text{(i) HCN/H}_3\text{O}^+}$ *A*. *A* is

a. *n*-heptanoic acid
b. 2-methyl hexanoic acid
c. *n*-heptane
d. 2-methyl hexane

14 Which of the following aldohexoses give the same osazone derivative?

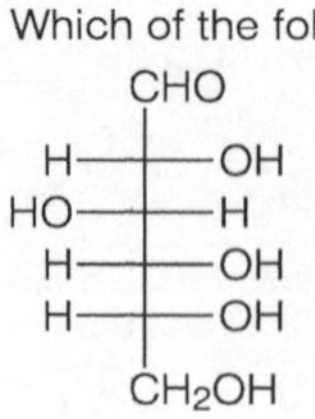

I.

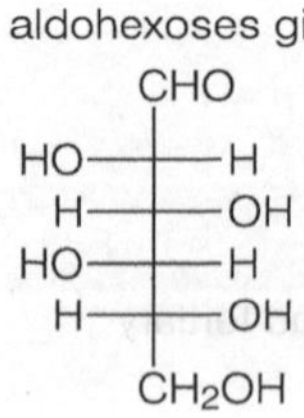

II.

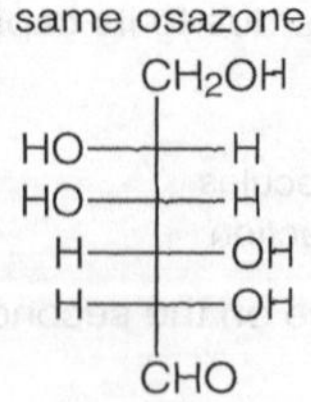

III.

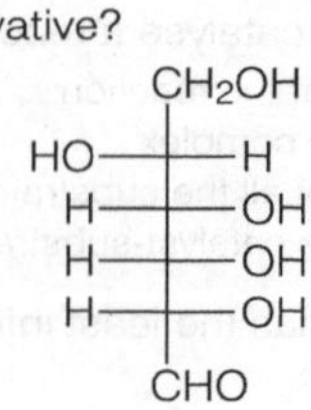

IV.

a. I and IV
b. I and III
c. II and III
d. III and IV

15 In the α-helix, the hydrogen bonds

a. are roughly parallel to the axis of the helix
b. are roughly perpendicular to the axis of the helix
c. occur mainly between electronegative atoms of the *R* groups
d. occur only between some of the amino acids of the helix

16 Which of the following amino acids has an amide side chain?

a. Aspartic acid b. Glutamic acid c. Asparagine d. Methionine

17 Which of the following statements about the conformation (secondary and tertiary structure) of proteins is false?

a. The four atoms of an amide linkage (—CO—NH—) lie in a plane as a result of resonance between the nitrogen and the carbonyl group
b. Electrostatic attraction occurs between basic and acidic side chains
c. The favoured conformation of a protein is always that which has the greatest number of hydrogen bonds
d. The presence of the amino acid proline has the effect of turning a corner in a protein

18 By adding SDS (sodium dodecyl sulphate) during the electrophoresis of proteins, it is possible to

a. determine a protein's isoelectric point
b. determine an enzyme's specific activity
c. determine the amino acid composition of the protein
d. preserve a protein's native structure and biological activity

19 Assume that a particular amino acid has an isoelectric point of 6.0. In a solution at pH 1.0, which of the following species will predominate?

a. $H_3N^+\overset{R}{\overset{|}{C}}HCO_2H$ b. $H_2N{-}\overset{R}{\overset{|}{C}}HCO_2H$ c. $H_3\overset{\oplus}{N}\overset{R}{\overset{|}{C}}HCO_2$ d. $H_2N\overset{R}{\overset{|}{C}}HCO_2$

20 Aspartic acid has the following structure $HOOC{-}\underset{\underset{\oplus NH_3}{|}}{CH}{-}CH_2COOH$

A

The pK_{a_1}, pK_{a_2} and pK_{a_3} of *A* respectively are 1.88, 3.65 and 9.60. Its pH at isoelectric point can be given by the expression

a. $\frac{pK_{a_1} + pK_{a_2}}{2}$ b. $\frac{pK_{a_1} + pK_{a_3}}{2}$

c. $\frac{pK_{a_2} + pK_{a_3}}{2}$ d. $\frac{pK_{a_1} + pK_{a_2} + pK_{a_3}}{3}$

21 Which of the following disaccharides is non-reducing sugar (does not react with Tollen's reagent)?

a.

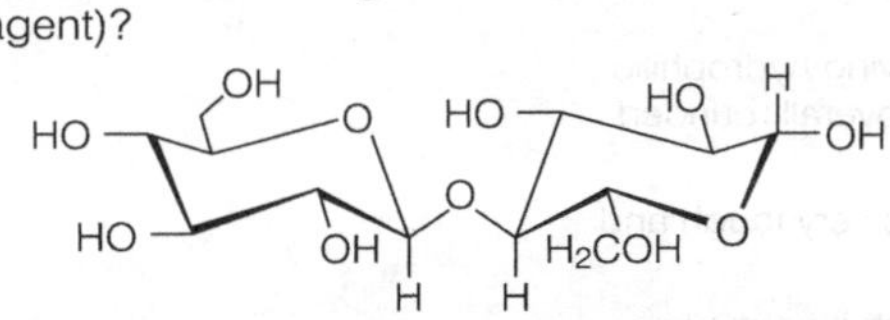

b.

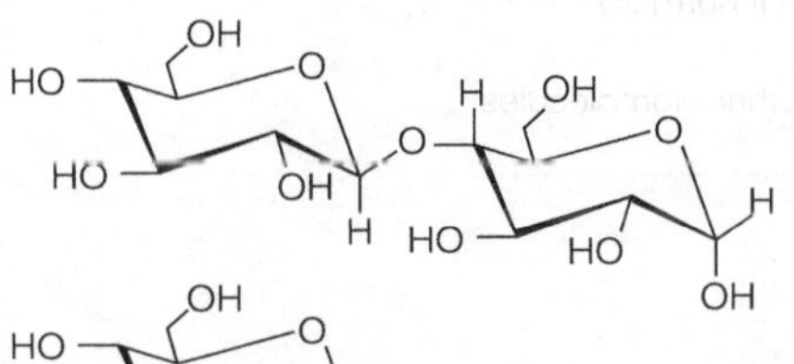

c.

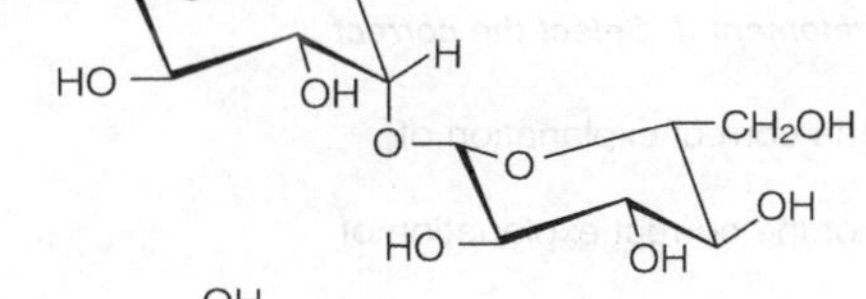

d.

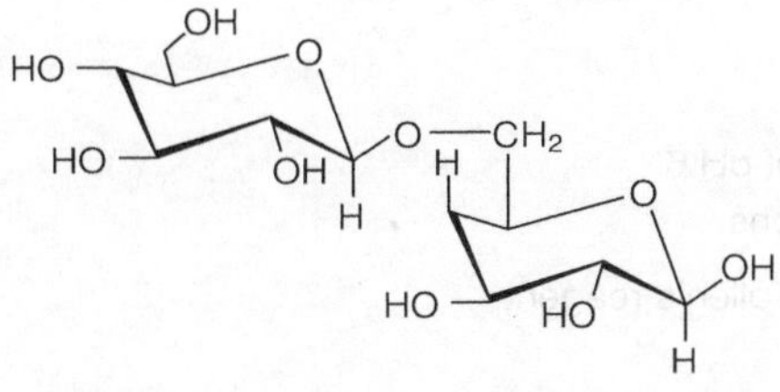

Working Space

22 Which of the following amino acids does not have an aromatic ring?

a. Tyrosine b. Histidine c. Tryptophan d. Proline

23 $\text{D-glucose} \xrightarrow[\text{EtONa}]{\text{MeNO}_2} \xrightarrow{\text{H}_2\text{SO}_4} \xrightarrow{\text{H}_2/\text{Ni}} X$

The final product X is

a. fructose b. mannose c. aldoheptose d. aldopentose

24 When octa-O-methyl D-cellobiose is hydrolysed by aqueous acid, two O-methylated glucose derivatives are formed.

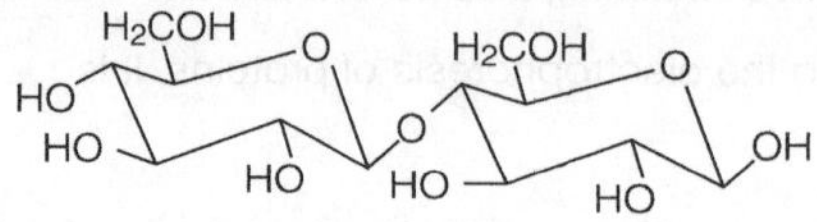

D-cellobiose

One is tetramethyl derivative and the other is trimethyl derivative. Why is a single methyl substituent lost in this process?

a. One methoxy group is lost by β-elimination
b. One methoxy group is an ester and the others are all ethers
c. One methoxy group is a part of an acetal, the other are all ethers
d. One glucose is an α-methyl glycoside; the other is a β-methyl glycoside

25 Periodic acid splits glucose and fructose into formic acid and CH_2O. Ratio of formic acid and CH_2O from glucose and fructose is

a. 5/1 and 4/2 b. 5/1 and 5/1
c. 4/2 and 4/2 d. 4/2 and 5/1

26 Reduction of hexose A (molecular formula $C_6H_{12}O_6$) with sodium borohydride gives compounds B and C. Compound B is optically inactive whereas compound C is optically active. Which of the following is compound A?

a. D-fructose b. D-glucose c. D-mannose d. .D-psicose

27 Which is/are correct pair(s) of matching?

	Protein	Property
a.	Native protein	A protein that is usually water soluble, having hydrophilic exterior and hydrophobic interior and an overall rounded shape.
b.	Fibrous protein	A protein that is usually soluble in water is very tough and has a long shape.
c.	Globular protein	A protein with the tertiary structure in which it normally occurs in living system.
d.	Conjugated protein	A protein that produces amino acids and other biomolecules or inorganic substances upon hydrolysis.

Statement Type

Direction (Q. Nos. 28 and 29) ***This section is based on Statement I and Statement II. Select the correct answer from the codes given below.***

a. Both Statement I and Statement II are correct and Statement II is the correct explanation of Statement I
b. Both Statement I and Statement II are correct but Statement II is not the correct explanation of Statement I
c. Statement I is correct but Statement II is incorrect
d. Statement II is correct but Statement I is incorrect

28 **Statement I** Aspartic acid migrates towards anode in a buffer of pH 6.

Statement II Aspartic acid exists as cation under given conditions.

29 **Statement I** Fructose has keto group hence, does not reduce Tollen's reagent.

Statement II Fructose is called fruit sugar.

SET - 2

Revisal Problems for JEE Advanced

Only One Option Correct Type

Working Space

Direction *(Q. Nos. 1-8) This section contains 8 multiple choice questions. Each question has four choices (a), (b), (c) and (d), out of which ONLY ONE option is correct.*

1 Which of the following is not a part of the primary protein structure?

a. The amino acid sequence
b. The disulphide linkages
c. The planar nature of the amide linkage
d. The conformers of the polypeptide backbone

2 Which of the following pairs are C-2 epimers as well as enantiomers?

a. D-glyceraldehyde and L-glyceraldehyde
b. D-erythrose and D-threose
c. D-ribose and D-arabinose
d. D-xylose and D-lyxose

3 Choose the response that provides the best match between the terms given and the structures shown.

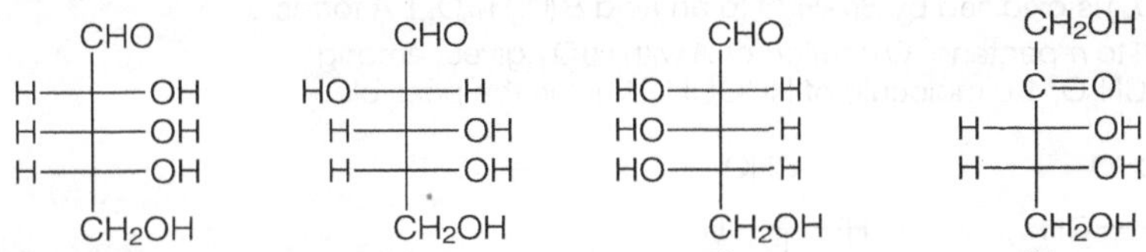

I. II. III. IV.

	Diastereomers	Enantiomers
a.	I, III and IV	I and III
b.	I and II	I and III
c.	I, II, and III	I and III
d.	I and IV	I and II

4 Which of the following acid(s) cannot form α-amino acid?

I. Benzoic acid
II. Glycolic acid
III. Phenyl acetic acid
IV. Formic acid

a. I and II
b. II and IV
c. I and IV
d. II and III

5 A D-aldohexose gives an optically inactive aldaric acid on oxidation with nitric acid. Ruff degradation of this aldohexose produces an aldopentose that is reduced by $NaBH_4$ to an optically active alditol. Which of the following is this hexose?

a. 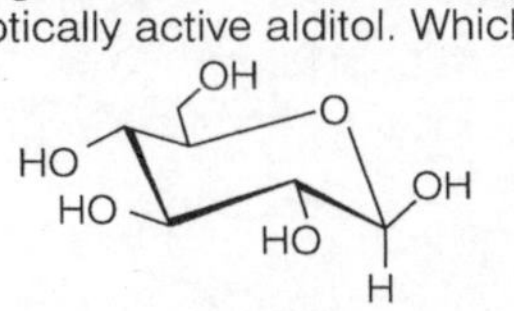

b. 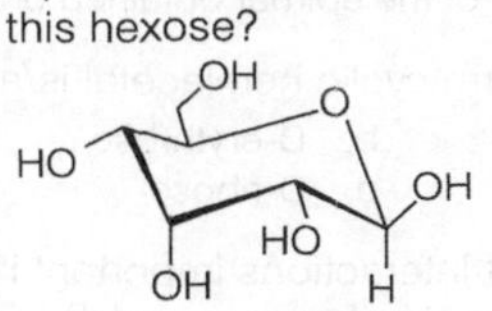

c. 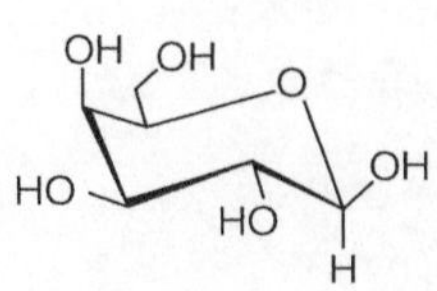

d. 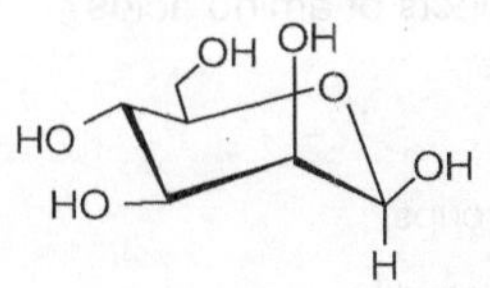

Working Space

6 Consider the following compounds.

I. 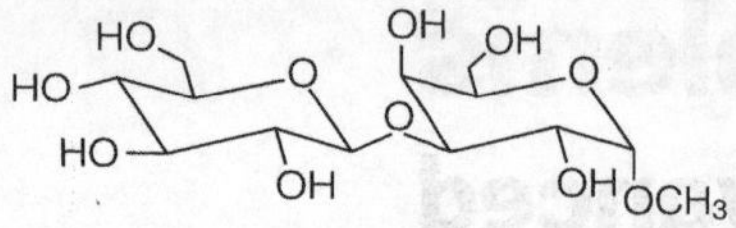

II. 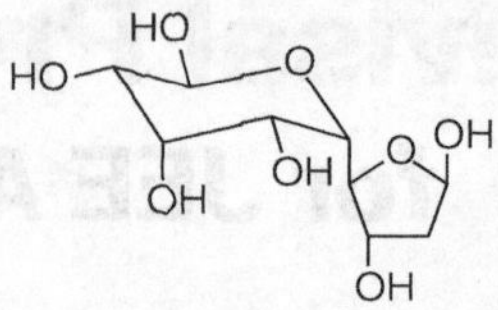

III. 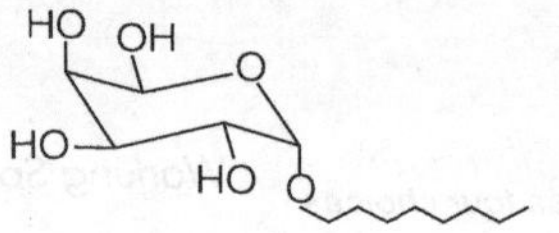

IV. 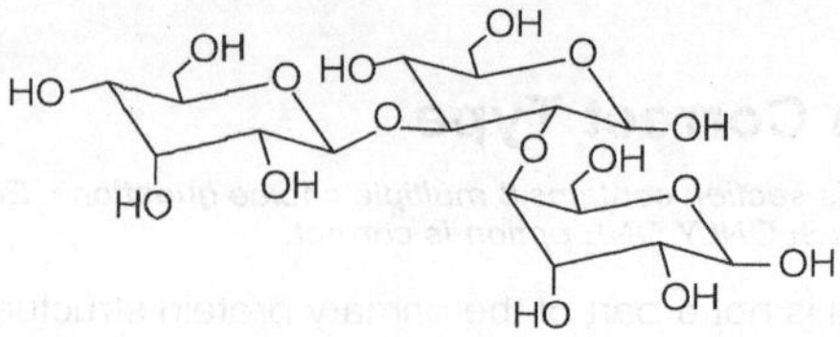

Which of the compounds (I-IV) depicted above has no hemiacetal linkage?

a. Compound I
b. Compound II
c. Compounds I and III
d. Compound IV

7 A hexapeptide has the composition Ala, Gly, Phe, Val_3. Both the N-terminal and C-terminal units are Val. Cleavage of the hexapeptide by chemotrypsin gives two different tripeptides, both having Val as the N-terminal group.

Among the products of random hydrolysis is a Aal-Val dipeptide fragment. What is the primary structure of the hexapeptide?

a. Val-Gly-Phe-Val-Ala-Val
b. Val-Ala-Phe-Val-Gly-Val
c. Val-Gly-Ala-Val-Phe-Val
d. Val-Phe-Val-Ala-Gly-Val

8 An organic compound $A (C_5H_{10}O_4)$ is oxidised by Br_2-H_2O to an acid $B (C_5H_{10}O_5)$. A forms a triacetate and it is reduced by HI to n-pentane. Oxidation of A with HIO_4 gives, among other products, 1.0 molecule of CH_2O, 1.0 molecule of HCOOH. What is the possible structure of A?

a. 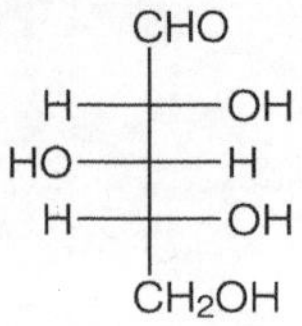
b. 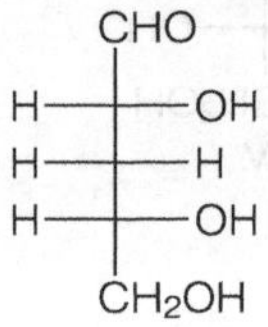
c. 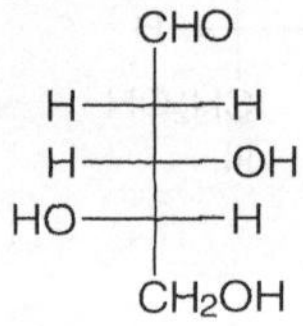
d. All of these

One or More than One Options Correct Type

Direction *(Q. Nos. 9-14) This section contains 6 multiple choice questions. Each question has four choices (a), (b), (c) and (d), out of which ONE or MORE THAN ONE are correct.*

9 Which is/are true regarding D-glucose and D-fructose?

a. Both forms same osazone with $C_6H_5NHNH_2$
b. Both are dextrorotatory sugars
c. Both forms a monocarboxylic acid with Br_2 -H_2O
d. Reduction product of D-glucose is one of the epimer obtained on reduction of D-fructose

10 The smallest aldose which is able to form cyclic hemiacetal is/are

a. D-glyceraldehyde
b. D-erythrose
c. D-threose
d. D-ribose

11 Which of the following are non-covalent interactions important in maintaining the secondary, tertiary and quarternary aspects of amino acids?

a. Hydrogen bonding along the backbone
b. Hydrogen bonding between R groups
c. Salt bridges between R groups
d. Hydrophobic interactions between R groups

12 The hemiacetal form of glucose is indicated by

a. reaction with $(CH_3—CO)_2O$
b. oxidation with Tollen's reagent
c. reduction with P/HI
d. glycoside formation

Working Space

13 Which statements is/are correct about the following reaction?

$$\text{D-glucose} \xrightarrow{\text{NaCN/HCN}} ?$$

a. The C-chain is increased by one C-atom
b. Two isomeric products, cyanohydrin and its C-2 epimeric cyanohydrin are formed
c. Epimers formed in products are in equal amounts
d. The presence of stereocentres in sugars causes their (C=O) groups to have diastereotopic faces that react at different rates, giving different amounts of diastereomers

14 Which of the following statements is/are correct about D, L of sugars and amino acids?

a. In sugars, D refers to dextrorotatory and L refers to laevorotatory
b. D and L-configurations are not used in case of amino acids
c. Serine is taken as the standard for D, L-configuration
d. In sugars, D and L shows relative configuration of —OH at the penultimate C-atom

Statement Type

Direction (Q. Nos. 15-17) ***This section is based on Statement I and Statement II. Select the correct answer from the codes given below.***

a. Both Statement I and Statement II are correct and Statement II is the correct explanation of Statement I
b. Both Statement I and Statement II are correct but Statement II is not the correct explanation of Statement I
c. Statement I is correct but Statement II is incorrect
d. Statement II is correct but Statement I is incorrect

15 **Statement I** Both α-D-glucose and β-D-glucose on oxidation with bromine water give the same gluconic acid.

Statement II α and β-forms of glucose are in equilibrium with its open chain structure in solution.

16 **Statement I** In α-helix structure, the protein molecules are coiled like spring.

Statement II In coiled form, entropy and hence stability is more.

17 **Statement I** Carboxypeptidase is an exopeptidase.

Statement II Carboxypeptidase cleaves N-terminal bond of peptide.

Comprehension Type

Direction (Q. Nos. 18-25) ***This section contains 3 paragraphs, each describing theory, experiments, data, etc. Eight questions related to the paragraphs have been given. Each question has only one correct answer among the four given options (a), (b), (c) and (d).***

Passage I

$$\underset{A}{\text{D-glucopyranose}} \xrightarrow[\text{HCl}]{\text{MeOH/}} B \xrightarrow[\text{Me}_2\text{SO}_4/\bar{\text{O}}\text{H}]{\text{Excess of}} C \xrightarrow{\text{Dil.HCl}} D \xrightarrow[\text{[O]}]{\text{HNO}_3} \text{2, 3-dimethoxy succinic acid, } E \text{ and 2, 3, 4-trimethoxy glutaric acid, } F$$

18 The incorrect statement about A is

a. It has a six-membered cyclic ring
b. It has a hemiacetalic linkage
c. It contains an acetalic linkage
d. All of these

19 Compound D is

a. 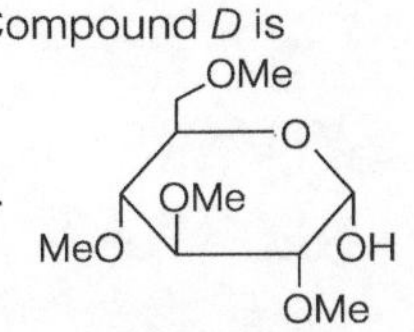

b. 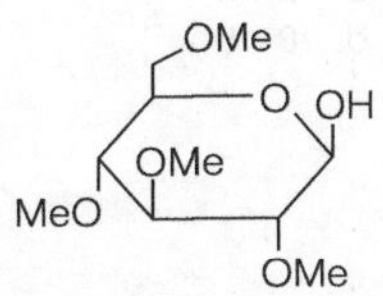

c. 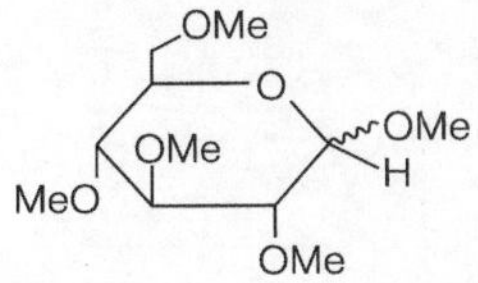

d. Both (a) and (b)

20 Which statement is correct about *E* and *F*? *Working Space*

a. *E* is obtained by the cleavage of C-4 and C-5 bond of compound *D*
b. *E* is obtained by the cleavage of C-5 and C-6 bond of compound *D*
c. *F* is obtained by the cleavage of C-4 and C-5 bond of compound *D*
d. *F* is obtained by the cleavage of C-3 and C-4 bond of compound *D*

Passage II

$$\underset{(I)}{\text{Tetrapeptide}} \xrightarrow{H_3O^+} \text{Four amino acids} \;\; \underset{\text{Glycine}}{A} \;\; \underset{\text{Alanine}}{B} \;\; \underset{\text{Valine}}{C} \;\; \underset{\text{Leucine}}{D}$$

$$\xrightarrow[\text{Sanger's method}]{\text{DNP method}} \underset{(II)}{\text{Tripeptide}} + B$$

$$\xrightarrow{\text{Carboxypeptidase}} \underset{(III)}{\text{Dipeptide}} + A \xrightarrow[\text{Edmann method}]{} C$$

21 The dipeptide III is

a. Val $\rightarrow$ Leu
b. Leu $\rightarrow$ Val
c. Gly $\rightarrow$ Ala
d. Ala $\rightarrow$ Gly

22 The tetrapeptide (I) is

a. Gly $\rightarrow$ Leu $\rightarrow$ Val $\rightarrow$ Ala
b. Ala $\rightarrow$ Val $\rightarrow$ Gly $\rightarrow$ Leu
c. Gly $\rightarrow$ Val $\rightarrow$ Leu $\rightarrow$ Ala
d. Ala $\rightarrow$ Val $\rightarrow$ Leu $\rightarrow$ Gly

Passage III

A solution of alanine hydrochloride, $[H_3N—CH(CH_3)—COOH]^+Cl^-$ is titrated with a solution of sodium hydroxide to produce a curve as follows:
K_1 and K_2 of alanine hydrochloride are 4.6×10^{-3} and 2.0×10^{-10} respectively.

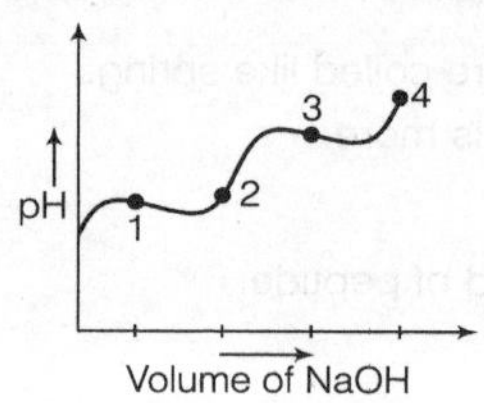

23 In the curve, at point 4 which of the molecule is found?

a. $H_3\overset{+}{N}—CH(CH_3)—COOH$
b. $H_3\overset{+}{N}—CH(CH_3)—COO^-$
c. $H_2N—CH(CH_3)—COOH$
d. $H_2N—CH(CH_3)—COO^-$

24 The pH at point 3 is

a. 2.34
b. 6.02
c. 9.07
d. 11.02

25 Which of the following is correct ratio of $\dfrac{H_3\overset{+}{N}—CH(CH_3)COO^-}{H_2N—CH(CH_3)COO^-}$ at pH 10?

a. 2 : 1
b. 5 : 8
c. 1 : 2
d. 8 : 5

Matching List Type

Direction (Q. Nos. 26-28) ***Choices for the correct combination of elements from Column I and Column II are given as options (a), (b), (c) and (d), out of which one is correct.***

26 Match the Column I with Column II and select the correct option from the codes given below.

	Column I (Mixture)		**Column II** (Reagent for distinction)
i.	Glucose + fructose	p.	Bromine water
ii.	Glucose + sucrose	q.	Tollen's reagent
iii.	Fructose + sucrose	r.	Fehling's solution
iv.	Benzaldehyde + acetaldehyde	s.	$I_2/\bar{O}H$

Codes

	i	ii	iii	iv		i	ii	iii	iv
a.	q,p	r,p	p,s	q	b.	r	r,p,q	r,s	p
c.	p	p,q,r	q,r	r,s	d.	q	r,s,q	r,s	s

27 Match the Column I with Column II and select the correct option from the codes given below.

	Column I (Group)		**Column II** (Compound)
i.	—NH_2 group	p.	Glycine
ii.	—OH group	q.	Picric acid
iii.	—COOH group	r.	Tyrosine
iv.	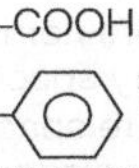 group	s.	Serine

Codes

	i	ii	iii	iv		i	ii	iii	iv
a.	q,r,p	s	p,s,r	r	b.	s	p	q	r
c.	q	p	q	s	d.	p,r,s	q,r,s	p,r,s	q,r

28 Match the Column I with Column II and select the correct option from the codes given below.

	Column I (Characteristics)		**Column II** (Organic compound)
i.	Lobry de Bruyn van Eikenstein rearrangement	p.	Sucrose
ii.	Ninhydrin test	q.	Glucose
iii.	Inversion of sugar	r.	Fructose
iv.	Decolouration of bromine water	s.	Alanine

Codes

	i	ii	iii	iv		i	ii	iii	iv
a.	q	r	s	p	b.	q,r	s	p	q
c.	r,q	p,s	r	q	d.	p	q	r	s

One Integer Value Correct Type

Direction (Q. Nos. 29-31) ***This section contains 3 questions. When worked out will result in an integer from 0 to 9 (both inclusive).***

29 Consider the following compounds. Glucose, fructose, mannose, ribose, arabinose, amylose, amylopectin, sucrose. How many pairs of such compounds are possible which form same osazone?

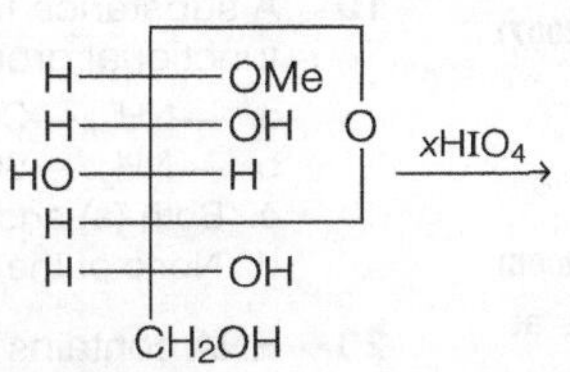

30 How many moles of HIO_4 are used in the above reaction?

31 Maltose $\xrightarrow{MeOH/HCl}$ $\xrightarrow{Ac_2O/NaOAc}$ $\xrightarrow{H_3\overset{\oplus}{O}}$ Product

How many moles of glucose are present in the product?

Working Space

JEE Main & AIEEE Archive

(Compilation of Last 13 Years' Questions)

1 Which one of the following bases is not present in DNA? **(JEE Main 2014)**

a. Quinoline b. Adenine c. Cytosine d. Thymine

2 Synthesis of each molecule of glucose in photosynthesis involves **(JEE Main 2013)**

a. 18 molecules of ATP b. 10 molecules of ATP
c. 8 molecules of ATP d. 6 molecules of ATP

3 Which of the following compounds can be detected by Molisch's test? **(AIEEE 2012)**

a. Nitro compounds b. Sugars
c. Amines d. Primary alcohols

4 Which one of the following statements is correct? **(AIEEE 2012)**

a. All amino acids except lysine are optically active
b. All amino acids are optically active
c. All amino acids except glycine are optically active
d. All amino acids except glutamic acids are optically active

5 The presence or absence of hydroxy group on which carbon atom of sugar differentiates RNA and DNA? **(AIEEE 2011)**

a. First b. Second c. Third d. Fourth

6 The change in the optical rotation of freshly prepared solution of glucose is known as **(AIEEE 2011)**

a. tautomerism b. racemisation
c. specific rotation d. mutarotation

7 Biuret test is not given by **(AIEEE 2010)**

a. carbohydrates b. polypeptides
c. urea d. proteins

8 The two functional groups present in a typical carbohydrate are **(AIEEE 2009)**

a. —OH and —COOH b. —CHO and —COOH
c. C$=$O and —OH d. —OH and —CHO

9 α-D-(+)-glucose and β-D-(+)-glucose are **(AIEEE 2008)**

a. conformers b. epimers
c. anomers d. enantiomers

10 The secondary structure of a protein refers to **(AIEEE 2007)**

a. α-helical backbone
b. hydrophobic interactions
c. sequence of α-amino acids
d. fixed configuration of the polypeptide backbone

11 The term anomers of glucose refers to **(AIEEE 2006)**

a. isomers of glucose that differ in configurations at carbons one and four (C-1 and C-4)
b. a mixture of D-glucose and L-glucose
c. enantiomers of glucose
d. isomers of glucose that differ in configuration at carbon one (C-1)

12 The pyrimidine bases present in DNA are **(AIEEE 2006)**

a. cytosine and adenine
b. cytosine and guanine
c. cytosine and thymine
d. cytosine and uracil

13 In both DNA and RNA, heterocyclic base and phosphate ester linkages are at **(AIEEE 2005)**

a. C_5' and C_1' respectively of the sugar molecule
b. C_1' and C_5' respectively of the sugar molecule
c. C_2' and C_5' respectively of the sugar molecule
d. C_5' and C_2' respectively of the sugar molecule

14 Identify the correct statement regarding enzymes. **(AIEEE 2004)**

a. Enzymes are specific biological catalysts that can normally function at very high temperatures ($T\sim$ 1000 K)
b. Enzymes are normally heterogeneous catalysts that are very specific in their action
c. Enzymes are specific biological catalysts that cannot be poisoned
d. Enzymes are specific biological catalysts that possess well defined active sites

15 Insulin production and its action in human body are responsible for the level of diabetes. This compound belongs to which of the following categories? **(AIEEE 2004)**

a. A coenzyme b. A hormone
c. An enzyme d. An antibiotic

16 Which base is present in RNA, but not in DNA? **(AIEEE 2004)**

a. Uracil b. Cytosine
c. Guanine d. Thymine

17 The reason for double helical structure of DNA is operation of **(AIEEE 2003)**

a. van der Waals' forces
b. dipole-dipole interaction
c. hydrogen bonding
d. electrostatic attractions

18 Complete hydrolysis of cellulose gives **(AIEEE 2003)**

a. D-fructose b. D-ribose
c. D-glucose d. L-glucose

19 A substance forms Zwitter ion. It can have functional groups **(AIEEE 2002)**

a. $—NH_2$, —COOH
b. $—NH_2$, $—SO_3H$
c. Both (a) and (b)
d. None of the above

20 RNA contains **(AIEEE 2002)**

a. ribose sugar and thymine
b. ribose sugar and uracil
c. deoxyribose sugar and uracil
d. deoxyribose sugar and thymine

JEE Advanced & IIT JEE Archive

(Compilation of Last 10 Years' Questions)

1 The total number of distinct naturally occurring amino acids obtained by complete acidic hydrolysis of the peptide shown below is **(2014 Adv., Integer Type)**

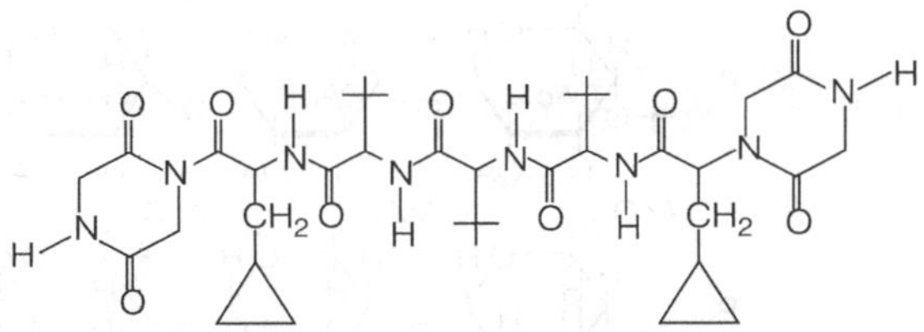

2 A tetrapeptide has —COOH group on alanine. This produces glycine (Gly), valine (Val), phenylalanine (Phe) and alanine (Ala) on complete hydrolysis. For this tetrapeptide, the number of possible sequences (primary structures) with $—NH_2$ group attached to a chiral centre is **(2013 Adv., Integer Type)**

3 The substituents R_1 and R_2 for nine peptides are listed in the table given below. How many of these peptides are positively charged at pH = 7.0 ? **(2012, Integer Type)**

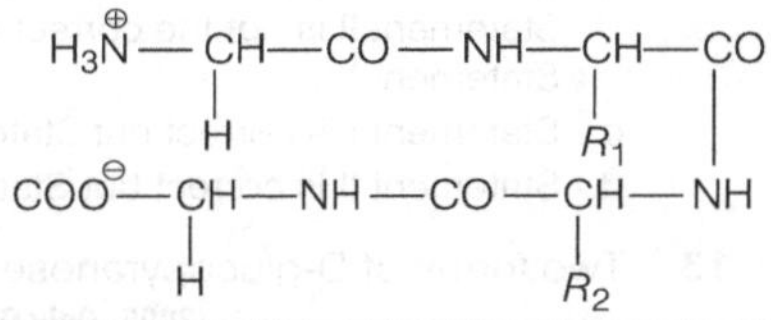

Peptide	R_1	R_2
I	H	H
II	H	CH_3
III	CH_2COOH	H
IV	CH_2CONH_2	$(CH_2)_4NH_2$
V	CH_2CONH_2	CH_2CONH_2
VI	$(CH_2)_4NH_2$	$(CH_2)_4NH_2$
VII	CH_2COOH	CH_2CONH_2
VIII	CH_2OH	$(CH_2)_4NH_2$
IX	$(CH_2)_4NH_2$	CH_3

4 When the following aldohexose exists in its D-configuration, the total number of stereoisomers in its pyranose form is **(2012, Integer Type)**

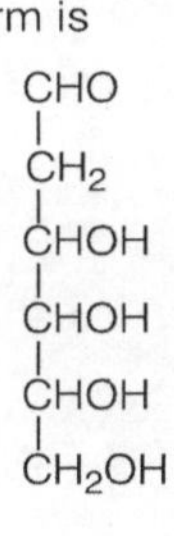

5 A decapeptide (mol. wt. 796) on complete hydrolysis gives glycine (mol. wt. 75), alanine and phenylalanine. Glycine contributes 47.0% to the total weight of the hydrolysed products. The number of glycine units present in the decapeptide is **(2011, Integer Type)**

6 The correct functional group X and the reagent/reaction conditions Y in the following scheme are **(2011, One or More than One Options Correct Type)**

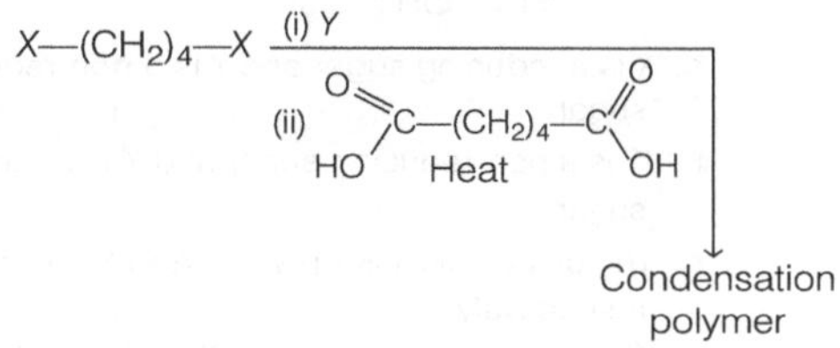

a. $X = COOCH_3$, $Y = H_2/Ni/Heat$
b. $X = CONH_2$, $Y = H_2/Ni/Heat$
c. $X = CONH_2$, $Y = Br_2/NaOH$
d. $X = CN$, $Y = H_2/Ni/Heat$

7 The following carbohydrate is

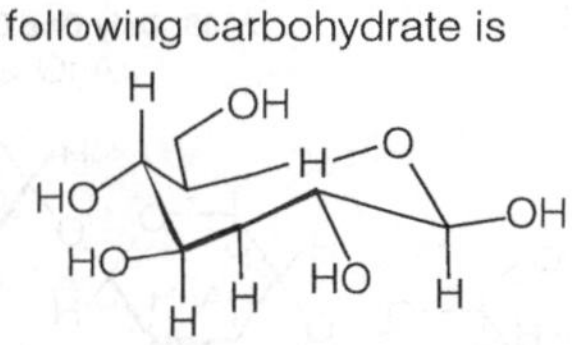

(2011, Only One Option Correct Type)

a. a ketohexose　　b. an aldohexose
c. an α-furanose　　d. an α-pyranose

8 The total number of basic groups in the following form of lysine is **(2010, Integer Type)**

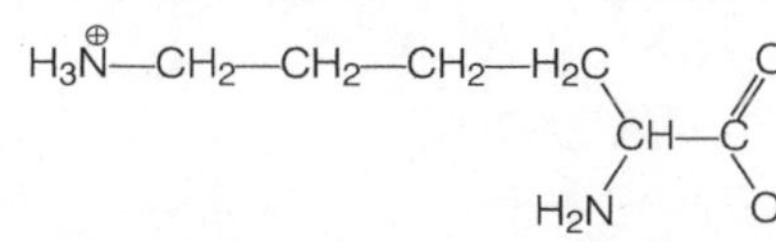

9 The correct statement about the following disaccharide is **(2010, Only One Options Correct Type)**

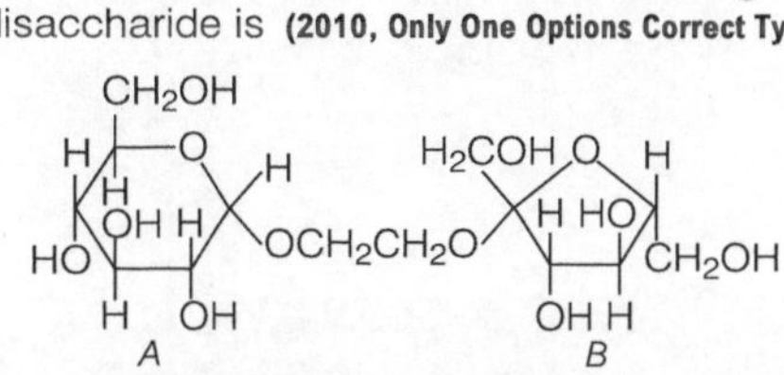

a. Ring A is pyranose with α-glycosidic linkage
b. Ring A is furanose with α-glycosidic linkage
c. Ring B is furanose with α-glycosidic linkage
d. Ring B is pyranose with β-glycosidic linkage

10 The correct statement(s) about the following sugars X and Y is(are)

(2009, One or More than One Options Correct Type)

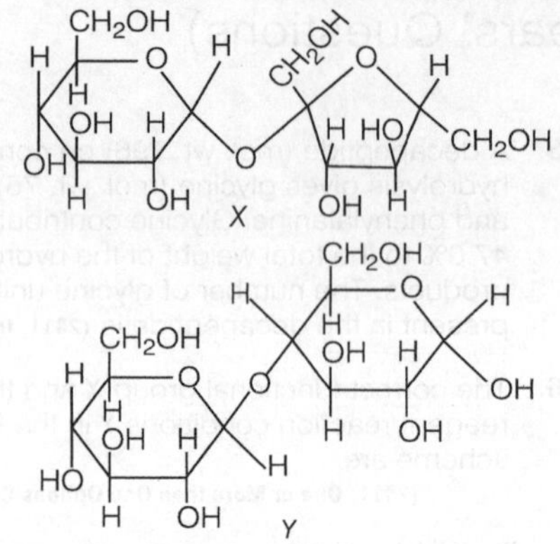

a. X is a reducing sugar and Y is a non-reducing sugar

b. X is a non-reducing sugar and Y is a reducing sugar

c. The glucosidic linkages in X and Y are α and β, respectively

d. The glucosidic linkages in X and Y are β and α, respectively

11 Cellulose upon acetylation with excess acetic anhydride/ H_2SO_4 (catalytic) gives cellulose triacetate whose structure is

(2008, Only One Option Correct Type)

a. 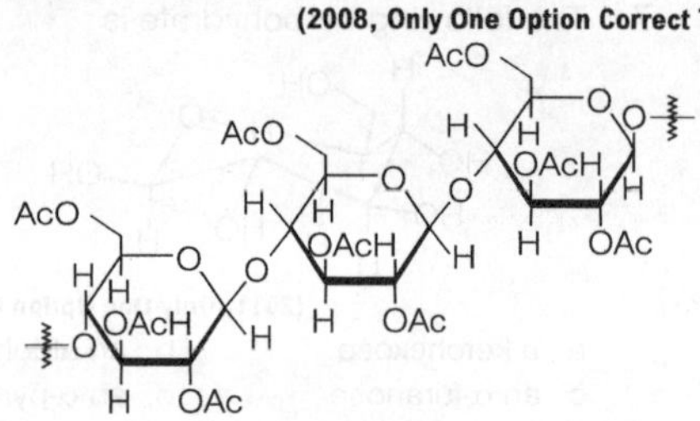

b. 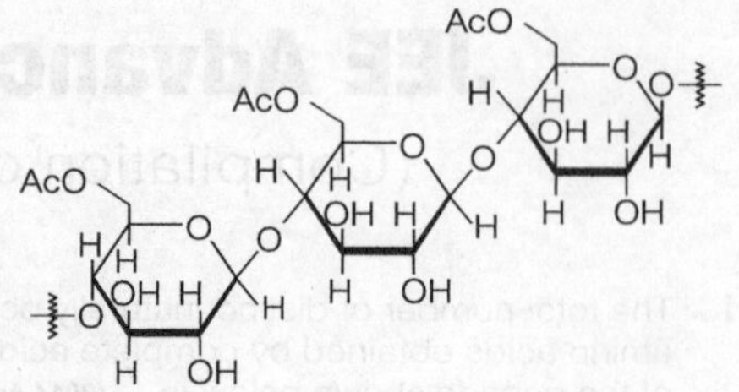

c. 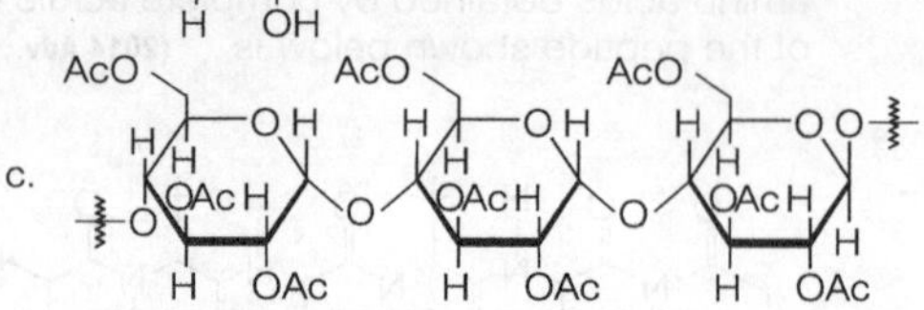

d. 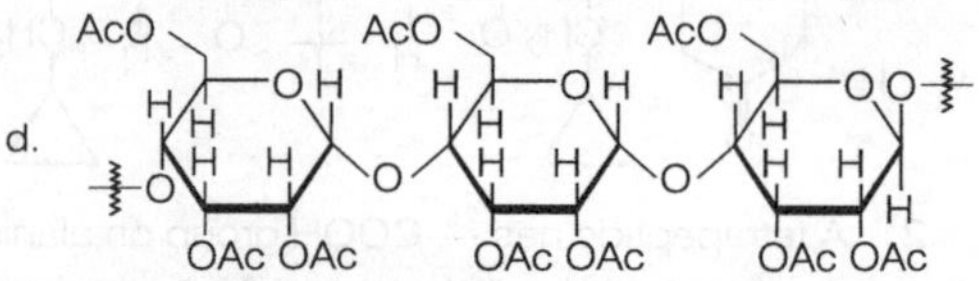

12 **Statement I** Glucose gives a reddish-brown precipitate with Fehling's solution.

Statement II Reaction of glucose with Fehling's solution gives CuO and gluconic acid.

(2007, Statement Type)

a. Both Statement I and Statement II are correct and Statement II is the correct explanation of Statement I

b. Both Statement I and Statement II are correct but Statement II is not the correct explanation of Statement I

c. Statement I is correct but Statement II is incorrect

d. Statement II is correct but Statement I is incorrect

13 Two forms of D-glucopyranose are called

(2005, Only One Option Correct Type)

a. enantiomers b. anomers

c. epimers d. diastereomers

6. Polymers

DPP
DAILY PRACTICE PROBLEMS

DPP-1 General Introduction and Classification of Polymers

Subjective Questions

Working Space

Direction (Q. Nos. 1-4) *These questions are subjective in nature, need to be solved completely on notebook.*

1 i. Show the repeat unit of polymers that would result from addition polymerisation of these monomers.

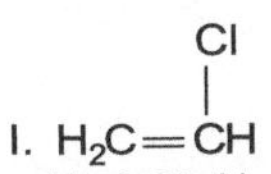

Vinyl chloride

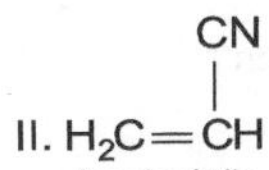

Acrylonitrile

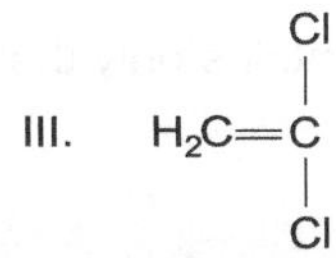

Vinylidene chloride

ii. Identify the monomer in the following polymeric structures.

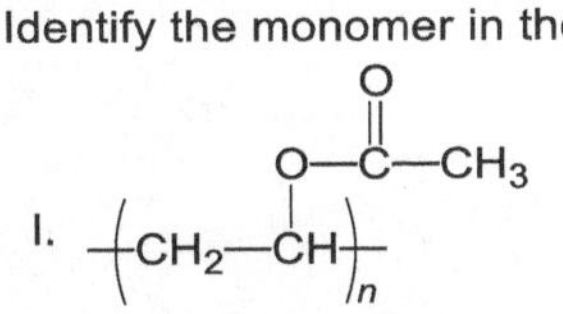

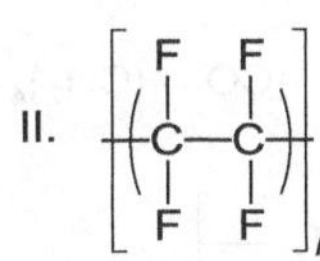

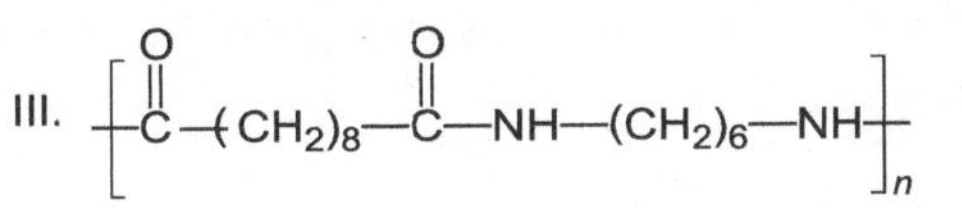

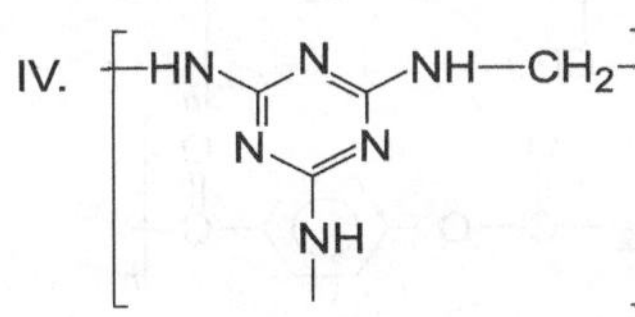

2 i. Why is teflon a highly thermally stable polymer?

ii. Explain, why teflon is linear and has no stereochemical complications even though it is prepared by free radical polymerisation?

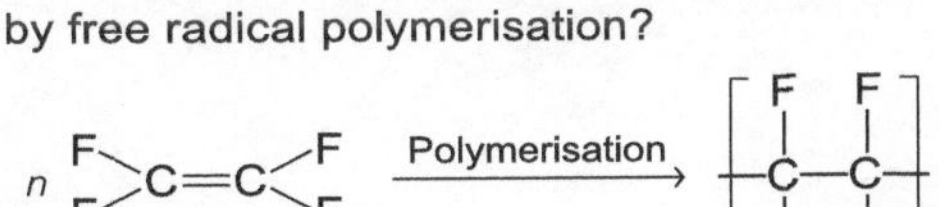

TFF Teflon

3 i. Write the IUPAC names of monomers of the following polymers:

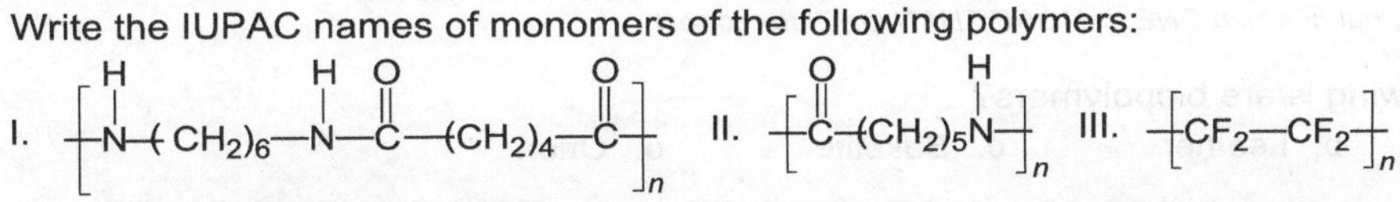

ii. Arrange the following polymers in increasing order of their intermolecular forces.

I. nylon-6 6, buna-S,polythene

II. nylon-6, neoprene, terylene, polyvinylchloride

Also justify your answer.

4 Explain, which of these monomers produces the more crystalline polymer?

i. 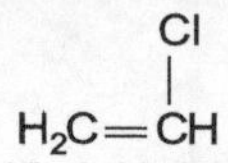
Vinyl chloride
or 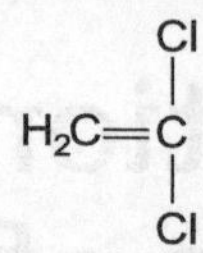 using radical polymerisation
Vinylidene chloride

ii. 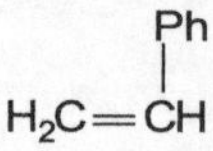 using radical polymerisation or coordination polymerisation
Styrene

Working Space

Only One Option Correct Type

Direction (Q. Nos. 5-7) *This section contains 3 multiple choice questions. Each question has four choices (a), (b), (c) and (d), out of which ONLY ONE is correct.*

5 Cellulose acetate is a

a. natural polymer
b. semisynthetic polymer
c. synthetic polymer
d. plasticiser

6 Which of the following sets contains only thermoplastics?

a. Glyptal, melmac, PAN
b. Polythene, bakelite, nylon-6
c. PVC, PMMA, polystyrene
d. Polypropylene, urea-formaldehyde, teflon

7 Which of the following is not correctly matched?

a. Nylon-6 6 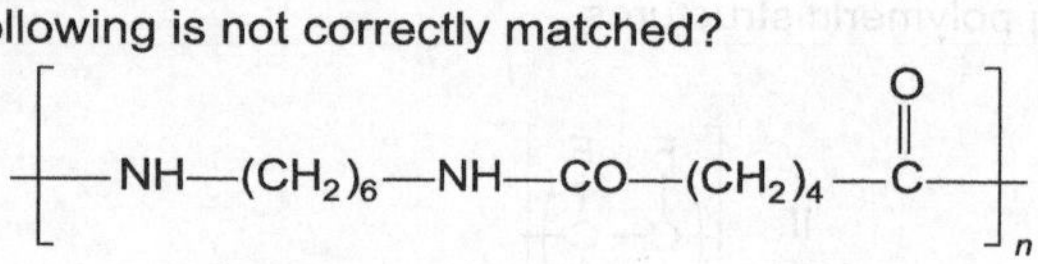

b. Neoprene 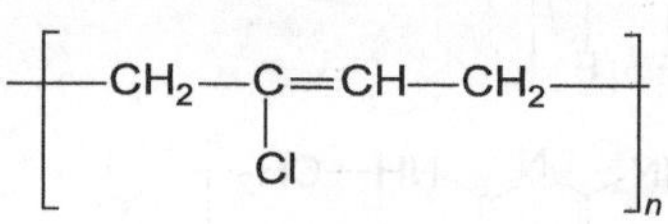

c. Terylene 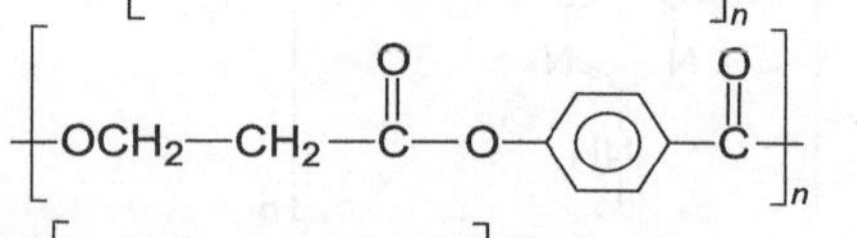

d. PMMA 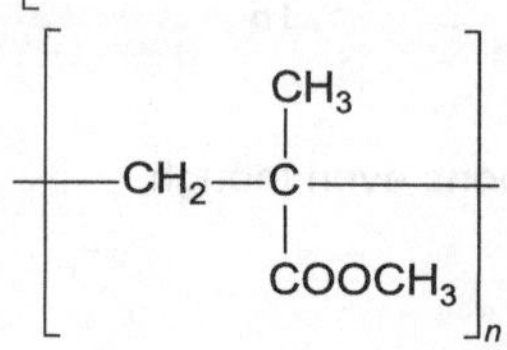

One or More than One Options Correct Type

Direction (Q. Nos. 8-10) *This section contains 3 multiple choice questions. Each question has four choices (a), (b), (c) and (d), out of which ONE or MORE THAN ONE are correct.*

8 Which of the following is/are biopolymers?

a. Nucleic acids
b. Leather
c. Bakelite
d. Orlon

9 Which of the following is/are not thermosetting polymers?

a. Bakelite
b. Polystyrene
c. PVC
d. Melmac

10 Which of the following is/are biodegradable polymers?

a. Nylon-6 6
b. PHBV
c. Nylon-2-nylon-6
d. Polychloroprene

Statement Type

Direction *(Q. Nos. 11 and 12) This section is based on Statement I and Statement II. Select the correct answer from the codes given below.*

a. Both Statement I and Statement II are correct and Statement II is the correct explanation of Statement I
b. Both Statement I and Statement II are correct but Statement II is not the correct explanation of Statement I
c. Statement I is correct but Statement II is incorrect
d. Statement II is correct but Statement I is incorrect

11 **Statement I** Nylon fibres are stronger than terylene fibres.
Statement II Intermolecular forces of attraction in terylene are H-bonding.

12 **Statement I** Bakelite is hard and has high melting point.
Statement II Intermolecular forces of attraction in it are H-bonding.

Matching List Type

Direction *(Q. Nos. 13-15) Choices for the correct combination of elements from Column I and Column II are given as options (a), (b), (c) and (d), out of which one is correct.*

13 Match the Column I with Column II and mark the correct option from the codes given below.

	Column I (Polymer)		Column II (Monomer)
i.	Resol	p.	Phenol + $CH_2{=}O$
ii.	PHBV	q.	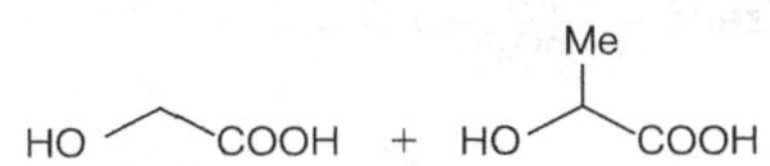
iii.	Polyglycolic and polylactic acid	r.	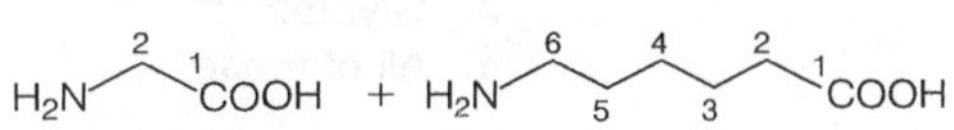
iv.	Nylon-2, 6	s.	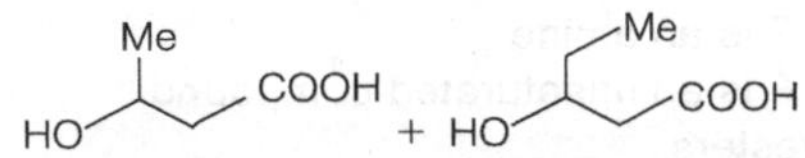

Codes

	i	ii	iii	iv
a.	s	q	p	r
b.	p	s	q	r
c.	r	p	s	q
d.	p	q	r	s

14 Match the Column I with Column II and mark the correct option from the codes given below.

	Column I		Column II
i.	Nylon-6 6	p.	Condensation polymerisation
ii.	Styrene	q.	Addition polymerisation
iii.	Nylon-6	r.	Homopolymer
iv.	Teflon	s.	Copolymer

Codes

	i	ii	iii	iv
a.	p,s	q,r	p,r	q,r
b.	q,r	p,q	r	s
c.	q,s	p,q	s,q	r,s
d.	s,r	q	s	p

Working Space

15 Match the Column I with Column II and mark the correct option from the codes given below.

	Column I (Polymer)		Column II (Type of polymer)
i.	Glyptal resin	p.	Polyamide
ii.	Perlon-L	q.	Copolymer
iii.	Nylon-6, 10	r.	Homopolymer
iv.	Perlon-U	s.	Polyester
v.	PHBV	t.	Polyurethane
		u.	Polycarbamate ester

Codes

	i	ii	iii	iv	v
a.	q	r	t	u	s
b.	s,q	p,r	p,q	q,t,u	q,s
c.	s	t	u	q	r
d.	t	s	u	r	q

Comprehension Type

***Direction** (Q. Nos. 16 and 17) This section contains a paragraph, describing theory, experiments, data, etc. Two questions related to the paragraph have been given. Each question has only one correct answer among the four given options (a), (b), (c) and (d).*

Passage

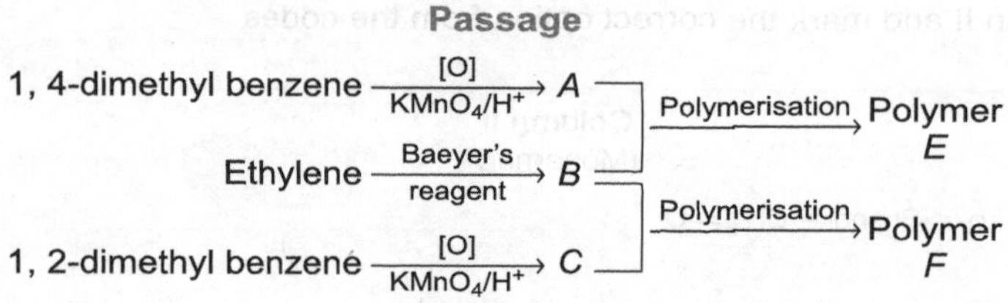

16 What is polymer *E*?

a. Dacron
b. Glyptal
c. Novolac
d. All of these

17 Select the correct statement about *E* and *F*.

a. *E* is a polyester while *F* is an amide
b. *E* is a polyester while *F* is an unsaturated compound
c. *E* and *F* both are polyesters
d. *E* and *F* are identical compounds

One Integer Value Correct Type

***Direction** (Q. Nos. 18 and 19) This section contains 2 questions. When worked out will result in an integer from 0 to 9 (both inclusive).*

18 How many of the following are condensation copolymers?
Nylon-6, nylon-6 6, dacron, glyptal, buna-S, ABS, neoprene, PHBV, perlon-U.

19 The molecular mass of the polymer is the integer multiple of molar mass of monomers (i.e. molar mass of polymer = $n \times$ (molar masses of monomers) in case of how many of the following polymers? Acrilon, vinyon, dynel, natural rubber, PCTFE, nylon-2, 6, polyurethane, PMMA, PHBV.

DPP-2 Mechanism of Polymerisation

Subjective Questions

Working Space

Direction (Q. Nos. 1-4) *These questions are subjective in nature, need to be solved completely on notebook.*

1 i. Free radical polymerisation is usually exothermic in nature. Explain.

ii. Free radical polymerisation of vinyl chloride can proceed in the following three ways:

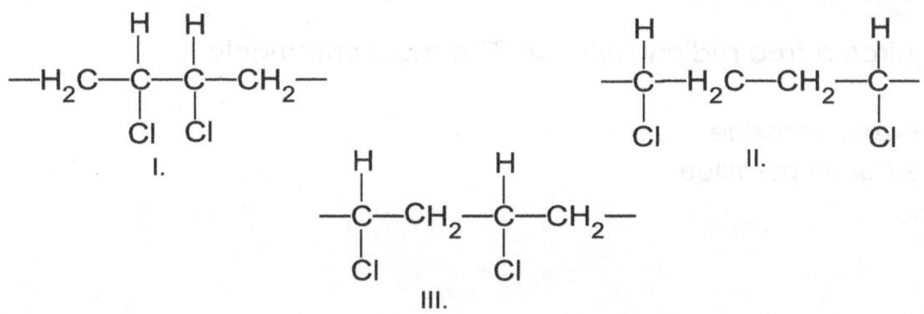

Out of the above three ways, only one type of polymer is formed. Which is the most preferred way of polymerisation and why? Provide mechanistic explanation.

2 i. Give the decreasing order of reactivities of the following monomers towards cationic addition polymerisation.

I. $MeCH{=}CH_2$ II. $PhCH{=}CH_2$

III. $CH_2{=}CH{-}COOMe$ IV. $CH_2{=}CH{-}Cl$

ii. Give the decreasing order of reactivities of the following monomers towards anionic addition polymerisation.

I. $Me{-}CH{=}CH_2$ II. $PhCH{=}CH_2$

III. $CH_2{=}CH{-}CN$ IV. $H_2C{=}CCl_2$

V. $H_2C{=}CF_2$ VI. $F_2C{=}CF_2$

3 Show, how each of the following polymer structures may be obtained from suitable monomers either by addition or condensation? More than one step may be required.

i. 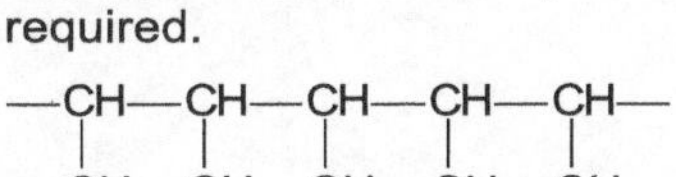

ii. 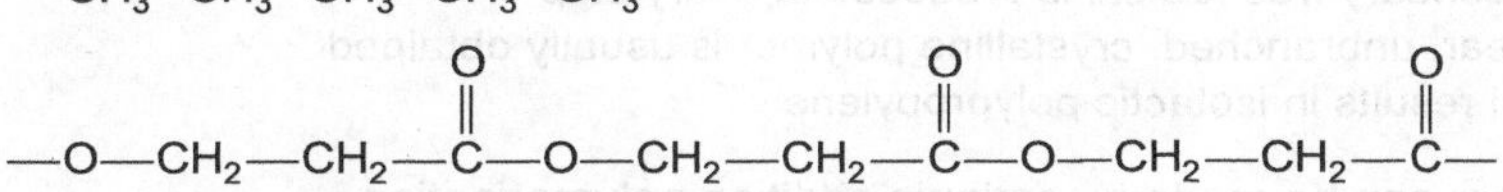

iii. 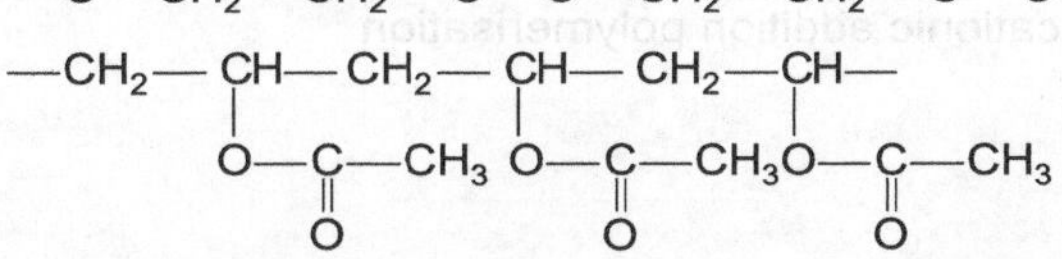

iv. 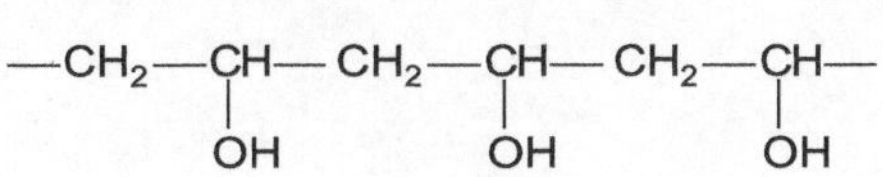

Working Space

4 The following two possible products can be formed in this reaction:

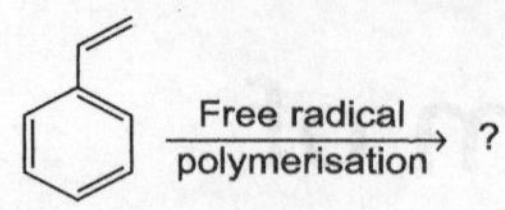

I. 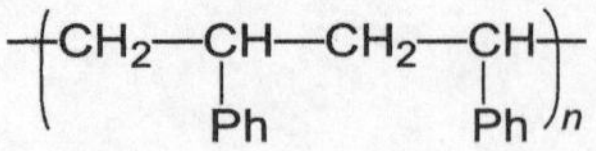

II. 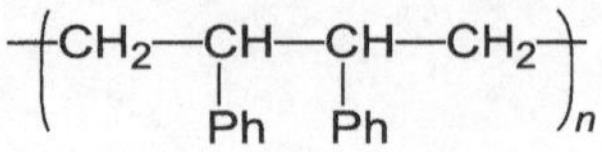

Which product is formed and why?

Only One Option Correct Type

***Direction** (Q. Nos. 5-9) This section contains 5 multiple choice questions. Each question has four choices (a), (b), (c) and (d), out of which ONLY ONE is correct.*

5 Free radical polymerisation requires a free radical initiator. The most commonly used free radical initiator is

a. Ph—CO—O—O—COPh, benzoyl peroxide
b. $(CH_3)_3C$—O—O—$C(CH_3)_3$, *tert*-butyl peroxide
c. C_6H_5—N ⟶ O, azoxybenzene
   ‖
   C_6H_5—N
d. CH_2N_2, diazomethane

6 Isotactic polypropylene polymer is one in which

a. all methyl groups are on one side of the extended chain. It is a highly crystalline, has high melting point and forms strong fibres
b. the methyl groups present alternate regularly from one side to the other
c. the methyl groups are distributed at random. It is a soft, elastic and rubbery material
d. None of the above

7 Which of the following monomers can undergoes radical, cationic as well as anionic polymerisation with equal ease?

a. $CH_3—C(CH_3)=CH_2$
b. $C_6H_5—CH=CH_2$
c. $CH_2=CH—CN$
d. $CH_2=CH_2$

8 What is not true regarding free radical polymerisation of propene?

a. Without proper control, atactic polypropylene is produced
b. During polymerisation, a secondary free radical is produced in every step
c. During polymerisation, a linear, unbranched, crystalline polymer is usually obtained
d. Use of *Ziegler-Natta* catalyst results in isotactic polypropylene

9 Which of the following polymers can be made by cationic addition polymerisation mechanism?

a. PVC
b. PP
c. HDPE
d. LDPE

Working Space

One or More than One Options Correct Type

Direction (Q. Nos. 10-12) *This section contains 3 multiple choice questions. Each question has four choices (a), (b), (c) and (d), out of which ONE or MORE THAN ONE are correct.*

10 Which of the following can be used as plasticisers?

a. Sodium hexametaphosphate
b. Di-*n*-butylphthalate
c. Tricresyl phosphate
d. Diethyl phthalate

11 Which monomer would polymerise in isotactic, syndiotactic, and atactic forms?

a. $CH_2{=}CCl_2$
b. $CH_3{-}CH{=}CH_2$
c. $Ph{-}CH{=}CH_2$
d. $CH_2{=}CH_2$

12 Polymerisation of buta-1, 3-diene by free radical mechanism gives

a. *trans*-1, 4-polybutadiene
b. *cis*-1, 4-polybutadiene
c. polyvinyl polyethene
d. polyallyl polyethene

Statement Type

Direction (Q. Nos. 13 and 14) *This section is based on Statement I and Statement II. Select the correct answer from the codes given below.*

a. Both Statement I and Statement II are correct and Statement II is the correct explanation of Statement I
b. Both Statement I and Statement II are correct but Statement II is not the correct explanation of Statement I
c. Statement I is correct but Statement II is incorrect
d. Statement II is correct but Statement I is incorrect

13 **Statement I** Polybutadiene is an example of step-growth polymer.
Statement II Copolymerisation of butadiene and acrylonitrile gives buna-N.

14 **Statement I** Buta-1, 3-diene is the monomer of gutta-percha.
Statement II Gutta-percha is formed through cationic addition polymerisation.

Matching List Type

Direction (Q. No. 15) *Choices for the correct combination of elements from Column I and Column II are given as options (a), (b), (c) and (d), out of which one is correct.*

15 Match the Column I with Column II and mark the correct option from the codes given below.

	Column I (Compound)		Column II (Characteristic property)
i.	Benzoquinone	p.	Free radical inhibitor
ii.	CCl_4 and CBr_4	q.	Chain transfer agents
iii.	Phenol	r.	*Ziegler-Natta* catalyst
iv.	$Et_3Al + TiCl_4$	s.	High density polyethylene

Codes

	i	ii	iii	iv		i	ii	iii	iv
a.	q	p	r	s	b.	s	p	q	r
c.	p	q	p	r,s	d.	r	p	s	q

Working Space

Comprehension Type

Direction *(Q. Nos. 16-18) This section contain a paragraph, describing theory, experiments, data, etc. Three questions related to the paragraph have been given. Each question has only one correct answer among the four given options (a), (b), (c) and (d).*

Passage

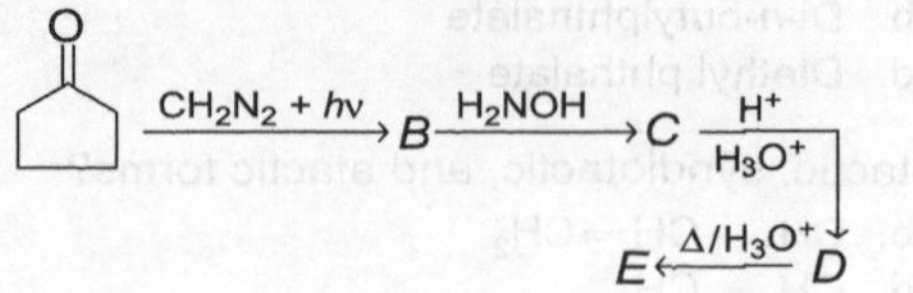

E is a macromolecule having a repeating unit.

16 What is the structure of D in this process?

a. 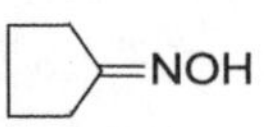

b. 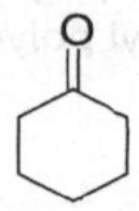

c. 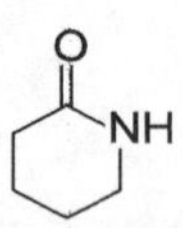

d. 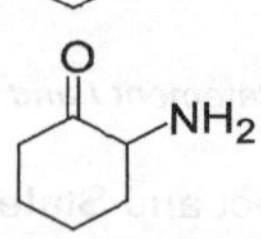

17 What type of polymerisation process is involve in the formation of E?

a. Addition homopolymerisation
b. Step growth copolymerisation
c. Step growth homopolymerisation
d. Chain growth copolymerisation

18 The intermediate(s) involved in the above reaction sequence is/are

a. 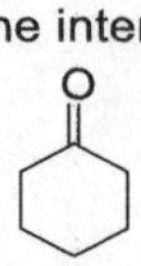

b. 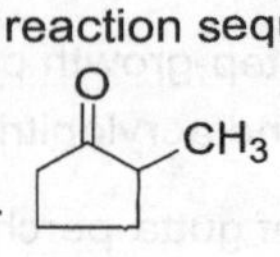

c. 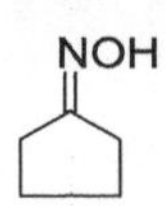

d. 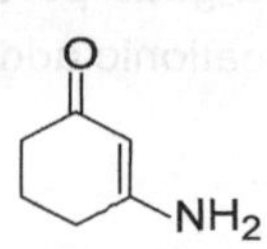

One Integer Value Correct Type

Direction (Q. Nos. 19 and 20) *This section contains 2 questions. When worked out will result in an integer from 0 to 9 (both inclusive).*

19 How many of the following are prepared by condensation polymerisation reaction?
Dacron, nylon-6 6, nylon-6, ABS, SBR, PE, bakelite, PMMA, kevlar

20 In the formation of how many of the following there is no loss of small molecules?
Orlon, PTFE, neoprene, bakelite, buna-S, melmac, vinyon, ABS, SAN.

DPP-3 Some Important Polymers (Preparation, Properties and Uses)

Subjective Questions

Working Space

Direction (Q. Nos. 1 and 2) *These questions are subjective in nature, need to be solved completely on notebook.*

1 Answer the following questions:

i. How does the number of carbons in diamine or diacid moiety affects the physical properties of nylons?

ii. Why does hole develop in nylon stocking when a drop of HCl is added to it?

iii. How irradiation of polyethylene with γ-rays bring about cross-linking?

2 A copolymer of ethene and vinyl chloride contains alternate monomers of each type. What is the mass percentage of vinyl chloride in this copolymer?

Only One Option Correct Type

Direction (Q. Nos. 3-9) *This section contains 7 multiple choice questions. Each question has four choices (a), (b), (c) and (d), out of which ONLY ONE is correct.*

3 Which pair of polymers have similar properties?

a. Nylon, PVC
b. PAN, PTFE
c. PCTFE, PTFE
d. Bakelite alkyl resin

4 Soft drink and baby feeding bottles are generally made of

a. polyurea
b. polyurethane
c. polyester
d. polystyrene

5 GRA is a copolymer of

a. butadiene and acrylonitrile
b. butadiene and adipic acid
c. chloroprene and acrylonitrile
d. chloroprene and adipic acid

6 Synthetic human hair wigs are made from a copolymer of vinyl chloride and acrylonitrile, which is called

a. PVC
b. polyacrylonitrile
c. cellulose
d. dynel

7 Terylene is a condensation polymer of ethylene glycol and

a. benzoic acid
b. phthalic acid
c. salicylic acid
d. terephthalic acid

Working Space

8 Gutta-percha is

a. *trans*-polyisoprene
b. non-elastic and softens to a plastic-like material on heating
c. used in underwater cables and golf balls
d. All of the above

9 If after complete ozonolysis of one mole of monomer of a synthetic polymer gives formaldehyde and $O{=}C(Cl){-}C(H){=}O$ as the only products, the polymer is

a. polyvinylchloride b. vinyon c. neoprene d. chloroprene

One or More than One Options Correct Type

Direction (Q. Nos. 10-12) *This section contains 3 multiple choice questions. Each question has four choices (a), (b), (c) and (d), out of which ONE or MORE THAN ONE are correct.*

10 Which of the following statement(s) is/are correct about nylon 66?

a. Nylon fibres have higher tensile strength than terylene fibres
b. Nylon fibres have lower tensile strength than terylene fibres
c. In nylon, there is strong intermolecular H-bonding, while in terylene, there is weak dipole-dipole interaction
d. In nylon, there is weak dipole-dipole interaction, while in terylene, there is strong intermolecular H-bonding

11 What is/are the correct regarding natural rubber?

a. It is *cis*-polyisoprene
b. It is amorphous in nature
c. It has syndiotactic stereochemistry
d. Molecules of rubber have coiled conformation

12 By which of the following reaction sequence can nylon-5 be prepared?

a. 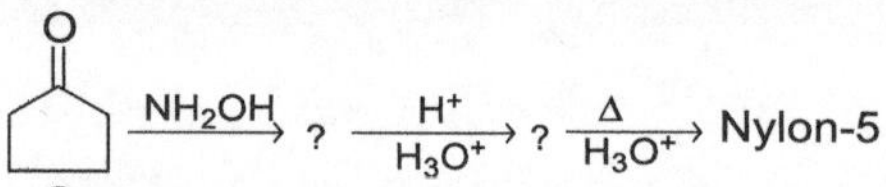

b. 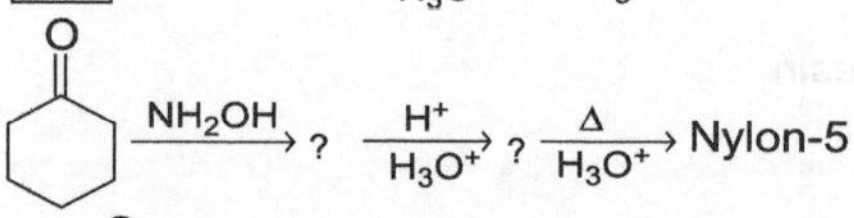

c. 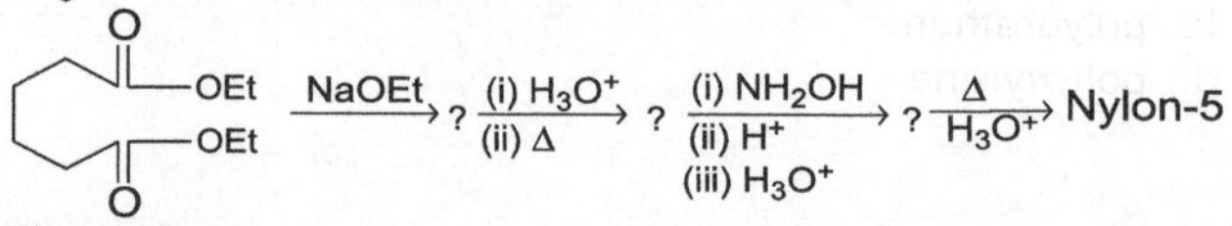

Diethyladipate

d. 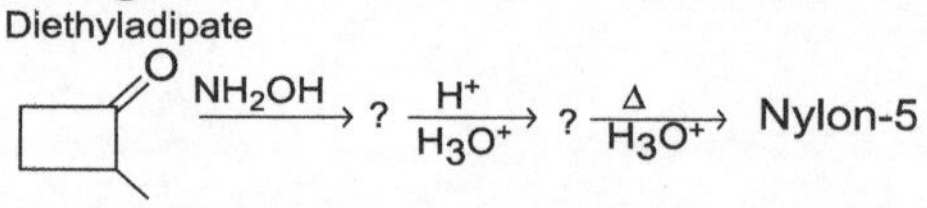

Statement Type

Direction (Q. Nos. 13 and 14) *This section is based on Statement I and Statement II. Select the correct answer from the codes given below.*

a. Both Statement I and Statement II are correct and Statement II is the correct explanation of Statement I
b. Both Statement I and Statement II are correct but Statement II is not the correct explanation of Statement I
c. Statement I is correct but Statement II is incorrect
d. Statement II is correct but Statement I is incorrect

Working Space

13 **Statement I** Plexiglas is the commercial name of PMMA.

Statement II It is used in making contact lens because it has an excellent light transmission property.

14 **Statement I** PUF (polyurethane foam) is spongy.

Statement II During the preparation of PUF, CO_2 is evolved, which forms bubbles that are trapped within the bulk of polymer. As it solidifies, it gives spongy product.

Matching List Type

Direction *(Q. Nos. 15 and 16) Choices for the correct combination of elements from Column I and Column II are given as options (a), (b), (c) and (d), out of which one is correct.*

15 Match the Column I with Column II and mark the correct option from the codes given below.

	Column I (Polymer)		Column II (Structure)
i.	Neoprene	p.	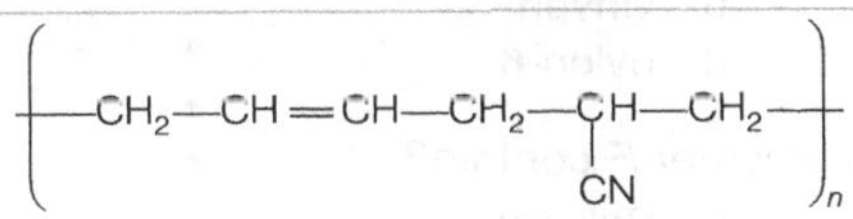
ii.	Gutta-percha	q.	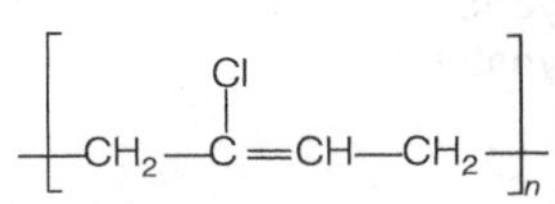
iii.	Buna-S	r.	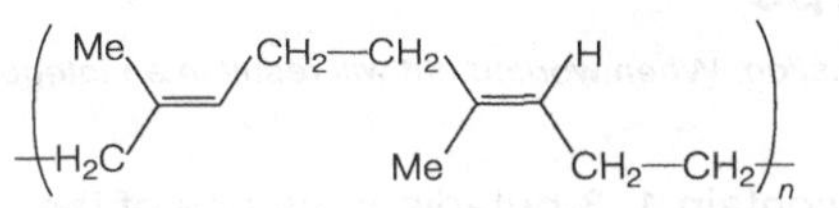
iv.	Buna-N	s.	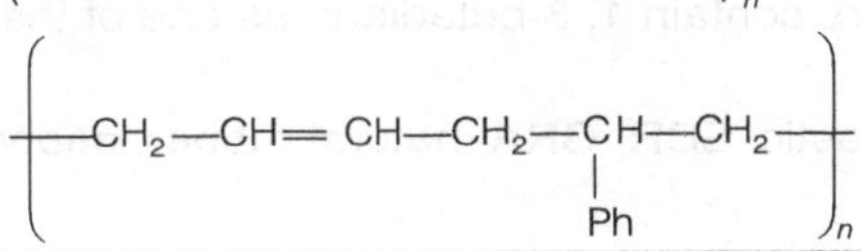

Codes

	i	ii	iii	iv
a.	p	s	q	r
b.	q	r	s	p
c.	s	p	q	r
d.	r	q	p	s

16 Match the Column I with Column II and mark the correct option from the codes given below.

	Column I (Polymer)		Column II (Step involve in preparation)
i.	Bakelite	p.	Electrophilic addition
ii.	SARAN	q.	Electrophilic substitution
iii.	Perlon-L (from cyclohexane)	r.	Beckmann rearrangement
iv.	Melmac	s.	Elimination of water
		t.	Nucleophilic addition

Codes

	i	ii	iii	iv
a.	q	s,p	r,s	t,q
b.	s,r	q	s	t
c.	t	p	q,s	r,t
d.	q,s	p	r,s,t	s,t

Comprehension Type

Working Space

Direction *(Q. Nos. 17-19) This section contains a paragraph, describing theory, experiments, data, etc. Three questions related to the paragraph have been given. Each question has only one correct answer among the four given options (a), (b), (c) and (d).*

Passage

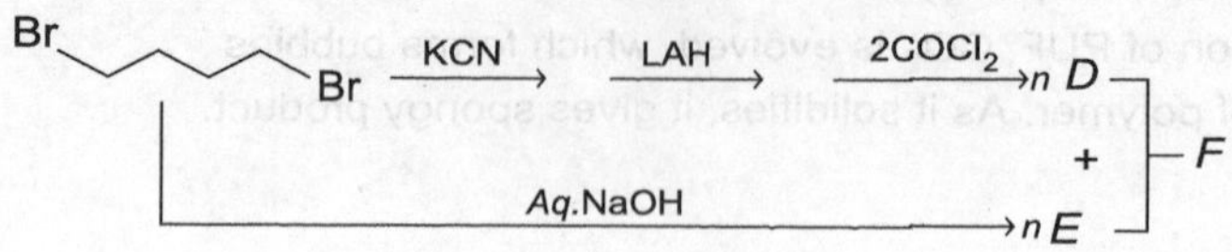

17 Compound *D* is

a. $O{=}C{=}N(CH_2)_6N{=}C{=}O$
b. $C{\equiv}\overset{+}{N}{-}(CH_2)_6\overset{+}{N}{\equiv}C$
c. $O{=}C{=}N(CH_2)_4N{=}C{=}O$
d. $C{\equiv}\overset{+}{N}{-}(CH_2)_4N{=}C{=}O$

18 The polymer *F* is

a. polyurethane
b. vinyon
c. perlon-L
d. nylon-6

19 Which of the following groups does polymer *F* contain?

a. Polyamide
b. Polyene
c. Polycarbamate ester
d. Polyester

One Integer Value Correct Type

Direction *(Q. No. 20) This section contains 1 question. When worked out will result in an integer from 0 to 9 (both inclusive).*

20 How many of the following polymers contain 1, 3-butadiene as one of the monomer?
Butyl rubber, nitrile rubber, ABS plastic, SBR, GRA, natural rubber, neoprene

SET - 1

Revisal Problems
for JEE Main

Only One Option Correct Type

Working Space

Direction (*Q. Nos.* 1-30) *This section contains* 30 *multiple choice questions. Each question has four choices (a), (b), (c) and (d), out of which ONLY ONE is correct.*

1 Cellulose has very high degree of hydrophilicity because of
- a. its amorphous nature
- b. crystalline nature
- c. presence of excessive voids in solid state
- d. presence of many hydroxyl groups on the polymer backbone

2 In the vulcanisation of rubber,
- a. sulphur reacts to form a new compound
- b. sulphur cross-links are introduced
- c. sulphur forms a very thin protective layer over rubber
- d. All of the above

3 Which one is not the chain growth polymer?
- a. Natural rubber
- b. Polythene
- c. Polypropylene
- d. Terylene

4 Which of the following is not a biodegradable polymer?
- a. PHBV
- b. Nylon-2, 6
- c. Polyglycolic and polylactic acid
- d. Perlon-U

5 Which percentage of sulphur is used in the vulcanisation of rubber?
- a. 3%
- b. 5%
- c. 30%
- d. 55%

6 Isotactic polyethenyl benzene has
- a. high crystallinity
- b. high density
- c. better mechanical strength
- d. All of the above

7 The interparticle forces between linear chains in nylon-6 6 are
- a. H-bonds
- b. covalent bonds
- c. ionic bonds
- d. coordinate bonds

8 Which of the following is an example of copolymer?
- a. Buna-S
- b. PAN
- c. Polythene
- d. PTFE

9 Which of the following fibre is not made of polyamides?
- a. Wool
- b. Natural silk
- c. Rayon
- d. Nylon

10 *Ziegler-Natta* catalysts
- a. are triethyl aluminium titanium tetrachloride complex $(C_2H_5)_3Al + TiCl_4$
- b. are used to prepare stereospecific addition polymers
- c. are employed to have stereochemical control of the polymerisation process
- d. All of the above

Working Space

11 What monomer would you use to prepare the following polymer?

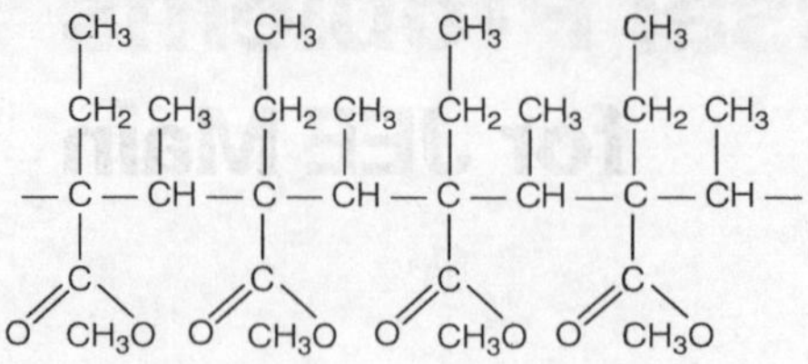

a. $CH_3-CH(COOCH_3)-CH_2-CH_3$
b. $CH_3-CH(COOCH_3)-CH=CH_2$
c. $CH_2=C(COOCH_3)-CH_2-CH_3$
d. $CH_3-CH=C(COOCH_3)-CH_2-CH_3$

12 Which one is not a step polymer?
a. Nylon-66 b. Nylon-6 c. Glyptal d. PMMA

13 $-[NH(CH_2)_6NHCO(CH_2)_4CO]_n-$ is
a. addition polymer
b. copolymer
c. homopolymer
d. thermosetting polymer

14 Which of the following polymer cannot be made by free radical addition polymerisation mechanism?
a. PE b. HDPE c. LDPE d. Teflon

15 Cellulose is a condensation polymer of
a. maltose
b. β-glucose
c. α-glucose
d. β-fructose

16 The turbidity of a polymer solution measures
a. light absorbed by the solution
b. light transmitted by the solution
c. light scattered by the solution
d. None of these

17 Which of the following is a step-growth polymer?
a. Polyacrylonitrile
b. Polyisoprene
c. Nylon
d. Polythene

18 Structures of some common polymers are given. Which is not correctly represented?
a. Teflon $-(CF_2-CF_2)_n-$
b. Neoprene

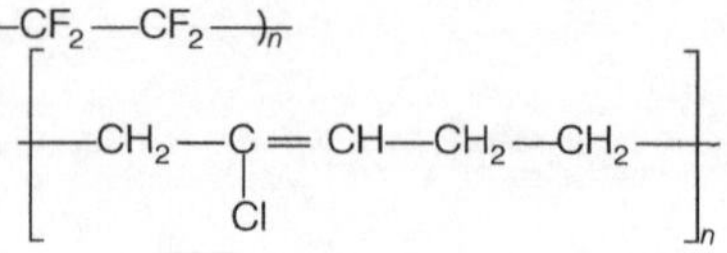

c. Terylene

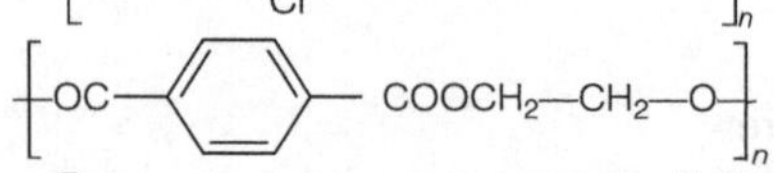

d. Nylon- 66

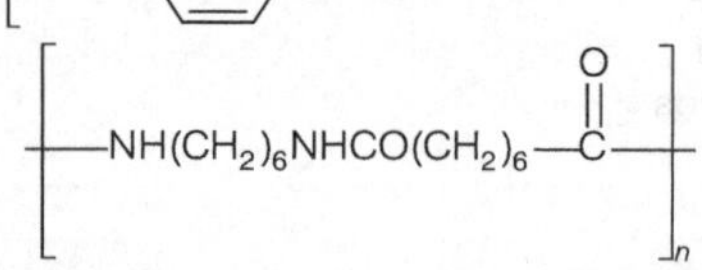

19 What is true regarding vulcanisation of rubber?
a. Rubber molecules are joined through S—S linkage at the ends
b. Rubber molecules are linked through S—S linkage at the various parts of polymer backbone
c. Vulcanisation makes rubber perfectly crystalline
d. Vulcanisation converts rubber into a thermosetting polymer

20 Which of the following statements is incorrect about phenol-formaldehyde resin?
a. Novolac or resol is a linear polymer and is used in the manufacture of adhesive
b. Bakelite is a cross-linked polymer and is used in making switches and plugs
c. Novolac is prepared when (P/F) (phenol/formaldehyde) ratio is greater than 1, whereas bakelite is prepared when (P/F) ratio is less than 1
d. Novolac is prepared when $P/F < 1$ and bakelite is prepared when $P/F > 1$

Working Space

21 The incorrect statement(s) regarding cellulose is/are

a. It is a polymer of D-glucose
b. It has β-1, 4-glucosidic linkage
c. It is used for making rayon fibre
d. It can be obtained by polymerisation of D-glucose

22 Bakelite is/are prepared by the reaction between

a. phenol and formaldehyde
b. ethylene glycol and dimethylphthalate
c. urea and formaldehyde
d. tetramethylene glycol and hexamethylene diamine.

23 Natural rubber is/are polymer of

a. butadiene
b. ethyne
c. styrene
d. isoprene

24 Polyurethane

a. have structural features of both an ester and an amide
b. have the formula $H_2N—\underset{\underset{O}{\|}}{C}—O—CH_2—CH_3$
c. are obtained from *p*-phenylene diisocyanate and ethylene glycol
d. All of the above

25 SBR (GRS. cold rubber) is obtained by free radical copolymerisation of

a. buta-1, 3-diene (70%) and 30% phenyl ethene
b. chloroprene and styrene
c. vinyl acetylene and styrene
d. isoprene and 1, 3-butadiene

26 Which of the following is/are true regarding teflon?

a. It is a linear, unbranched polymer of tetrafluoroethylene
b. It has very high thermal stability
c. polymer molecules are associated by strong dipole-dipole attraction
d. All of the above

27 What is/are true regarding free radical polymerisation of ethylene?

a. Polymerisation is invoked by either peroxide initiator or radiation
b. Polymerisation is exothermic in nature
c. Molecular weight gain occur in propagation step of polymerisation
d. All of the above

28 Which statement(s) is/are incorrect about the following polymer?

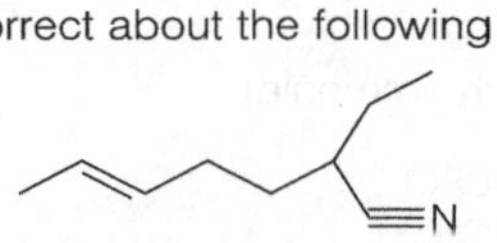

a. Its monomers are buta-1, 3-diene and $CH_2{=}CH—CN$
b. Its monomer is $CH_2{=}CH—CH{=}CH—CH{=}CH—C{\equiv}N$
c. It is a addition homopolymer
d. It is a condensation copolymer

29 $\underset{\text{Amine}}{X} + \underset{\text{Acid}}{Y} \longrightarrow$ Nylon-6, 10

Acid *Y* used here is/are

a. Adipic acid
b. Sebacic acid
c. Oxalic acid
d. Octanoic acid

30 The reagent used to obtain PAN from acetylene is/are

a. KCN
b. AgCN, C_2H_5OH
c. HCN, CuCl/HCl
d. None of these

SET - 2

Revisal Problems
for JEE Advanced

Only One Option Correct Type

Direction *(Q. Nos. 1-12) This section contains 12 multiple choice questions. Each question has four choices (a), (b), (c) and (d), out of which ONLY ONE is correct.*

Working Space

1 The chemical name of melamine is

a. 2, 4-diamino-1,3,5-triazine
b. 2-amino-1,3,5-triazine
c. 2,4,6-triamino-1,3,5 triazine
d. 1,3,5-triamino-2,4,6-triazine

2 Which of the following is the biodegradable polymer of polyamide class?

a. Nylon-6 6
b. Nylon-2-nylon-6
c. Dextran
d. PHBV

3 Which of the following sets contains only copolymers?

a. SBR, glyptal, nylon-6 6
b. Polythene, polyester, PVC
c. Nylon-6, butyl rubber, neoprene
d. Melmac, bakelite, teflon

4 Which is used for the formation of nylon-6 6?

a. Sulphurous acid
b. Adipic acid
c. Sulphur hexafluoride
d. Phthalic acid

5 $\{CH_2—CH(—COOH)\}_n \xrightarrow{NaOH} X$

Polypropanoic acid
(high molecular weight)

The true statement about X is

a. It has intermolecular forces stronger than parent species
b. It is a mixture of several compounds
c. It is more cross-linked than the parent species
d. All of the above

6 Identify X.

$$CHCl_3 \xrightarrow[HF]{SbF_3} \xrightarrow{1070\ K} X + \text{gas (form white fumes with ammonia)}$$

a. $F_2C=CF_2$
b. $ClFC=CFCl$
c. $F_2C=CFCl$
d. $F_2C=CCl_2$

7 $R_2C=O \xrightarrow[\text{(ii) Methanol, } \Delta \text{ (iii) Peroxide}]{\text{(i) } HCN/H_3O^+} X$

X is

a. PMMA if $R = Me$
b. PAN if $R = E^+$
c. PEMA if $R = H$
d. All of these

8 Atactic polymethyl-2-methyl propenoate is not an elastomer. This is because of the presence of

a. linear structure of the polymer
b. weak intermolecular forces of attraction
c. presence of polar group
d. All of the above

Working Space

9 When a diamine is treated with some diacid under suitable conditions, substance X is formed. Under what conditions, substance X behaves as an elastomer?

a. By increasing crystallinity of the substance
b. By introducing cross-links and branching
c. By cooling the substance
d. Substance X is a fibre and cannot behave as an elastomer

10 The formation of polyethylene from calcium carbide takes place as follows:

$$CaC_2 + 2H_2O \longrightarrow Ca(OH)_2 + C_2H_2$$
$$C_2H_2 + H_2 \longrightarrow C_2H_4$$
$$nC_2H_4 \longrightarrow -\!\!(CH_2-CH_2)\!\!-_n$$

The amount of polyethylene obtained from 64 kg of CaC_2 is

a. 14 kg b. 7 kg c. 21 kg d. 28 kg

11 Bakelite is a condensation copolymer of phenol and formaldehyde. The initial step between the two compounds is an example of

a. free radical reaction
b. aldol condensation
c. aromatic nucleophilic substitution
d. aromatic electrophilic substitution

12 Vulcanisation makes rubber

I. more elastic
II. soluble in inorganic solvent
III. crystalline
IV. more stiff

The correct option is

a. I and IV b. I and III c. II and III d. II, III and IV

One or More than One Options Correct Type

Direction (Q. Nos. 13-18) *This section contains 6 multiple choice questions. Each question has four choices (a), (b), (c) and (d), out of which ONE or MORE THAN ONE are correct.*

13 Which of the following polymers can be made by anionic addition polymerisation mechanism?

a. PVC b. PAN c. Teflon d. PP

14 Which of the following are polycarbamate ester polymers?

a. Polyurethane b. Perlon-U c. Melmac d. Saran

15 Which of the following processes can be used to prepare polystyrene?

a. Anionic b. Cationic c. Free radicals d. *Ziegler-Natta*

16 In which case, the polymer is/are not chiral?

a. n [2-ethylpropene] $\xrightarrow{\text{Polymerisation}}$
b. n [Propene] $\xrightarrow{\text{Polymerisation}}$
c. n [2-methyl buta-1,3-diene] $\xrightarrow{\text{Polymerisation}}$
d. n [buta-1,3-diene] $\xrightarrow{\text{Polymerisation}}$

17 Which is/are true regarding rayon?

a. It is pure regenerated cellulose
b. It is obtained by dissolving wood pulp in alkaline CS_2
c. It is obtained by passing Na salt of cellulose xanthate through spinneret into aqueous $NaHNO_3$ solution
d. It is extracted as fibres of cellulose

18 Nylon-5,10 can be prepared by

a. $H_2N(CH_2)_5NH_2$ + (sebacic acid)
b. reduced form of pentan-1,5-dinitrile + $HOOC(CH_2)_8COOH$
c. $H_2N(CH_2)_5NH_2 + HOOC(CH_2)_{10}COOH$
d. $H_2N(CH_2)_{10}NH_2 + HOOC(CH_2)_3COOH$

Working Space

Comprehension Type

***Direction** (Q. Nos. 19-22) This section contains 2 paragraphs, each describing theory, experiments, data, etc. Four questions related to the paragraphs have been given. Each question has only one correct answer among the four given options (a), (b), (c) and (d).*

Passage I

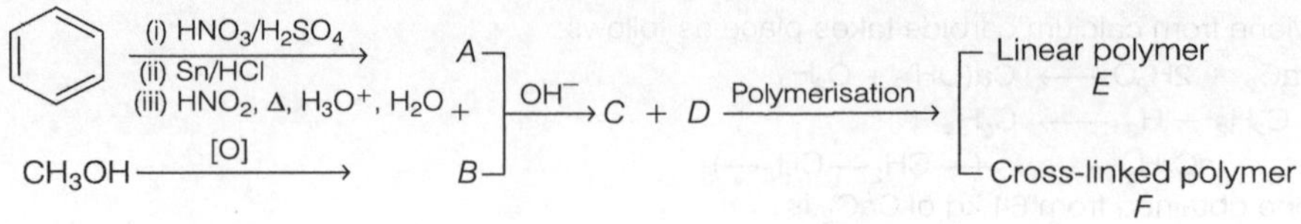

19 The linear polymer E is formed when

a. $\frac{P}{F}\left(\frac{\text{Phenol}}{\text{Formaldehyde}}\right)=1$

b. $\frac{P}{F}>1$

c. $\frac{P}{F}<1$

d. None of these

20 Which of the following statement(s) is/are correct about the polymer F?

a. It is thermoplastic polymer
b. It is thermosetting polymer
c. It is used in the manufacture of adhesive
d. It is used in the manufacture of switches and plugs

Passage II

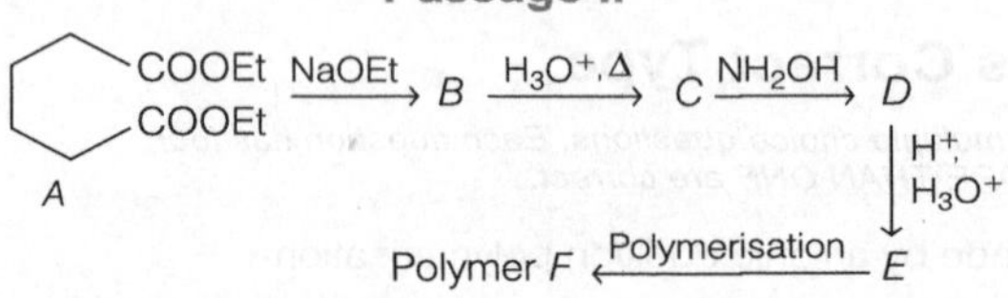

21 The conversion A to B is called

a. Claisen ester condensation
b. Dieckmann reaction
c. intramolecular Claisen ester condensation
d. Claisen-Schmidt reaction

22 Polymer F is

a. nylon-6
b. nylon-5
c. nylon-6 6
d. nylon 5,5

Statement Type

***Direction** (Q. Nos. 23-25) This section is based on Statement I and Statement II. Select the correct answer from the codes given below.*

a. Both Statement I and Statement II are correct and Statement II is the correct explanation of Statement I
b. Both Statement I and Statement II are correct but Statement II is not the correct explanation of Statement I
c. Statement I is correct but Statement II is incorrect
d. Statement II is correct but Statement I is incorrect

23 **Statement I** Rayon is a semisynthetic polymer and is taken as a better choice than cotton fabric.

Statement II Mechanical and aesthetic properties of cellulose can be improved by acetylation.

24 **Statement I** Polyvinyl alcohol is obtained by polymerisation of vinyl alcohol.

Statement II Keto form of vinyl alcohol is more stable.

25 **Statement I** HDPE is prepared by free radical addition polymerisation.

Statement II HDPE consists of linear chains.

Working Space

Matching List Type

Direction *(Q. Nos. 26 and 27) Choices for the correct combination of elements from Column I and Column II are given as options (a), (b), (c) and (d), out of which one is correct.*

26 Match the Column I with Column II and mark the correct option from the codes given below.

	Column I (Polymers and plasticisers)		Column II (Property)
i.	*o*-cresyl phosphate	p.	Thermosetting
ii.	Bakelite	q.	Thermoplastic
iii.	Resol or novolac	r.	Natural polymer
iv.	Cellulose	s.	Synthetic polymer

Codes

	i	ii	iii	iv		i	ii	iii	iv
a.	q	p,s	p,s	r,s	b.	s	p	q	r
c.	q	s,p	r	q	d.	p	q,s	r	p

27 Match the Column I with Column II and mark the correct option from the codes given below.

	Column I		Column II
i.	HDPE	p.	An ester
ii.	Polypropene	q.	Reduction in % *s*-character
iii.	PVC	r.	Free radical addition polymerisation
iv.	Dacron	s.	Homopolymer

Codes

	i	ii	iii	iv		i	ii	iii	iv
a.	q,s	q,r,s	q,s	p	b.	p	q,r	q	s
c.	q	r,s	q	p	d.	p	q,r	q	p

One Integer Value Correct Type

Direction *(Q. Nos. 28 and 29) This section contains 2 questions. When worked out will result in an integer from 0 to 9 (both inclusive).*

28 How many of the following are the examples of addition homopolymers?
ABS, SAN, SBR, neoprene, PMMA, PVC, PCTFE, glyptal, teflon

29 What is the number of carbon atoms in the ketone used to prepare ε-aminovaleric acid, a monomer for nylon category?

JEE Archive

(Compilation of Last 13 Years' Questions of JEE Main, AIEEE, JEE Advanced & IIT JEE)

1 Which one is classified as a condensation polymer? **(JEE Main 2014)**

a. Dacron b. Neoprene
c. Teflon d. Acrylonitrile

2 The species which can best serve as an initiator for the cationic polymerisation is **(AIEEE 2012)**

a. $LiAlH_4$ b. HNO_3
c. $AlCl_3$ d. BaLi

3 Thermosetting polymer, bakelite is formed by the reaction of phenol with **(AIEEE 2011)**

a. CH_3CH_2CHO b. CH_3CHO
c. HCHO d. HCOOH

4 The polymer containing strong intermolecular forces, e.g. hydrogen bonding, is **(AIEEE 2010)**

a. teflon b. nylon-66
c. polystyrene d. natural rubber

5 Buna-N synthetic rubber is a copolymer of **(AIEEE 2009)**

a. $H_2C=CH-\overset{Cl}{\overset{|}{C}}=CH_2$ and $H_2C=CH-CH=CH_2$
b. $H_2C=CH-CH=CH_2$ and $H_5C_6-CH=CH_2$
c. $H_2C=CH-CN$ and $H_2C=CH-CH=CH_2$
d. $H_2C=CH-CN$ and $H_2C-\underset{CH_3}{\underset{|}{C}}=CH_2$

6 Bakelite is obtained from phenol by reacting with **(AIEEE 2008)**

a. $(CH_2OH)_2$ b. CH_3CHO
c. CH_3COCH_3 d. HCHO

7 Which of the following is fully fluorinated polymer? **(AIEEE 2005)**

a. PVC b. Thiokol
c. Teflon d. Neoprene

8 Which of the following is a polyamide? **(AIEEE 2005)**

a. Bakelite b. Terylene
c. Nylon-6 6 d. Teflon

9 Nylon threads are made up of **(AIEEE 2003)**

a. polyvinyl polymer
b. polyester polymer
c. polyamide polymer
d. polyethylene polymer

10 Monomers are converted to polymer by **(AIEEE 2002)**

a. hydrolysis of monomers
b. condensation reaction between monomers
c. protonation of monomers
d. None of the above

11 Among cellulose, poly (vinyl chloride), nylon and natural rubber, the polymer in which the intermolecular force of attraction is weakest is **(2009, Only One Option Correct Type)**

a. nylon
b. poly (vinyl chloride)
c. cellulose
d. natural rubber

12 Match the chemical substances in Column I with type of polymers/type of bond in Column II. **(2007, Matching Type)**

	Column I		Column II
i.	Cellulose	p.	Natural polymer
ii.	Nylon-6 6	q.	Synthetic polymer
iii.	Protein	r.	Amide linkage
iv.	Sucrose	s.	Glycoside linkage

Codes

	i	ii	iii	iv
a.	s	p	q,s	q,r
b.	p,s	q,r	p,r	s
c.	s	p,s	q	r
d.	p	q	s	r,s

13 Monomer *A* of a polymer on ozonolysis yields two moles of HCHO and one mole of CH_3COCHO. **(2005, Subjective Type)**

i. Deduce the structure of *A*.
ii. Draw the structures of all *cis*-forms of polymer of compound *A*.

Answers *with* Explanations

Answers

1. Aryl Halides and Phenols

DPP-1 Methods of Preparation of Aryl Halides

5. *a* **6.** *c* **7.** *c* **8.** *b* **9.** *d* **10.** *d* **11.** *c* **12.** *a,b* **13.** *a,b,d*
14. *a,c,d* **15.** *d* **16.** 2

DPP-2 Physical Properties and Electrophilic Substitution in Aryl Halides

4. *a* **5.** *b* **6.** *c* **7.** *c* **8.** *a* **9.** *a* **10.** *c* **11.** *c* **12.** *d* **13.** *a,b*
14. *a,b* **15.** *a* **16.** *d*

DPP-3 Chemical Reactions of Aryl Halides (Other than Electrophilic Substitution)

7. *c* **8.** *d* **9.** *d* **10.** *c* **11.** *a* **12.** *d* **13.** *d* **14.** *b* **15.** *a,b,c,d*
16. *a,b,c* **17.** *a,b,c* **18.** *a* **19.** *b* **20.** *a* **21.** (i) 3 (ii) 2

DPP-4 Methods of Preparation of Phenols

2. *b* **3.** *d* **4.** *c* **5.** *c* **6.** *c* **7.** *b* **8.** *b* **9.** *b*
10. *a,c,d* **11.** *a,c,d* **12.** *b,c,d* **13.** *a,c,d* **14.** *a,d* **15.** *a,c,d* **16.** *c* **17.** *b* **18.** *d* **19.** *a*
20. 4

DPP-5 Physical Properties, Acidity and Electrophilic Substitution Reactions of Phenols

2. *b* **3.** *c* **4.** *b* **5.** *a,b,c,d* **6.** *a,b,c* **7.** *b,c,d* **8.** *a,b,c,d* **9.** *a,d* **10.** *a,c,d*
11. *a* **12.** *d* **13.** *a* **14.** *c* **15.** *b* **16.** *b* **17.** *c* **18.** *a* **19.** *b*
20. (i) 5 (ii) 4 (iii) 8

DPP-6 Chemical Properties of Phenols (Other than Electrophilic Substitution)

4. *d* **5.** *d* **6.** *a* **7.** *d* **8.** *c* **9.** *c* **10.** *c* **11.** *c* **12.** *a,c,d* **13.** *b,d*
14. *c,d* **15.** *c* **16.** *c* **17.** *c* **18.** *b* **19.** *c* **20.** *d*

Revisal Problem (JEE Main)

1. *c* **2.** *c* **3.** *b* **4.** *d* **5.** *a* **6.** *b* **7.** *a* **8.** *a* **9.** *b* **10.** *d*
11. *b* **12.** *a* **13.** *d* **14.** *c* **15.** *a* **16.** *d* **17.** *b* **18.** *c* **19.** *c* **20.** *b*
21. *c* **22.** *d* **23.** *b* **24.** *c* **25.** *d* **26.** *b* **27.** *b* **28.** *c* **29.** *a* **30.** *d*

Revisal Problems (JEE Advanced)

1. *c* **2.** *a* **3.** *b* **4.** *a* **5.** *d* **6.** *c* **7.** *a* **8.** *b* **9.** *a* **10.** *c*
11. *d* **12.** *b,c,d* **13.** *a,b,d* **14.** *a,b* **15.** *b,c* **16.** *a,d* **17.** *a,c* **18.** *a,b,c,d* **19.** *a,c* **20.** *a,b*
21. *a,c* **22.** *a* **23.** *b* **24.** *d* **25.** *a* **26.** *c* **27.** *b* **28.** *a* **29.** *a* **30.** *b*
31. *a* **32.** *c* **33.** 3 **34.** 4 **35.** 4

JEE Main & AIEEE Archive

1. *c* **2.** *a* **3.** *b* **4.** *d* **5.** *d* **6.** *b* **7.** *c* **8.** *b* **9.** *b* **10.** *a*
11. *b* **12.** *c*

JEE Advanced & IIT JEE Archive

1. *d* **2.** *b,d* **3.** *b* **4.** *a,b,c* **5.** *a* **6.** *c* **7.** *c* **8.** *c* **9.** *b*
10. *a,b,c*

2. Aromatic Aldehydes, Ketones and Acids

DPP-1 Methods of Preparation of Carbonyl Compounds

2. *a* **3.** *b* **4.** *d* **5.** *c* **6.** *b* **7.** *d* **8.** *a* **9.** *c* **10.** *a* **11.** *a*
12. *c* **13.** *b,d* **14.** *b,c* **15.** *b* **16.** *b* **17.** *c* **18.** *b* **19.** *c* **20.** 3

DPP-2 Properties of Aromatic Aldehydes

1. (i) (a) PhCOOH, (b) $PhCH_2OH$, (c) $PhCH_2OH$, (d) $PhCH_3$, (e) Pinacol or hydrobenzoin (f) $PhCH_3$
4. *a* **5.** *a* **6.** *d* **7.** *c* **8.** *c* **9.** *c* **10.** *b,d* **11.** *a,b,c,d* **12.** *d* **13.** *a*
14. *a* **15.** *b* **16.** *d* **17.** *a* **18.** *b* **19.** *a* **20.** 4

DPP-3 Properties of Aromatic Ketones

6. *c* **7.** *b* **8.** *a* **9.** *c* **10.** *b* **11.** *c* **12.** *a* **13.** *b* **14.** *c* **15.** *a*
16. *a* **17.** *b* **18.** *a* **19.** *b*

DPP-4 Combined Properties of Aromatic Aldehydes and Ketones

3. *c* **4.** *c* **5.** *b* **6.** *c* **7.** *a* **8.** *b* **9.** *a* **10.** *c,d* **11.** *a,b,d*
12. *a,b,c* **13.** *d* **14.** *c* **15.** *c* **16.** *c* **17.** *c* **18.** *a* **19.** *b* **20.** *a*

DPP-5 Preparation and Properties of Aromatic Carboxylic Acids

4. *c* **5.** *d* **6.** *a* **7.** *a* **8.** *c* **9.** *c* **10.** *b* **11.** *c* **12.** *d*
13. *a,b,d* **14.** *b,c* **15.** *a,d* **16.** *a* **17.** *a* **18.** *d* **19.** *c* **20.** *b*

DPP-6 Acidity of Carboxylic Acids

1. *c* **2.** *a* **3.** *b* **4.** *b* **5.** *a* **6.** *b* **7.** *b* **8.** *b* **9.** *c* **10.** *b*
11. *b* **12.** *d* **13.** *d* **14.** *b* **15.** *a,c* **16.** *c,d* **17.** 6 **18.** 2 **19.** 1 **20.** 6

Revisal Problems (JEE Main)

1. *d* **2.** *b* **3.** *b* **4.** *b* **5.** *c* **6.** *a* **7.** *b* **8.** *b* **9.** *b* **10.** *a*
11. *a* **12.** *d* **13.** *c* **14.** *c* **15.** *a* **16.** *d* **17.** *b* **18.** *b* **19.** *b* **20.** *c*
21. *d* **22.** *b* **23.** *a* **24.** *b* **25.** *a* **26.** *b* **27.** *a* **28.** *b* **29.** *a*

Revisal Problems (JEE Advanced)

1. *c* **2.** *c* **3.** *c* **4.** *c* **5.** *b* **6.** *c* **7.** *b* **8.** *d* **9.** *a,d* **10.** *a,c*
11. *a,b* **12.** *a,c* **13.** *a,c* **14.** *b* **15.** *a,b,c* **16.** *a* **17.** *d* **18.** *a* **19.** *a* **20.** *d*
21. *c* **22.** *b* **23.** *c* **24.** *b* **25.** *b* **26.** *b* **27.** *d* **28.** *a* **29.** *b* **30.** *c*
31. 5 **32.** 5

JEE Main & AIEEE Archive

1. *a* **2.** *c* **3.** *d*

JEE Advanced & IIT JEE Archive

1. *c* **2.** *a* **3.** *c* **4.** *a*

3. Aliphatic Nitrogen Containing Compounds

DPP-1 Classification, Nomenclature and Isomerism of Amines

1. (i) Both are amide groups (ii) 3° amine, 2° amine and 2° amide
   (iii) 3° amine (iv) 3° amine, imine, amide
2. (i) Dimethyl ammonium carbonate
   (ii) N-ethylcyclopentyl amine
   (iii) N,N-dicyclopropyl cyclopropanamine
   (iv) 2-amino pyrimidine

5. *c* 6. *c* 7. *b* 8. *b* 9. *b* 10. *c* 11. *d* 12. *c* 13. *a* 14. *c*
15. *a,b,c* 16. *a,,c,d* 17. *b,c,d* 18. *a* 19. 2 20. 2

DPP-2 General Methods of Preparation of 1° Amines and Other than 1° Amines

4. *a* 5. *b* 6. *b* 7. *a* 8. *a* 9. *d* 10. *d* 11. *a* 12. *c* 13. *b,d*
14. *a,b,c,d* 15. *b,c* 16. *b,c,d* 17. *b,c* 18. *a* 19. *c* 20. *c* 21. 3 22. 3 23. 5

DPP-3 Physical Properties and Basicity of Amines

1. *a* 2. *a* 3. *a* 4. *d* 5. *d* 6. *d* 7. *a* 8. *c* 9. *d* 10. *a*
11. *c* 12. *b* 13. *b* 16. *c,d* 17. *b,c* 18. *a,b,c* 19. *b* 20. 5

DPP-4 Distinction and Separation of 1°, 2°, and 3° Amines

4. *d* 5. *c* 6. *c* 7. *b* 8. *a* 9. *b* 10. *c* 11. *c,d* 12. *a,b,c* 13. *a,b*
14. *a* 15. *b* 16. *d* 17. *a* 18. *c* 19. 6 20. 2

DPP-5 Chemical Properties of Aliphatic Amines

3. *b* 4. *b* 5. *c* 6. *c* 7. *b* 8. *b* 9. *a* 10. *d* 11. *c* 12. *a*
13. *d* 14. *a,b* 15. *b,d* 16. *a,b,c* 17. *a,b,d* 18. *a,b,c* 19. *a,b,d* 20. *a* 21. 4

Revisal Problems (JEE Main)

1. *a* 2. *c* 3. *b* 4. *a* 5. *a* 6. *a* 7. *b* 8. *a* 9. *d* 10. *b*
11. *a* 12. *a* 13. *a* 14. *a* 15. *a* 16. *d* 17. *c* 18. *b* 19. *d* 20. *d*
21. *d* 22. *d* 23. *c* 24. *c* 25. *b* 26. *c* 27. *c* 28. *b* 29. *b* 30. *a*

Revisal Problems (JEE Advanced)

1. *b* 2. *d* 3. *a* 4. *c* 5. *c* 6. *a* 7. *d* 8. *b* 9. *b* 10. *d*
11. *b* 12. *a,b,d* 13. *b,c,d* 14. *a,c,d* 15. *b,d* 16. *c,d* 17. *b,d* 18. *a,c* 19. *b,c* 20. *a*
21. *d* 22. *a* 23. *d* 24. *c* 25. *a* 26. *b* 27. *a* 28. *a* 29. *c* 30. *b*
31. *c* 32. *a* 33. 2 34. 8

JEE Main & AIEEE Archive

1. *d* 2. *a* 3. *c* 4. *a* 5. *c* 6. *c* 7. *c* 8. *b* 9. *c* 10. *c*

JEE Advanced & IIT JEE Archive

1. *a* 2. *d*

4. Aromatic Nitro Compounds, Anilines and Diazonium Salts

DPP-1 Preparation and Properties of Aromatic Nitro Compounds

5. *c* 6. *b* 7. *c* 8. *d* 9. *d* 10. *c* 11. *b,c* 12. *a,c* 13. *a,c* 14. *b,c*
15. *d* 16. *b* 17. *b* 18. *b* 19. 9 20. 4

DPP-2 Preparation Methods and Basicity of Anilines

3. *a*	4. *b*	5. *d*	6. *d*	7. *c*	8. *d*	9. *d*	10. *a*	11. *c*	12. *c*
13. *a*	14. *b,c,d*	15. *a,b*	16. *b*	17. *d*	18. *b*	19. 3	20. 3		

DPP-3 Chemical Properties of Anilines

3. *b*	4. *d*	5. *c*	6. *c*	7. *d*	8. *a*	9. *c*	10. *d*	11. *c*	12. *c*
13. *a,b*	14. *a,b,c,d*	15. *a,c,d*	16. *a,b,c*	17. *c*	18. *d*	19. *a*	20. *a*	21. 4	22. 4

DPP-4 Preparation and Properties of Diazonium Salts

3. *d*	4. *b*	5. *a*	6. *d*	7. *d*	8. *a*	9. *b*	10. *a,c,d*	11. *a,c,d*
12. *a,b,c*	13. *a*	14. *a*	15. *a*	16. *a*	17. *c*	18. 9		

Revisal Problems (JEE Main)

1. *b*	2. *d*	3. *a*	4. *c*	5. *a*	6. *b*	7. *a*	8. *b*	9. *b*	10. *a*
11. *d*	12. *b*	13. *b*	14. *a*	15. *a*	16. *d*	17. *b*	18. *b*	19. *d*	20. *b*
21. *a*	22. *a*	23. *d*	24. *a*	25. *b*	26. *a*	27. *a*			

Revisal Problems(JEE Advanced)

1. *c*	2. *a*	3. *c*	4. *d*	5. *a*	6. *a*	7. *c*	8. *a*	9. *b*	10. *a*
11. *b*	12. *d*	13. *a*	14. *a,b,d*	15. *b,c*	16. *a,b*	17. *a,b*	18. *a,b,c*	19. *a,b,c*	
20. *a,b*	21. *a,c,d*	22. *c*	23. *a*	24. *a*	25. *b*	26. *d*	27. *c*	28. *d*	29. *b*
30. *c*	31. *a*	32. 4	33. 4						

JEE Main & AIEEE Archive

1. *b*	2. *c*	3. *a*	4. *b*	5. *d*	6. *d*	7. *d*

JEE Advanced & IIT JEE Archive

1. *a*	2. *c*	3. *a*	4. *d*	5. *b*	6. *b*

5. Biomolecules

DPP-1 Introduction and Classification of Carbohydrates

5. *a*	6. *c*	7. *c*	8. *b*	9. *d*	10. *b,c*	11. *b,d*	12. *b*	13. *d*	14. *a*
15. *a*	16. *b*	17. 9	18. 4						

DPP-2 Preparation and Properties of Monosaccharides

6. *b*	7. *c*	8. *a*	9. *a*	10. *a*	11. *b*	12. *b*	13. *c,d*	14. *a,d*
15. *a,b,c*	16. *a*	17. *a*	18. *c*	19. *c*	20. *c,d*	21. 4	22. 7	

DPP-3 Some Important Oligos and Polysaccharides

5. *a*	6. *c*	7. *c*	8. *a*	9. *a,b,d*	10. *a,b,c,d*	11. *a*	12. *d*	13. *b*	14. *a*
15. *a*	16. 9								

DPP-4 Classification and Physical Properties of Amino Acids

6. *d*	7. *a*	8. *a*	9. *b*	10. *b*	11. *a*	12. *d*	13. *b,c,d*	14. *b,c*	15. *b*
16. *c*	17. *c*	18. *a*	19. 3	20. 5					

DPP-5 Peptides and Proteins

2. *d*	3. *c*	4. *a*	5. *c*	6. *a,b,c,d*	7. *a*	8. *b*	9. *c*	10. *b*	11. *d*
12. 4	13. 6	14. *a*	15. *a*	16. *b*					

DPP-6 Other Biomolecules (DNA, RNA, Vitamins, Hormones, etc.)

3. *d* **4.** *c* **5.** *c* **6.** *a* **7.** *d* **8.** *b* **9.** *c* **10.** *a* **11.** *c* **12.** b
13. d **14.** *b* **15.** *a,c* **16.** *a,b,c* **17.** *b* **18.** *b* **19.** *b* **20.** *c*

Revisal Problems (JEE Main)

1. *d* **2.** *c* **3.** *c* **4.** *a* **5.** *d* **6.** *a* **7.** *a* **8.** *b* **9.** *c* **10.** *a*
11. *a* **12.** *b* **13.** *b* **14.** *b* **15.** *a* **16.** *c* **17.** *d* **18.** *d* **19.** *a* **20.** *a*
21. *c* **22.** *d* **23.** *c* **24.** *c* **25.** *a* **26.** *d* **27.** *d* **28.** *c* **29.** *d*

Revisal Problems(JEE Advanced)

1. *d* **2.** *a* **3.** *b* **4.** *c* **5.** *c* **6.** *a* **7.** *a* **8.** *b* **9.** *a,d* **10.** *b,c*
11. *a,b,c,d* **12.** *a,d* **13.** *a,b,d* **14.** *c,d* **15.** *a* **16.** *c* **17.** *c* **18.** *c* **19.** *d* **20.** *a*
21. *a* **22.** *d* **23.** *d* **24.** *c* **25.** *c* **26.** *c* **27.** *d* **28.** *b* **29.** 4 **30.** 2
31. 2

JEE Main & AIEEE Archive

1. *a* **2.** *a* **3.** *b* **4.** *c* **5.** *b* **6.** *d* **7.** *a* **8.** *c* **9.** *c* **10.** *a*
11. *d* **12.** *c* **13.** *b* **14.** *d* **15.** *b* **16.** *a* **17.** *c* **18.** *c* **19.** *c* **20.** *b*

JEE Advanced & IIT JEE Archive

1. 4 **2.** 4 **3.** 4 **4.** 8 **5.** 6 **6.** *a,b,c,d* **7.** *b* **8.** 2 **9.** *a* **10.** *b,c*
11. *a* **12.** *c* **13.** *b*

6. Polymers

DPP-1 General Introduction and Classification of Polymers

5. *b* **6.** *c* **7.** *c* **8.** *a,b* **9.** *b,c* **10.** *b,c* **11.** *c* **12.** *c* **13.** *b*
14. *a* **15.** *b* **16.** *a* **17.** *c* **18.** 5 **19.** 6

DPP-2 Mechanism and Polymerisation

5. *a* **6.** *a* **7.** *b* **8.** *c* **9.** *b* **10.** *b,c*
11. *b,c* **12.** *a,b,c* **13.** *d* **14.** *d* **15.** *c* **16.** *c* **17.** *c* **18.** *a* **19.** 5 **20.** 7

DPP-3 Some Important Polymers (Preparation, Properties and Uses)

3. *c* **4.** *d* **5.** *a* **6.** *d* **7.** *d* **8.** *d* **9.** *c* **10.** *a,c*
11. *a,b,d* **12.** *a,c* **13.** *a* **14.** *a* **15.** *b* **16.** *d* **17.** *a* **18.** *a* **19.** *c* **20.** 5

Revisal Problems (JEE Main)

1. *d* **2.** *b* **3.** *d* **4.** *d* **5.** *b* **6.** *d* **7.** *a* **8.** *a* **9.** *c* **10.** *d*
11. *d* **12.** *d* **13.** *b* **14.** *d* **15.** *b* **16.** *c* **17.** *c* **18.** *b* **19.** *b* **20.** *d*
21. *d* **22.** *a* **23.** *d* **24.** *c* **25.** *a* **26.** *d* **27.** *d* **28.** *a* **29.** *b* **30.** *c*

Revisal Problems (JEE Advanced)

1. *c* **2.** *b* **3.** *a* **4.** *b* **5.** *a* **6.** *a* **7.** *a* **8.** *c* **9.** *b* **10.** *d*
11. *d* **12.** *a* **13.** *a,b,c* **14.** *a,b* **15.** *a,b,c,d* **16.** *a,c,d* **17.** *a,c,d* **18.** *a,b* **19.** *b* **20.** *b,d*
21. *b,c* **22.** *b* **23.** *a* **24.** *d* **25.** *d* **26.** *a* **27.** *a* **28.** 5 **29.** 5

JEE Archive

1. *a* **2.** *c* **3.** *c* **4.** *b* **5.** *c* **6.** *d* **7.** *c* **8.** *c* **9.** *c* **10.** *b*
11. *d* **12.** *b*

1. Aryl Halides and Phenols

DPP-1 Methods of Preparation of Aryl Halides

1. (i)

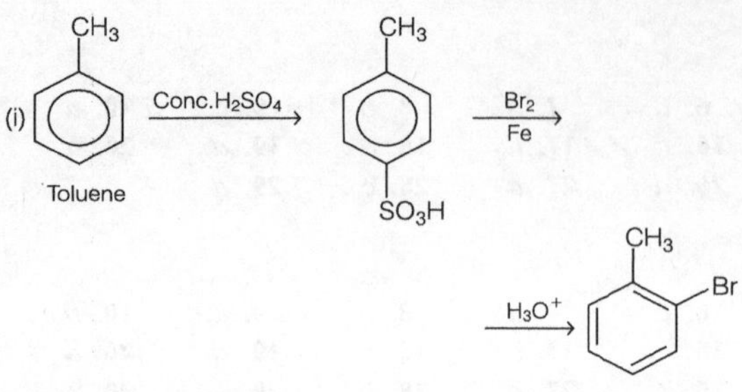

(ii) Bromine is *o*/*p* directing group, so direct bromination cannot be used for this synthesis.

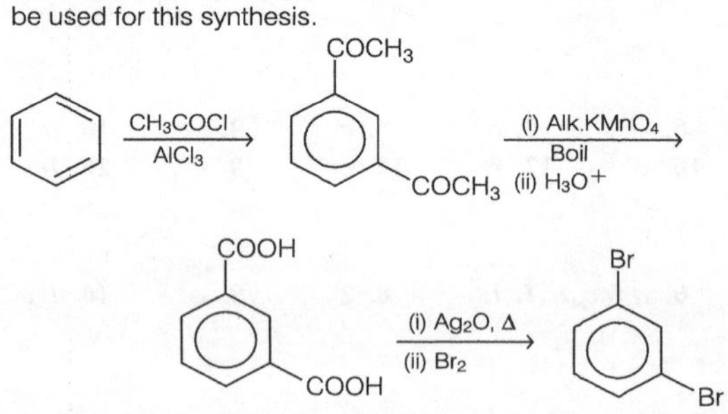

m-dibromobenzene

(iii)

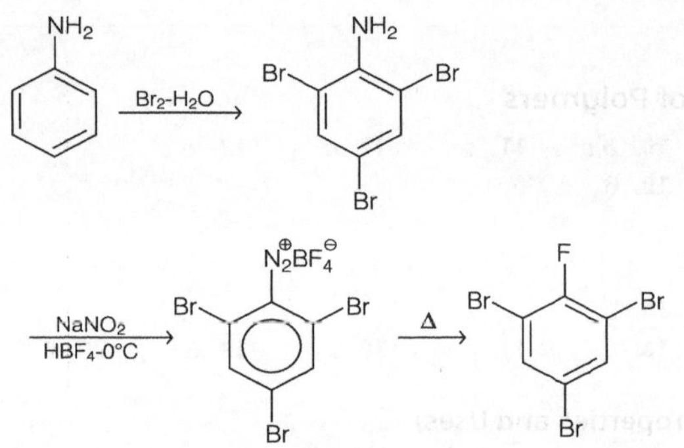

(iv)

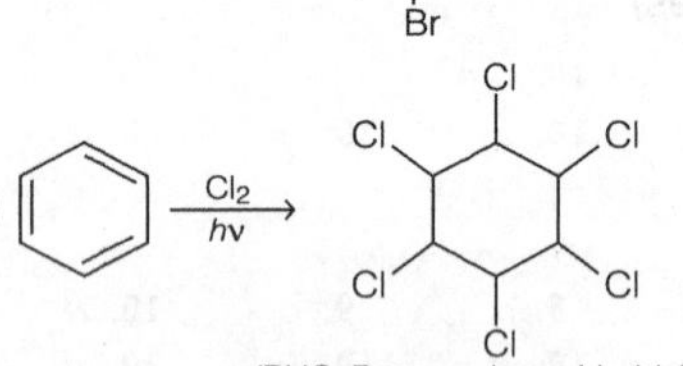

(BHC, Benzene hexachloride)

2. (i)

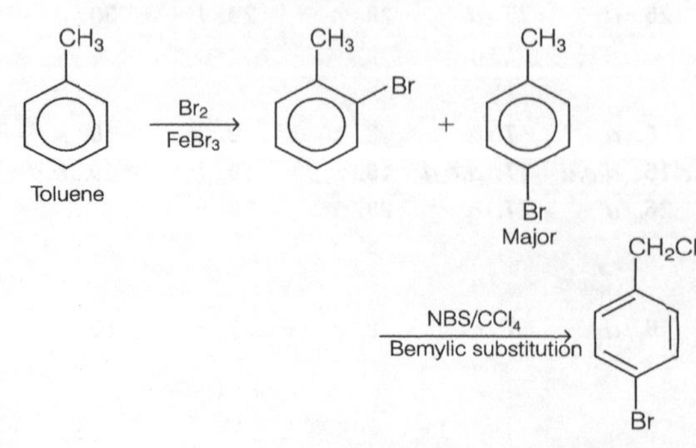

(ii)

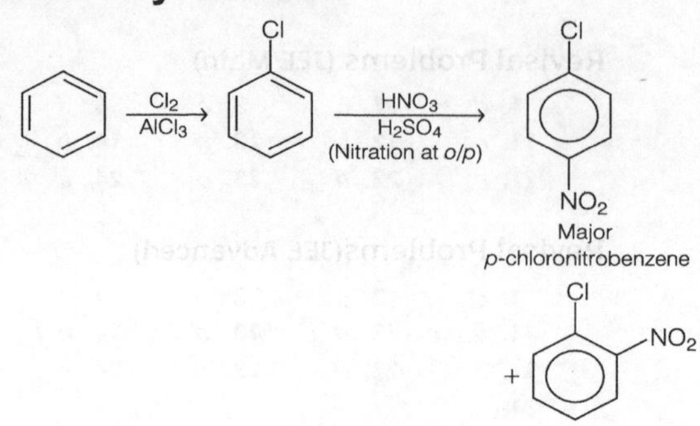

(iii)

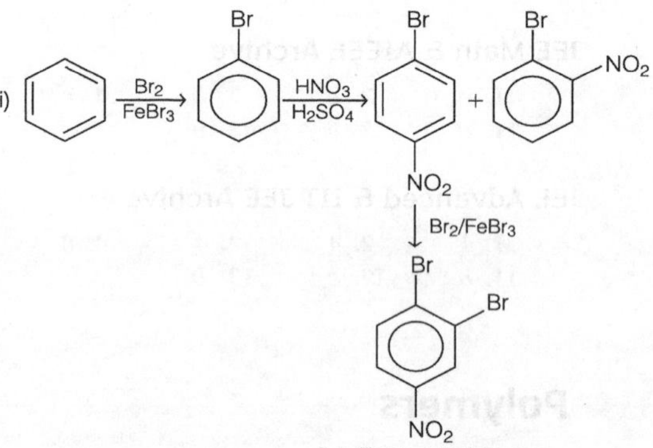

3,4-dibromonitrobenzene

(iv)

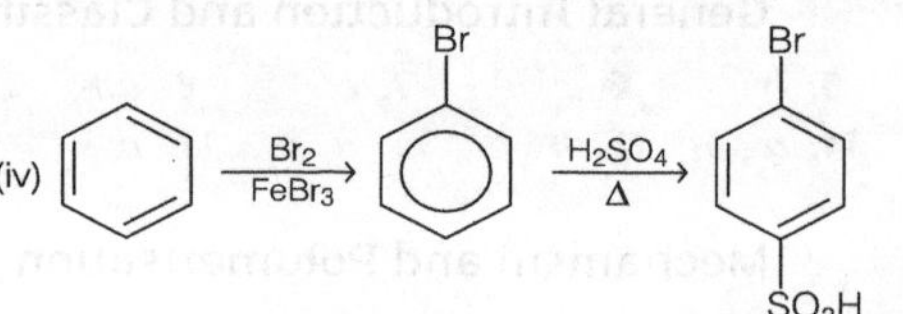

p-bromobenzene sulphonic acid

(v)

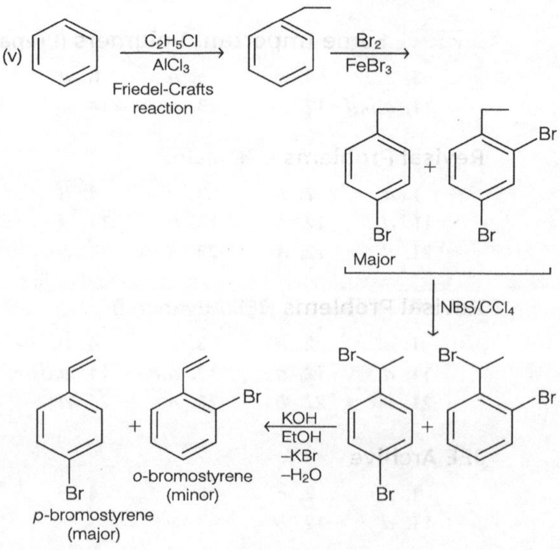

3. (i) 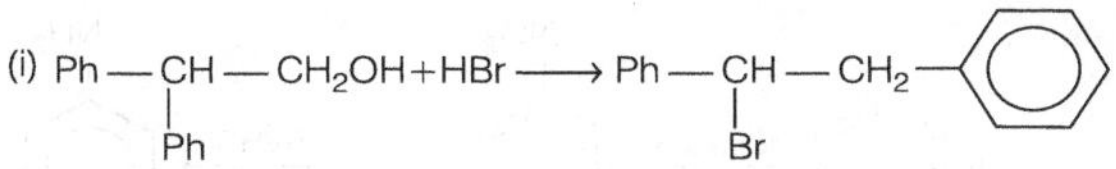

Mechanism

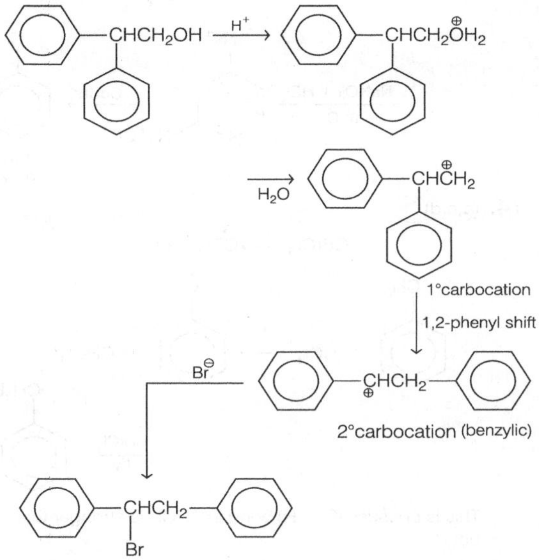

(ii) 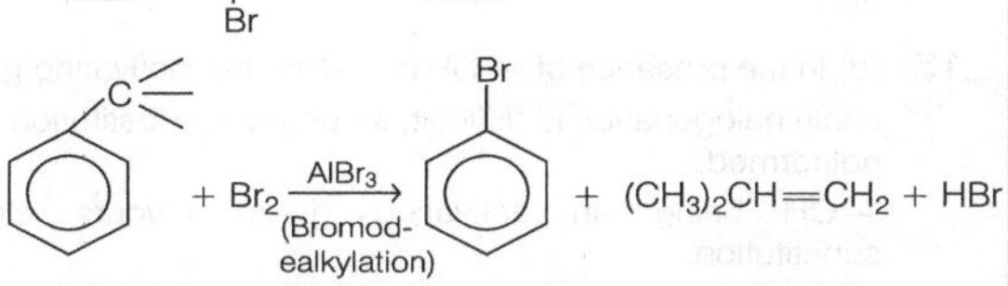

(The Friedel-Crafts alkylation can be reversed especially when a 3° alkyl group such as Me_3C is present). Dealkylation is affected with $AlBr_3$ by transfer of Me_3C^+ to another arene used as solvent or at high temperature by protonation as H^+ and loss of Me_3C as $Me_2C{=}CH_2$.

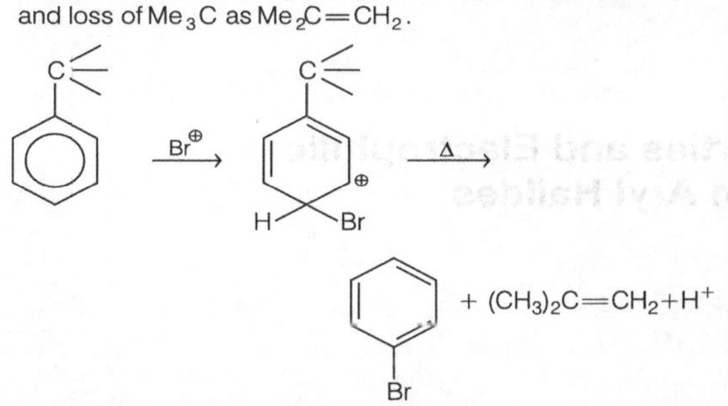

4. (i) Presence of Lewis acid catalyst like $FeBr_3$, $AlCl_3$, etc., is necessary for the **nuclear** substitution, so in absence of it, no nucleophilic **nuclear** substitution takes place.

(ii) Unlike sulphonation, bromination of benzene does not show kinetic isotopic effect.

5. (a) HNO_3 oxidises the by-product HI (i.e. I^-) to I_2, otherwise it reduces the product iodobenzene back to benzene.

6. (c)

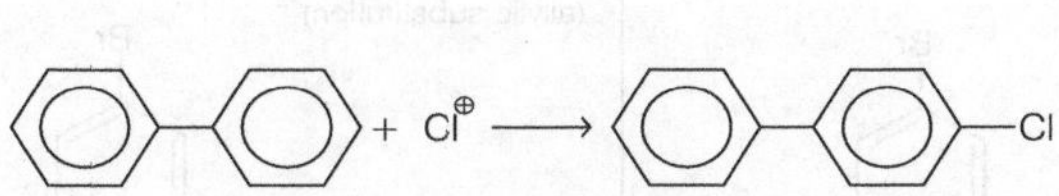

($\overset{+}{C}l$ attacks at the *para*-position)

7. (c) Since, —Br and —NO_2 are at *meta* positions and CH_3 is a *o/p* directing group, so it is first oxidised with $K_2Cr_2O_7/H_2SO_4$, Δ to carboxylic acid, which is *meta* directing.

The second step is nitration, not bromination because —NO_2 is also a *meta* directing group but Br is an *o/p* directing group.

8. (b)

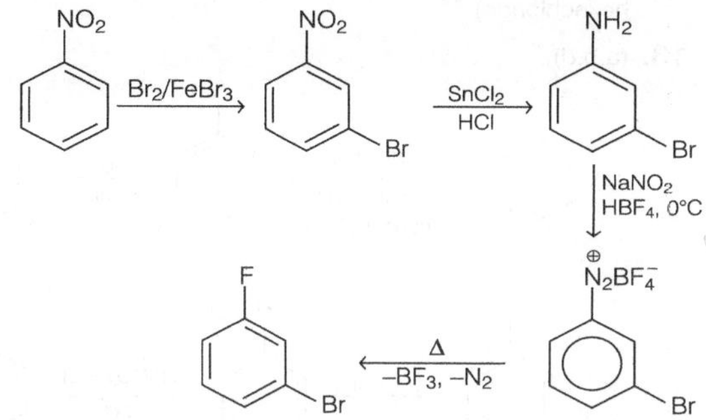

9. (d)

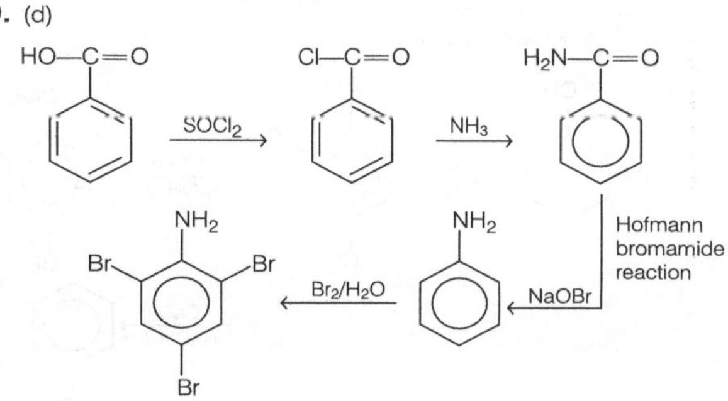

10. (d) Degree of unsaturation in

$$A = \frac{C \times 2 - H + 2}{2} = \frac{18 - 10 + 2}{2} = 5$$

Oxidation product of *C* is benzoic acid, so *A* must have a benzene ring (with four degree of unsaturation, one ring and three C=C)

A reacts with 1 mole of Br_2, so there is a C=C bond in the side chain. Oxidation product is also CH_3COOH. Hence, *A* has C=C bond in conjugation with benzene.

Thus,

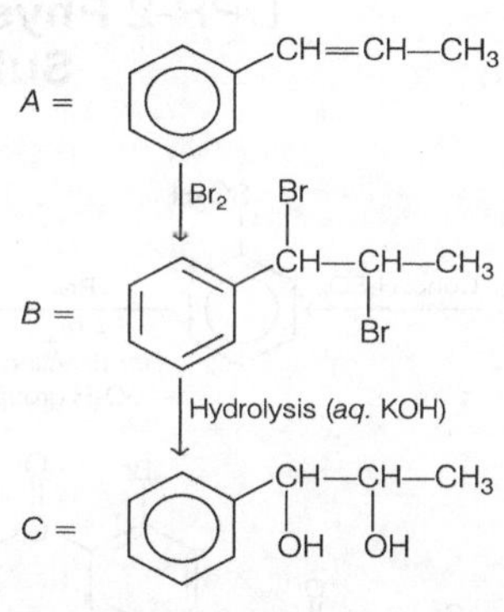

11. (c) (i)→(s), (ii)→(r), (iii)→(q), (iv)→(p)

In Sandmeyer reaction, diazonium salt is converted into arylhalide by treating it with respective Cu (I) salt. In Gattermann reaction diazonium salt with corresponding halogen acid in the presence of coppor powder to convert it into respective halide.

Rasching process is used to prepare chlorobenzene from benzene. Formation of diazonium salt by treatment with HNO_2 is called diazotisation.

12. (a,b) Benzene diazonium chloride decomposes on heating to give chlorobenzene along with the evolution of nitrogen. It also gives chlorobenzene when treated with CuCl.

Reaction of benzene with C_6H_5COCl in the presence of $AlCl_3$ gives benzophenone while with $Cl_2(h\nu)$ it gives BHC (benzene hexachloride).

13. (a,b,d)

(a) 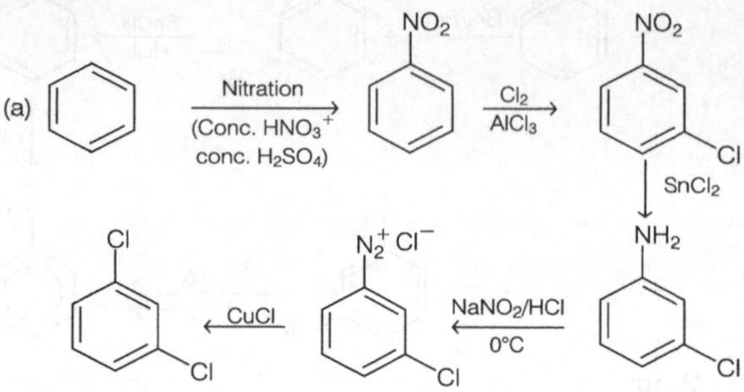

1,3-dichloro benzene

(b) 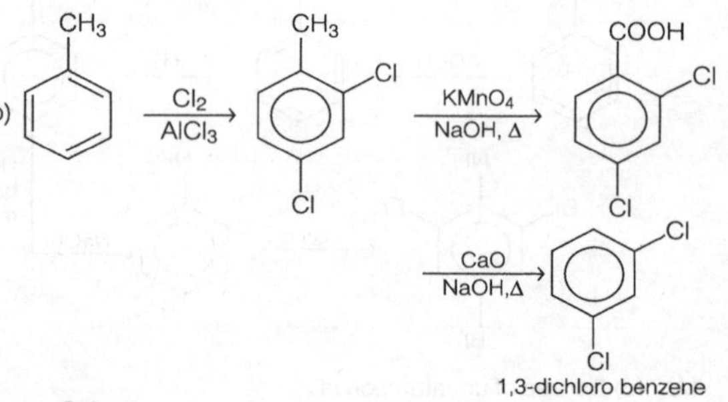

(c) 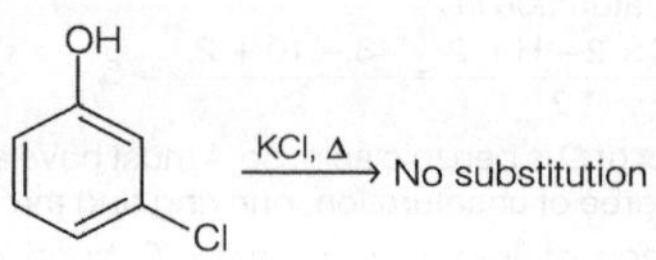

(d) 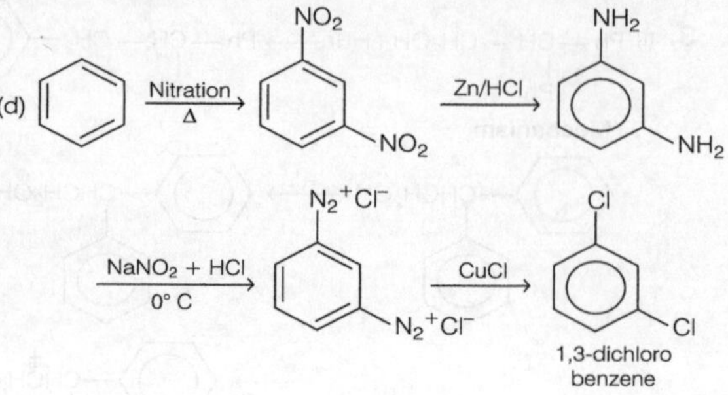

14. (a,c,d)

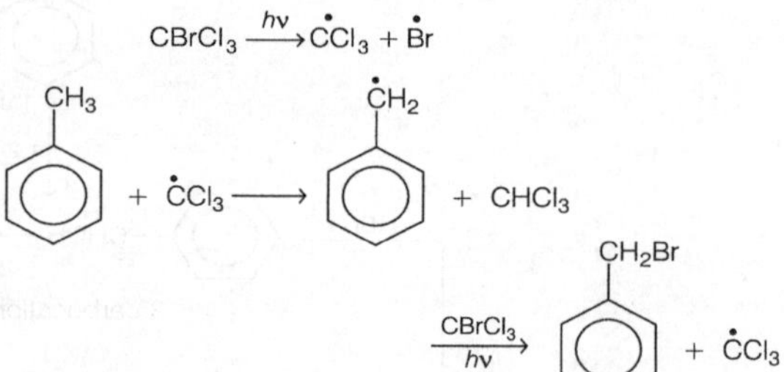

This is because (C—H) bond in $CHCl_3$ is stronger than H—Br bond.

15. (d) In the presence of —OH or —NH_2 like activating group, side chain halogenation is difficult, so benzylic substitution product is not formed.

—OH being an activating group favours electrophilic substitution.

16. (2) Benzoic acid is formed when *X* is treated with Zn dust/$KMnO_4$, so *X* contains —OH and CH_3 group. On bromination Br atom is also inserted. So, it contains only two functional groups, i.e. —OH and —Br. (CH_3 is a substituent, not a functional group).

DPP-2 **Physical Properties and Electrophilic Substitution in Aryl Halides**

1. (a)

(i) 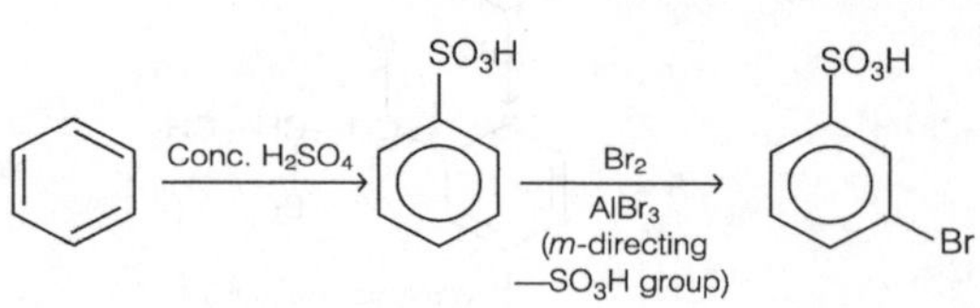

(ii) 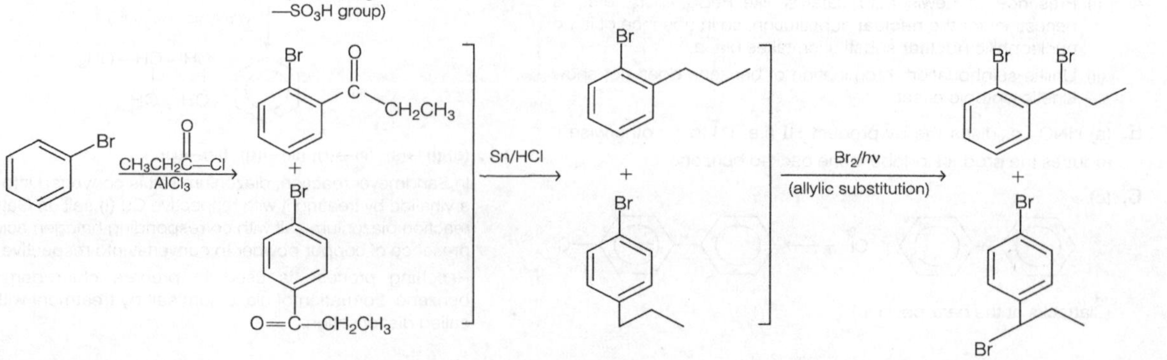

(iii) 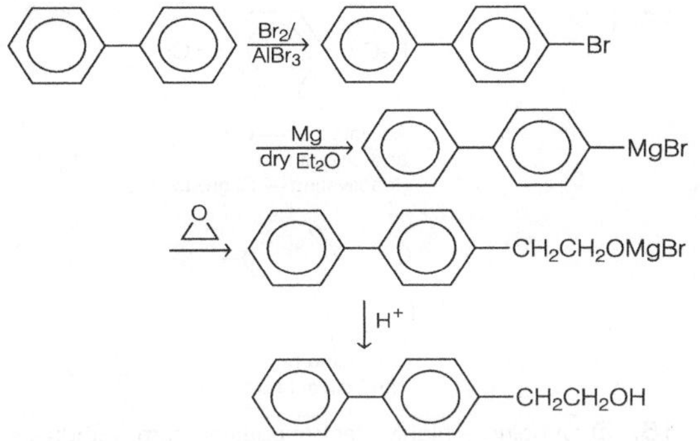

(b) The structures of the three isomeric dichlorobenzenes are

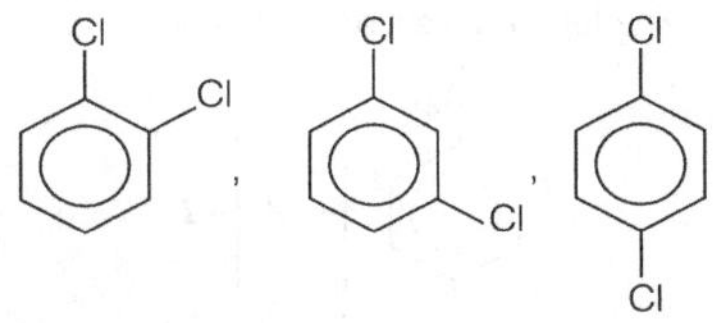

Out of these 1,4-dichlorobenzene is most symmetrical hence, have greater intercrystalline attraction and thus, higher melting point.

2. (i) 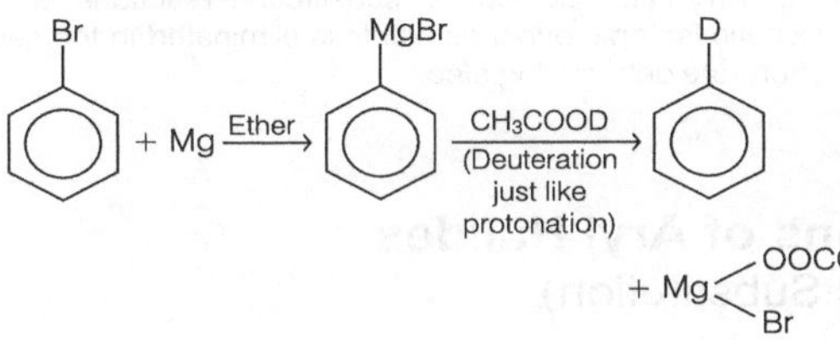

(ii) $AlCl_3 + CH_3CH_2CH_2Cl \longrightarrow CH_3CH_2\overset{\oplus}{C}H_2 + AlCl_4^{\ominus}$

1°carbocation

$\downarrow$ 1,2-hydride shift

$CH_3\overset{\oplus}{C}HCH_3$

2°carbocation (more stable)

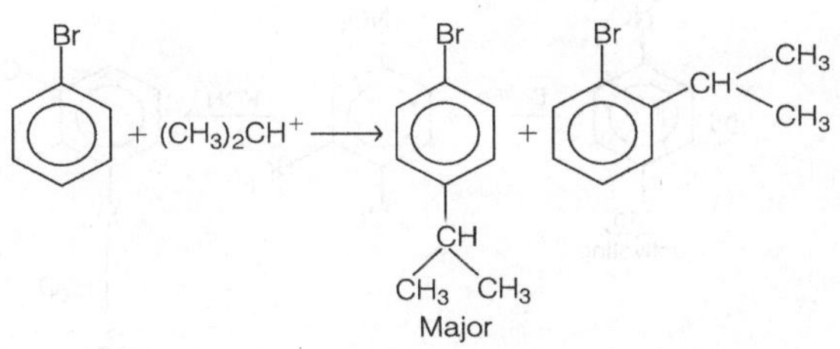

(iii) 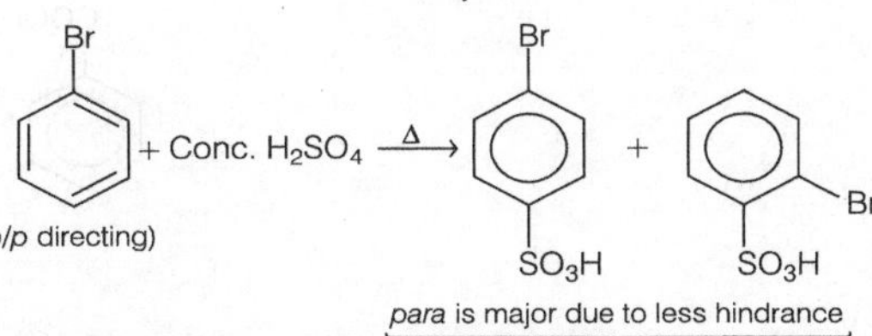

(*o/p* directing)

para is major due to less hindrance

p-bromo- *o*-bromo-benzene sulphonic acid

(iv) No reaction

(v) No reaction (because of the some double bond character in C—Br bond.)

3. Greater the reactivity of the substituent for *m*-director, greater is the $\left(\frac{m}{o+p}\right)$ ratio.

Thus, (i) II > III > IV > I.

(ii) Reactivity of halogens is controlled by $-I$ effect but orientation is controlled by $+R$-effect.

Reactivity order is I > Br > Cl > F

($—CX_3$) is *m*-director but ($—CH_3$) is *o*/*p* director.

Thus, the order is IV > II > III > I.

4. (a) $$\mu = 2\mu_{C-X}\cos\frac{\theta}{2}$$

where

$\mu_{(C-X)}$ = vector moment of side ($C-X$)

$\therefore$ $$\mu \propto \cos\frac{\theta}{2}$$

or $$\mu \propto \frac{1}{\theta}$$

In case of *ortho* isomer, θ is least, so μ is maximum. Further for Br-atoms, repulsion is maximum due to its larger size as compared to Cl atom, so angle increases hence, μ decreases.

The order of correct μ is

$$\mu(a) > \mu(d) > \mu(b) > \mu(c)$$

Thus, dipole moment is highest for *o*-dichlorobenzene.

5. (b) The given data suggests that the ease of nitration is more at *o*/*p* positions but not very high. It means the group *X* is *o*/*p* directing but deactivating. This condition matches with halogens like —Cl, —Br, etc.

6. (c) This condition is true only in case I and III.

e.g.

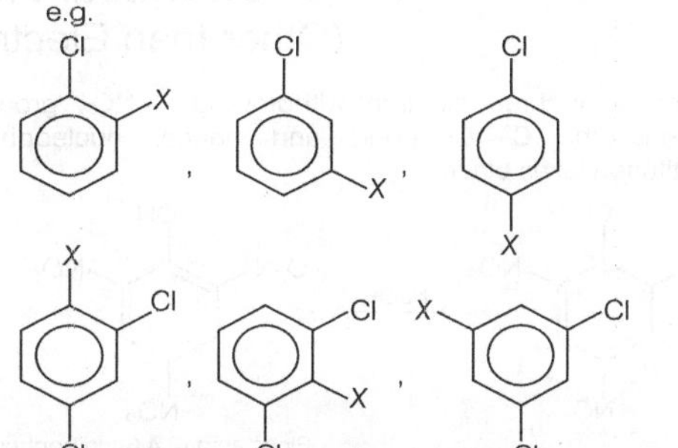

(Three isomers)

(Three isomers)

7. (c) $—CH_3$ being an activating group decides the position of incoming electrophile. So, E^+ adds at position *ortho* to it.

8. (a) 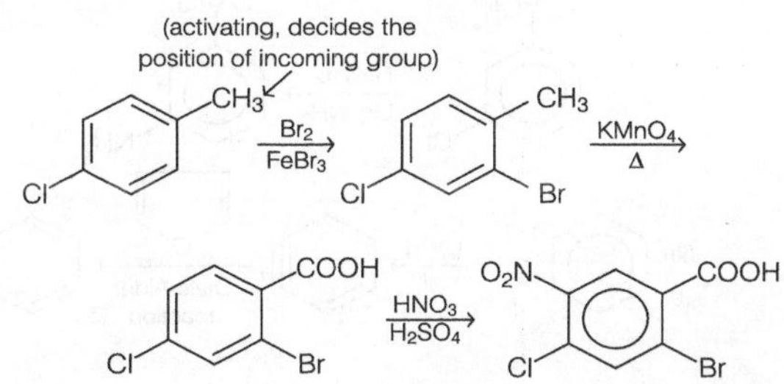

9. (a) Electrophilic substitution takes place *ortho* to (—OH) group, not *ortho* to (—Ö—) group. Since, it is deactivated by the strong e^- withdrawing group ($—COO^-$ and $\overset{+}{N}H_3$) in another ring.

10. (c)

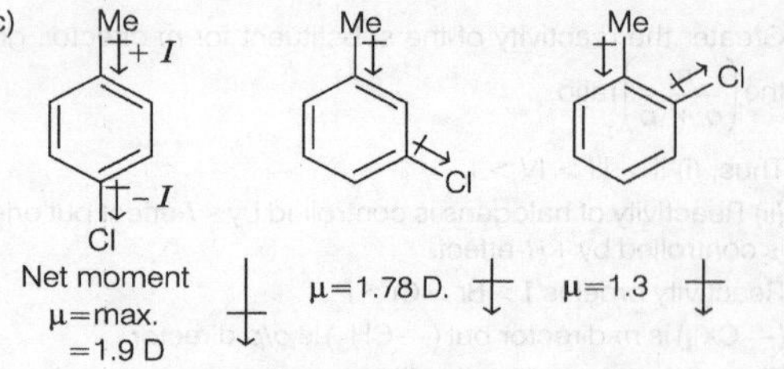

Thus, the order of μ is $p > m > o$.

So, X with medium value of μ should be *m*-chlorotoluene.

11. (c) —OMe is *o*/*p* directing group. Further due to large size of *t*-butyl group, *para*-substitution predominates to give less hindered structure.

12. (d) In this reaction, the least hindered position of the anilide ring is sulphonated.

13. (a,b)

(a) Grignard reagent when treated with CH_3COOD undergoes neutralisation to give deuterobenzene.

(b) $C_6H_5F + C_6H_5Li \xrightarrow[-LiF]{} $ Ph—Ph (biphenyl)

(c) Chlorobenzene does not give Friedel-Crafts reaction (as cleavage of C—Cl bond is difficult).

(d) Chlorine (—Cl) is *o*/*p* directing group, so *ortho* or *para* (not *meta*) dichlorobenzene is formed.

14. (a,b)

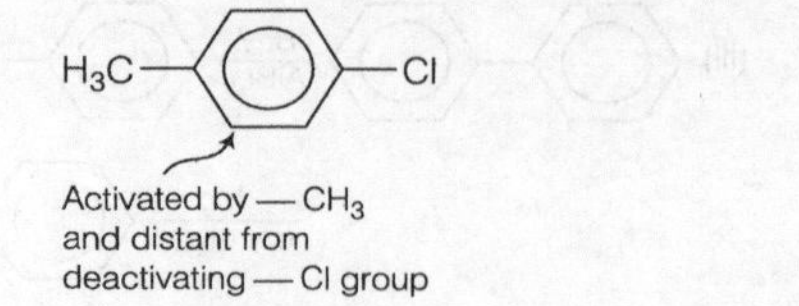

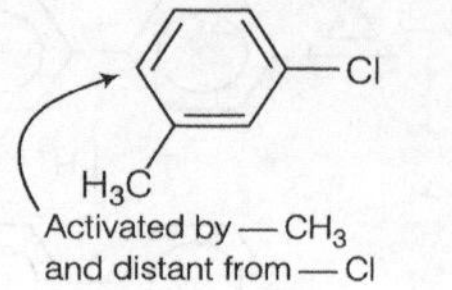

15. (a) Cl being a highly electronegative atom, withdraws electrons from the ring and deactivates it for electrophilic substitution. However, $+R$ effect makes it *o*/*p* directing.

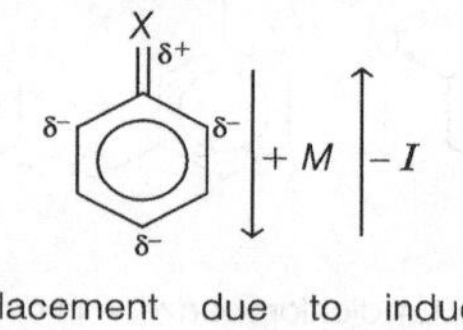

(Electron displacement due to inductive effect is not compensated by resonance effect).

16. (d) Aromatic nucleophilic substitution reactions do not show kinetic isotopic effect as halide is eliminated in the second, fast non-rate determining step.

DPP-3 **Chemical Reactions of Aryl Halides**
(Other than Electrophilic Substitution)

1. (i) Presence of three electron withdrawing $—NO_2$ groups weakens the C—Cl bond and hence, nucleophilic substitution takes place.

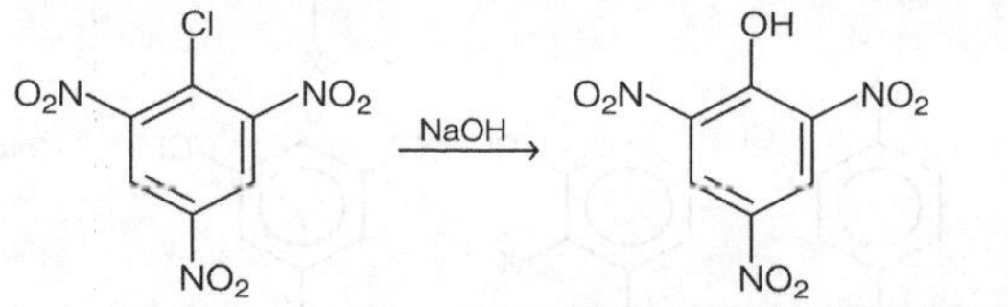

Picric acid (2,4,6-trinitrophenol)

(ii) CF_3 is also an electron withdrawing group so facilitates nucleophilic substitution.

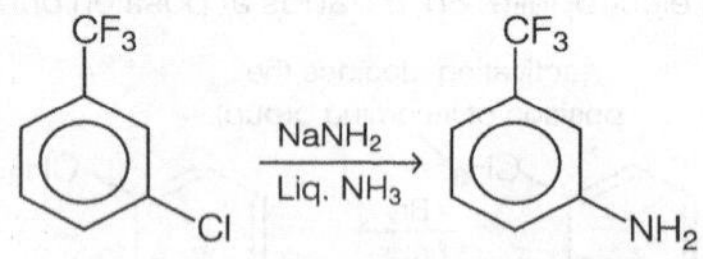

(iii)

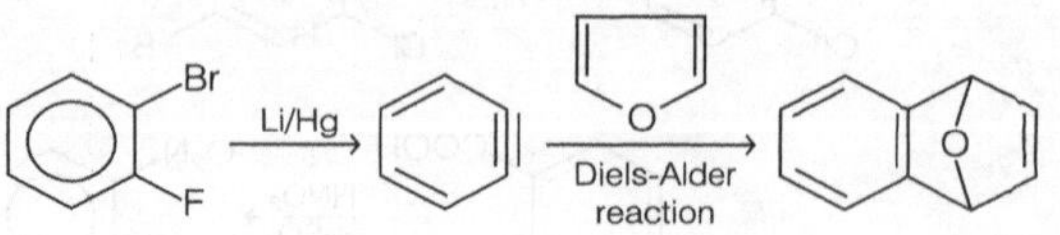

(iv)

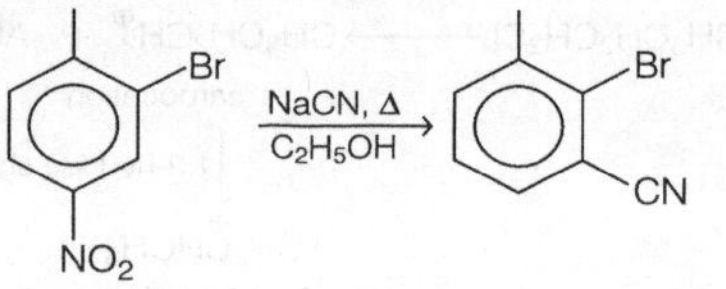

(v)

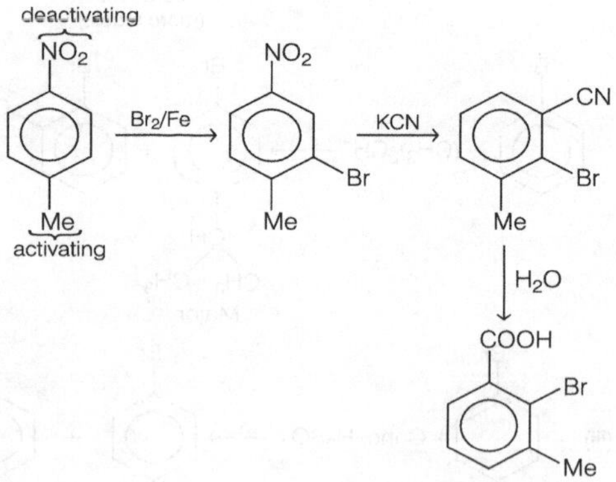

2. (i) 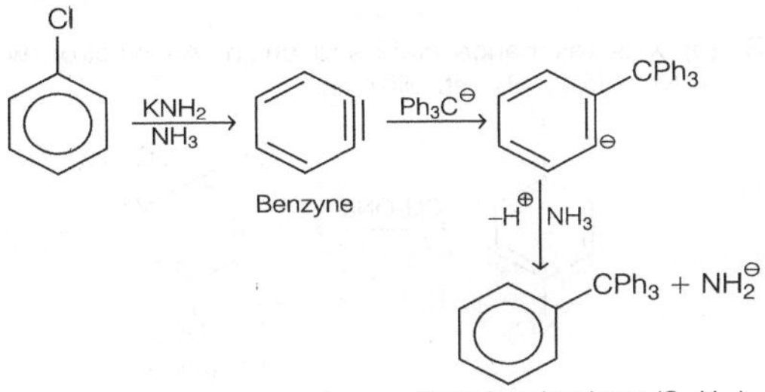

Tetraphenylmethane ($C_{25}H_{20}$)

Here, KNH_2 is used to generate benzyne intermediate which combines with $Ph_3C^{\ominus}K^{\oplus}$ to give the final product.

(ii) Exchange of H from solvent (NH_3) to the intermediate anion is faster than breaking of the (C—F) bond to form benzyne because the (C—F) bond is very strong. D/H exchange eliminate the possibility of E2 type mechanism.

3. Aromatic nucleophilic substitution reactions are favoured by electron withdrawing groups at *o*- and *p*-positions. More the electron withdrawing groups at these positions, faster is the aromatic nucleophilic substitution.

Thus,

(i) The decreasing order towards ArS_N reaction is
2, 4, 6-trinitrochlorobenzene > 2, 4-dinitrochlorobenzene > p-NO_2—C_6H_4—Cl > PhCl
[∵ NO_2 is a electron withdrawing group]

(ii) p-HO_3S—C_6H_4F > p-NO_2—C_6H_4F > p-HOOC—C_6H_4F > PhF
[∵ $-R$ effect of —SO_3H > —NO_2 = —COOH but $-I$ of —NO_2 > —COOH]

(iii) o-$NO_2C_6H_4Cl$ > p-$NO_2C_6H_4Cl$ > m-$NO_2C_6H_4Cl$ > PhCl
∵ Net electron withdrawing effect of — NO_2 group
o-NO_2 > p-NO_2 > m-NO_2.
(Although at *o* and *p*-positions, $-R$ effect is equal but $-I$ effect at *o*-position is slightly greater than that at *p*-position. At *m*-position only $-I$ effect is operative).

4. Presence of electron withdrawing group weakens the C—X bond and thus, makes the substituent more reactive towards nucleophilic substitution.

(i) More the number of electron withdrawing — NO_2 group, more is the reactivity towards CH_3O^-.

II is more reactive because a — NO_2 group at *para* position activate more for S_NAr due to electron withdrawing resonance effect than two nitro at *meta* position by $-I$ effect.

So, the correct order is

I < III < II

(iii) Cl(II) is *para* to —NO_2, closest to —NO_2, most reactive in S_NAr. I is also activated by —NO_2 by its resonance effect, it is 2nd most reactive. Cl(III) is not accelerated by electron withdrawing resonance effect of —NO_2 as well as it is very far away from —NO_2, just little accelerated for S_NAr. Hence, the overall reactivity of labelled Cl in a S_NAr reaction is

III < I < II

5. (i) Aromatic nucleophilic substitution (addition-elimination) takes place in the ring which contains strong electron withdrawing groups at *o* or/and *p* to the eliminating group.

(a) 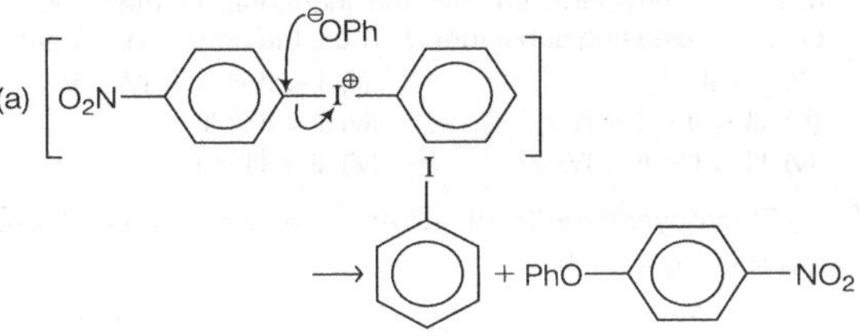

(b) 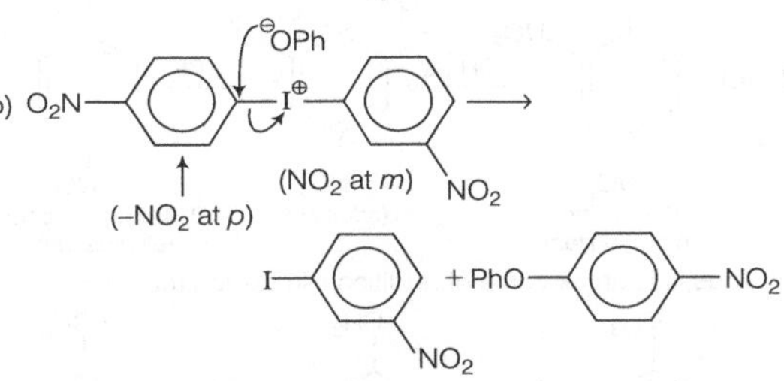

(c) 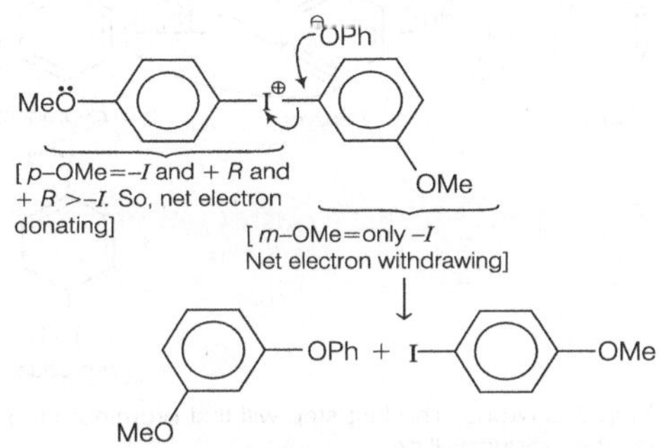

(d) 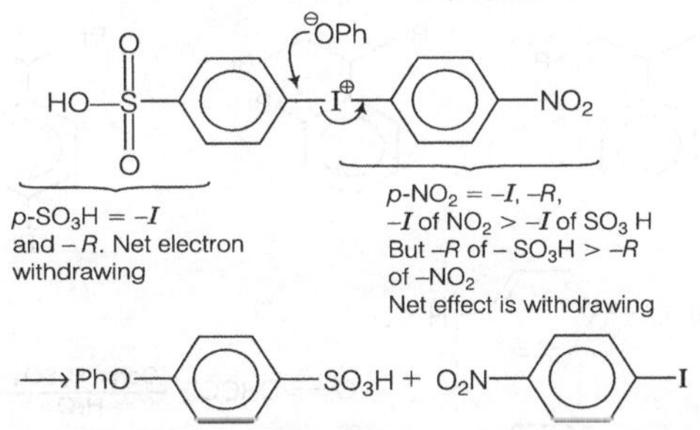

(ii) (a) When treated with $AgNO_3$, benzyl iodide gives a yellow ppt of AgI but *m*-iodotoluene does not (because in iodotoluene C—I bond acquires some double bond character due to resonance).

(b) When treated with NaOMe/$AgNO_3$, *p*-nitrobromobenzene gives yellow ppt of AgBr but *p*-bromoanisole does not give this reaction. *p*-nitrobromobenzene gives addition elimination (ArS_N) reaction due to the presence of electron withdrawing —NO_2 group.

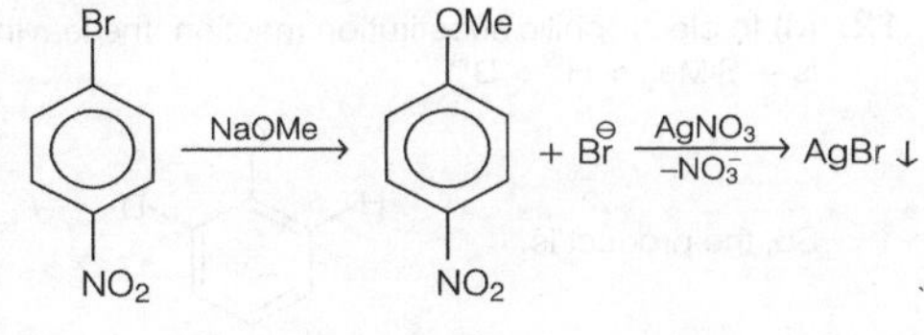

6. Rate of S_NAr (aromatic nucleophilic substitution) increases by the presence of electron withdrawing groups as a substituent while rate of electrophilic substitution increases in the presence of electron releasing substituents. Thus, the correct order are

(i) I < II
(ii) I < II < III < IV < V
(iii) III < II < I < IV
(iv) I < II < III
(v) III < I < II < IV
(vi) II < III < I

7. (c) Chloropyridine will undergo ArS_N reaction so the —Cl group is replaced by —OMe.

8. (d)

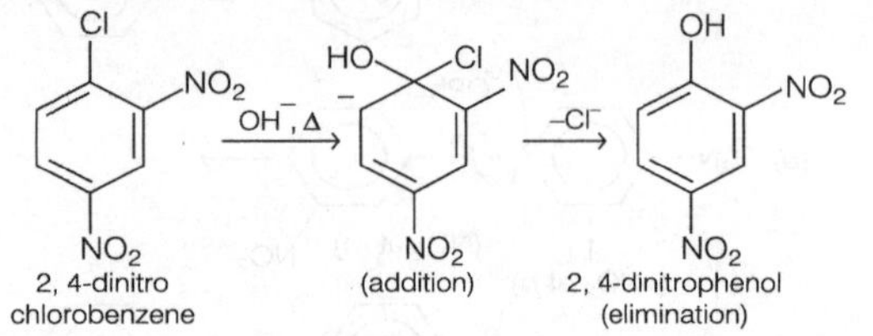

Thus, it is an example of addition elimination reaction.

9. (d)

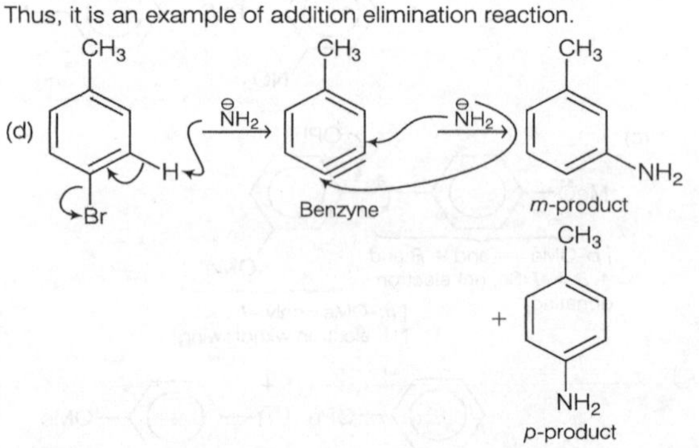

10. (c) Step 3 is wrong. The last step will first brominate the double bond. The product will be

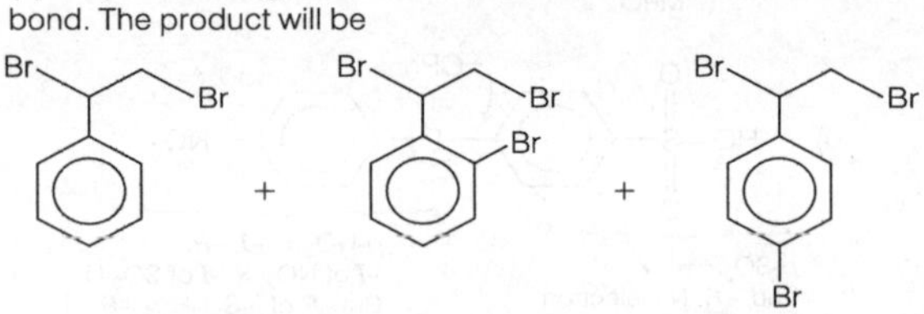

11. (a)

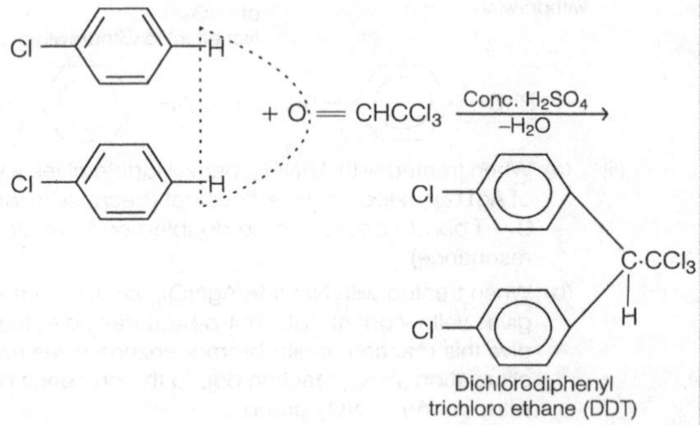

12. (d) In electrophilic substitution reaction, the leaving group ability is $—SiMe_3 > H^{\oplus} > D^{\oplus}$.

So, the product is

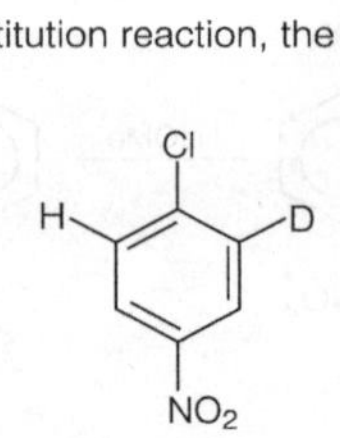

13. (d) X is resonance stabilised anion. An electron withdrawing $—NO_2$ group also stabilises it.

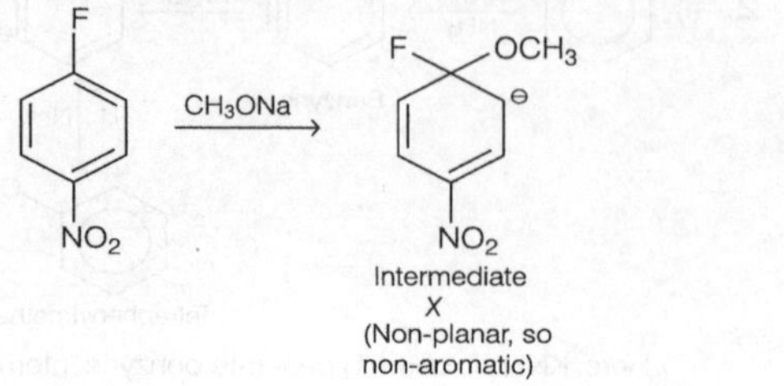

14. (b)

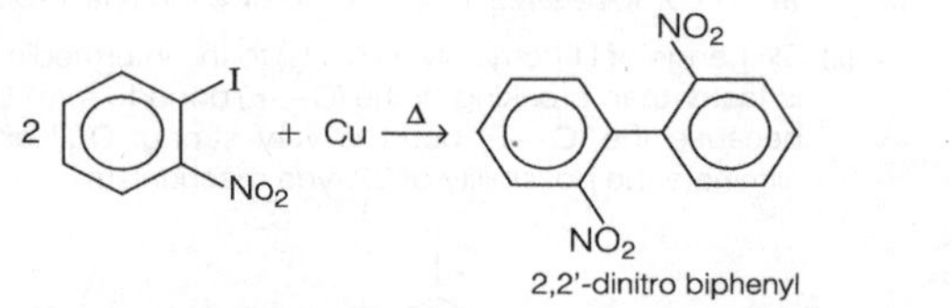

In A, free rotation about the single bond is possible and each ring has a vertical plane of symmetry. Hence, it does not show optical isomerism.

The diphenyl is sterically hindered because of bulky *ortho* substituents, therefore phenyl ring cannot be coplanar, and the energy barrier for rotation of C′—C1′ σ-bond is very high for interconversion of enantiomers.

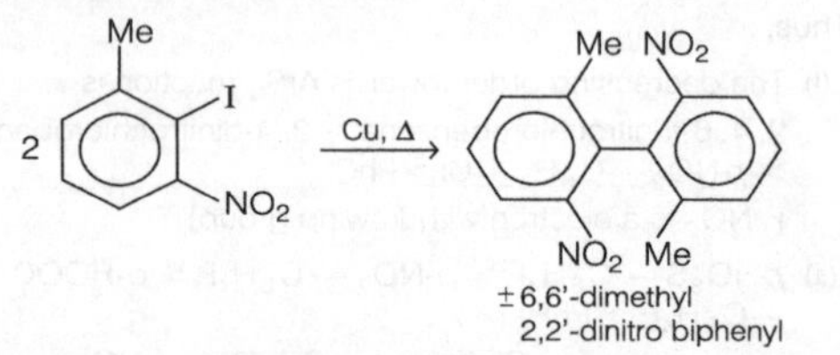

15. (a,b,c,d) ArS_N reactions are favoured by the presence of electron withdrawing groups (like $—NO_2$, — F, —Cl, —CN, etc.,) but disfavoured by electron releasing groups (R, – OMe etc.). Thus, substitution of Cl^- or NO_2^- or CN^- at position 1 or 2 favours the substitution (S_NAr) reaction but substitution by — OMe (methoxy group) at these positions discourage this reaction.

16. (a,b,c) Product II is Cine substitution and product I is direct substitution . Reaction is ArS_N (elimination - addition).

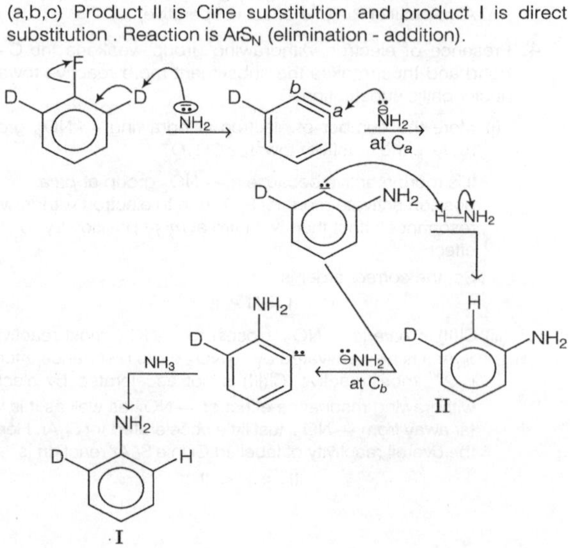

17. (a,b,c)

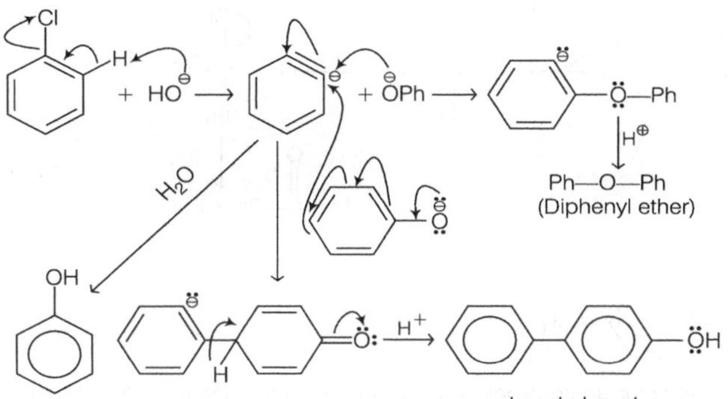

Phenol   *p*-phenyl phenol

18. (a) (i)→(p,q,s), (ii)→(p,q,r,s), (iii)→(q,r,s)

All the given reactions are elimination-addition reactions.

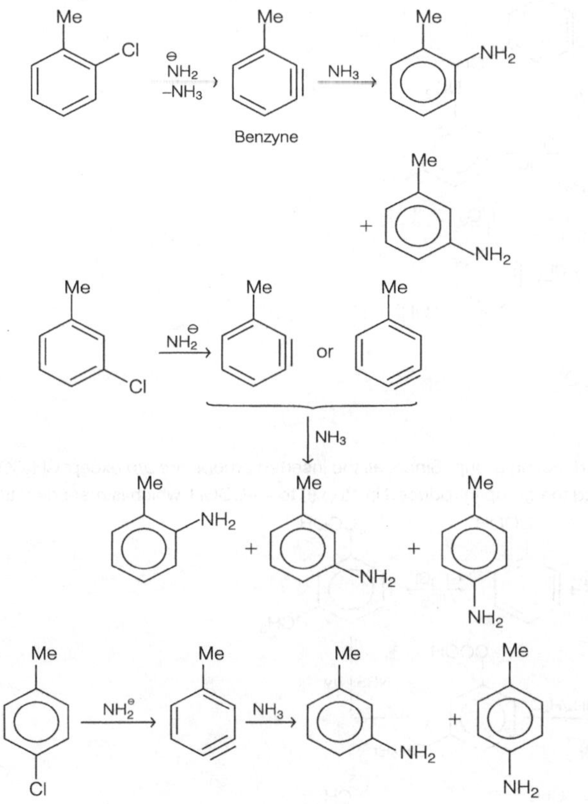

19. (b) Chlorobenzene is less reactive than an aliphatic chloride towards S_N reaction. This is because of resonance of aromatic ring with chloro group to impart partial double bond character to C—Cl bond.

Chlorine has both $-I$ and $+R$ effect. So, the net effect is electron withdrawing.

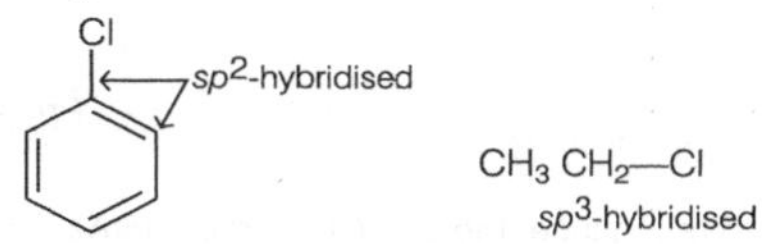

20. (a) The net effect (from $-I$ and $+R$) of chloride is electron withdrawing on benzene ring, so it increases the reactivity of the ring towards S_NAr reaction.

21. (i) (3)

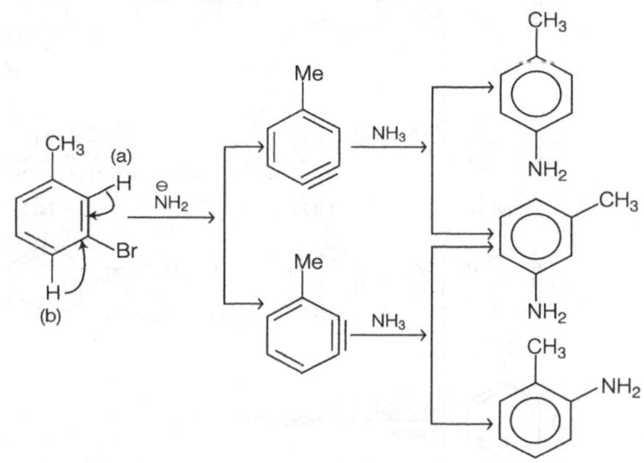

(ii) (2)

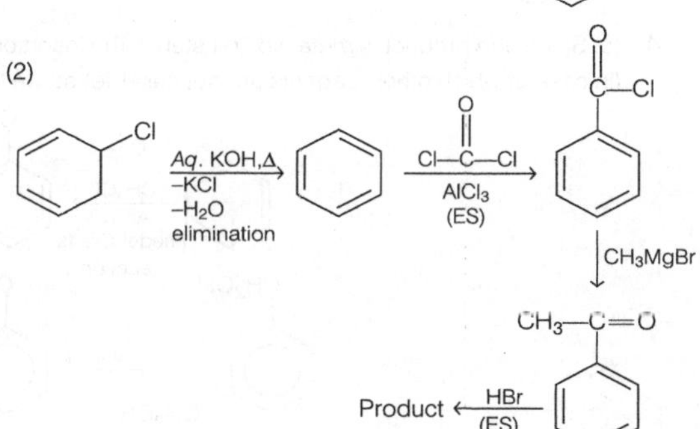

Methods of Preparation of Phenols

1. (i)

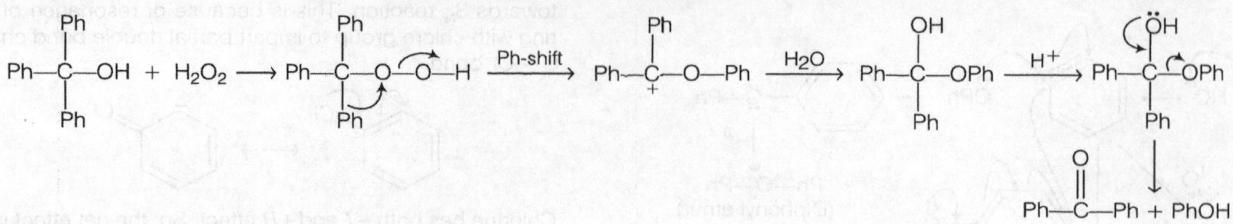

(ii) Similar to step (i), the products formed in this case are

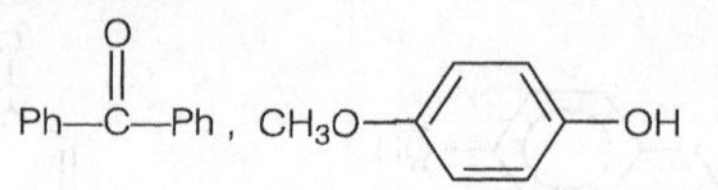

2. (b) Due to resonance, the C—Cl bond acquires some double bond character, so does not cleave by the $\overline{O}H$ at temperature 50 – 60°C. For occurrence of the reaction, the required temperature is atleast 600°C.

3. (d)

(a) 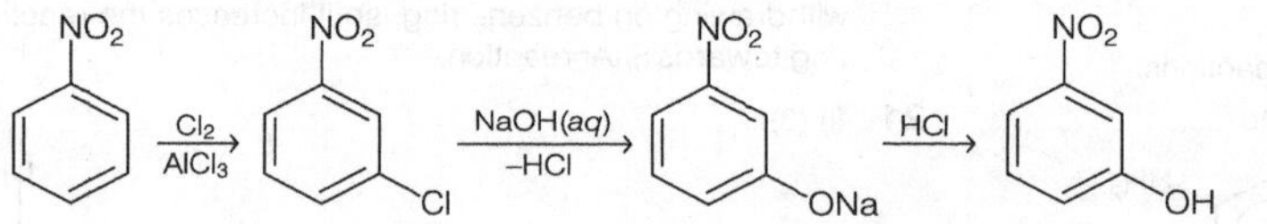

(b) 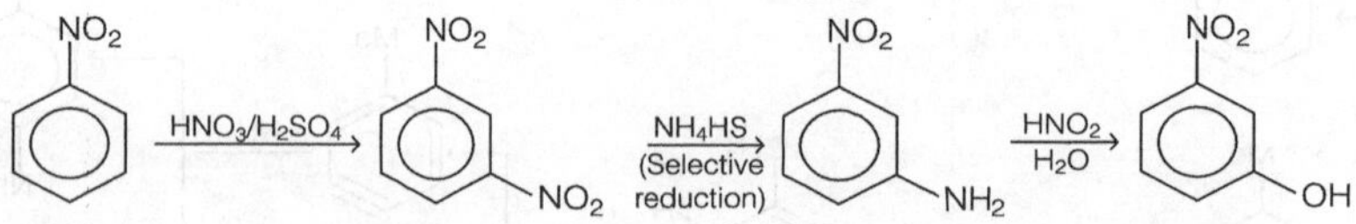

(c) 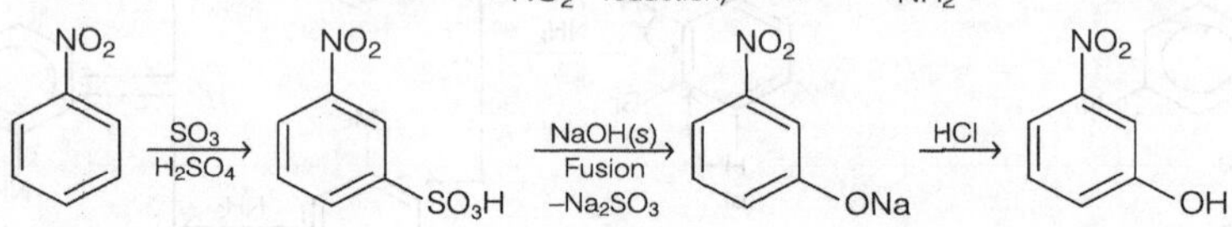

(d) 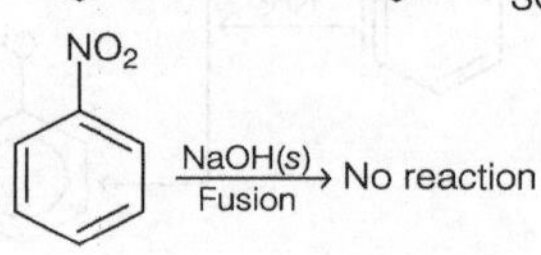

4. (c) Since, the product is *meta*, so first step is the insertion of a *meta* directing group. Since, all the inserted groups are *o/p* except CH_3COCl (in case of which other reagents are not suitable) so we then oxidised the group introduced in step (i) to — COOH, which is *meta* directing.

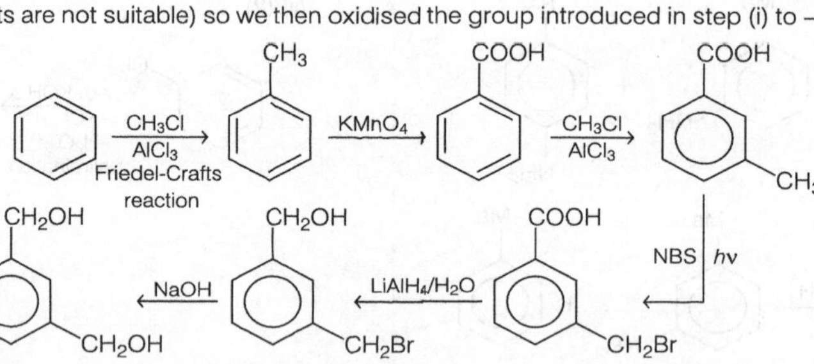

5. (c) 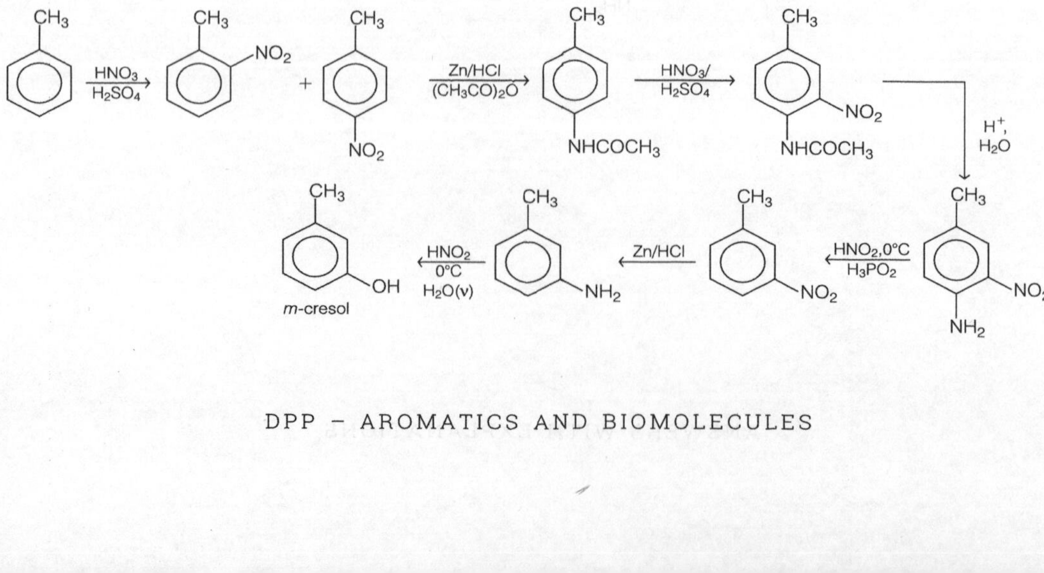

6. (c)

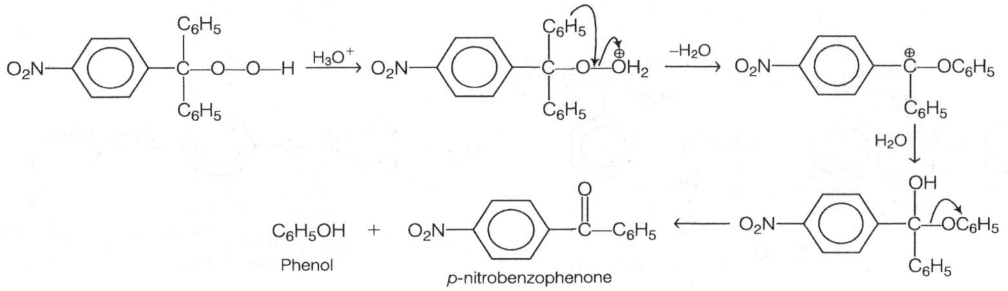

7. (b)

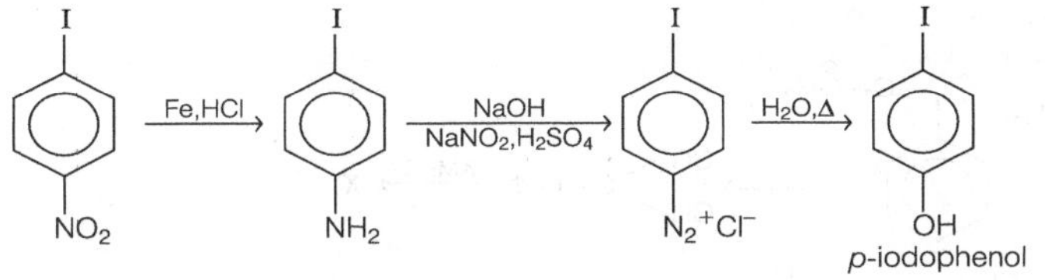

8. (b)

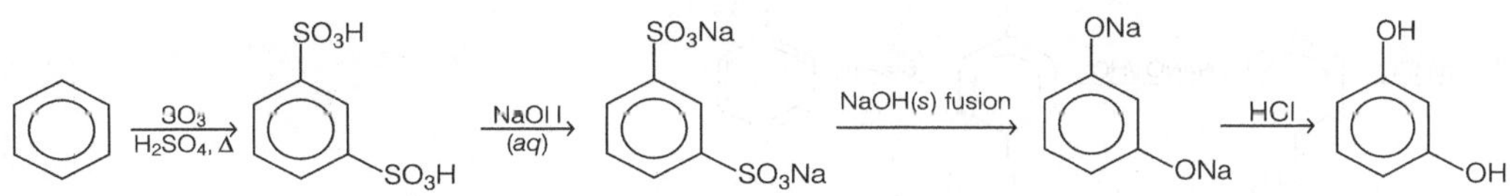

9. (b)

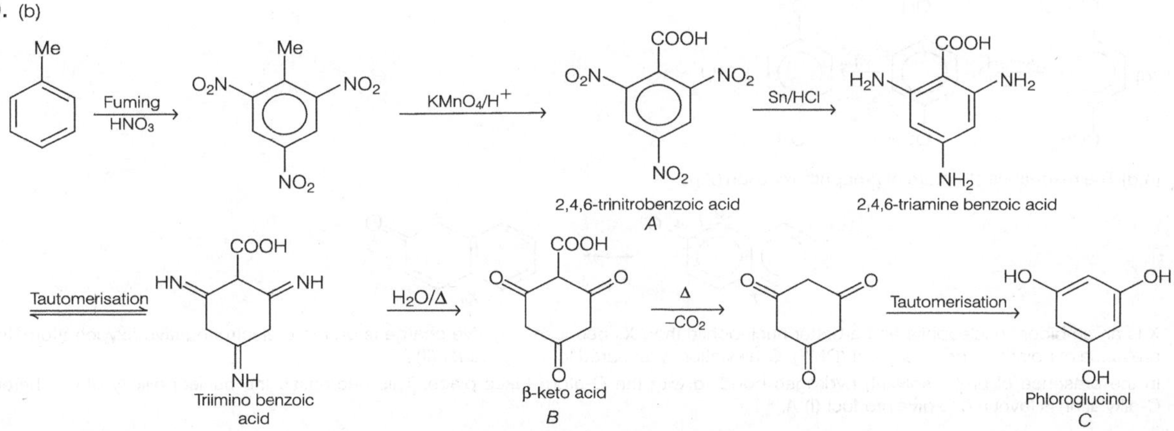

10. (a,c,d)

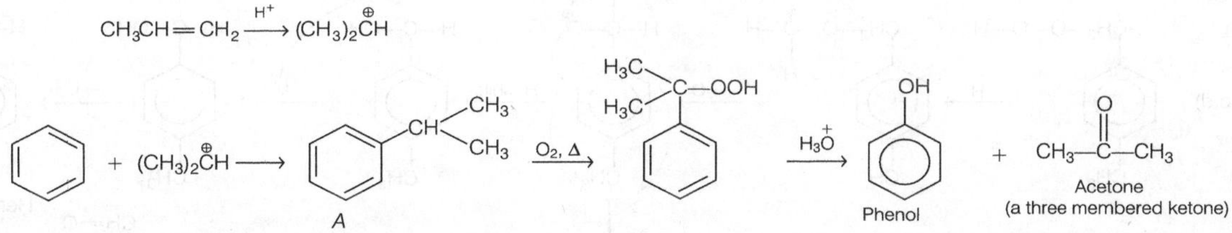

11. (a,c,d) Except (b), all the given methods are used to prepare phenol. In this case, $—NO_2$ is not diazotised.

12. (b,c,d) Hydrolysis of chlorobenzene with *aq*. KOH does not give phenol because $\bar{O}H$ is not able to break C—Cl bond, which becomes more stable due to resonance.
(Other given reactions give phenol).

13. (a,c,d)

(a) 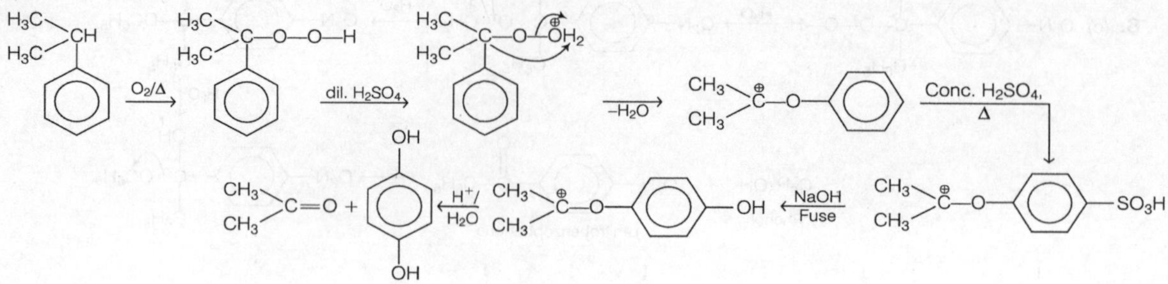

(b) 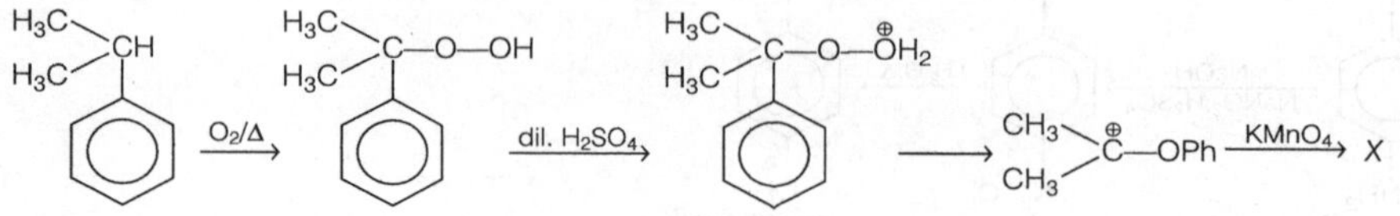

(c) 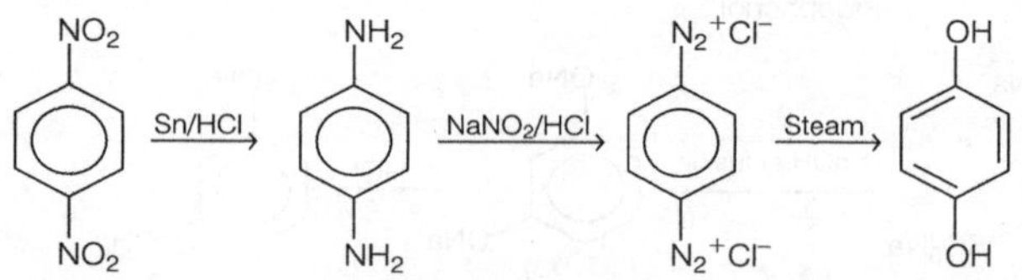

(d) 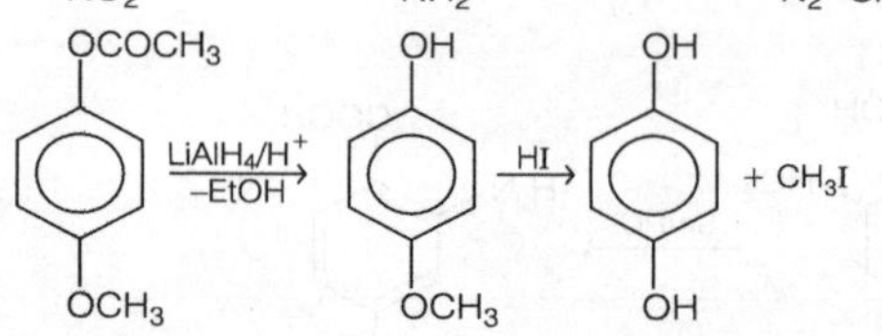

14. (a,d) The resonance structure of 2-naphthoxide ion (*A*) is

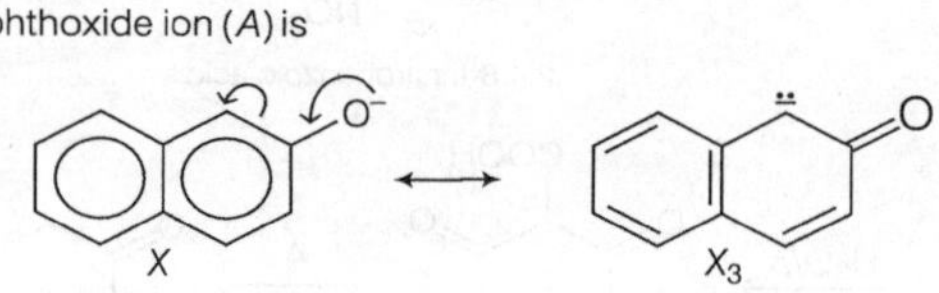

X is an ambident nucleophile and a better nucleophile than X_3 because negative charge is on more electronegative oxygen atom. In the presence of polar non-protic solvent (DMF), O-alkylation is favoured to give product (II) *B*.

In the presence of protic solvent, hydrogen bonding with the O-atom takes place. This decreases the nucleophilicity of O. Therefore, C-alkylation is favoured to give product (I) *A*.

15. (a,c,d)

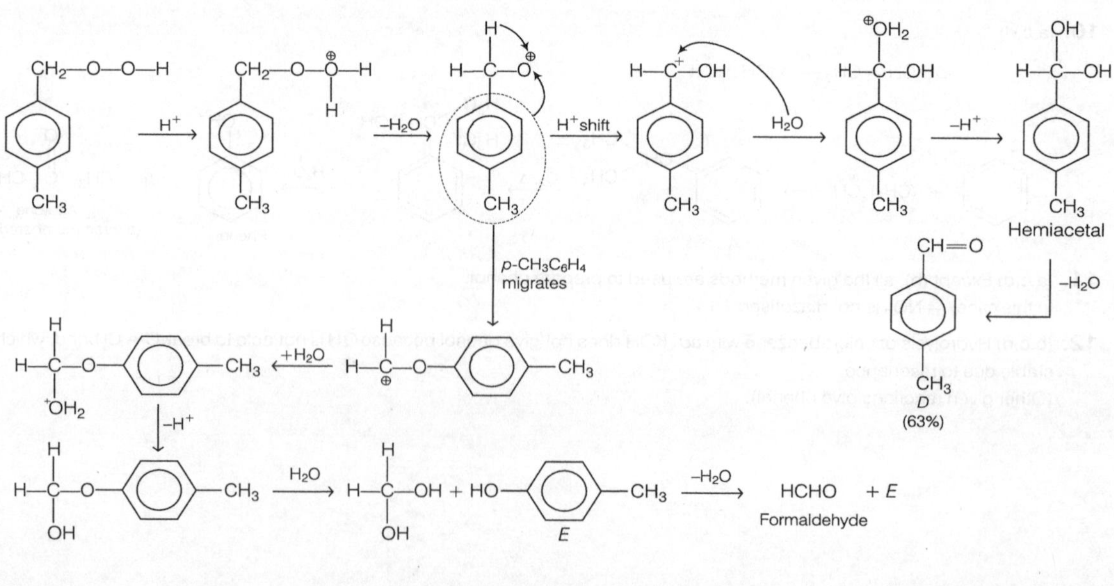

Explanations for Q. Nos. (16-18)

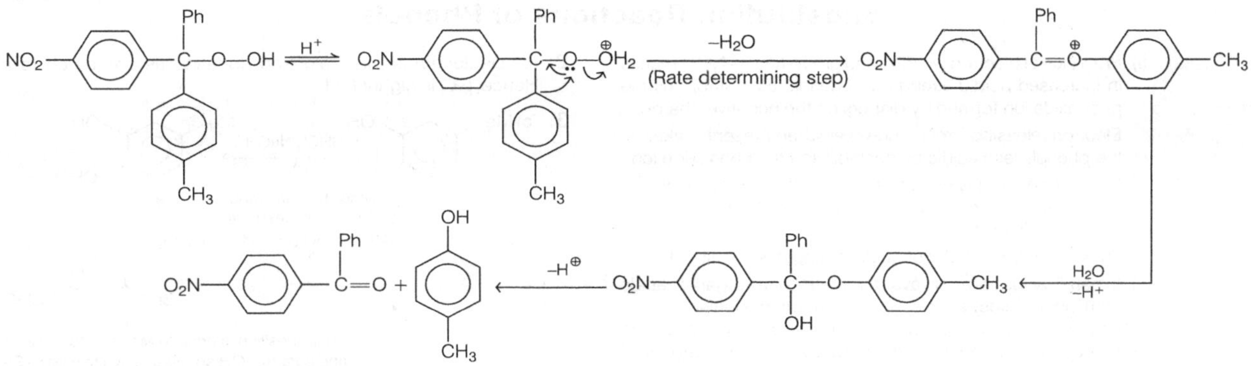

16. (c) **17.** (b)

18. (d) When $(CH_3)_2CHOOH$ is hydrolysed, acetone is obtained as the product.

19. (a) (i) → (p,s), (ii) → (p), (iii) → (q), (iv) → (p)

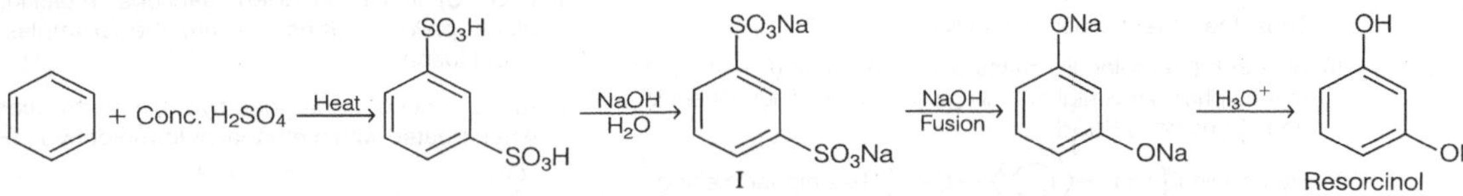

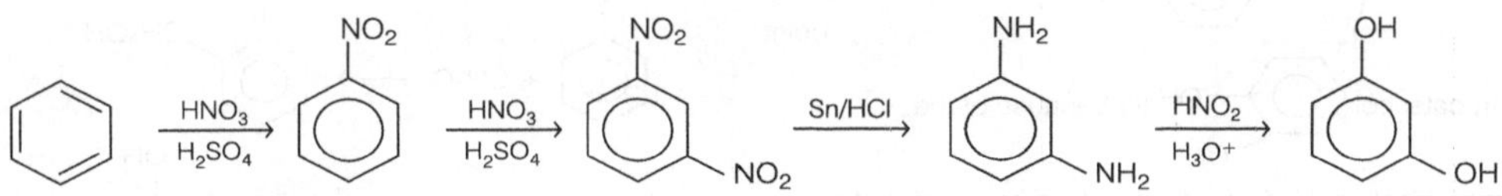

(b)

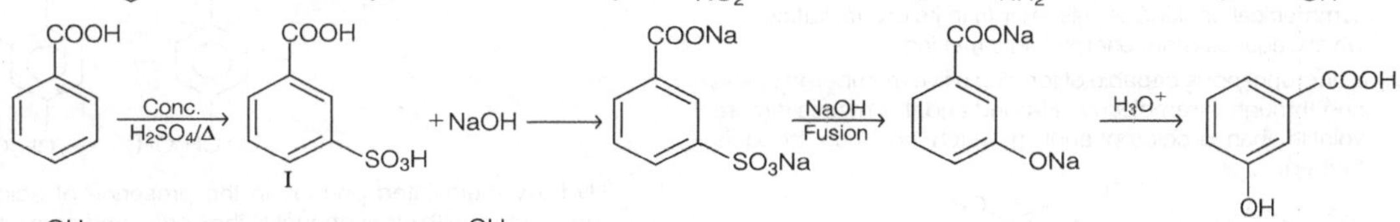

(iii)

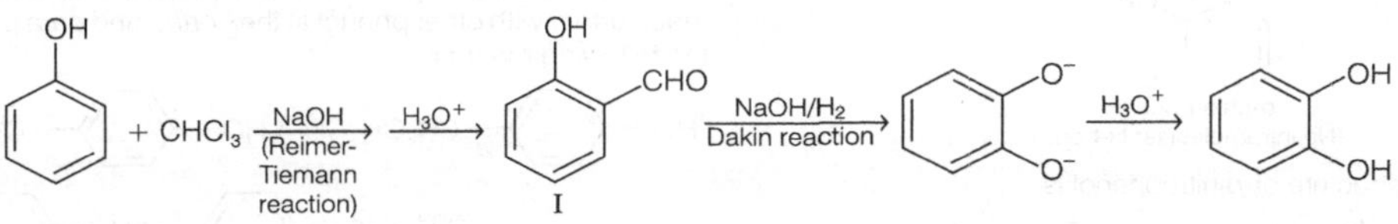

(iv)

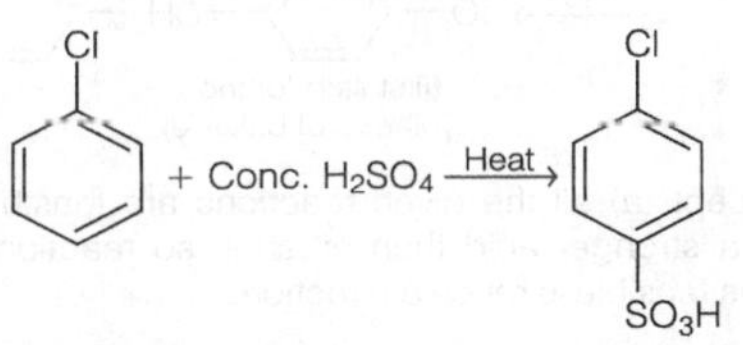

p-chlorosulphonic acid formed above, on fusion with NaOH gives the desired product as

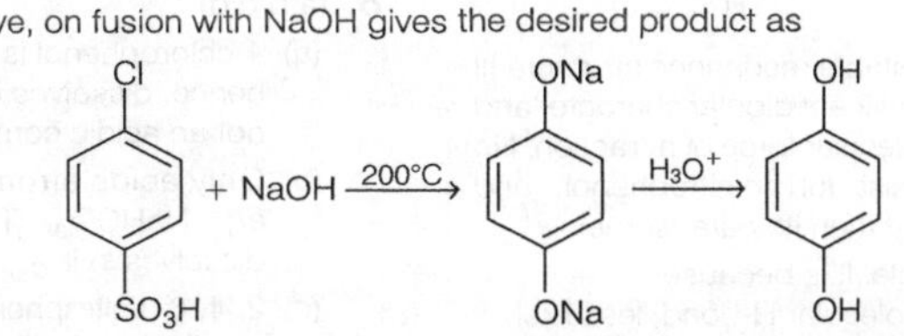

20. (4)

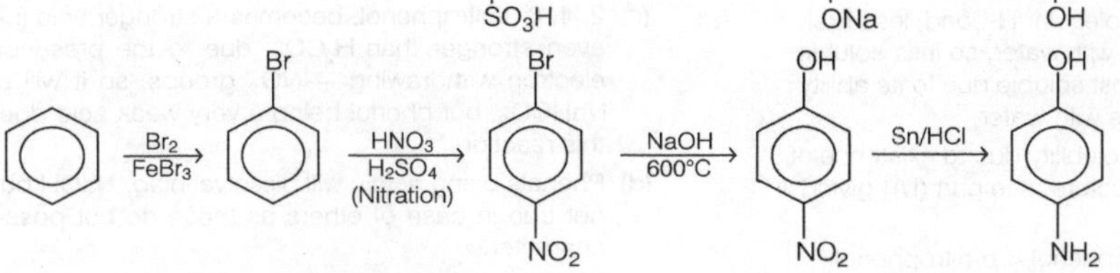

DPP-5 Physical Properties, Acidity and Electrophilic Substitution Reactions of Phenols

1. (a) (i) Presence of electron withdrawing group (like —NO_2) results in increased acidic strength of phenols as it stabilises the phenoxide ion formed by dispersing the negative charge.

Electron releasing groups,however when present makes the phenols less acidic by destabilising the phenoxide ion.

Since H-bonding is present in case of III so removal of H^+ is easy from I as compared to III. In case of II, the operative effect is only $-I$, not $-R$.

Thus, the correct order of acidic strength is II < III < I.

(ii) Mineral acids are always stronger than organic acids (carboxylic acids) as they undergo complete ionisation. Organic acids (carboxylic acids) are stronger acids as compared to phenols or alcohols due to more stability of carboxylate ion. Thus, the order of acidity is IV < III < I < II.

(iii) Same as (i), the order of acidity is II < IV < I < III.

(iv) More the number of electron withdrawing Cl groups, more the dispersal of charge and more is the stability of phenoxide ion, making the respective phenol more acidic. Thus, the correct order of acidity is I < II < III.

(b) (i) Although the molecular mass of benzenethiol (Ph—SH) is higher, phenol has higher boiling point. It is because there is no H-bonding in PhSH.

(ii) Hydroquinol 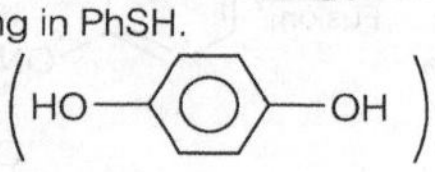 I has higher melting point than catechol 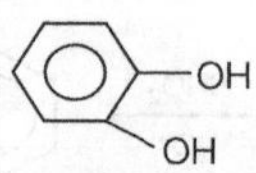 (II) because of the symmetrical packing of *p*-isomer (I) in its crystal lattice which requires more energy for its melting.

(iii) *o*-fluorophenol is capable of forming a five membered stable ring through intramolecular H-bond and this makes it more volatile than *o*-chlorophenol in which no such bond is formed.

o-fluorophenol   *o*-chlorophenol (No intramolecular H-bond)

(iv) The resonating structure of *p*-nitrophenol is

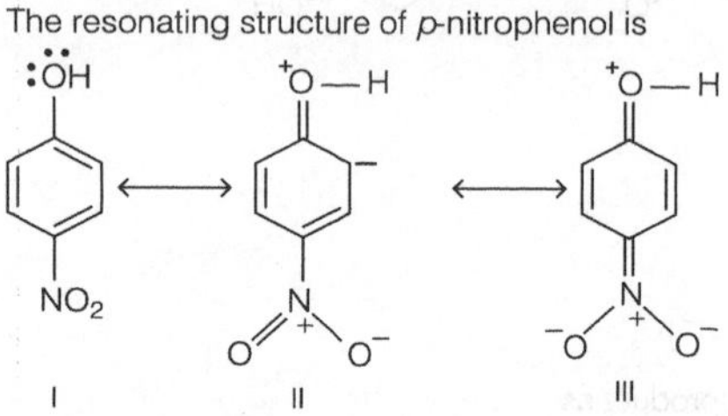

I   II   III

Due to the presence of a very stable resonance structure III in *p*-nitrophenol, it acquires significant dipolar character and develops a very strong intermolecular force of attraction. No such resonance structure exist for *m*-nitrophenol, and hence, it has lower boiling point than its *para*-isomer.

(v) *ortho*-nitrophenol is least soluble. It is because *ortho*-nitrophenol forms intramolecular H-bond, less likely to form intermolecular H-bond with water, so less soluble. *Meta* nitrophenol is second most soluble due to its ability to form intermolecular H-bonds with water.

Para-nitrophenol has highest solubility due to existence of a stable dipolar resonance structure [see part (iv)] giving it a large polar character.

Order : *o*-nitrophenol < *m*-nitrophenol < *p*-nitrophenol

2. (b) Deuterium is more electron donating as compared to H. Hence, pK_a is higher for it.

3. (c)

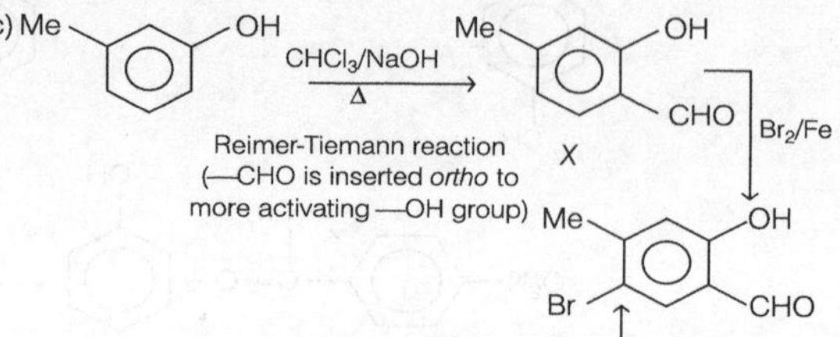

(This position is *ortho* to Me *meta* to — CHO and *para* to—OH so more active towards ESAr)

4. (b) With increase in the size of halogen electrophiles, steric hindrance increases, hence number of halogen substitution decreases.

5. (a,b,c,d) In all the given reactions, attacking species is an electrophile, so these all are the examples of electrophilic substitution.

6. (a,b,c) Phenol undergoes hydroxy methylation at *o*/*p* position when treated with methanal or formaldehyde (HCHO).

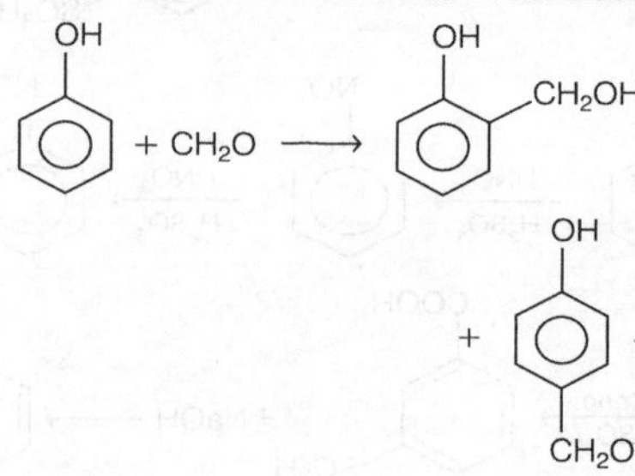

Hydroxy methylated phenol, in the presence of acid catalyst react further with other phenol at their *ortho* and *para* positions in the following manner.

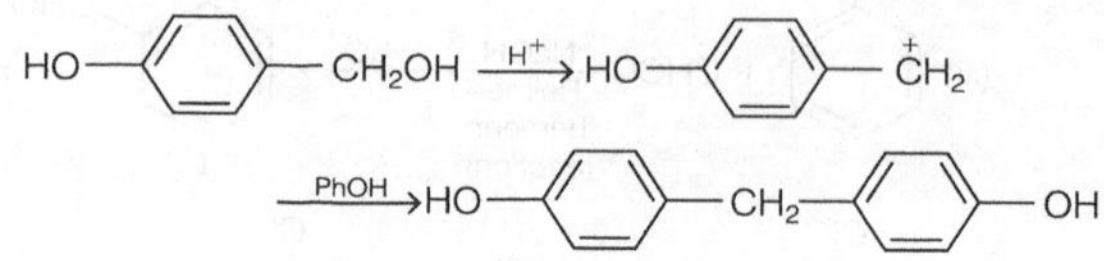

(first step for the synthesis of bakelite)

7. (b,c,d) Except (a) all the given reactions are feasible. Since, phenol is a stronger acid than ethanol, so reaction given in option (a) is feasible is reverse direction.

8. (a,b,c,d)

(a) 4-chlorophenol is acidic because of the + I-effect of Cl and hence, dissolves in NaOH but 4-chloro-1-methyl benzene is not an acidic compound, so does not react with NaOH.

(b) Only acids stronger than carbonic acid dissolve in *aq*. $NaHCO_3$. That's why only 4-methyl benzoic acid dissolves in it.

(c) 2, 4, 6-trinitrophenol, becomes a stronger acid (pK_a = 0.38) even stronger than H_2CO_3, due to the presence of three electron withdrawing —NO_2 groups, so it will dissolve in $NaHCO_3$, but phenol being a very weak acid does not give this reaction.

(d) Phenols being acidic will dissolve in *aq*. NaOH but similar is not true in case of ethers as these do not possess acidic character.

9. (a,d)

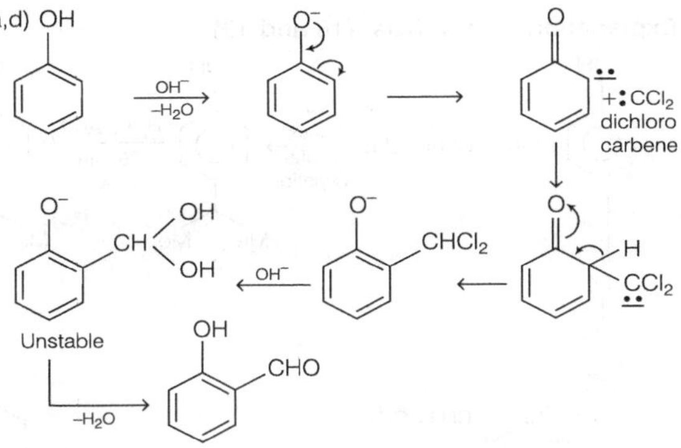

10. (a,c,d)

(a) Phenol gives white ppt of 2, 4, 6-tribromophenol, which is a visible change.

(b) No visible change is observed in Reimer-Teimann reaction.

(c) HCl will be evolved, and a dye is formed which is a visible change.

(d) Libermann's nitroso test of phenol, so a visible change appears.

11. (a)

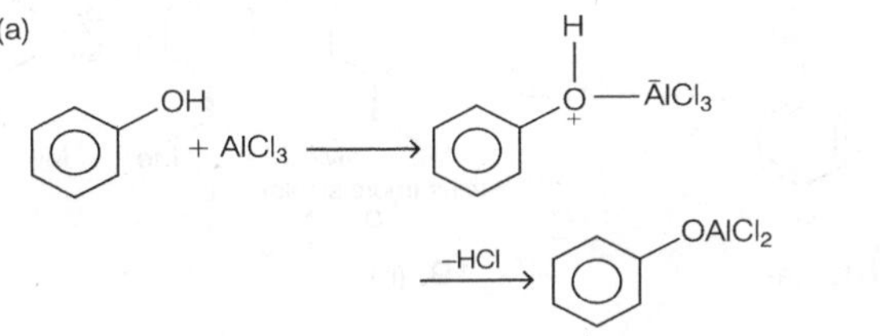

12. (d)

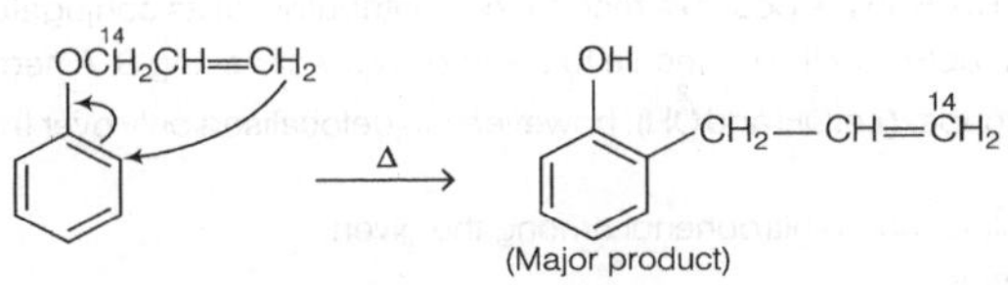

This is an example of Claisen rearrangement.

13. (a) (i) → (p,q,s) (ii) → (p,q,s) (iii) → (p,q,r) (iv) → (p,q,r)

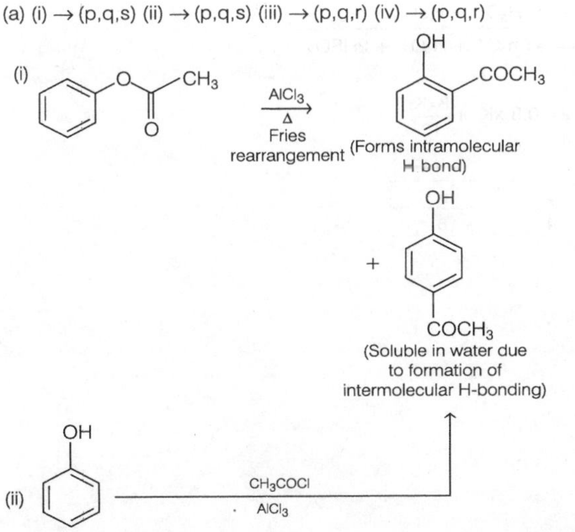

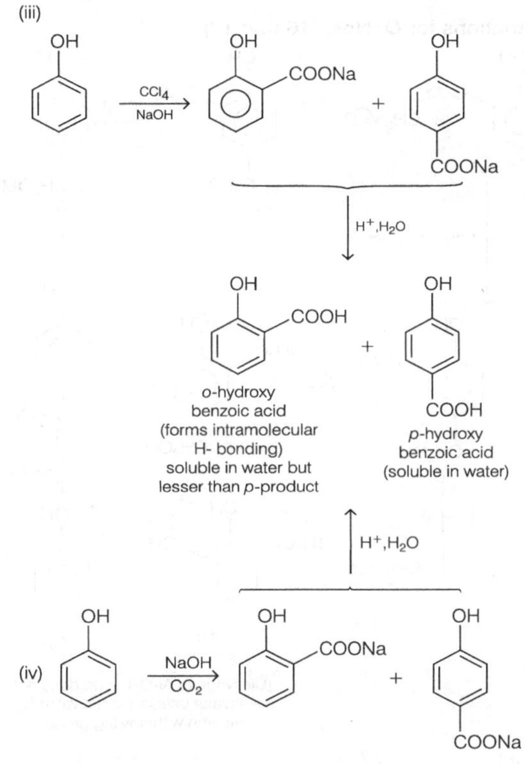

14. (c)

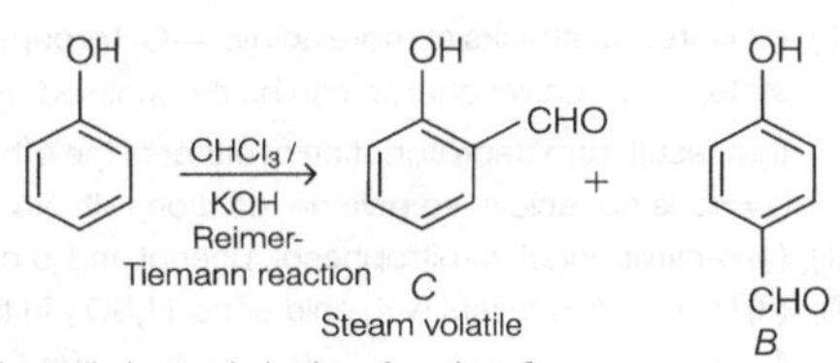

[$CHCl_3$ = trihalogen derivative of methane]

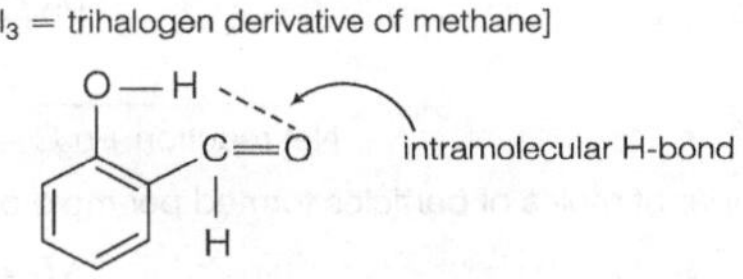

So, carbonyl group in this case does not react with Schiff's reagent.

15. (b)

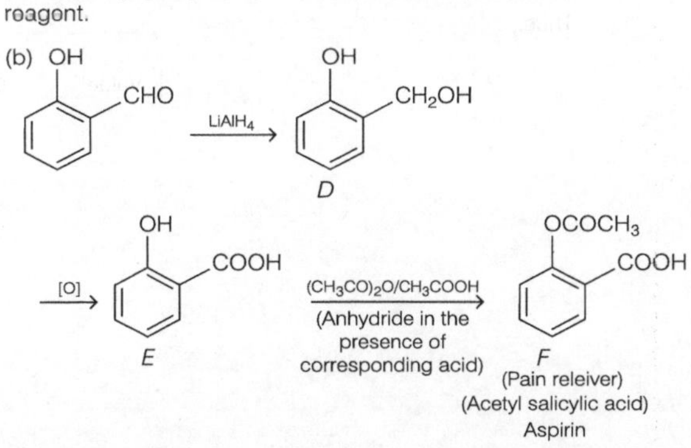

Explanations for Q. Nos. (16 and 17)

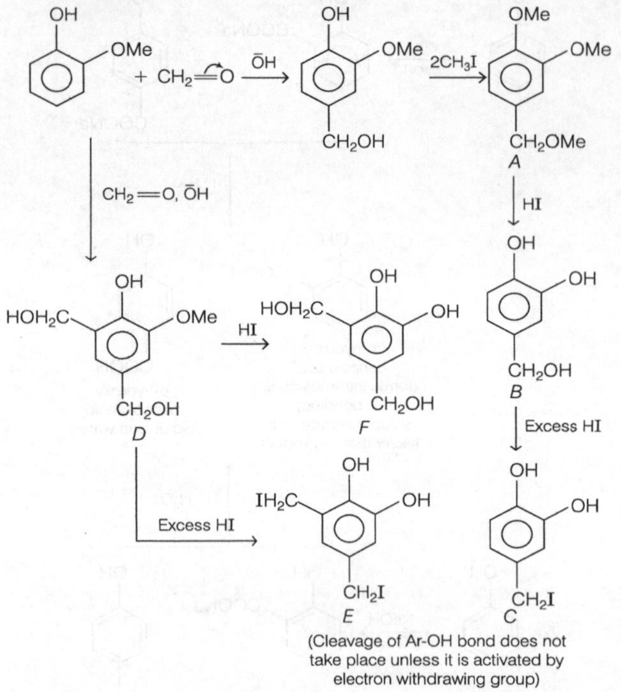

Explanations for Q. Nos. (18 and 19)

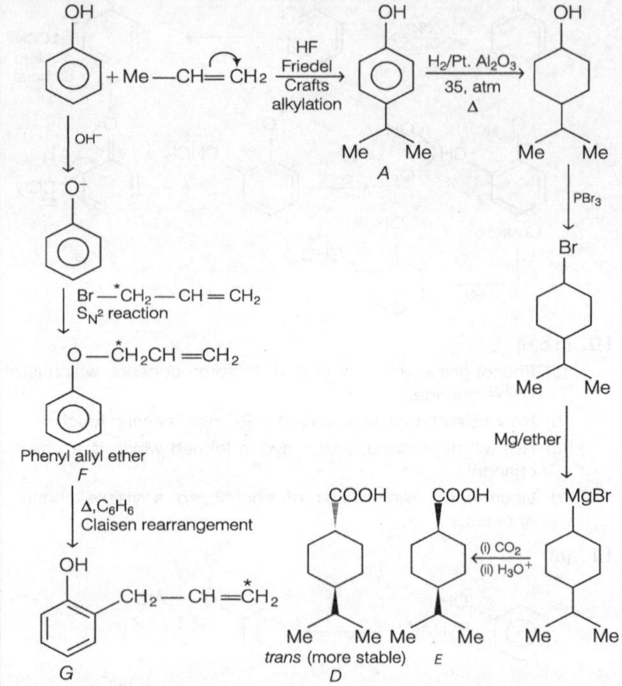

16. (b) **17.** (c) **18.** (a) **19.** (b)

20. (i) (5) Here, Na attacks at more acidic —OH group, which is marked as 5. In the possible resonance contributors of its conjugate base in this state, the negative charge can be delocalised over the entire π structure of the molecule to the ring *A* as well as ring *B*. A negative charge that result from deprotonation of either of the other two phenol groups, (i.e. $\overset{1}{O}H$ and $\overset{2}{O}H$), however can delocalised only over the one ring. C=C is not acidic, so give no reaction with Na.

(ii) (4) *o*-nitrophenol, *m*-nitrophenol, phenol and *p*-cresol are less acidic than *p*-nitrophenol among the given.

(iii) (8) Ph_3C—OH reacts with cold conc. H_2SO_4 in the following manner

1. $$Ph_3C{-}OH + H_2SO_4 \longrightarrow Ph_3C^+ + H_2O + HSO_4^-$$
2. $$H_2SO_4 + H_2O \longrightarrow H_3O^+ + HSO_4^-$$

Net reaction $$Ph_3C{-}OH + 2H_2SO_4 \longrightarrow Ph_3C^+ + H_3O^+ + 2HSO_4^-$$

Number of moles of particles formed per mole of solute = 4

$\therefore$ $$\Delta T_f = iK_f \cdot m = 4x \times 0.2 = 0.8\ xK \text{ or } \frac{8xK}{10}$$

So, the value of $z = 8$

Thus,

	x	y	z
Value	5	4	8

DPP-6 **Chemical Properties of Phenols**
(Other than Electrophilic Substitution)

1. (i) (a)

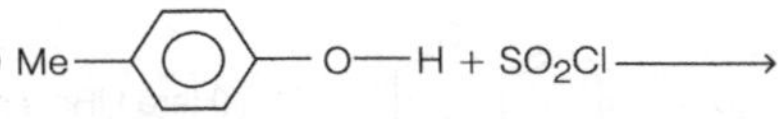

(Cleavage of Ar—O bond does not take place unless, it is activated by electron withdrawing group.)

(b)

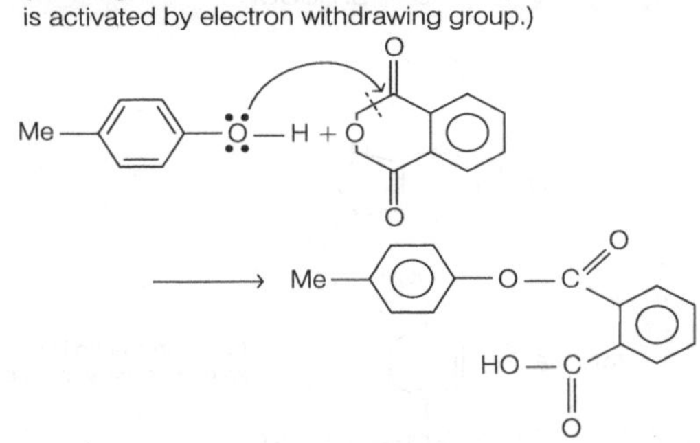

o-tolyl hydrogen phthalate

(c)

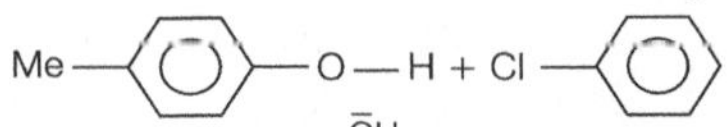

$\xrightarrow{\bar{O}H}$ No reaction (due to the absence of electron withdrawing group in ArCl)

(d)

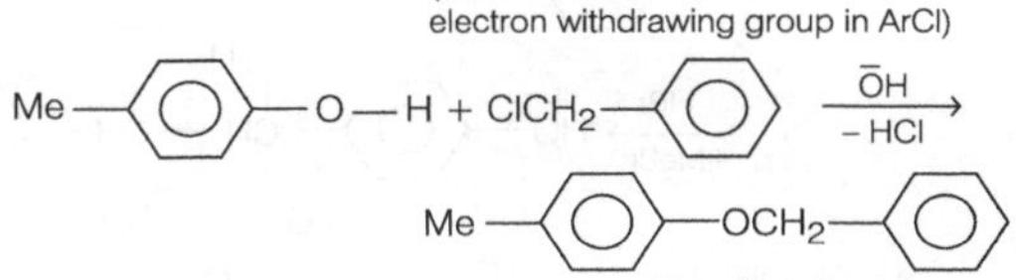

Benzyl-*p*-tolylether

(e)

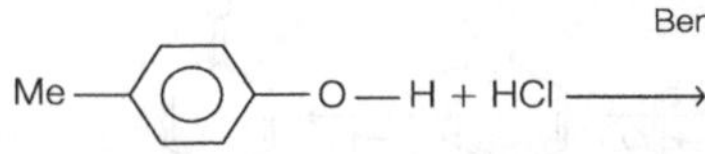

No reaction

(Cleavage of Ar—O bond does not take place with HCl.)

(ii) (a)

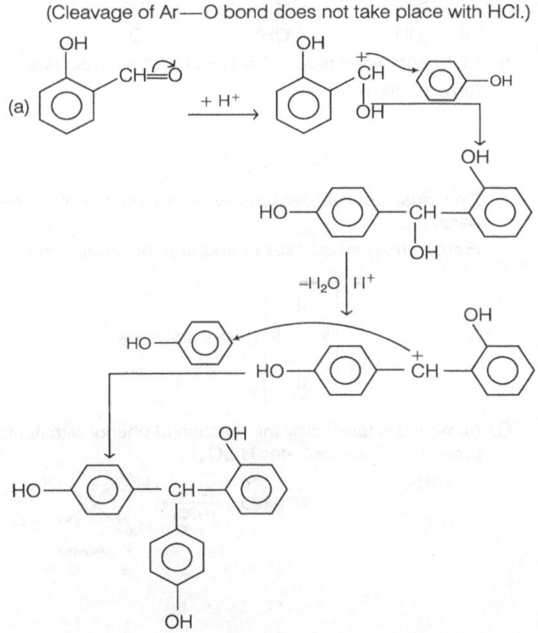

A

(b)

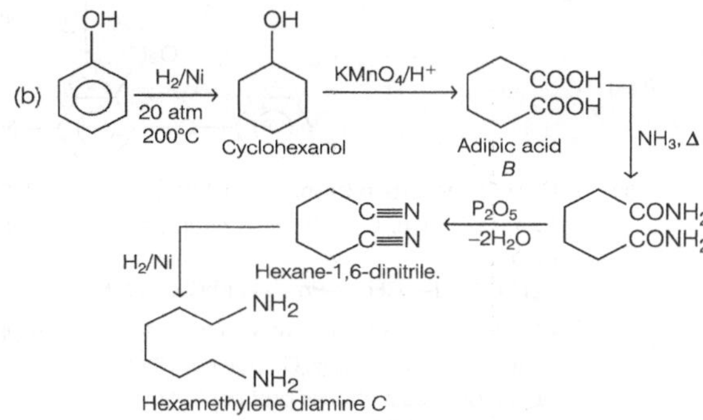

(c)

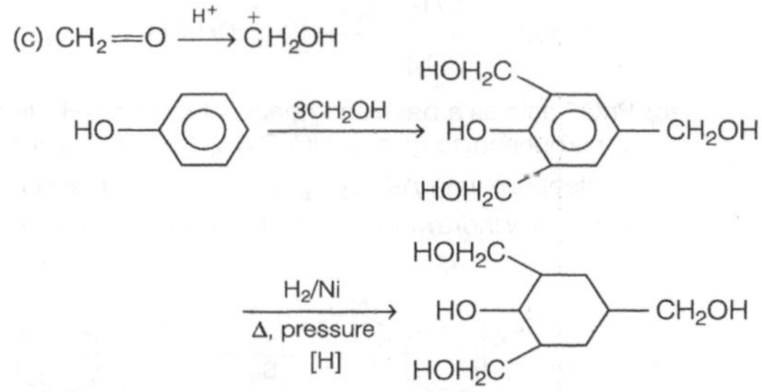

(d)

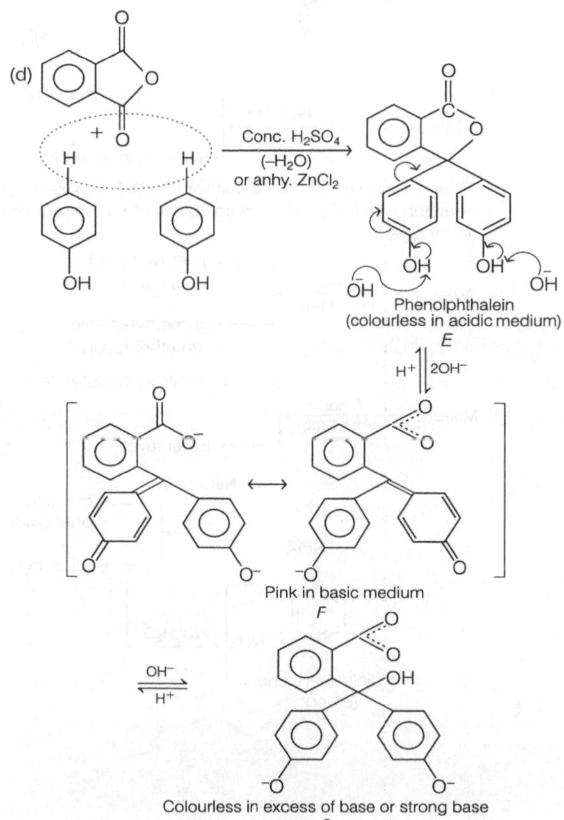

2. (i)

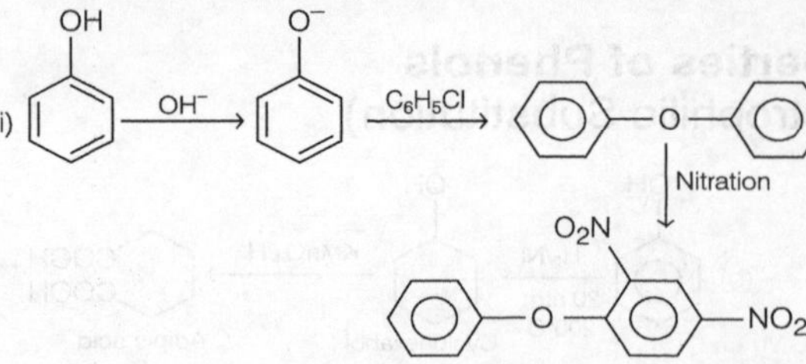

(ii) (a) $C_2H_5O^-$ acts as a base, so abstracts H^+ from PhOH to form PhO^- ion. $C_2H_5O^-$ is a stronger nucleophile than PhO^- so,

$$C_2H_5O^- + I—CH_2—Ph \longrightarrow PhCH_2OC_2H_5$$

(b) CH_3O^- acts as a base. It abstracts H^+ from *p*-nitrophenol to form *p*-$NO_2—C_6H_4O^-$. Since, CH_3O^- is a stronger nucleophile so it attacks preferentially.

$$CH_3O^- + \underset{CH_2I}{\overset{CH_3}{|}} \xrightarrow{S_N2} CH_3OCH_2CH_3$$

(c) PhO^- acts as a base and hence, abstracts a H^+ from *p*-nitrophenol to give *p*-$NO_2C_6H_5O^-$. PhO^- is stronger nucleophile than *p*-$NO_2C_6H_5O^-$. Moreover, strong electron withdrawing —NO_2 in (*A*) favours ArS_N reaction.

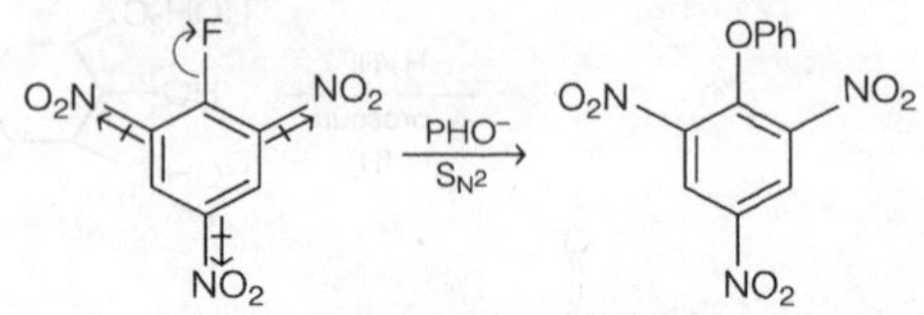

3. (i)

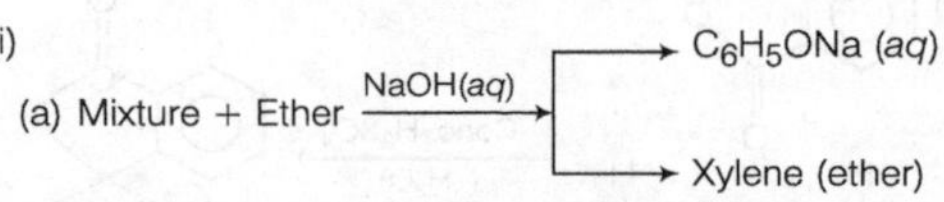

Phenol being acidic forms salt with NaOH but xylene give no reaction with it. Further, it is soluble in ether, so remain in ether layer.

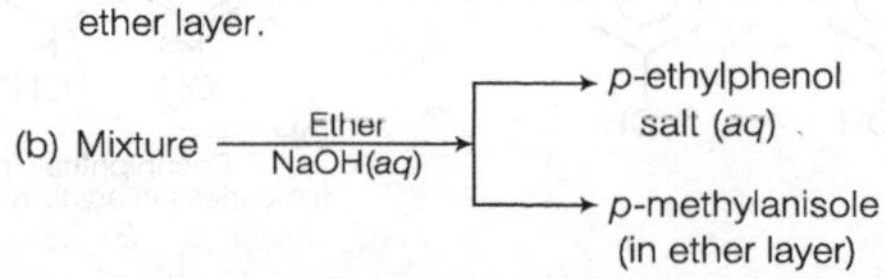

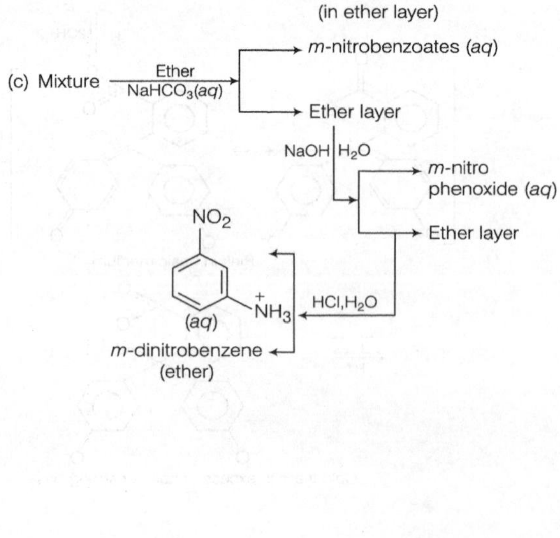

(ii) (a) $A =$

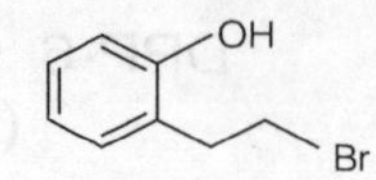

(Only aliphatic –OH group reacts to HBr)

$B =$

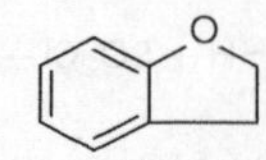

(Here HBr is eliminated)

(b) $C =$

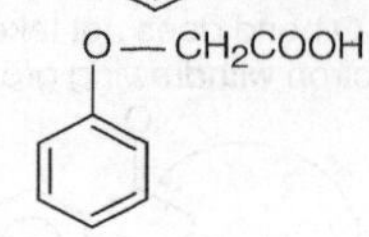

$D =$

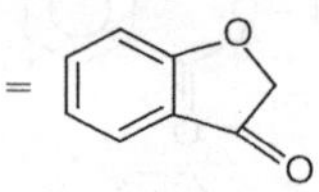

(c) $E =$

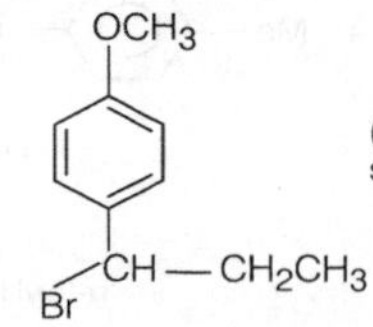

(∵ Double bond is more reactive so preferentially attacked by HBr)

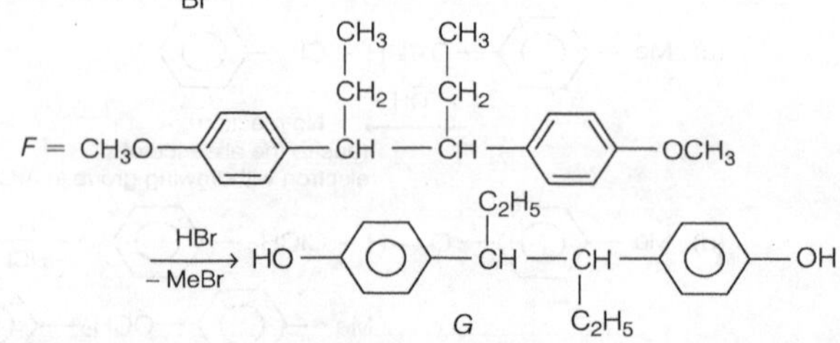

4. (d)

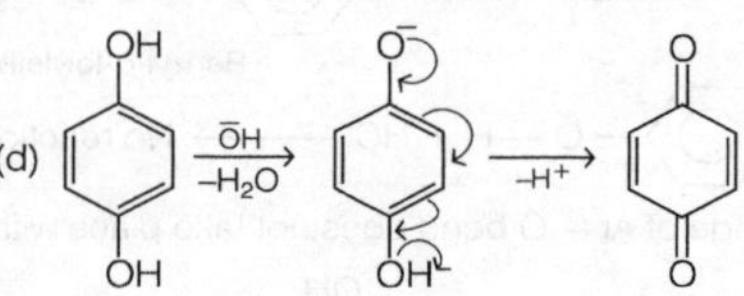

5. (d) Phenol when reduced with H_2/Ni forms cyclohexanol (an aliphatic alcohol).

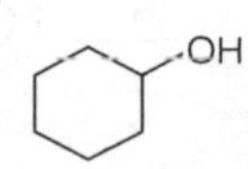

Zinc dust causes deoxygenation so phenol is converted into benzene.

Fremy's reagent oxidises phenol in *p*-benzoquinone.

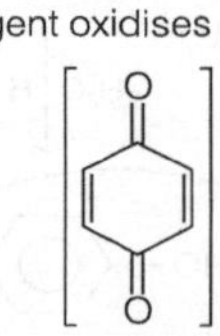

p-benzoquinone

6. (a) Acetal is formed by the reaction of phenol with acetone in the presence of concentrated H_2SO_4.

$$(H_3C)_2C=O + 2C_6H_5OH \xrightarrow[-H_2O]{\text{Conc. } H_2SO_4} (H_3C)_2C(OC_6H_5)_2$$

Acetal

7. (d)

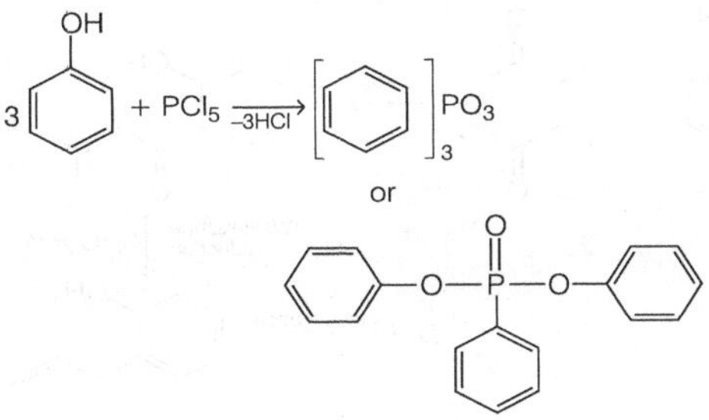

Triphenyl phosphate
(Main product)

Some PhCl is also formed but in very less amount.

8. (c)

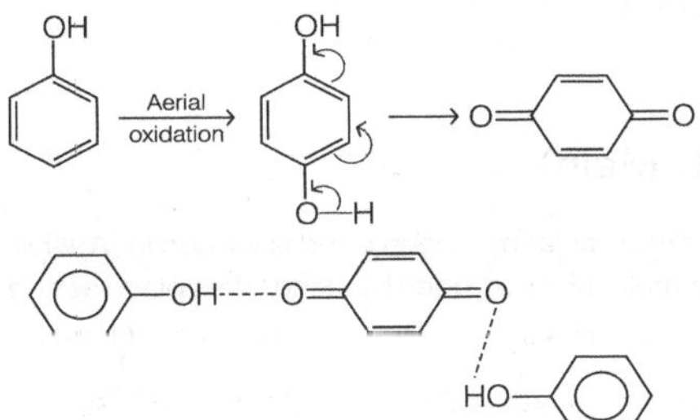

Phenoquinone
or
Quinhydrone (coloured form)

9. (c) The reaction takes place by S_N1 mechanism, i.e. involves the formation of carbocation (benzyl carbocation) intermediate, which is stabilised by the presence of electron releasing group like —OMe or —Me (electron releasing effect of OMe is more than Me) at *o*/*p* positions but not at *meta*.

So, the order of reactivity is III > II > I > IV

10. (c)

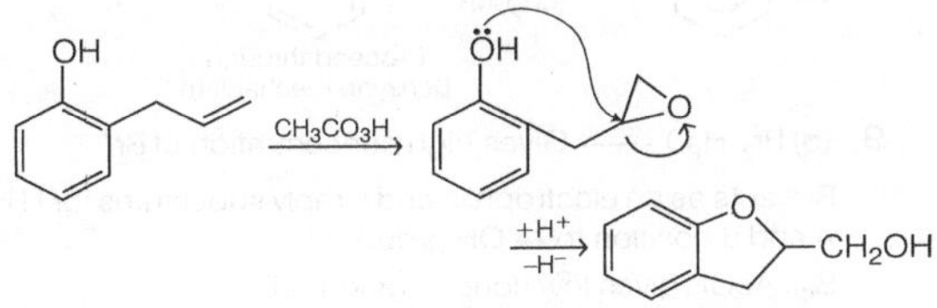

11. (c)

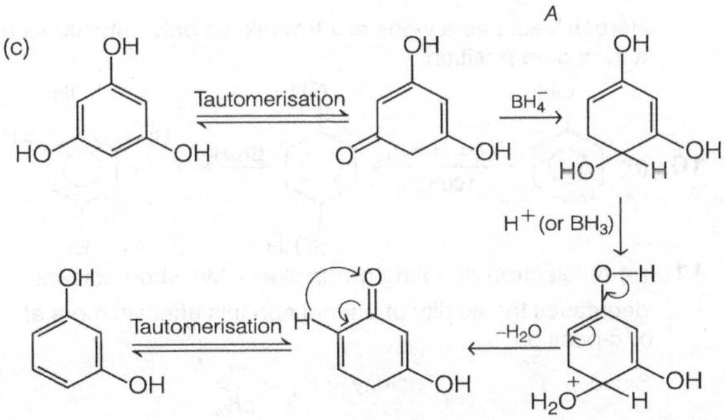

12. (a,c,d) Phenol is a stronger acid than water but weaker than carbonic acid (H_2CO_3), hence, it gives no reaction with $NaHCO_3$.

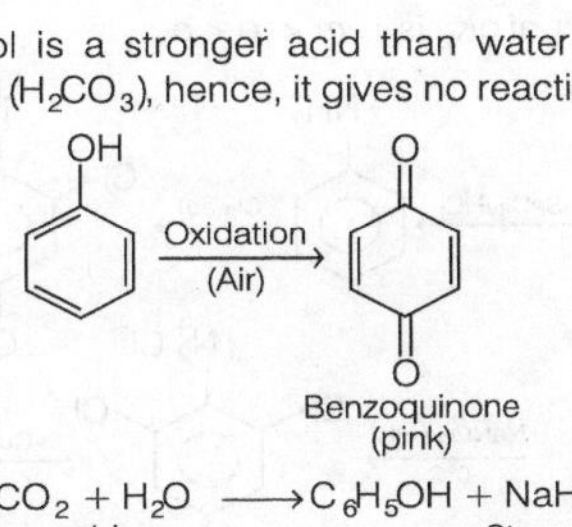

Benzoquinone
(pink)

$$\underset{\text{Strong acid}}{C_6H_5ONa} + CO_2 + H_2O \longrightarrow C_6H_5OH + \underset{\text{Strong base}}{NaHCO_3}$$

13. (b,d)

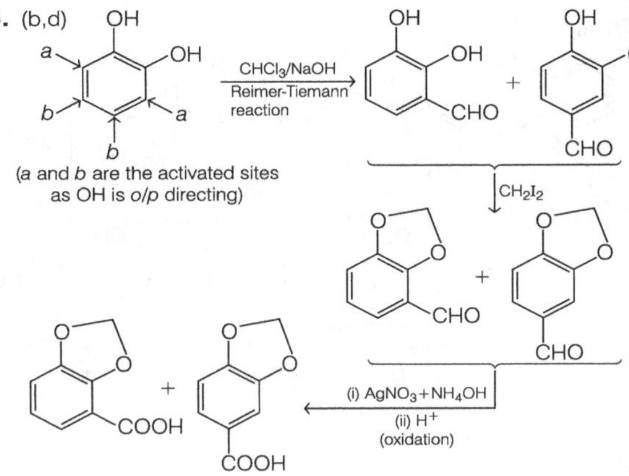

14. (c,d)

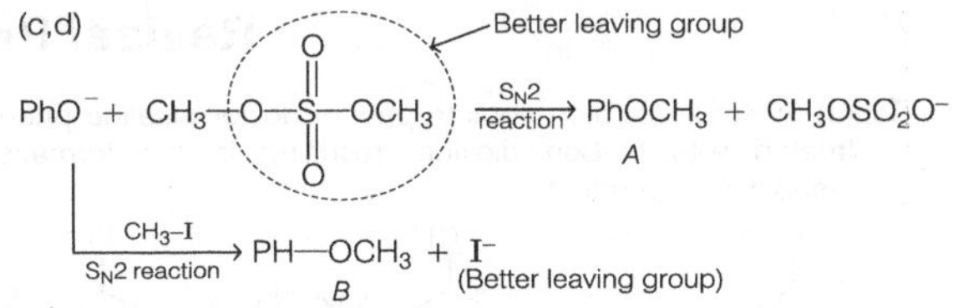

15. (c) Benzoic acid is a stronger acid than carbonic acid (H_2CO_3) but phenol is a weaker acid (or stronger base) than it, so former will react with it but later does not and hence, these two can be separated by using it.

Explanation for Q. Nos. (16 and 17)

Vanillin gives colour with $FeCl_3$, which shows presence of a phenolic group.

Reaction with Tollen's reagent $[Ag(NH_3)_2]^+$ suggests the presence of a —CHO group in it.

Since, it reacts with only one mole of MeI, so it contains only one —OH group.

Vanillin is not steam distilled which suggests that —OH and —CHO groups are not in *o*-position.

$$C_8H_8O_3-[CHO—OH—C_6H_4] = OCH_2$$

Thus, it also has a OCH_3 group.

The reactions are as follows:

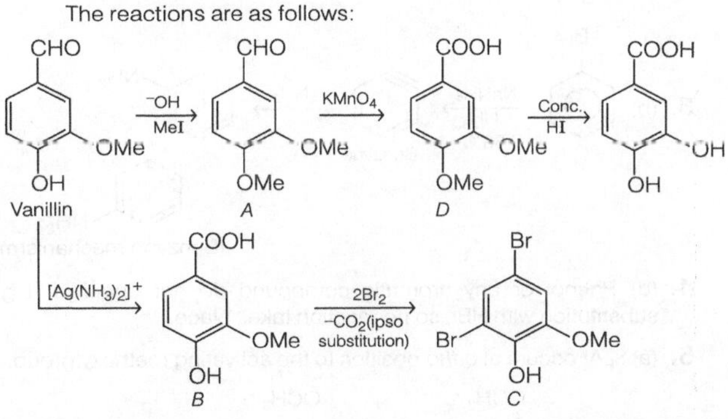

16. (c) **17.** (c)

18. (b) No effervescence with Na indicates that no acidic hydrogen is present in *A*. However when hydrolysed, it gives phenolic group and aceto group as *B* gives both $FeCl_3$ test and iodoform test.

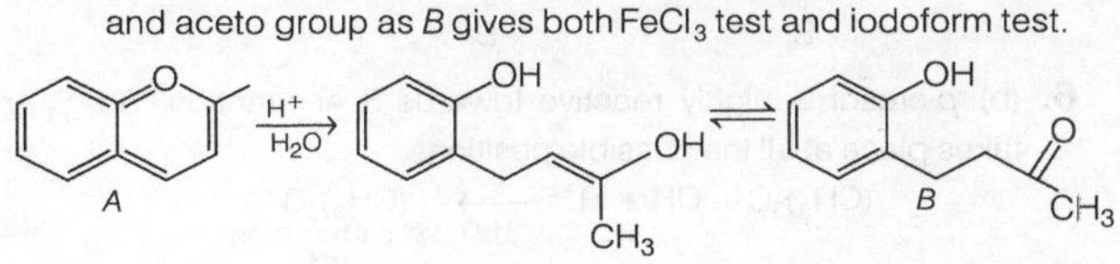

19. (c)

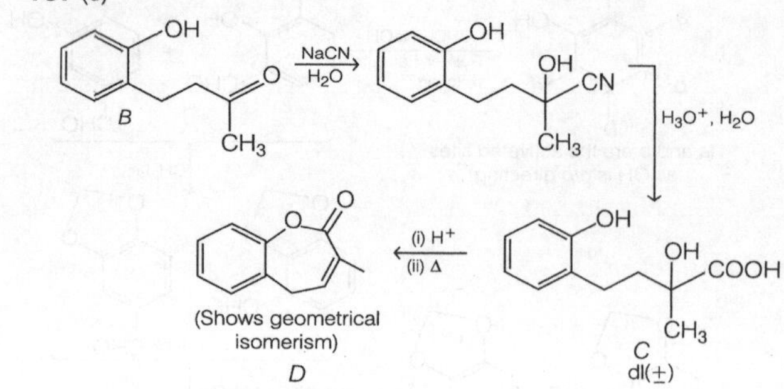

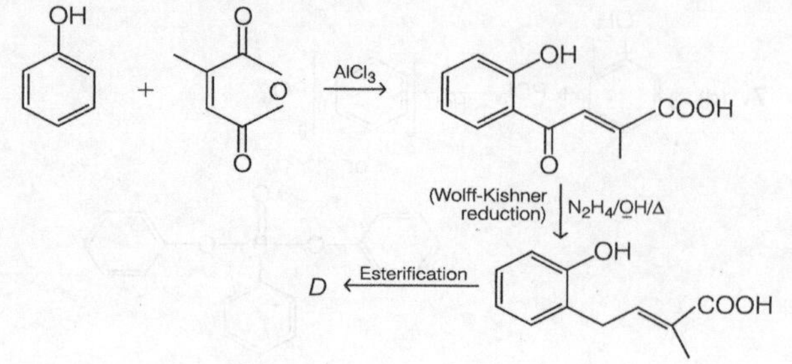

20. (d) B on reduction with $NaBH_4$ gives a pair of enantiomers (racemic mixture), as a new chiral centre is formed from planar sp^2 carbon.

Revisal Problems (JEE Main)

1. (c) In Kolbe reaction sodium phenoxide (or alkaline phenol) is treated with carbon dioxide, resulting in the formation of *o*-substituted product.

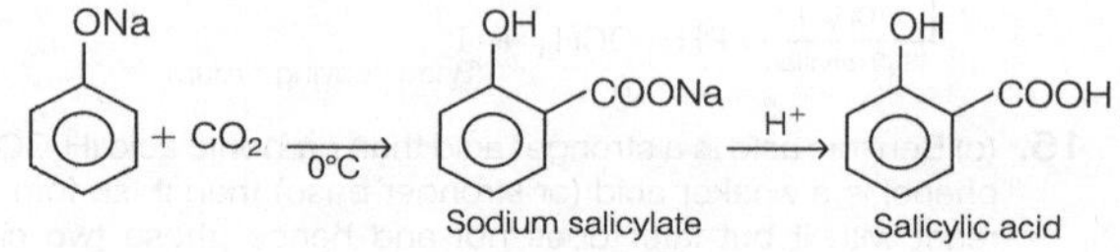

2. (c) This is an example of Claisen rearrangement.

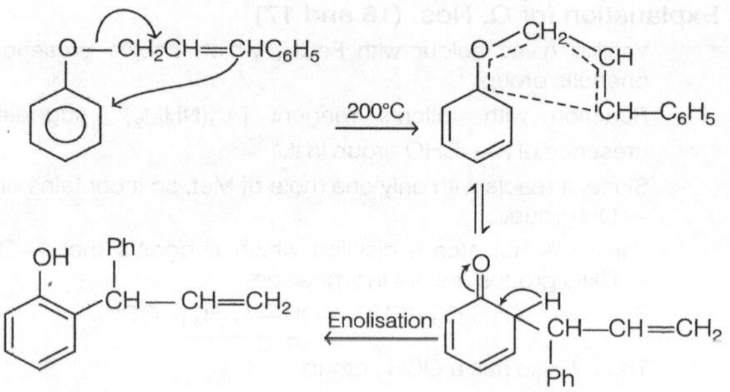

3. (b)

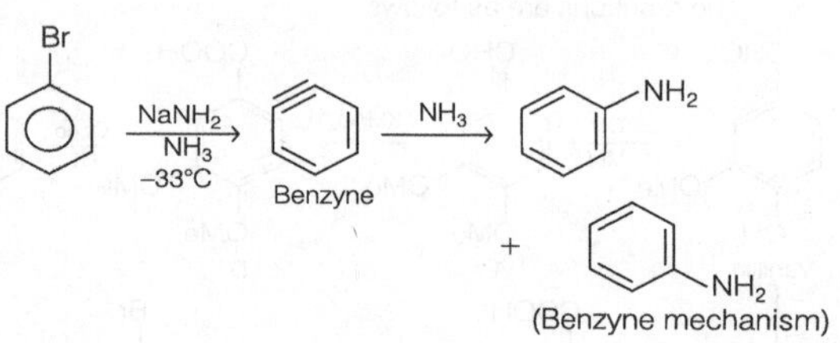

4. (d) Phenol or any aromatic compound do not undergo ring substitution with HBr, so no reaction takes place.

5. (a) S_EAr occurs at *ortho* position to the activating methoxy group.

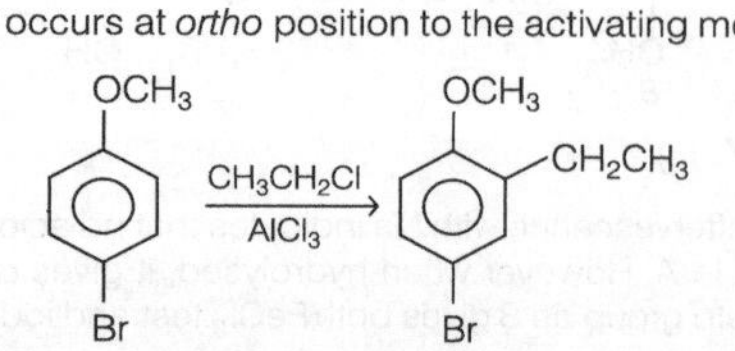

6. (b) *p*-cresol is highly reactive towards S_EAr reaction. So, S_EAr takes place at all the possible positions.

$$(CH_3)_3C{-}OH + H^+ \xrightarrow[-H_2O]{} \underset{\text{3° carbocation } (E^+)}{(CH_3)_3C^+}$$

7. (a) Intramolecular hydrogen bond is formed in (A) while in case of B, intermolecular hydrogen bonding takes place as shown below.

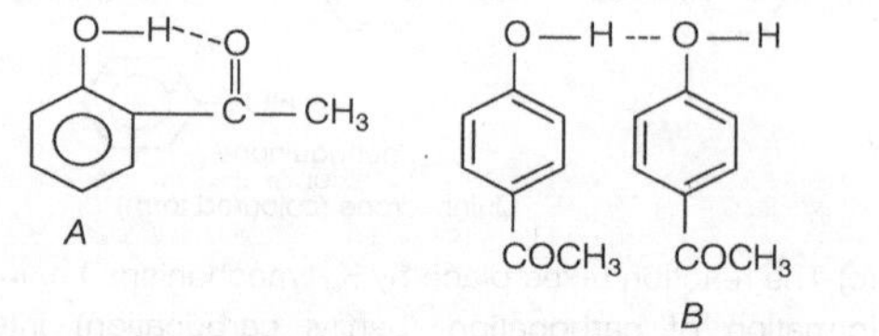

Thus, A is more volatile or has a lower boiling point than B.

8. (a)

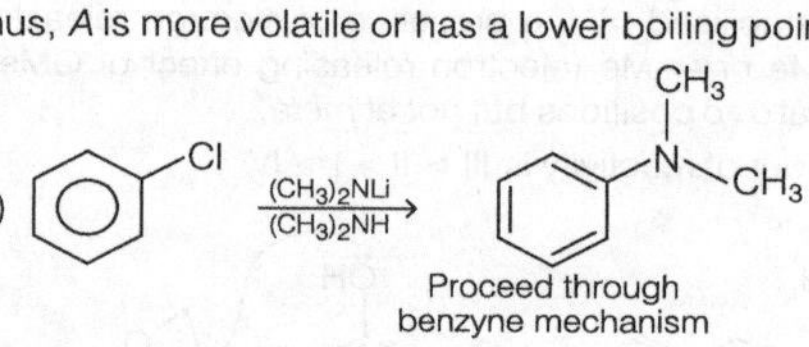

9. (b) $Br_2/H_2O \longrightarrow$ Gives high concentration of Br^+.

Br^+ acts as an electrophile and simply substitutes for H in any of 2, 4 and 6 position to —OH group.

$Br_2/AcOH$ gives low concentration of Br^+.

Here Br^+ acts as a weak electrophile so only substitutes at *ortho* and/or *para* position.

10. (d)

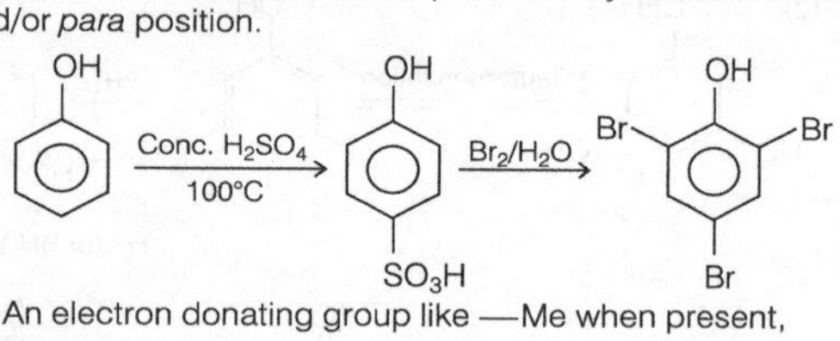

11. (b) An electron donating group like —Me when present, decreases the acidity of phenol and this effect is more at *o*/*p*-position.

Since, $\text{acidity} \propto K_a \propto \dfrac{1}{pK_a}$

So, the order of pK_a is $m < p < o$.

12. (a)

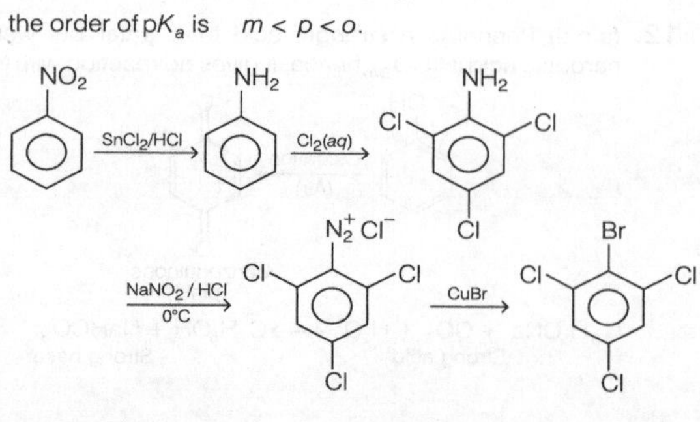

13. (d)

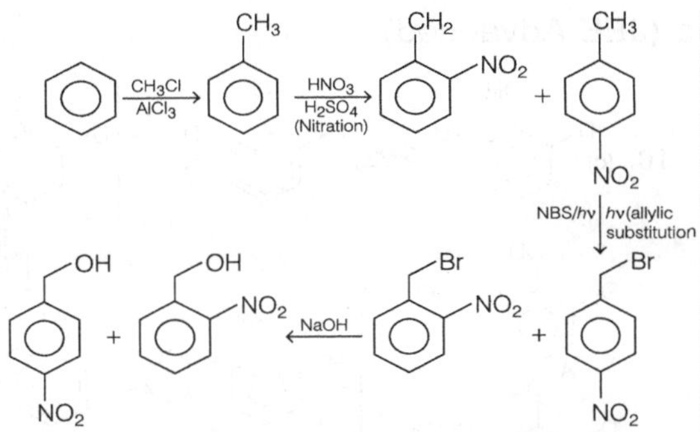

14. (c) Dil. $AgNO_3$ is used for this rearrangement.

15. (a) Electron withdrawing groups like $\rangle C{=}O$ etc., decreases the rate of ESAr, (i.e. reaction with $Br_2/AlCl_3$).

Thus, the order of reactivity towards $Br_2/AlCl_3$ is

$$III < I < II < IV$$

16. (d) Presence of electron withdrawing $—NO_2$ makes the phenol more acidic as it stabilises the phenoxide intermediate. The effect is more at *o*/*p*-position.

Aliphatic alcohols are less acidic as compared to phenol due to low stability of their anion formed.

Thus, the order of acidic strength is

$$3 > 1 > 4 > 2$$

17. (b) Here *A* is a *m*-derivative while *B* is *ortho*.

18. (c)

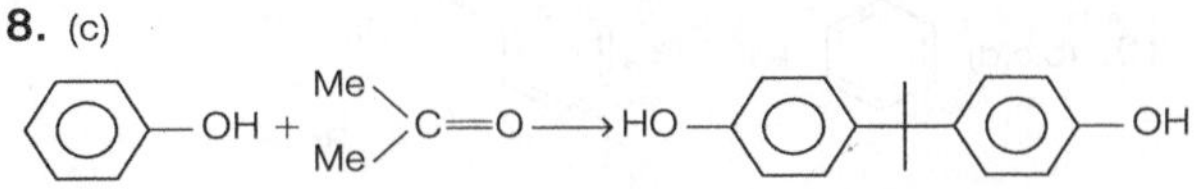

19. (c) *X* must have a —OH group and the $—CH_3$ substituent must be at *meta*-position so that *ortho* and *para*-position gets brominated to give tribromo derivative.

20. (b) $pK_a \propto \frac{1}{K_a}$

Thus, more the number of electrons donating Me groups on the ring, lesser is the acidic character and hence, more is the pK_a value.

21. (c)

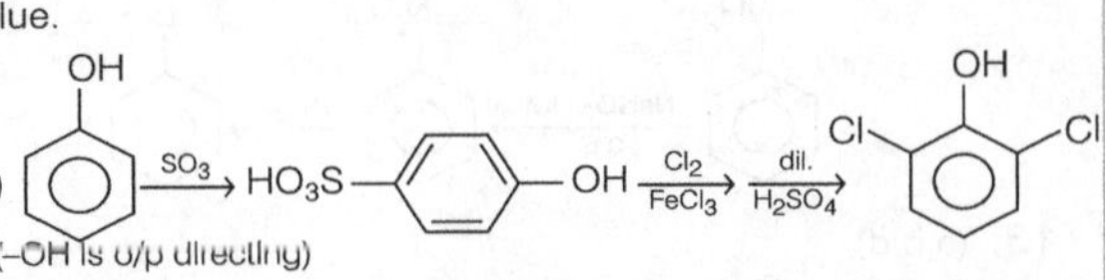

22. (d) Ethylation followed by chlorination is the correct way for the given transformation. If chlorination is done first, it will give deactivated chlorobenzene, which will not react efficiently in ethylation.

23. (b)

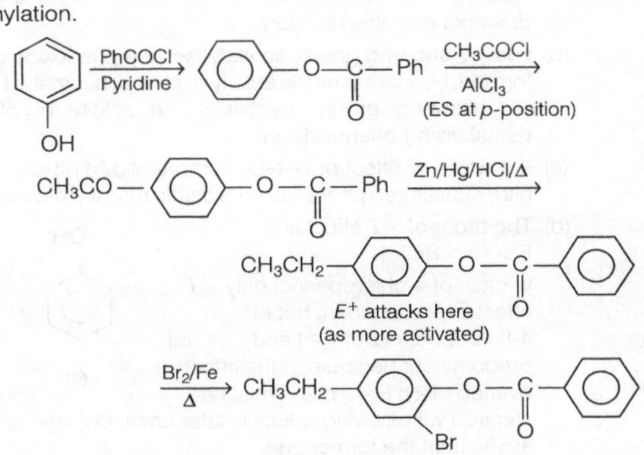

24. (c)

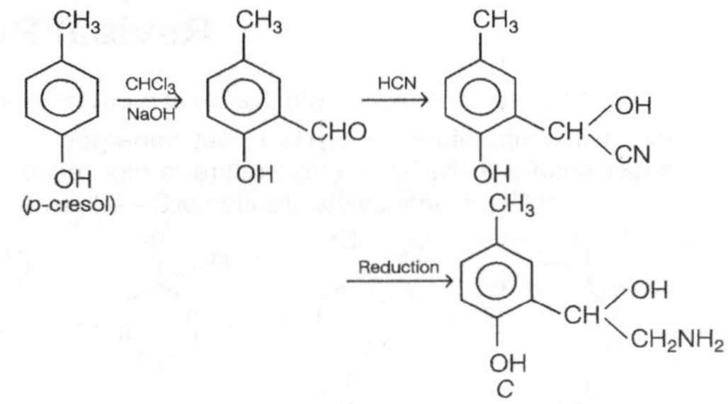

25. (d)

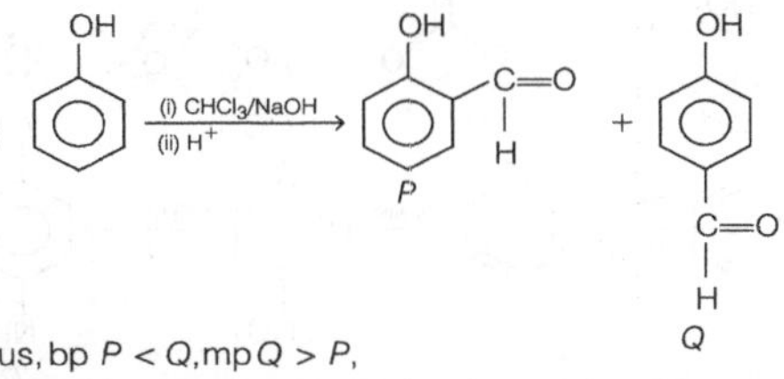

Thus, bp $P < Q$, mp $Q > P$,

Solubility in water $P < Q$, Acidic strength $Q > P$.

26. (b) Solubility in NaOH suggests the presence of phenolic (Ph—OH) group. Whereas, no reaction with it shows the absence of this group.

Formation of white ppt of tribromo derivative with bromine water indicates that the methyl group is present at *meta*-position. So, *A* is 3-methyl phenol. Its isomer which also gives tribromo derivative in this reaction is anisole. So (*C*) is anisole.

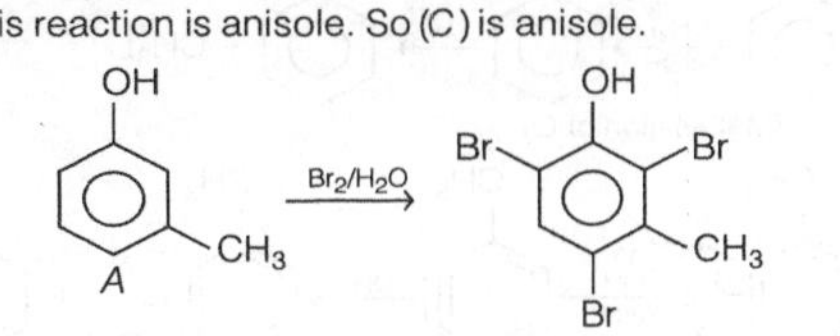

2, 4, 6-tribromo-3-methyl phenol
B

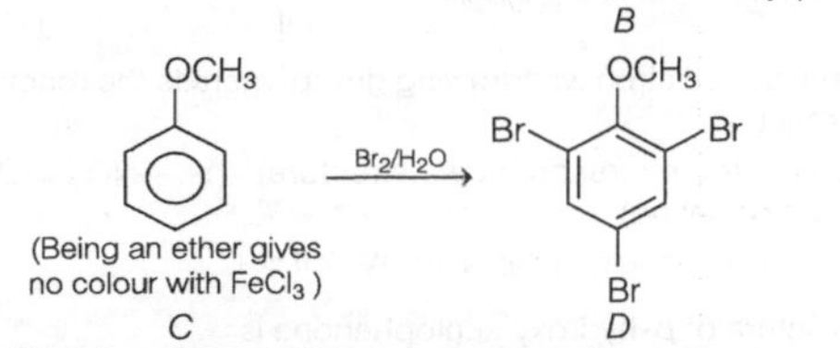

Thus, *B* and *D* are different compounds.

27. (b) II is most reactive among the given as it contain two $—NO_2$ groups at *ortho* and *para*-positions. IV is least reactive due to lack of resonance activation by $—NO_2$ group from *meta*-position.

I being less sterically hundred is more reactive than the III. Further greater electrophilicity at α—C also makes it more reactive.

28. (c)

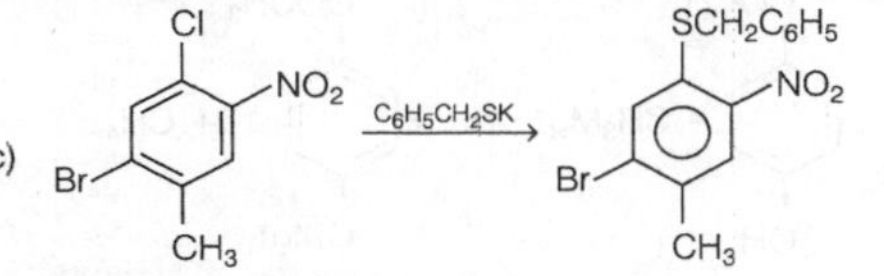

Presence of $—NO_2$ group (electron withdrawing) activates the Cl to undergo S_NAr reaction.

29. (a)

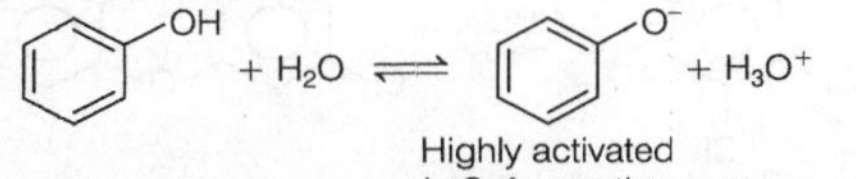

Phenoxide ion undergoes trichlorination even in the absence of Lewis acid catalyst.

30. (d) *p*-nitrofluoro benzene is more reactive in S_NAr reaction than *p*-nitro chlorobenzene due to less steric hindrance and increased electrophilicity at α—C in later case.

Revisal Problems (JEE Advanced)

1. (c) C_6H_5Cl gives no reaction with NaOH at room temperature (or mild temperature) or C_2H_5OH but undergoes nucleophilic substitution with NaOH if temperature is high enough. This is because of the resonance stabilisation of C—Cl bond.

2. (a)

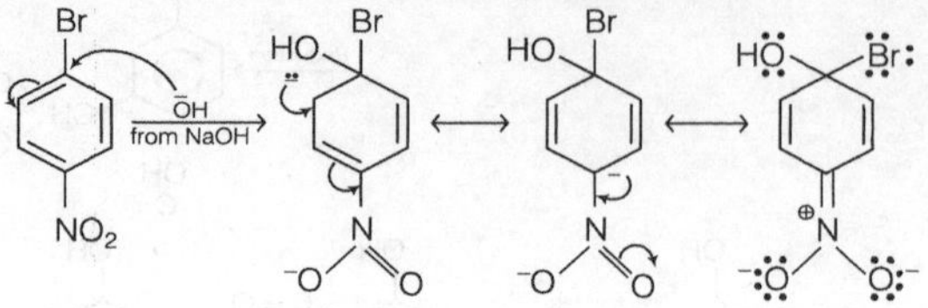

3. (b)

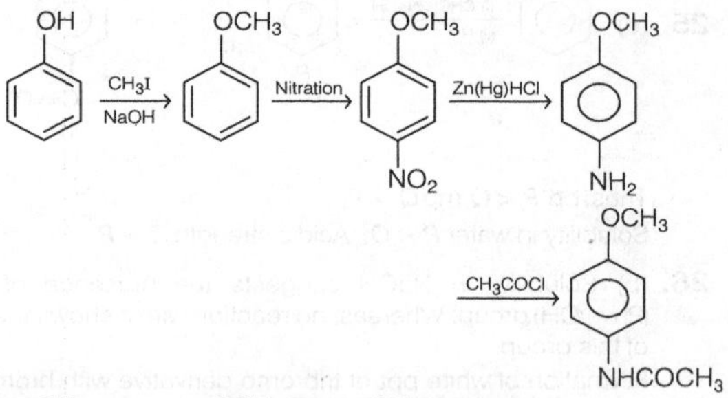

4. (a)

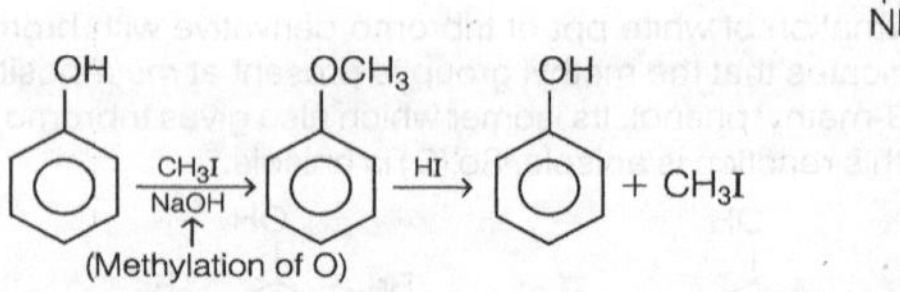

5. (d)

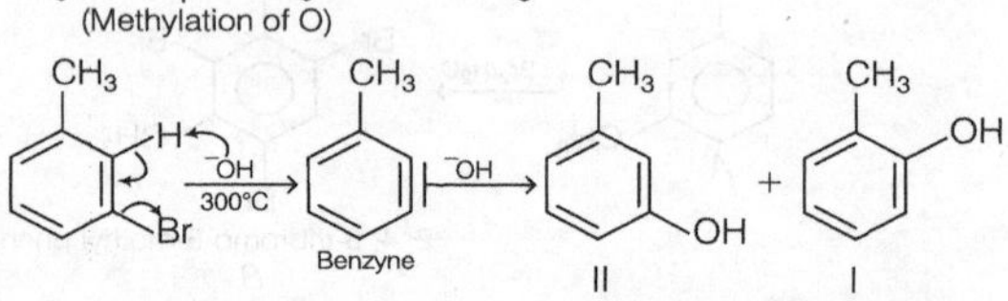

6. (c) More the electron withdrawing group, more is the reactivity for ArS_N reaction.

—SO_3H (more resonance structure) > —NO_2 > Cl > Me (Electron donating)

Thus, the decreasing order is III > IV > II > I.

7. (a) Structure of *p*-hydroxy acetophenone is

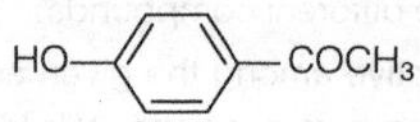

It contains an acidic —OH group. So, neutralises with Grignard reagent (methyl magnesium iodide) to give methane and —OMgI at —OH group.

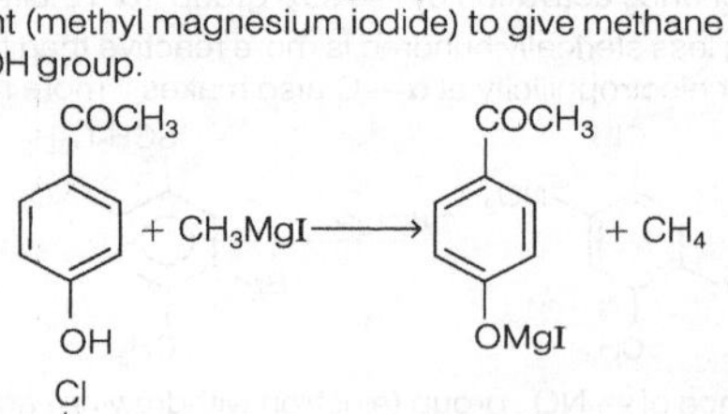

8. (b)

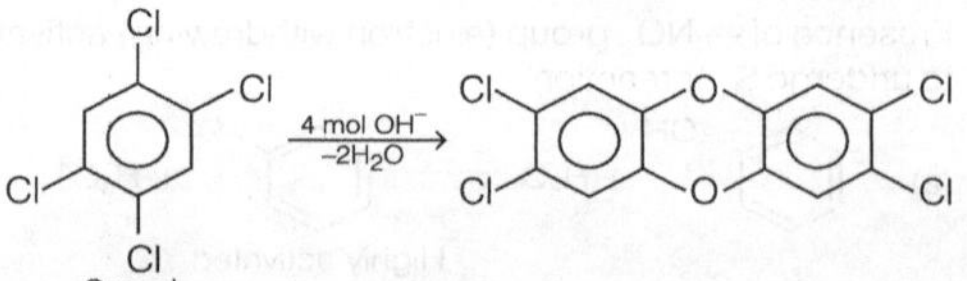

(—Cl being an electron withdrawing group activates the other Cl to undergo nucleophilic substitution).

9. (a) This is because F is *o*/*p* directing but highly deactivating group and *para* product is much more than *ortho*.

10. (c)

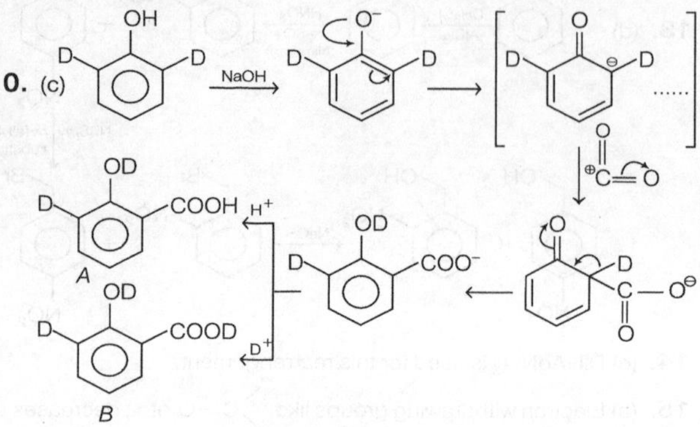

11. (d)

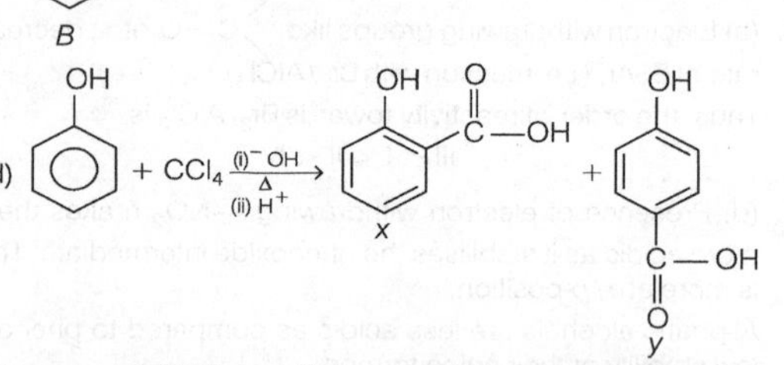

x is more volatile than *y* (due to formation of intramolecular H-bonding). *y* is more soluble (as can easily form intermolecular H-bond with the molecules of water). *x* is more acidic than *y*.

12. (b,c,d)

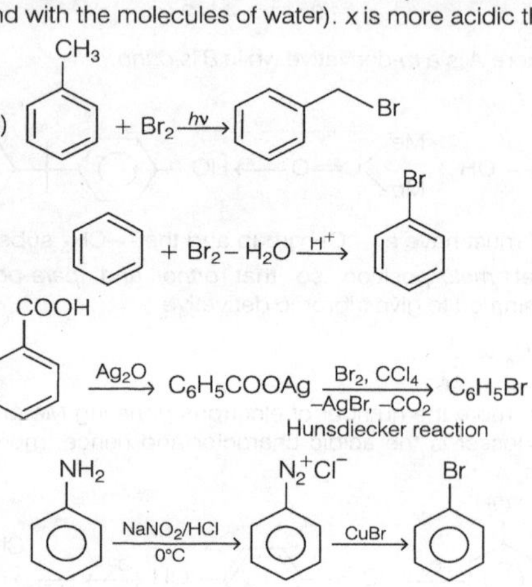

13. (a,b,d)

(a) 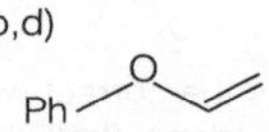 being an unsaturated compound decolourises Br_2 in CCl_4, but 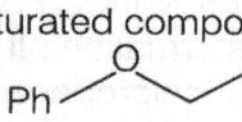 being saturated does not give this reaction.

(b) F is $-I$ showing group, so stabilises the phenoxide group ion formed, i.e. increases the acidity of phenol but methyl being a $+I$ showing group decreases the acidity of phenol by destabilising phenoxide ion.

(c) $-I$ and $-R$ effect of —NO_2 is more at *o*/*p* position, so *para* nitro substituted phenol is more acidic than its *m*-counter part.

(d) The order of $-I$ effect is $F > Cl > Br > I$.

In case of 4-chlorophenol only $-I$ effect in predominant but in 4-fluorophenol both $-I$ and $+R$ are predominant because of the effective overlap of $2p$ F – $2p$ C. Thus, net electron withdrawing is less in later case. So, later is less acidic than the former one.

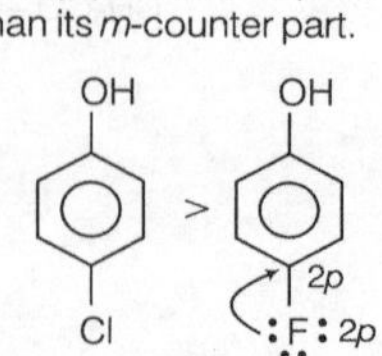

14. (a,b)

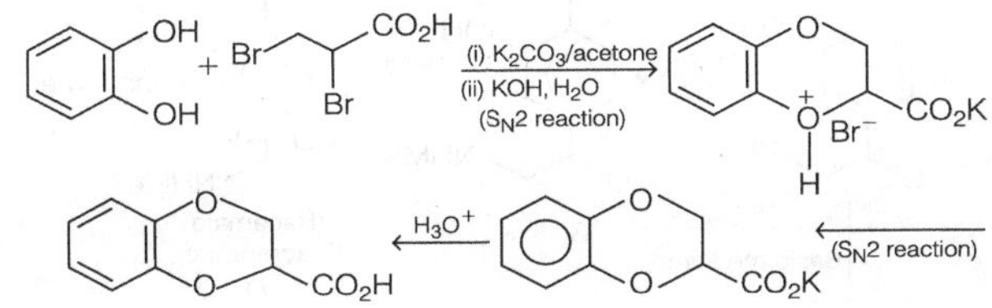

15. (b,c) Since, F is *para* to —NO_2, so it will be substituted first. After its substitution by —OMe, the ring becomes more deactive towards S_NAr reaction and hence, further substitution of remaining Cl will become difficult.

16. (a,d) At lower temperature I (*ortho*) product is the major product because the reaction is under kinetic or rate control.

When temperature is high i.e. 100°C, the reaction is under thermodynamical or equilibrium control, leading to the formation of *para* product. *Para* product is more stable than *ortho*.

17. (a,c) In case (*I*) —NO_2 group is free to involve in resonance, so it weakens the C—Cl bond and makes it more reactive towards S_NAr reactions.

In case (II), —NO_2 group is not in resonance with the ring due to steric inhibition to resonance, hence, it does not activate the ring to undergo S_NAr reaction.

18. (a,b,c,d)

(a)

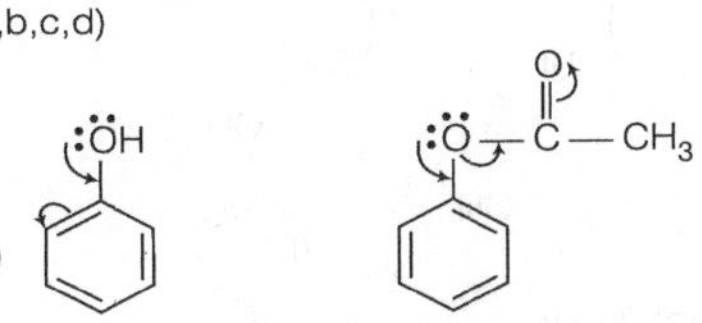

(+*R* and –*I* effect) (Cross-conjugation)

II

Due to cross-conjugation in (II) electron donating power of oxygen to the ring decreases and hence, it is less reactive than phenol.

(b) PhO^- is more basic and is a stronger nucleophile than,

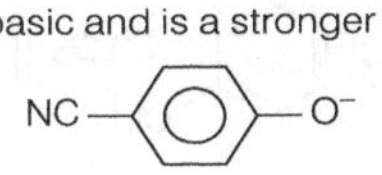

so it will react faster with $PhCH_2Cl$.

(c)

(i)

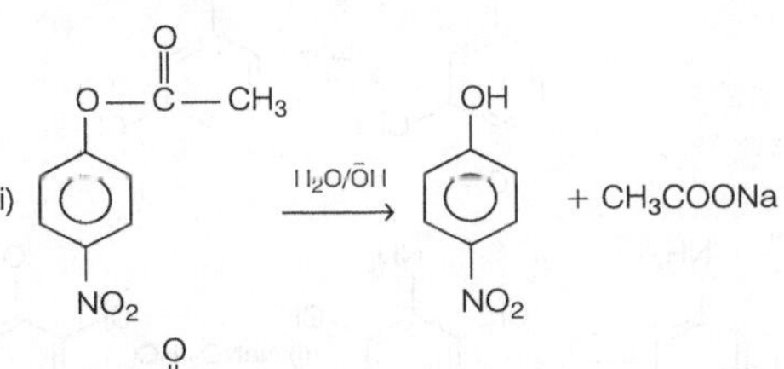

(ii)

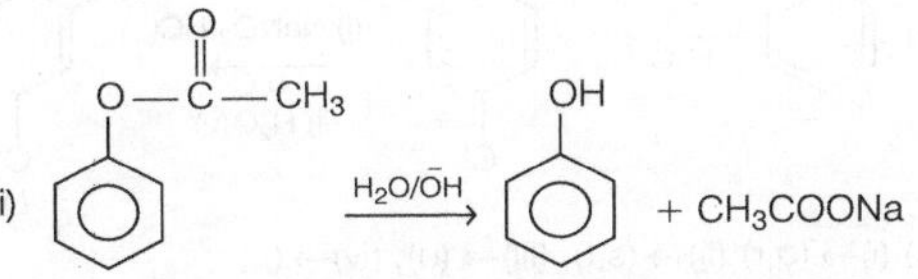

The electron withdrawing effect of (p-NO_2) group (i.e. – *R* and – *I*) from O, makes p-NO_2—$C_6H_4O^-$ a weaker base and a better leaving group thus reaction (i) is faster.

(d) In the presence of an electron withdrawing group like —NO_2, the phenol becomes less basic (more acidic), so its reactivity decreases. Thus, PhOH undergoes esterification faster than *p*-nitrophenol.

19. (a,c)

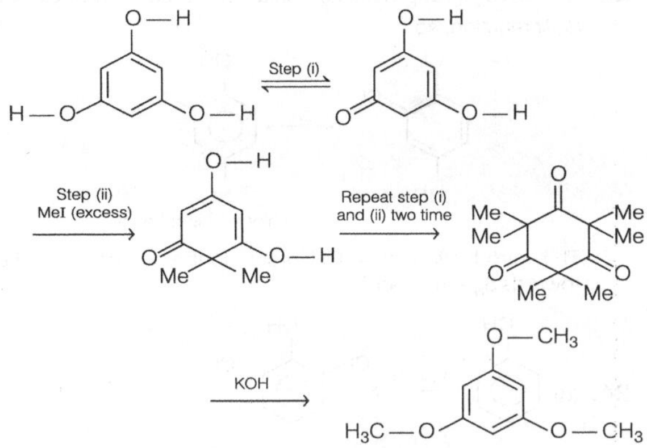

20. (a,b)

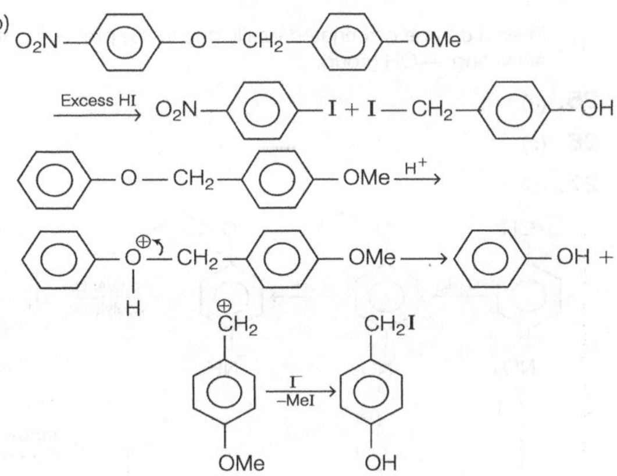

Benzyl carbocation

21. (a,c)

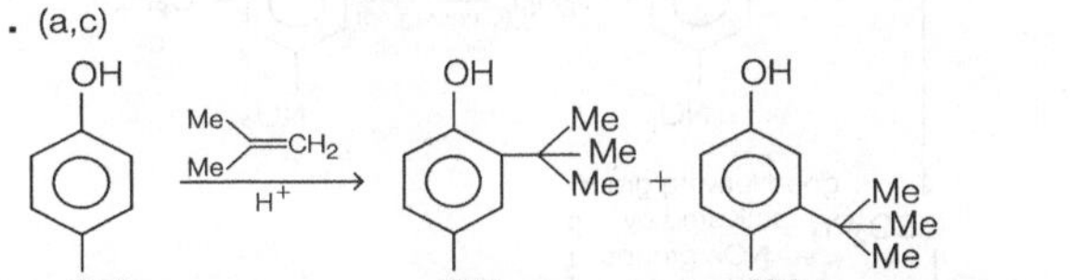

p-methoxy phenol

Mixture of 2- and 3- *t*-butyl-4-methoxy phenol or butylated hydroxy anisole (BHA)

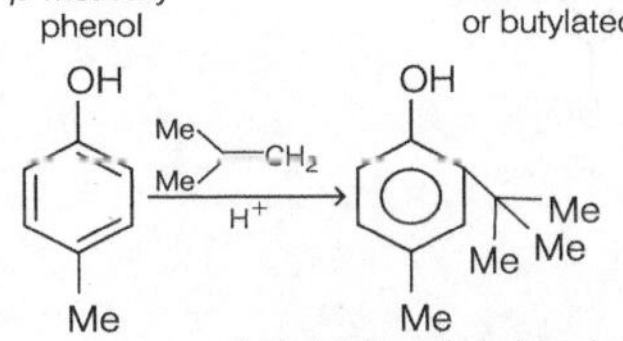

2-*t*-butyl-4-methyl phenol
(BHT, butylated hydroxy toluene)
B

BHA and BHT are the antioxidants, i.e. preserve the food by the action of oxygen.

22. (a) Phenolic ester when treated with Lewis acid like $AlCl_3$ undergo Fries rearrangement.

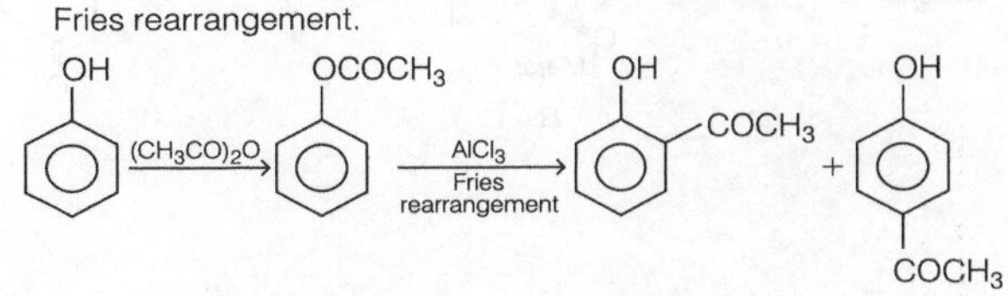

23. (b) —NO_2 group activates *para* Cl—more than *m*-Cl, towards S_NAr reaction, so

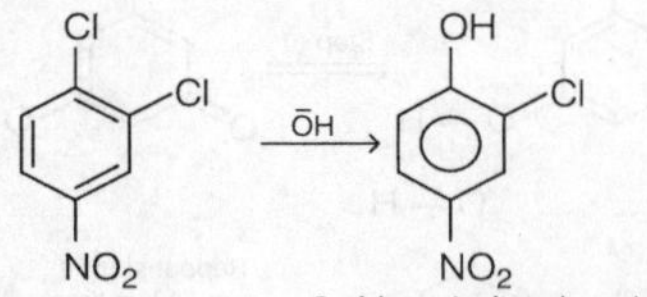

2-chloro-4-nitrophenol

The introduction of —OH group further deactivates the ring towards S_NAr reaction.

24. (d)

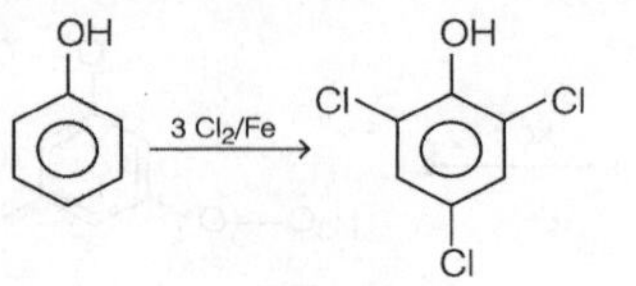

Phenol can be chlorinated easily due to the presence of activating —OH group.

25. (a)

26. (c)

27. (b)

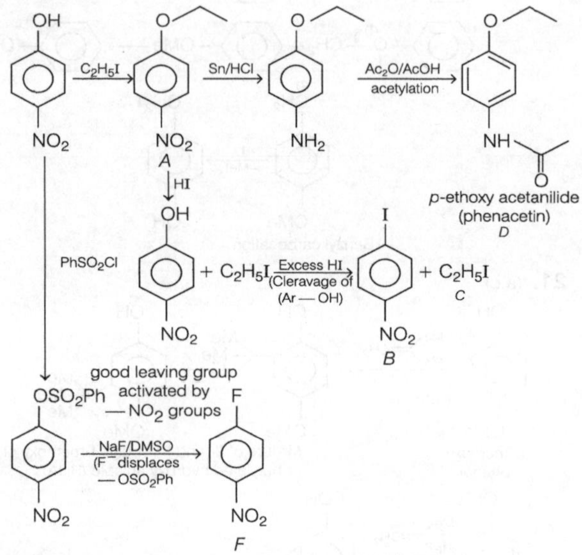

28. (a)

29. (a)

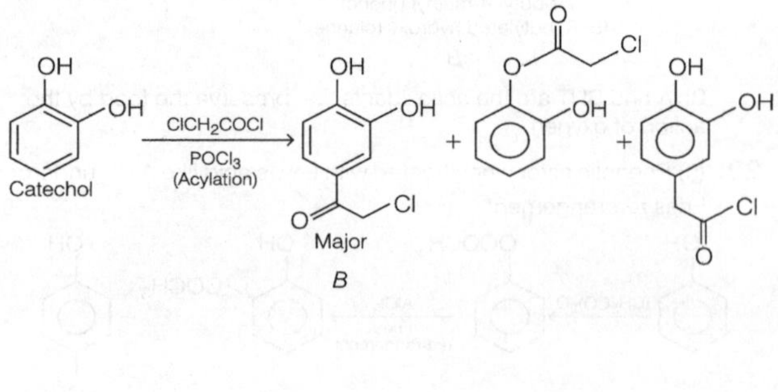

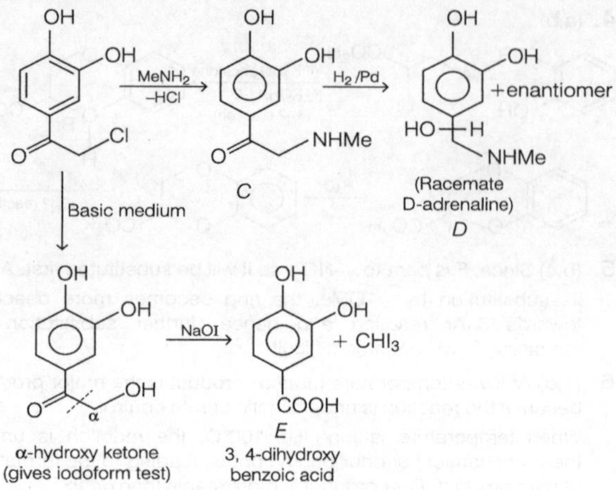

30. (b) (i) → (p,q), (ii)→ (q), (iii)→ (r), (iv)→ (s)

(p)

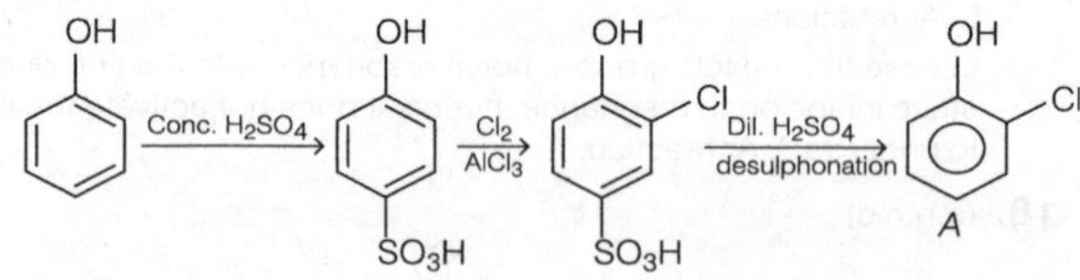

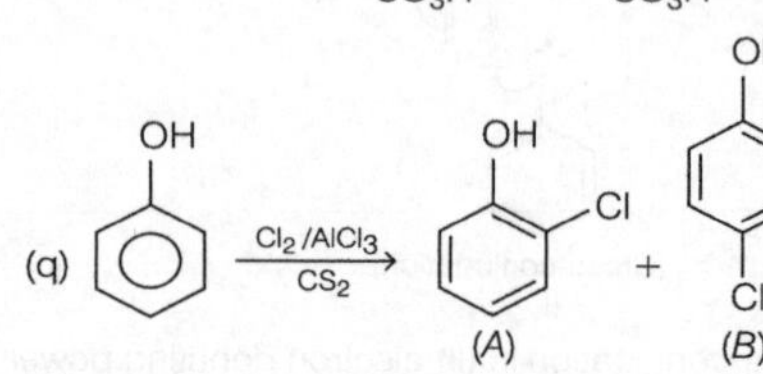

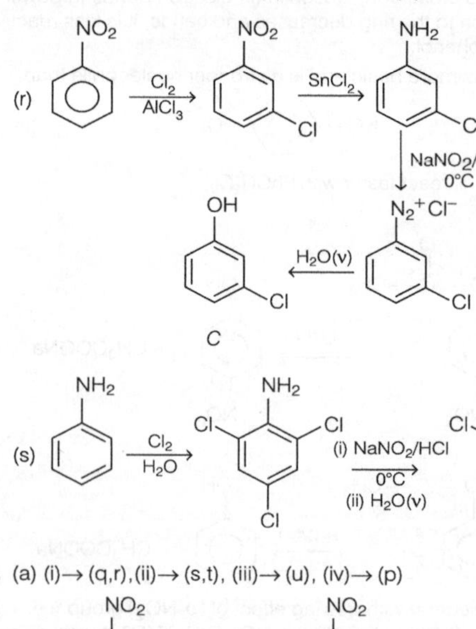

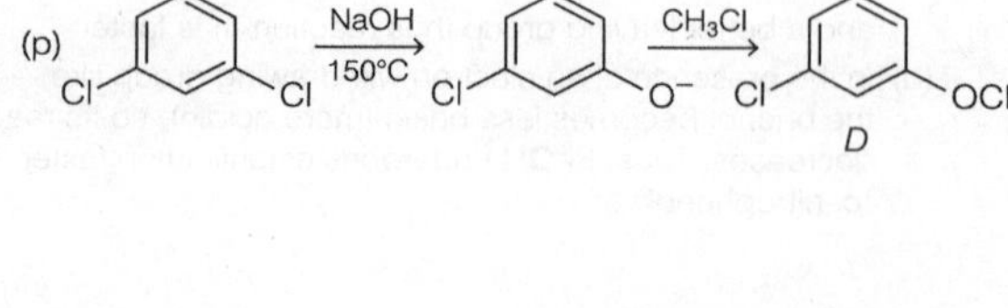

31. (a) (i)→ (q,r), (ii)→ (s,t), (iii)→ (u), (iv)→ (p)

(q)

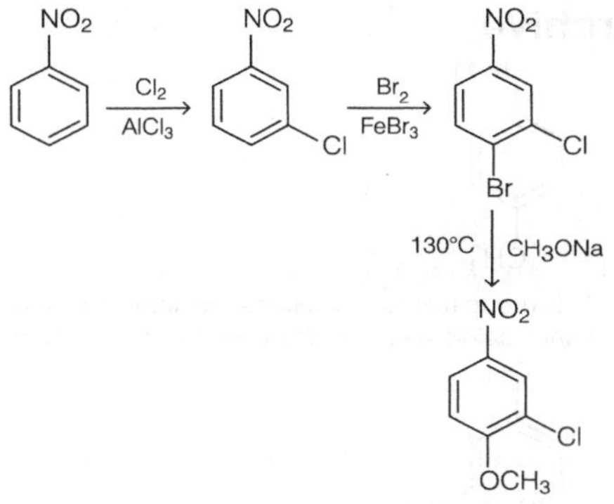

A

(r)

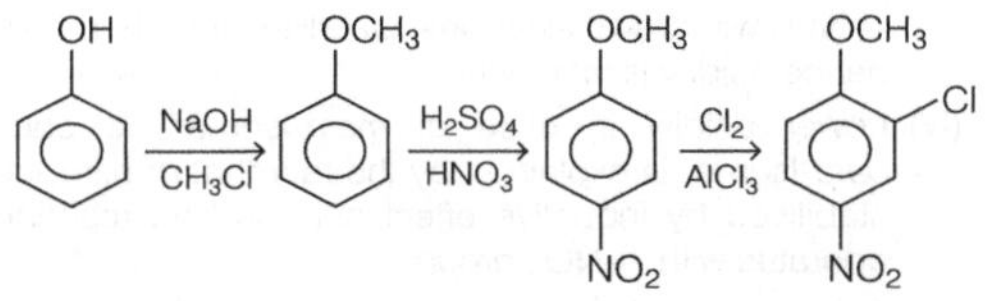

A

(s)

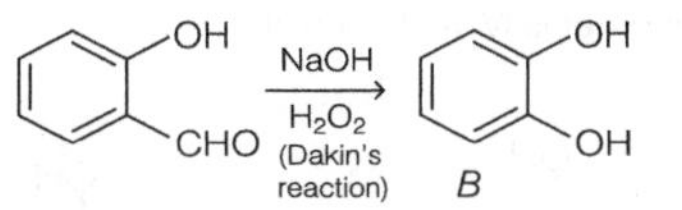

B

(t)

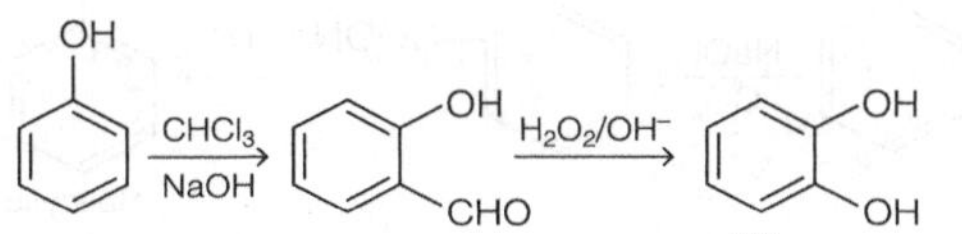

(*B*)

(u)

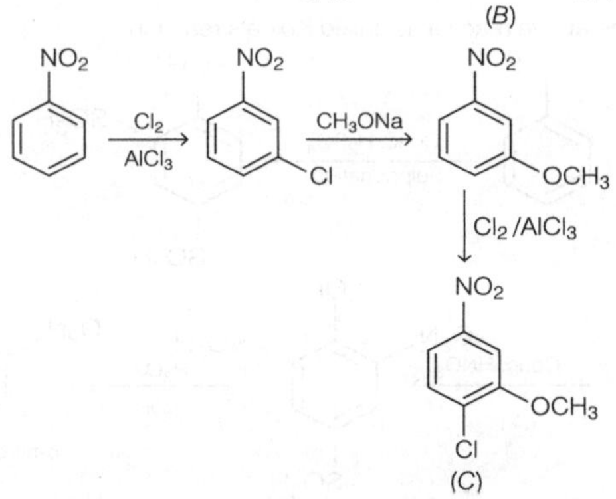

(*C*)

32. (c) (i) → (q,r), (ii) → (p,s), (iii) → (q), (iv) → (p)

(i)

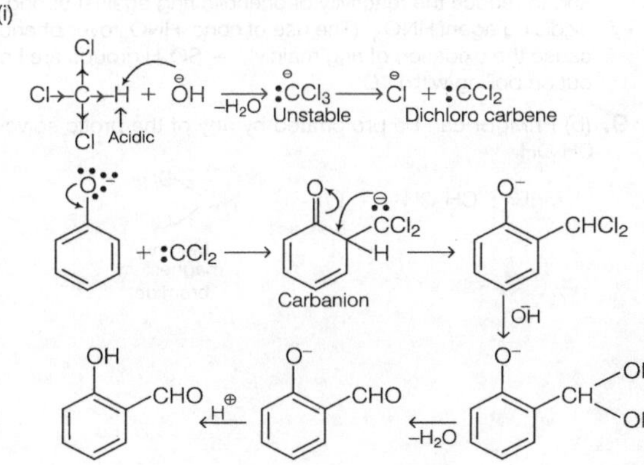

(ii)

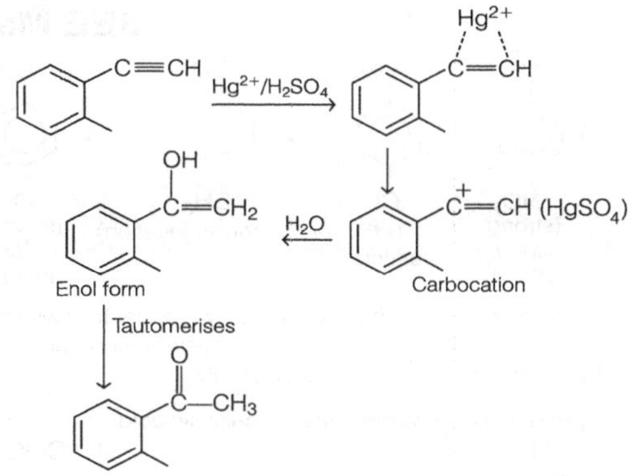

(iii)

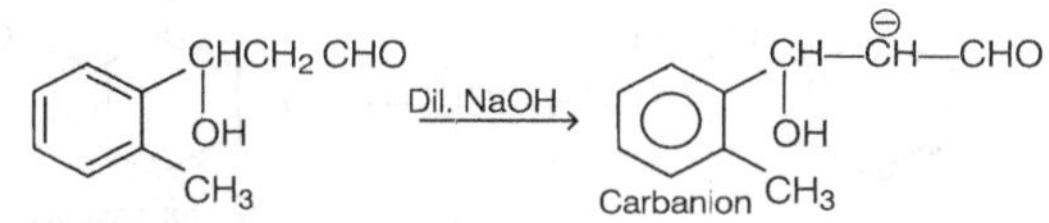

(iv)

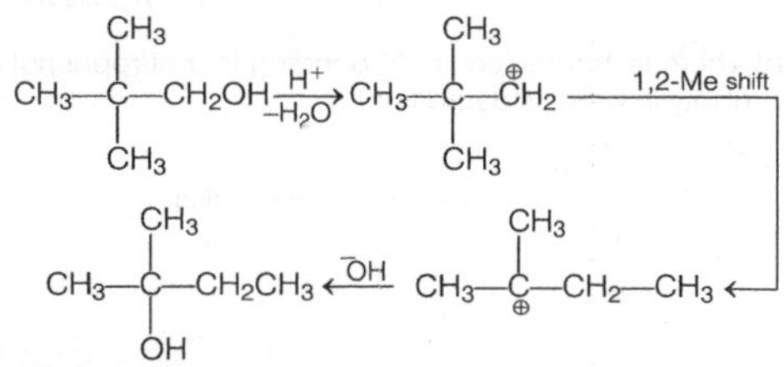

33. (3) Salol, oil of wintergreen (methyl salicylate)

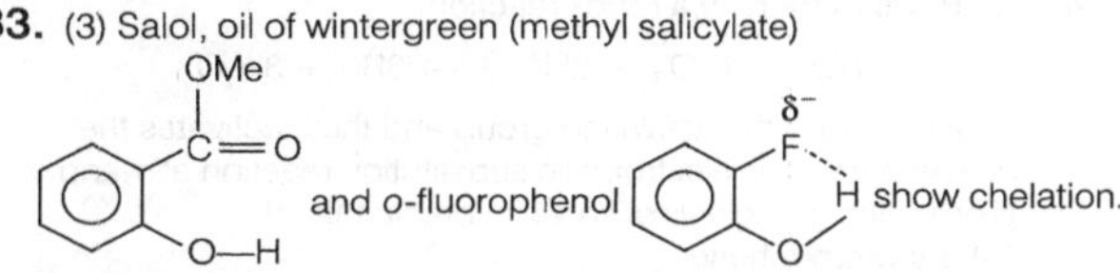

and *o*-fluorophenol show chelation.

34. (4) Because of the presence of four allylic sites, four products are possible.

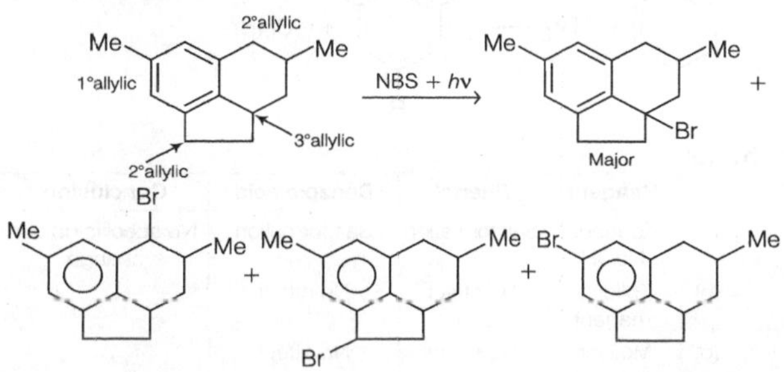

35. (4) *A* gives effervescence with Na, so it must has an —OH group.

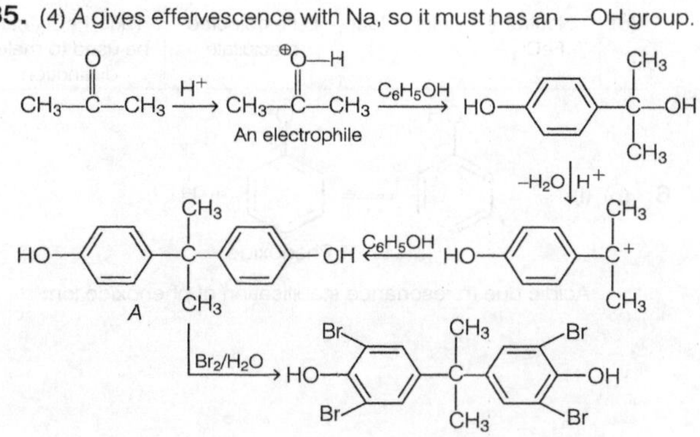

JEE Main & AIEEE Archive

1. (c) 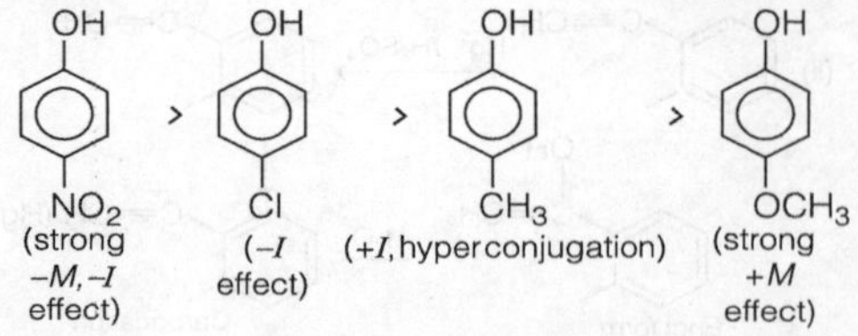

Electron releasing group decreases while electron withdrawing group increases acidic strength by destabilising and stabilising the phenoxide ion formed respectively.

2. (a) Aspirin is acetyl derivative of salicylic acid.

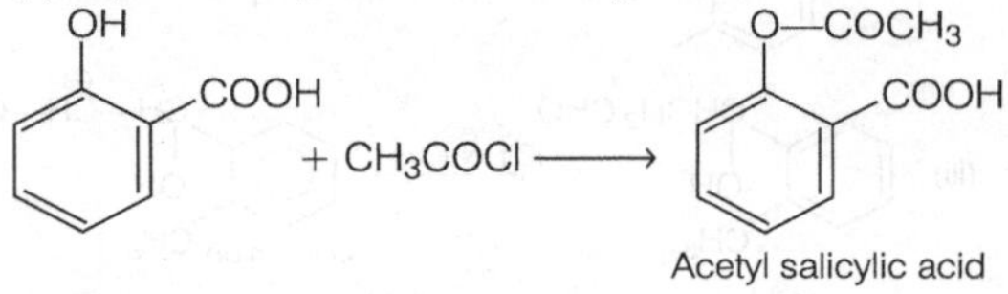

Acetyl salicylic acid

3. (b) There is intramolecular H-bonding in *o*-nitrophenol and thus, solubility in water is decreased.

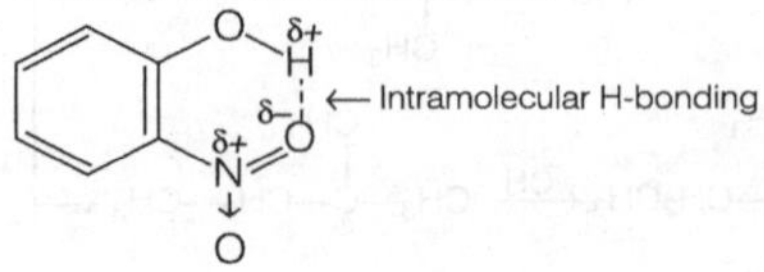

4. (d) Br_2 is formed by a redox reaction.

$$5Br^- + BrO_3^- + 6H^+ \longrightarrow 3Br_2 + 3H_2O$$

—OH group is the activating group and thus, activates the benzene towards electrophilic substitution reaction at *o*-and *p*-positions giving yellowish white precipitate of 2,4,6-tribromophenol.

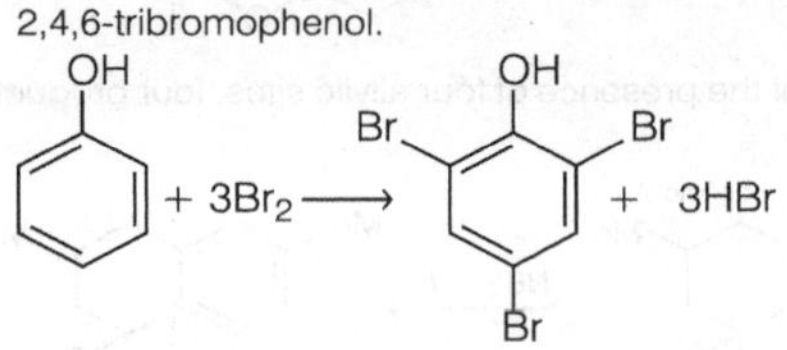

5. (d)

	Reagent	Phenol	Benzoic acid	Conclusion
(a)	Aqueous NaOH	Salt formation	Salt formation	No specific colour change
(b)	Tollen's reagent	No effect	No effect	
(c)	Molisch reagent	No effect	No effect	
(d)	Neutral $FeCl_3$	Violet colour	Buff- coloured precipitate	Thus, $FeCl_3$ can be used to make distinction

6. (b) (I) 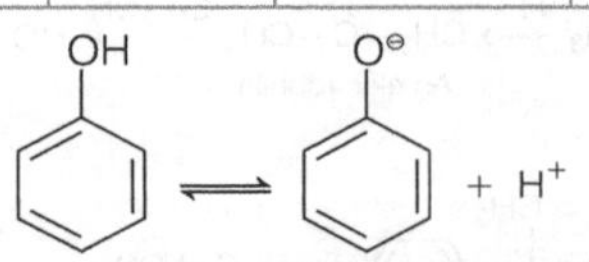

Acidic due to resonance stabilisation of phenoxide ion.

(II) 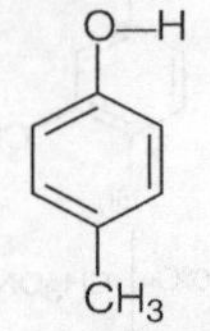

—CH_3 group releases electrons, destabilises the phenoxide ion, thus acidic nature is decreased hence II < I (acidic).

(III) 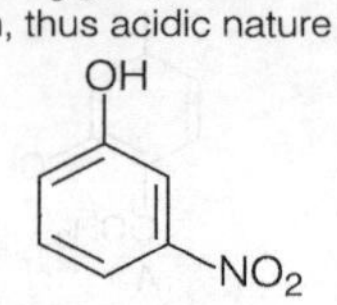

—NO_2 withdraws electrons, stabilises the phenoxide ion, hence acidity is increased.

(IV) Lower acidity in case of *meta*-isomer as compared to *para*-isomer is explained by the fact that *m*-nitrophenoxide is stabilised by inductive effect only and no resonance effect operates with —NO_2 group.

Thus, *m*-nitrophenol (III) < *p*-nitrophenol (acidic) (IV)

Thus, correct order is IV > III > I > II.

7. (c)

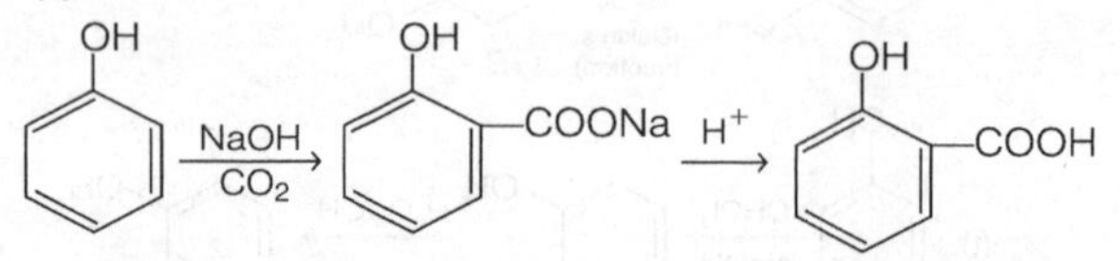

The above reaction is called Kolbe's reaction.

8. (b) 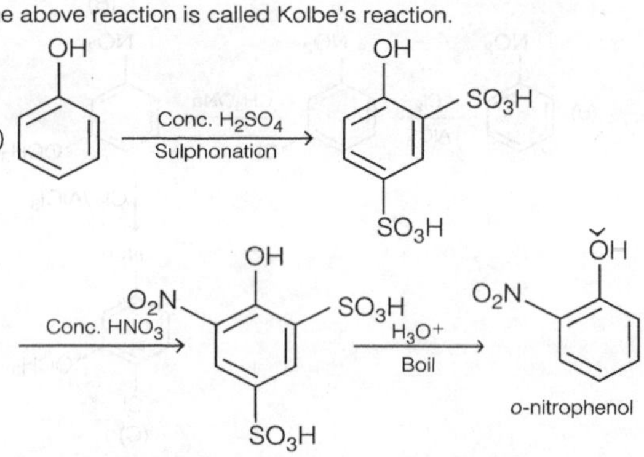

First sulphonation is the means to block *para* and *ortho* position and to reduce the reactivity of phenolic ring against strong oxidising agent HNO_3. (The use of conc. HNO_3 over phenol cause the oxidation of ring mainly). — SO_3H groups are knocked out on boiling with H_2O.

9. (b) PhMgBr can be protonated by any of the protic solvent, e.g. CH_3OH.

$$\text{Ph MgBr} + CH_3OH \longrightarrow \underset{\text{Benzene}}{\text{Ph—H}} + \underset{\substack{\text{Methoxy}\\\text{magnesium}\\\text{bromide}}}{\text{Mg}\begin{matrix}\text{Br}\\\text{OCH}_3\end{matrix}}$$

10. (a)

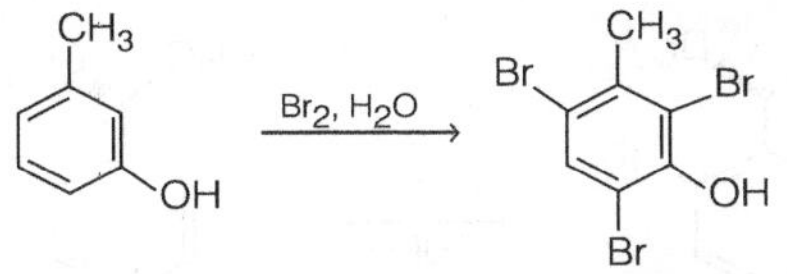

Tribromo derivative

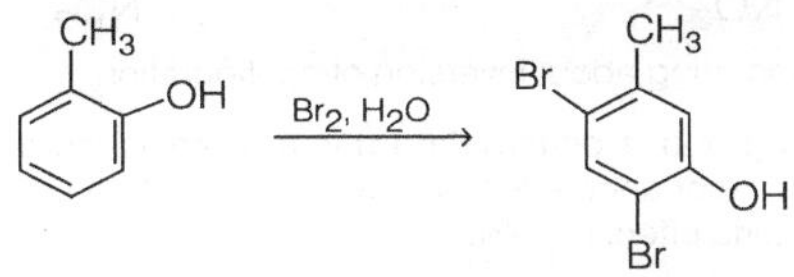

Dibromoderivative

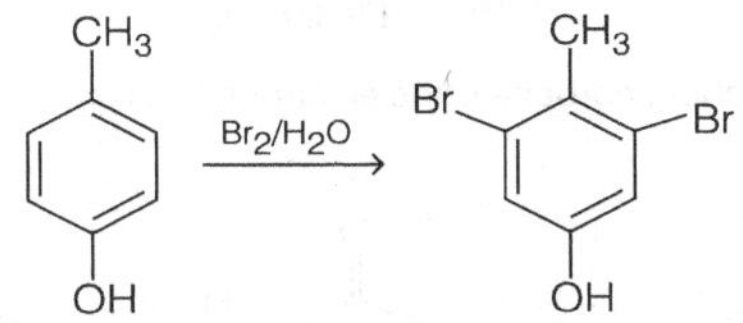

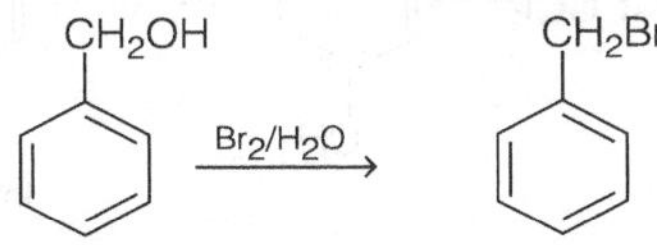

m-cresol due to phenoxide ion in H_2O solvent, gives tribromo derivative at all *ortho and para* positions.

11. (b)

$$OH^- + CHCl_3 \rightleftharpoons H_2O + CCl_3^- \longrightarrow :CCl_2 + Cl^-$$

[from Na^+OH^-]

Dichloro carbene E^+

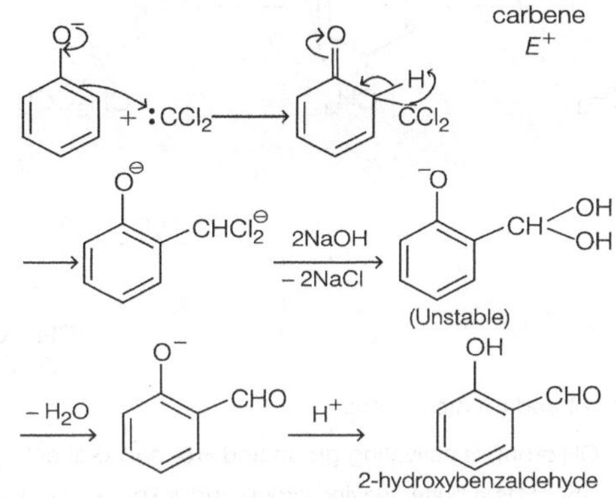

12. (c) Picric acid is 2, 4, 6-trinitrophenol.

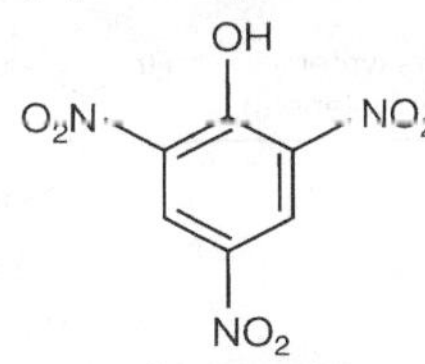

Strongly acidic

JEE Advanced & IIT JEE Archive

1. (d) In basic (alkaline) solution naphthol exists as naphthoxide ion which is a strong *o*, *p*-directing group.

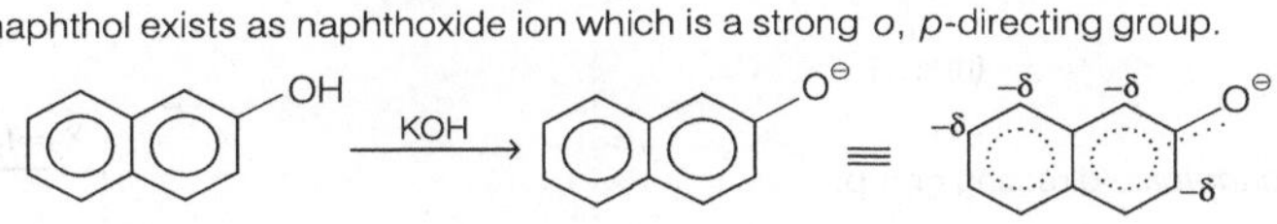

Thus, formation of dye can be shown as

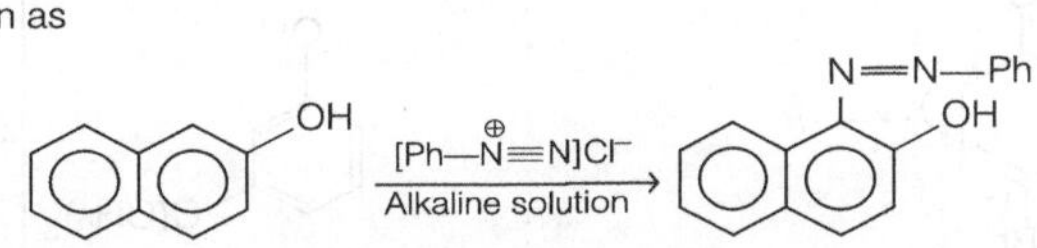

Thus, (d) is the correct choice.

2. (b,d) Phenolic compounds in alkaline solution react with chloroform ($CHCl_3$) at a temperature lower than that of $CHCl_3$ to form *ortho*-isomer as the major product (due to greater stability resulting from intramolecular hydrogen bonding).

$$HO^- + H{-}CCl_3 \rightleftharpoons H_2O + \overset{\ominus}{C}Cl_3 \xrightarrow{-Cl^-} :CCl_2$$

Dichloro carbene

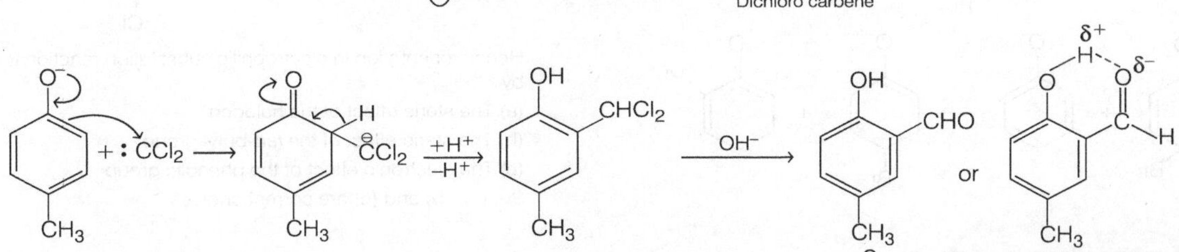

(Major as stable due to intramolecular H-bonding).

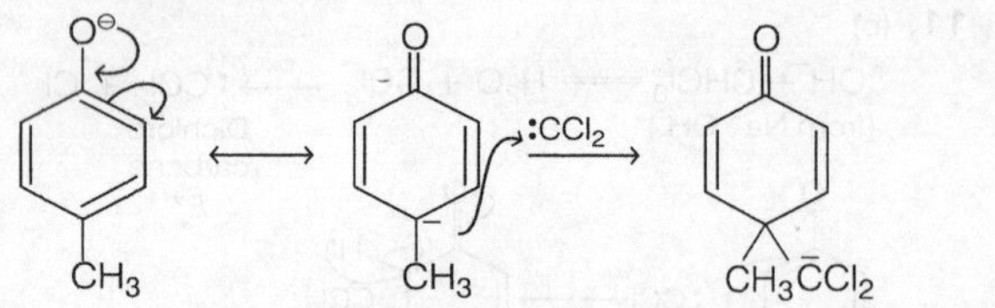

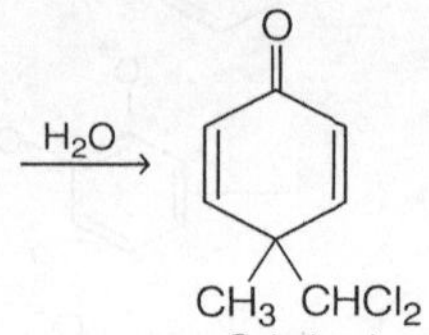

Q (minor)

Thus, (b) and (d) are correct.

3. (b) —OH group is activating group and is *o*- and *p*-directing. Also —SO_3H is a better leaving group and is knocked out by Br^-.

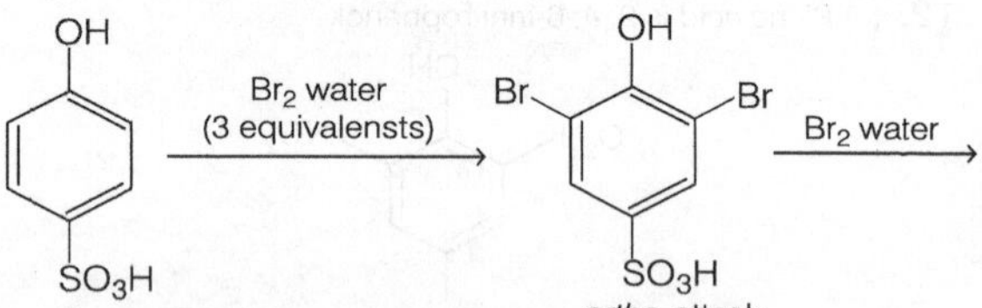

ortho-attack

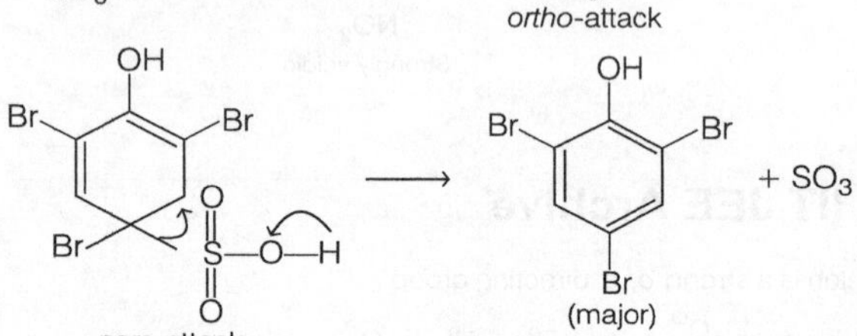

para-attack

4. (a,b,c) —OH in phenol is *ortho/para* directing group.

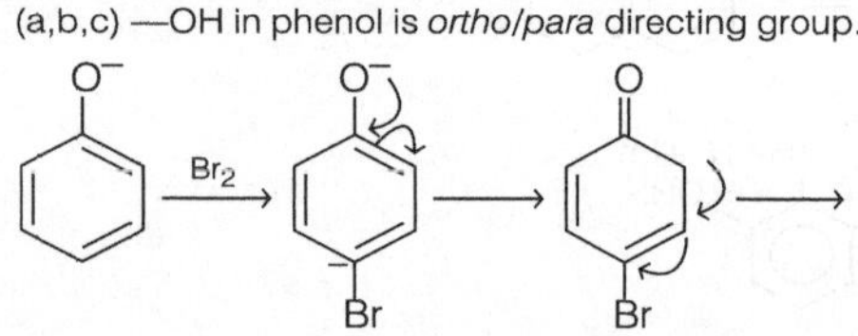

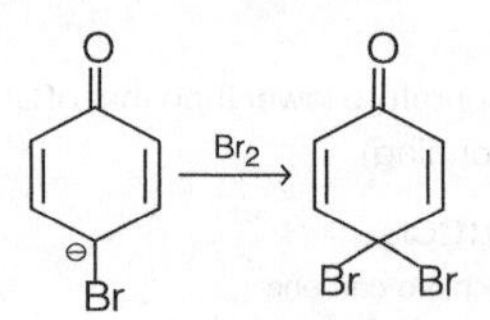

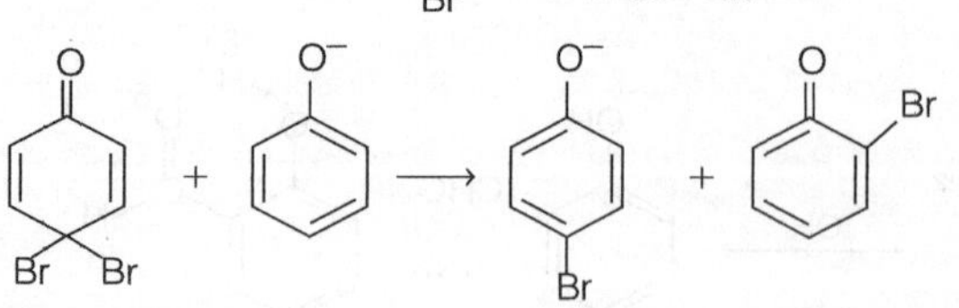

5. (a)

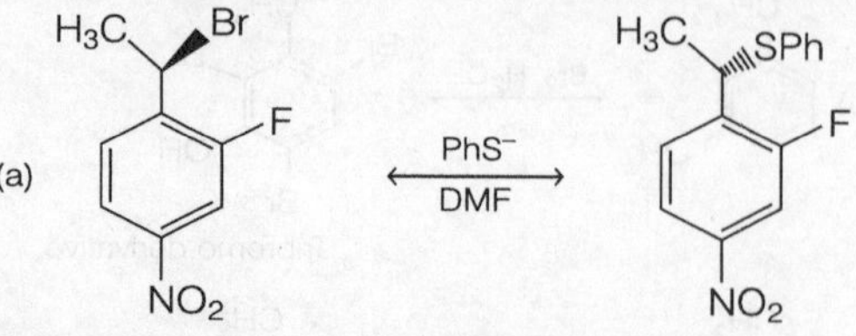

S_N2 reaction bring about inversion of configuration.

6. (c) Bromo group is deactivating due to dominance of inductive effect over resonance effect. However, orientation is determined by mesomeric effect of —Br.

7. (c) $CHCl_3 + NaOH \longrightarrow \underset{\text{Electrophile}}{:CCl_2}$ (Dichloro carbene)

8. (c) Dichlorocarbene is the electrophile as shown above.

9. (b)

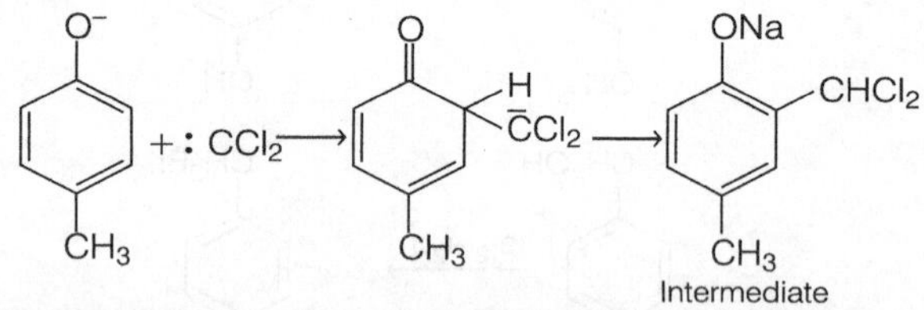

10. (a,b,c) Steric effect of halogens are as follows $Cl_2 < Br_2 < I_2$
Electronic effect of phenolic group directs the approaching electrophile towards *ortho* and *para* positions.
Tertiary butyl group has large size so it causes steric effect around aromatic nucleus.
On the basis of above factors the products of the given reactions are as follows:

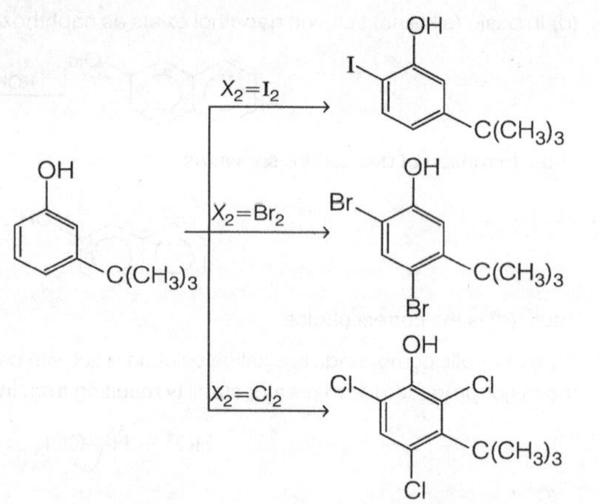

Hence, orientation in electrophilic substitution reaction is decided by

(a) The steric effect of the halogen
(b) The steric effect of the *tert*-butyl group
(c) The electronic effect of the phenolic group

So, (a), (b) and (c) are correct choices.

2. Aromatic Aldehydes, Ketones and Acids

DPP-1 Methods of Preparation of Aromatic Carbonyl Compounds

1. (i)

(a)

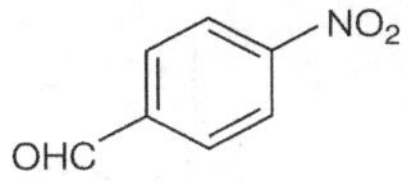

(b)

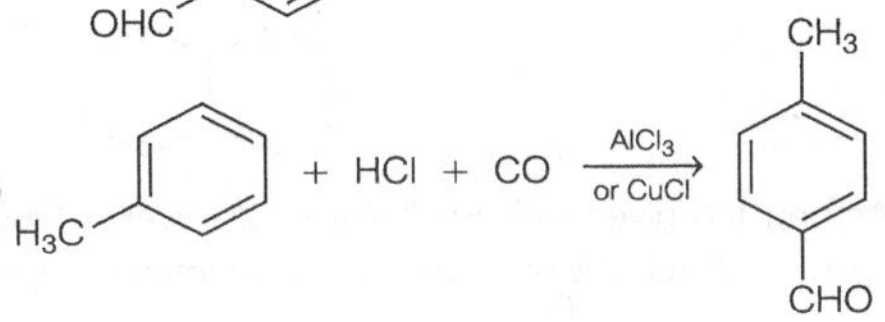

(c) SeO_2

(d)

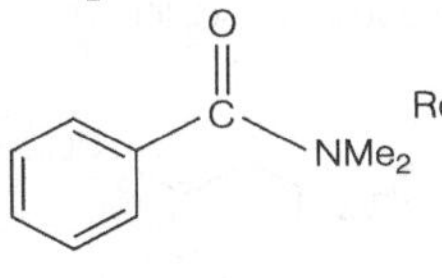

Reagent = $Li(OEt)_2AlH_2$

(ii)

(a)

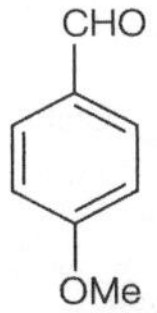

(b)

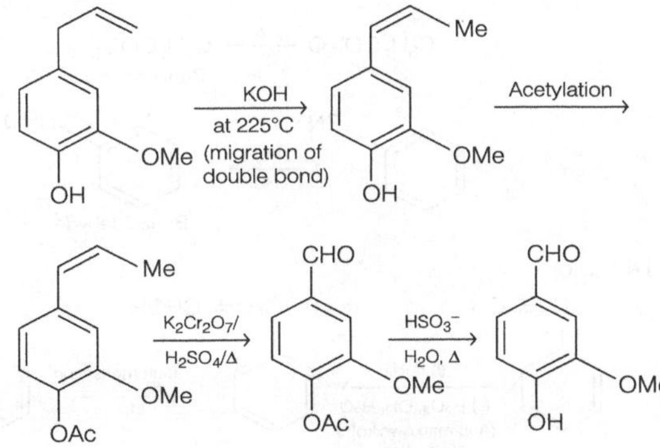

(c)

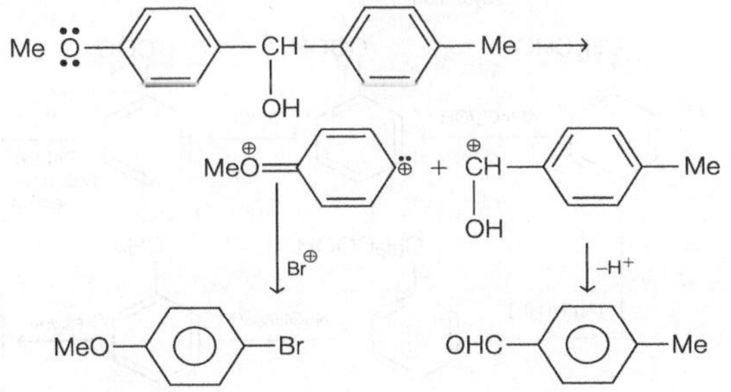

(d)

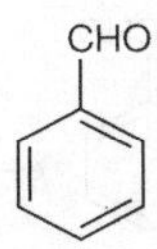

A

(iii)

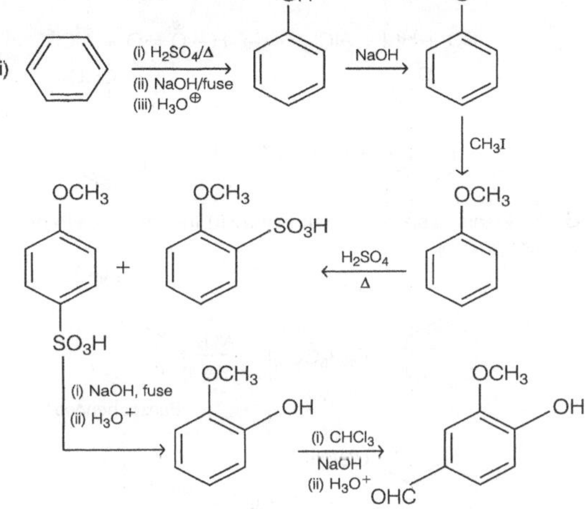

2. (a)

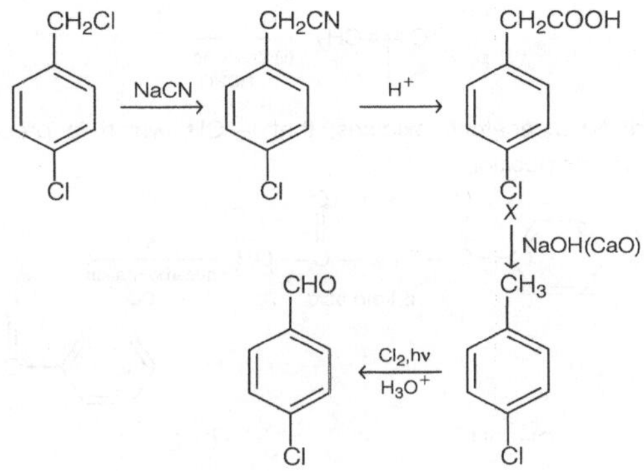

3. (b)

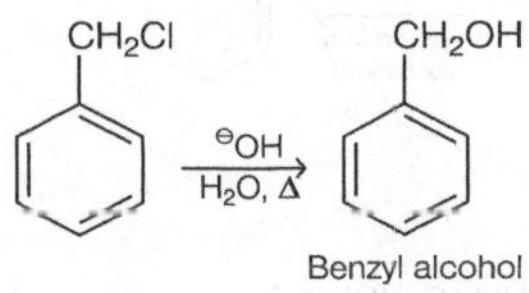

Benzyl alcohol

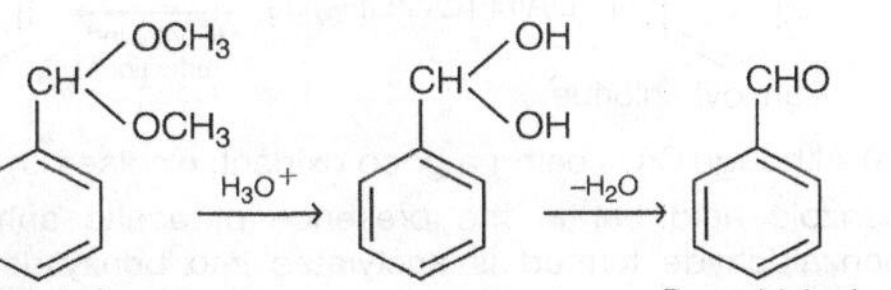

Acetal of benzaldehyde

Benzaldehyde

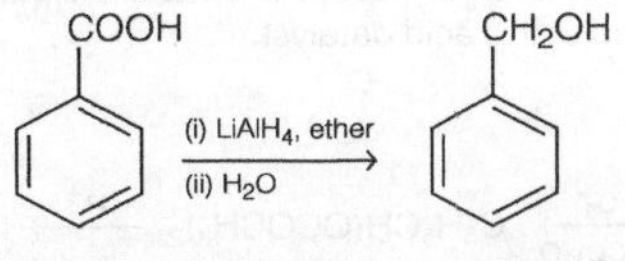

4. (d) Here oxidation occurs at α-position thus, 2° benzylic carbon is oxidised to ketone.

5. (c) $CO + DCl + AlCl_3 \longrightarrow \overset{\oplus}{C}DO + AlCl_4^-$

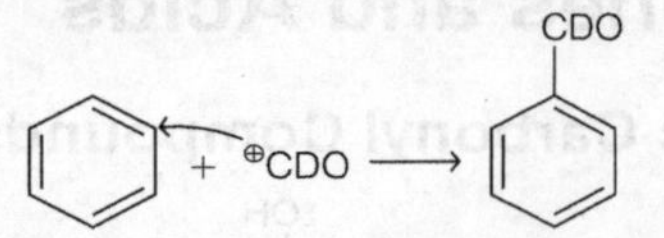

$CO + HCl + AlCl_3 \longrightarrow H-\overset{\oplus}{C}=O + AlCl_4^{\ominus}$

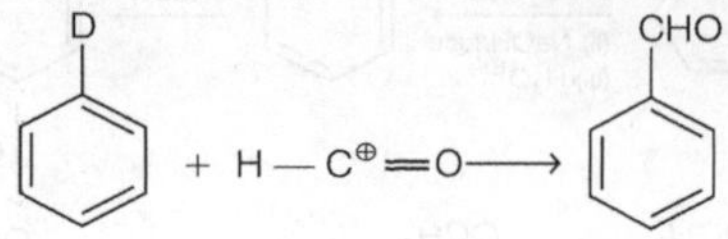

6. (b) In this case the oxidation may further proceed to benzoic acid and phenol.

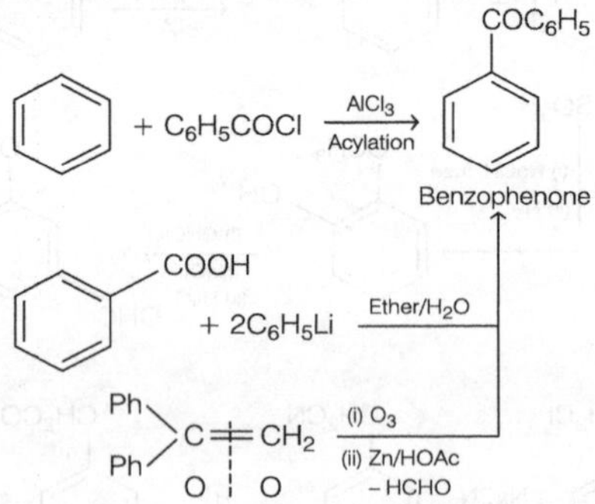

7. (d) MnO_2/acetone oxidises, that —OH, which is present at benzylic position.

8. (a)

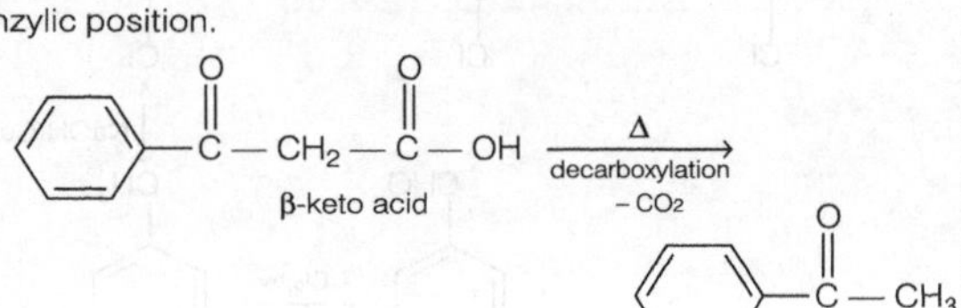

9. (c)

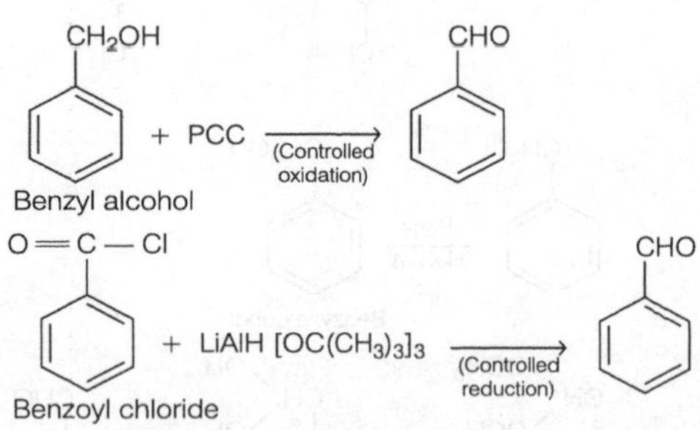

10. (a) Although CrO_3 being a good oxidant, oxidises C_6H_5CHO into benzoic acid but in the presence of acetic anhydride, the benzaldehyde formed is acetylated into benzylidene acetate, which prevents it from further oxidation. After the completion of reaction, the unreacted CrO_3 is first removed and the benzylidene acetate is hydrolysed using acid catalyst.

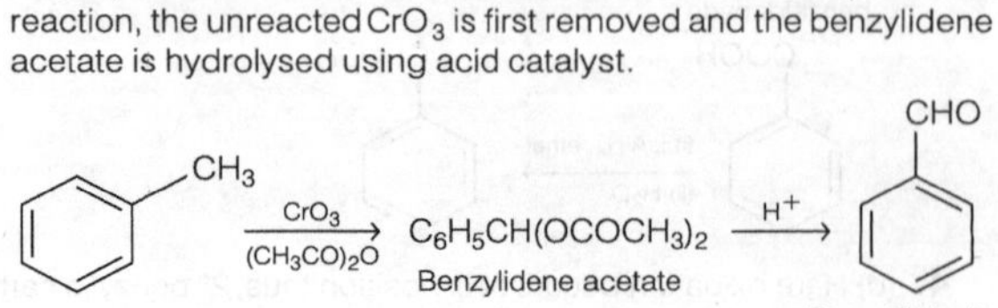

11. (a)

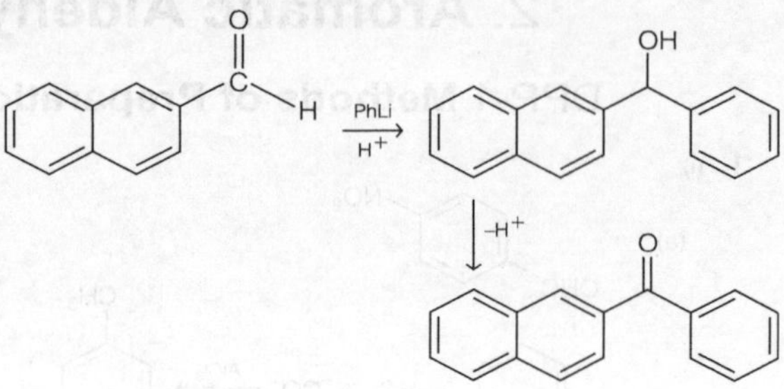

12. (c) Among the given only $SnCl_2/HCl$ selectively reduced —CN group to —CHO. DIBAL-H (diisobutyl aluminium hydride) and LBAH, also reduce $-\overset{O}{\overset{\|}{C}}-Cl$ group into —CHO.

13. (b,d)

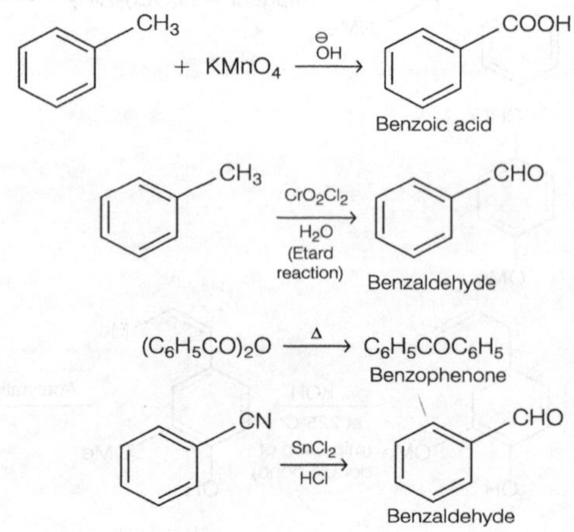

14. (b,c)

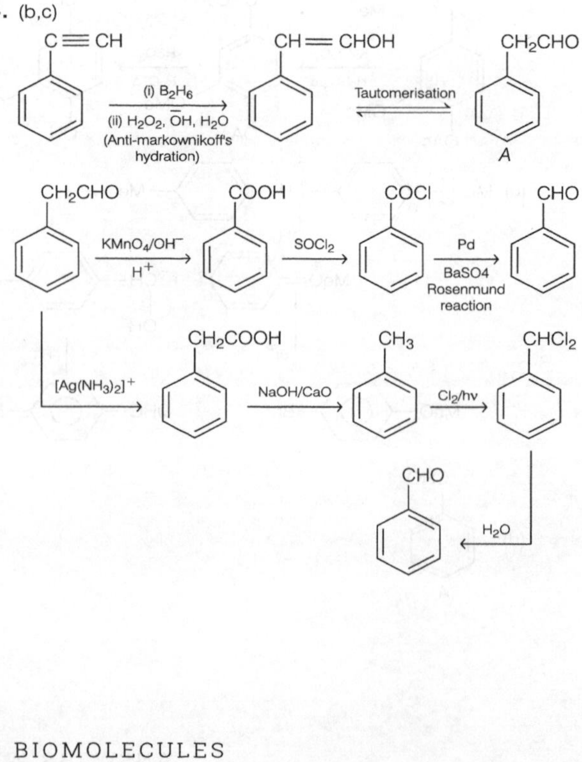

Explanations for Q. Nos. (15 and 16)

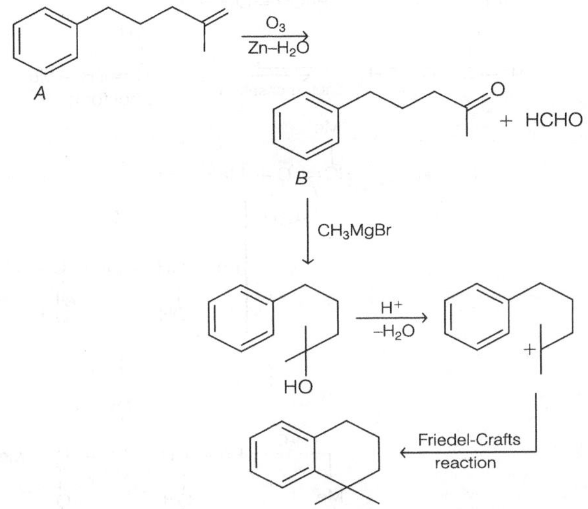

15. (b) **16.** (b)

17. (c) Gattermann's Koch synthesis involves formylation of benzene (which is a modified form of Friedel-Crafts reaction). Presence of electron withdrawing —NO_2 group deactivates the ring to such an extent that it does not undergo formylation. So only product formed, even if nitrobenzene is used as a solvent, is benzaldehyde.

18. (b) MnO_2 acts as oxidising agent in the presence of H_2SO_4, so it oxidises —CH_3 group attached with benzene nuclei to —CHO group.

Since, benzyl alcohol is an intermediate in this process, so the reaction becomes faster if benzyl alcohol is taken instead of toluene.

19. (c) (i) → (r,s), (ii) → (q,t), (iii) → (p,r,s), (iv) → (p)

(i)

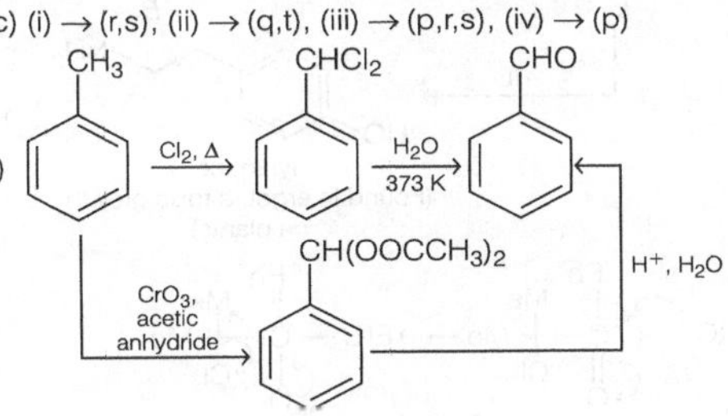

(ii)

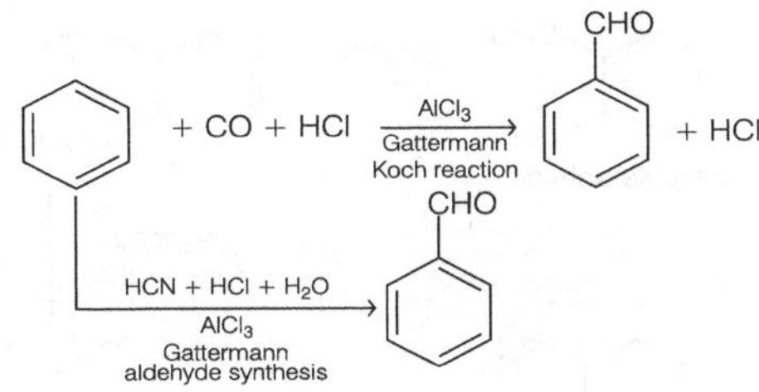

(iii)

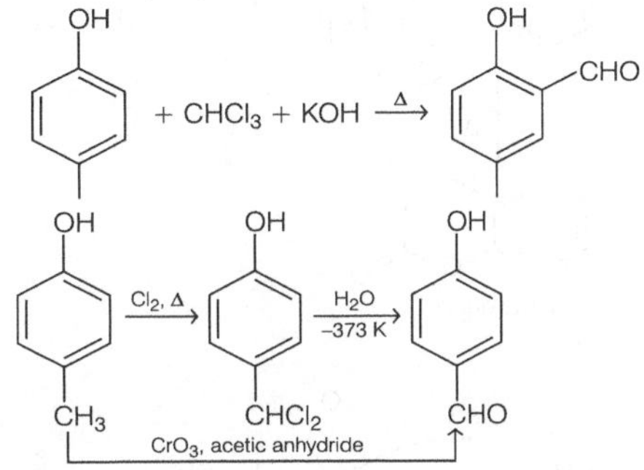

(iv)

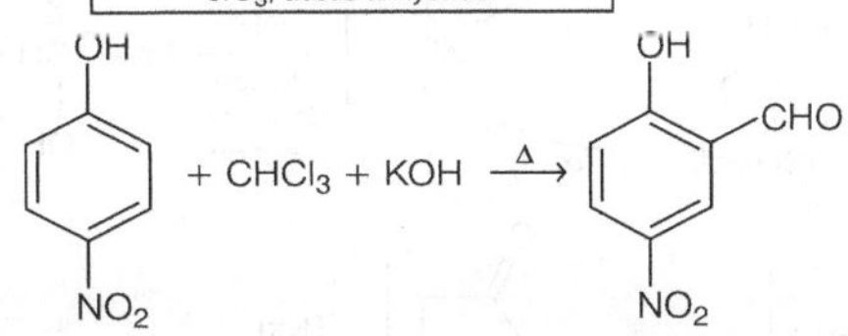

20. (3)

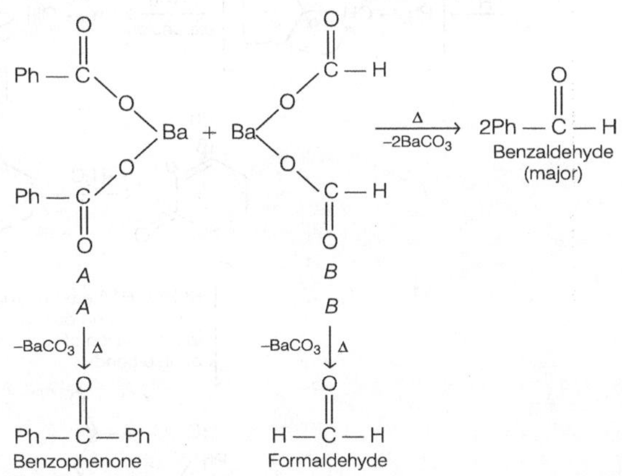

DPP-2 **Properties of Aromatic Aldehydes**

1. (i) (a) PhCOOH, (b) $PhCH_2OH$ (c) $PhCH_2OH$, (d) $PhCH_3$ (e) Pinacol or hydrobenzoin 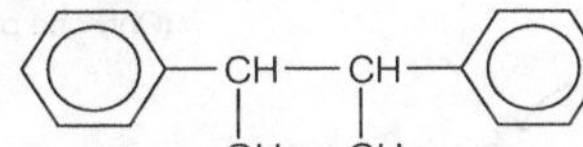 , (f) $PhCH_3$

(ii)

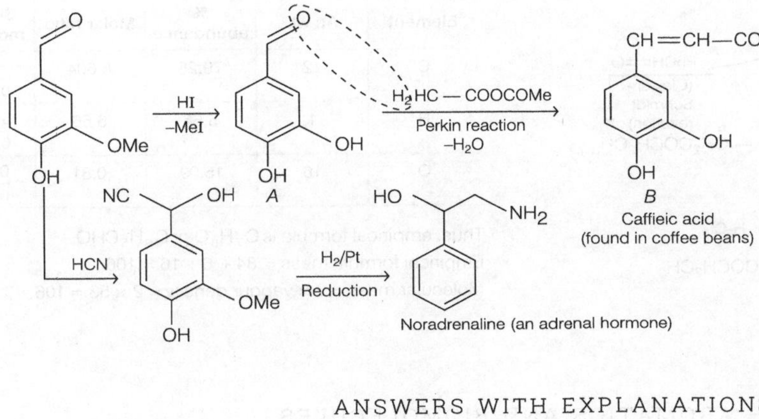

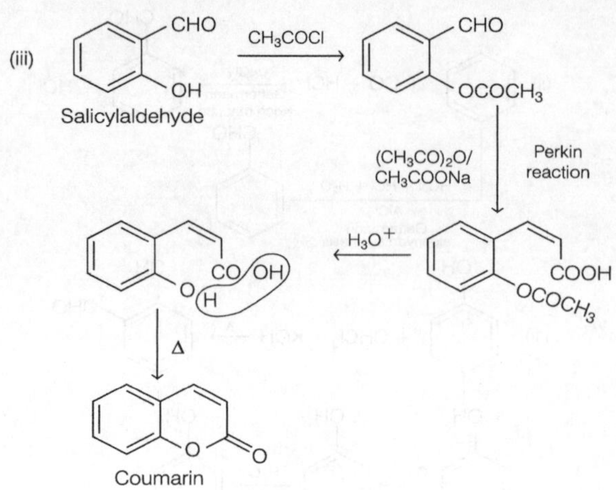

2. (i)

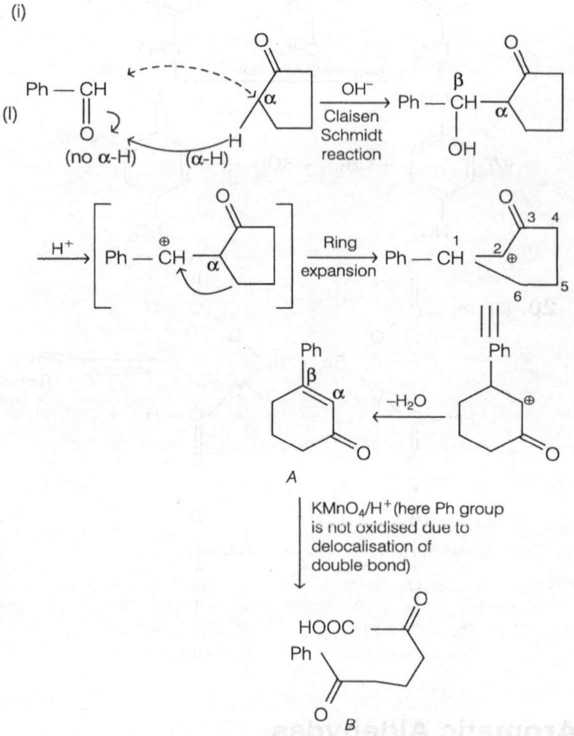

6-phenyl-2, 6-dioxohexanoic acid
(Only one product)

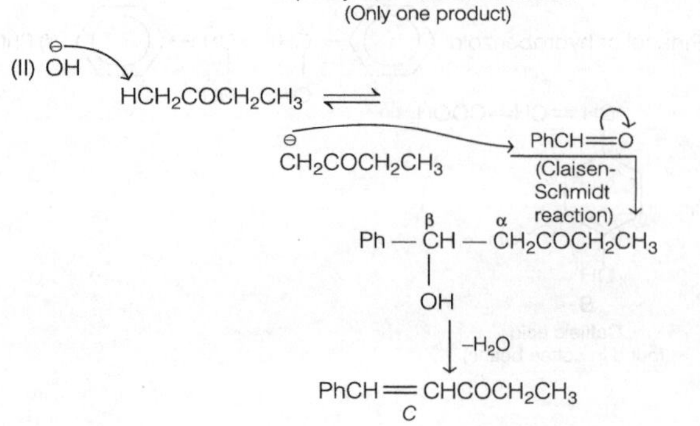

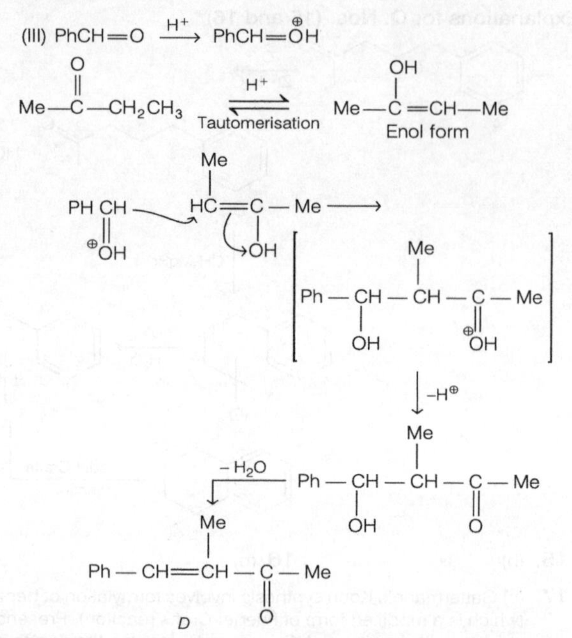

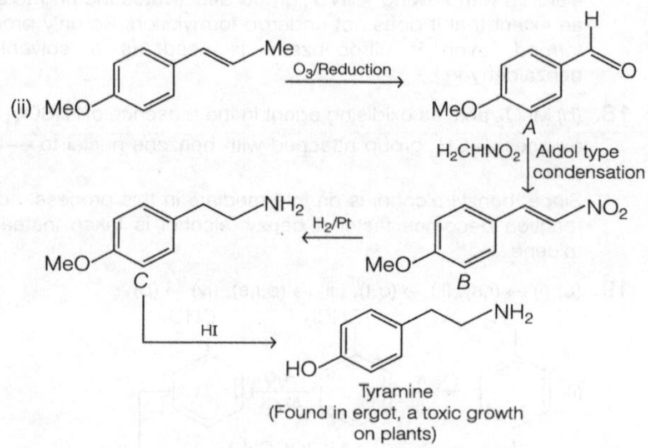

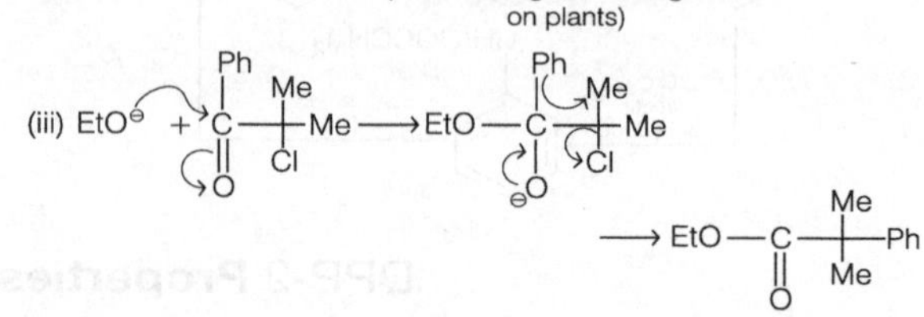

3. Finding empirical formula.

Element	At. wt.	% abundance	Molar ratio	Simplest molar ratio
C	12	79.25	6.604	$\frac{6.6}{0.81} = 8$
H	1	5.66	5.66	$\frac{5.66}{0.81} = 6$
O	16	15.09	0.81	$\frac{0.81}{0.81} = 1$

Thus, empirical formula is C_7H_6O or C_6H_5CHO

Empirical formula mass = 84 + 6 + 16 = 106.

Molecular mass = 2 × vapour density = 2 × 53 = 106

Degree of unsaturation = 5 (4 due to benzene and one due to aldehyde).

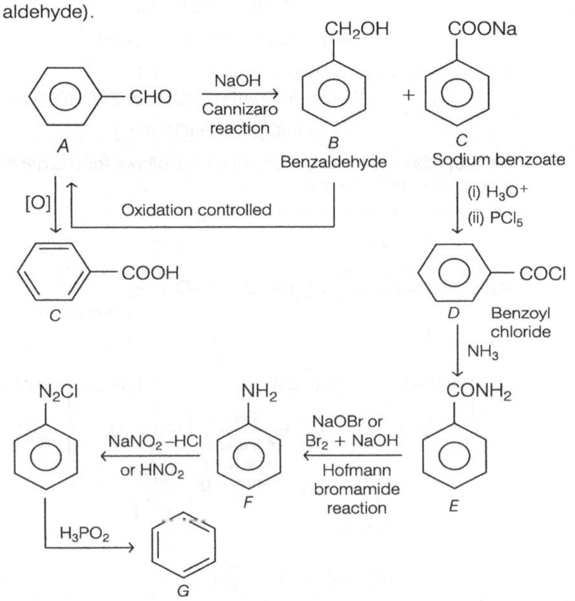

4. (a) Because of the presence of —CHO group, acetal formation takes place.

5. (a) Since, CN^- attacks at planar sp^2- hybridised carbon, the attack is possible from both the sides and hence, a racemic mixture is obtained.

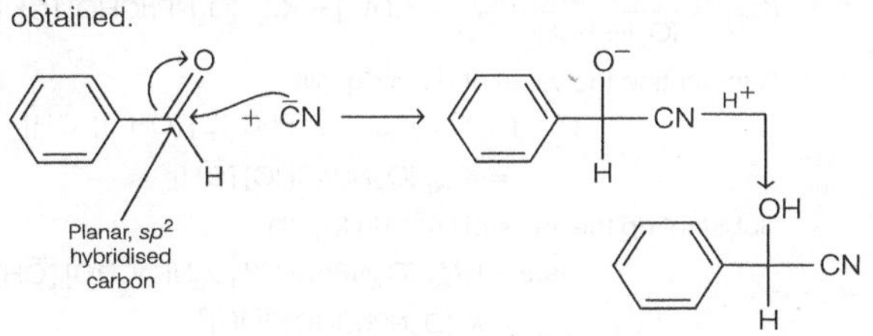

6. (d) When treated with concentrated alkali solution, —CHO group is oxidised to acid as well as reduced to alcohol (i.e. disproportionate) as

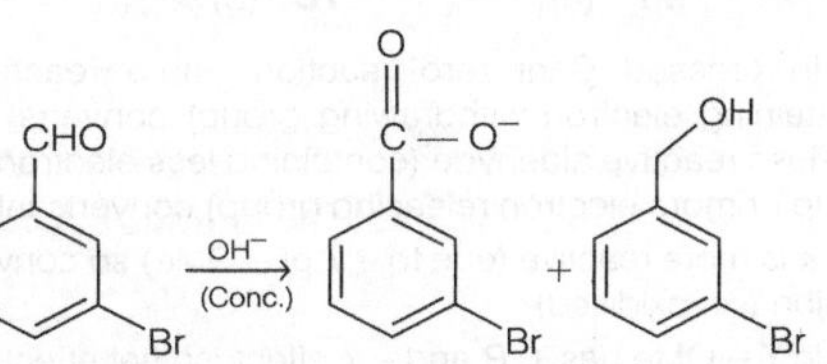

7. (c) Presence of electron withdrawing group increases the reactivity of carbonyl group towards nucleophilic addition reactions as they make the carbonyl carbon more electron deficient (good electrophile)

Electron releasing groups work in an opposite manner, so if present, reduces the activity of carbonyl group towards nucleophilic addition reactions.

Thus, the correct sequence of decreasing order of reactivity is

II > IV > I > III

8. (c) $\underset{1\,\text{mol}}{\text{HCHO}} + \underset{1\,\text{mol}}{\text{PhCHO}} \xrightarrow{\bar{O}H} \underset{1\,\text{mol}}{\text{HCOONa}} + \underset{1\,\text{mol}}{\text{PhCH}_2\text{OH}}$

Since, 1 mole of PhCHO remains, so

$$\underset{\substack{2\,\text{mol}\\1\,\text{mol}}}{2\text{PhCHO}} \xrightarrow{OH^-} \underset{\substack{1\,\text{mol}\\\frac{1}{2}\,\text{mol}}}{\text{PhCH}_2\text{OH}} + \underset{\substack{1\,\text{mol}\\\frac{1}{2}\,\text{mol}}}{\text{PhCOONa}}$$

Thus, HCOONa produced = 1 mol

PhCOONa produced = 0.5 mol

and $PhCH_2OH$ produced = 1 + 0.5 = 1.5 mol

9. (c)

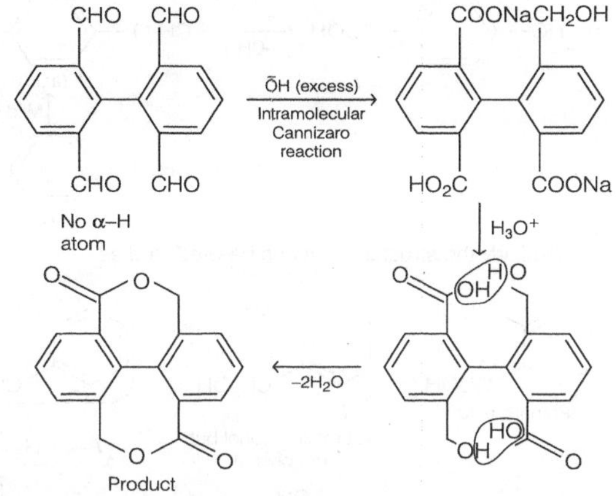

10. (b,d) Benzaldehyde reduces Tollen's reagent and forms adduct with $NaHSO_3$ but gives no reaction with Fehling solution and forms C_6H_5—CH=N—CH(C_6H_5)—N=CH—C_6H_5 as a product with ammonia.

11. (a,b,c,d)

(a) Phenol undergoes condensation with benzaldehyde to form the given product. The reaction takes place at *para* position.

(b) In the presence of HCl/Zn, the >C=O group gets reduced to —CH_2.

(c)

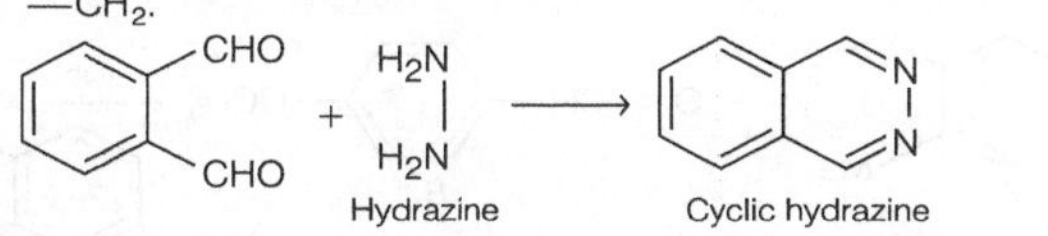

(d)

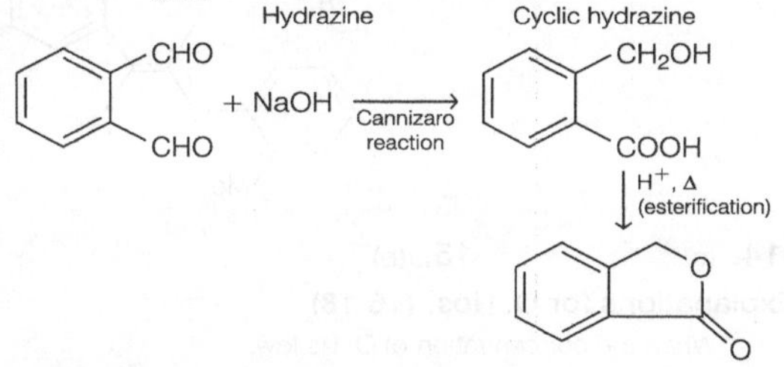

12. (d) Ammoniacal solution of cupric tartarate, i.e., Fehling solution oxidises —CHO group into —COOH but not in case of benzaldehyde.

13. (a) It is an example of Claisen-Schmidt reaction.

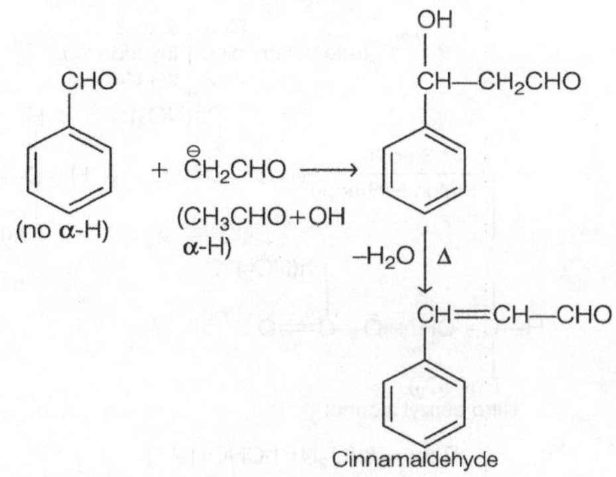

Explanations for Q. Nos. (14 and 15)

To predict the structure of C, we have to proceed in reverse direction.

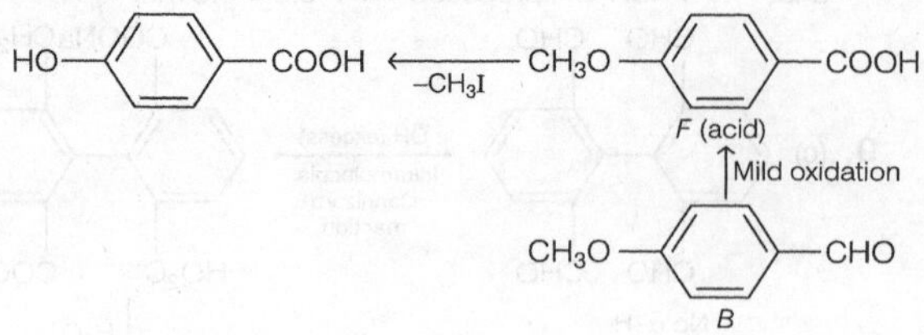

Similarly the structure of A can be predicted as

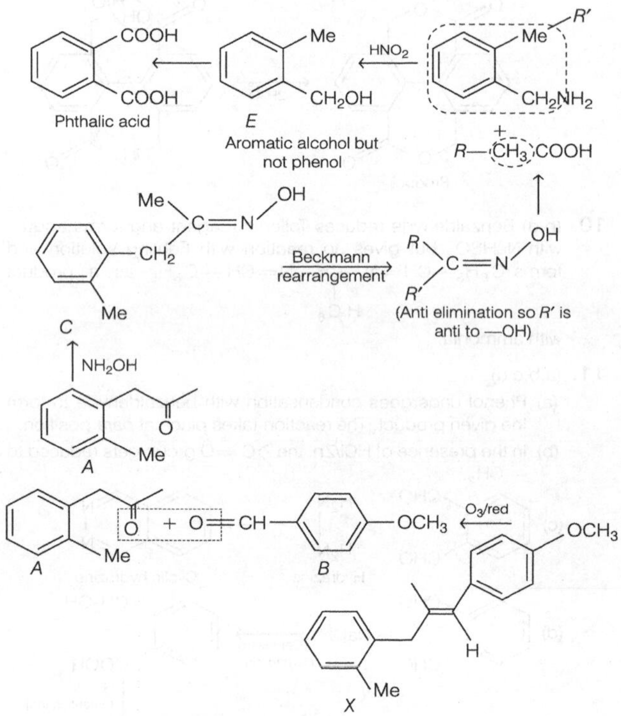

14. (a) **15.** (b)

Explanations for Q. Nos. (16-18)

When the concentration of $\bar{O}H$ is low,

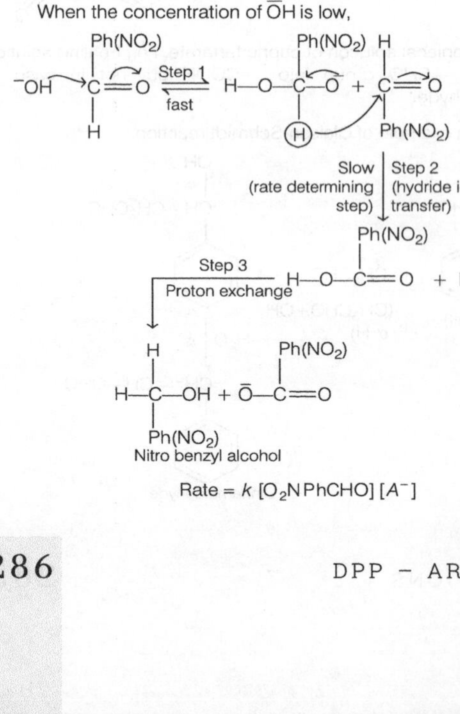

$$\text{Rate} = k\,[O_2NPhCHO]\,[A^-] \qquad \ldots\text{(i)}$$

$$K_{eq} = \frac{[A^-]}{[O_2NPhCHO]\,[\bar{O}H]} \qquad (\bar{A} = \text{anion})$$

or
$$[A^-] = K_{eq}\,[O_2NPhCHO]\,[\bar{O}H]$$

Substituting the value of $(\bar{A})$ in Eq. (i) we get

$$\text{Rate} = k\,[O_2NPhCHO]\,[O_2NPhCHO]\,[OH^-]$$
$$= k\,[O_2NPhCHO]^2\,[\bar{O}H]$$

Hence, this reaction is bimolecular and follows third order kinetics when concentration of $\bar{O}H$ is low.

Where the concentration of $\bar{O}H$ is high,

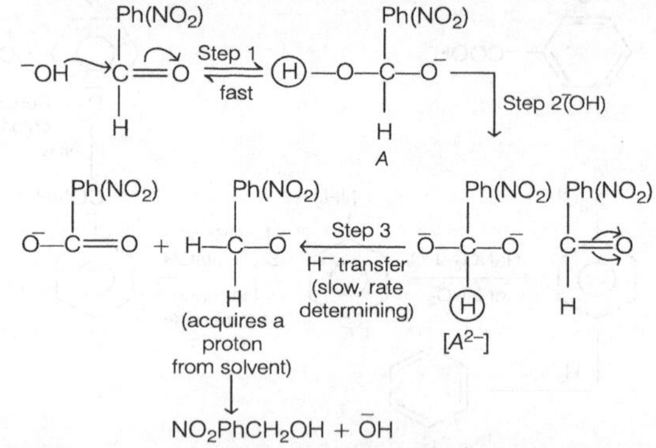

$$\text{Rate} = k\,[NO_2PhCHO]\,[A^{2-}] \qquad \ldots\text{(ii)}$$

$$K_{eq_2} = \frac{[A^{2-}]}{[A^-]\,[OH^-]}, \qquad [A^{2-}] = K_{eq_1}[A^-]\,[OH^-] \qquad \ldots\text{(iii)}$$

$$K_{eq_1} = \frac{[A^-]}{[O_2NPhCHO]\,[\bar{O}H]}, \qquad [A^-] = K_{eq_2}[O_2NPhCHO]\,[\bar{O}H]$$

Substituting the value of A^- in Eq. (iii)

$$[A^{2-}] = K_{eq_2}K_{eq_1}\,[O_2NPhCHO]\,[OH^-]\,[^-OH]$$
$$= K'_{eq}\,[O_2NPhCHO]\,[\bar{O}H]^2$$

Substituting the value of $[A^{2-}]$ in Eq. (ii)

$$\text{Rate} = kK'_{eq}\,[O_2NPhCHO]\,[O_2NPhCHO]\,[\bar{O}H]^2$$
$$= k''[O_2NPhCHO]^2[\bar{O}H]^2$$

Thus, in this condition the reaction is bimolecular and follows fourth order kinetics.

16. (d) **17.** (a) **18.** (b)

19. (a) In crossed Cannizaro reaction, more reactive aldehyde (containing electron withdrawing group) converts into acid ion and less reactive aldehyde (containing less electron withdrawing group or more electron releasing group) converts into alcohol.

(i) X is more reactive (due to $-I$ of $-$ OMe) so converts into acid ion (i.e. oxidised).

In Y – OMe has $+R$ and $-I$ effect, so net effect is electron releasing, so converts into alcohol i.e. reduced.

So, **(i)→ (p,s)**

(ii) Y is more reactive (less electron donating $+I$ effect of three Me groups) than X (more electron donating (i – Pr) group, one hyperconjugative structure, hyperconjugative effect is more powerful than $+I$ effect).

So, Y converts into acid ion, i.e. oxidises while X is reduced

So, **(ii)→ (q,r)**

(iii) X is more reactive (less electron donating, one hyper conjugative structure) than Y (two hyperconjugative structures). So, X is oxidised while Y is reduced.

So, **(iii)→ (p,s)**

(iv) Y is more reactive (Et group is less electron releasing than Me) so converts into acid ion i.e. oxidises while X is reduced.

So, **(iv) → (q, r)**

20. (4) Four products are formed during this reaction.

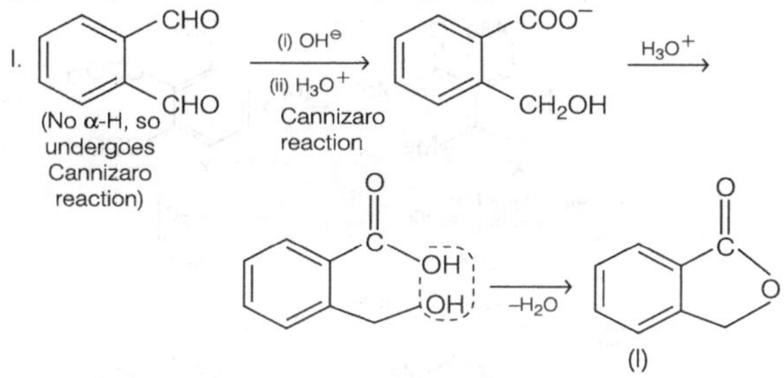

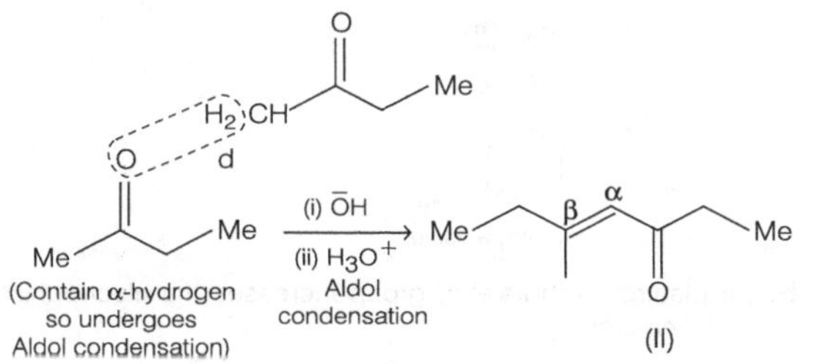

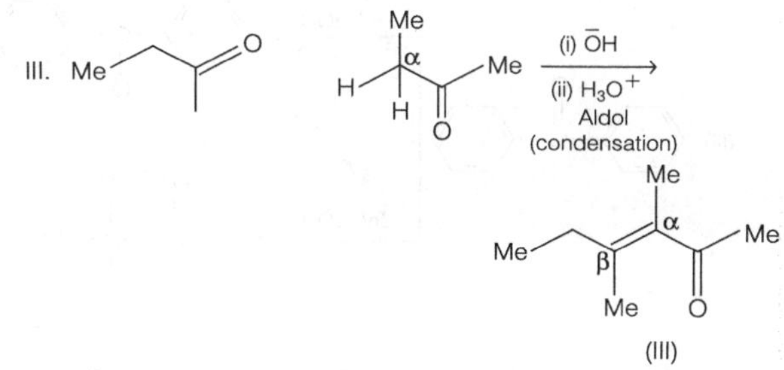

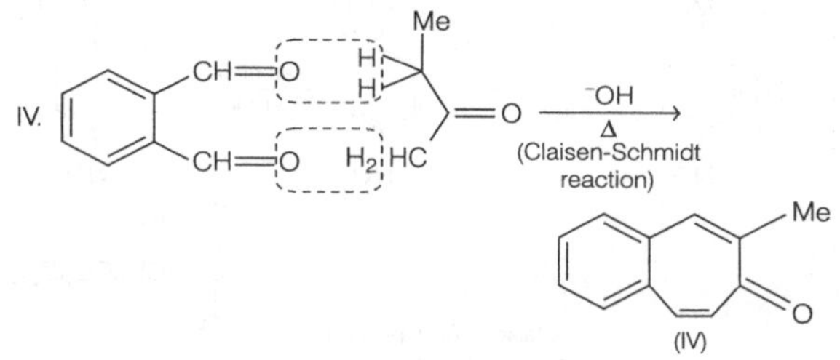

DPP-3 **Properties of Aromatic Ketones**

1. (i)

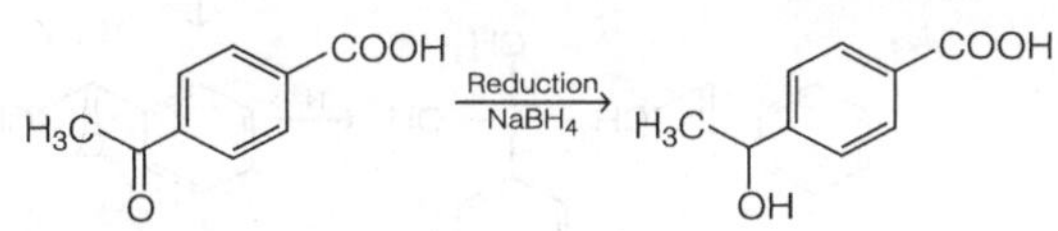

Note $LiAlH_4$ *cannot be used for this purpose.]*

(ii)

(a) SeO_2 converts active methylene group to $>C=O$ (methyl group α to $>C=O$).

(b) Ph—CH(CN)—CH_2COPh (Here conjugate addition of HCN takes place)

(c) $Ph-\overset{O}{\overset{\|}{C}}-Ph \xrightarrow[NaOH]{Zn} Ph_2CHOH$

(d)

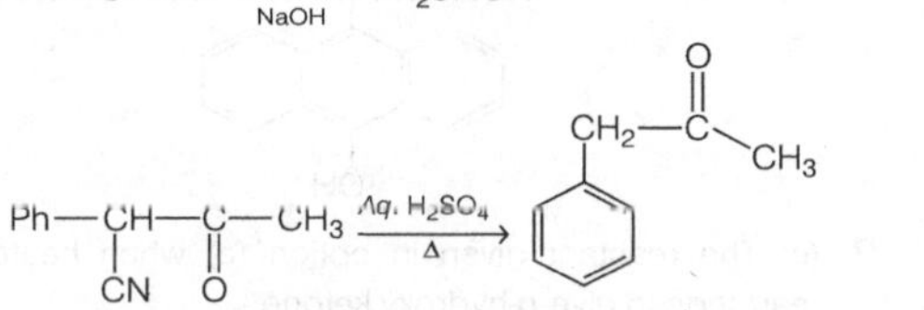

[As β-keto acids readily undergo decarboxylation on heating.]

(e) Zn - Hg/HCl converts $>C=O$ group to $—CH_2$. (Clemmensen's reduction).

(f) $A = Ph-\overset{O}{\overset{\|}{C}}-CH=NOH$

$B = Ph-\overset{O}{\overset{\|}{C}}-CN$

$C = Ph-\overset{O}{\overset{\|}{C}}-COOH$

2. From the reverse process, we get the structures of *A*, *B* and *C* as follows

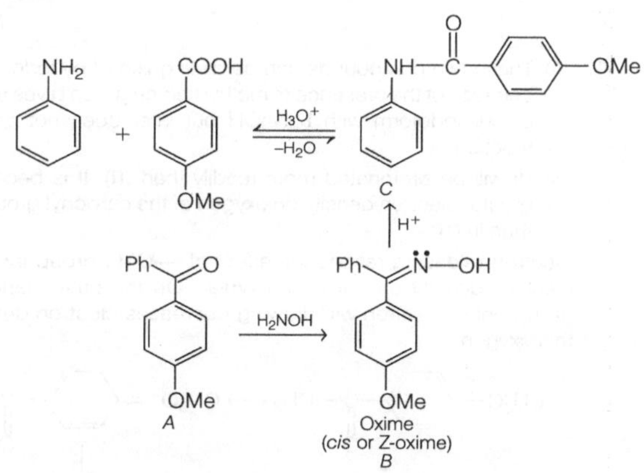

3. (i) Here 1, 4-addition takes place. The obtained product tautomerises to give the final product.

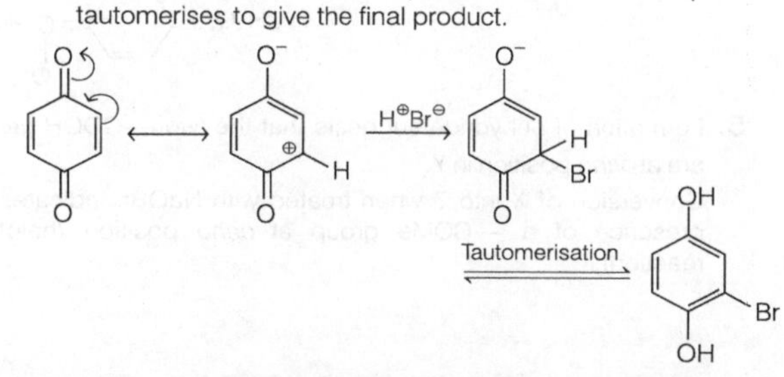

(ii) SeO_2 oxidises active methylene group to carbonyl group. So, product is PhCOCHO.

(iii) 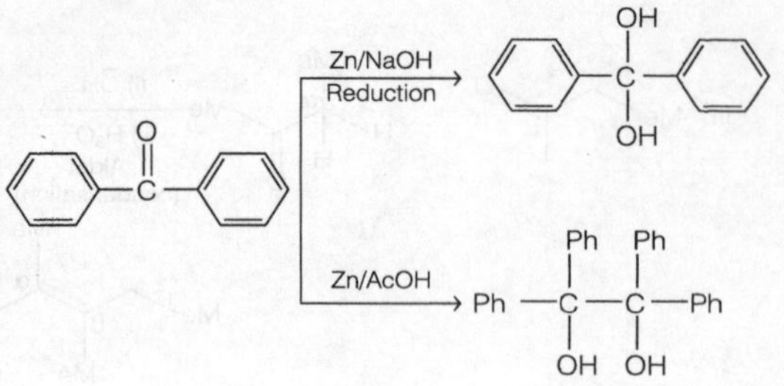

Pinacol (Benzopinacol)

(iv) 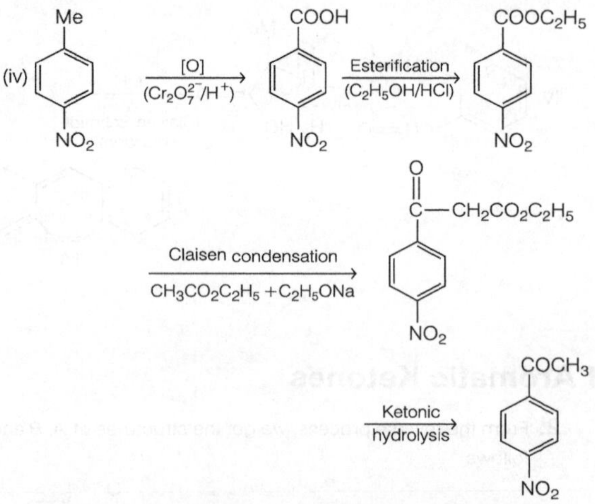

4. (i) These two compounds can be distinguished by iodoform. (I) because of the presence of methyl ketone group gives a yellow colour iodoform with I_2/NaOH but later does not give this reaction.

(ii) (I) will be protonated more readily than (II). It is because of greater electron density on oxygen of the carbonyl group in (I) than in (II).

Electron releasing resonance effect of —OMe group increases electron density on carbonyl oxygen. On the other hand, NO_2 group being electron withdrawing decreases electron density at this oxygen.

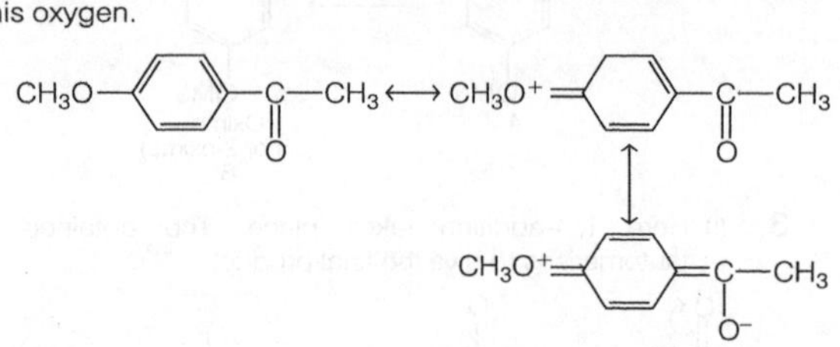

5. Formation of anhydride suggests that the two —COOH groups are at *ortho* position in *Y*.

Conversion of *X* into Z when treated with NaOBr, indicates the presence of a —COMe group at *ortho* position (haloform reaction).

Thus, the possible structure of *X* is

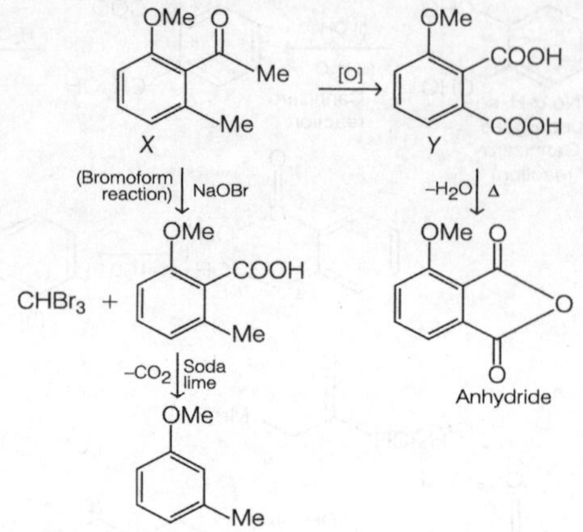

m-methylanisole

6. (c) Electron withdrawing group increases the electrophilicity of carboxyl group.

7. (b)

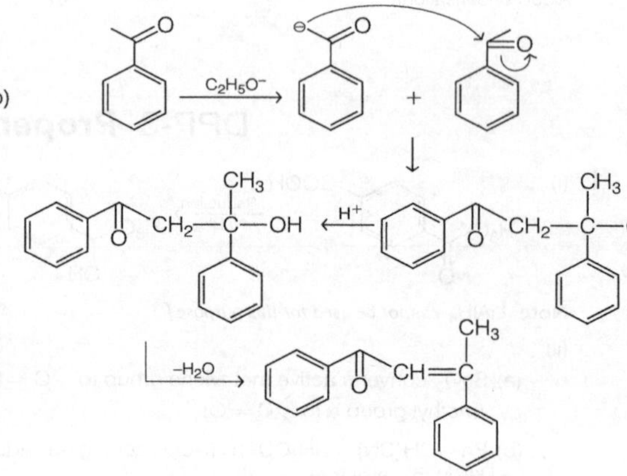

8. (a) First $\bar{O}H$ attacks at α-carbon forming a carbanion, which rearranges to give

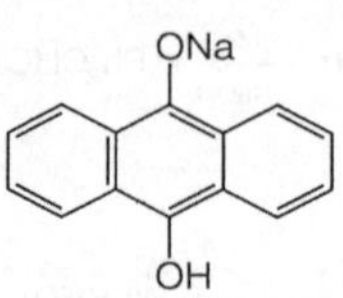

9. (c) The reactant given in option (c) when heated with $AlCl_3$ rearrange to give *p*-hydroxy ketone.

10. (b)

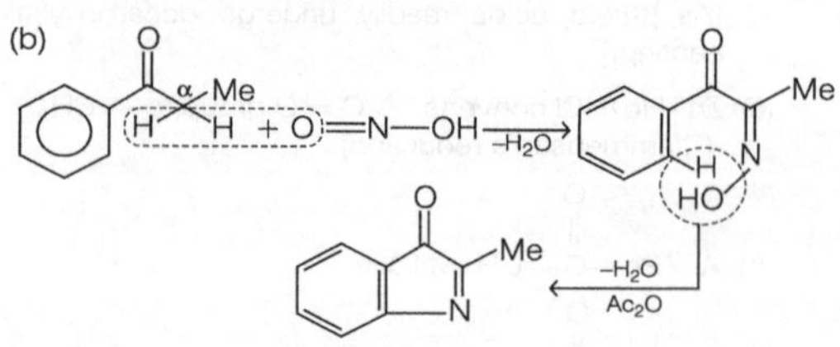

11. (c) Here reaction occurs at carbonyl carbon, so —COOH group can be protected by esterification with CH_3CH_2OH. (as only one —COOH group is present).

12. (a)

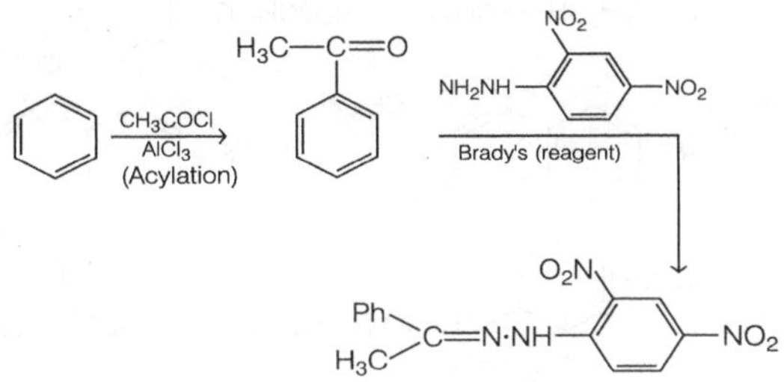

13. (b)

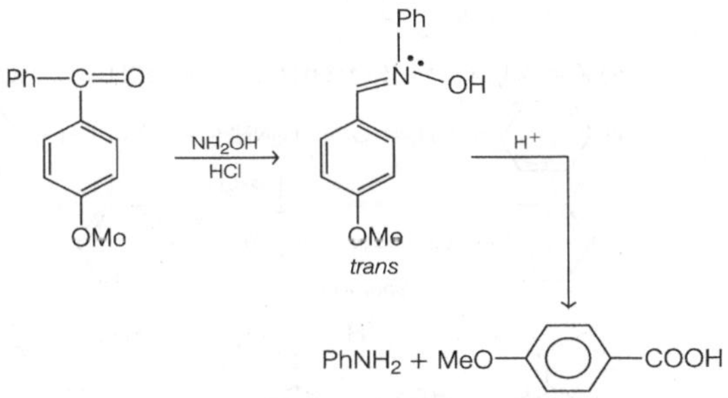

14. (c)

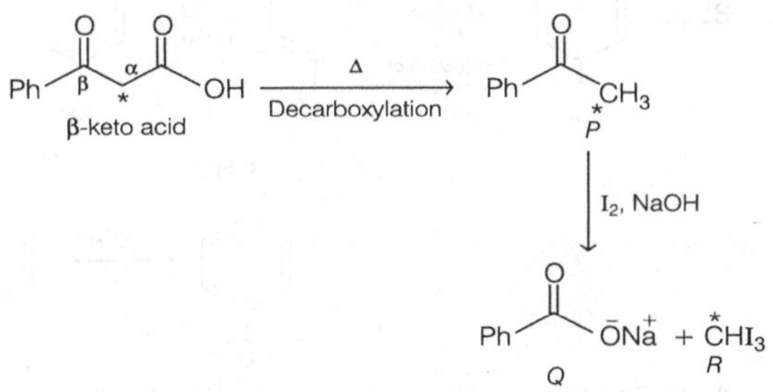

15. (a)

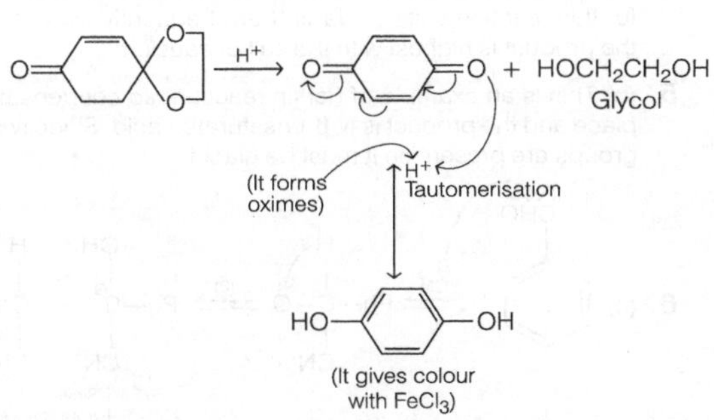

16. (a) The given reactions suggest that the compound *A* is acetophenone.

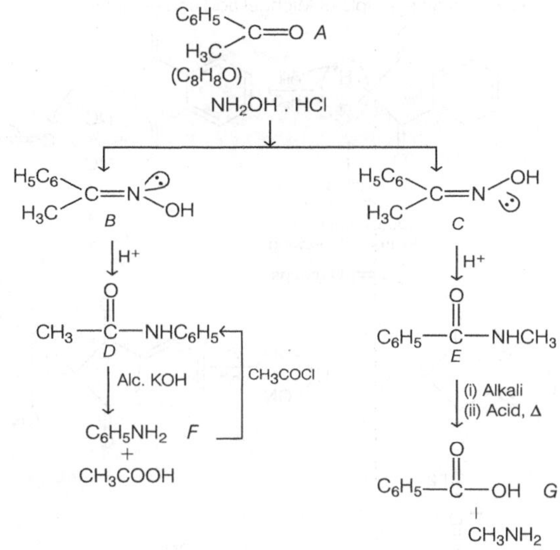

Explanations for Q. Nos. (17-19)

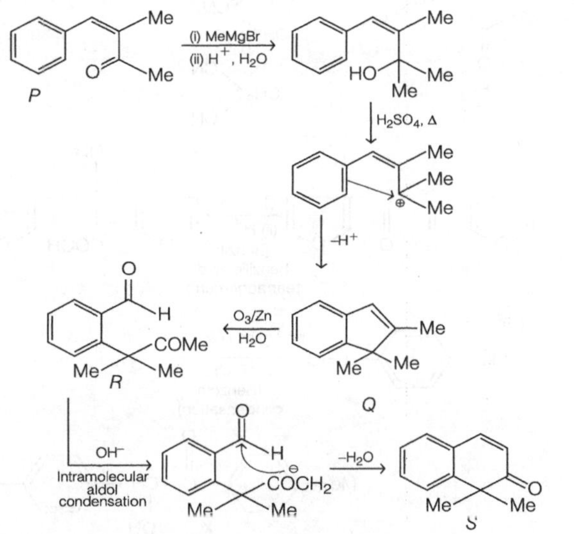

17. (b)

18. (a)

19. (b)

DPP-4 Combined Properties of Aromatic Aldehydes and Ketones

1. (i) It is an example of Michael addition reaction.

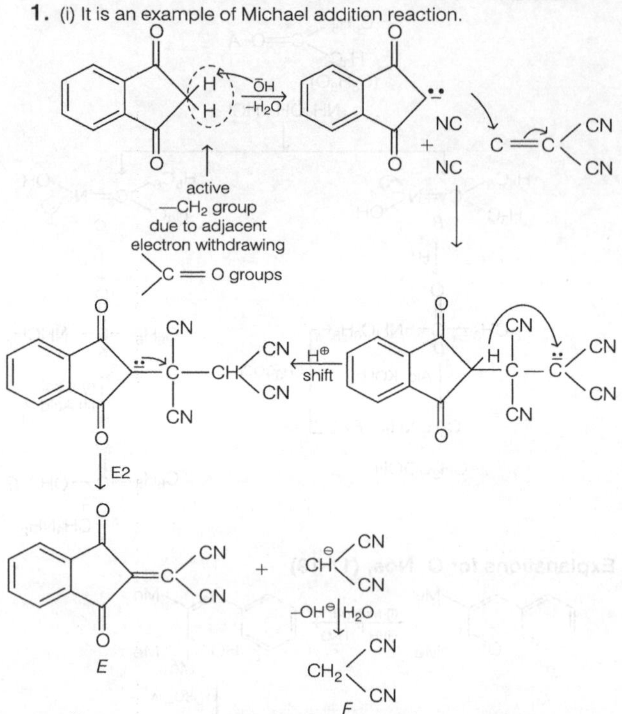

(ii)

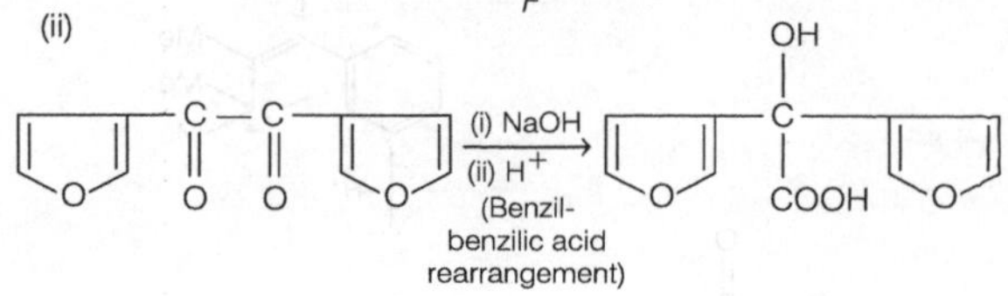

(iii)

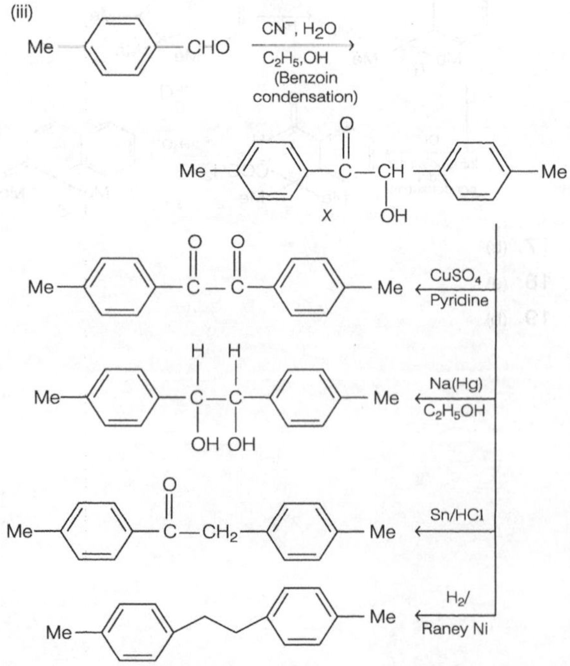

2. (i)

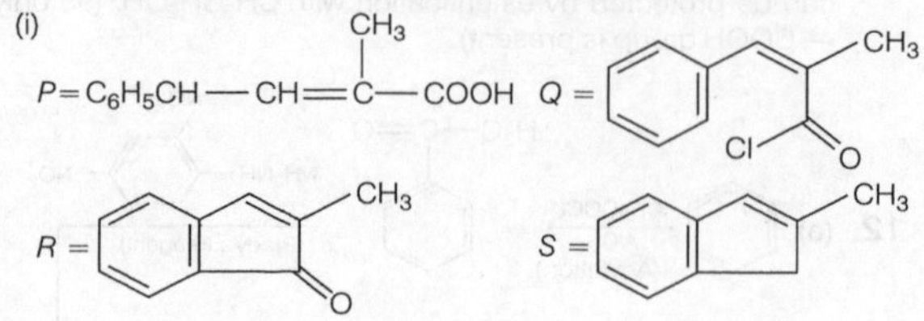

(ii)

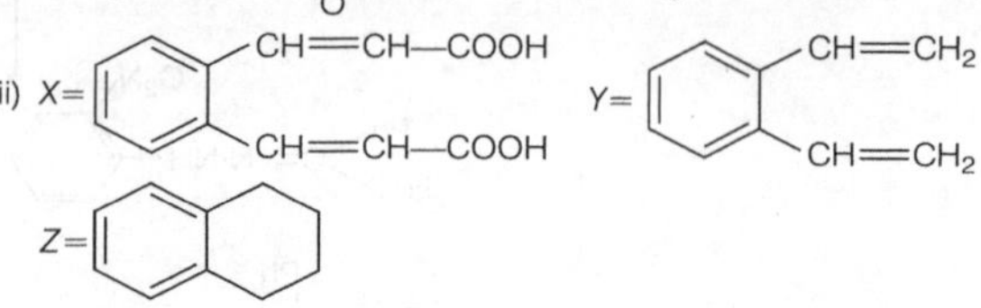

(iii) $A = CH_2COOEt/Zn$; then H^+/Δ $B = PhCH{=}CHCOOEt$

(iv)

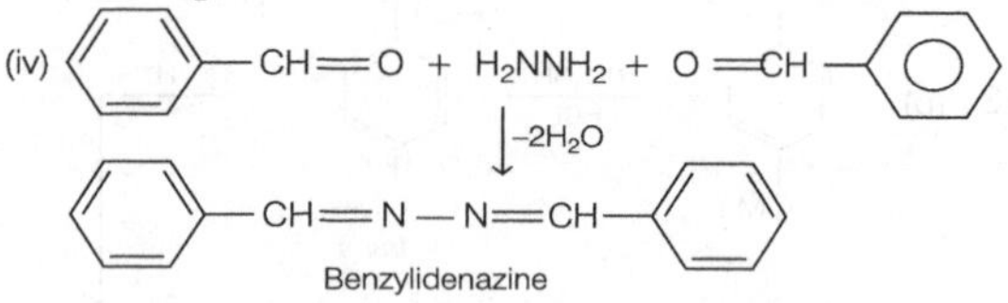

3. (c)

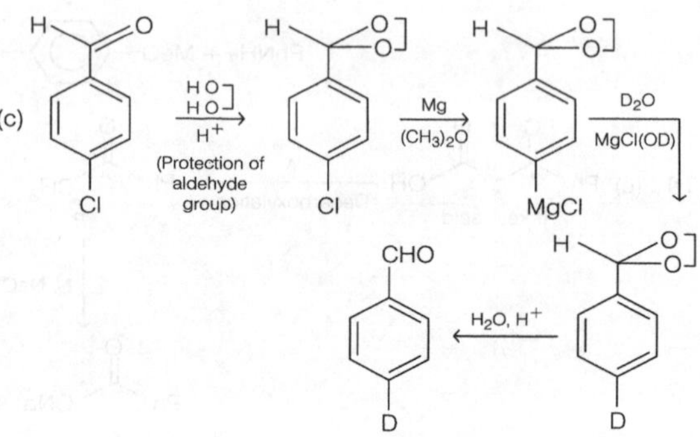

4. (c) The given reaction is an example of Perkin reaction. Although for this reaction salts of Na or K are frequently used, the yield of the product is highest with the salt of caesium.

5. (b) This is an example of Perkin reaction, so condensation takes place and the product is α, β-unsaturated acid. Since two —CHO groups are present so it must be diacid.

6. (c)

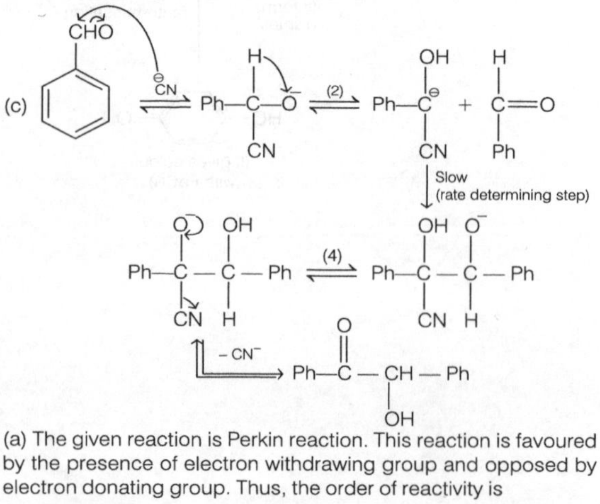

7. (a) The given reaction is Perkin reaction. This reaction is favoured by the presence of electron withdrawing group and opposed by electron donating group. Thus, the order of reactivity is

II > IV > I > III

$(-I, -R)$ $(-I)$ $(+R, -I$, so net electron releasing)

8. (b) It is an esterification reaction, so the product formed is

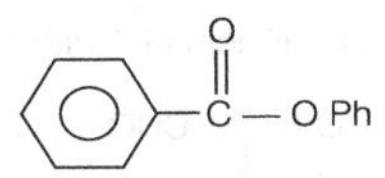

(an ester)

9. (a)

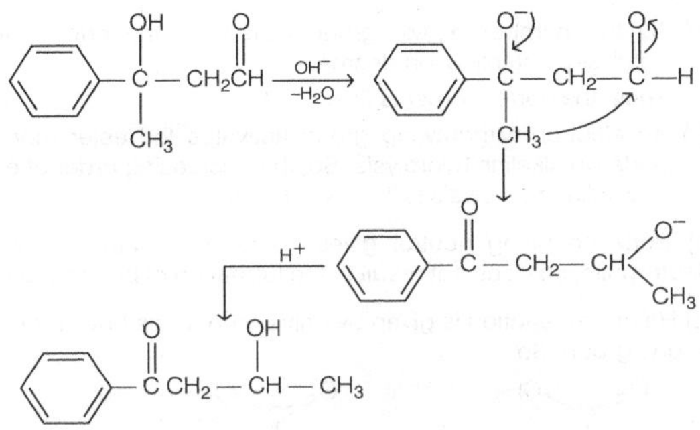

10. (c,d) (c) The product of this reaction is 2° alcohol.

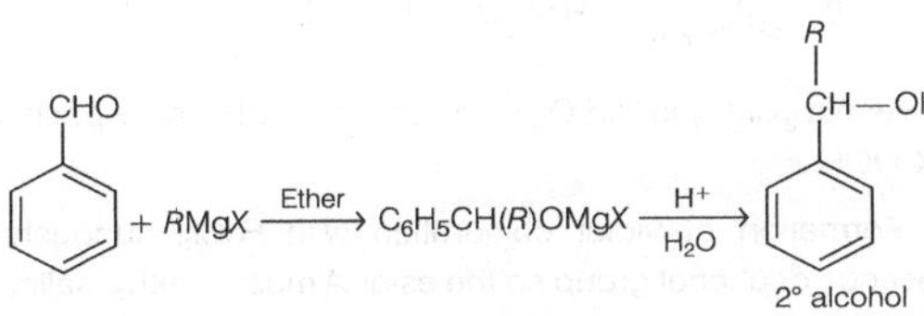

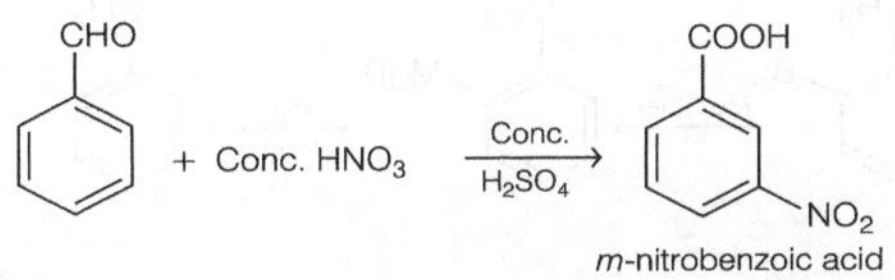

This is because nitric acid is a good oxidising agent, thus also oxidises the —CHO group into —COOH.

11. (a,b,d)

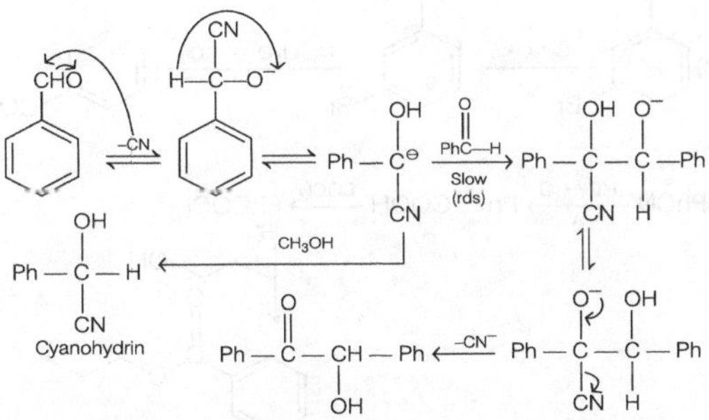

12. (a,b,c) If electron withdrawing group is present at the benzene nuclei of benzaldehyde, then it becomes more reactive towards Parkin condensation (—OMe at *meta* has only-*I* effect).

13. (d) In the given case ring *A* is more reactive due to lack of any electron releasing —OMe group at *para* position. (+ *R* and – *I* effect).

At *meta* position, —OMe has only – *I* effect, so ring *A* is more reactive.

Explanations for Q. Nos. (14 and 15)

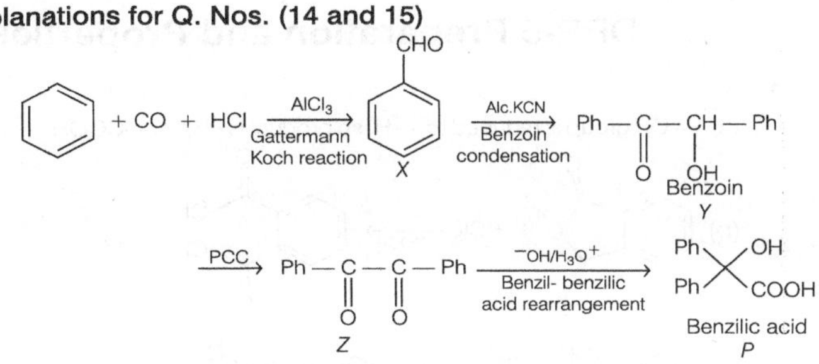

14. (c) **15.** (c)

Explanations for Q. Nos. (16-18)

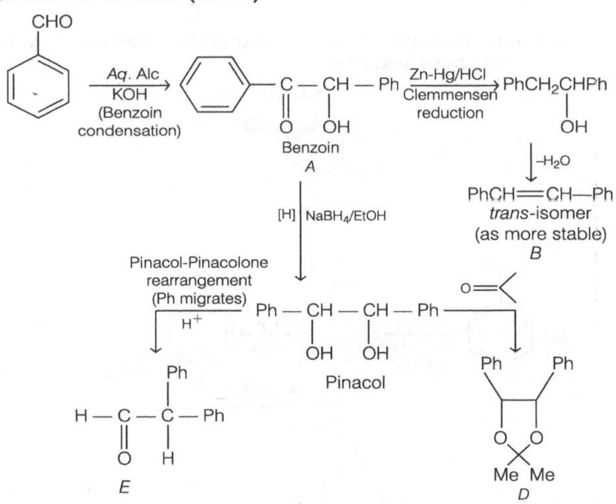

16. (c) **17.** (c) **18.** (a)

19. (b) (i)→ (p), (ii) → (r,s), (iii) → (p), (iv) → (r,s)

In the mixed benzoin condensation, CN^- adds to more reactive aldehyde (containing electron withdrawing group) and after the removal of CN^-, the same aldehyde group is converted to C$=$O group and the other aldehyde group is reduced to —CHOH group.

(i) Presence of Me group deactivates the ring, so *A* is more reactive and hence $X = (=\text{O})$ and $Y = (\text{—OH})$

(ii) Since – *I* effect of Me at *para* is less than at *ortho*, so *B* is more reactive and $X = (\text{—OH})$ and $Y = (=\text{O})$

(iii) *A* is more reactive as OMe at *meta* has only – *I* effect. So $X = (=\text{O})$ and $Y = (\text{—OH})$

(iv) *B* is more reactive, so $X = (\text{—OH})$ and $Y = (=\text{O})$.

20. (a) (i) → (q), (ii) → (q,r), (iii) → (p,s), (iv) → (p)

In Benzil-benzilic acid rearrangement reaction, $\overline{\text{O}}\text{H}$ reacts at the more reactive (C$=$O) group attached to benzene ring containing electron withdrawing group and the same group is migrated.

(i) ^{2}C$=$O is more reactive than ^{1}C$=$O (+ *R* and – *I* effects of OMe), so $\overline{\text{O}}\text{H}$ attacks at this and Ph group migrates.

(ii) (^{2}C$=$O) is more reactive (only – *I* effect of OMe) than (^{1}C$=$O) (+ *R* and – *I* effects of OMe) so *m*-MeO-C_6H_4- group migrates.

(iii) (^{1}C$=$O) is more reactive (less e^- donating Et group with two hyperconjugative structures) than ^{2}C$=$O so *p*-Et C_6H_4 group migrates.

(iv) (^{1}C$=$O) is more reactive (less e^- donating *i*-Pr group) than (^{2}C$=$O) (more e^- donating Et group) so *p*-*i* PrC_6H_4 group migrates.

DPP-5 Preparation and Properties of Aromatic Carboxylic Acids

1. (i)

(a) —OH group is reduced to —H, so product is $Ph_2CH—COOH$.

(b) 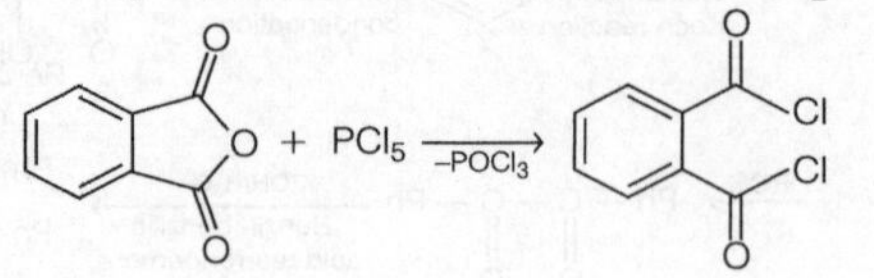

(c) 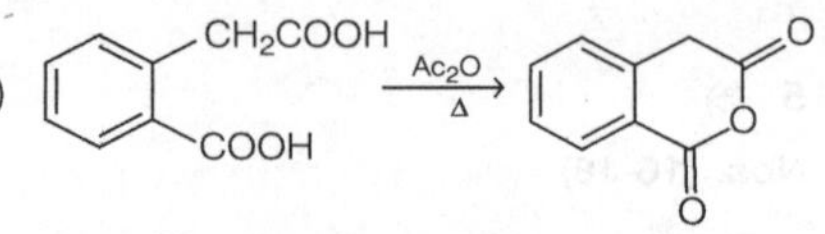

(d) —Br group occupy a position *ortho* to —OH and *meta* to —COOH so product is

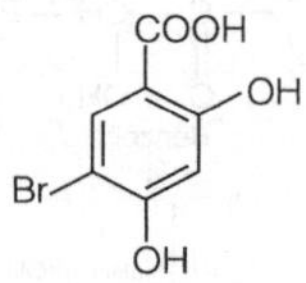

(ii)

(a) 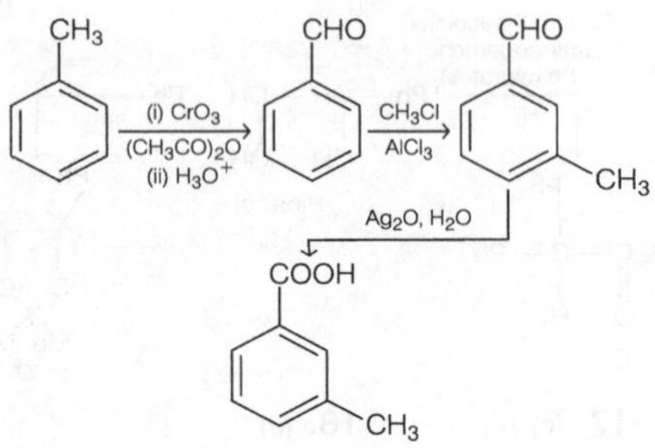

(b) 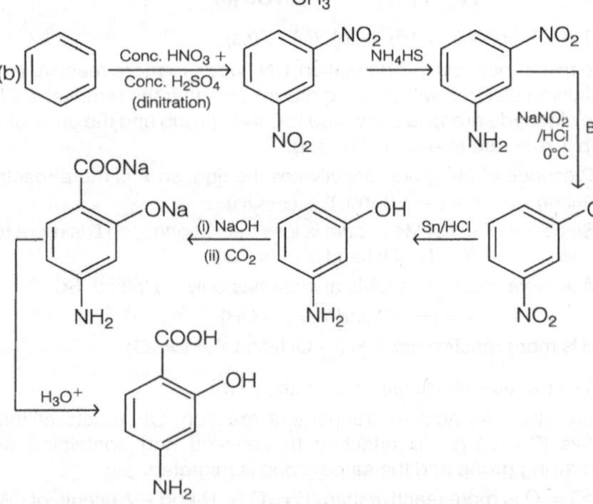

2. Formation of only benzoic acid suggests that only one carbon chain is attached directly with benzene nuclei. Thus, the possible structure of *B* is

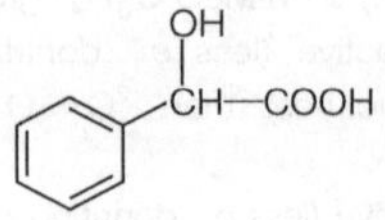

B

This *B* is obtained by the treatment of *A* with *aq*. NaOH followed by acidification, so

A = 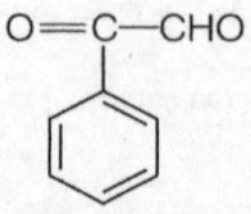

3. (i) As the number of Me group increases, the rate of acid catalysed esterification decreases.
Thus, the correct order is III < II < I.

(ii) An electron withdrawing group activates the ester more to undergo alkaline hydrolysis. So, the decreasing order of ease of alkaline hydrolysis is III > IV > I > II.

4. (c) Benzene being neutral gives no reaction with very weak electrophile, so does not result in the formation of benzoic acid.

5. (d) Haloform reaction is given by only compounds having methyl ketone group. So

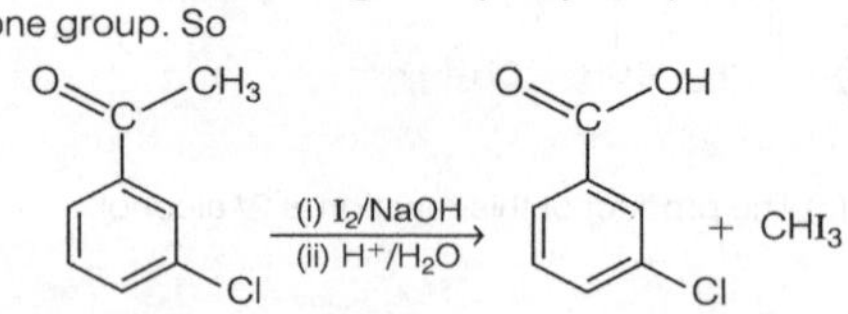

6. (a) As $Na_2Cr_2O_7$(in H_2SO_4) oxidises only CH_2 at α-position in —COOH.

7. (a) Formation of violet colouration with $FeCl_3$ suggests the presence of phenol group so the ester *A* must be ethyl salicylate.

8. (c) 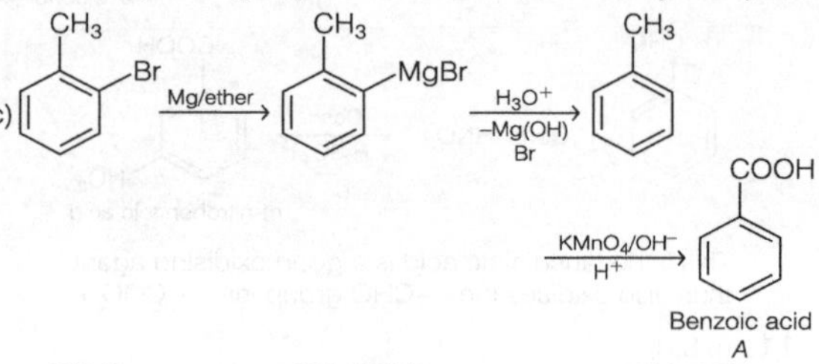

9. (c) 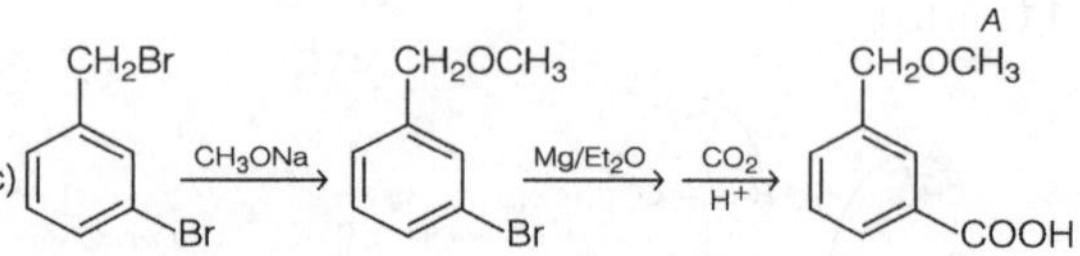

10. (b)

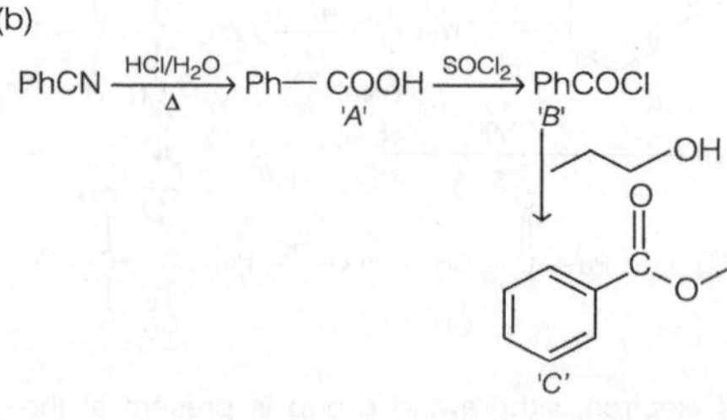

11. (c) The given structure is of aspirin, which is the acetylation product of salicylic acid.

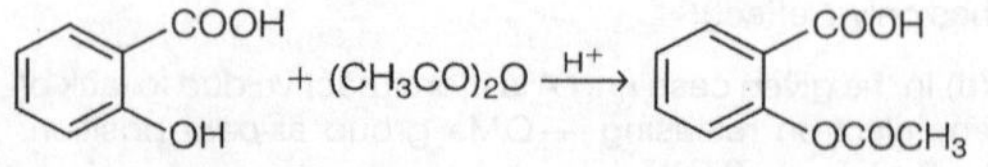

12. (d) Since ethylene glycol is a bifunctional compound, so esterification take place at both the —OH group giving (d) as the major product.

13. (a,b,d)

(a) $KMnO_4$ in alkaline medium oxidises the carbon chain attached directly with benzene nuclei into —COOH group.

(b) 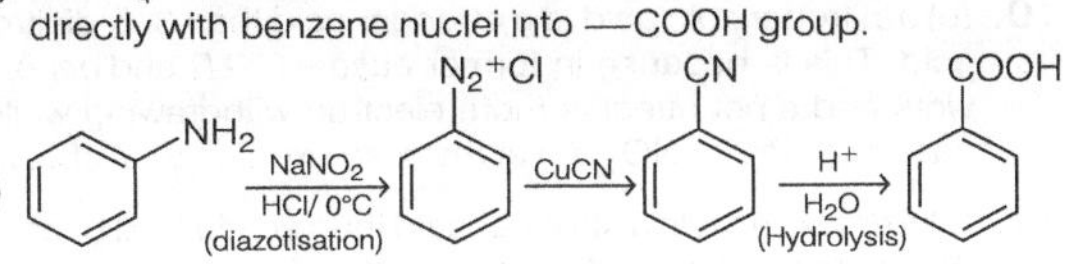

(c) 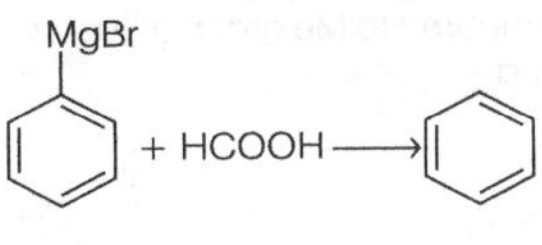

(d) 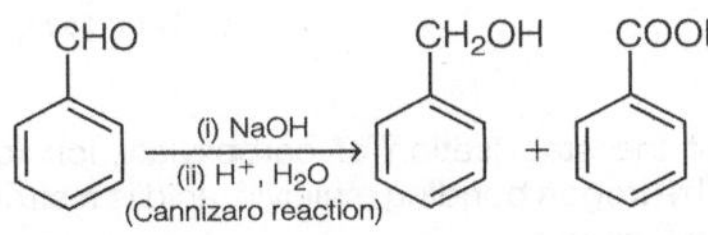

14. (b,c)

(a) Benzoic acid is slightly more acidic than ethanoic acid due to electron withdrawing nature of phenyl group.

(b) Phenyl free radical is very less likely, so Kolbe's reaction does not occur.

(c) Due to hydrophobic phenyl ring, it is less soluble than ethanoic acid. Its boiling point is higher because of its higher molecular mass and hence more van der Waals' forces.

(d) Since, benzoic acid is a stronger acid than carbonic acid, so it forms salt with $NaHCO_3$.

15. (a,d)

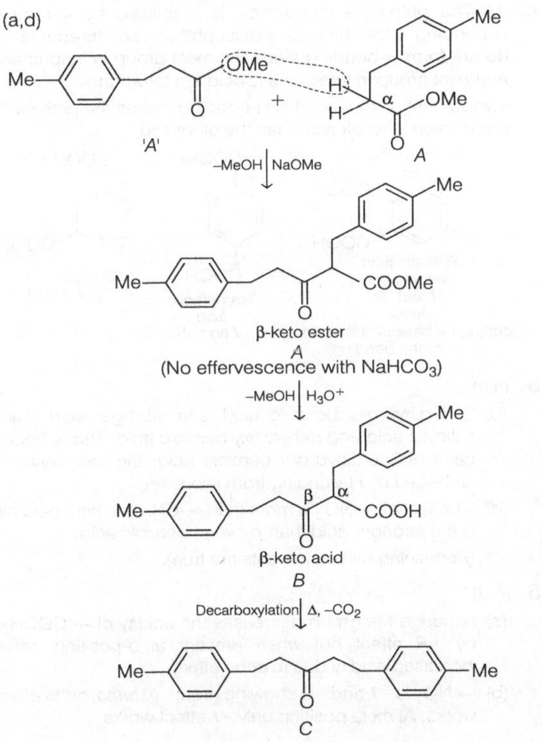

1, 5-bis (p-tolyl) pentan-3-one (gives no iodoform with NaOI)

16. (a) —OH group is strongly activating group, hence, substitution takes place readily at *o*-and *p*-positions and this causes elimination of —COOH group.

17. (a) Although acid (I) is sterically hindered, but formation of acylium ion as intermediate minimizing the strain and making the reaction feasible.

The acylium ion is stabilised due to delocalisation of π-electrons.

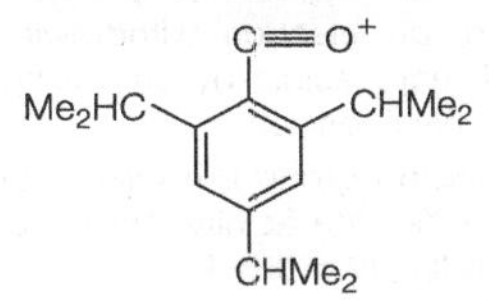

Explanations for Q. Nos. (18-20)

(I)

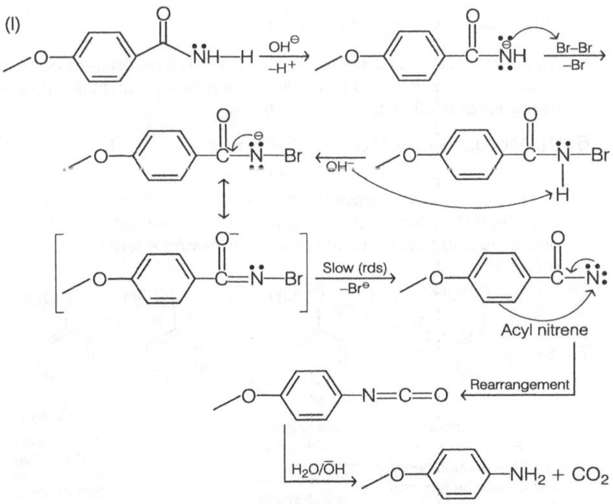

This reaction is called Hofmann bromamide reaction.

(II)

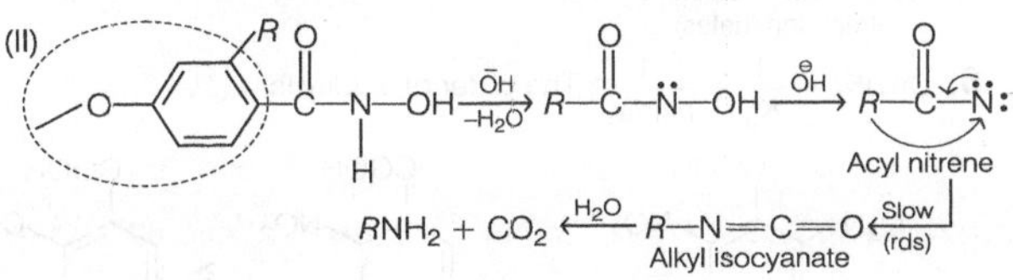

This is Lossen rearrangement reaction.

(III)

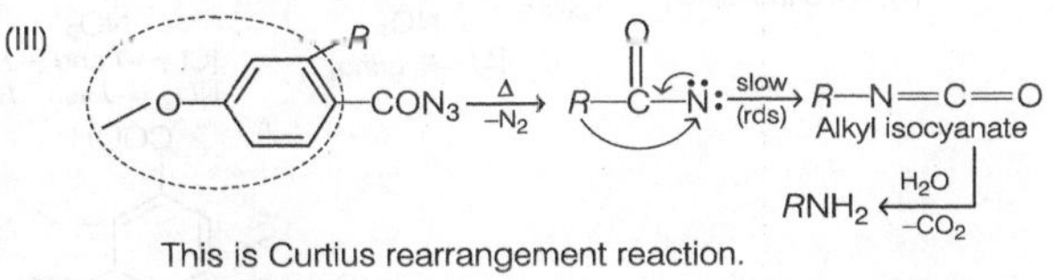

This is Curtius rearrangement reaction.

(IV)

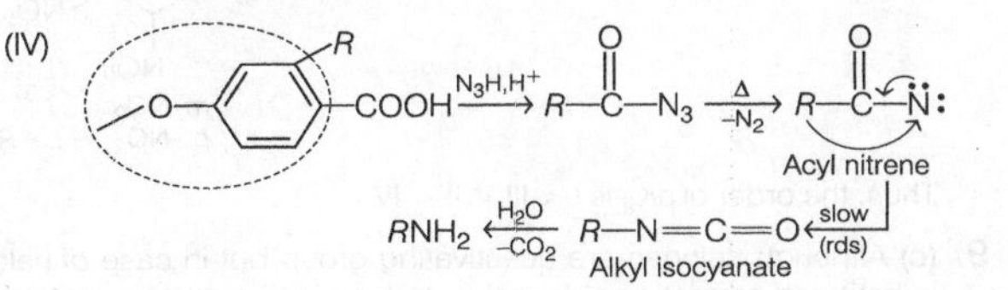

18. (d)

19. (c)

20. (b)

DPP-6 Acidity of Carboxylic Acids

1. (c) Acidity $\propto K_a$ and acidity increases in the presence of more electron withdrawing group.

2. (a) During the reaction of carboxylic acid with sodium bicarbonate, all the CO_2 comes from the $NaHCO_3$ but not from the acid.

3. (b) Presence of an electron withdrawing group makes the carboxylic acid more acidic by stabilising carboxylate ion. Its effect is more at *para* position.

An electron releasing group like OMe (due to $+R$ and $-I$ and $+R > -I$) decreases the acidity. Thus the decreasing order of acidic strength is II > IV > III > I.

4. (b) $K_a \propto \dfrac{1}{K_b}$ and $K_a \propto$ acidity.

Thus, the order of K_b is IV > III > I > II.

5. (a) All the given acids have steric inhibition of resonance but it is greater in I than in II and III and in II, it is greater than in III. Hence, the order of acidic strength is I > II > III.

6. (b) Although $-I$ effect order of halogen is

$$-F > -Cl > -Br > -I$$

but net electron withdrawing effect is Cl > F > Br > I. It is due to the effective overlap of $2p$ of F with $2p$ of C so $+R$ effect becomes operative and hence, net electron withdrawing effect ($-I$) and $+R$ of Cl > F.

7. (b) 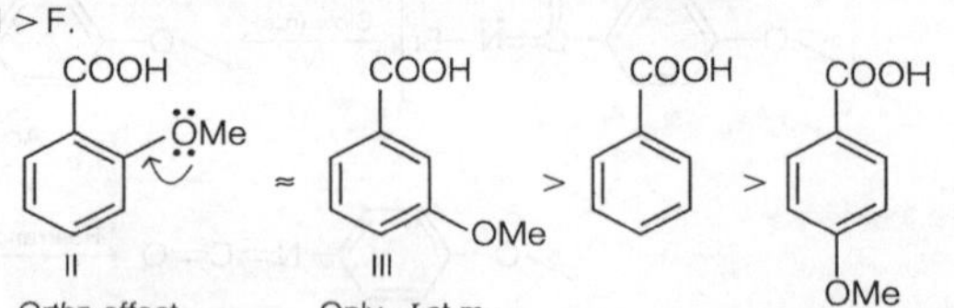

II — *Ortho* effect (If $-R$ and $+R$ effects of –OMe is considered, net should be electron donating and should be weaker than I, so *ortho* effect dominates)

III — Only $-I$ at *m*– of –OMe group. Net electron withdrawing

$-I$ at *para* and $+R$ of OMe — Net electron donating

8. (b) $pK_a \propto \dfrac{1}{K_a} \propto \dfrac{1}{\text{acidity}}$. The order of acidity is

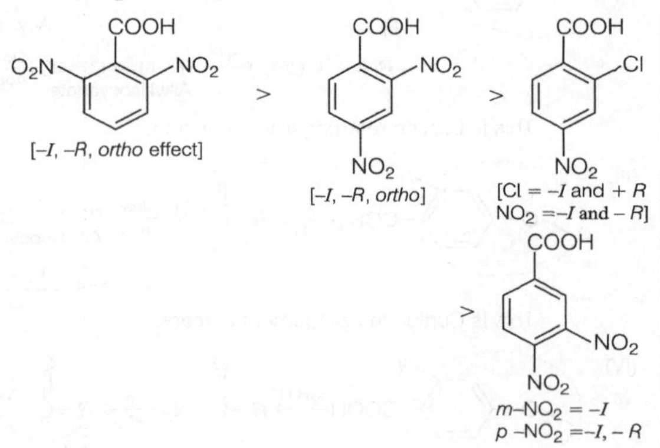

Thus, the order of pK_a is I < III < II < IV.

9. (c) Although halogen are deactivating group but in case of halo substituted aromatic acids, only $-I$ effect is considered and this effect is maximum at *o*- and minimum at *para*-position.

$$pK_a \propto \frac{1}{\text{acidity}}$$

Since, the order of acidity of given halo substituted acids is

$$\text{(i)} > \text{(ii)} > \text{(iii)} > \text{(iv)}$$

so, the order of pK_a value is $r < s < p < q$

10. (b) *o*-nitrobenzoic acid is a stronger acid than 3, 5-dinitrobenzoic acid. This is because in former case $-I$, $-R$ and *ortho* effect all work and a net effect is more electron withdrawing while in later case both the —NO_2 are at *meta*, so exert only $-I$ effect.

11. (b) Benzoic acid is a stronger acid than acetic acid because of the presence of $-I$ and $+R$ showing phenyl group (net effect is slight electron donating) as compared to Me group (hyperconjugation), so more electron donating.

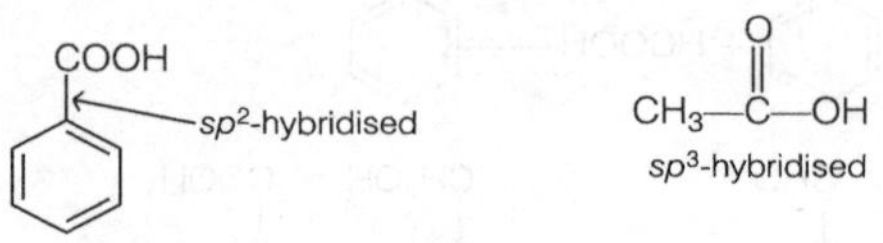

12. (d) Because of the stabilisation of carboxylate ion formed by intramolecular hydrogen bonding, salicylic acid is a stronger acid than its *m* and *p*-isomers.

13. (d) Formic acid (HCOOH) is a stronger acid than benzoic acid (PhCOOH). This is because here the e^- donating resonance effect ($+R$) of Ph group outweighs the electron attracting inductive effect, i.e. $+R > -I$.

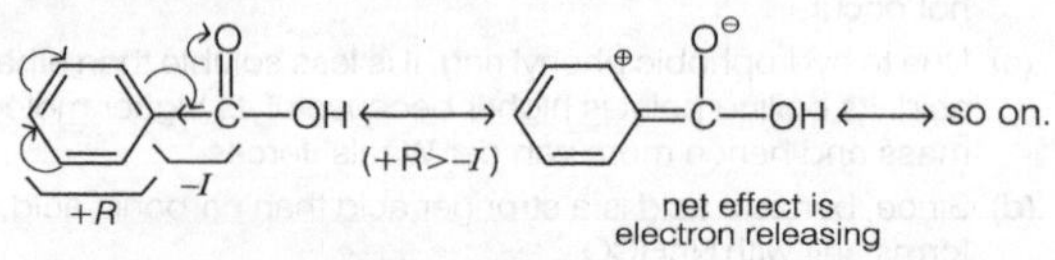

14. (b) The phthalate monoanion is stabilised by intramolecular H-bonding, while the anions of isophthalic and terephthalic acids do not form H-bonding since the *meta* group in isophthalic acid and *para* group in terephthalic acid are far apart.

Involvement of second H in H-bonding makes the phthalate monoanion, a weak acid than the other two.

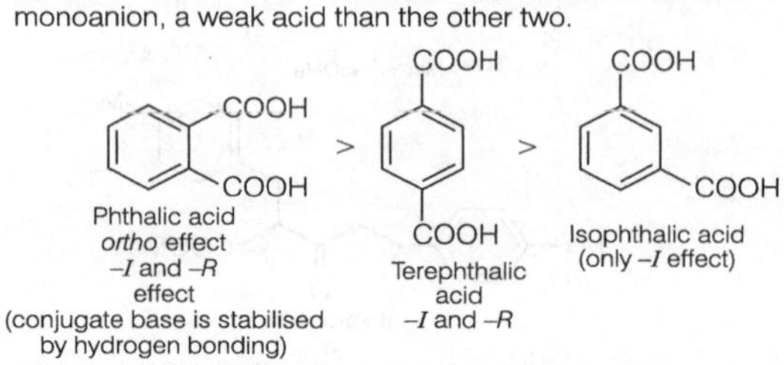

15. (a,c)

(b) 2, 6-dihydroxy benzoic acid is a stronger acid than both salicylic acid and *m*-hydroxy benzoic acid. This is because in case of 2, 6-dihydroxy benzoic acid, the carboxylate ion is stabilised by H-bonding from two sides.

(d) $-I$ effect of —NO_2 is more than —CN, so *p*-nitro benzoic acid is the stronger acid than *p*-cyano benzoic acid.

(Remaining two statements are true).

16. (c,d)

(a) Although Me group decreases the acidity of —COOH group by $+R$ effect but when present at *o*-position, results in increased acidity due to *ortho* effect.

(b) —NO_2 is $-I$ and $-R$ showing group. At *ortho*, *ortho* effect also works. At *meta* position only $-I$ effect works.

(c) *o*-methoxy benzoic acid is most acidic among the given due to *ortho* effect, *m*-methoxy also has the similar acidity due to its $-I$ effect. *p*-methoxy benzoic acid is although less acidic than benzoic acid due to net effect of $+R$ and $-I$ (i.e. net electron releasing).

(d) The correct order is

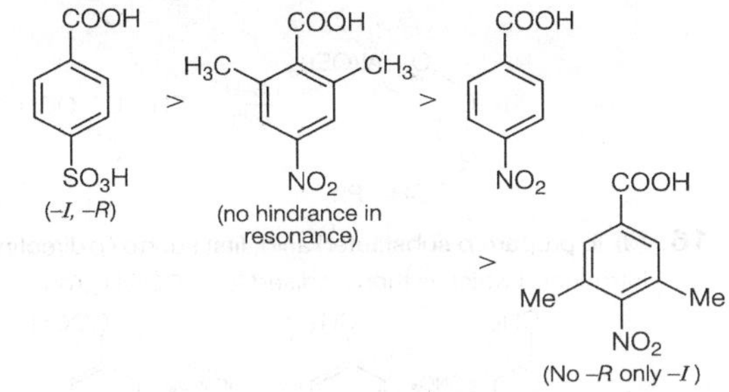

17. (6) Out of the given I, II, III, IV, VII and VIII are most acidic than benzoic acid so react with NaOH with a faster rate than benzoic acid.

18. (2) —OH and —COOH both the groups react with Na metal and per group 1 mole of Na is utilised and 0.5 mole of H_2 is released. So, with 4 moles of Na, H_2 released is $4 \times 0.5 = 2$ mole.

19. (1) Na will attack at the most acidic position and a carboxylic acid is usually more acidic as compared to phenol.

20. (6) Since, 1 mole of $NaHCO_3$ reacts with 1 mole of —COOH. Here (in product *Y*) 6—COOH groups are present, so 6 moles of $NaHCO_3$ are required.

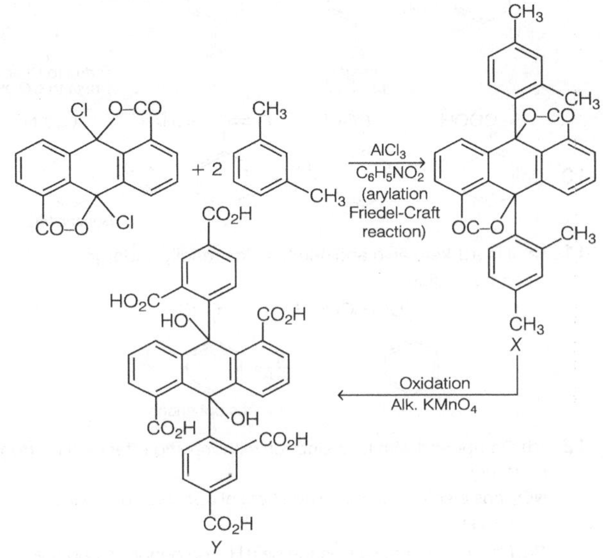

Revisal Problems (JEE Main)

1. (d) Alkaline $KMnO_4$ oxidises the carbon chain attached directly with benzene nuclei into —COOH group.

2. (b)

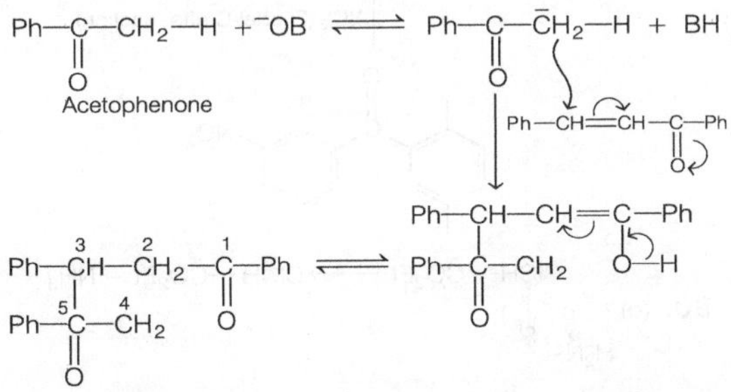

3. (b) The aceto group (CH_3CO—group) in phenyl methyl ketone has $-R$ effect just like —COOH, so gives *meta* product on nitration like benzoic acid.

4. (b) In the presence of acid, acetophenone reacts with H_2O^{18} to give unstable gem-diol, which dehydrated to give O^{18} incorporated acetophenone.

5. (c)

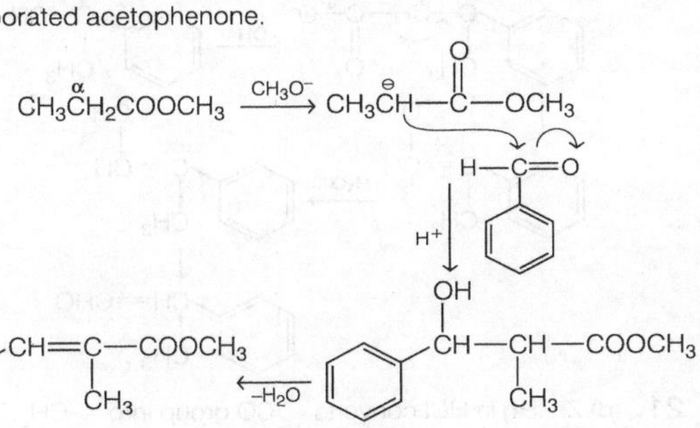

6. (a) Presence of an electron withdrawing group makes the carbonyl group more reactive towards nucleophilic addition reaction by decreasing electron density at carbonyl carbon. (i.e. K_{eq} is more).

7. (b) An aldehyde is more reactive towards nucleophilic addition reaction, so $C_6H_5CH_2CHO$ undergo hydration to a greater extent.

8. (b)

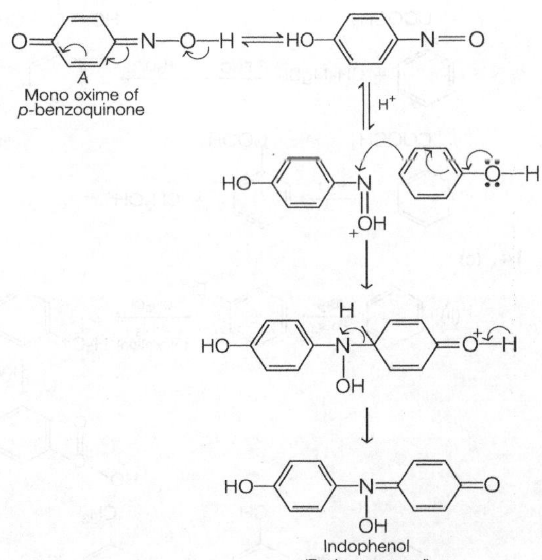

9. (b)

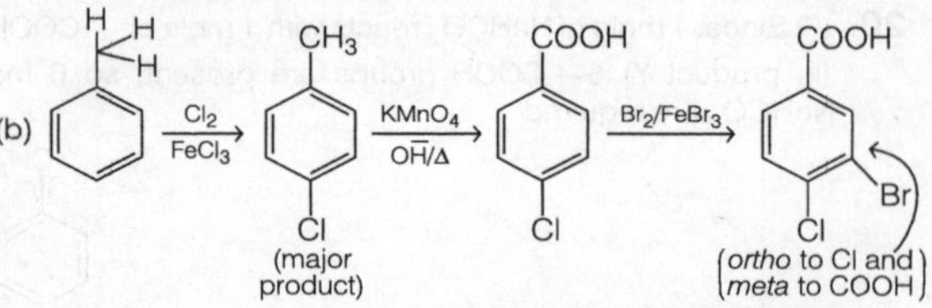

10. (a)

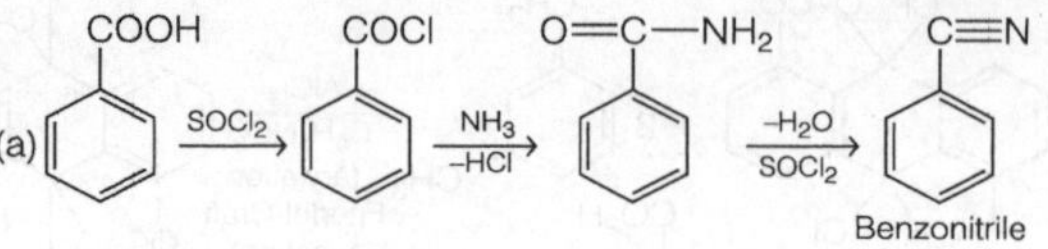

11. (a) It is a β-keto acid and β-keto acids readily undergo decarboxylation.

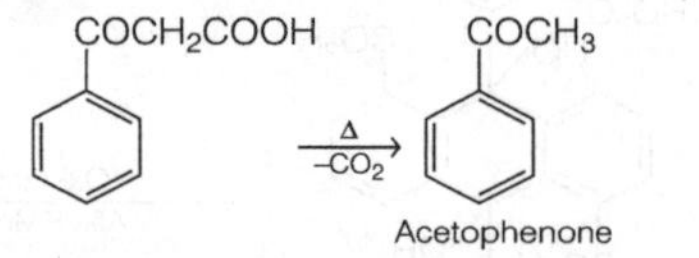

12. (d) Compound with less electron withdrawing effect will be good H^- donor.

NO_2 has electron withdrawing effect at *o*, *m*- and *p*- positions.

So, Ph—C(H)(O⁻)—O⁻ dianions is the best H^- ion donor among the given.

13. (c)

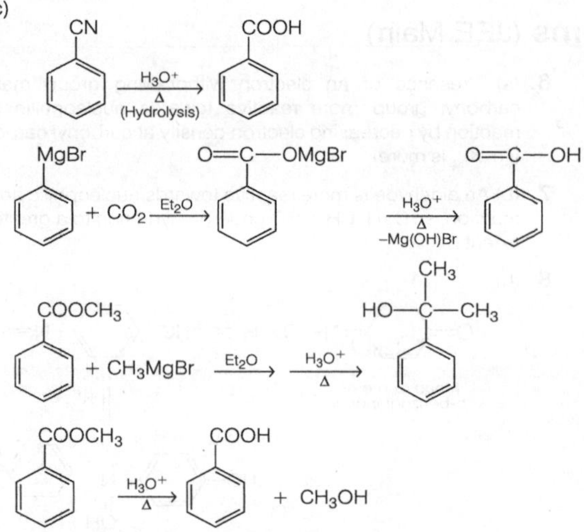

14. (c)

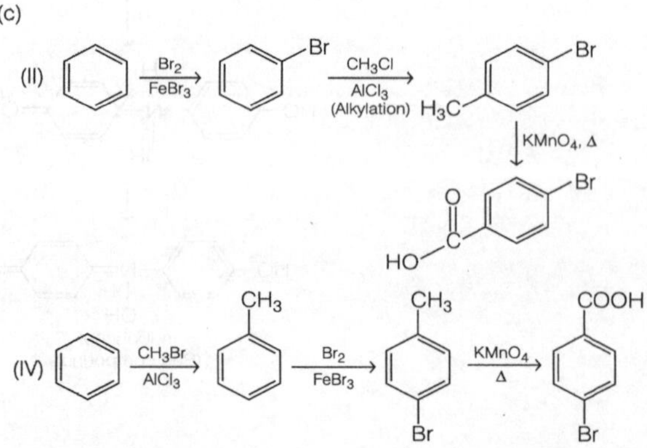

15. (a) Tischenko reaction on benzaldehyde in the presence of anhy. $Al(OEt)_3$ gives benzyl benzoate.

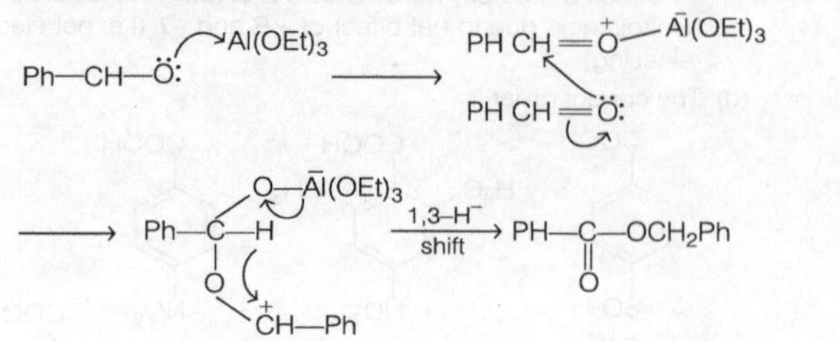

16. (d) To prepare *p*-substituted acid, first such *o*/*p* directing group is introduced which is then oxidised to —COOH group.

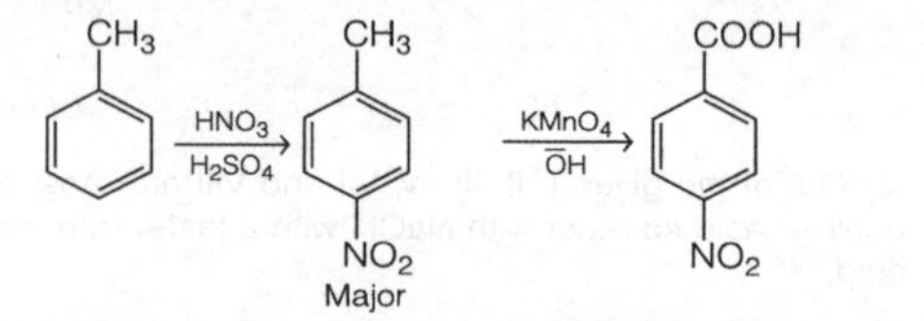

17. (b) To prepare *p*-product, first a *o*/*p* directing group like alkyl group $(CH_3)C$—, CH_3 etc., has to be introduced and then acylation is done to obtain *p-tert*-butyl acetophenone.

18. (b) Here $\overset{\ominus}{O}H$ attacks at CHO of C-1, not at C=O of C-2. All other given statements are true.

19. (b)

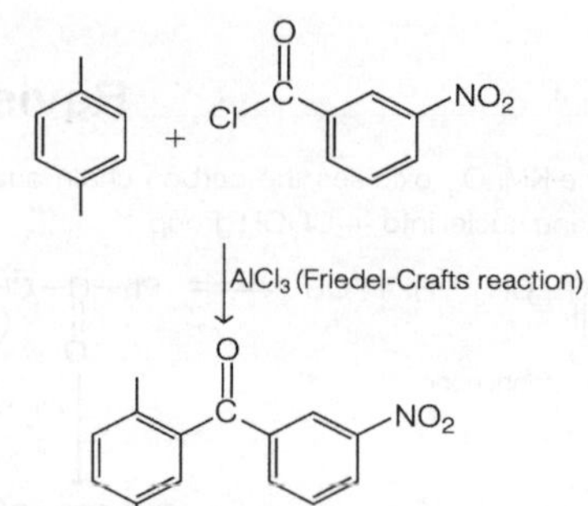

20. (c)

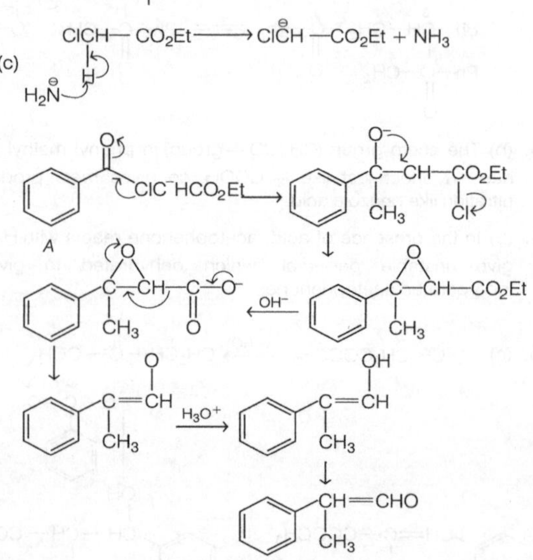

21. (d) Zn-Hg in HCl converts >CO group into $—CH_2$. This reaction is called Clemmensen reduction.

22. (b) Degree of unsaturation in

$$A = \frac{(2n_c + 2) - n_H}{2} = \frac{(2 \times 7 + 2) - 6}{2} = \frac{16 - 6}{2} = 5°$$

5 degree of unsaturation shows the presence of a benzene nuclei (4 degree of unsaturation) and one —CHO group (1-degree of unsaturation). Thus, *A* is benzaldehyde.

It forms oxime as

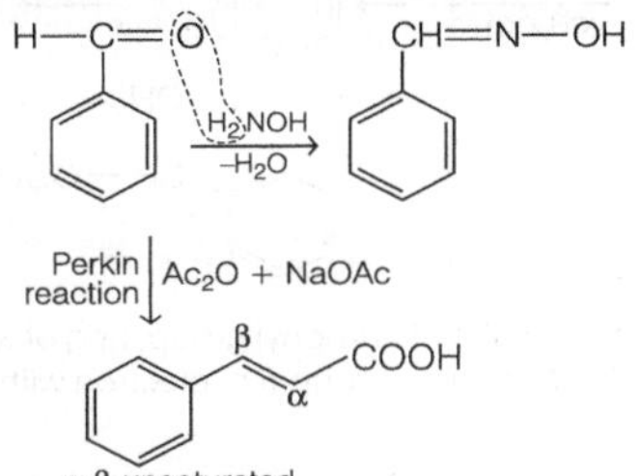

α,β-unsaturated acid (cinnamic acid)
B

Because of the presence of unsaturation, i.e. double bond, it decolourises $KMnO_4$. Further gives effervescence with $NaHCO_3$ due to the presence of —COOH group.

23. (a) Oxidation of *A* introducing four oxygen indicating that there are two alkyl groups on benzene ring which on oxidation giving benzenedicarboxylic acid. This establishes that the two alkyl groups are none other than two methyl group. Also, *B* on heating loses a molecule of water and forms C, which is an anhydride, indicates that the two carboxylic groups are on adjacent carbons of benzene ring. Hence, *A*, *B* and *C* are

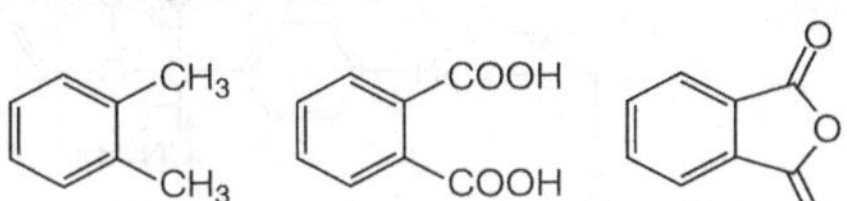

A *B* *C*

24. (b)

(a) A trimerisation followed by enolisation reaction.

(b) Repeated Aldol condensation, but the final product is

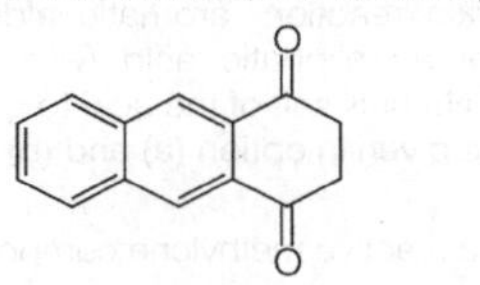

(c)

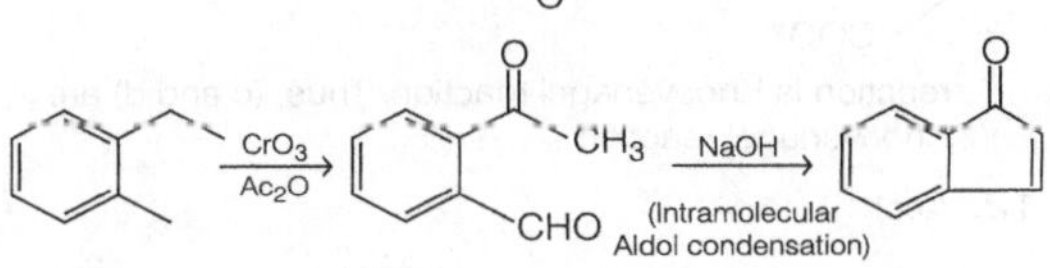

(d)

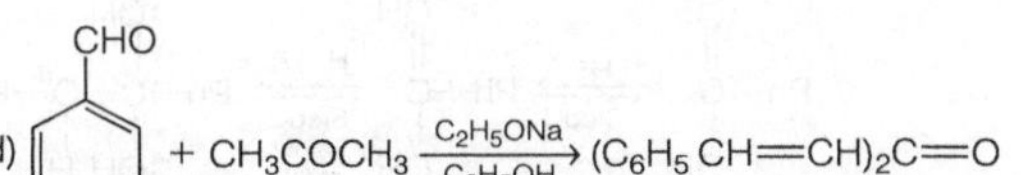

Repeated Aldol condensation

25. (a) As —CH_3 group has a strong + *I* effect and —OCH_3 has a weak –*I* and strong +*R* (net electron releasing) so they increase the electron density on oxygen atom and makes the O—H bond stronger.

On the other hand, —NO_2 has a strong –*I* and –*R* effect, so, it withdraws electron from Ph as well as oxygen atom of —OH and facilitates the removal of a proton.

Thus, the order of esterification is

$$I > II > III > IV$$

26. (b) (i) → (q), (ii) → (p), (iii) → (s), (iv) → (r)

Allylic bromination takes place by free radical mechanism.

Claisen condensation involves the formation of carbanion. In Fries rearrangement acyl (—$COCH_3$) group migrates while in Woff rearrangement alkyl group migrates.

27. (a) (i) → (s), (ii) → (p), (iii) → (q), (iv) → (r)

(i). Aldol condensation

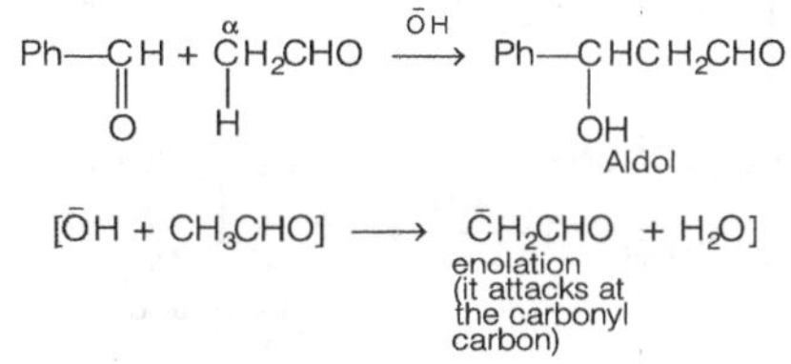

(ii) Cannizaro reaction

Transfer of H^- ion is involved.

(iii) In Reformatsky reaction, organo zinc compound is used.

(iv) In Benzoin condensation, CN^- acts as a catalyst.

28. (b) In the presence of —NO_2, e^- density on the —OH decreases, and hence, intramolecular electrophilic attack becomes difficult, so ring closure does not take place.

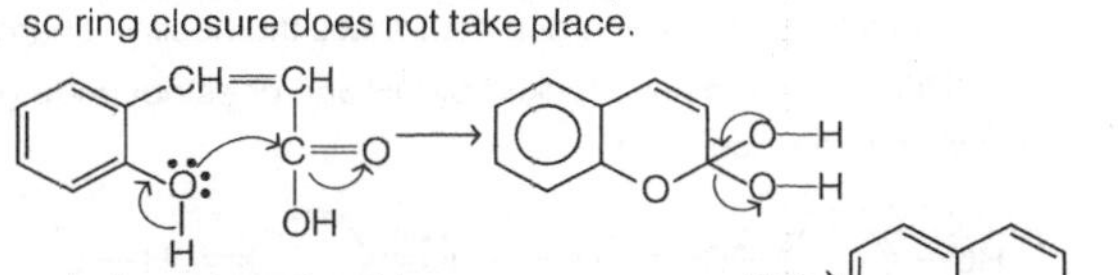

o-hydroxy cinnamic acid

Coumarin

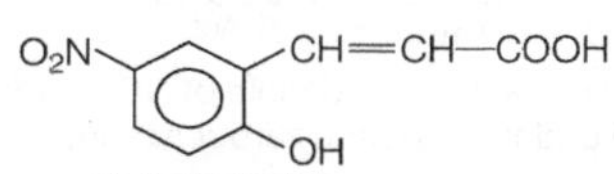

No lactonisation

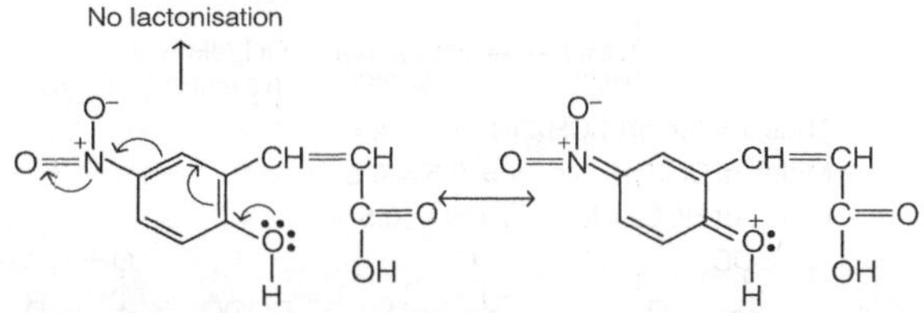

29. (a) Because of the presence of acidic hydrogen at carbonyl carbon, it (oxime of benzaldehyde) readily loses water (prior to rearrangement) to give benzonitrile instead of normal N-substituted amide.

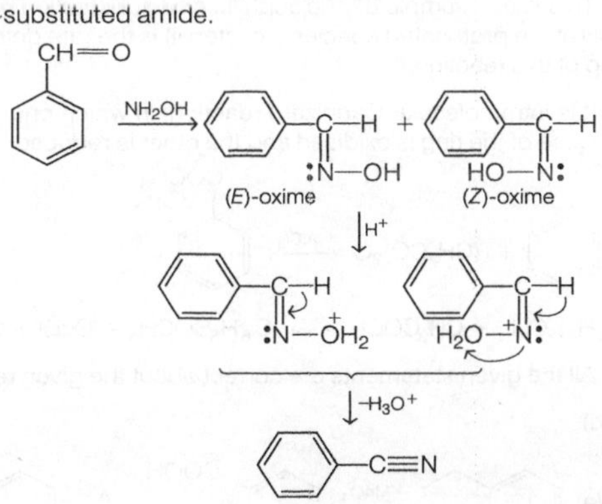

Benzoritrile

Revisal Problems (JEE Advanced)

1. (c)

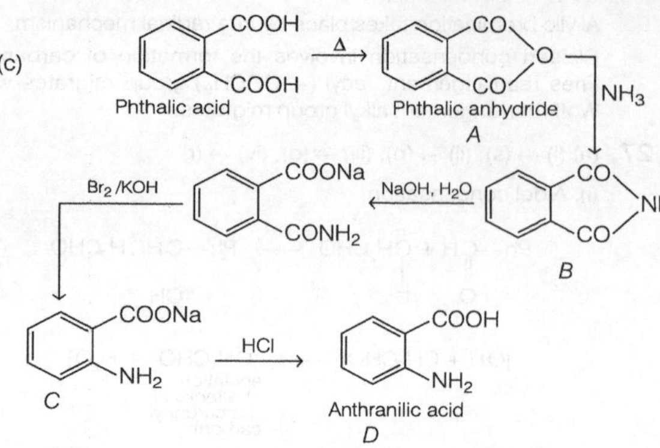

2. (c)

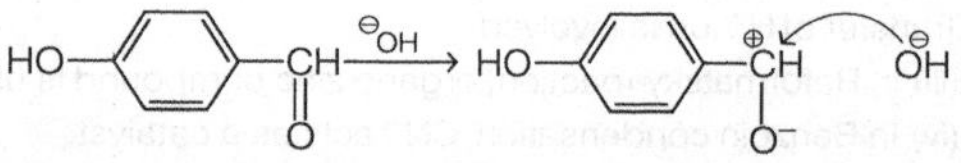

As the positive charge increases on the C-atom of CHO group, nucleophile $\bar{O}H$ will readily attack to it but the resonance effect of electron releasing —OH group neutralises the positive charge of —CHO, hence it is not attacked by $\bar{O}H$ and does not undergo Cannizaro reaction.

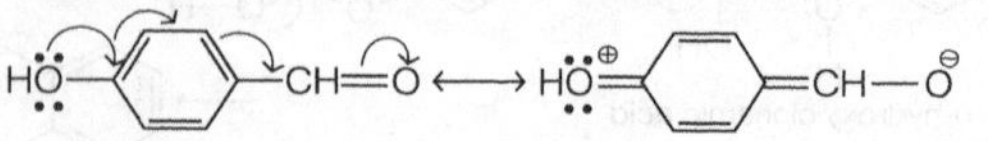

3. (c) $\underset{1\,mol}{HCHO} + \underset{1\,mol}{PhCHO} \xrightarrow{\bar{O}H} \underset{1\,mol}{PhCH_2OH} + \underset{1\,mol}{HCOONa}$

Since, 2 moles of HCHO is taken, so one mole of HCHO remains after the reaction. It also undergoes Cannizaro reaction.

$$\underset{1\,mol}{HCHO} \xrightarrow{\bar{O}H} \underset{0.5\,mol}{HCOONa} + \underset{0.5\,mol}{CH_3OH}$$

Thus, moles of $PhCH_2OH$ i.e. $x = 1$

Moles of CH_3OH i.e. $y = 0.5 = 0.5$

Moles of HCOONa, i.e. $z = 1 + 0.5 = 1.5$

4. (c)

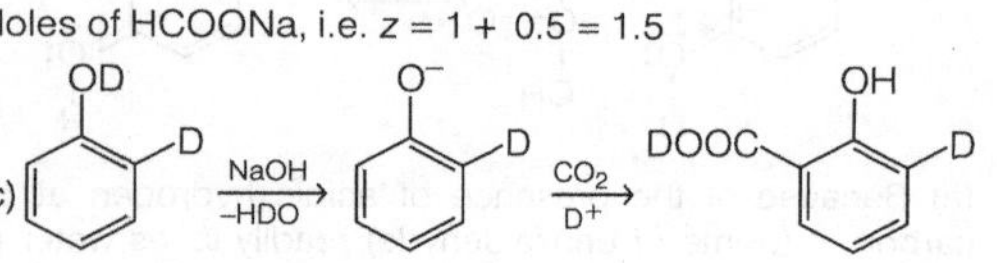

5. (b) This is an example of nucleophilic acyl substitution. Attack of ROH at the protonated species (i.e. step ii) is the rate determining step of this reaction.

6. (c) It is intramolecular Cannizaro reaction, in which one group of one side of the ring is oxidised and the other is reduced.

7. (b)

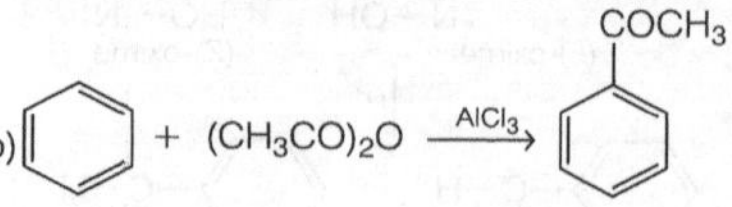

$(C_6H_5)_2CuLi + CH_3COCl \longrightarrow C_6H_5COCH_3 + CuCl + C_6H_5Li$

8. (d) All the given statements are correct about the given reaction.

9. (a,d)

(a)

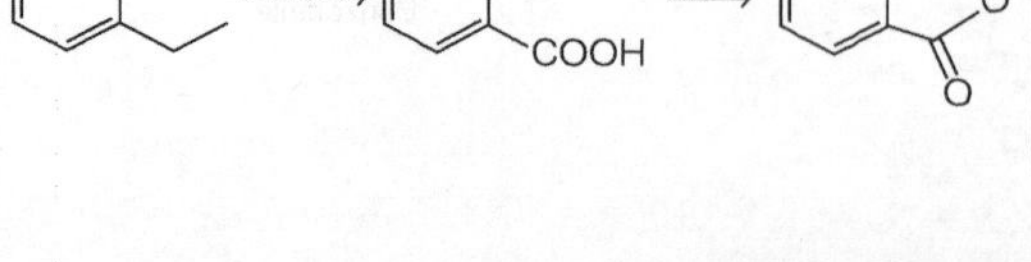

(b,c) Since —CHO or —NO_2 deactivates the ring, so electrophile does not attack the ring.

(d)

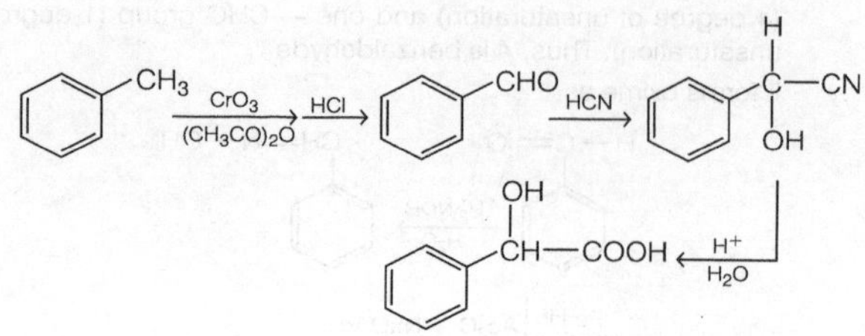

10. (a,c) CN^- attacks at that carbonyl group, ring of which is activated by a $-I$ and $-R$ showing group (i.e. electron withdrawing group).

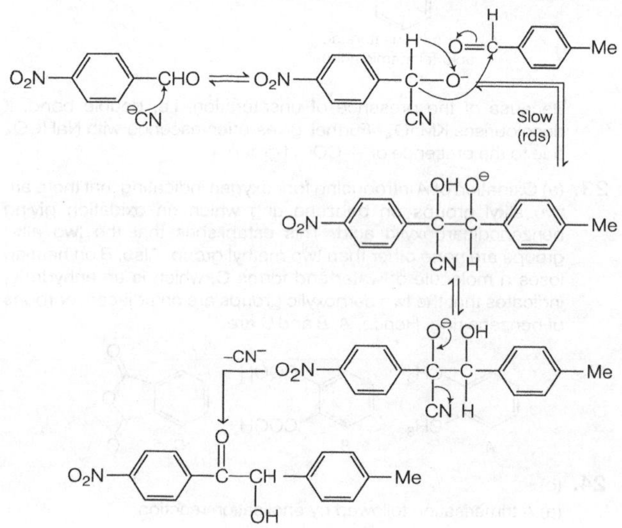

11. (a,b) In Perkin reaction, aromatic aldehydes are heated with anhydride of an aliphatic acid (with two α-H atoms) in the presence of Na or K salt of the acid to give α, β-unsaturated acid. So, reactions given in option (a) and (b) are Perkin reactions.

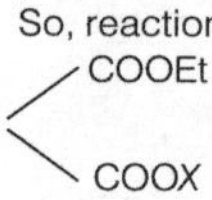 is a active methylene compound, so in its case the reaction is Knoevenagel reaction. Thus, (c and d) are Knoevenagel reaction.

12. (a,c)

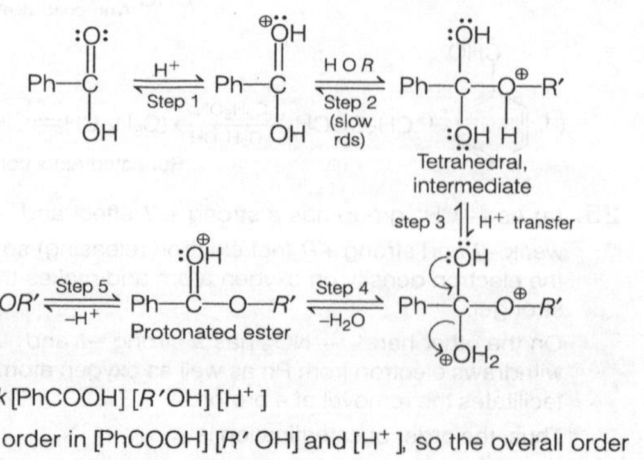

Rate = k[PhCOOH] [R'OH] [H^+]

i.e. first order in [PhCOOH] [R'OH] and [H^+], so the overall order = 3 (third order reaction).

13. (a,c)

(a)

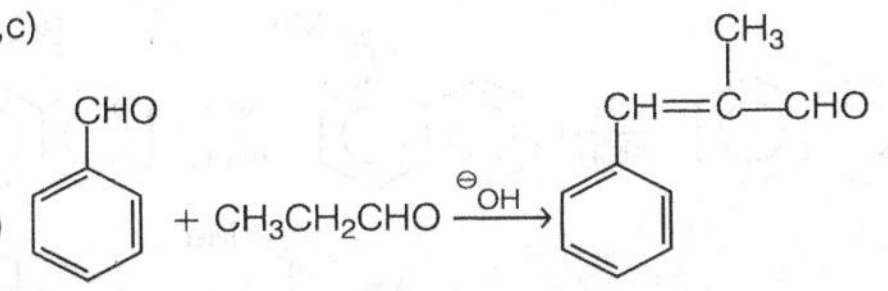

(b) Perkin condensation, correct reaction.

(c) Here cyanohydrin is formed, benzoin is formed with NaCN.

(d) Zn reduces CHO into —CH(OH)—

14. (b) When $R = C_2H_5$

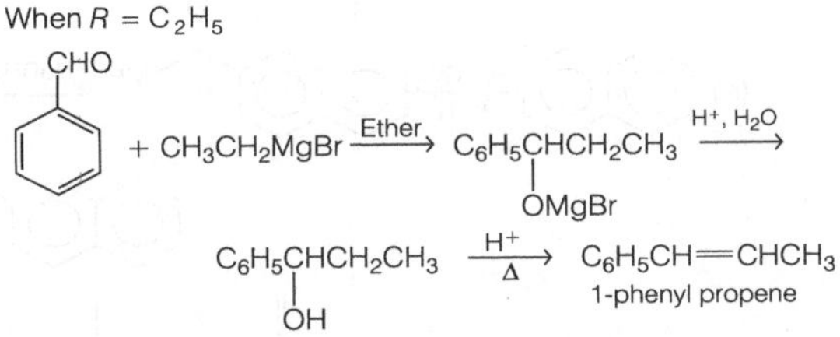

When $R = CH_3$

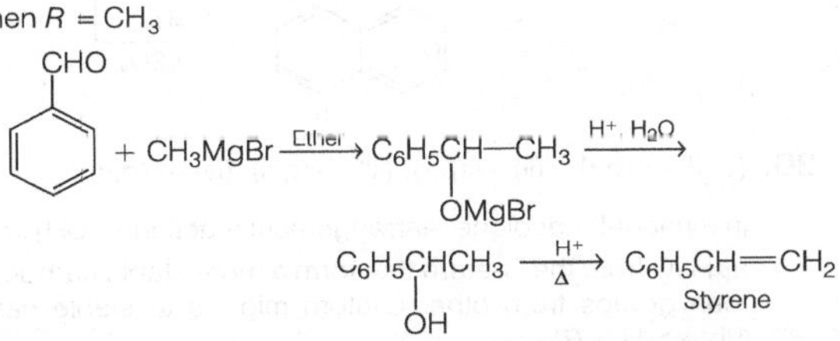

15. (a,b,c)

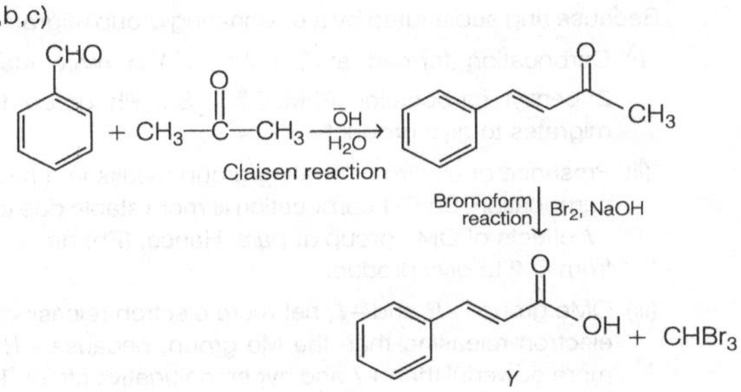

Because of the presence of a double bond (unsaturation) *Y* decolourises Br_2 water and 1% alkaline $KMnO_4$. Being a strong acid, it reacts with all Na, NaOH and $NaHCO_3$.

16. (a) Dichromate is a good oxidising agent, so directly oxidises CH_2OH group into —COOH *via* the formation of —CHO as intermediate.

17. (d) 2,6-$Me_2C_6H_3COMe$ does not give iodoform with I_2/NaOH (although it contains a —COMe group). This is because it cannot form the necessary tetrehedral intermediate due to strong steric interaction offered by *o*-methyl groups.

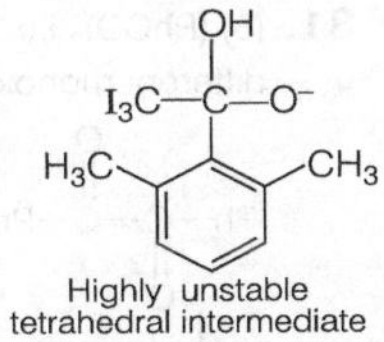

Highly unstable tetrahedral intermediate

18. (a) (E)-oxime of $Ph_2CHCOCH_3$ does not undergo normal Beckmann rearrangement in the presence of conc. H_2SO_4, due to the formation of a highly stable diphenylmethyl cation instead of usual migration of anti-group.

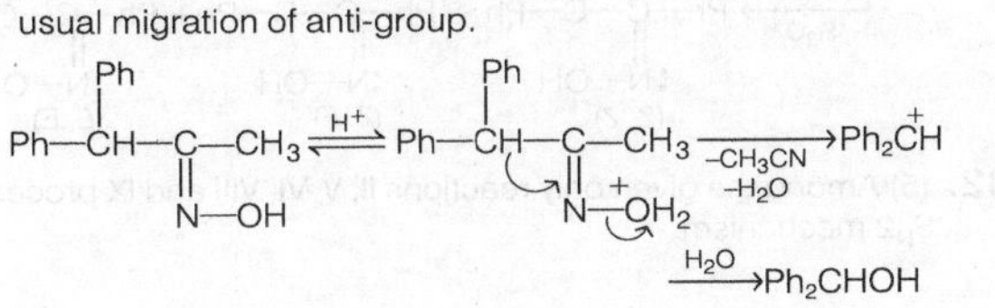

Explanations for Q. Nos. (19-21)

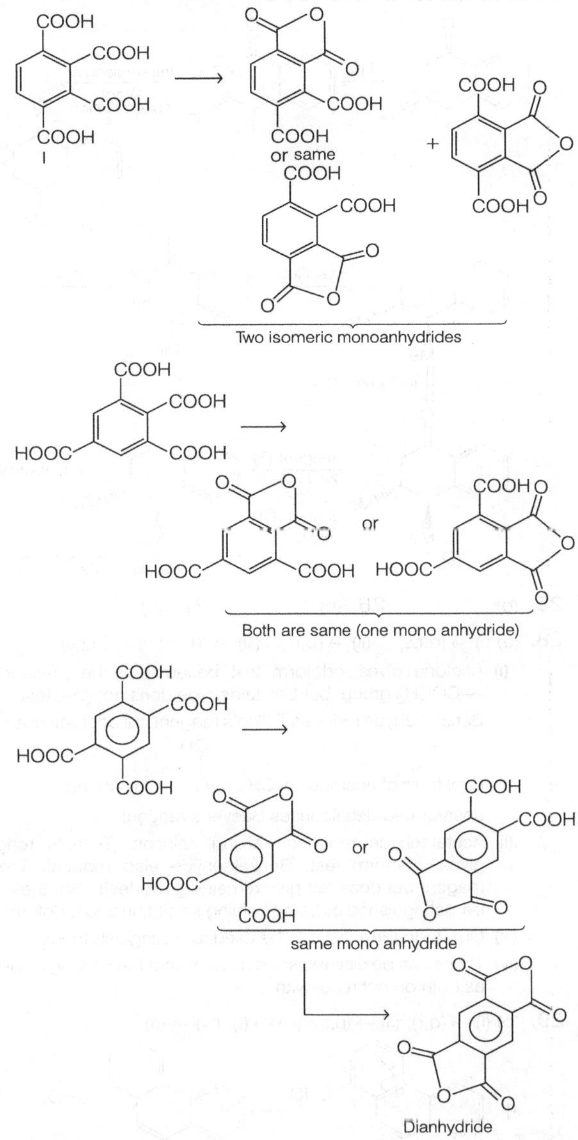

19. (a)

20. (d)

21. (c)

Explanations for Q.Nos. (22-24)

The given reaction is called Friedel-Crafts acylation, so the reagents used are $CH_3COCl/AlCl_3$. $LiAlH_4$ reduces amide into amine.

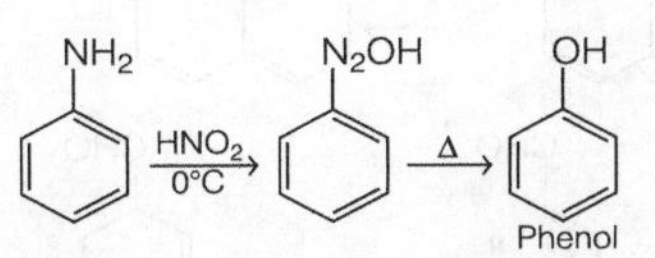

22. (b) **23.** (c) **24.** (b)

Explanations for Q. Nos. (25-27)

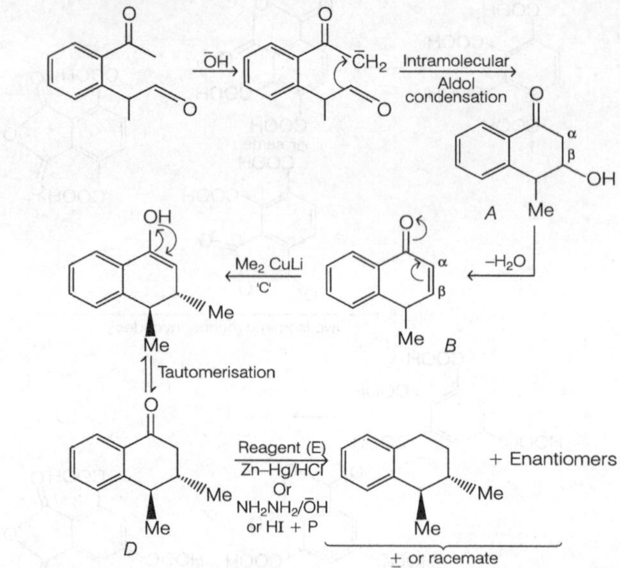

25. (b) **26.** (b) **27.** (d)

28. (a) (i) → (q,r,s), (ii) → (p,r), (iii) → (r), (iv) → (q,r,s)

(i) Acetone gives iodoform test because of the presence of $-COCH_3$ group, but benzaldehyde does not give test.
Benzaldehyde reduces Tollen's reagent but acetone does not.
Enol form of acetone i.e. $CH_2=\overset{OH}{\overset{|}{C}}-CH_3$ being unsaturated decolourises Baeyer's reagent.

(ii) Acetaldehyde reduces Fehling solution, Tollen's reagent, gives iodoform test. Benzaldehyde also reduces Tollen's reagent but does not give remaining two tests. So, these can be distinguished by using Fehling's solution and iodoform test.

(iii) Only iodo form test can be used to distinguish them.

(iv) These can be distinguished by all except the Fehling's solution as both do not react with it.

29. (b) (i) → (q,r), (ii) → (p,s), (iii) → (t), (iv) → (u)

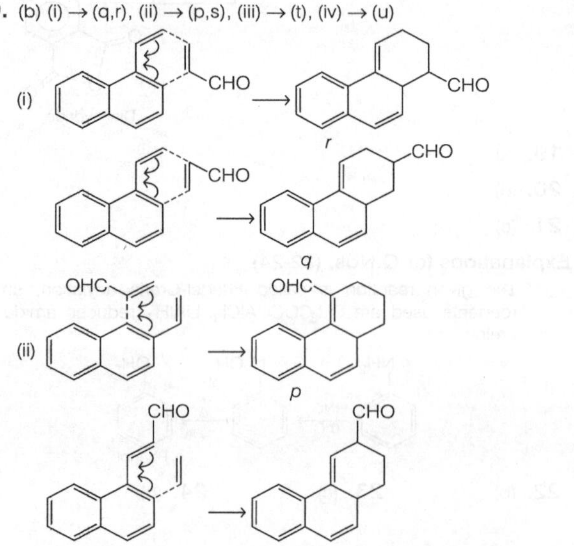

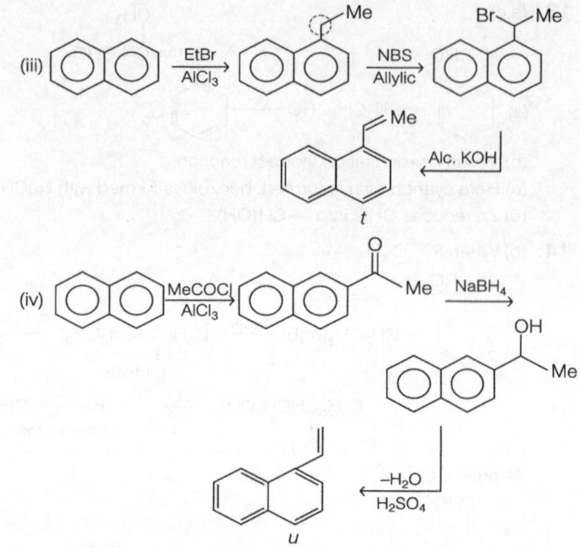

30. (c) (i) → (p,r) (ii) → (p,r), (iii) → (q,s), (iv) → (q,s)

In Pinacol-Pinacolone rearrangement reaction —OH group is lost as H_2O from the C-atom that form a more stable carbocation and then groups from other C-atom migrate to stable carbocation (Ph >> H > *R*).

Because ring substituted by a e^- donating group migrates readily.

(i) Carbocation formed at C-1 (Ph_2C^+) is more stable than 2° benzyl carbocation $PhMeC^{\oplus}$. So, Ph group from C-2 migrates to give product.

(ii) Presence of electron donating group results in a more stable carbocation. So C-1 carbocation is more stable due to $+R$ and $-I$ effects of OMe group at *para*. Hence, (Ph) group migrates from C-2 to give product.

(iii) OMe group ($+R$ and $-I$, net more electron releasing) is more electron releasing than the Me group, because $+R$ effect is more powerful than $+I$ and hyperconjugative effect. Thus, C-2 carbocation is more stable and the group *X* migrates to C-2 to give product.

(iv) C-2 carbocation is more stable (diphenyl$C^{\oplus}$ with $+I$ and hyper conjugation of Me group at *para*) than $PhMeC^{\oplus}$ (with $-I$ effect of Cl at *para*). So, the group *X* migrates from C-1 to form product.

31. (5) $(PhCO)_2$ i.e. Benzil can form three different dioximes and two different monoximes as

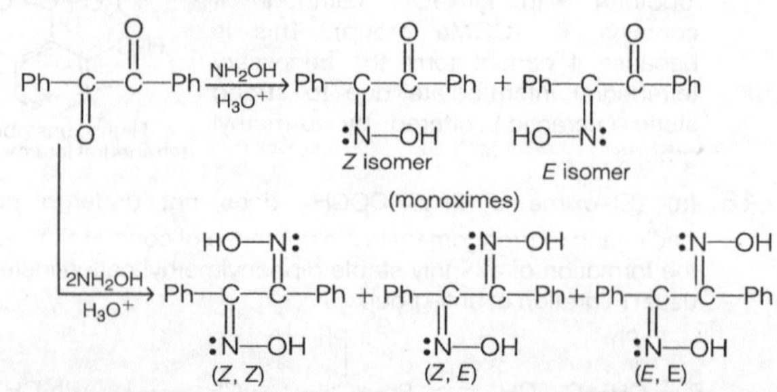

32. (5) Among the given only reactions II, V, VI, VIII and IX proceed by S_N2 mechanism.

JEE Main & AIEEE Archive

1. (a)

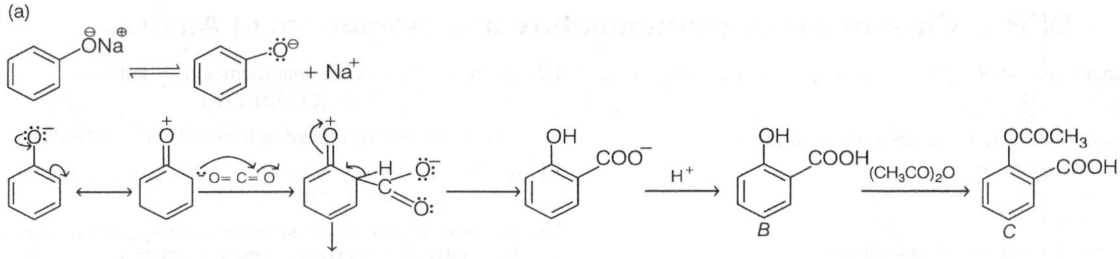

2. (c)

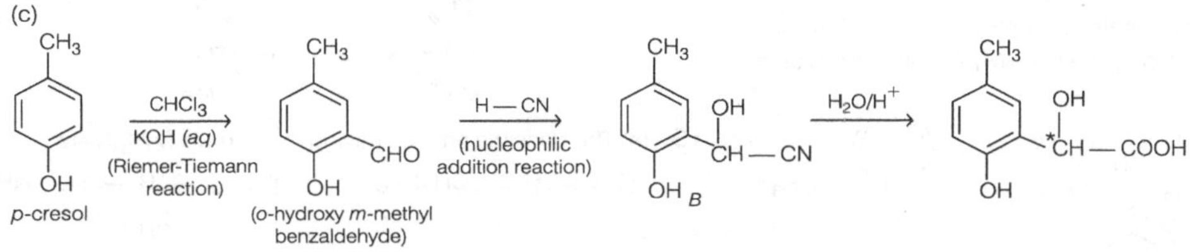

3. (d)

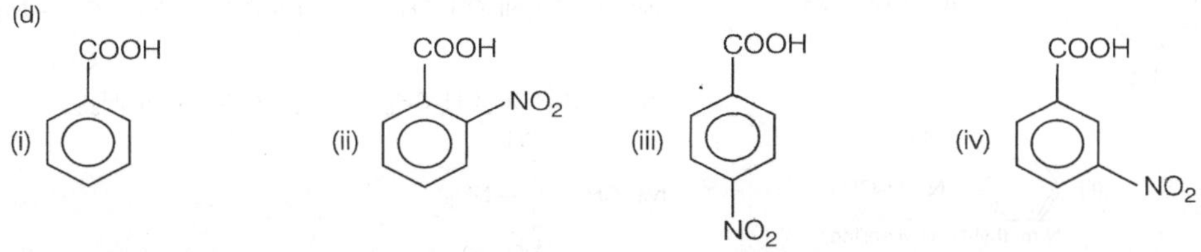

$-NO_2$ group at any position shows electron withdrawing effect, thus acidic strength is increased. But *o* -nitro benzoate ion is stabilised by intramolecular H-bonding like forces, hence its acidic strength is maximum. Thus, the order of acidic strength is (ii) > (iii) > (iv) > (i).
The effect is more pronounced at *para* position than *meta*.

JEE Advanced & IIT JEE Archive

Explanations for Q. Nos. (1 and 2)

The first step of reaction is Perkin's condensation.

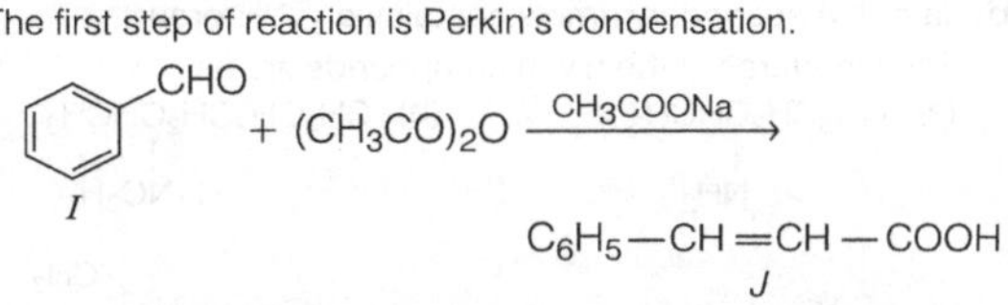

J being a carboxylic acid gives effervescence with $NaHCO_3$. Also, *J* has olefinic bond, it will decolourise Baeyer's reagent.

In the second step, *J* on treatment with $H_2 / Pd / C$ undergo hydrogenation at olefinic bond only as :

$$J + H_2 / Pd \longrightarrow C_6H_5—CH_2—CH_2—COOH$$

The hydrogenated acid, on treatment with $SOCl_2$ gives acid chloride.

$$C_6H_5—CH_2—CH_2—COOH + SOCl_2 \longrightarrow C_6H_5—CH_2—CH_2—COCl + HCl + SO_2$$

In the final step, acid chloride formed above undergoes intramolecular Friedel-Crafts acylation as

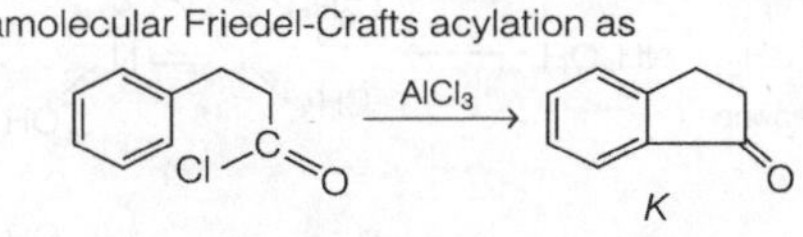

1. (c)

2. (a)

3. (c)

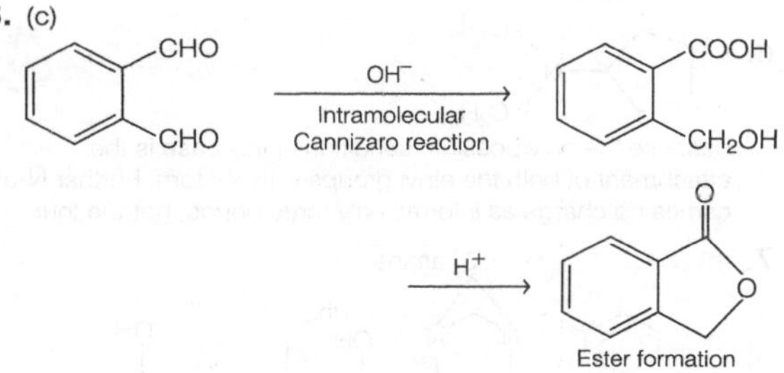

4. (a)

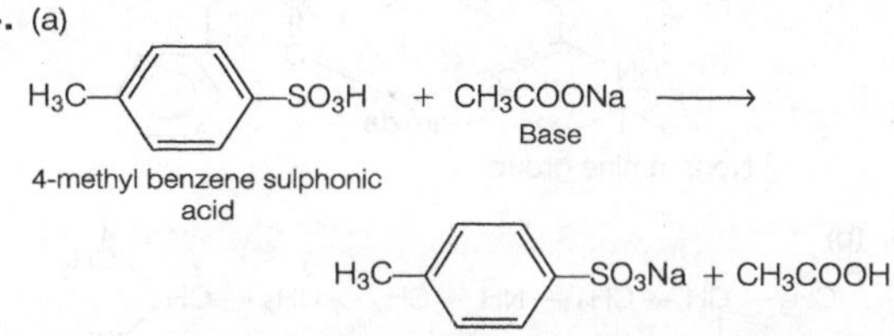

3. Aliphatic Nitrogen Containing Compounds

DPP-1 Classification, Nomenclature and Isomerism of Amines

1. 1° amines have —NH_2 group, 2° have >NH group and 3° have >N group, —C(=O)—NH— group is an amide group.
(i) Both are amide groups
(ii) 3° amine, 2° amine and 2° amide
(iii) 3° amine
(iv) 3° amine, imine, amide (substituted)

2. (i) Dimethyl ammonium carbonate (ii) N-ethylcyclopentylamine
(iii) N, N-dicyclopropyl cyclopropanamine (iv) 2-amino pyrimidine

3.

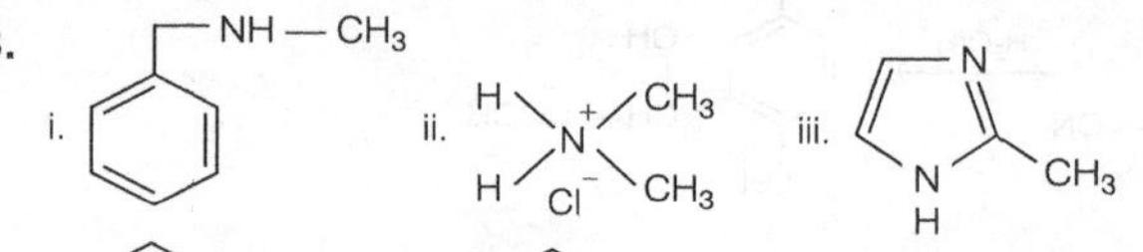

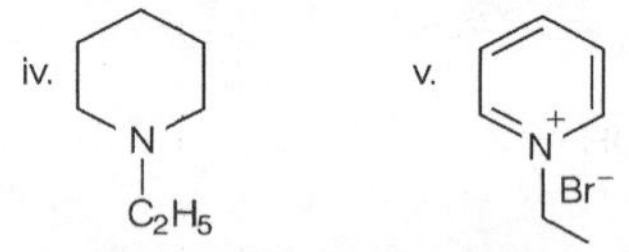

vi. $(C_2H_5)_4N^+OH^-$

4. (i) CH_3—CH(CH_3) $NHCH_3$
N-methyl-2-aminopropane

(ii)

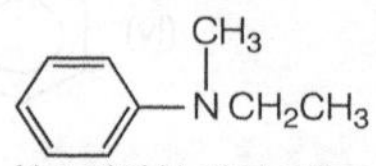

N-methyl-N-ethyl aniline

(iii)

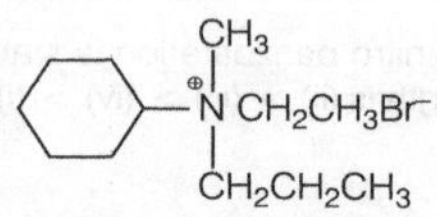

N-methyl-N-ethyl-N-propyl cyclohexyl ammonium bromide

5. (c) In this molecule, there are 3 bond pairs and a lone pair, so hybridisation is sp^3 because of the involvement of 4 pairs of electrons.
However, presence of a lone pair, changes its geometry from tetrahedral to pyramidal.

6. (c)

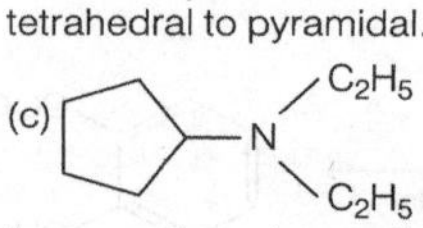

because the only possible condition in this case is the attachment of both the ethyl groups with N-atom. Further N-atom carries no charge as it forms only three bonds, not the four.

7. (b)

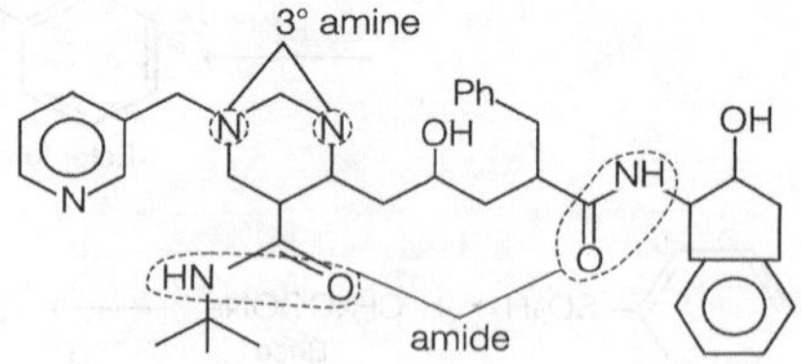

8. (b) CH_3—CH(CH_3)—CH_2—NH—$\overset{1}{C}H_2$—$\overset{2}{C}H_2$—$\overset{3}{C}H$($\overset{4}{C}H_3$)CH_3

3-methyl –N– (2-methylpropyl)–1–butanamine

9. (b) Tertiary (3°) amines have >N group, but no charge. Charge is present only in case of quarternary ammonium salts.

10. (c) The structure of tetramethylammonium iodide is

$$(CH_3)_3N^+CH_3I^-$$

Thus, it can also be named as trimethylaminemethiodide.

11. (d) Only quarternary ammonium salt because of hindrance are capable of resolution into enantiomers.

12. (c) The N-atom of ethylmethylamine undergoes fast inversion, so it cannot be resolved under normal conditions.

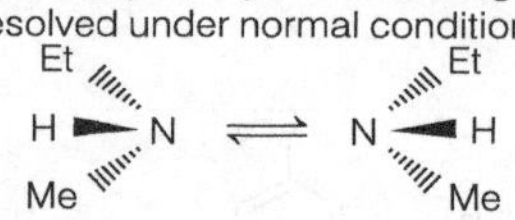

13. (a) The possible amines having the formula $C_4H_{11}N$ are
(i) $CH_3CH_2CH_2CH_2NH_2$ (ii) CH_3—CH(CH_3)—CH_2—NH_2
(iii) $CH_3CH_2NHCH_2CH_3$ (iv) CH_3—CH_2—CH(CH_3)—NH_2
(v) $CH_3NH\ CH_2CH_2CH_3$ (vi) $CH_3CH_2N(CH_3)_2$
(vii) CH_3—C(CH_3)$_2$—NH_2

Out of the above (i), (ii), (iv) and (vii) are 1° amines and (vi) is a tertiary (3°) amine.

14. (c) α, β-unsaturated amines are called enamines. Thus, only ii is an enamine.

15. (a,b,c) Although compounds given in (a), (b) and (c) also has chiral centre but they undergo fast inversion due to the small energy barrier between the two arrangements (enantiomers).

16. (a,c,d) A secondary amine contains a >NH group.
The structures of the given compounds are as
(a) $CH_3CH_2CH(NH_2)CH_3$ (b) $CH_3CH_2CH_2CH(HNC_2H_5)CH_3$

(c) (d)

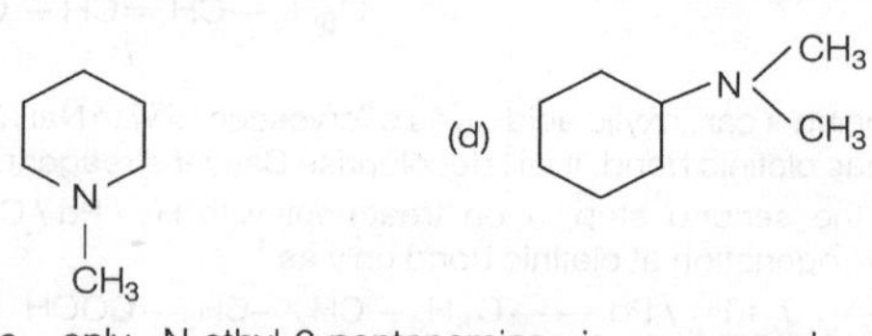

Thus, only N-ethyl-2-pentanamine is a secondary amine. 2-butanamine is a 1° amine while (c) and (d) are 3° amines.

17. (b,c,d) If *R* is asymmetric like acetophenone, benzaldehyde and acetaldehyde, the oxime formed is also asymmetric and exhibits geometrical isomerism. e.g.

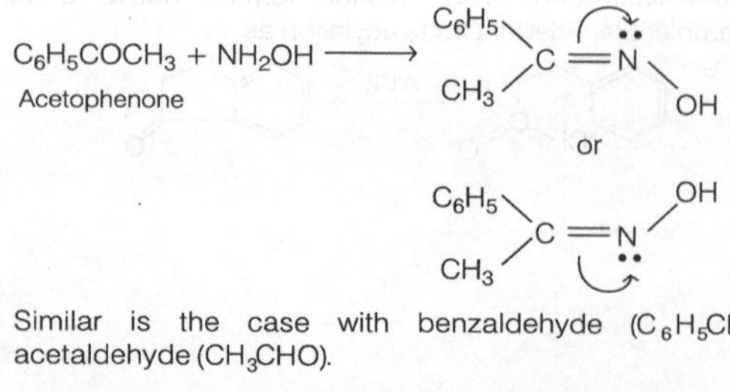

Similar is the case with benzaldehyde (C_6H_5CHO) and acetaldehyde (CH_3CHO).

18. (a)

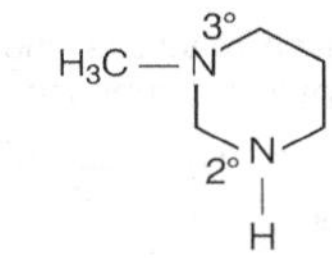

It contains a 3° and a 2° amino group so, it is classified as 3° (tertiary) as well as 2° (secondary) amines.

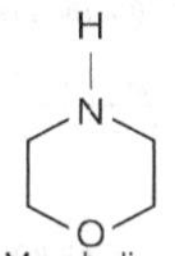

Morpholine

It contains a >NH group. So, classified as a 2° (secondary) amine.

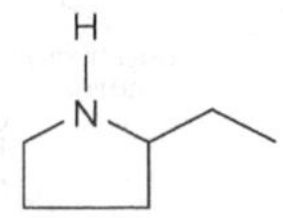

2-ethylpyrolidine

It also contains only a secondary amine group so, it is a secondary amine.

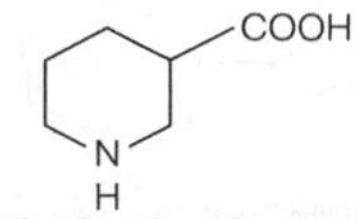

Piperidine-3-carboxylic acid

It is also a secondary (2°) amine.

19. (2) The possible structure is

$$\overset{1}{CH_3}\ \underset{NH_2}{\overset{2}{CH}}\ \underset{CH_3}{\overset{3}{CH}}\ \underset{CH_3}{\overset{4}{CH}}\ \overset{5}{CH_3}$$

Thus, $x = 2$

(Here, NH_2 group can occupy only 2 second position because according to IUPAC nomenclature NH_2 group got preference as compared to R group).

20. (2) The compound in which all the four valencies of carbon are satisfied by four different groups are generally considered as optically active. The possible optically active 1°amine of $C_5H_{13}N$ are n-$C_3H_7C(CH_3)H\ NH_2$, $C_2H_5C(H)(CH_3)CH_2NH_2$.

DPP-2 General Methods of Preparation of 1° Amines and Other than 1° Amines

1. (i) $A = RCH_2NH_3^+$

When treated with a base, A gives RCH_2NH_2

$$RCH_2NH_3^+ \xrightarrow[-H_2O]{OH^-} \underset{1^\circ \text{ amine}}{RCH_2NH_2}$$

(ii)

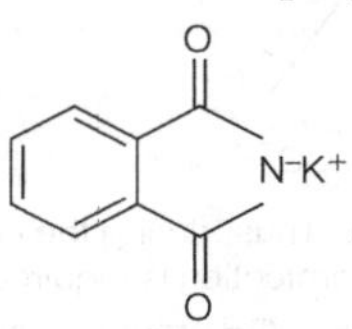

acts as a nucleophile. This reaction takes place by S_N2 mechanism.

2. (i).

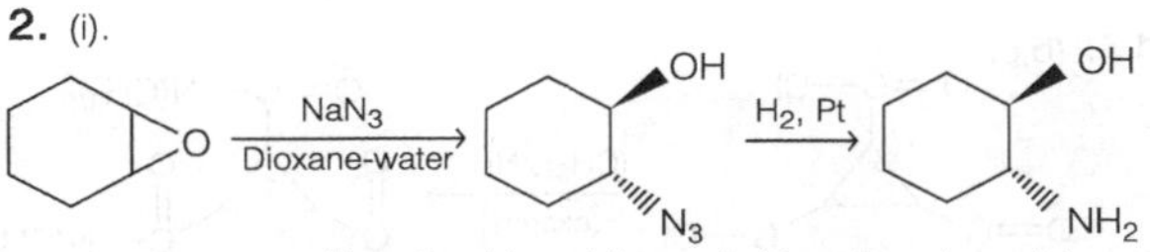

Trans-2- azidocyclohexanol *A*   *Trans*-2- aminocyclohexanol *B*

(ii)

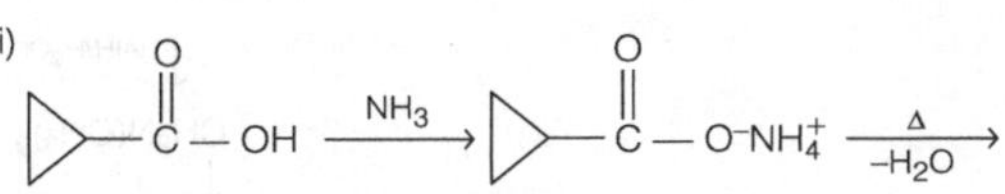

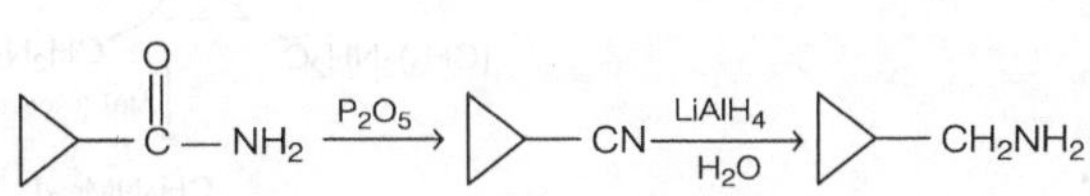

C   *D*   *E*

3. (i)

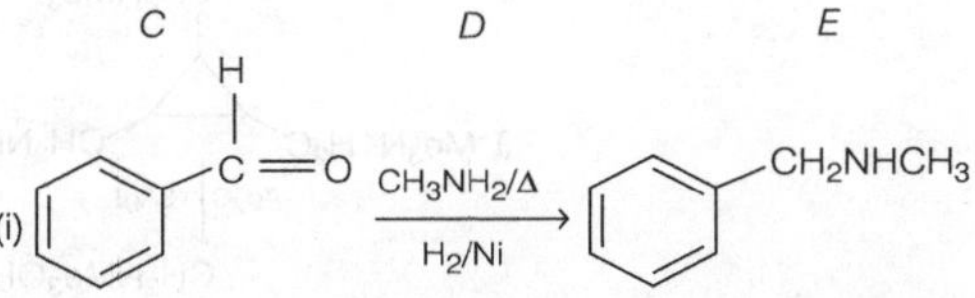

(ii)

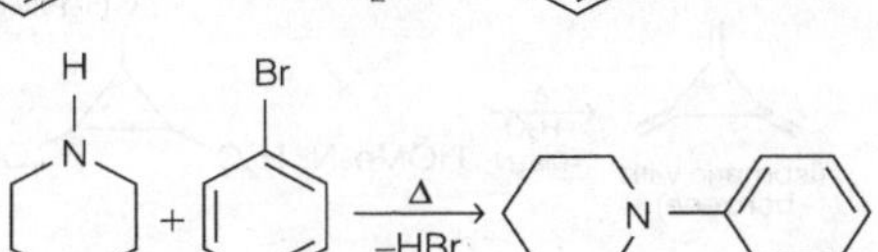

(iii)

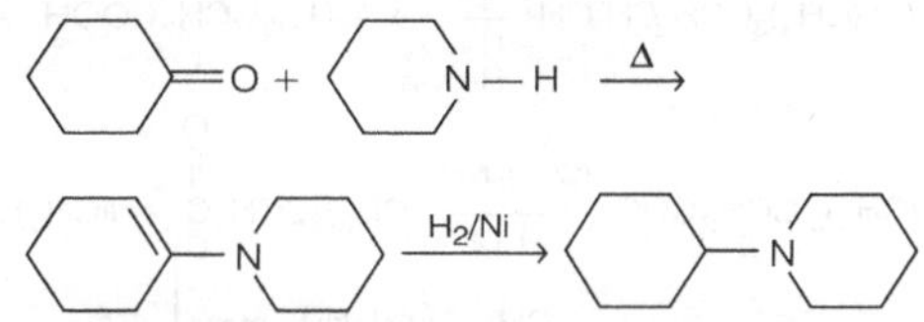

4. (a) Here only Cl group is replaced by H, so the product is

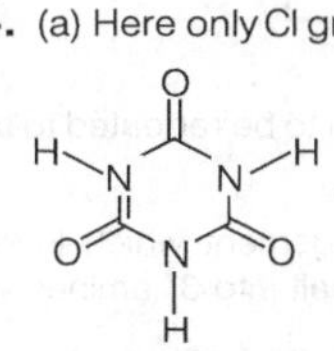

with 12σ and 3π bond i.e. σ/π ratio is 4 : 1

5. (b) $Ph—\underset{}{\overset{OH}{CH}}—\overset{O}{\overset{\|}{C}}—NH_2 \xrightarrow[KOH]{Br_2}$

$$\underset{\text{Unstable}}{Ph—\overset{OH}{CH}—NH_2} \xrightarrow{-NH_3} \underset{\text{Benzaldehyde}}{Ph—\overset{O}{\overset{\|}{C}}—H}$$

6. (b)

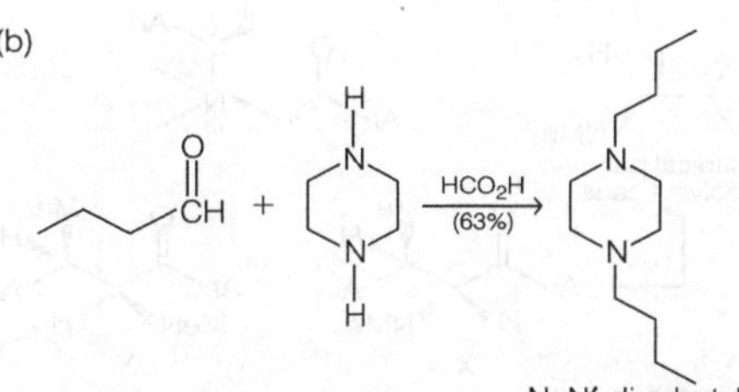

N, N′-di n-butyl piperadine

7. (a)

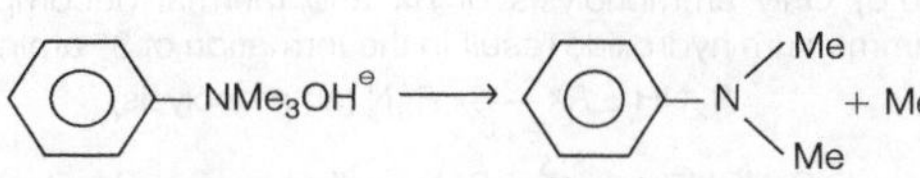

3° amine
N, N-dimethyl aniline

8 (a) This is an example of Leuckart reaction.

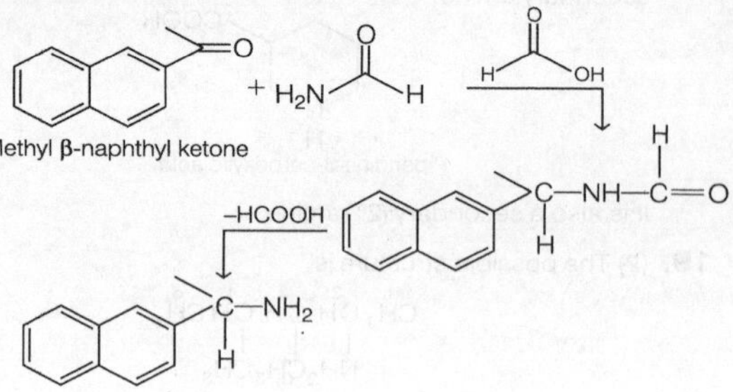

9 (d)

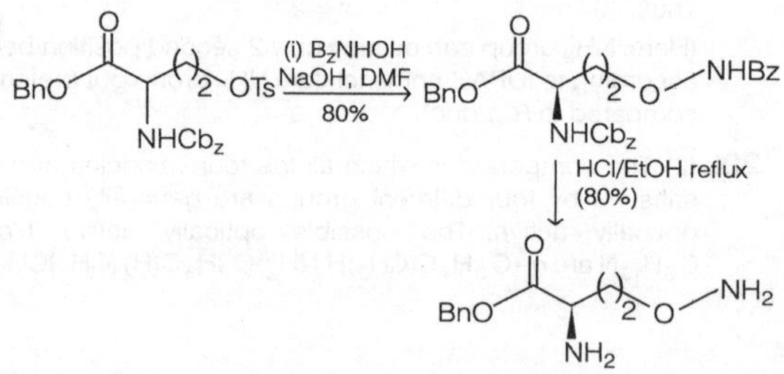

(–OTs being a good leaving reagent is readily replaced by $-\overset{16}{O}NHB_Z$. Both the $-NHB_Z$ groups are hydrolysed to give 1° amine).

10. (d)

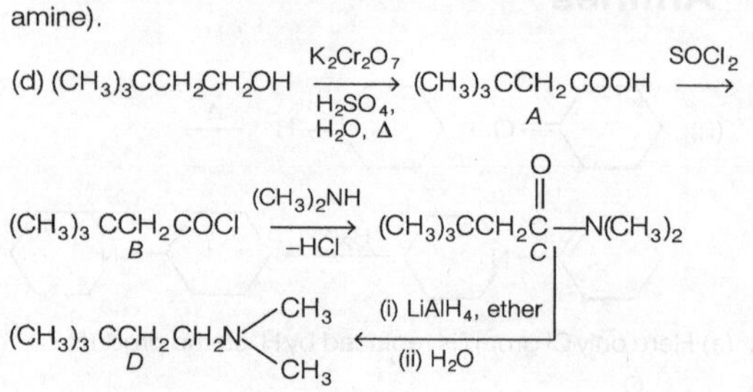

11. (a) It would require Hofmann's elimination to be repeated to three times (which is the highest).

12. (c) It is an example of Stevens rearrangement which involves conversion of a quaternary ammonium salt into 3° amines while with a base like $\bar{O}H$ or RLi (R) etc.

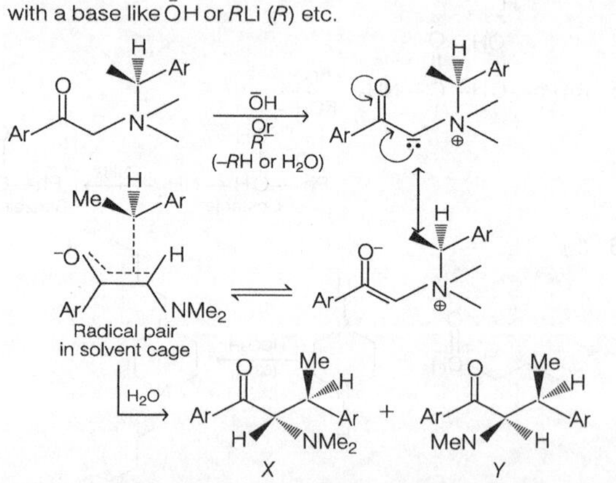

13. (b,d) Only ammonolysis of RX and thermal decomposition of ammonium hydroxide result in the formation of 3° amines as

$$R_2NH + RX \xrightarrow[-HX]{} R_3N \text{ (ammonolysis)}$$

$$\underset{\text{(Quarternary salt)}}{R_4NOH^-} \xrightarrow{\Delta} R_3N + \text{alkene} + \underset{\text{(Thermal decomposition)}}{H_2O}$$

14. (a,b,c,d) In all the given reactions, the amine formed has one carbon less than the parent compound.

$$RCONH_2 \xrightarrow[\text{Hofmann rearrangement}]{Br_2/NaOH} R\text{—}NH_2 \xrightarrow[\text{rearrangement}]{\text{Lossen's}} RNH_2$$

$$RCOCl \xrightarrow{NaN_3} RCON_3 \xrightarrow[\Delta]{2NaOH} RNH_2$$

(Curtius rearrangement)

$$RCOOH \xrightarrow[\text{Schmidt reaction}]{NaN_3/H_2SO_4\text{(conc.)}} RNH_2$$

15. (b,c)

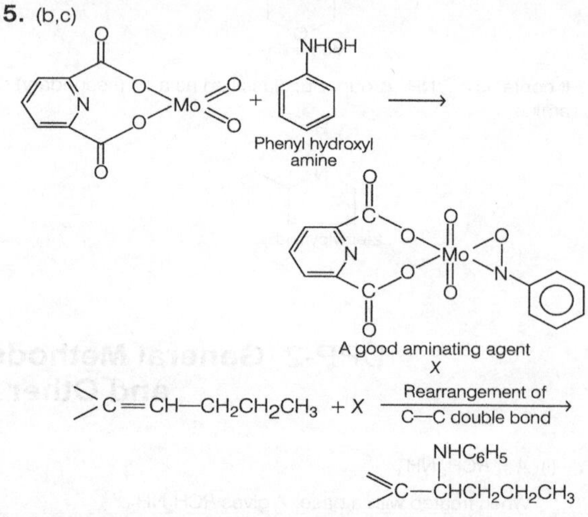

16. (b,c,d) A is

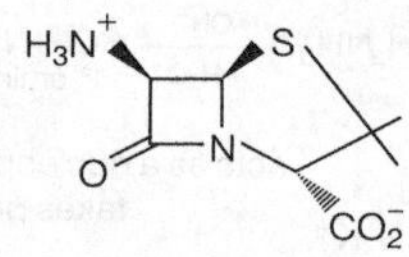

It is the deoxylation on product of **I**. Thus, during the conversion of **I** into A by the given method, no protection is required.

If R = H product is ampicillin and if R = OH product is amoxicillin.

17. (b,c)

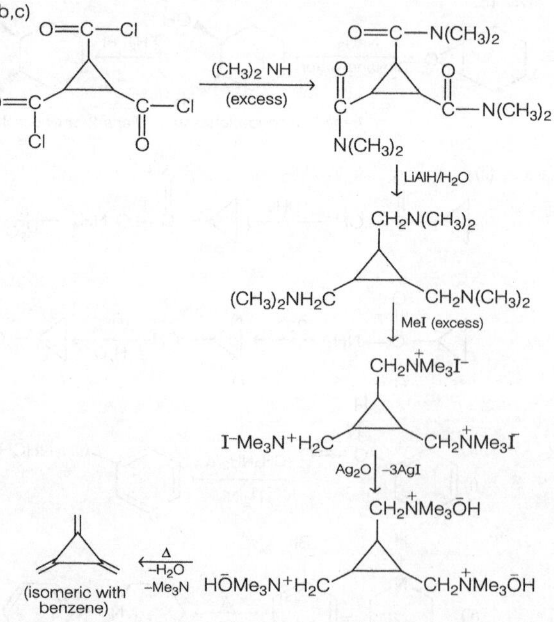

Degree of unsaturation = 4

(3 due to double bond and one due to ring)

Being unsaturated decolourised Br_2 water

Does not undergo ES reactions.

18. (a) $(CH_3)_3CNH_2$ although is a 1° amine but it cannot be prepared by the Gabriel phthalimide reaction because it would require an S_N2 reaction on a tertiary alkyl halide in the first step. In such a condition elimination takes place instead of substitution and alkene is the main product.

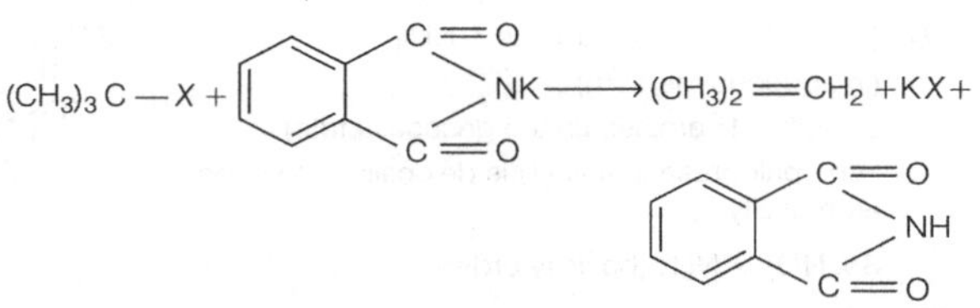

19. (c) All the given statements are true, except II. Because both the rearrangements involve complete retention of configuration in this step.

20. (c) Curtius method converts —COCl group into —NH_2 when it is treated with NaN_3 followed by heating (evolution of N_2) and hydrolysis.

21. (3) $\underset{\substack{1\,mol \\ =60\,g}}{CH_3COOH} \xrightarrow{N_3H,\ \Delta} \underset{\substack{1\,mol \\ =31\,g}}{CH_3NH_2} + N_2 + CO_2$

31 g of 1° amine require acid = 60 g

1.55 g of this amine require acid = $\frac{60}{31} \times 1.55 = 3$ g

22. (3) Only RNO_2, RCN and 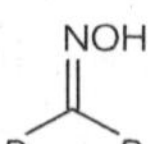 , get reduced by LAH to 1° amine, but other given are not.

23. (5)

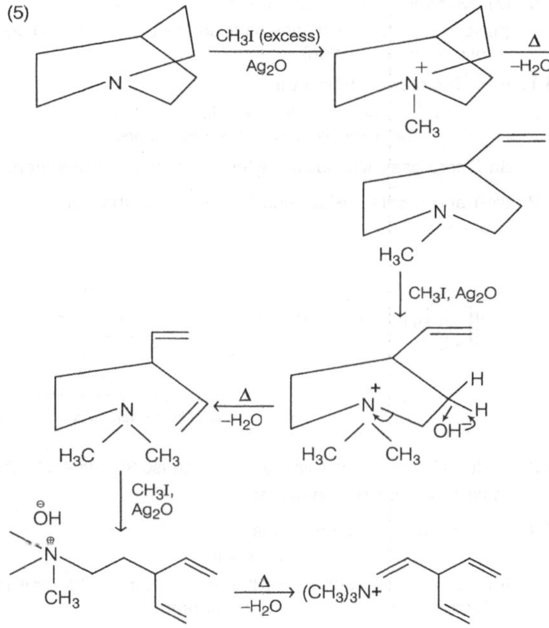

So, $X = 3$

Similarly, 2 times HEM occur in case of

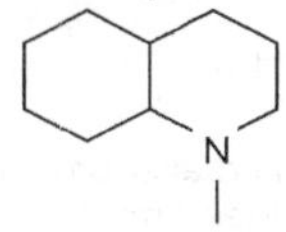

So, $Y = 2$

Total $X + Y = 3 + 2 = 5$

DPP-3 Physical Properties and Basicity of Amines

1. (a) Boiling point decreases with branching due to decrease in surface area. That is why trimethyl amine has the lowest boiling point among the given.

2. (a) More acidic or electron deficient centre is readily attacked by NaOH. Thus,

$$(CH_3)_4N^+I^- \xrightarrow{NaOH} (CH_3)_4N^+OH^- \xrightarrow[-CH_3OH]{\Delta} (CH_3)_3N$$

3. (a) More the electron density on the N-atom, more is the basicity of the compound. In case of $(H_2N)_2C{=}NH$ electron density is more so, it is the strongest base.

4. (d) After losing a proton (d) (ammonium salt) is converted in a 2° amine, which is stabilised by conjugation.

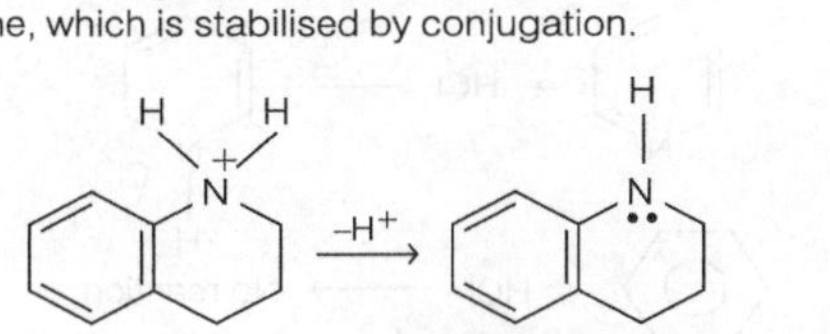

Most stable conjugate base

5. (d) Four CH_3 groups attached with the N increases the electron density of N and $\bar{O}H$ also involve in this enhancement, so it is the strongest base in aqueous solution.

Because of ionic nature of this compound $\bar{O}H$ is much free and the most effective to donate the pair of electron.

6. (d) 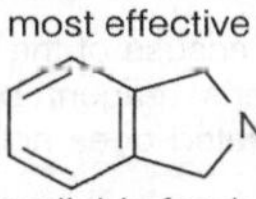 N—H is the strongest base because lone pair of N are not in conjugation with unsaturation, so available for donation.

IV is the weakest base as lone pair of N are in conjugation with double bond as well as electron withdrawing >C=O group is present at nearby position which results in decreased electron density.

7. (a) ∵ *a* is 1° amine, *b* is in 2° amine and *c* is in 3° amine.

8. (c) More is the electronegativity difference, higher the polarity and hence, the dipole moment.

Thus, correct order of dipole moment is I < III < II.

9. (d) pK_b value is log inverse of K_b value i.e. $pK_b (= -\log K_b)$ so it is maximum for the weakest base and presence of two electron withdrawing phenyl group, makes the NH of (d) electron deficient, so ease of attracting H^+ decreases, i.e. pK_b value increases.

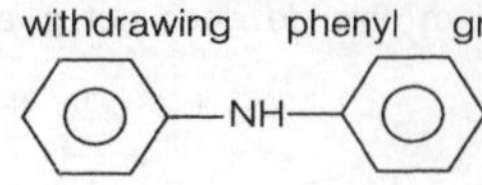

10. (a) Pyrimidine < Pyridine < Imidazole.

(have two N-atoms with lone pairs and lone pair of one N does not involve in delocalisation).

11. (c) ∵ The order of basicity is

IV > II > III > I

2° amine 1° amine 3° amine aniline

and more basic site has more tendency to get protonated.

12. (b) I and II sets are correct. In case of III, the correct order of basicity is

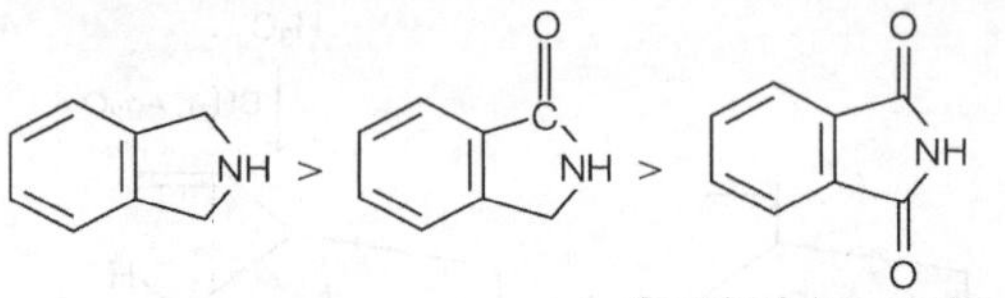

(least basic because lone pair of N are delocalised with >C=O)

13. (b) It is the product of this reaction because 3° amine ($Et_3\ddot{N}$) being most basic gets protonated easily.

14. The correct order of basicity is

III > I > II

II is least basic because of the involvement of its lone pair in delocalisation with two >C=O groups.

15. (i)

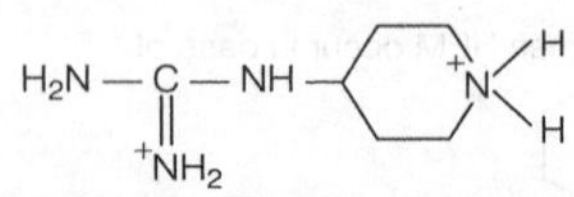

(ii) This is because the conjugate acid formed after protonation contains hydrogen bonding, which stabilises it more.

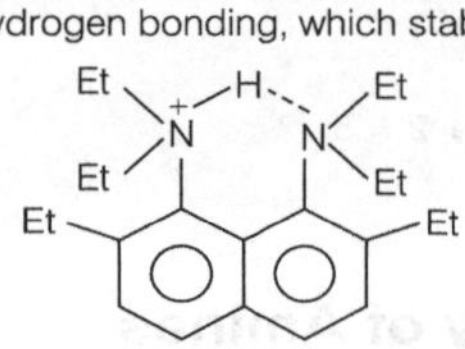

16. (c,d) IV is more basic as lone pair of electron is not a part of delocalisation. In III and II no delocalisation of lone pair of electrons on $\ddot{N}H_2$ or $\ddot{N}$. In I delocalisation of lone pair of e^- of takes place. The order of basic character is IV > III > II > I.

17. (b,c) Aromatic amines are usually weaker bases, so cannot neutralise the acid although the more acidic group like $—SO_3H$ is neutralised and dipolar ion is formed in case of it.

Aliphatic α-amino acid also forms dipolar ion as the amine is sufficiently basic.

18. (a,b,c) 2° > 1° > 3° amines > NH_3

(in aqueous medium)

3° > 2° > 1° amines (in the gaseous state)

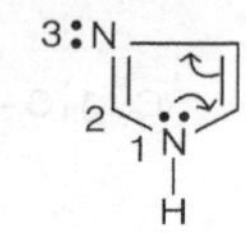

Lone pair present at N (1) is delocalised to impart aromaticity

So, N(3) > N(1) (basicity order).

19. (b) (i) → (q), (ii) → (s), (iii) → (p), (iv) → (r)

More the pK_b value, weaker is the base. Delocalisation of lone pair on N decreases its basicity. sp^2-hybridised N is less basic than corresponding sp^3-hybridised N-atom. $—C_2H_5$ has higher $+I$-effect than $—CH_3$.

Thus, the order of basicity is

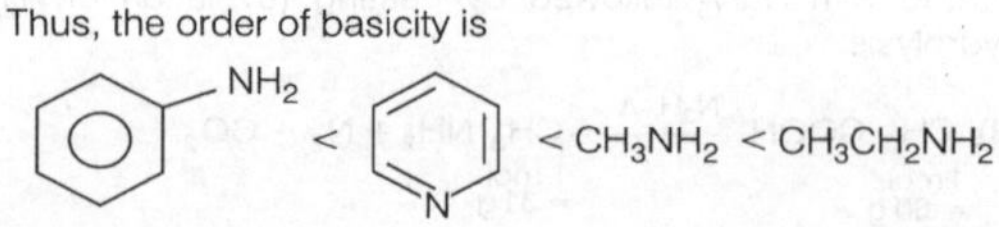

and corresponding pK_b values are as

20. (5) Due to resonance N – 5 carries fractional negative charge.

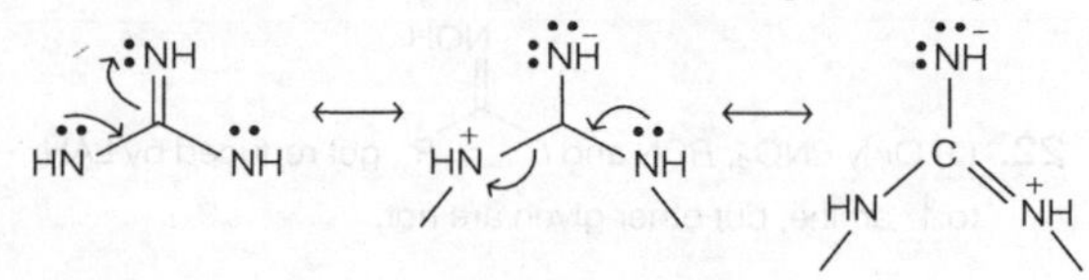

Also the protonated species will be resonance stabilised and have three equivalent resonating structures.

DPP-4 Distinction and Separation of 1°, 2° and 3° Amines

1. (i) Allyl amine ($CH_2{=}CH—CH_2NH_2$) because of the presence of unsaturation decolourises Baeyer's reagent but propyl amine ($CH_3CH_2CH_2NH_2$) being saturated does not give this reaction.

$$CH_2{=}CHCH_2NH_2 + H_2O + [O] \xrightarrow[1\%\ KOH]{KMnO_4} \underset{OH}{CH_2}—\underset{OH}{CH}—CH_2NH_2$$

Colourless

(ii) Tetrapropyl ammonium chloride when treated with $AgNO_3$ gives white precipitate of AgCl while tetrapropyl ammonium hydroxide gives no reaction with it.

$(n-Pr)_4N^+Cl^- + AgNO_3 \longrightarrow AgCl + (nPr)_4N^+NO_3^-$

(iii) Tripropyl ammonium chloride reacts with Na evolving H_2 gas whereas tetrapropylammonium chloride does not give this reaction.

(iv) When di and triethyl amines are treated with nitric acid, the former one form yellow coloured oily nitrosamine whereas the later gives no reaction.

(v) Cyclohexylamine ($C_6H_{11}NH_2$) is a primary amine, so it gives carbylamine test.

$$C_6H_{11}NH_2 + CHCl_3 + 5KOH \longrightarrow C_6H_{11}NC + K_2CO_3 + 3KCl + H_2O$$

Piperidine is not a primary amine, so does not give this reaction.

(vi) Pyridine is basic in nature, so forms salt with hydrochloric acid. Benzene being neutral does not give this reaction.

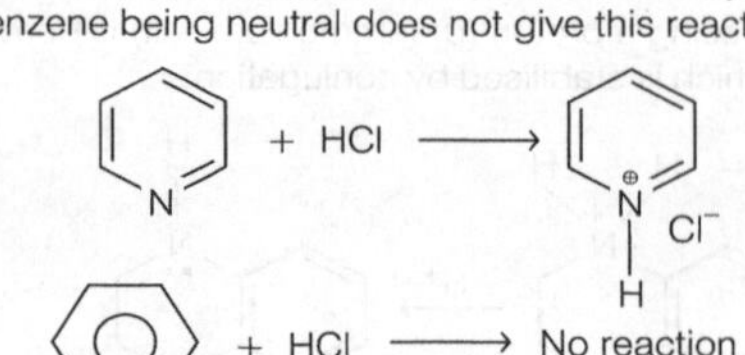

2. The amine *B* must be a secondary amine as the product obtained with benzene sulphonyl chloride is insoluble in aqueous sodium hydroxide.

The acid *C* must be formic acid as it reduces Tollen's reagent to give silver mirror.

So, the reaction involved and *A* are as

$$\underset{A}{H-\overset{\overset{O}{\|}}{C}-\overset{\overset{CH_3}{|}}{\underset{\underset{CH_3}{|}}{N}}} \xrightarrow{\text{Hydrolysis}} \underset{C}{H-\overset{\overset{O}{\|}}{C}-OH} + \underset{B\ 2^\circ \text{ amine}}{CH_3NHCH_3}$$

$$CH_3NHCH_3 \xrightarrow{C_6H_5SO_2Cl} \underset{\text{Insoluble in alkali}}{C_6H_5SO_2N(CH_3)_2} + HCl$$

$$\underset{C}{HCOOH} + Ag_2O \longrightarrow H_2O + CO_2 + Ag\downarrow$$

3. Since, the compound gives a foul smelling gas on treatment with $CHCl_3$ and alc. KOH, the compound must be a primary amine.

Further, the compound on treatment with $NaNO_2/HCl$ at 0°C produces a colourless gas, so it must be an aliphatic primary amine.

$$RNH_2 + HNO_2 \longrightarrow ROH + N_2 + H_2O$$

$$\text{Amount of gas liberated} = \frac{112}{22400} = \frac{1}{200}\ \text{mol}$$

$$\therefore \text{Amount of compound } RNH_2 = \frac{1}{200}$$

$$\frac{0.295}{M} = \frac{1}{200}$$

$$M = 59\ \text{g mol}^{-1}$$

The molar mass of alkyl group *R* is 59 – 16 = 43 g. Thus, *R* must be C_3H_7 (mass of NH_2 = 14 + 216 g)

Since, the compound obtained after distillation gives iodoform with NaOI, so it must be a 2-ol with $CH_3\underset{OH}{\underset{|}{C}H}$ group.

Hence, volatile nitrogen compound is $CH_3CH(NH_2)CH_3$.

4. (d) 2° amines (compound having $>$NH group) form oily nitrosoamine with nitrous acid.

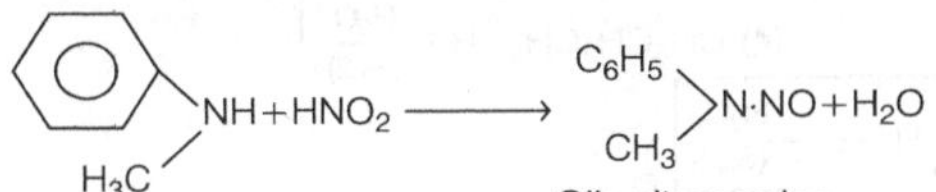

Oily nitrosamine

5. (c) This is the property of 1° amine (i.e. compound having —NH_2) group. Thus, 2,2-dimethylpropylamine gives this reaction.

$$CH_3-\overset{\overset{CH_3}{|}}{\underset{\underset{CH_3}{|}}{C}}-CH_2-NH_2 + CHCl_3 + 3KOH \longrightarrow (CH_3)_3CCH_2NC + 3KCl + 3H_2O$$

6. (c) These are the characteristics of secondary amines and $C_6H_5NHCH_3$ is a secondary amine.

7. (b) One of the nitrogen is secondary while other is tertiary. After forming sulphonamide at secondary nitrogen, tertiary nitrogen remain unreacted which later reacts with H_2SO_4. Hence $(CH_3)_2NCH_2CH_2NHCH_3$ satisfy the given set of conditions.

8. (a) 1° amine forms N-alkyl benzene sulphonamide with $C_6H_5SO_2Cl$, which is soluble in alkali.

2° amine forms N,N-dialkyl benzene sulphonamide, which is insoluble in alkali as well as acid.

3° amines do not react with $C_6H_5SO_2Cl$ and hence are soluble in acid. Thus, *A*, *B* and *C* here represent 1°, 2° and 3° amines respectively.

9. (b) Formation of phthalic acid on oxidation with $KMnO_4/\bar{O}H$ suggests the presence of two carbon chains at the benzene nucleus at 1,4 positions. further formatin of a liquid with oxamic ester that gets decomposed by KOH shows that, it is a secondary (2° amine). Hence, its possible structure is

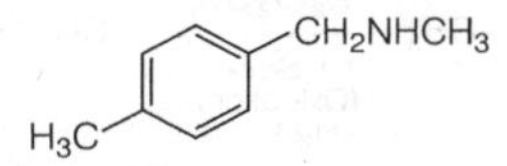

10. (c) 1° amines gives N_2 (inert gas) with HNO_2 and forms isocyanate with $HgCl_2/CS_2$ and a solid (N, N′-dialkyl oxamide) with diethyl oxalate.

Thus, it can be

(a) $CH_3\underset{NH_2}{\underset{|}{C}H}CH_3$ (optically inactive)

(b) $RCl \xrightarrow{AgCN} RNC \xrightarrow{\text{Reduction}} \underset{2^\circ \text{ amine}}{RNH\,CH_3}$

(c) $R-\overset{\overset{O}{\|}}{C}-N_3 \xrightarrow[-Na_2CO_3]{OH^-} \underset{1^\circ \text{ amine}}{RNH_2}$

(c) $R-\overset{\overset{O}{\|}}{C}-NH_2$ (compound with —NH_2 at carbonyl carbon)

11. (c,d) With $KMnO_4$, 1° amines give ketone, 2° gives tetra alkyl hydrazine and 3° gives no reaction. Thus,

$X = (CH_3)_3\,N$, $Z = (CH_3)\underset{H}{\underset{|}{N}}CH_2\,CH_2$

$Z = CH_3CH_2\,CH_2\,NH_2$

Thus, *X* forms amine oxide with Caro's acid (H_2SO_5) and *Y* being a 1° amine gives smell of mustard oil when treated with $CS_2/HgCl_2$.

12. (a,b,c) The possible structure of 3° alcohol is $(CH_3)_3C$—OH, so structure of amine is $(CH_3)_3CNH_2$.

Being 1° amine, it will give all the reactions except Hofmann bromamide reaction, which is given by 1° amides.

13. (a,b)

$$\underset{\substack{(2^\circ \text{ amine as 2 mol}\\ \text{of MeI are added})}}{C_6H_{15}N} + MeI \xrightarrow[-HI]{} C_6H_{14}N{<}^{Me}$$

$$C_6H_{14}N{<}^{Me} \xrightarrow{MeI} C_6H_{14}\overset{+}{N}{<}^{Me}_{Me}\ I^- \xrightarrow{KOH} C_8H_{20}N^+OH^- \text{ or } C_6H_{14}\overset{+}{N}{<}^{Me}_{Me}\ OH^-$$

$$\xrightarrow{\Delta} CH_2{=}\overset{\overset{CH_3}{|}}{C}-CH_3 + (CH_3)_2\,NEt + H_2O$$

Thus, possible structure of *A* is

$$CH_3-\overset{\overset{CH_3}{|}}{C}H\,CH_2NHCH_2CH_3 \text{ or } CH_3-\overset{\overset{CH_3}{|}}{\underset{\underset{CH_3}{|}}{C}}-NH-CH_2CH_3$$

14. (a) Formation of racemate suggest that *A* is an optically active compound.

$$A \xrightarrow{HBr} \underset{\text{Optically active}}{R-Br} \xrightarrow{NH_3} \underset{\substack{\text{Optically}\\ \text{active } 1^\circ\\ \text{amine } C}}{C_4H_{11}N} \xrightarrow[-N_2]{HNO_2} \underset{\text{Alcohol}}{D}$$

Thus, the possible structure of optically active 1° amine with molecular formula $C_4H_{11}N$ is $CH_3CH_2-\underset{NH_2}{\underset{|}{C}H}-CH_3$ (as only in it a chiral C is present).

15. (b)

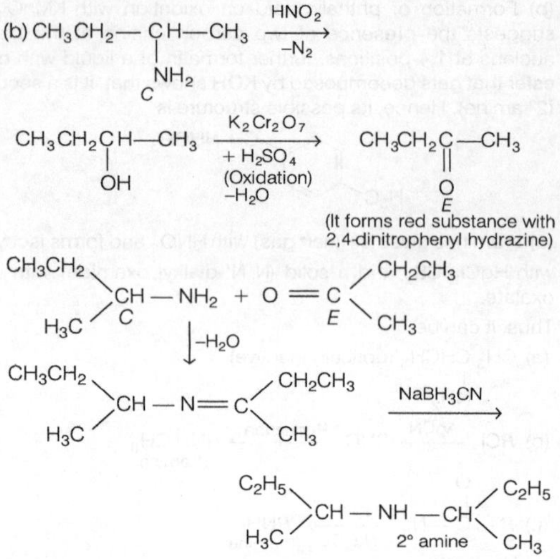

16. (d)

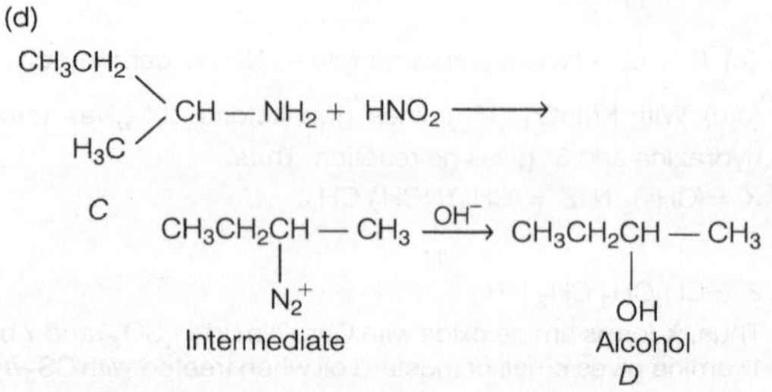

17. (a) Primary amines give all the tests given in Column II. So, ethylamine also gives all these tests.

(i) $\underset{\text{or } C_2H_5NH_2 \text{ 1° or ethylamine}}{R-NH_2} + \underset{\text{Hinsberg's reagent}}{C_6H_5SO_2Cl} \xrightarrow{-HCl} \underset{\text{(soluble in alkali)}}{RNHSO_2C_6H_5}$

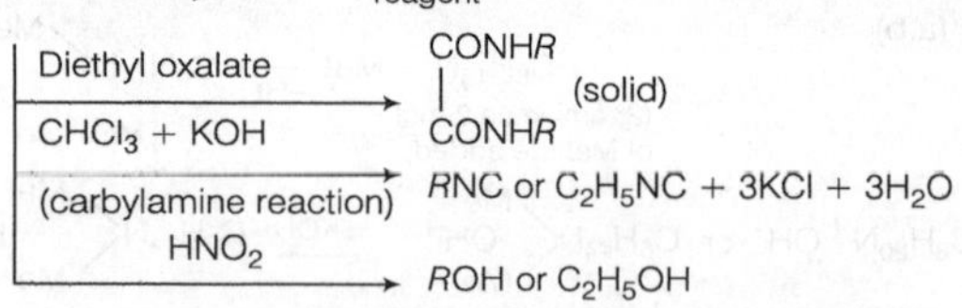

(ii) 2° amines react with all reagents except $CHCl_3/KOH$, i.e. do not give carbylamine reaction. They form alkali insoluble sulphonamide with $C_6H_5SO_2Cl$ (Hinsberg's reagent) and a liquid product with oxalate and HNO_2.

(iii) 3° amines react only with HNO_2

$$\underset{\text{3° amine}}{R_3N} + HNO_2 \longrightarrow R_3NH^+ NO_2^-$$

18. (c) Primary 1° amines (compound having —NH_2 group) give Hofmann mustard oil reaction, i.e. they first produce dithiocarbamic acid with CS_2 which is decomposed by $HgCl_2$ into alkyl isothiocyanate, with smell of mustard oil.

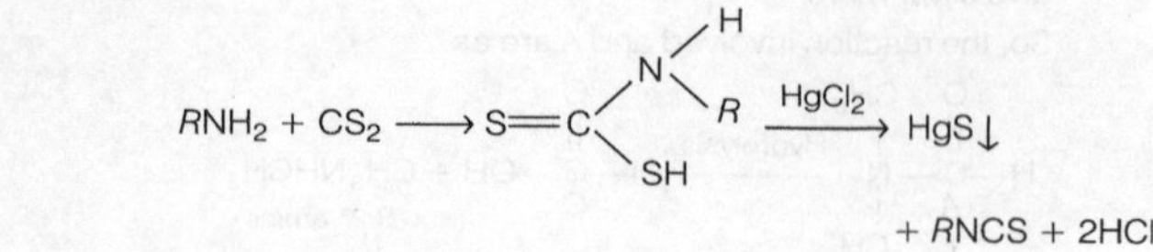

2° (secondary amines) produce alkali insoluble sulphonamide with p-toluene sulphonyl chloride.

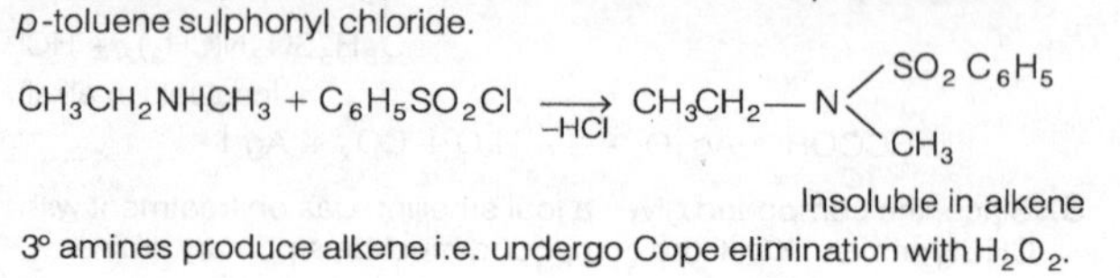

3° amines produce alkene i.e. undergo Cope elimination with H_2O_2.

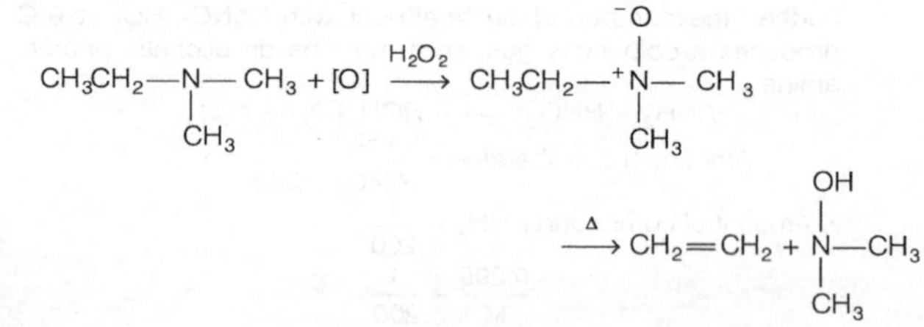

19. (6)

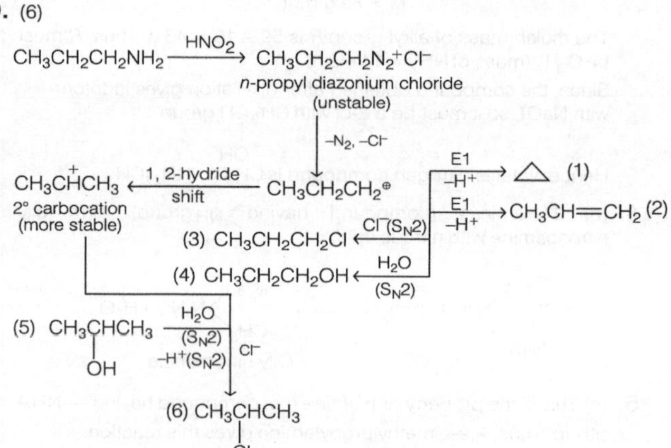

20. (2) 2° amines forms tetraalkyl hydrazine with $KMnO_4$.

$$2\,R_2NH \xrightarrow{[O]} R_2N-NR_2$$

So, difference in no. of carbon atoms is multiple of 2 as no. of C-atoms becomes double.

DPP-5 Chemical Properties of Aliphatic Amines

1. Formation of ethanol with HNO_2 suggests that 'Z' is ethylamine ($C_2H_5NH_2$).

Since, Z is obtained by Hofmann bromamide degradation (reaction with Br_2/NaOH) of Y, so Y must be an amide of formula

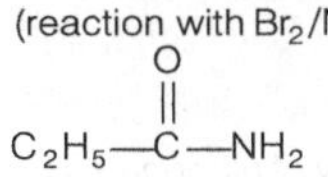

Data also supports this formula

Element	% age	At · wt.	Molar ratio	Simplest mole ratio
C	49.13%	12	4.09	3
H	9.59%	1	9.59	7
N	19.18%	14	1.37	1
O	22.1%	16	1.38	1

Thus, formula is C_3H_7NO or $C_2H_5CONH_2$.

Since, X is a chlorine compound. So,

X Is 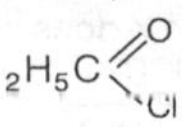

C_2H_5COCl when treated with strong ammonia forms $C_2H_5CONH_2$ (propanamide).

2. (i)

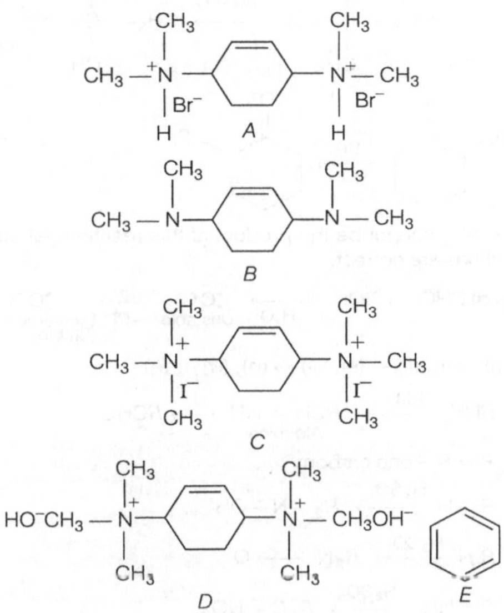

(ii)

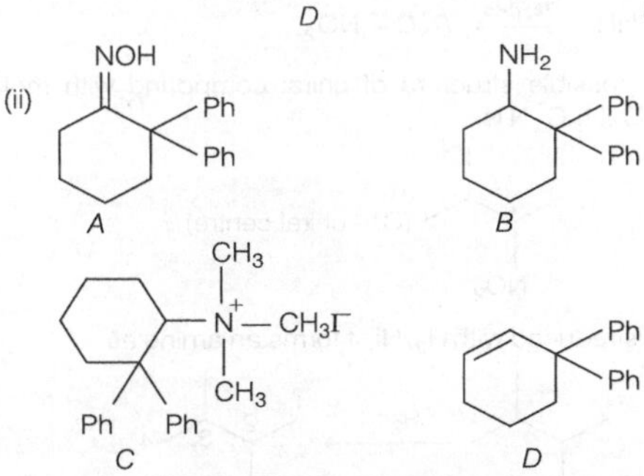

3. (b)

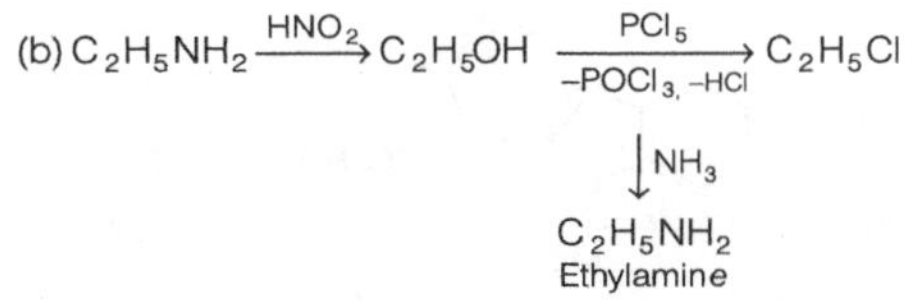

4. (b)

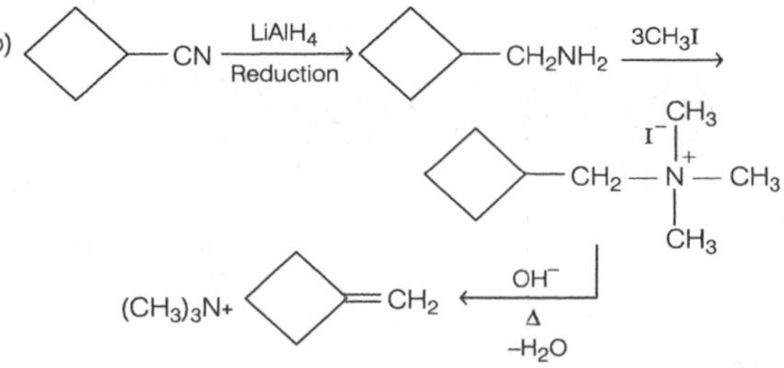

5. (c) $C_2H_5NH_2 + \underset{A}{C_6H_5CHO} \longrightarrow$

$$C_2H_5N = CHC_6H_5 + H_2O$$

$$NH_2CONH_2 + \underset{\substack{\text{Hydrazine}\\ B}}{NH_2—NH_2} \longrightarrow NH_2NHCONH_2 + NH_3$$

$$C_2H_5NH_2 + \underset{C}{NOCl} \longrightarrow C_2H_5Cl + H_2O + N_2$$

6. (c) This is because of the formation of a stable carbocation.

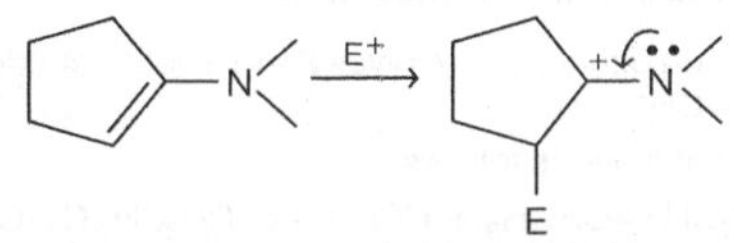

Stable carbocation

7. (b)

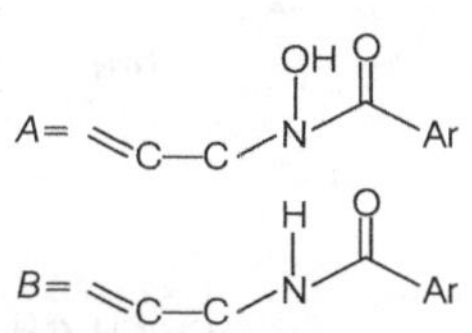

Thus, B is the deoxygenated product of A.

8. (b)

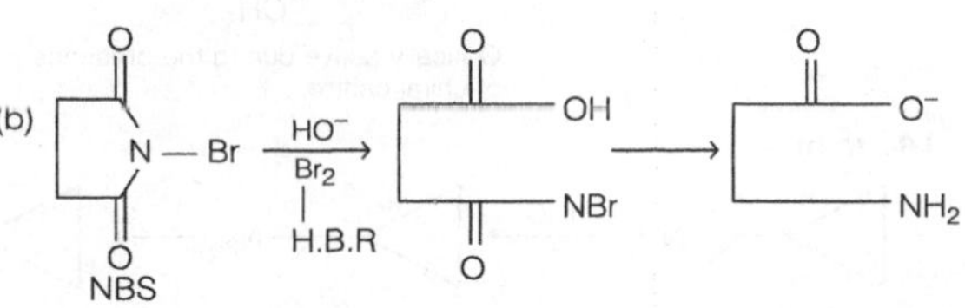

9. (a) I. $\underset{\text{Ethylamine}}{CH_3CH_2NH_2} \xrightarrow{NaNO_2 + HCl} CH_3CH_2OH + N_2 + H_2O + NaCl$

II. $NH_2CONH_2 \xrightarrow[2]{NaNO_2/\ HCl} 2N_2 + 2H_2O + CO_2 + NaCl$

III. $CH_2CONH_2 \xrightarrow{NaNO_2 + HCl} CH_3COOH + N_2 + H_2O + NaCl$

IV. $C_6H_5CH_2NH_2 \xrightarrow{NaNO_2 + HCl} C_6H_5CH_2OH + H_2O + NaCl + N_2$

10. (d) The configuration of both the substituent remain the same. Thus, product is

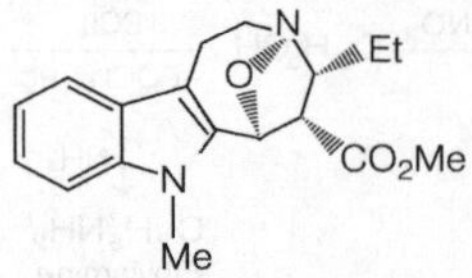

11. (c)

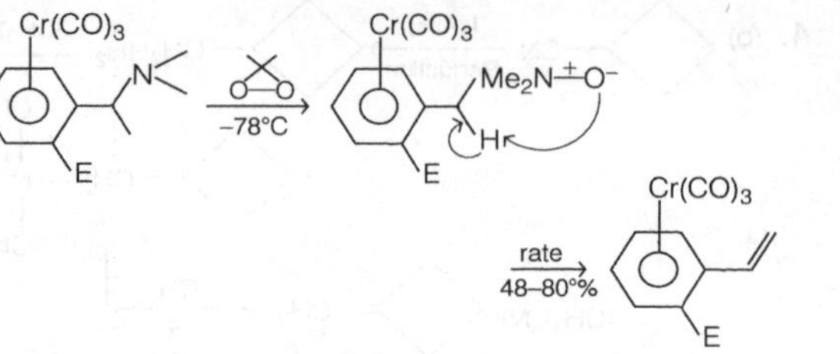

12. (a) The reactive species is the nitrosonium ion (electrophile) and the most likely site of attack will be that with the highest electron density.

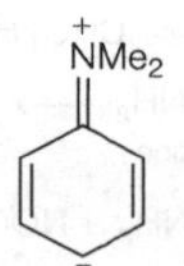

Such kind of delocalisation is not found in 2,6-dimethyl derivative due to steric inhibition of resonance.

13. (d) Since, on ozonolysis, *A* gives 2 moles of acetaldehyde, so *A* is $CH_3CH{=}CHCH_3$

The reactions are as follows

$$CH_3CH{=}CHCH_3 + HCl \longrightarrow \underset{B}{CH_3CH_2\underset{|}{C}HCH_3}\ (Cl)$$

$$\xrightarrow{NH_3} CH_3CH_2-\underset{\underset{\text{C (2-aminobutane)}}{NH_2}}{CH}-CH_3$$

$$\underset{D}{CH_3CH_2\overset{*}{C}H(OH)-CH_3} \xleftarrow{HNO_2}$$

Optically active due to the presence of chiral centre

14. (a,b)

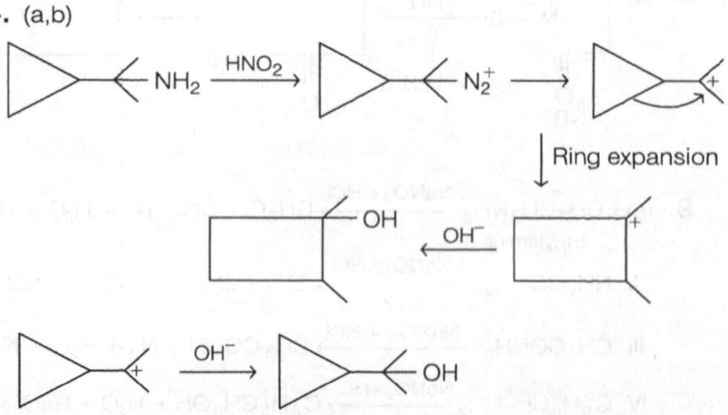

15. (b,d) There is initially ring opening of epoxides by amines.

$$H_3C-NH_2 + \underset{\text{(epoxide, O)}}{CH_2-CH_2} \longrightarrow \underset{\beta\text{-amino alcohol } A}{H_3CNHCH_2CH_2OH}$$

B (β-amino alcohol) is less reactive because of stabilisation by intramolecular hydrogen bonding between the lone pair of electron on nitrogen and the hydrogen of the hydroxyl group.

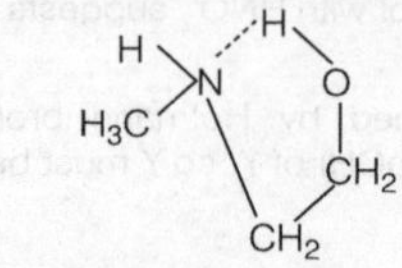

H-bonding also lowers the nucleophilicity of nitrogen in *B*.

16. (a,b,c)

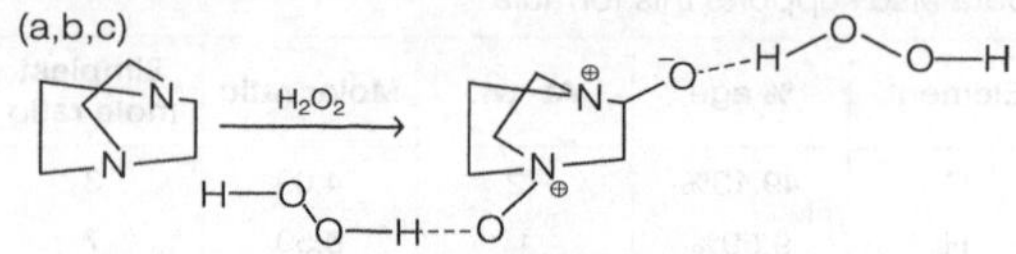

1, 4-diazobicyclo [2, 2, 2] octane
1, 4-dioxide diperhydrate

17. (a,b,d) Presence of α-H atom is the main condition for exhibiting tautomerism.

$(CH_3)_3CNH_2$ does not contain any α-H atom, so it is also absent in product also hence *Y* does not show tautomerism.

$$(CH_3)_3CNH_2 \xrightarrow[-N_2,-H_2O]{NOCl} (CH_3)_3C-Cl \xrightarrow{AgNO_2} (CH_3)_3C-NO_2$$

18. (a,b,c)

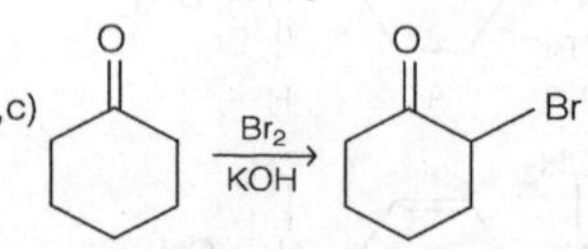

$C_6H_{11}NH_2$ cannot be the product of this reaction. All other given reactions are correct.

19. (a,b,d) $CHCl_3 + {}^{-}OH \xrightarrow{-H_2O} \underset{\text{Unstable}}{:\bar{C}Cl_3} \xrightarrow{-Cl^-} \underset{\text{Dichlorocarbene (Stable)}}{:CCl_2}$

20. (a) (i) → (r), (ii) → (s), (iii) → (p), (iv) → (q)

(i) $RNH_2 \xrightarrow{KMnO_4} \underset{\text{Aldimine}}{RCH{=}NH} \xrightarrow{H^+} RCHO$

$R = R$ – one carbon

(ii) $R_2NH \xrightarrow{H_2SO_5} R_2-N-OH$

(iii) $R_3N \xrightarrow{H_2SO_5} R_3N \longrightarrow O$

(iv) $R_3CNH_2 \xrightarrow{H_2SO_5} R_3C-NO_2$

21. (4) The possible structure of chiral compound with molecular formula $C_5H_{11}O_2N$ is

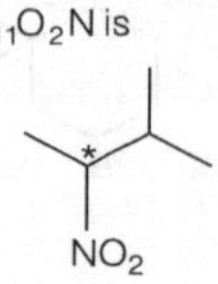

(C*– chiral centre)

When it is reduced with H_2/Ni, it forms an amine as

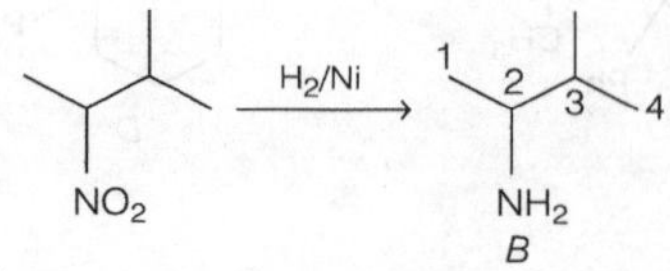

B

Thus, four carbon atoms are present in the main chain of chiral *B*

Revisal Problems (JEE Main)

1. (a) N_2 is the best leaving group among the given.

2. (c) $CH_3CH_2Cl \xrightarrow{KCN} CH_3CH_2CN \xrightarrow[C_2H_5OH]{Na} CH_3CH_2CH_2NH_2$

Thus, KCN (potassium cyanide) is the source of nitrogen in this reaction.

3. (b) $\underset{R}{C_6H_5-CH(CH_3)-CONH_2} \xrightarrow[\text{Hofmann bromamide degradation}]{NaOH + Br_2} \underset{R}{C_6H_5-CH(CH_3)-NH_2}$

4. (a)

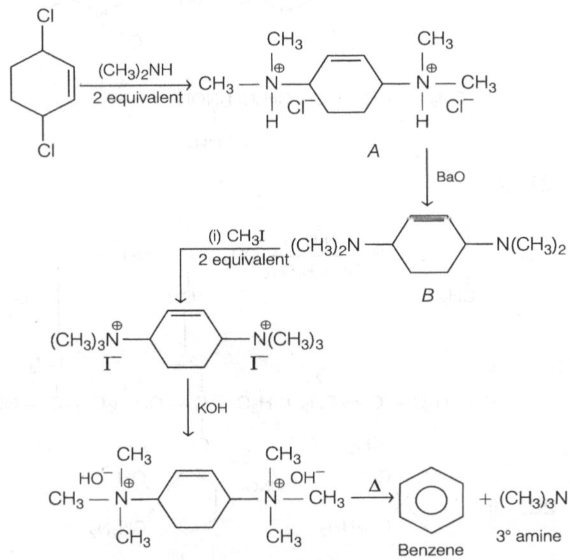

5. (a) Primary amine reacts with chloroform ($CHCl_3$) and KOH to give bad smelling isocyanide. (Carbylamine test).

6. (a)

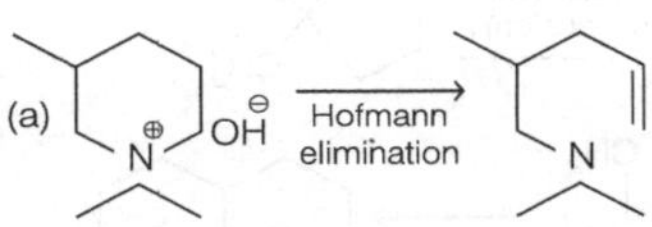

(as less substituted alkene is the main product.)

7. (b)

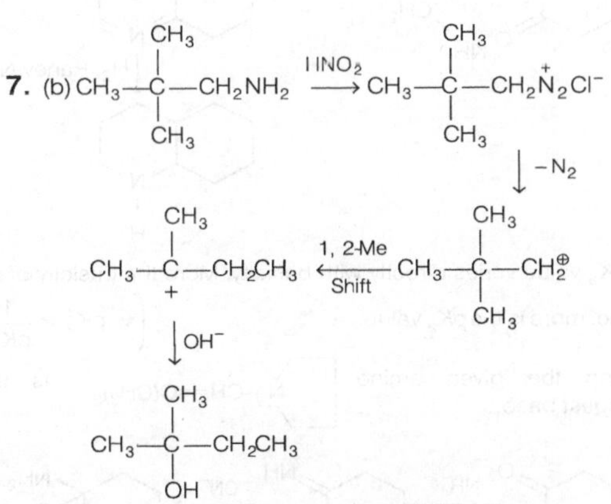

8. (a) $\underset{}{CH_3CH_2NH_2} \xrightarrow{3CH_3I} \underset{B}{CH_3CH_2\overset{\oplus}{N}(CH_3)_3\,\overset{\ominus}{I}} \xrightarrow[H_2O]{Ag_2O} \underset{C}{CH_3CH_2\overset{\oplus}{N}(CH_3)_3\,\overset{\ominus}{O}H} \xrightarrow{\Delta} (CH_3)_3N + CH_2{=}CH_2 + H_2O$

9. (d)

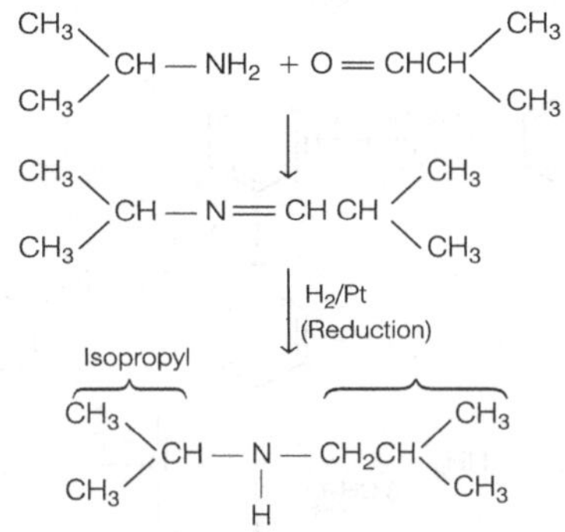

Isobutyl isopropylamine

10. (b)

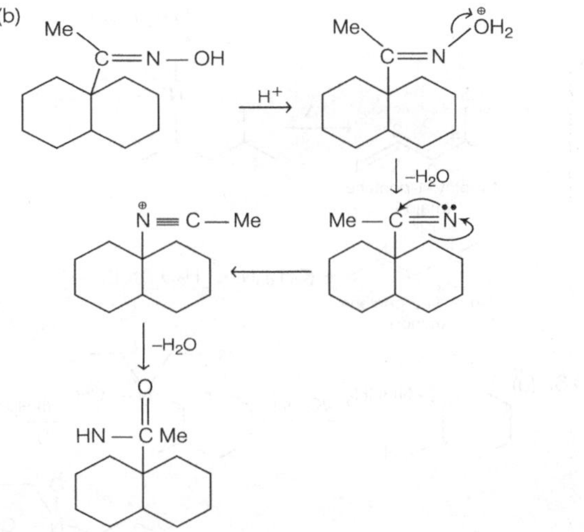

Since, it involves the conversion of ketoxime into amide in acidic media which then hydrolysed into 1° amine, so it is an example of Beckmann rearrangement.

11. (a) This is because in case of it lone pair of N are available on N for protonation.

$\ddot{N}CH_2-\overset{O}{\overset{\|}{C}}-OCH_3$

(Readily protonated)

12. (a) Metal/acid or $LiAlH_4$ in ether are generally used for the reduction $-NO_2$ group into $-NH_2$.

13. (a) Boiling point increases with increase in molecular mass but decreases with branching. Thus, the order of boiling point is

n – butylamine(II) (no - branching) > diethylamine (III) > ethyldimethylamine (I) (maximum branching)

14. (a)

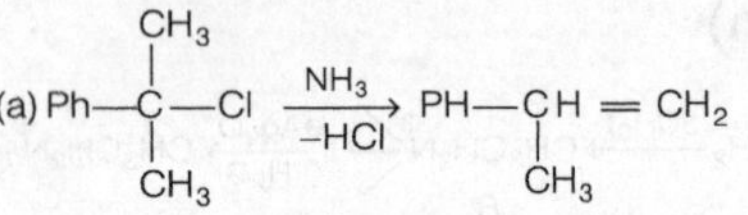

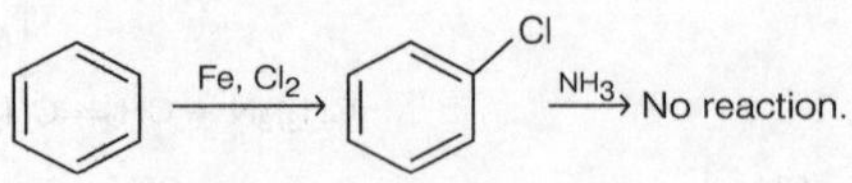

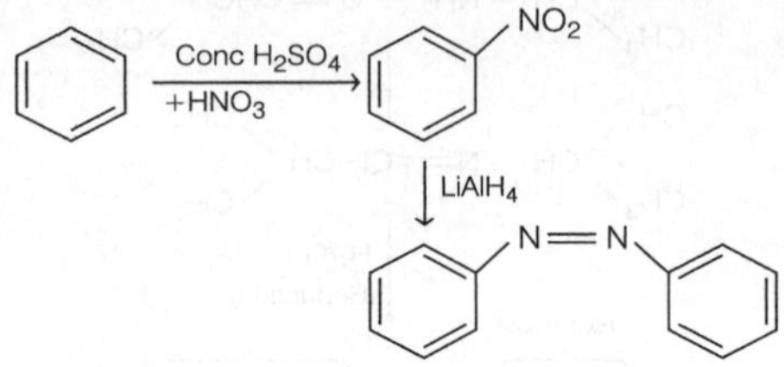

15. (a)

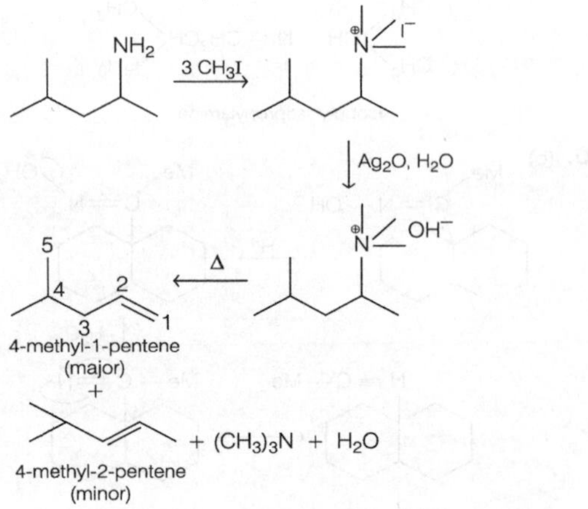

16. (d)

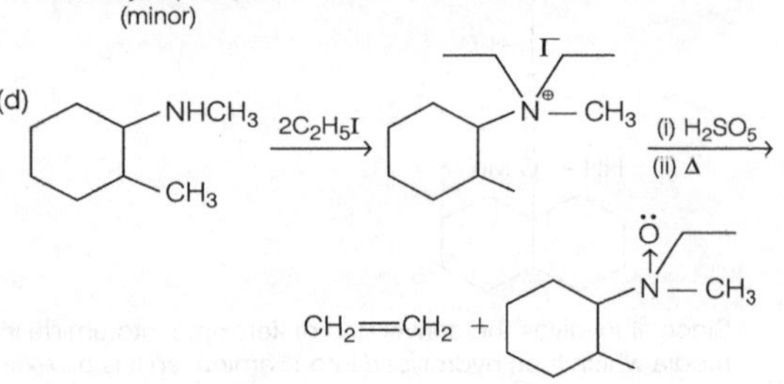

17. (c)

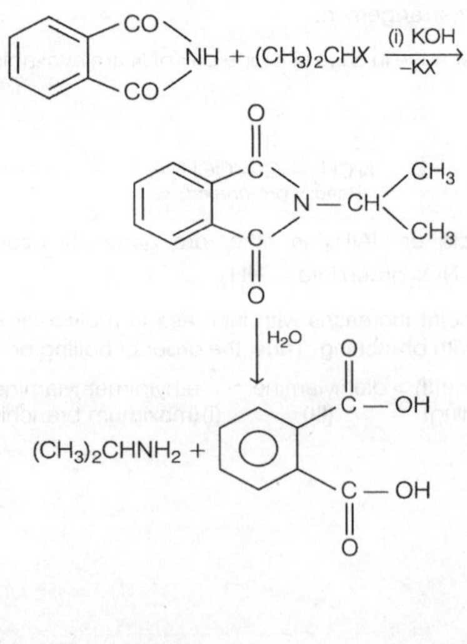

18. (b)

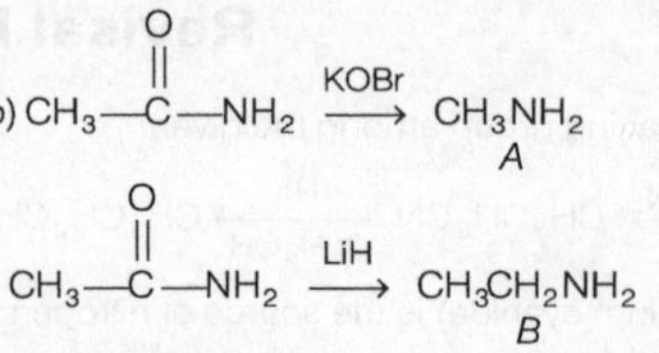

A and B belong to same homologous series, so they are homologues.

19. (d) Because here lone pair of N are more available for donation.

20. (d)

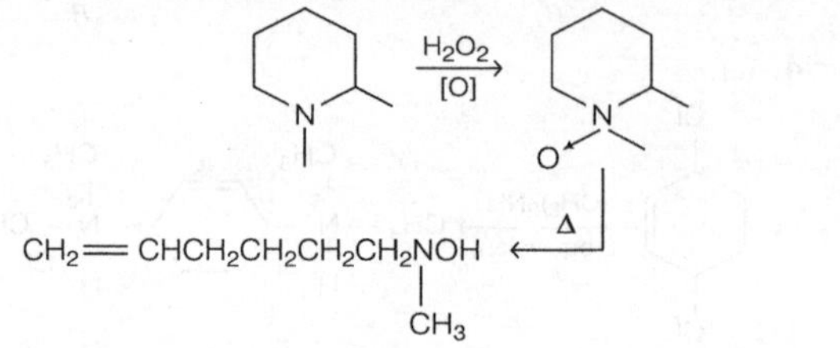

21. (d)

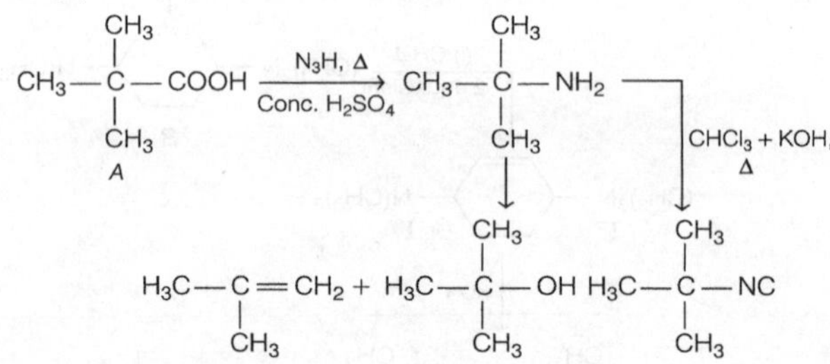

22. (d)

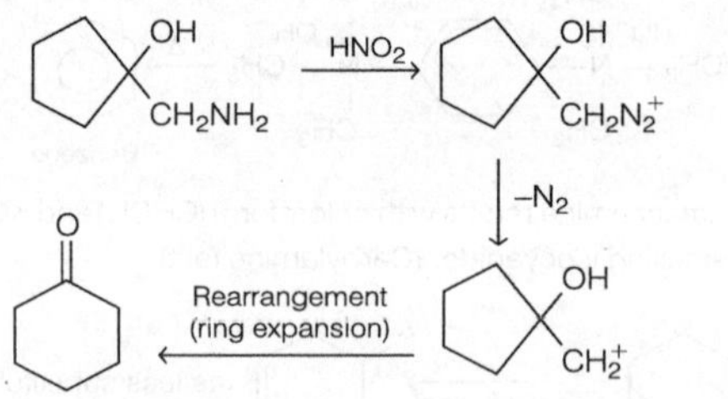

23. (c)

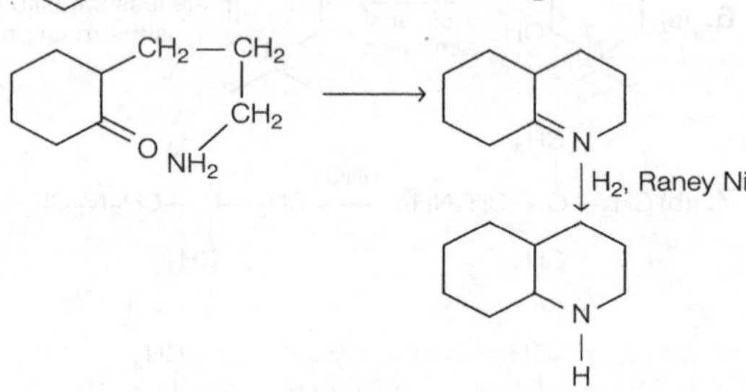

24. (c) pK_a value varies directly with basicity. More the basicity of an amine, more is the pK_a value. $\left(\because pK_a \propto \frac{1}{pK_b}\right)$

Among the given amine 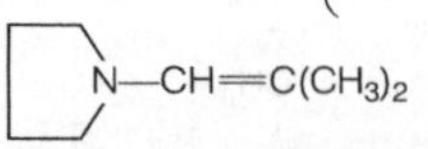 is the strongest base.

25. (b)

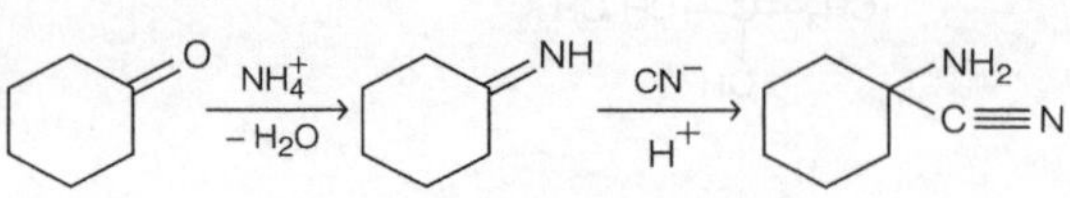

26. (c) $LiBH_3CN$ is a good reducing agent, so preferred for the reduction of imine to amine.

27. (c)

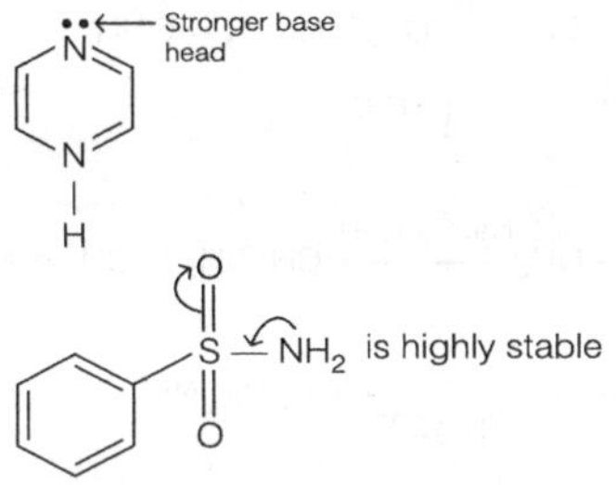

28. (b) This is an example of cope elimination. Thus, the product is

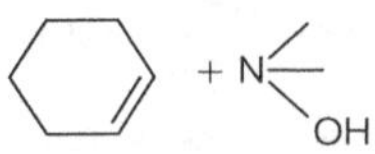

29. (b) To separate dibutylamine $(C_4H_9)_2NH$ and anisole, first the mixture is dissolved in ether, and then amine (being basic) is extracted with 10% *aq*. HCl.

30. (a)

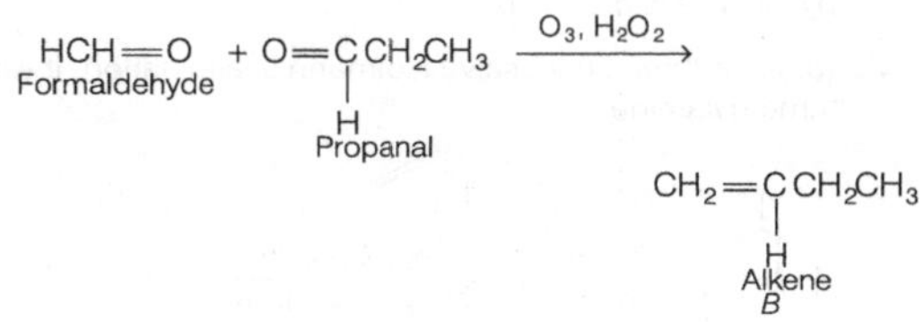

Since, the alkene has linear structure, so amine also has linear chain.

$-NH_2$ group should be at second position so that *A* is optically active

Thus, $A = CH_3CH(NH_2)-CH_2CH_3$

Revisal Problems (JEE Advanced)

1. (b) Among cyanides, imines and amines, the amines are most basic because of the more availability of lone pair for protonation. In case of cyanides, this lone pair is involved in bonding between C and N, so cyanides are least basic.

2. (d)

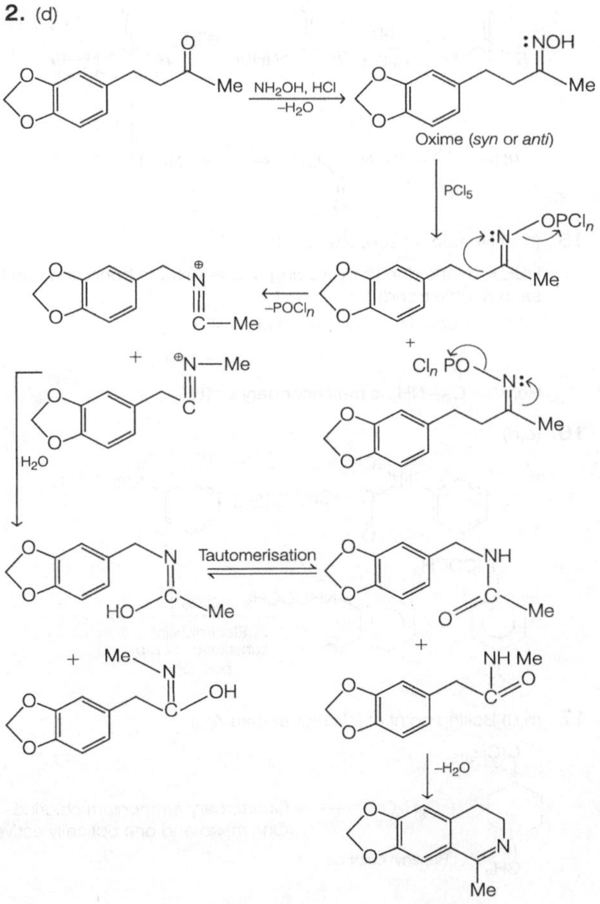

3. (a)

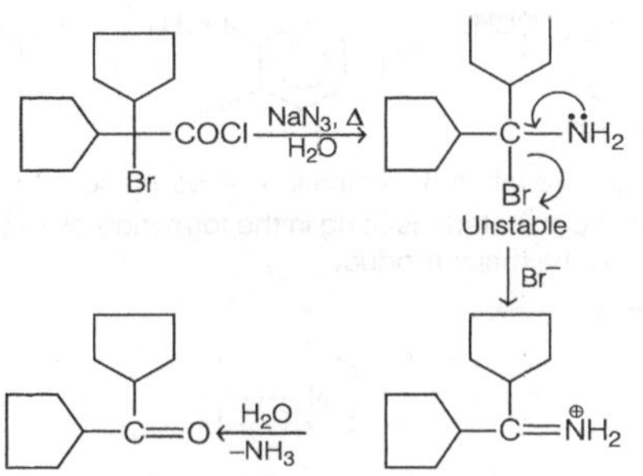

4. (c) $CH_2N_2 \xrightarrow{\Delta} :CH_2 + N_2$
Carbene

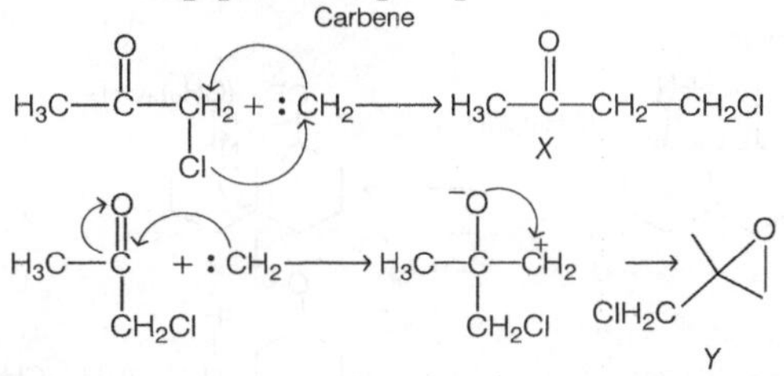

5. (c)

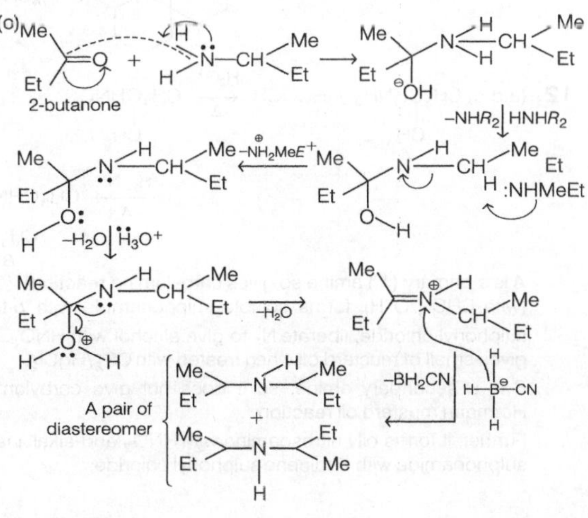

6. (a) The correct order is I < II < III

pK_a (11.3) (11.2) (9.28)

Thus, substituted morpholine is least basic whereas substituted pyrrolidine is most basic.

7. (d) After three successive Hofmann's elimination, it would lose a trimethylamine.

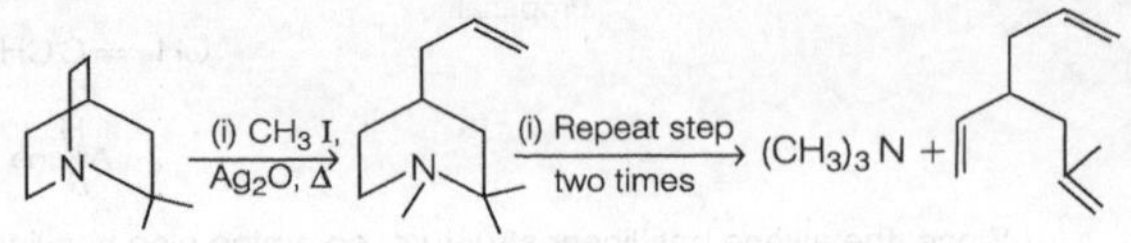

8. (b)

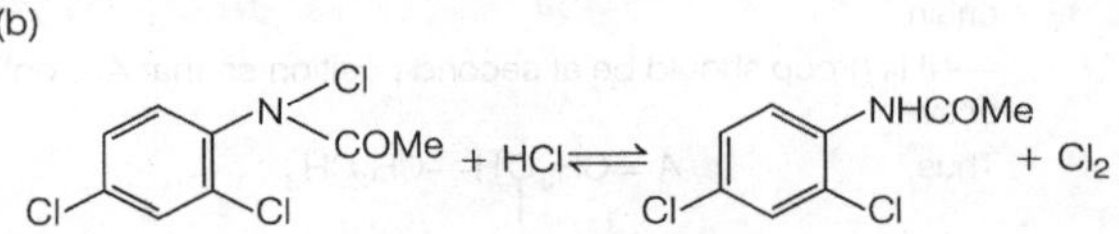

N-chloro-2, 4-dichloro acetanilide   I.

An intermolecular rearrangement takes place. Cl_2 becomes free. Anisole is more reactive for further electrophilic substitution, chlorination than molecule I and $—OCH_3$ is *o*/*p*-directing group.

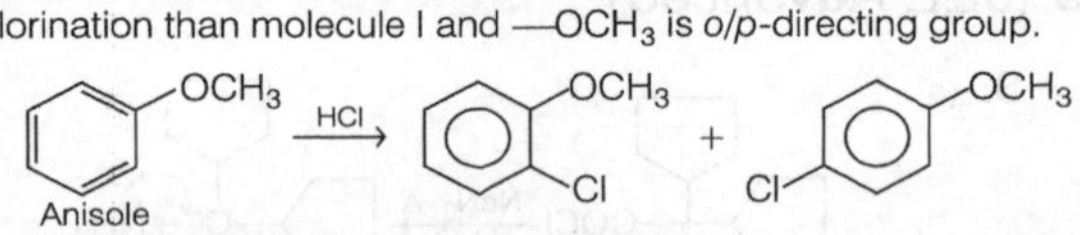

Anisole

9. (b) In this reaction β-elimination takes place. The most acidic proton is eliminated resulting in the formation of least substituted alkene as the major product.

10. (d) (LDA)

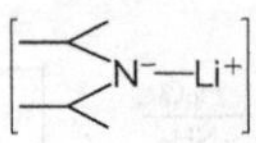

is the strongest base as it has a full negative charge over the N atom which is not a part of any resonating structure.

11. (b)

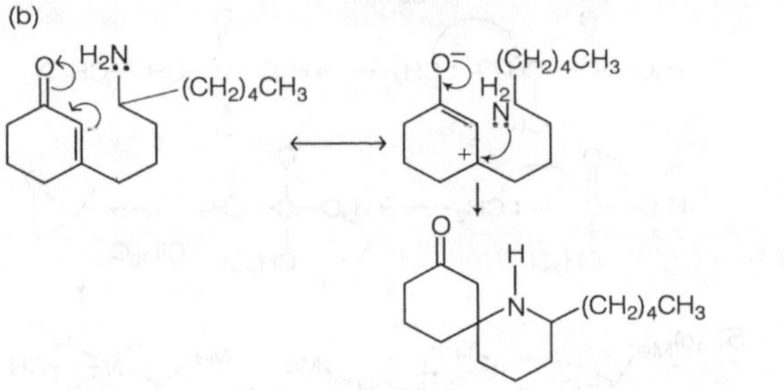

12. (a,b,d)

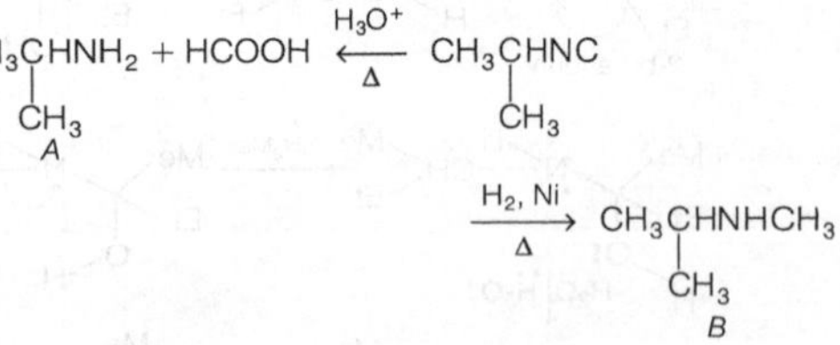

A is a primary (1°) amine so gives carbylamine reaction (with $CHCl_3/O^-H$), forms soluble sulphonamide with *p*-toluene sulphonyl chloride, liberate N_2 to give alcohol with HNO_2. *A* also gives smell of mustard oil when treated with $CS_2/HgCl_2$.

B is a secondary amine, so it does not give carbylamine or Hofmann mustard oil reaction.

Further, it forms oily nitrosoamine with HNO_2 and alkali insoluble sulphonamide with *p*-toluene sulphonyl chloride.

13. (b,c,d)

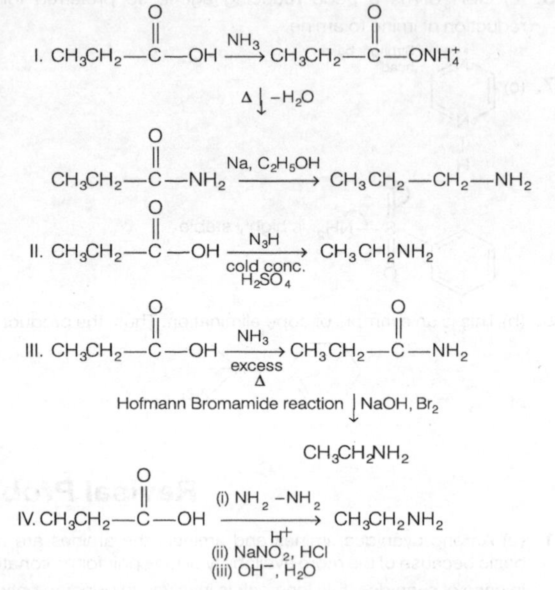

14. (a,c,d)

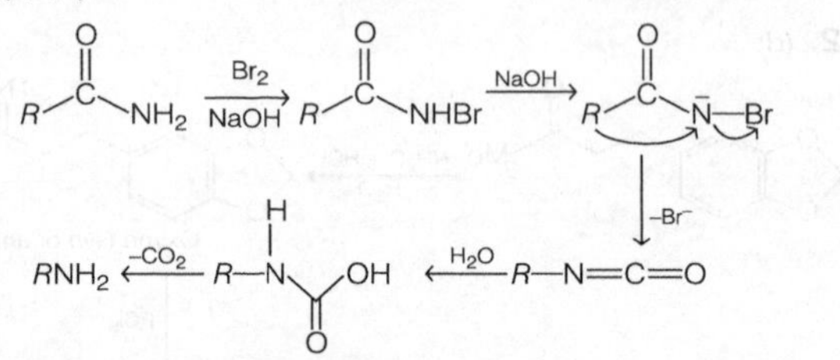

15. (b,d) Number of substituent = 2.

DBE = 6 (one due to cyclic ring, 2 due to triple bond and one for each double bond).

Total number of C in the parent chain = 6.

Here, $—\overset{\overset{O}{\|}}{C}—NH_2$ is the secondary suffix.

16. (c,d)

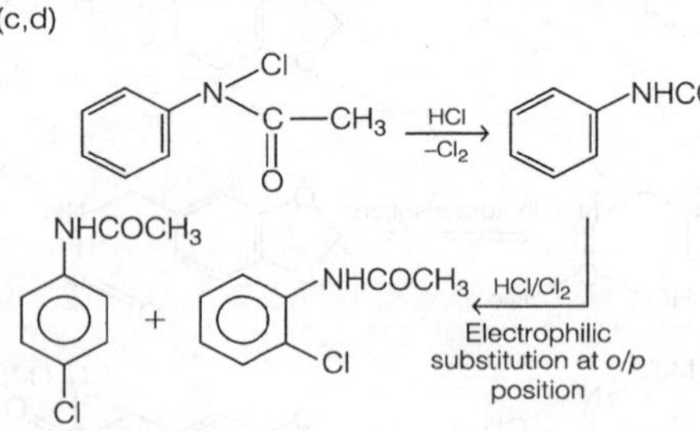

17. (b,d) Boiling point of *B* is higher than *A*.

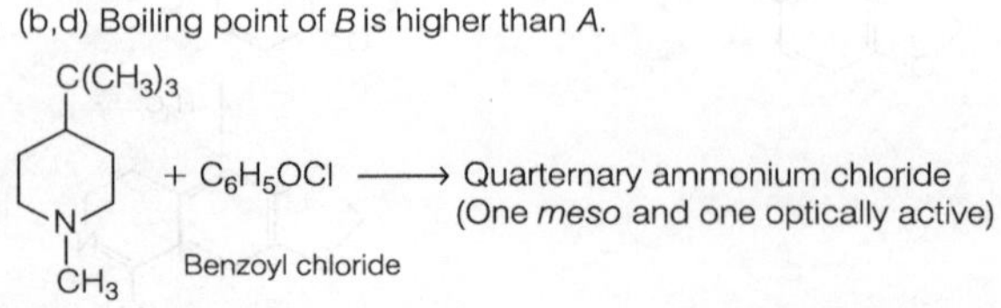

18. (a,c) Alkyl cyanide is reduced into amine by a two steps reduction process through imine intermediate as

$$RCH_2CN \xrightarrow{H_2/Ni} \underset{\text{Imine}}{R-CH_2CH{=}NH} \xrightarrow{H_2/Ni} R-CH_2CH_2NH_2$$

During reduction process, some intermediate imine undergo nucleophilic attack by the product amine, which result in formation of secondary amine as :

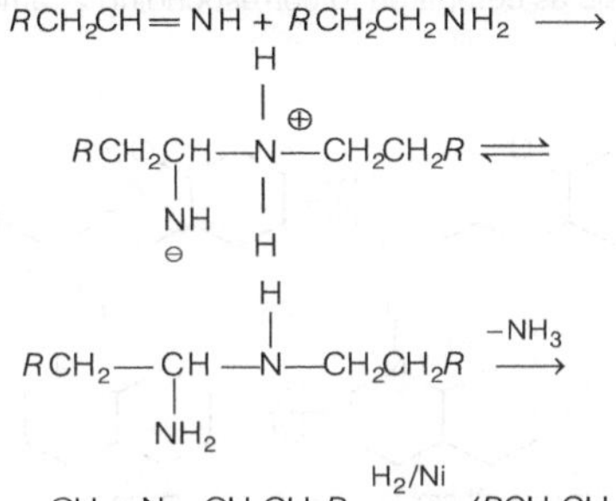

$$RCH_2-CH{=}N-CH_2CH_2R \xrightarrow{H_2/Ni} (RCH_2CH_2)_2NH$$

19. (b,c) Plane *X* is better because of the involvement of reductive amination.

Further, plane *Y* involves the elimination of HBr from 2° alkyl halide, so yield is lesser.

20. (a)

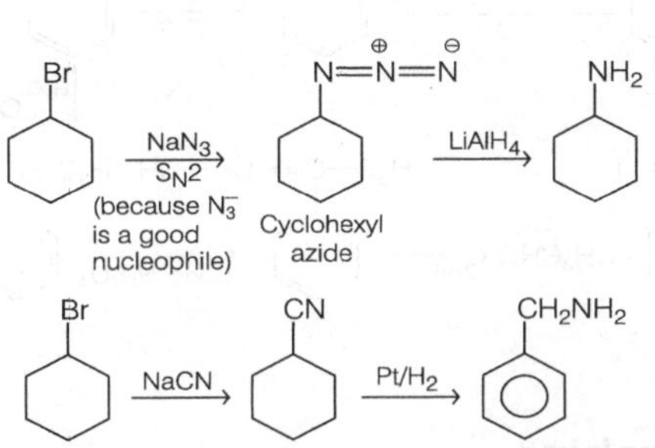

21. (d) All the given can be used as organic solvent *A*.

22. (a) Amine is in the form of its salt with acid.

$$RNH_2 + HCl \longrightarrow \underset{\text{In aqueous layer } C}{RNH_3^+Cl^-}$$

23. (d) Aqueous layer contains NaCl while amine goes in etheral layer.

24. (c) Since *B* is an optically inactive compound with formula $C_6H_{14}O$ so its possible structure is

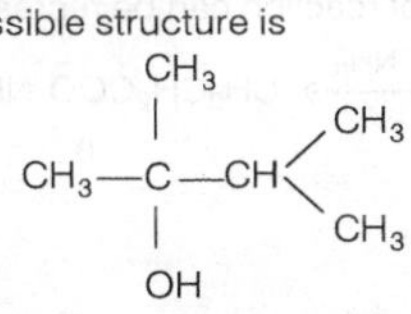

A, although is optically active, so its structure on the basis of *B* is

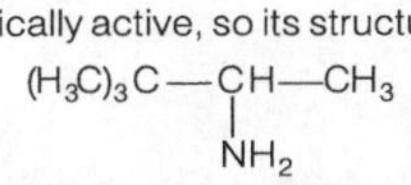

Formation of single alkenic product on treatment with CH_3I/Ag_2O, Δ, also support this structure of amine.

25. (a)

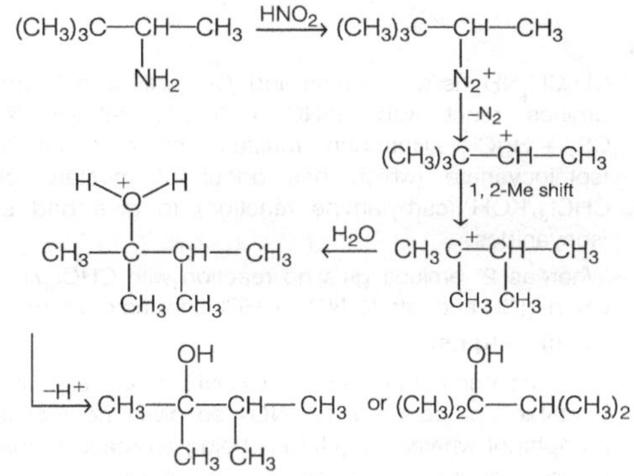

26. (b)

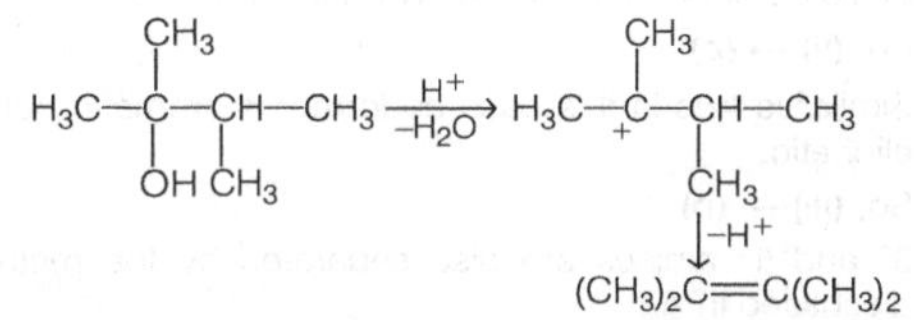

27. (a) (i) → (p,q), (ii) → (p,q,r,s), (iii) → (p,q), (iv) → (r)

Curtius degradation

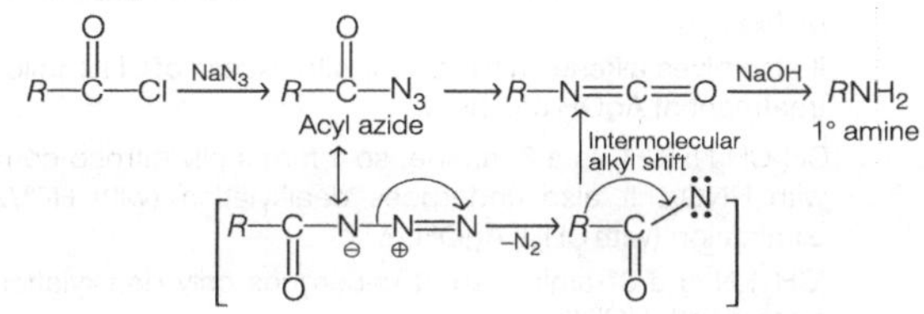

The shift takes place with complete retention of configuration in the migrating alkyl group.

Hofmann degradation

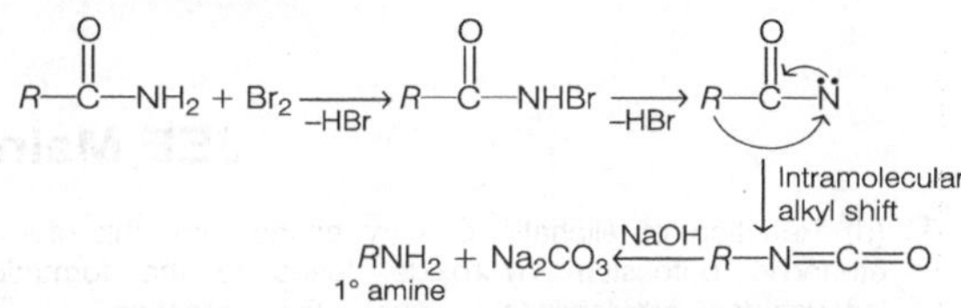

It also occurs with retention of configuration.

Schmidt reaction

$$R-\overset{O}{\overset{\|}{C}}-OH \xrightarrow[\Delta]{NaN_3+\text{Conc. }H_2SO_4} \underset{1^\circ\text{ amine}}{RNH_2} + N_2 + CO_2$$

(Release of $-CO_2$ is called decarboxylation.)

Gabriel syntheis It converts *RX* into 1° amine.

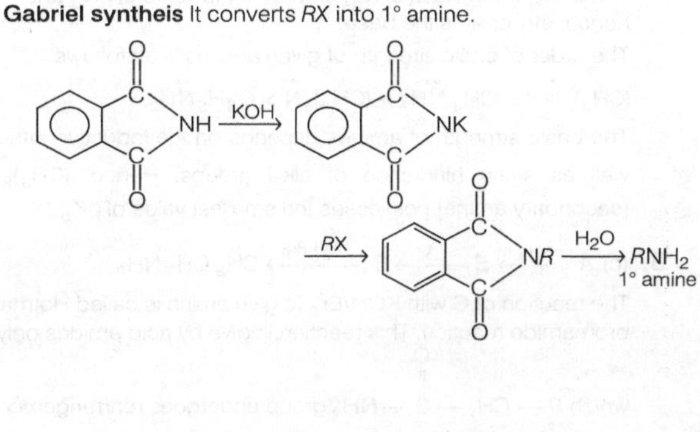

28. (a)

(i) $CH_3CH_2NH_2$ is a 1° amine and $(CH_3)_2NH$ is a 2° amine. 1° amines react with $NaNO_2 + HCl$ to release N_2 gas, $CS_2 + HgCl_2$ (Hofmann mustard oil reaction) to give isothiocyanate (which has odour of mustard oil) and $CHCl_3/KOH$ (carbylamine reaction) to give bad smelling isocyanides.

Whereas 2° amines give no reaction with $CHCl_3/KOH$ and $CS_2/HgCl_2$ and with $NaNO_2 + HCl$, it forms oily nitrosamine.

So, **(i) → (q,r,s)**

(ii) Both are primary amines but CH_3NH_2 being aliphatic forms unstable compound with HNO_2 so gives no reaction with β-naphthol whereas $C_6H_5NH_2$ being aromatic forms stable diazo compound and gives dye with β-naphthol.

So, **(ii) → (p)**

(iii) Similar is true in this case as former is aromatic but later is aliphatic.

So, **(iii) → (p)**

(iv) 3° and 1° amines are also separated by the methods as discussed in *A*.

29. (c) (i) → (r,s), (ii) → (p,r,s), (iii) → (s), (iv) → (q)

1° amines undergo dealkylation when treated with HCl followed by heating.

It also gives alkene on treatment with excess of CH_3I, followed by treatment of AgOH and heat.

$CH_3CH_2NHCH_3$ is a 2° amine, so it forms oily nitroso compound with HNO_2. It also undergoes dealkylation (with HCl/Δ) and elimination (with CH_3I, AgOH, Δ).

$(CH_3)_3N$ is a 3° amine, so, it undergoes only dealkylation when treated with HCl/Δ.

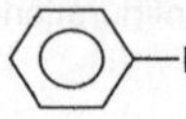—NH_2 being aromatic form diazonium salt with $NaNO_2$, HCl

30. (b) Both the statements are true, $NaBH_3CN$ being a good reducing agent reduces imine into 1° amine.

31. (c) $NaBH_3CN$ reduces imine to 1° amine. However, it is less reactive as compared to $NaBH_4$.

32. (a) Because of over crowding due to $+I$-effect of three alkyl groups, strain increases in case of 3° amines, hence, these are less basic as compared to corresponding 2° amines in aqueous solution.

33. (2)

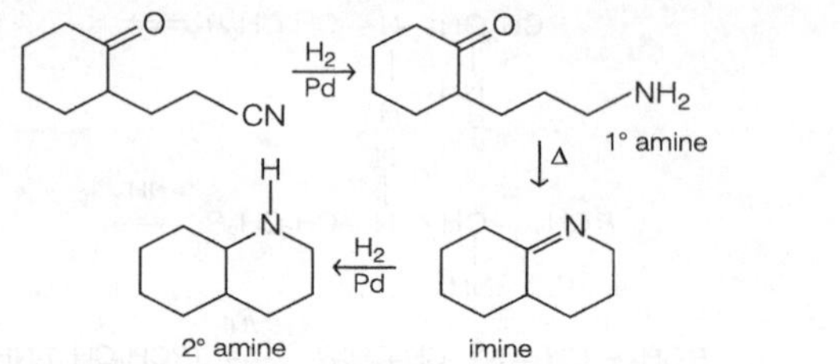

34. (8)

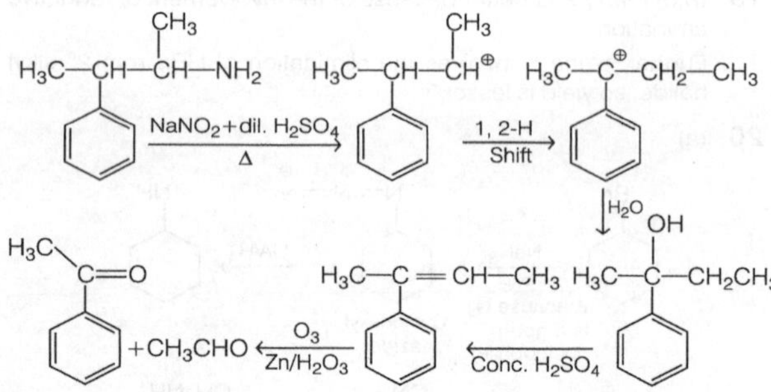

JEE Main & AIEEE Archive

1. (d) Reaction of aliphatic primary amine with chloroform and ethanolic potassium hydroxide, leads to the formation of isocyanides/carbylamines. Thus, this reaction is called carbylamine reaction.

$$R-NH_2 + CHCl_3 + KOH \xrightarrow{\Delta} \underset{\text{Alkylic isocyanide}}{R-NC} + 3KCl + 3H_2O$$

2. (a) It is known that, $pK_b = -\log K_b$

Thus, larger the value of K_b, smaller is the value of pK_b and hence, stronger is the base.

The order of basic strength of given amines is as follows

$(CH_3)_2\ddot{N}H > CH_3\ddot{N}H_2 > (CH_3)_3\ddot{N} > C_6H_5\ddot{N}H_2$

The basic strength of amines depends on the inductive effect as well as steric hindrance of alkyl groups. Hence, $(CH_3)_2\ddot{N}H$ [secondary amine] possesses the smallest value of pK_b.

3. (c) $A \xrightarrow{NH_3} B \xrightarrow{\Delta} C \xrightarrow{KOH/Br_2} CH_3CH_2NH_2$

The reaction of C with KOH/Br_2 to give amine is called Hofmann bromamide reaction. This reaction is give by acid amides only in which $R-CH_2-\overset{O}{\overset{\|}{C}}-NH_2$ group undergoes rearrangement along with the loss of CO_2 molecule. Thus, the compound C must be acid amide with three carbon atoms.

Hence, the compound C is $CH_3-CH_2-\overset{O}{\overset{\|}{C}}-NH_2$.

Since, all the options shows that *A* is an acid and it forms acid amide on reaction with NH_3. Thus, acid must contain three carbon atoms. Hence, the compound *A* is CH_3CH_2COOH. The complete series of reaction can be represented as

$$\underset{A}{CH_3CH_2COOH} \xrightarrow{NH_3} \underset{B}{CH_3CH_2CO\bar{O}\,\overset{+}{N}H_4} \xrightarrow{\Delta} \underset{C}{CH_3CH_2-\overset{O}{\overset{\|}{C}}-NH_2} \xrightarrow[\text{Hofmann bromamide reaction}]{KOH/Br_2} CH_3CH_2NH_2$$

4. (a) Methyl isocyanate $CH_3-N=C=O$ (MIC) gas was leaked from the storage tank of the union carbide plant in Bhopal gas tragedy.

5. (c) During acetylation, one H-atom with atomic mass 1 amu of $—NH_2$ group is replaced by an acetyl group CH_3CO [molecular mass = 43 u]. Thus,

$$—NH_2 + CH_3COCl \longrightarrow —NHCOCH_3 + HCl$$

The above equation suggests that, the acylation of each $—NH_2$ group increases the mass by 42 u [43 – 1]. If the molecular mass of the organic compound is 180 u while that of the acylated product is 390 u, then the increase in the mass due to acylation is given by

$$390 - 180 = 210 \text{ u}$$

Hence, the number of $—NH_2$ groups $= \dfrac{210 \text{ u}}{42 \text{ u}} = 5$

6. (c) In aqueous solution, basicity order is

Dimethylamine > methylamine > trimethylamine > aniline

This order depends upon the inductive effect and steric hindrances of alkyl groups.

7. (c) $CH_3CH_2NH_2 + CHCl_3 + 3KOH \xrightarrow{\text{(alc)}}$

$$\underset{A}{CH_3CH_2—N \overset{\rightarrow}{=} C} + 3KCl + \underset{D}{3H_2O}$$

The above reaction is called carbylamine reaction in which isocyanides or carbylamines are produced.

8. (b) Wurtz reaction is used to prepare alkanes from alkyl halides.

$$2R—X + 2Na \xrightarrow{\text{Dry ether}} R—R + 2NaX$$

9. (c)

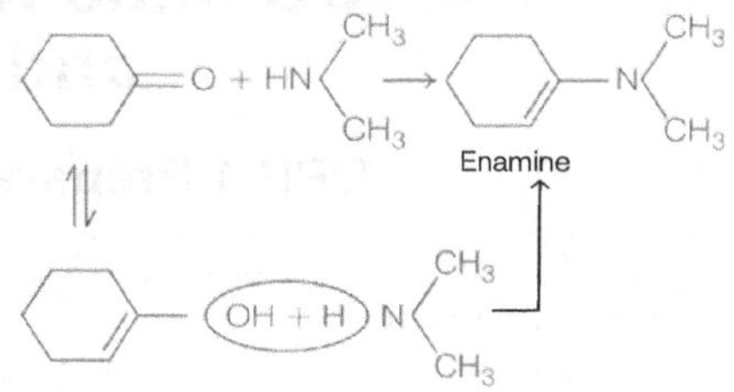

10. (c) The basicity of an amine in aqueous solution primarily depends upon the stability of the ammonium cation or the conjugate acid formed by accepting the proton from water.

$$RNH_2 + H_2O \rightleftharpoons \underset{\text{[ammonium cation]}}{R\overset{+}{N}H_3} + OH^-$$

The stability of the ammonium cation depends upon

(i) $+I$ effect of alkyl groups

(ii) Extent of H-bonding with water molecules

(iii) Steric effects of the alkyl groups

On the basis of above factors, the basic strength of amine is as follows

$$(CH_3)_2NH > CH_3NH_2 > NH_3$$

($+I$ effect of $—CH_3$ group predominates)

JEE Advanced & IIT JEE Archive

1. (a) (i) → (p,q,s,t), (ii) → (p,s,t), (iii) → (p), (iv) → (r)

i.	$CH_3CH_2CH_2CN$:	Gives amine with Pd -C / H_2 Gives aldehyde with $SnCl_2$ / HCl Gives amide with diisobutyl- aluminium hydride. Gives carboxylic acid on alkaline hydrolysis.
ii.	$CH_3CH_2OCOCH_3$: Ester	Reduced to alcohol with Pd -C / H_2 Reduced with diisobutylaluminium hydride into aldehyde. Undergo alkaline hydrolysis.
iii.	$CH_3CH = CHCH_2OH$:	Reduced to butanol when treated with Pd -C / H_2 Not reduced with $SnCl_2$, diisobutyl aluminium hydride.
iv.	$CH_3CH_2CH_2CH_2NH_2$:	A primary amine, gives carbylamine test.

2. (d) $CH_3NH_2 + CHCl_3 + KOH \longrightarrow \underset{\text{Isocyanide}}{CH_3—\overset{+}{N} \equiv \overset{-}{C}} + KCl + H_2O$

4. Aromatic Nitro Compounds, Anilines and Diazonium Salts

DPP-1 Preparation and Properties of Aromatic Nitro Compounds

1. (i) *p*-product is the major product.

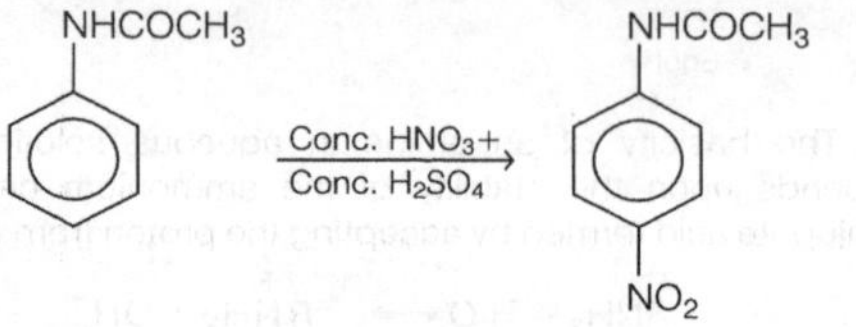

(ii) CH_3COO^- is a *meta* directing group so product is

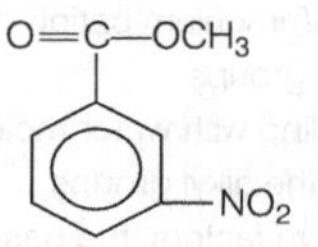

(iii) Alkoxy is an *o* / *p* directing group and *p*-is the major product. Thus, the product is

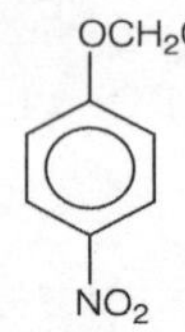

(iv) Electrophilic substitution takes place at position *ortho* to Me group. Thus, product is

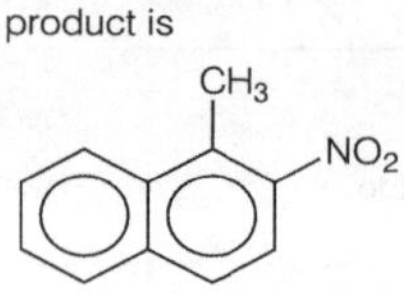

2. (i) $PhNO_2 \xrightarrow{Zn + aq.\ NH_4Cl} PhNHOH$
N–phenyl hydroxylamine

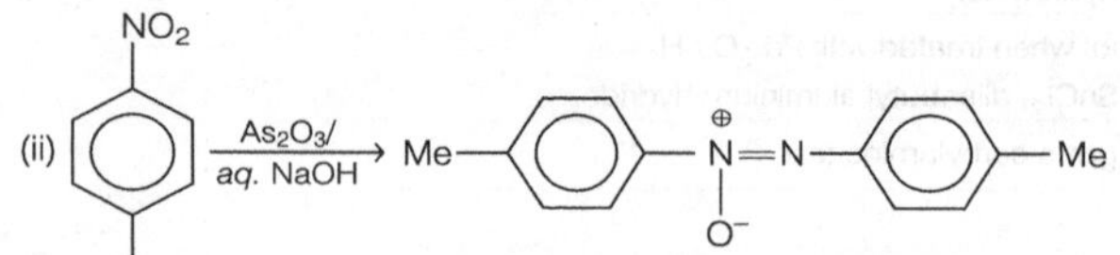

p-methylazoxybenzene

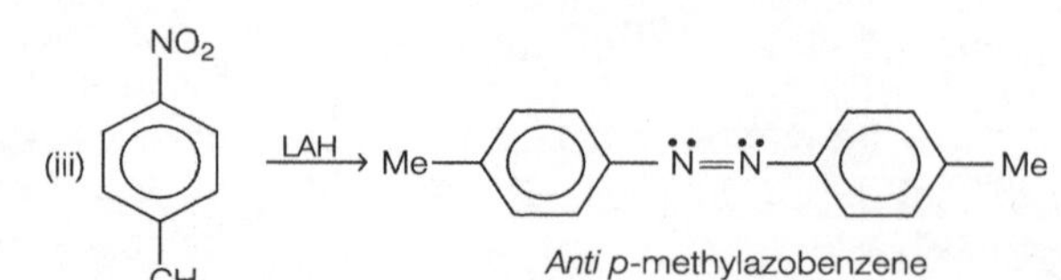

Anti p-methylazobenzene
+

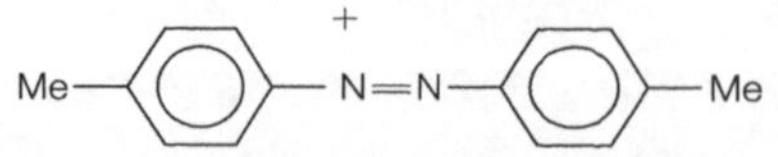

Syn p-methylazobenzene

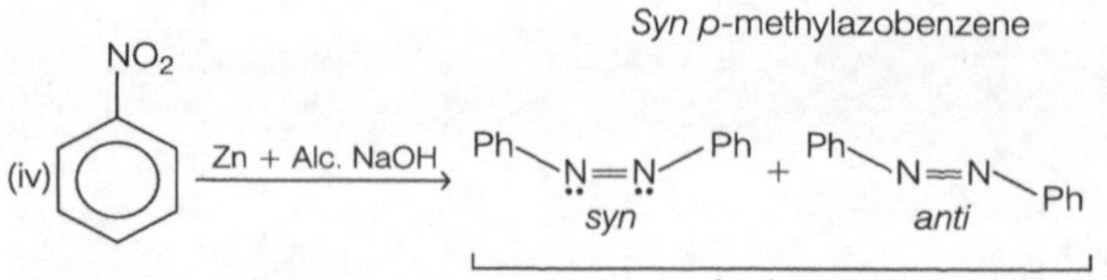

Azobenzene

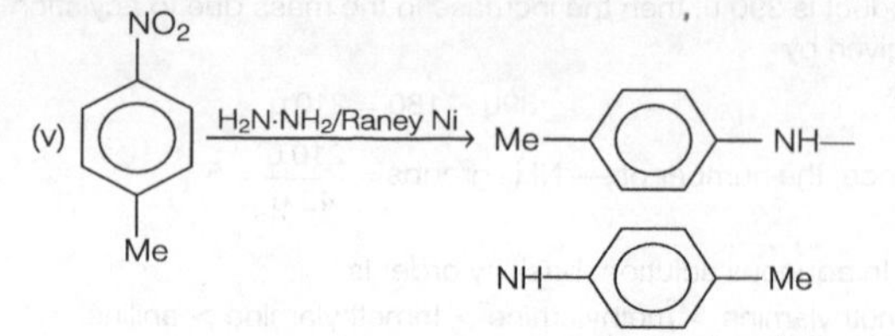

p-methyl hydrazobenzene

3. (i) $Br_2 + AlCl_3 \longrightarrow AlCl_3Br^{\ominus} + Br^{\oplus}$

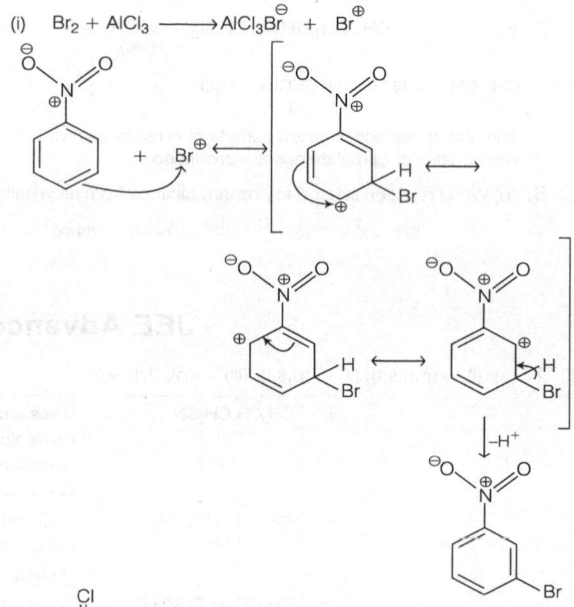

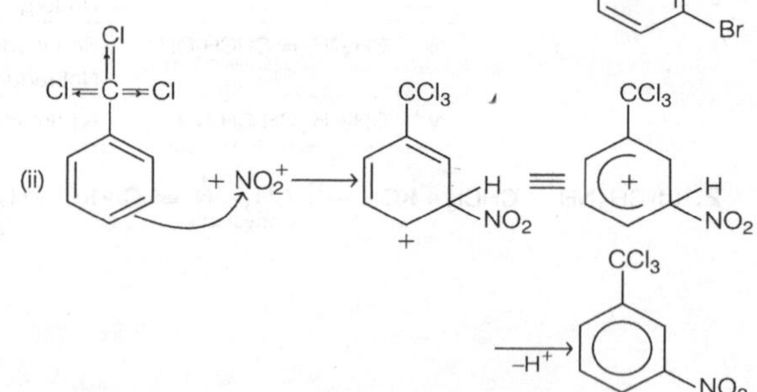

$[HNO_3 + H_2SO_4 \longrightarrow NO_2^+ + H_2O + HSO_4^-]$
Electrophile

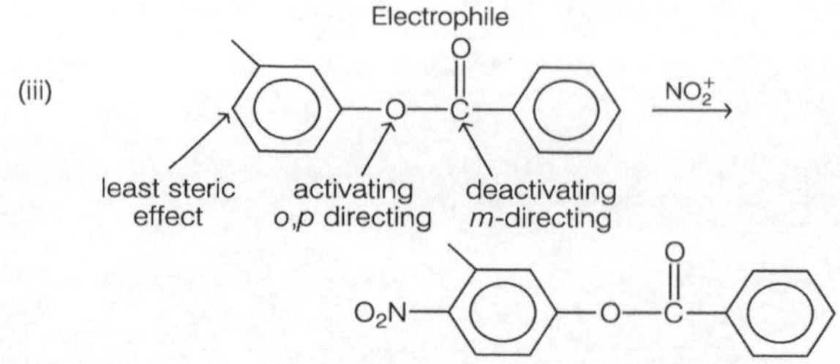

4. Sn/HCl reduces —NO_2 group into NH_2. Presence of which makes ring I more reactive towards oxidation.

Thus, the product is

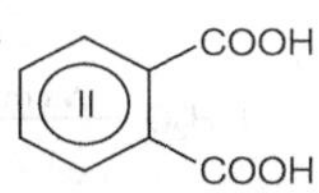

In case II direct oxidation result in the formation of

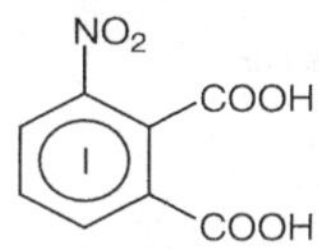

because —NO_2 group deactivates the ring towards oxidation.

5. (c) Formation of nitronium ion (NO_2^+) require some energy and is the slowest step of the reaction. Hence, it is expected as a rate determining step.

Protonation is usually fast, so cannot be the rate determining step and formation of arenium ion 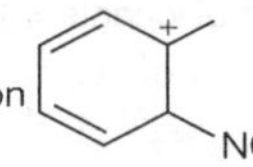 cannot be the rate determining step because the reaction is independent of the concentration of toluene.

6. (b)

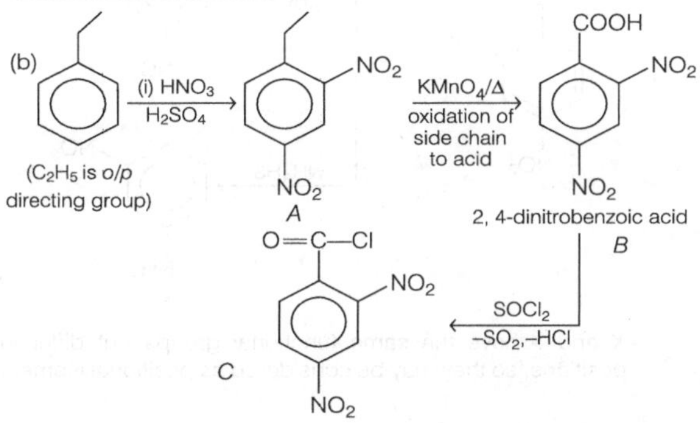

7. (c)

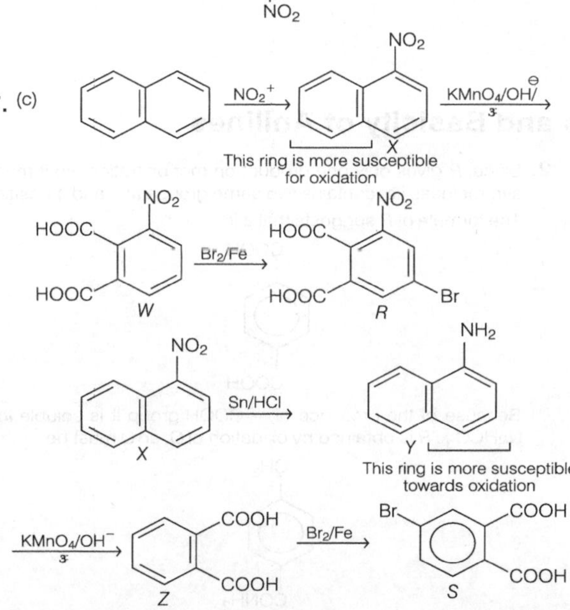

8. (d) —NO_2 is a highly deactivating group and deactivates the benzene nucleus to such an extent that it does not undergo Friedel-Craft's reaction if —NO_2 group is present (i.e. in case of nitrobenzene).

9. (d) Although bromination of nitrobenzene is more difficult than nitration of bromobenzene (i.e. $r_2 > r_1$) but in case of second electrophilic substitution $r_4 > r_3$.

10. (c)

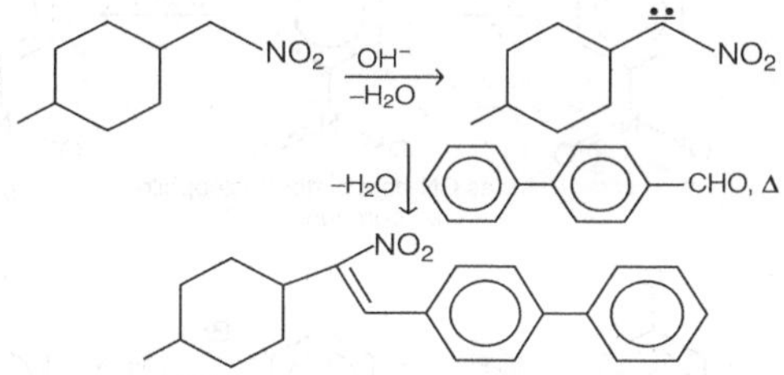

11. (b,c) $PhNO_2 \xrightarrow{LAH} Ph—N=N—Ph$ (Azobenzene)

$PhNO_2 \xrightarrow{Al\text{-}Hg/H_2O} PhNHOH$ (phenyl hydroxyl amine)

12. (a,c) Experimentally, it was found that nitration of benzene is a zero order reaction. It can be achieved by using concentrated nitric acid alone also. NO_2BF_4 salt is also a source of nitronium ion (NO_2^+) so able to cause nitration.

Presence of H_2SO_4 increases the rate of generation of NO_2^+ ion.

13. (a,c) (a)

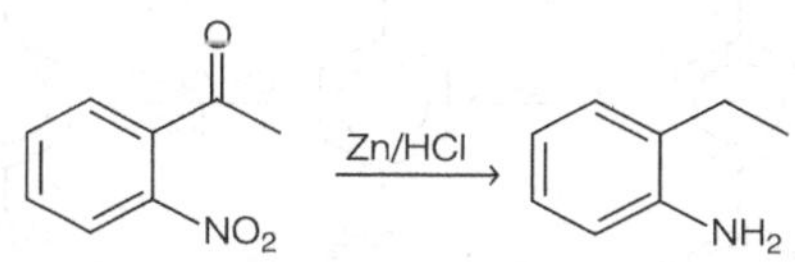

(Because Zn/HCl will reduce both —NO_2 as well as keto group).

(b)

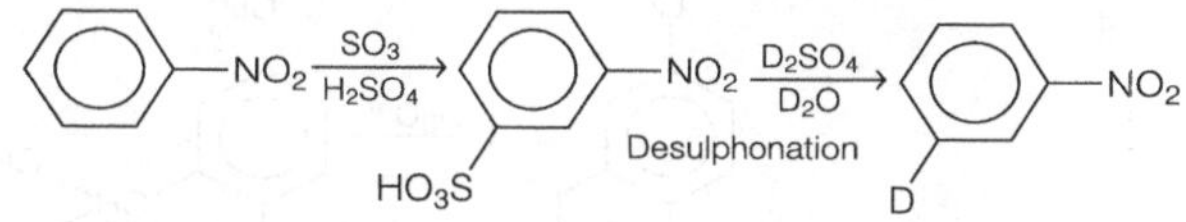

(c) In the presence of—NO_2 group, Grignard synthesis does not occur i.e. no reaction takes place.

(d) In the final step, diazonium group will be replaced by deuterium to give the product shown.

14. (b,c) The given reactions are the examples of nucleophilic substitution. In the substituted benzene containing an e^- with drawing group S_N reaction is possible only at *o* or *p*-position with respect to electron withdrawing group.

S_N reaction never takes place at the *m*-position with respect to electron withdrawing group because an electron deficient centre cannot be generated at the *m*-position.

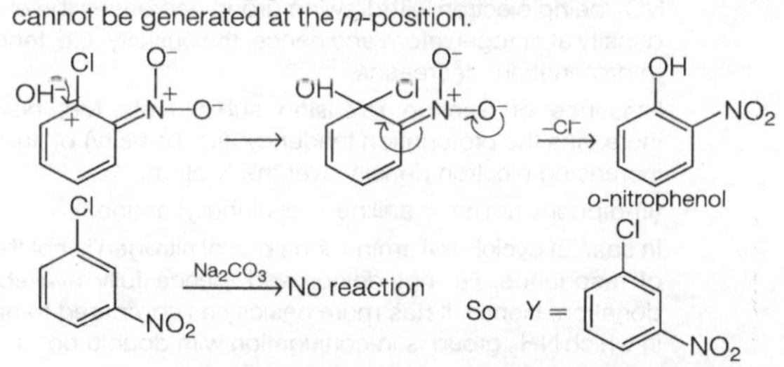

15. (d) Nitrobenzene because of the presence of highly deactivating —NO_2 group does not undergo Friedel-Craft reaction, so reaction takes place with chloroethane in the presence of $AlCl_3$. However, —NO_2 group is a *meta* directing group in EAS reaction because e^- density is more at these positions.

16. (b) Presence of two nitro groups on the benzene ring deactivates the aromatic ring to such a level that introduction of third nitro group becomes impractical, thus, trinitrobenzene cannot be prepared by direct nitration of benzene.

If three —NO_2 groups are present steric hindrance is more.

17. (b)

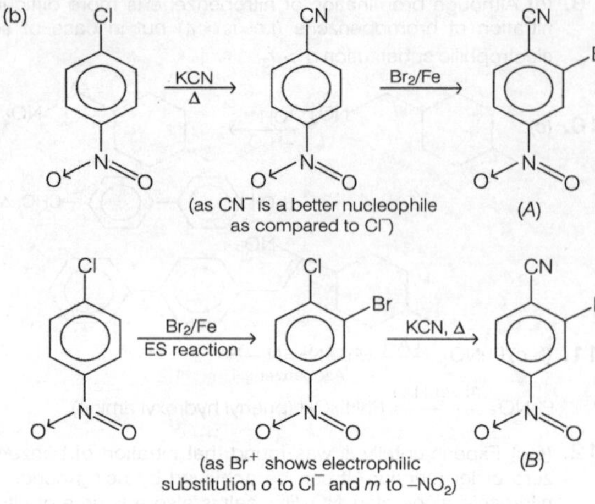

Thus, *A* and *B* are identical molecules.

18. (b)

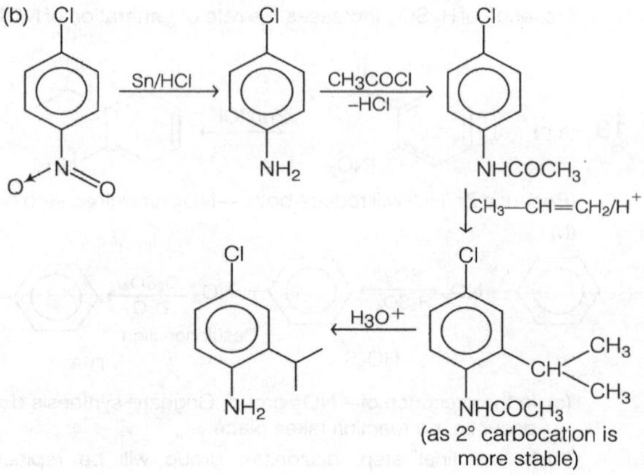

19. (9)

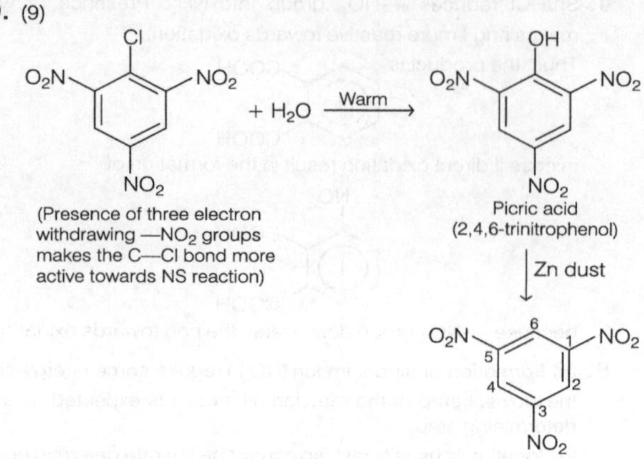

Sum of positions of locants = 1 + 3 + 5 = 9

20. (4)

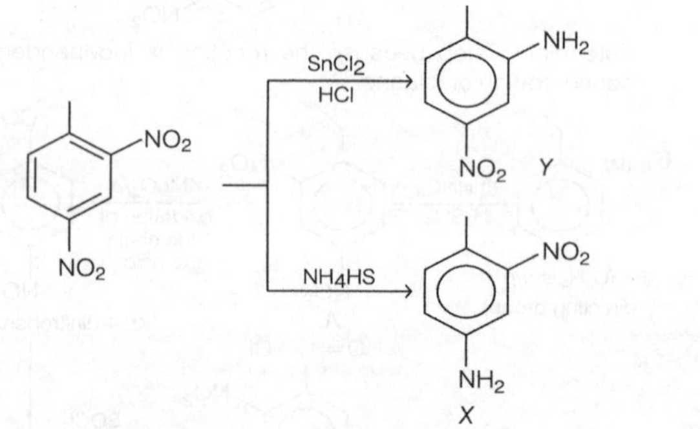

X and *Y* have the same functional groups but differ in their positions, so they may be considered as positional isomers.

DPP-2 Preparation Methods and Basicity of Anilines

1. (a) (i) *p*-nitroaniline < aniline < *p*-methylaniline

NO_2 being electron withdrawing group decreases the electron density at nitrogen atom and hence, the basicity, (i.e. tendency to protonation) decreases.

Presence of electron releasing substituents Me, however, increases the protonation tendency. (i.e. basicity) of aniline by increasing electron density over the N-atom.

(ii) diphenylamine < aniline < cyclohexyl amine.

In case of cyclohexyl amine, lone pair of nitrogen is not the part of resonance, i.e. not delocalised, hence fully available for donation. Hence, it has more basicity as compared to amines in which NH_2 group is in conjugation with double bond.

In case of diphenylamine, the lone pair of nitrogen are much less available for protonation, hence, it is least basic among the given.

(b) (i) I is more basic as lone pair of N is not a part of conjugation and is more available for protonation.

(ii) I is more basic as amides are less basic because lone pair of N is involve in resonance with $>C=O$.

2. Since, *R* gives only one product on mononitration, so it must be symmetrical, i.e. contains two same groups at 1 and 4 positions. The formula of *R* suggests that it is

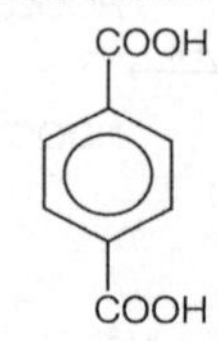

Because of the presence of —COOH group it is soluble in *aq*. $NaHCO_3$. *R* is obtained by oxidation of *Q*, so *Q* must be

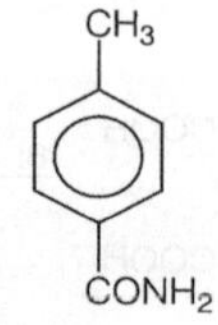

This compound being neutral is insoluble in dilute acid and base.

The reactions are as follows.

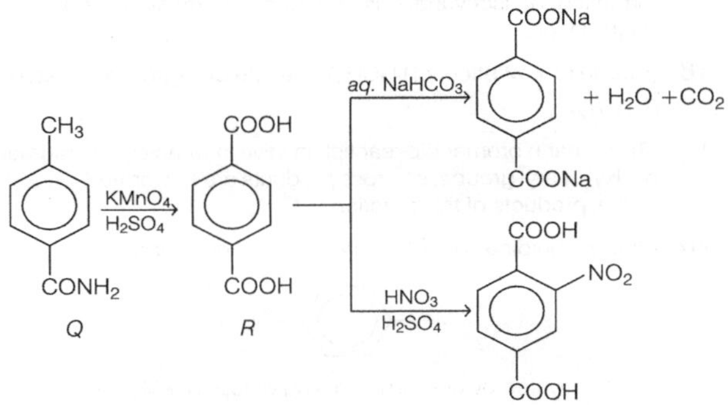

Product

3. (a) Since, the reaction involves conversion of an N, N-dialkylated aniline into *o*/*p*-alkylated aniline, so it is an example of Hofmann Martius rearrangement.

4. (b) Amine group (—NH_2) attached with CH_2 group is more basic, hence, first attacked by the HCl molecule.

5. (d) In acetanilide N is attached directly with electron withdrawing >C=O group along with phenyl group. So, lone pair is less available.

6. (d) When the migratory group is aryl (as in the given case), the rate of degradation is increased by the presence of electron releasing substituents in the aromatic ring.

7. (c)

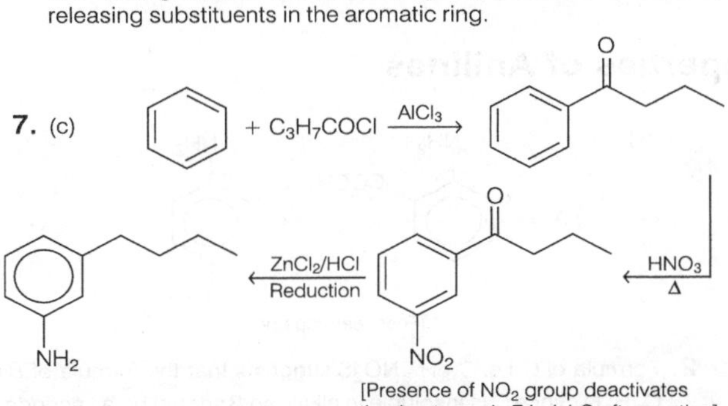

[Presence of NO_2 group deactivates the ring towards Friedel-Crafts reaction]

8. (d) In this structure, there is steric inhibition to resonance due to lack of planarity of nitrogen with aromatic ring. This occurs due to steric crowding by the three bulky adjacent groups.

9. (d)

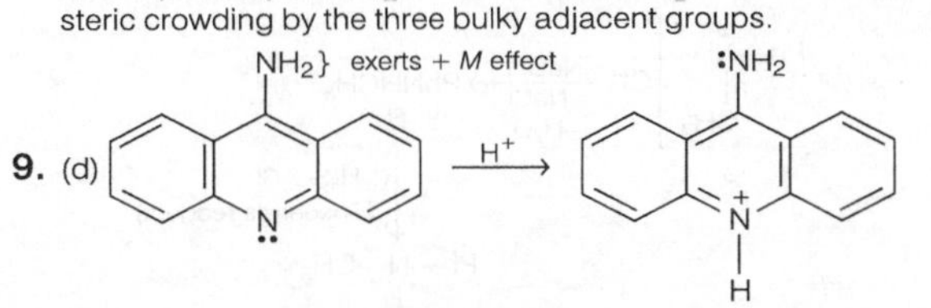

This lone pair does not delocalised in the ring.

(Since, lone pair of N is not delocalised and NH_2 group exerts + *M* effect, so it is most reactive towards protonation, i.e. most basic).

10. (a) —CCl_3 has electron withdrawing effect, so decreases the electron density of N. The effect is more at *p*-position so the order of basic strength is

Aniline > *p* > *m* > *o*

11. (c) Alkyl amines are more basic than aryl amines or amides due to delocalisation of lone pair of N in later case or in case of amides.

Electron withdrawing group like NO_2 also result in decreased electron density over N-atom of amine group and hence, makes the aniline less basic.

Thus, the order of basicity is

$$IV < III < I < II$$

12. (c) A strong base has more tendency of protonation. Pyridine

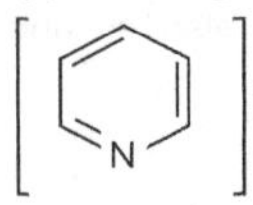

is stronger base as compared to pyrrole

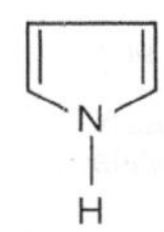

, so its protonated form readily loses proton to pyridine.

Hence, it is a feasible reaction.

In other reactions, less basic site gets protonated which is not possible because of the less availability of lone pair for protonation.

13. (a) Compound *B* when hydrolysed gives equimolar quantities of *p*-chloroaniline and benzoic acid, so it must be a *para*-chloro substituted secondary amide.

Solid *A* on reaction with PCl_5 gives an amide so it must be an oxime of formula.

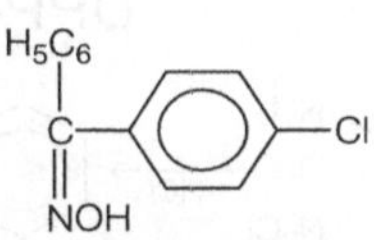

Ketoxime (*A*)

This ketoxime undergoes Beckmann rearrangement in the presence of PCl_5 (in ether) to give respective amide *B* as

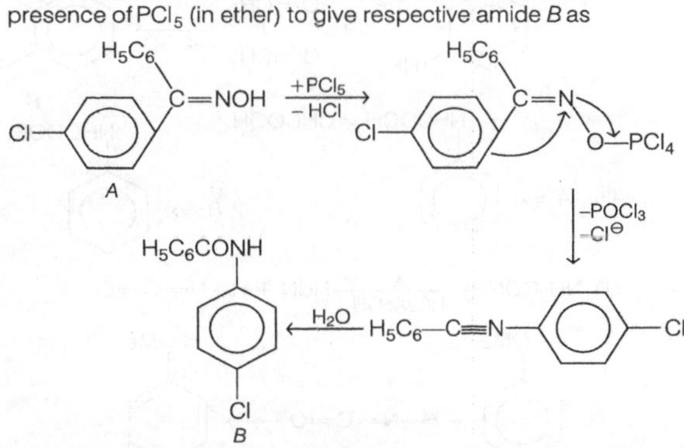

14. (b,c,d) (a) —OH group of phenol does not undergo nucleophilic substitution reaction, so reaction given in option (a) is not feasible.

(b) At high temperature, chlorobenzene reacts with ammonia to give aniline *via* nucleophilic substitution reaction.

(c) It occurs by benzyne mechanism.

(d)

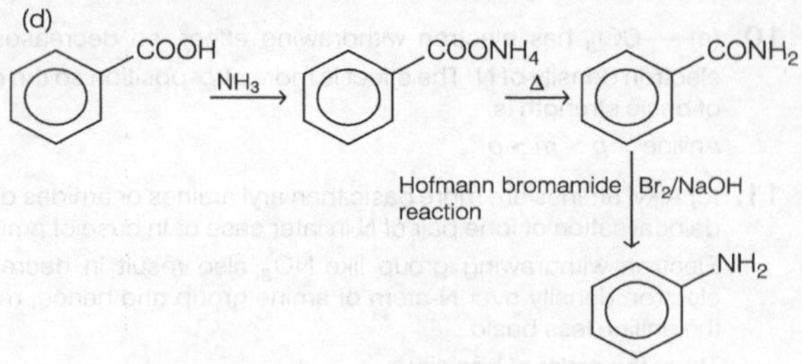

15. (a,b) Lone pair of N-atom of pyridine does not involve in delocalisation, so more available for protonation, hence it is more basic as compared to aniline.

Presence of electron releasing Me group also enhance the basicity of amines.

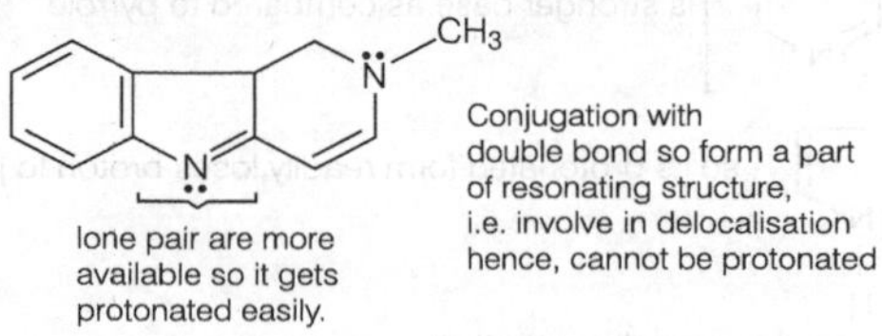

16. (b) Both are the correct statements.

N, N-diethylaniline is more basic as compared to aniline. This is because of the presence of bulky ethyl groups on nitrogen which destroy planarity and hence, resonance of nitrogen lone pair with aromatic ring. Consequently basic strength increases.

17. (d) Slowest step in the rate determining step of a reaction thus, formation of isocyanate is the rate determining step of this degradation.

18. (b) Due to + *M* effect of N $(CH_3)_2$ the rate of migration is faster i.e. $R_1 < R_2$.

19. (3) Hofmann bromanide reaction involve intramolecular migration of alkyl or aryl groups, so cross products are not formed. Thus, I, V and IX products of this reaction.

20. (3) (I) *p*-toluidine cannot be prepared, since arylhalide

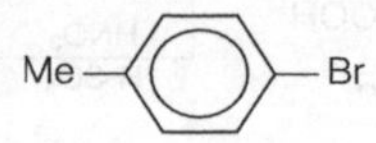

is not activated by electron donating group (like Me group). Aromatic nucleophilic substitution (ArS_N) takes place only when the benzene ring is activated by electron withdrawing group present at *o*/*p*-positions.

(II) Since —NO_2 group is present at *m*-position so ArS_N does not take place.

(III) Vinylamine is also not prepared by this method because vinylhalides are unreactive towards S_N2 reactions.

(IV) *p*-nitroaniline and *o*-nitroaniline (V) because their respective halide due to the presence of electron withdrawing —NO_2 group at *o*/*p*-positions undergo ArS_N reaction and hence can be prepared by Gabriel method.

(V) Allyl halide can also undergo S_N2 reaction, so allylamine can also be prepared by Gabriel method.

DPP-3 Chemical Properties of Anilines

1. (i) $A \Rightarrow$

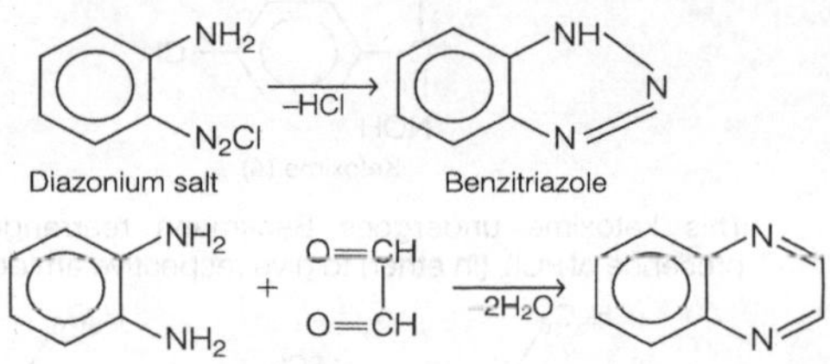

(ii) $A \Rightarrow$ $B \Rightarrow$

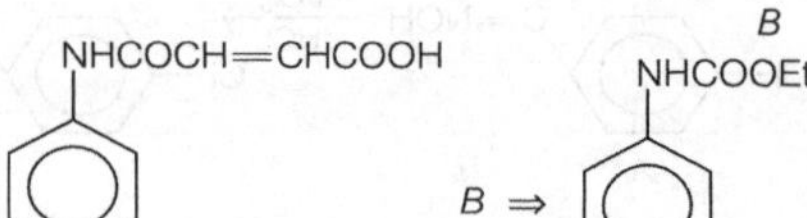

(iii) $NH_2CONH_2 \xrightarrow[HCl/AcOH]{\Delta} HCN + H—N=C=O$

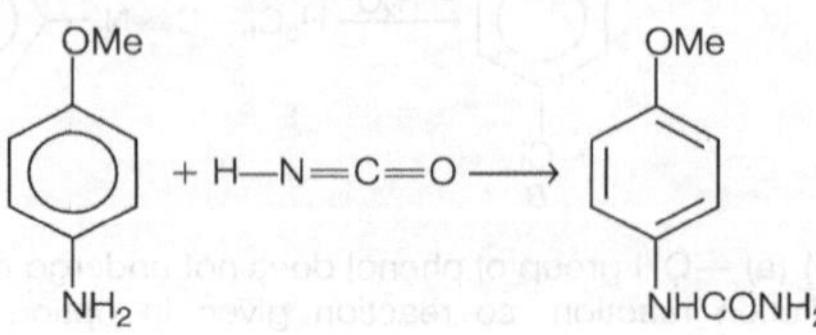

(iv) *A*

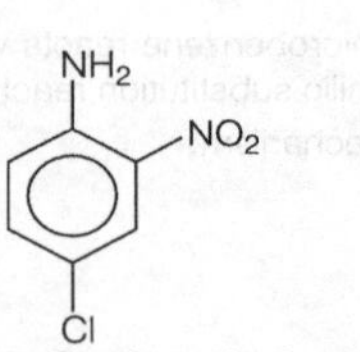

Only —Cl *ortho* to nitro is activated.

(v)

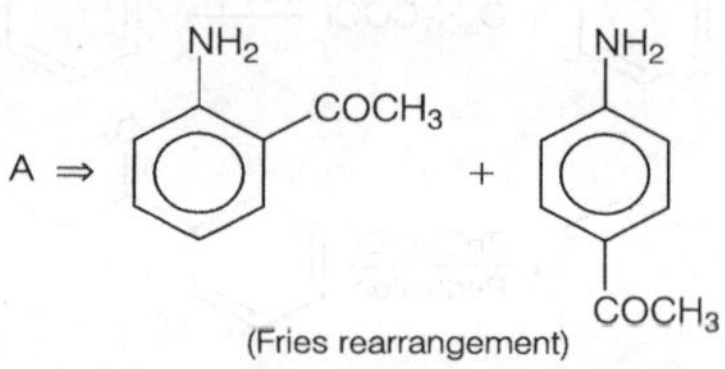

(Fries rearrangement)

2. Formula of *C* i.e. $C_{13}H_{13}NO_2S$ suggests that the formula of *B* is C_7H_9N. Since *C* is insoluble in alkali, so *B* should be a secondary amine of formula $C_6H_5NHCH_3$.

A is acidic to litmus and gives chlorine free compound on treatment with NaOH. So, it must be chloride salt of amine i.e. *A* is

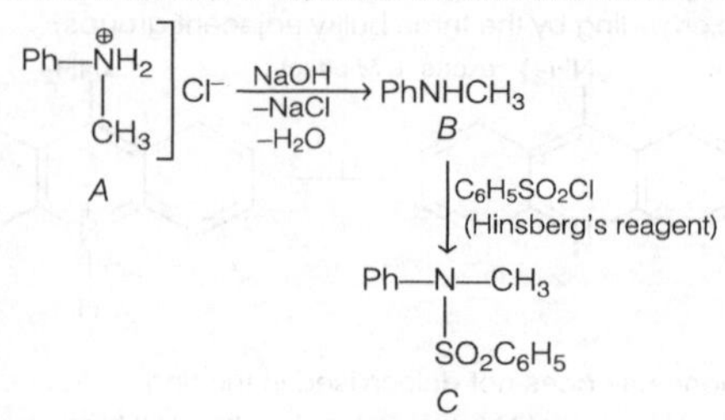

[(C) is insoluble in alkali due to the absence of H attached directly with N.]

3. (b)

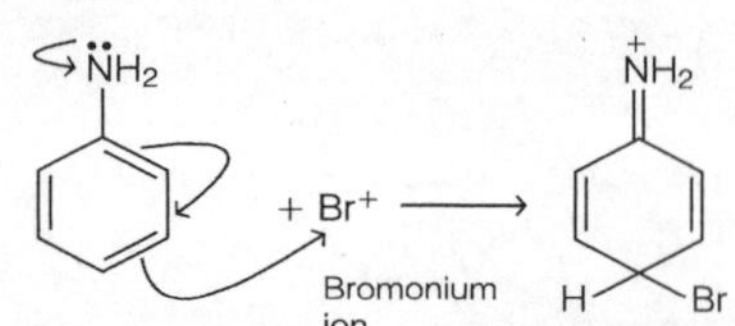

4. (d)

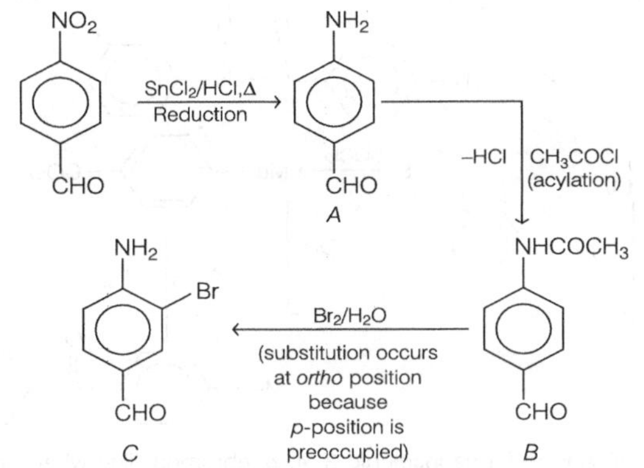

5. (c)

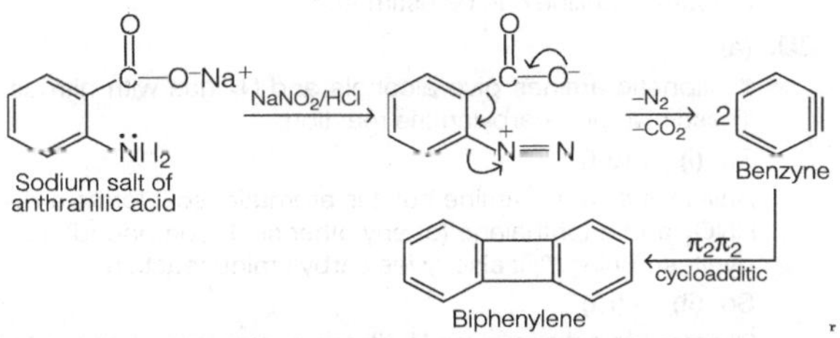

6. (c) This is because of the protonation of amine group which makes it *meta* directing.

7 (d)

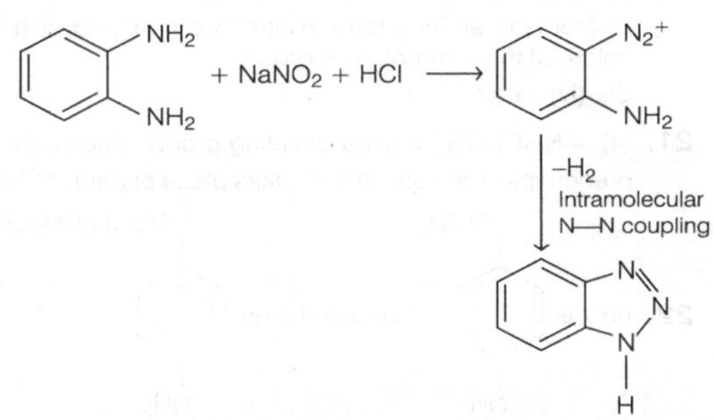

8. (a) In case of pyrrole electrophilic aromatic substitution takes place at C-2 and C-5 position.

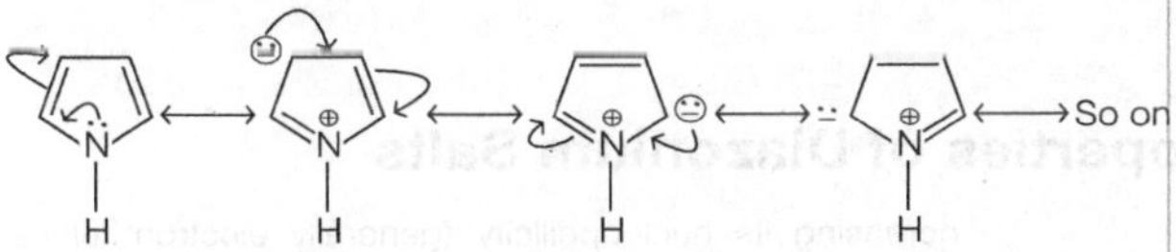

9. (c) $AgNO_3$ gives white ppt of AgCl with anilinium hydrochloride because of the presence of free chlorine in it but *p*-chloroaniline gives no reaction with it. Hence, $AgNO_3$ can be used to distinguish them.

10. (d) Formation of single product in electrophilic substitution reaction indicates the symmetrical structure, (i.e. *p*-substitution) of *A*. Formation of dye stuff with $NaNO_2$/HCl (0°C) and 2-naphthol suggests that it is a aromatic primary amine. Thus, possible structure of *A* is

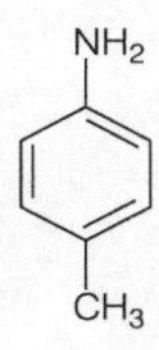

11. (c)

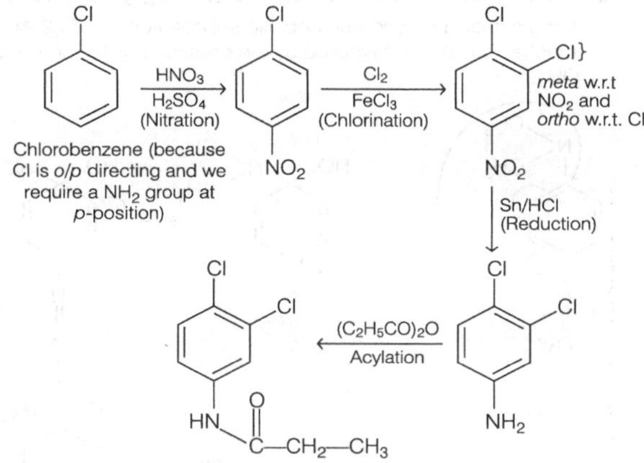

12. (c) In acidic medium, NH_2 group accepts a proton and forms anilinium ion. The position *ortho* to it is more activated. But in basic medium, OH loses its proton to form phenoxide ion. The position *ortho* to O^- is more activating for coupling in this condition.

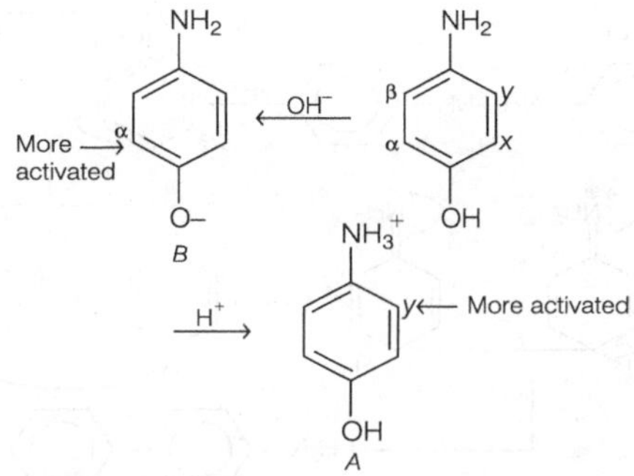

13. (a,b)

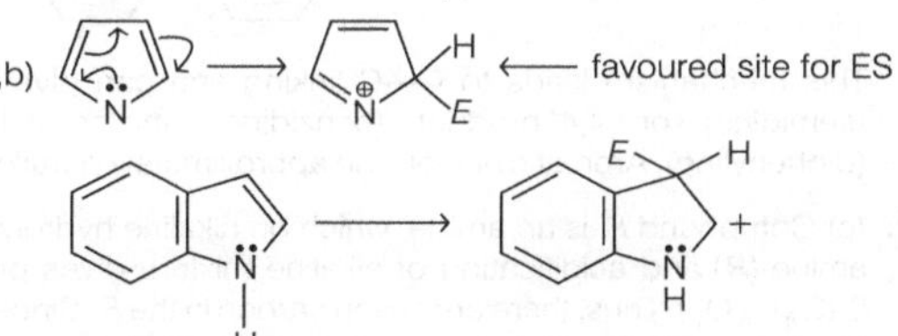

Here, *E* = electrophile

14. (a,b,c,d)

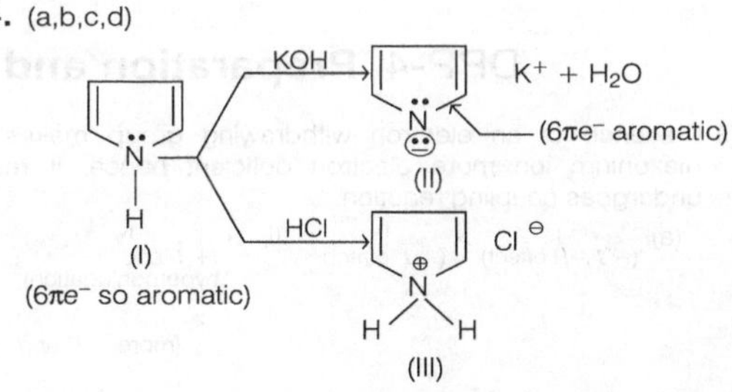

Thus, only one of the three is non-aromatic.

15. (a,c,d) —N=O is a $-M$ and $-I$ showing group, so favours bimolecular aromatic nucleophilic substitution (i.e. S_N2 Ar). This mechanism is also favoured by the presence of strong base e.g. OH^-.

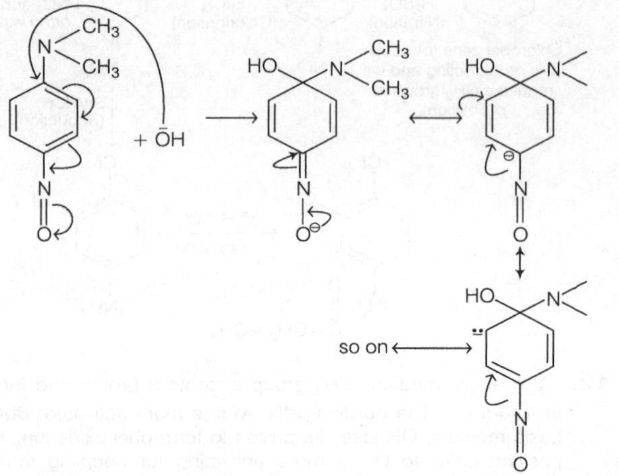

16. (a,b,c)

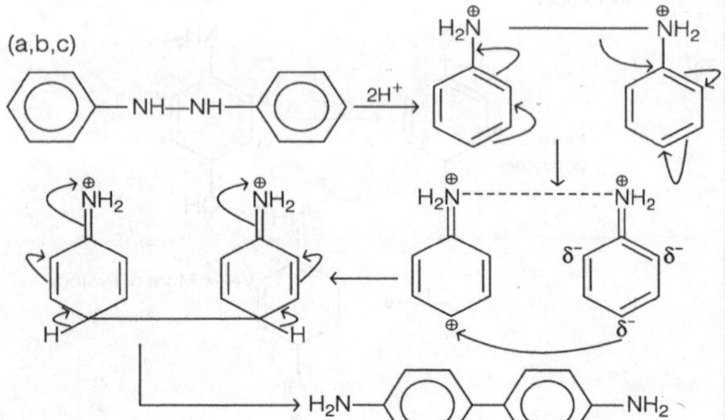

This mechanism leads to C—C linking and can give rise to 4 (semidine) or 4,4′-product (benzidine) and/or 2,4-product (diphenyline). Aromatic nuclei lie in approximately parallel planes.

17. (c) Compound *P* is an amide, which on alkaline hydrolysis gives amine (*R*) and acidification of alkaline mixture gives precipitate *S* ($C_8H_8O_2$). Thus, there are seven carbon in the *R*. Since, *R* gives tribromide with Br_2-H_2O, it must be *meta*-toluidine.

Thus,

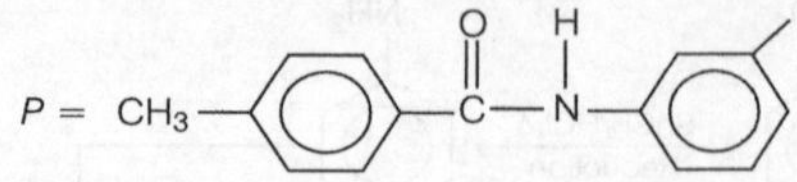

18. (d)

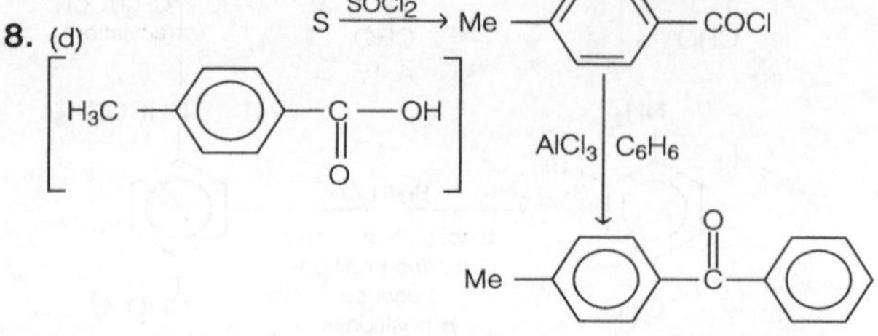

19. (a) Aniline forms insoluble 2, 4, 6-tribromoaniline when treated with bromine water (Br_2-H_2O). Hence, by weighing this precipitate, amount of aniline can be estimated.

20. (a)

1° aliphatic amines give alcohols and N_2 gas with nitrous acid. These also give carbylamine reaction.

So, **(i) → (q,t)**

Aniline is also a 1° amine but it is aromatic, so it forms a dye with HNO_2 and naphthalene (or any other such compound), i.e. gives dye test. Being 1°, it also gives carbylamine reaction.

So, **(ii) → (r,t)**

2° aromatic amines gives N-nitroso compound when treated with HNO_2, which gives Liebermann's nitroso reaction.

So, **(iii) → (p)**

3° aromatic amines form *p*-nitroso compound with HNO_2. The colour of this compound is green.

So, **(iv) → (s)**

21. (4) —$NHCOCH_3$ is a *o/p* directing group. Since *ortho* position is preoccupied so substitution takes place at *para* (4th) position.

22. (4) *X* is 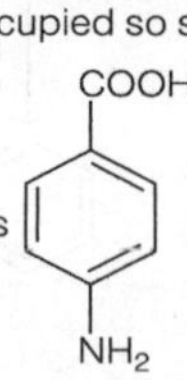 . To convert it into 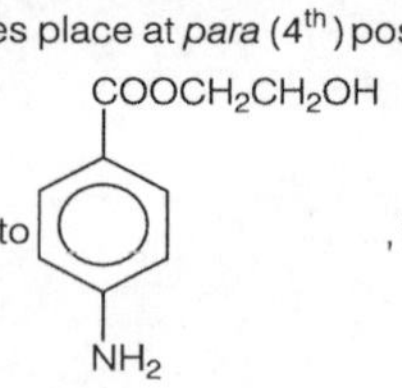 , it should be first treated with NaOH followed by treatment with 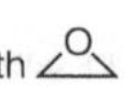 and H^+.

DPP-4 Preparation and Properties of Diazonium Salts

1. (i) Presence of an electron withdrawing group makes the diazonium ion more electron deficient hence, it readily undergoes coupling reaction.

(a) I ($-I, -R$ effect) > II ($-I$ effect) > III, > IV ($+I$ and hyperconjugation) > V (more $+R$ and $-I$)

(b) I ($-I, -R$ effects) > II ($-I$ only) > IV ($-I$ but $-I$ at $m > p$ and $-I$ of $-NO_2$ > $-I$ of Cl) > III ($-I$ at p) > V

(ii) In the coupling reaction, the compound coupling with the diazonium ion behaves as nucleophile and the group increasing its nucleophilicity (generally electron releasing groups) makes it more reactive for coupling reaction. Thus,

(a) I > II > III > IV > V

∵ The order of electron releasing tendency is

—$\ddot{N}H_2$ > $\ddot{O}H$ > —Me > —Cl > —NO_2

(b) V > I >> II > III > IV

∵ Order of electron releasing tendency is

—$\ddot{O}$Me > —Me (three hyperconjugative structure and $+I$) > —CH_2CH_3 (2 HC structure and $+I$) > Me_2CH— (One HC structure and $+I$) > Me_3C— (No HC, only $+I$)

2. *C* is formed when the diazotisation product of *B* is treated with CuCN/KCN, hence, it should contain a cyanide group, at benzene nucleus. Thus, possible formulae of *C* may be

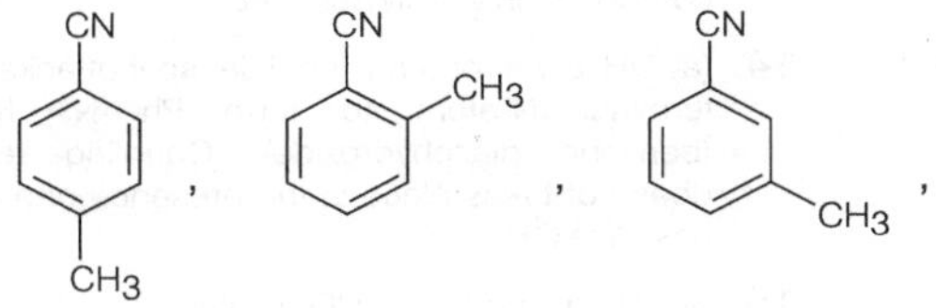

C on hydrolysis (followed by oxidation) gives *D* with equivalent mass 83. *D* gives only one mono substitution product, so it must have symmetrical structure, i.e. groups are present at 1,4 positions

Hence, the correct structure of *C* is

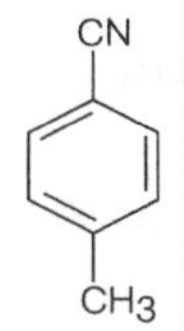

On this basis *A* should be a nitro compound of structure

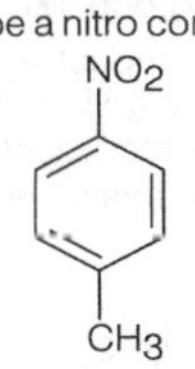

The reaction involved are as follows

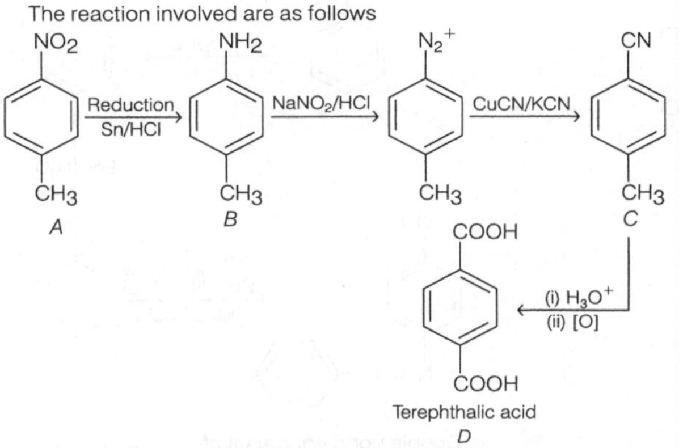

3. (d) Diazo group cannot be replaced directly by aceto (CH_3CO—) group. Remaining all groups (—CN, —Br and —F) can be brought directly. Hence, among the given, 4-bromoaceto phenone cannot be prepared directly from corresponding diazonium salt.

4. (b)

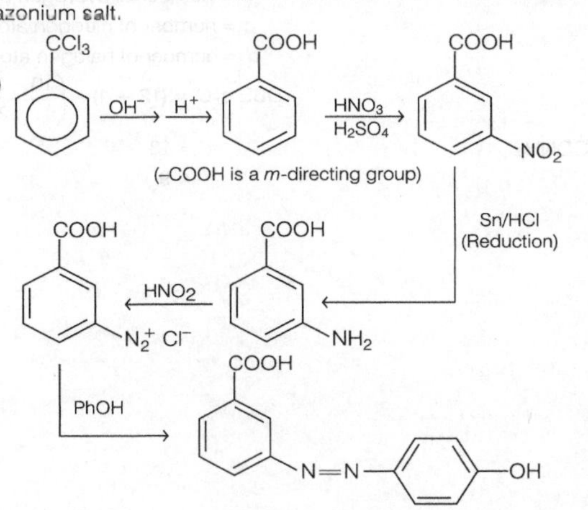

para-coupling

5. (a) In 2, 6-$Me_2C_6H_3NMe_2$, —NMe_2 group cannot coplanar with the benzene ring because of steric interaction offered by *o*-methyl group. Due to steric inhibition of resonance, coupling reaction fails with benzene diazonium cation. With phenol, the reaction takes place in alkaline medium.

Mesitylene is a weaker compound (nucleophile) as compare to phenol or aromatic amines, so gives no coupling with $PhN_2^+Cl^-$ [but with 2,4,6-trinitrobenzene diazonium cation coupling takes place.]

6. (d)

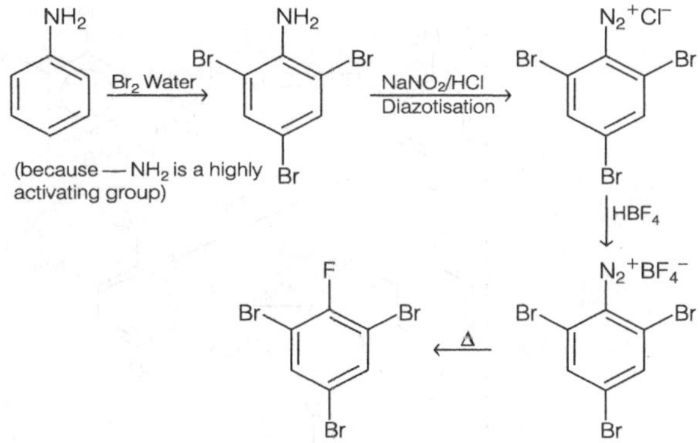

2, 4, 6-tribromo fluorobenzene

7. (d)

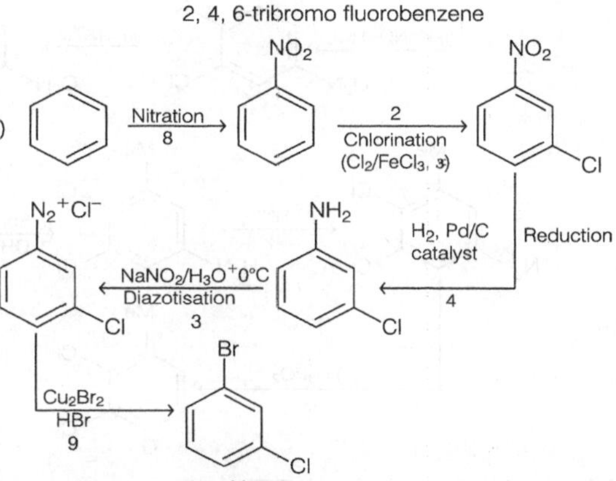

m-chloro bromobenzene

8. (a) Here Sn/HCl reduces —NO_2 group to NH_2 while $NaNO_2$/HCl causes its diazotisation. The diazotised compound couples with phenol to

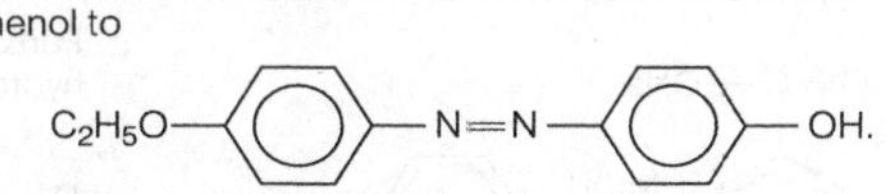

This compound after following steps 6 to 9 gives back *X*.

9. (b) For the coupling reaction, diazonium ion should contain some electron withdrawing groups and the other compound should have electron releasing groups, i.e. should behave like a good nucleophile.

Thus, *A* should be

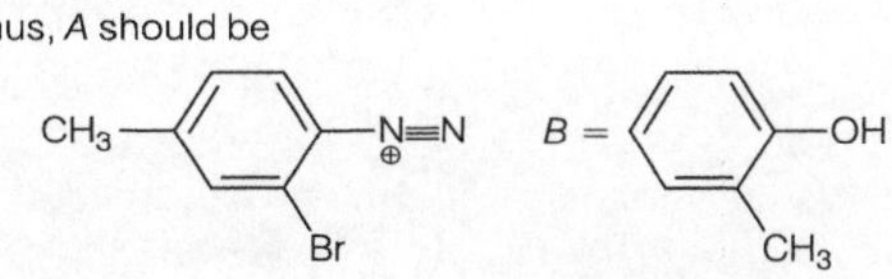

Formation of 3-bromo-4-amino toluene and 2-methyl-4-amino phenol also support these structures for *A* and *B*.

10. (a,c,d) (a) Diazocoupling with phenol takes place at position *para* to —OH group, so correct.

(b) Here also coupling takes place at position *para* to —OH group, so incorrect.

(c) *Ortho* coupling takes place with β-naphthol, so correct.

(d) In case of α-naphthol, the coupling takes place at *para* position, so correct.

11. (a,c,d)

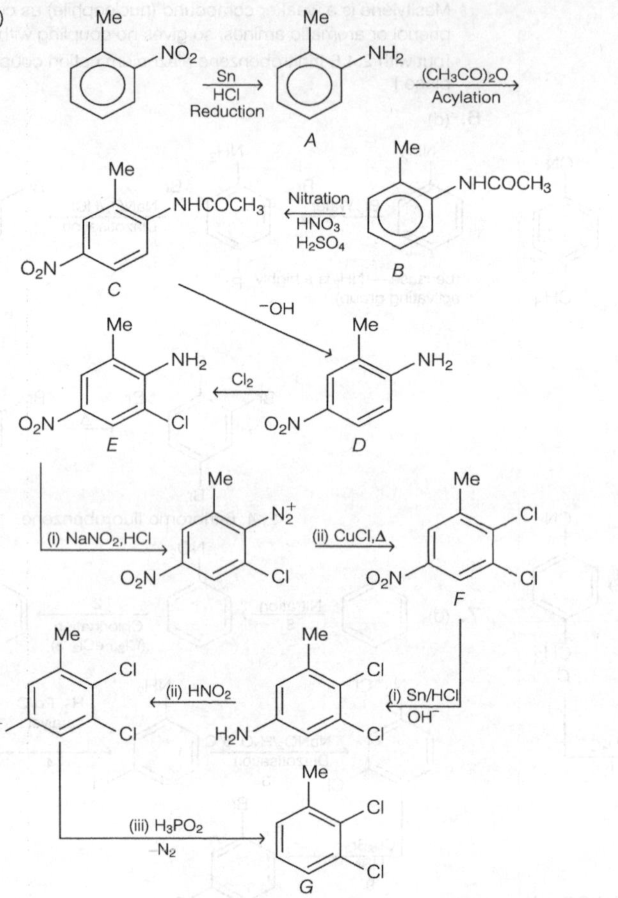

12. (a,b,c)

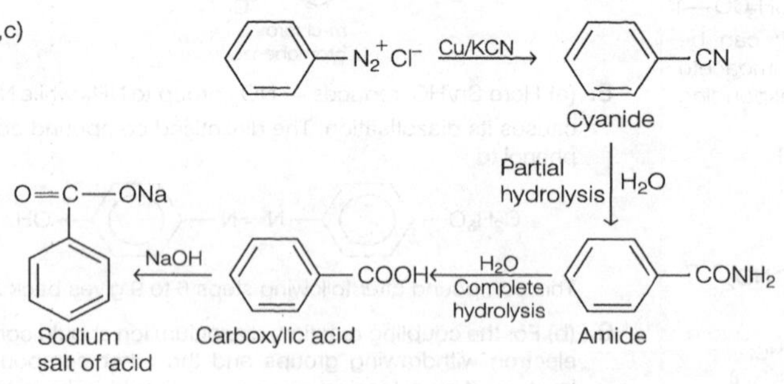

13. (a) During the coupling reaction, the two aromatic rings are in conjugation *via* a azo group, which is a strong chromophore and hence, makes the compound intensely coloured.

14. (a) $\overline{O}H$ is a strong nucleophile, so it attacks at the terminal N-atom to form Ph—N=N—OH (benzene diazohydroxide). Coupling reaction does not takes place in the presence of a strong base like $\overline{O}H$.

15. (a) Nitro group (—NO_2) intensify the positive charge on diazo group and makes it to couple even with less active anisole.

16. (a) In diazocoupling reaction, diazonium ion acts as electrophile and presence of electron withdrawing group (like —NO_2) enhance its electrophilicity.

17. (c) Basic group, like —NH_2 containing compound gives diazocoupling in acidic medium i.e. at a pH less than 7 (pH < 7) but acidic group like —OH, —COOH etc., containing compound gives this reaction in basic medium (i.e. pH > 7).

18. (9)

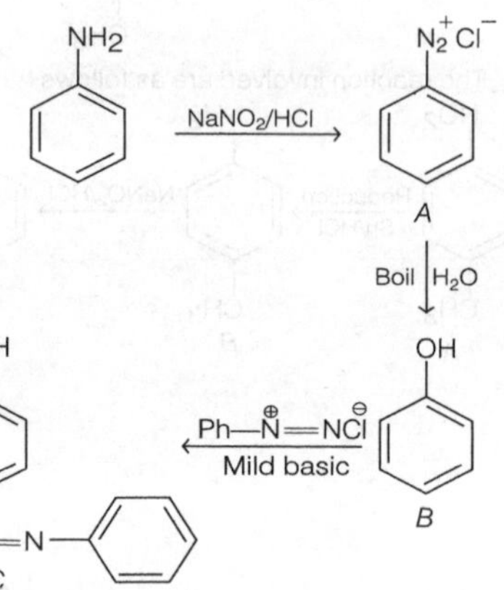

Double bond equivalent of

$$C = (a+1) - \left(\frac{b-c+d}{2}\right)$$

Here, a = number of carbon atoms

b = number of hydrogen atoms

c = number of nitrogen atoms

d = number of halogen atoms.

$$\therefore \quad \text{DBE of } C = (12+1) - \left(\frac{10-2+0}{2}\right)$$

$$= 13 - 4$$

$$= 9$$

Revisal Problems (JEE Main)

1. (b)

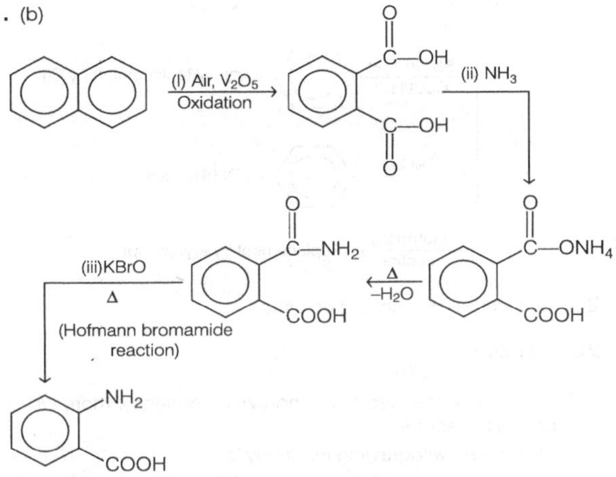

A

2. (d) In the diazocoupling reaction, the diazonium ion behaves as electrophile, so presence of an electron withdrawing group like $—NO_2$ makes it more electrophilic and hence, more reactive towards this reaction. Thus, S = $—NO_2$ or —CHO etc. The other compound taking part in this reaction should be electron rich (i.e. behave like a nucleophile) or in other words, it should contain electron donating group (like $—OCH_3, CH_3, C_2H_5$ etc.)

3. (a) First step of this reaction is biprotonation of NH groups. Thus, intermediate is

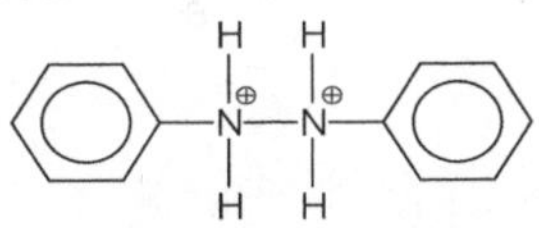

4. (c) Secondary amines (compound containing —NH— bond) gives only N-nitroso amine when treated with HNO_2.

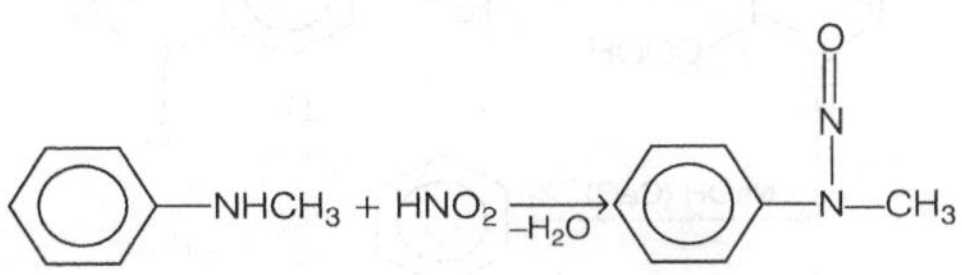

5. (a)

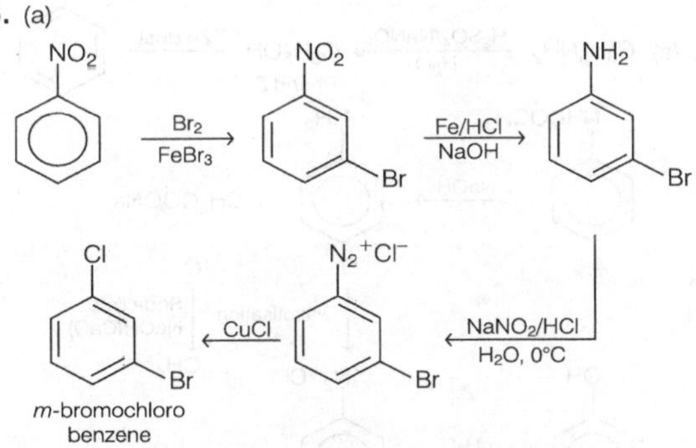

m-bromochloro benzene

6. (b) Formation of two nitro derivatives suggests the presence of two $—CH_3$ groups at successive position (i.e. it is *ortho*-xylene)

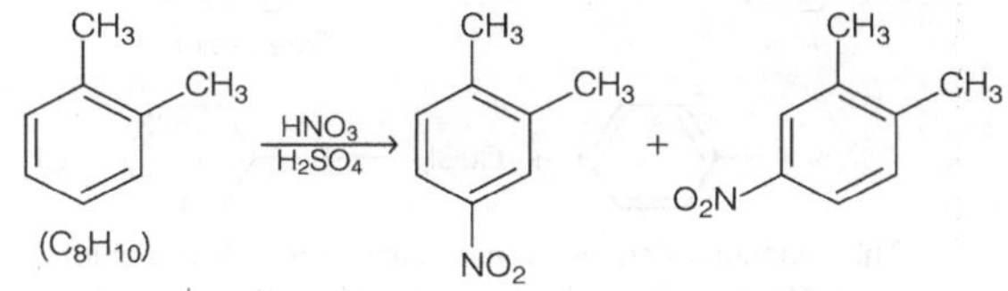

7. (a) Being more basic than water, methylamine accepts a proton from water liberating $\bar{O}H$ ion

$$CH_3\ddot{N}H_2 + H—OH \longrightarrow CH_3—\overset{+}{N}H_3 + \bar{O}H$$

The $\bar{O}H$ ions combine with Fe^{3+} ion present in H_2O to form a brown ppt. of hydrated ferric oxide.

$$FeCl_3 \longrightarrow Fe^{3+} + 3Cl^-$$

$$2Fe^{3+} + 6OH^- \longrightarrow \underset{\text{Brown ppt}}{2Fe(OH)_3 \text{ or } Fe_2O_3 \cdot 3H_2O}$$

8. (b)

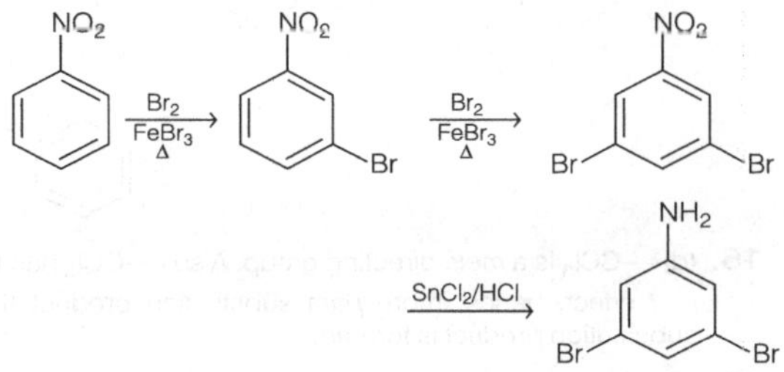

3,5 -dibromoaniline

9. (b)

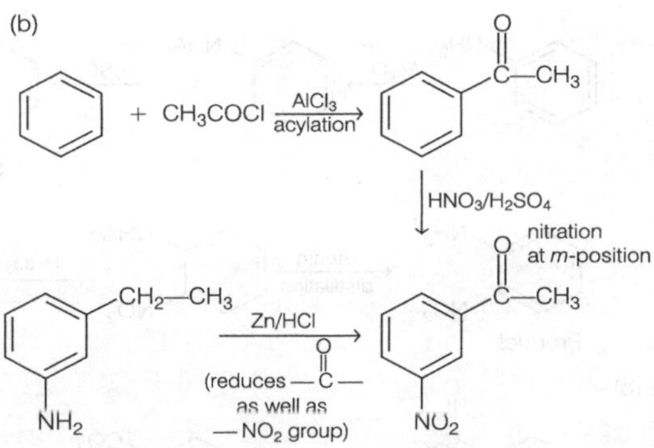

10. (a) Being steam volatile, aniline is separated by steam distillation.

11. (d) —O— is electron releasing group. So activate the ring attached to it more and hence, electrophilic substitution takes place *para* to it (—O—).

12. (b)

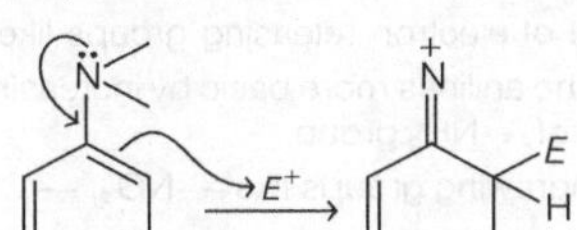

13. (b) $CuCl + Cl^- \longrightarrow CuCl_2^-$

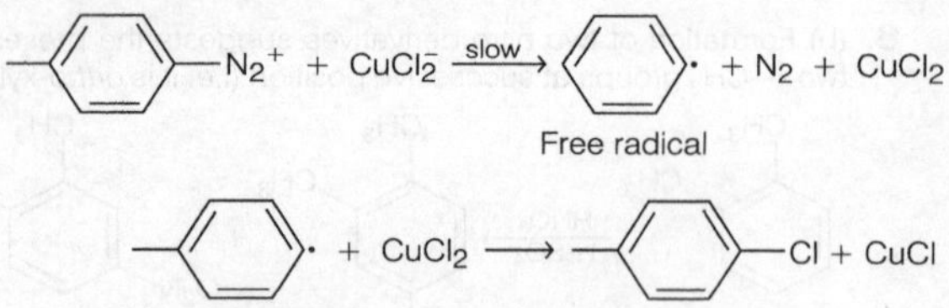

This reaction of conversion of diazonium salt into chlorobenzene by treating it with CuCl/HCl is called Sandmeyer reaction.

14. (a) *Para*-product is formed with α-naphthol.

15. (a) Basic strength depends upon the availability of lone pair of N for protonation.

If the lone pair of N are in conjugation with = bond, the availability of lone pair for protonation and hence. basicity decreases. Thus, order of basicity is

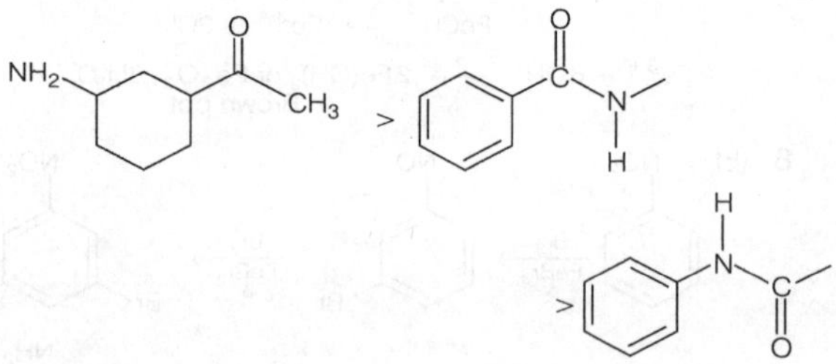

16. (d) —CCl_3 is a *meta* directing group. Also, —CCl_3 has strong $-I$-effect, hence, more *para* substitution product than *ortho* substitution product is formed.

17. (b)

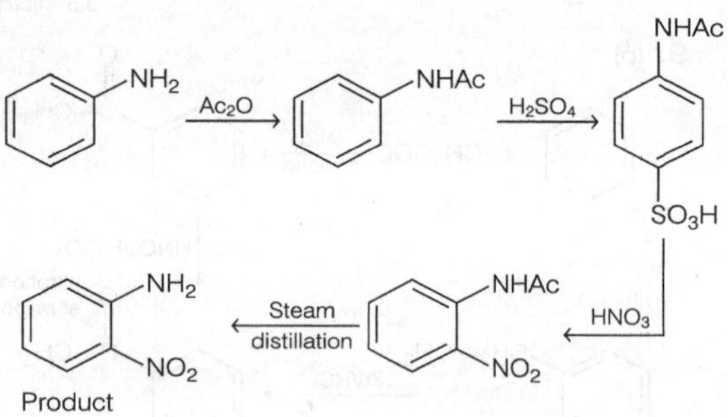

18. (b)

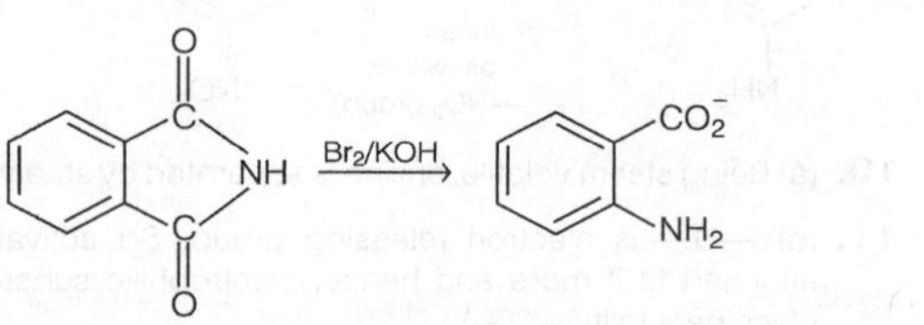

19. (d) Presence of electron releasing groups like —CH_3, —OCH_3 etc., makes the anilines more basic by increasing electron density over N-atom of —NH_2 group.

Electron withdrawing groups like —NO_2, —Cl etc., makes them less basic.

Thus, the order of basicity is

$$\underset{(-R-I)}{\text{III}} < \underset{(-I-R)}{\text{IV}} < \underset{(+I)}{\text{I}} < \underset{\substack{(+R \text{ and } -I \\ \text{net effect } +R)}}{\text{II}}$$

20. (b)

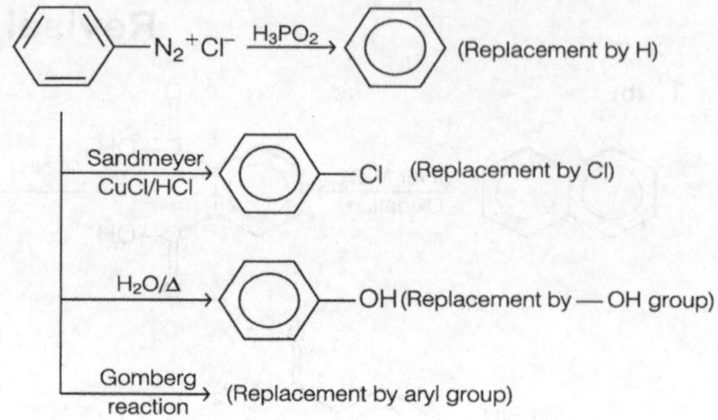

21. (a) Here, H_3PO_2 should be used.

22. (a) Basicity $\propto \dfrac{1}{pK_b}$

Thus, more the electron withdrawing tendency, more will be the pK_b of the aniline.

Order of e^- withdrawing tendency is

$$\underset{(e^- \text{ releasing})}{CH_3} < H < CF_3 < NO_2$$

23. (d)

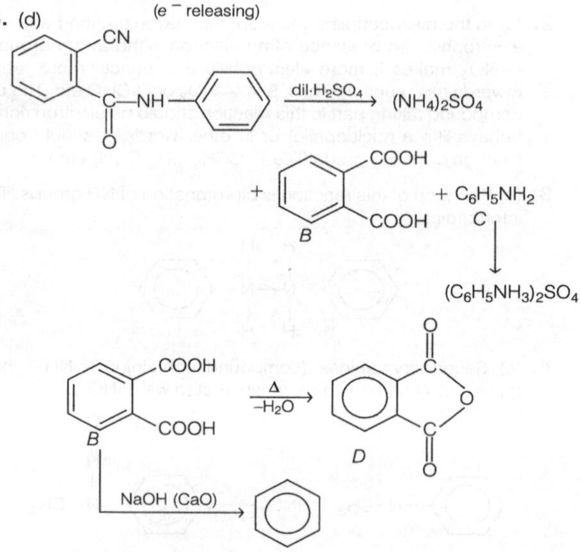

24. (a)

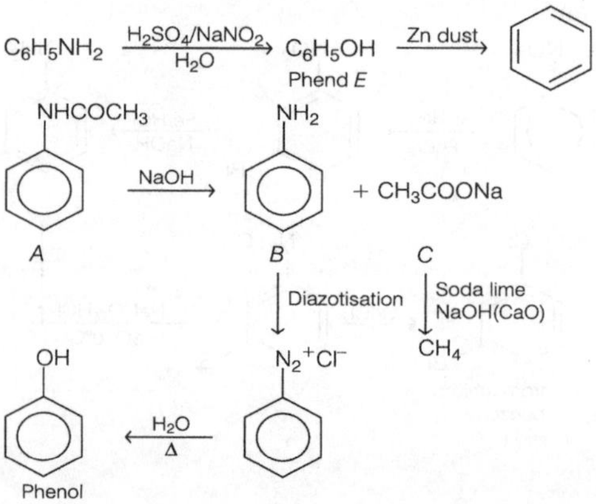

25. (b)

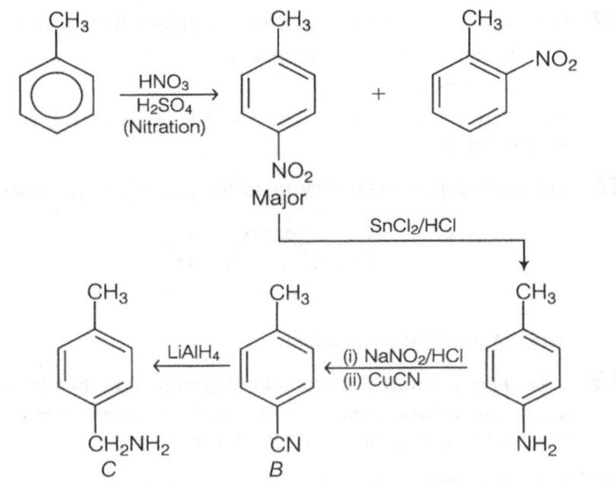

26. (a) $\overset{\oplus}{PhN_2}\overset{\ominus}{Br} \xrightarrow[\text{(from alkali)}]{OH^-}$ Ph—N=N—OH (Benzene diazohydroxide) $\longrightarrow$ Ph + N_2 + OH
Phenyl free radical

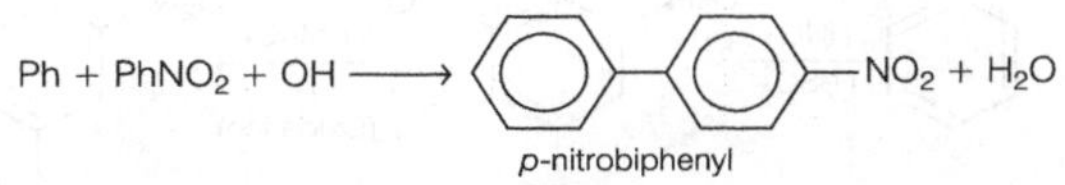

This is an example of Gomberg reaction.

27. (a) In the coupling reaction, the compound coupled with $PhN_2^{\oplus}$ behaves like a nucleophile and the compound with substituent having more electron releasing tendency has more nucleophilicity.

Since, aniline is more electron donating as compared to phenol, it coupled readily with $PhN_2^{\oplus}$.

Revisal Problem (JEE Advanced)

1. (c) In acidic medium aniline derivative gets protonated to give anilinium ion, which is *meta* directing, so *meta* it is the major product.

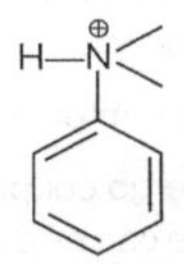

2. (a) Phenol undergoes coupling reaction in the presence of a base. The diazonium salt is less reactive for this reaction but presence of electron withdrawing group like —NO_2 makes it sufficiently reactive to undergo coupling reaction with phenol.

3. (c) Here σ complex formation is the rate determining step, so

$$C_6H_6 = C_6D_6 = C_6T_6$$

Presence of electron donating group increases the rate while that of electron withdrawing group (like Br) decreases the rate for electrophilic substitution (nitration) reactions.

4. (d)

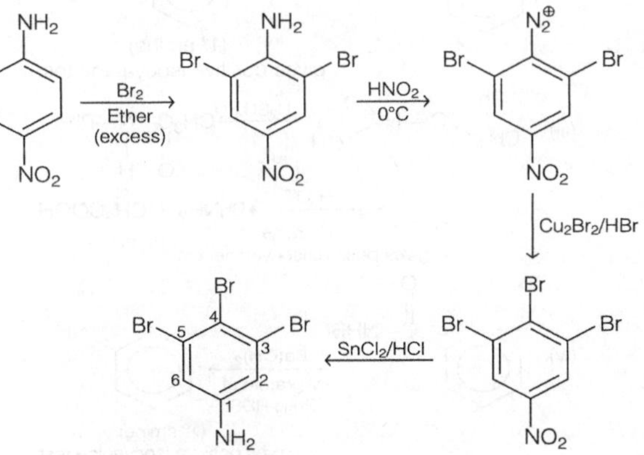

3, 4, 5-tri bromoaniline

5. (a)

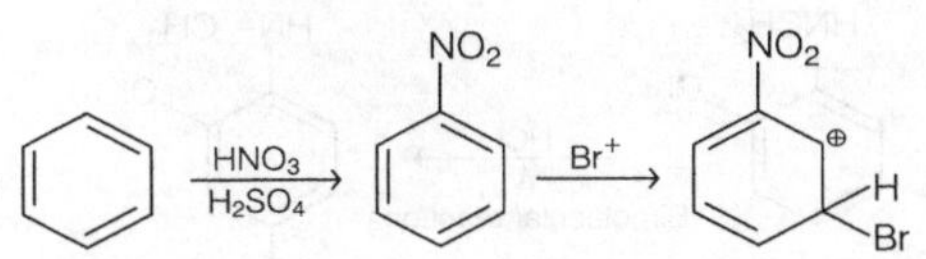

6. (a) More the basic character, (i.e. more the availability of lone pair over N for protonation), more is the value of basicity constant. Thus, pyridine is more basic as compared to aniline. Presence of two —Me group over N also enhance its protonation tendency. Thus, the order of K_b is

$$I < II < III$$

7. (c)

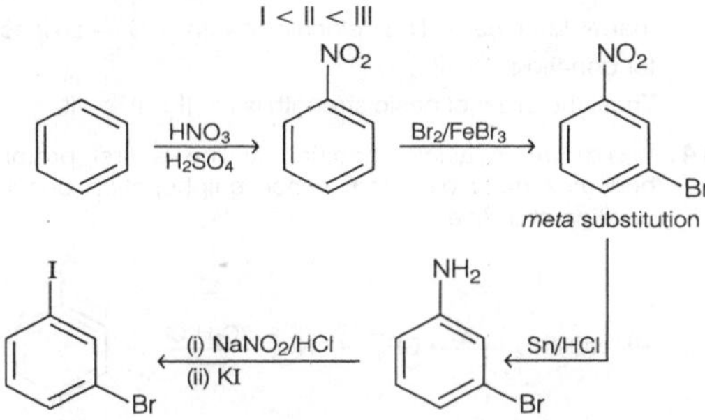

8. (a) In charge transfer complex, the electron donor should have low ionisation potential and high electron density while electron acceptor should have a very low electron density.

9. (b)

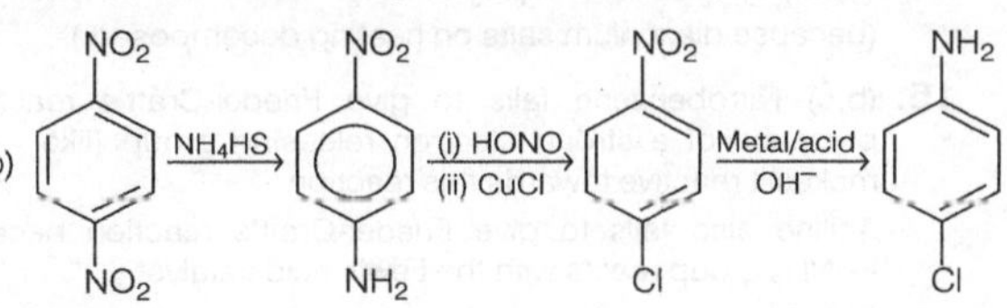

Thus, Q = CuCl

10. (a) R = Fe/HCl, then $\overline{O}H$

11. (b)

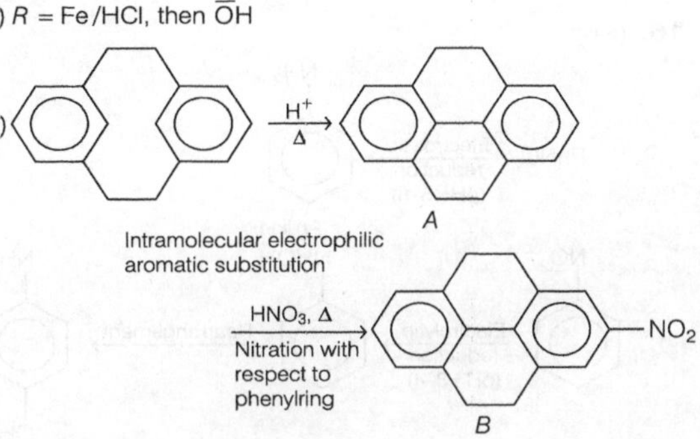

12. (d)

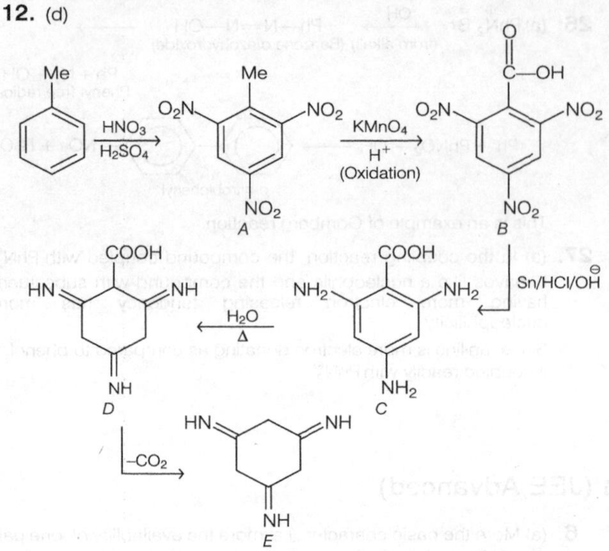

13. (a) Presence of bulkier groups at *ortho* positions restrict the delocalisation of lone pair of N with the ring so, N (II) is more basic than N (IV). Lone pair of N (III) is delocalised with —C(=O)— so least available for donation

Thus, the order of basic strength is I > II > IV > III

14. (a,b,d) (a) In acidic medium, aniline is first protonated and becomes *meta* directing, hence sulphonation occurs at *meta* position of aniline.

(b) 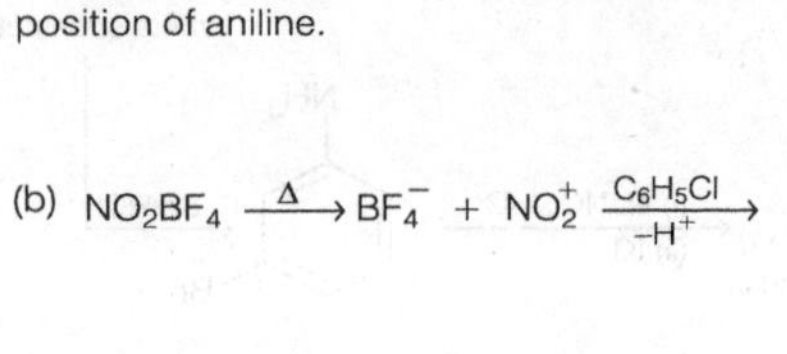

(c) $KMnO_4$ will oxidise also —NO to —NO_2.

(d) $C_6H_5N_2Cl \xrightarrow{\Delta} C_6H_5Cl + N_2$

(because diazonium salts on heating decomposed.)

15. (b,c) Nitrobenzene fails to give Friedel-Craft's reaction but presence of a strong electron releasing group (like —OCH_3) makes it reactive towards this reaction.

Aniline also fails to give Friedel-Craft's reaction because its —NH_2 group reacts with the Lewis acid catalyst.

Chlorobenzene, although with difficulty, but undergoes Friedel-Craft's reaction.

16. (a,b)

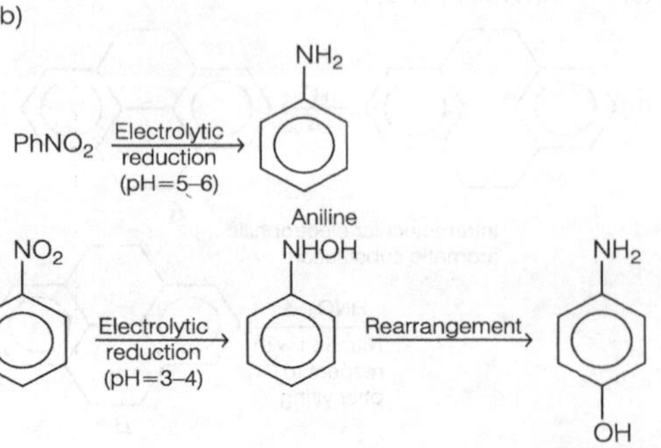

17. (a,b) (a) Precipitate of 2, 4, 6-tribromoaniline would be formed.

(b) A coloured dye will be formed

(c) No reaction.

(d) Acetylation would occur at —NH_2 but will not produce any visible effect.

18. (a,b,c) Except benzaldehyde, all the given give *p*-benzoquinone

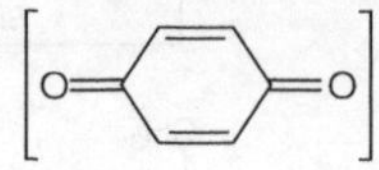

when subjected to oxidation.

19. (a,b,c) NO_2 being a deactivating group, deactivates the ring to such an extent that it does not undergo further nitration (electrophilic substitution). Other given reactions are correct.

20. (a,b) II is more basic than IV because of the presence of two electron releasing Me groups at N.

Similar is true in case of III and I (III > I).

21. (a,c,d) Presence of electron withdrawing group or conjugation (with N) makes it less basic but presence of electron releasing group enhance the basicity of amines.

Thus, the order of decreasing basicity is

$$PhCH_2NH_2 > PhNH_2 > PhNHPh$$

other given orders are correct.

22. (c) Chloronium ion is not formed rather chloride ion is formed.

23. (a) Diazonium salts undergo coupling reaction with β-naphthol to give coloured dye, hence can be used for identification of aniline.

24. (a) (i) → (p,r,s), (ii) → (q,r), (iii) → (p,r,s), (iv) → (p,r,s)

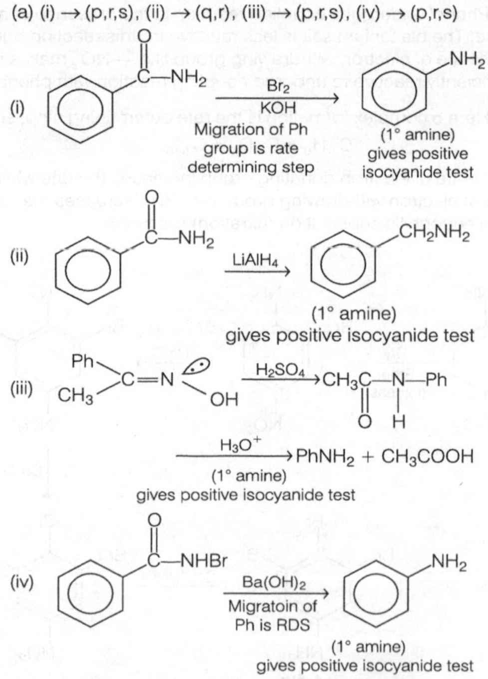

25. (b) (i) → (p,s), (ii) → (q,r), (iii) → (p,q), (iv) → (q,r)

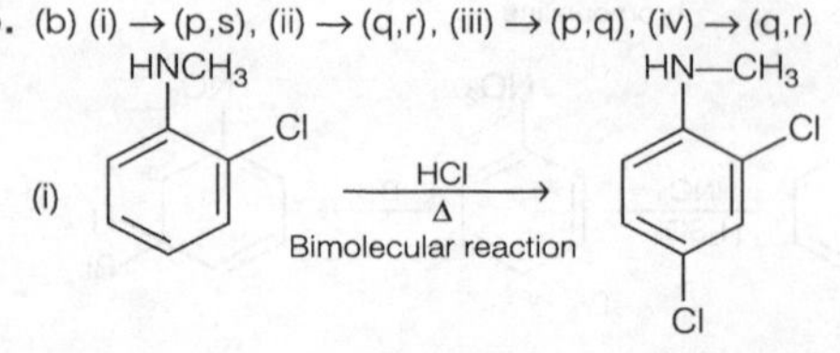

(ii) 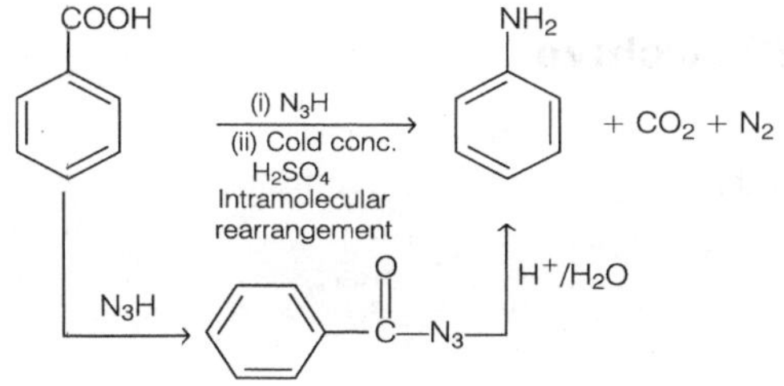

(iii) 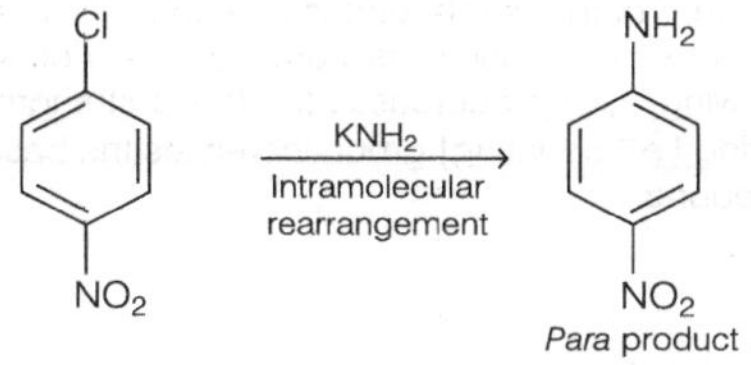

(iv) 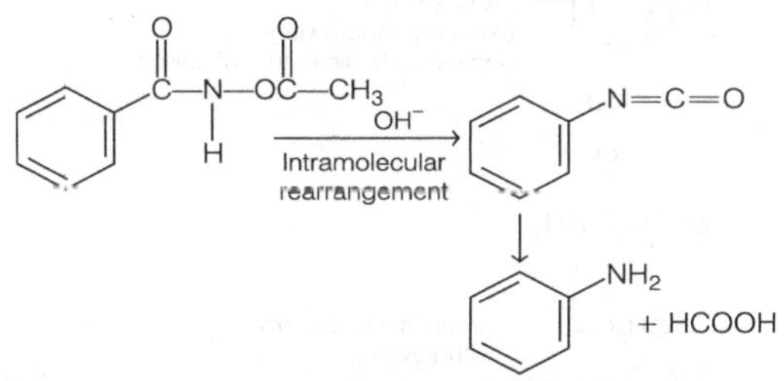

26. (d) Nitro compound does not reacts with Grignard reagent.

27. (c) Since, C_2 is a diazo compound, so C_4 should be a cyanide which on hydrolysis gives —COOH groups.

Thus, 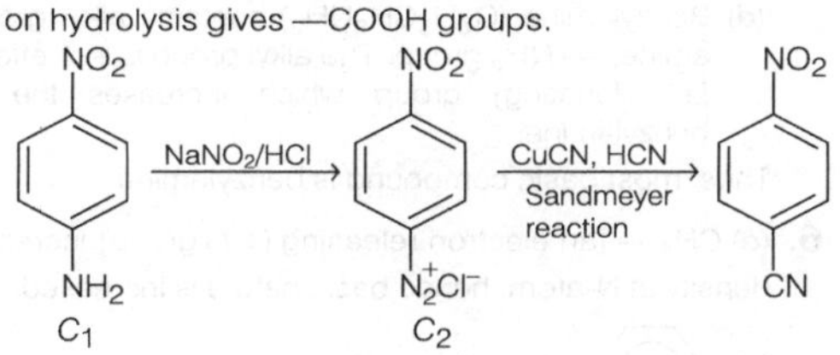

28. (d) 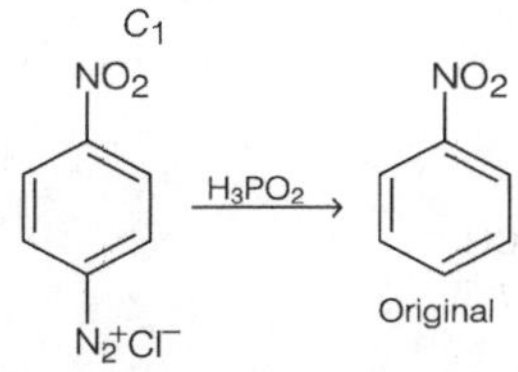

Explanations for Q. Nos. (29-31)

Compound *P* is 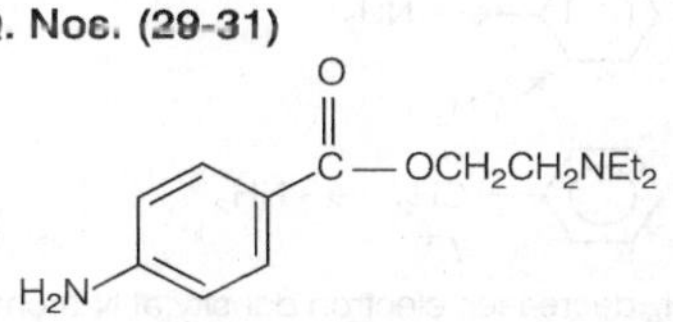

Diazonium salt of *P* will form a highly coloured dye with β-naphthol.

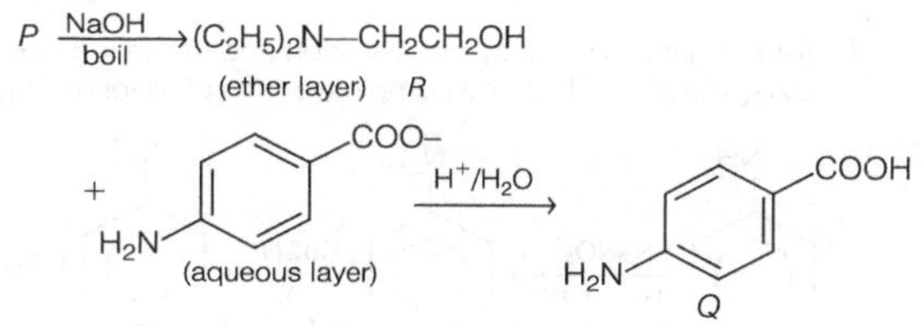

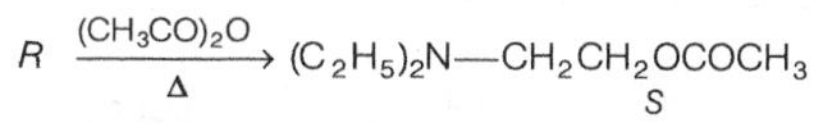

29. (b) **30.** (c) **31.** (a)

32. (4) Presence of electron releasing group makes aniline more basic. Further pyridine is more basic than aniline.

33. (4)

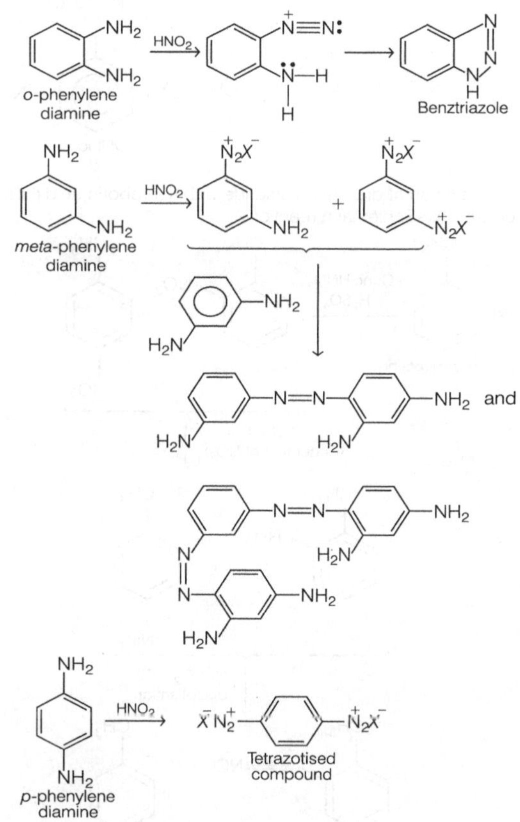

JEE Main & AIEEE Archive

1. (b) Formation of *A* is by diazotisation and formation of *B* from *A* is by S_N reaction. Thus, the complete series of reaction is given by

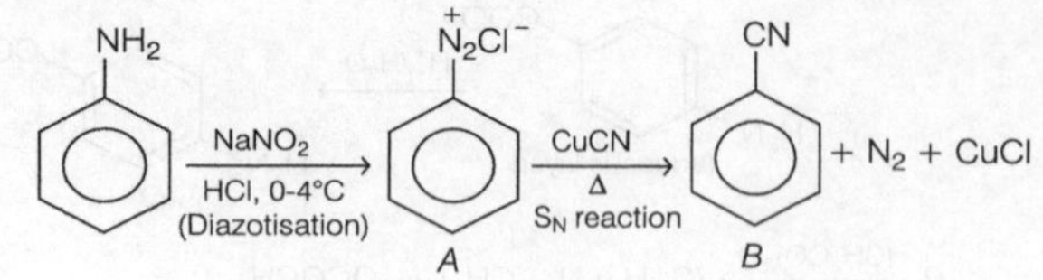

2. (c)

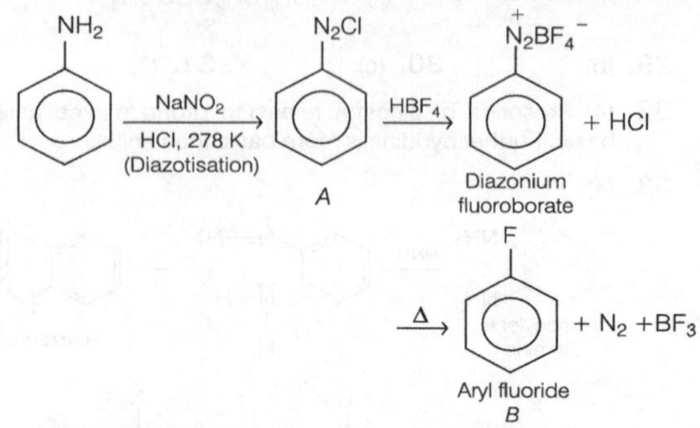

The reaction of diazonium chloride with fluoroboric acid (HBF_4) is called Balz-Schiemann reaction.

3. (a)

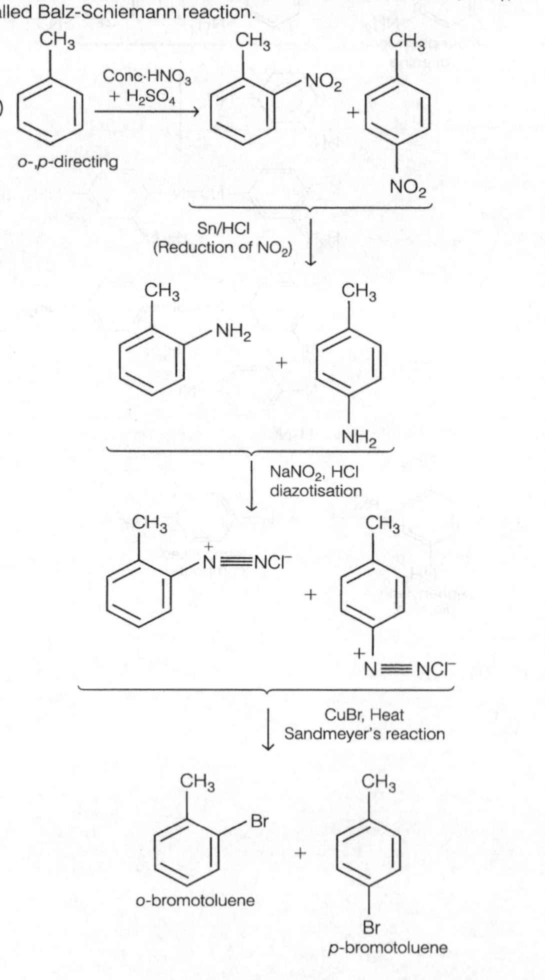

4. (b)

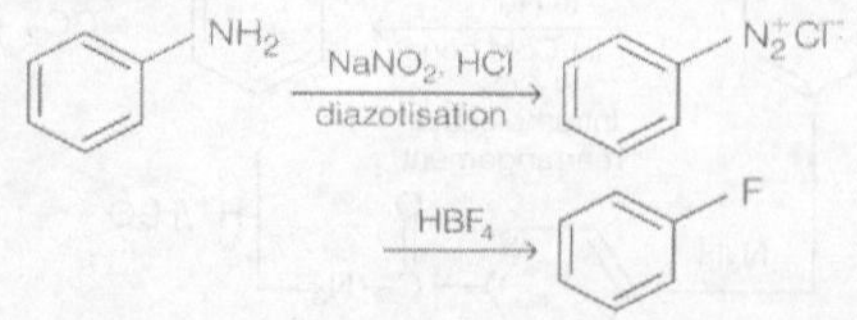

5. (d) Basic nature of the compound is related to their tendency to donate their lone pair of electrons more readily. – I effect exerting [e^- withdrawing] group decreases the basic strength while + I effect exerting [e^- donating] group increases the basic strength of the compound.

(a)

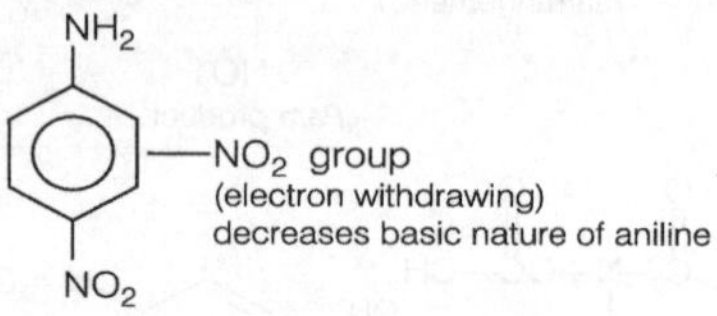

(b)

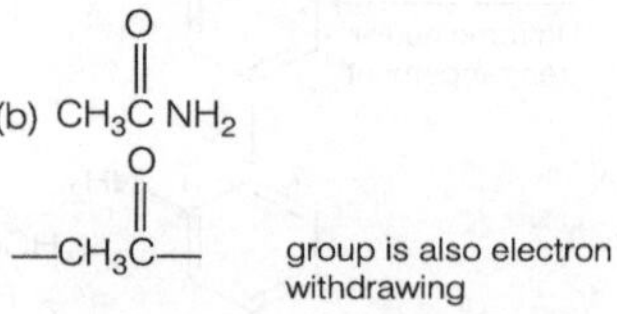

(c)

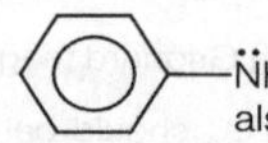

phenyl group is also electron withdrawing.

(d) Benzylamine ($C_6H_5CH_2NH_2$) contains alkyl group linked to amine, — NH_2 group. The alkyl group is + I effect exerting [e^- donating] group which increases the basicity of benzylamine.

Thus, most basic compound is benzylamine.

6. (d) CH_3— [an electron releasing (+ I) group] increases electron density at N-atom, hence basic nature is increased.

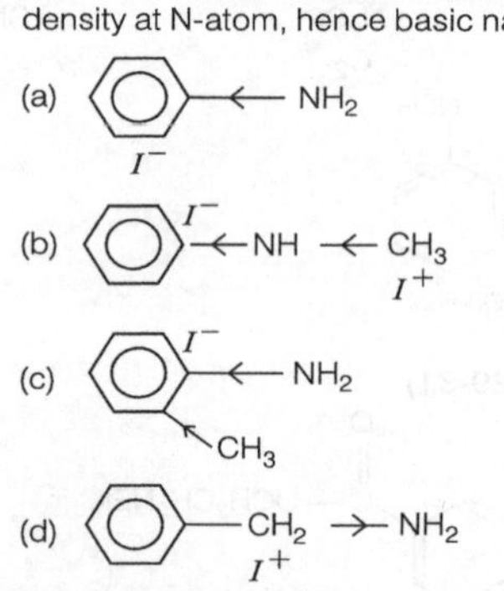

C_6H_5 decreases electron density at N-atom due to delocalisation of e^- of NH_2 with πe^- of benzene. Thus, basic nature is decreased.

Hence, (d) is the strongest base.

7. (d) It is carbylamine reaction.

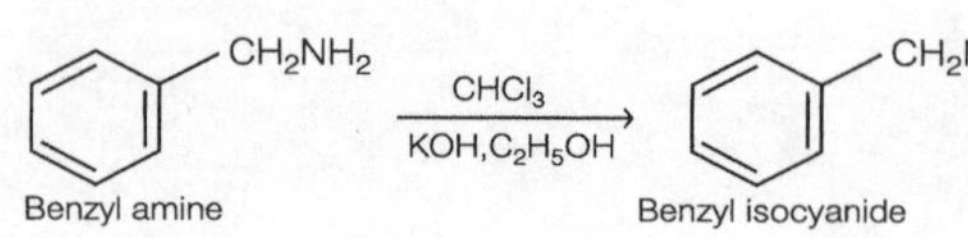

Carbylamine reaction (test for 1° amine)

JEE Advanced & IIT JEE Archive

1. (a) NH_2 is more nucleophilic hence, NH_2 group will undergo reaction faster than $CONH_2$.

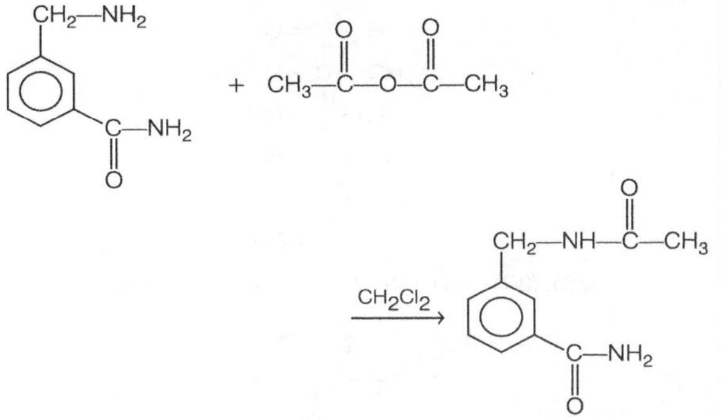

2. (c) As we know, benzenediazonium salt forms brilliant coloured dye with β-naphthol, the compound under consideration must be *p*-toluidine (c) as it is a primary aromatic amine.

Primary aromatic amine, on treatment with $NaNO_2$ in dil. HCl forms the corresponding diazonium chloride salt.

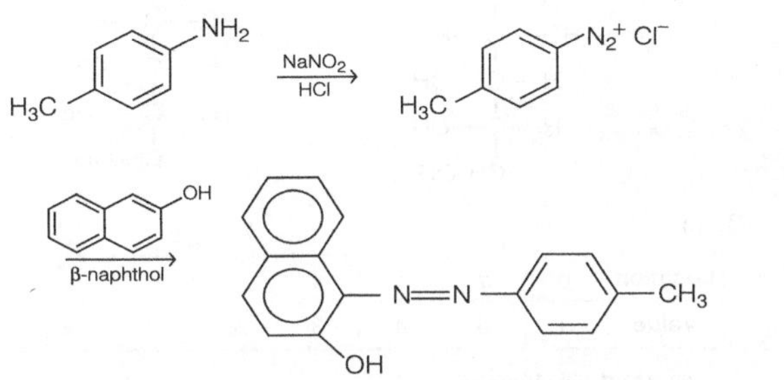

3. (a) It is the first step of Gabriel's phthalimide synthesis. The hydrogen bonded to nitrogen is sufficiently acidic due to two α-carbonyl groups.

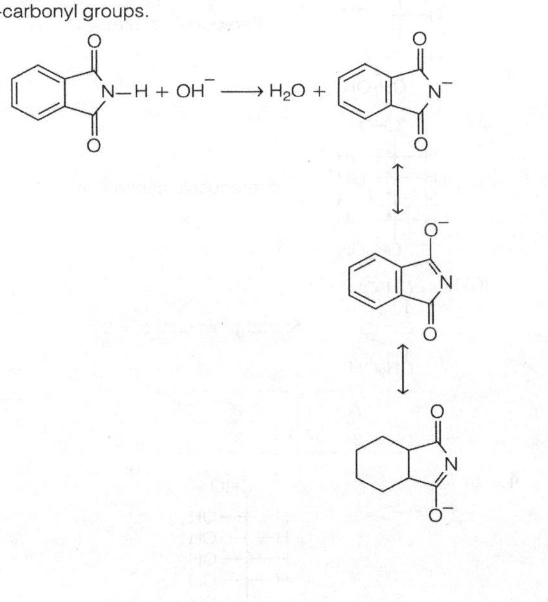

The conjugate base formed above acts as nucleophile in the subsequent step of reaction. As shown above, the nucleophile exists in three resonating forms, one may think of oxygen being the donor atom in the nucleophilic attack. However, nitrogen acts as donor as it is better donor than oxygen.

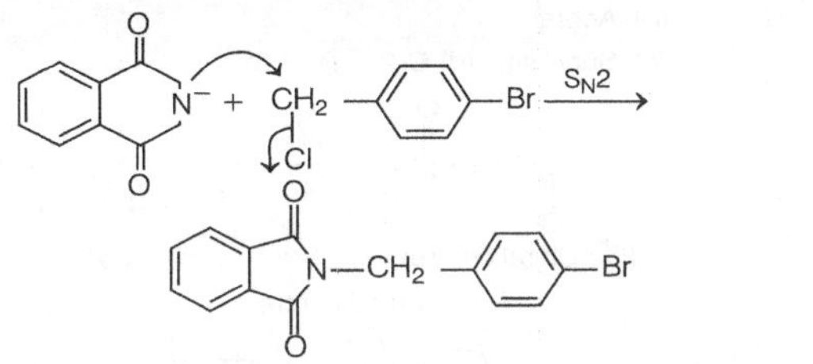

Bromine is not substituted in the above reaction as it is in resonance with benzene ring giving partial double bond character to C—Br bond, hence difficult to break.

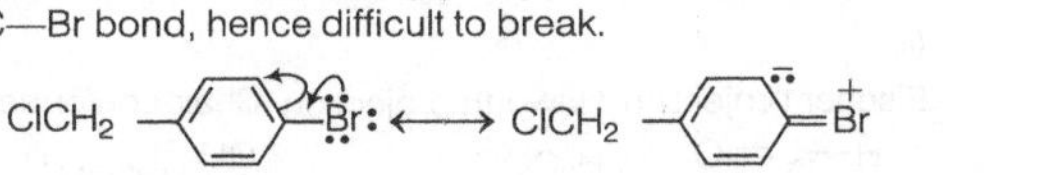

4. (d) $C_6H_5NH_2 + NaNO_2/HCl \xrightarrow{0°C} C_6H_5N_2^+Cl^-$

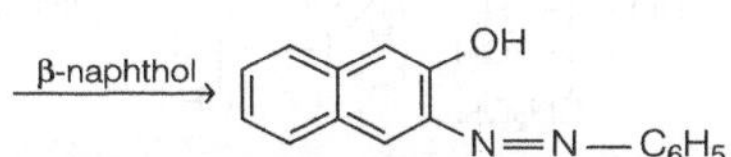

Coloured dye (orange or red)

5. (b) Sodium fusion extract gives Prussian blue colouration, when nitrogen and carbon both are present in the compound.

Phenolic group and salt of carboxylic acid gives $FeCl_3$ test.

Chloride salt gives white precipitate of AgCl on treatment with $AgNO_3$.

Hydrazone formation occur effectively at pH = 4.5. The reaction proceeds in that condition only when H^+ concentration is just sufficient to activate the following enolization.

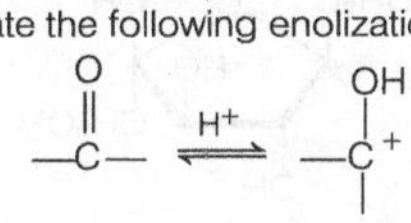

As H^+ concentration rises sufficiently, a large number of molecules of hydrazine gets converted into hydrazinium ion which is not nucleophilic and reaction becomes impossible. Further low concentration of H^+ (in the case of 2,4-dinitrophenylhydrazinium bromide) is not effective to proceed elimination.

6. (b)

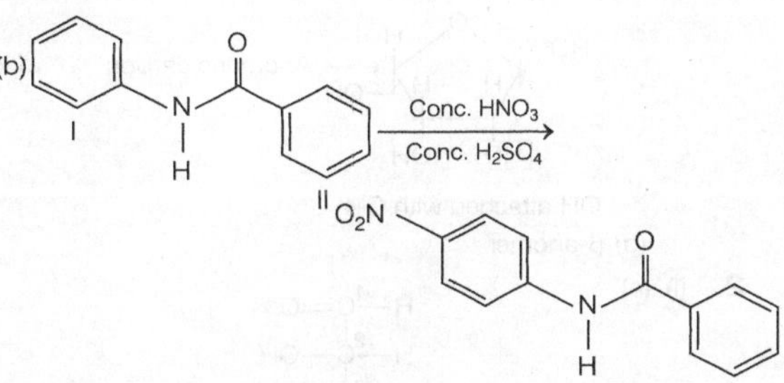

Ring-I is activated while ring-II is deactivated towards electrophilic aromatic substitution reaction.

5. Biomolecules

DPP-1 Introduction and Classification of Carbohydrates

1. (i) (a) Diastereomers (the non-superimposable forms which are not related as mirror images)

(b) Acetal

(c) Sucrose (d) C-2. (e) 5 to 1

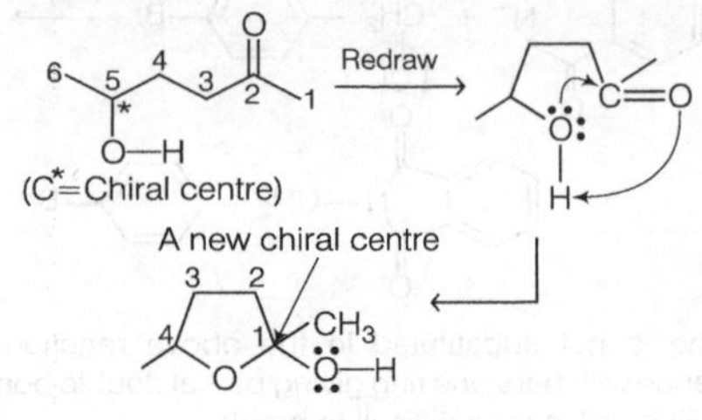

(ii)

Fischer projection Haworth projection Chair conformation

(a)

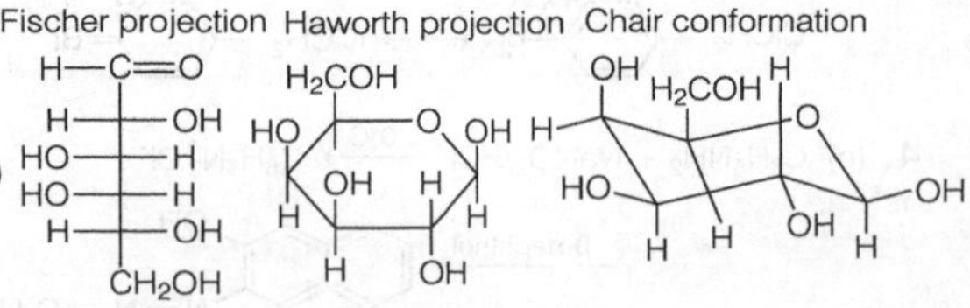

(b)

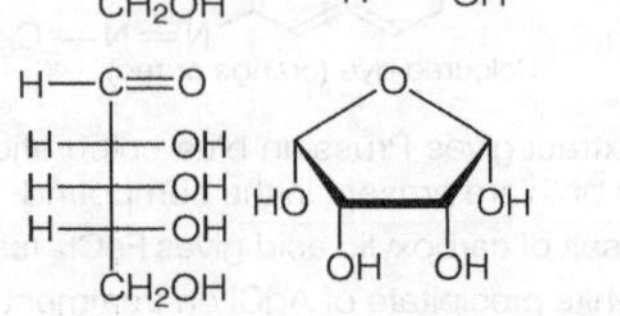

Furanoses do not have chair conformation

(c)

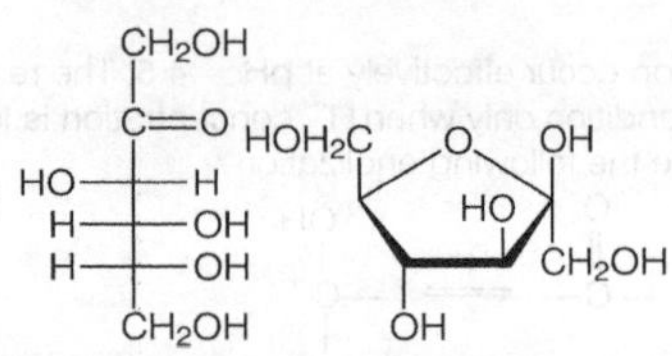

Furanoses do not have chair conformation

Note *Hydrogens can be omitted from Haworth projections to avoid cluttering but do not draw a stick instead of* H*-atom. As in all bond, line or stick structures, a stick that is only attached to one other atom represents a Me group.*

(iii) (a) Ketose

(b) to (c)

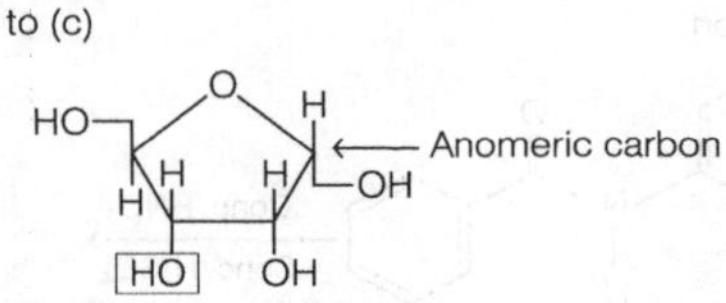

OH attached with C-4

(d) β-anomer

2. (i) (a)

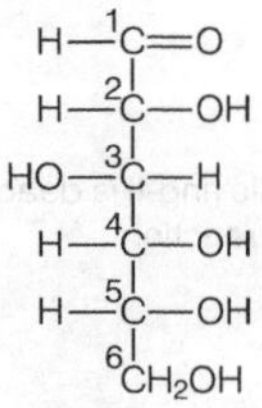

(2*R*), (3*S*), (4*R*),(5*R*) -6-pentahydroxyhexanal

(b)

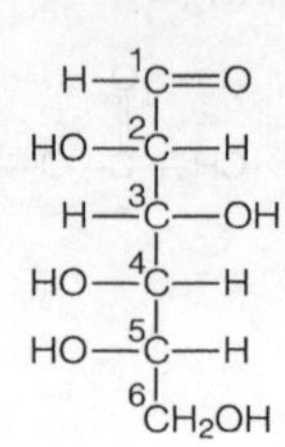

(2*S*), (3*R*), (4*S*), (5*S*) -6-pentahydroxyhexanal

(c)

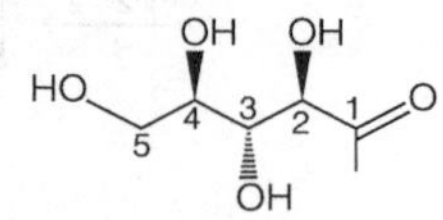

(2*R*), (3*R*), (4*R*) -5-tetrahydroxy pentanal

(ii)

(a)

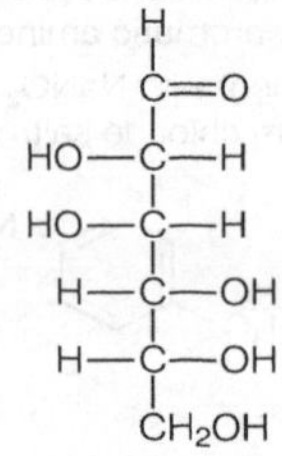

(b)

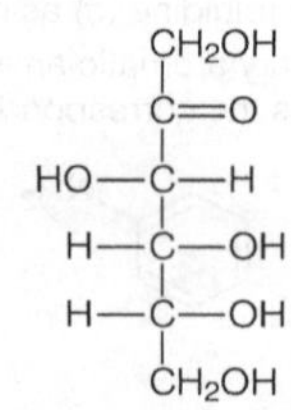

3. (i)

Location	*p*	*q*	*r*	*s*	*t*	*u*	*v*	*w*
Value	3	5	4	3	2	3	2	4

(ii) (A) 2-ketohexose

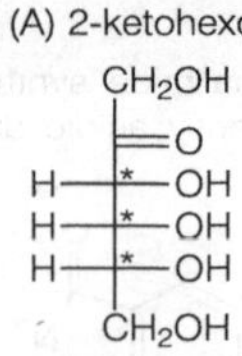

Stereogenic centre = 3 = *A*

(B)

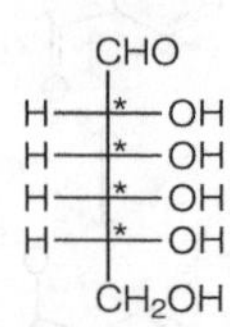

Stereogenic centre = 4 = *B*

(C)

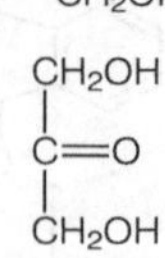

Stereogenic centre = 0 = *C*

A	*B*	*C*
3	4	0

4. (i)

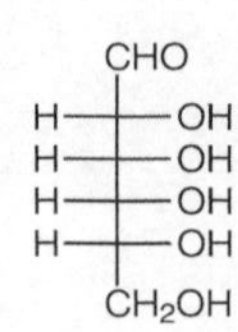

Since in this cyclic structure, only OH group in an axial position is the one at C-3. Thus, this sugar is the C-3 epimer of D-glucose. The substituent at the anomeric carbon is in the β-position. Thus, the name of the sugar is propyl β-D-alloside or propyl-β-D-allopyranoside.

Since, the sugar is an acetal hence, it is a non-reducing sugar.

(ii)

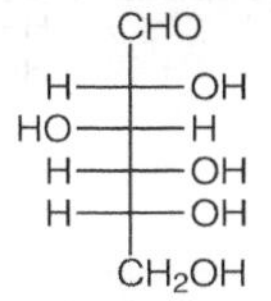

The substituent at the anomeric carbon is in the α-position. So, its name is methyl α-D-glucopyranoside.

It is also an acetal, hence, it is also a non-reducing sugar.

(iii)

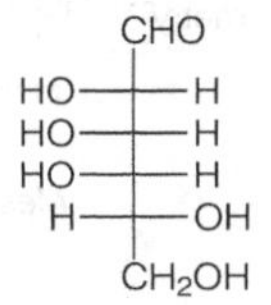

The name of the given compound is α-D-talopyranoside. Since, it is a hemiacetal, so reducing in nature.

(iv)

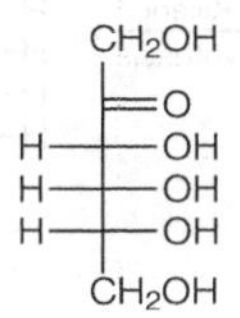

The name of the given compound is ethyl β -D-psicopyranoside.

Since, it is a acetal hence, it is non-reducing in nature.

5. (a) Because of its five membered cyclic structure, it is furanose. It has lack of an oxygen atom, so it is pentulose (i.e. five carbon ketone with lack of O at one position).

6. (c) Furanose means a cyclic five membered ring with an O-atom. Hexa shows presence of six carbon atoms. β means OH groups is towards the O-atom. Thus, the correct structure of β-ketohexafuranose is

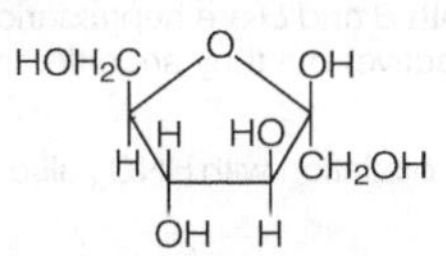

7. (c) Structure given at left hand side is α-form and that at right hand side is β-form.

These two are non-superimposable non-mirror images (i.e. are diastereomers).

Conversion of α form to β or *vice-versa* is called mutarotation.

8. (b)

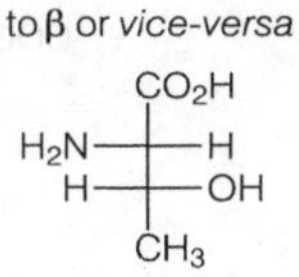

For C_2, order of priority

$$NH_2 > COOH > —CH(OH)CH_3 > —H$$

So, 2*S*

For C_3, order of priority

$$—OH > —CH(NH_2)CO_2H > —CH_3 > —H$$

So 3*R*.

9. (d) Compounds differing in configuration at C-1 are called anomers while that differing in configuration at any one carbon other than C-1 are called epimers.

Thus, I and III are anomers and I and II are epimers.

10. (b,c) Epimers differ in the configuration of any carbon other than C-1 position. (b) and (c) are C-2 epimers.

11. (b,d) β form of glucose is more stable because all the bigger groups occupy equatorial positions in it. Since, α and β-forms are non-superimposable non-mirror images, so they are related as diastereomers, not as enantiomers.

12. (b) Anomers differ in configuration of anomeric carbon (i.e. C_1).

13. (d) The possible keto forms of the given enol are

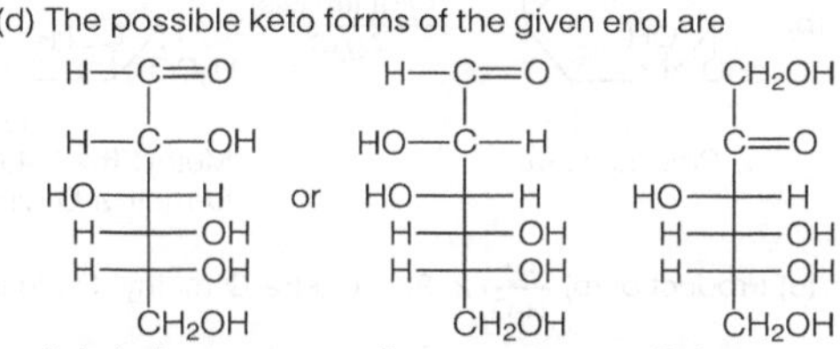

D (+) glucose — D-mannose — D-fructose

14. (a) Structure of β-D-glucopyranose is

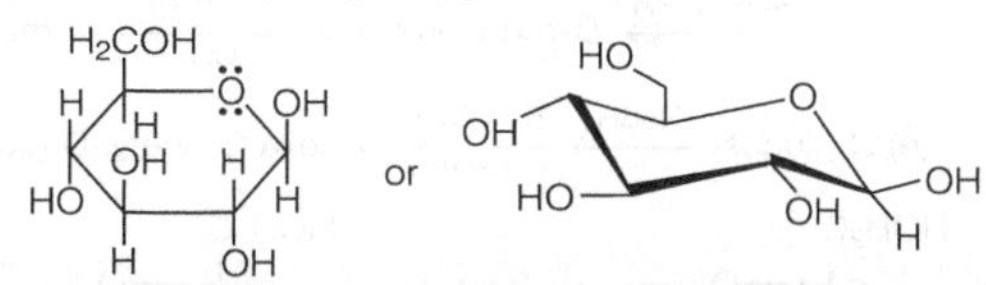

Since, all the ring substituents are at equatorial positions, so, it is more stable and hence, abundant aldohexose.

15. (a) D-fructose cannot be used as a sweetener in hot food as high temperature shift the pyranose ⇌ furanose equilibrium towards less sweet furanose form.

16. (b) (i) → (r,s), (ii) → (r,s), (iii) → (p,q), (iv) → (p,q)

Glucose is an aldehyde while fructose is a ketone, although both have the same molecular formula. Thus, these are related as functional isomers.

LDBAVE transformation involves interconversion of glucose into fructose or mannose or *vice-versa*.

Mannose is an anomer of glucose, so it is also an aldehyde. Thus, it is also functional isomer of fructose.

α and β (cyclic) forms of glucose are related as diastereomers (i.e. non superimposable non-mirror images).

17. (9)

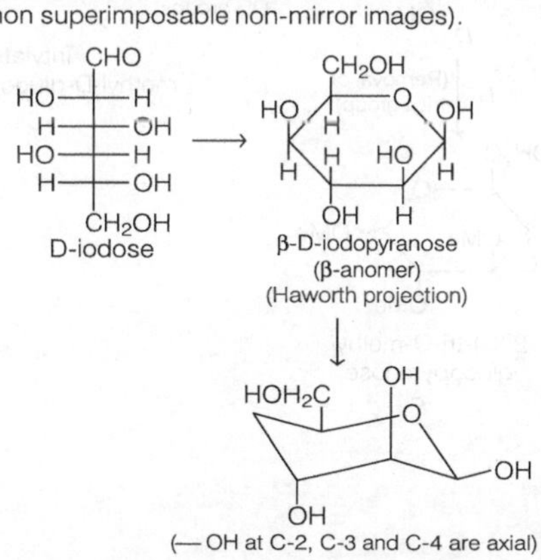

D-iodose → β-D-iodopyranose (β-anomer) (Haworth projection)

(—OH at C-2, C-3 and C-4 are axial)

∴ The sum of locations of axial —OH groups

$$= 2 + 3 + 4 = 9$$

18. (4) Only the pairs of (allose, atrose), (glucose, mannose), (gulose, indose) and (galactose, talose) contain C-2 epimers.

DPP-2 Preparation and Properties of Monosaccharides

1. (i) (a)

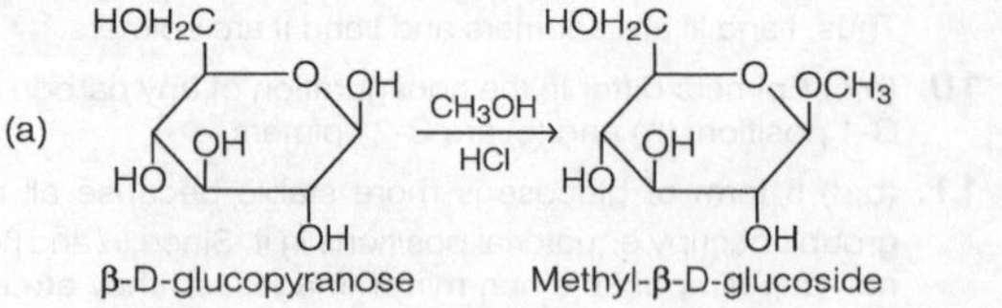

(b)

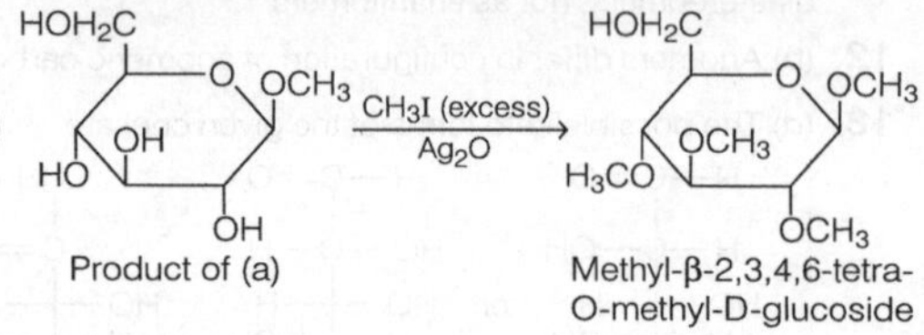

(c) Product of (b) $\xrightarrow[H_2O]{H^+}$ 2, 3, 4, 6-tetra-O-methyl-D-glucose

(d) D-glucose $\xrightarrow[H_2O]{Br_2}$ D-gluconic acid $\xrightarrow{\text{Pyridine}}$ Epimers

$\xrightarrow{\text{Separation}}$ D-mannonic acid $\xrightarrow[CO_2]{Na(Hg)}$ D-mannose

(e) D-glucose $\xrightarrow{NaBH_4}$ $\xrightarrow[\text{Excess}]{(CH_3CO)_2O}$ Hexa-O-acetyl-D-glucitol

(ii)

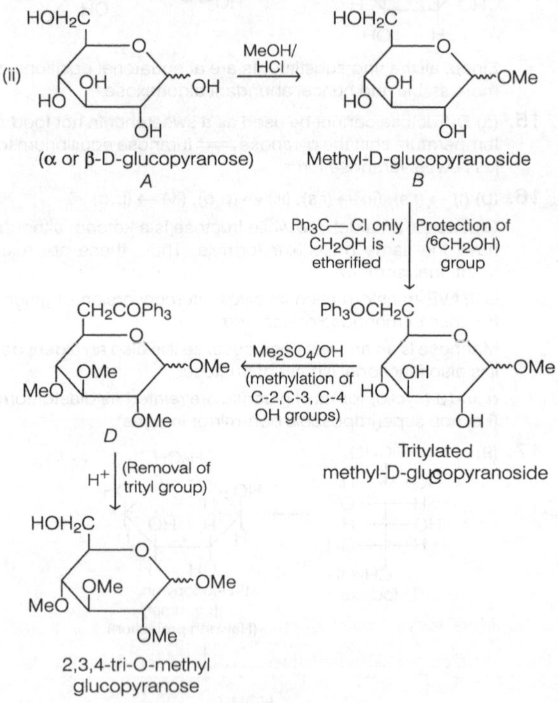

2. (i) (a)

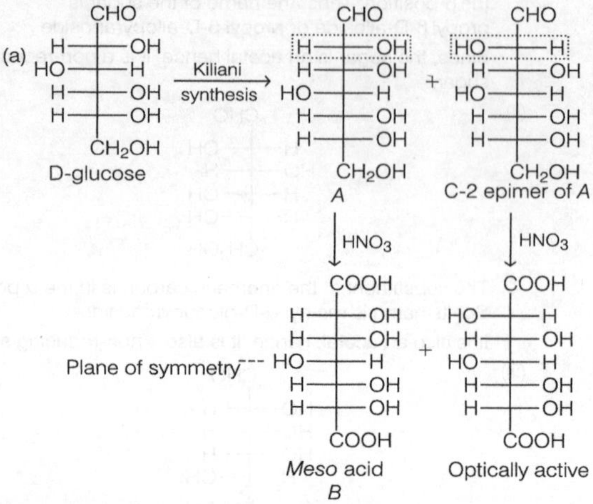

(b)

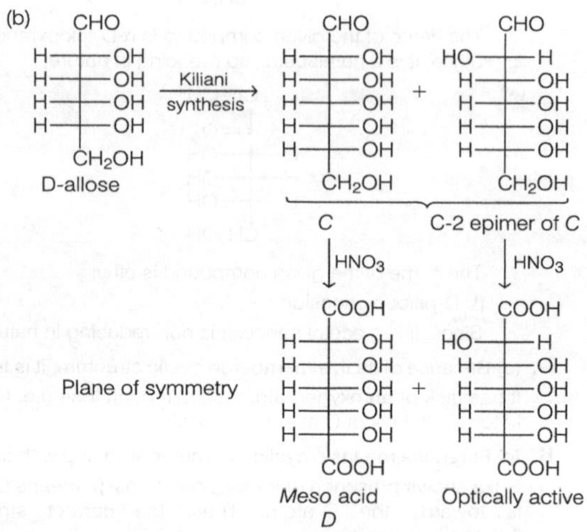

Although both *B* and *D* are heptaldaric acid, both are *meso* (optically inactive) but they are different.

(ii)

(a) L-gulose on oxidation with HNO_3 also give D-glucaric acid as

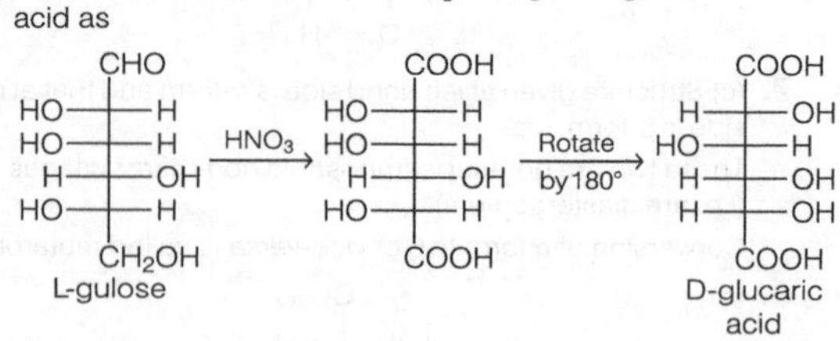

(b) D-altrose and D-talose on oxidation give same aldaric acid.

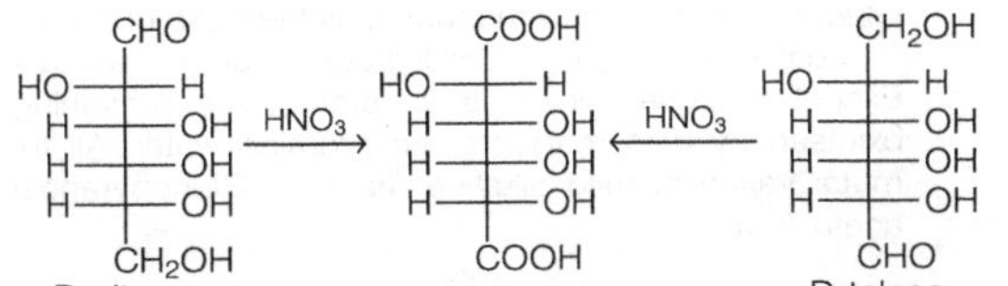

D-altrose D-talose (Rotated upside down)

3. (i) (a) D-lyxose
(b) D-tagatose
(c) D-talose

(ii) (a) D-ribose, D-arabinose, D-xylose and D-lyxose are all enantiomers with their corresponding L-forms.
(b) D-ribose and D-arabinose (since, differ in the configuration of C-2) gives the same osazone.

Similarly, D-xylose and D-lyxose also give the same osazone.

4. (i)

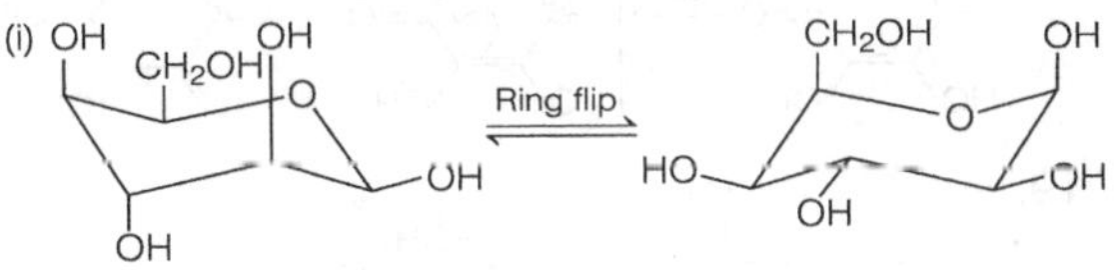

β-D-iodopyranose (Less stable) (More stable)

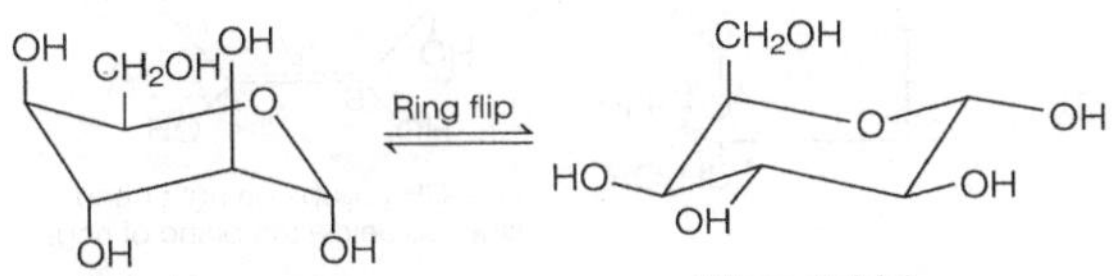

α-D-iodopyranose (Less stable) (More stable)

(ii) α-form is more stable as compared to β because in it only one substituent —OH is at axial position in its more stable chair conformation.

(iii) In case of β-anomer of iodose, the axial hydroxyl groups on C_1 and C_6 are close enough for five membered ring to form anhydro.

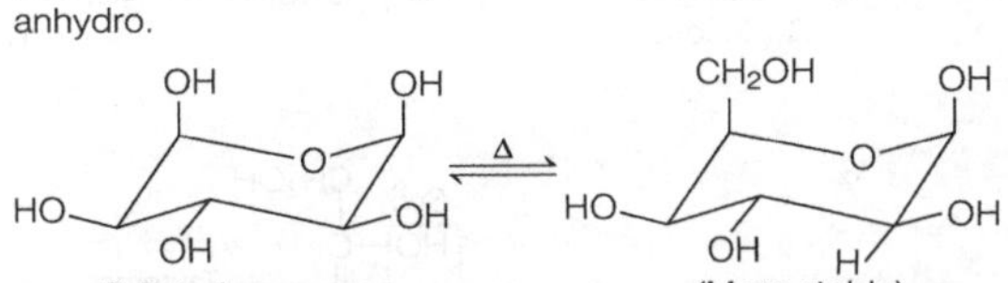

β-D-iodopyranose (More stable)

However, in case of D-glucopyranose, the —OH group of C_1 and C_6 occupy equatorial position in its most stable chair conformation and thus, too far away for a ring to form.

5. (i)

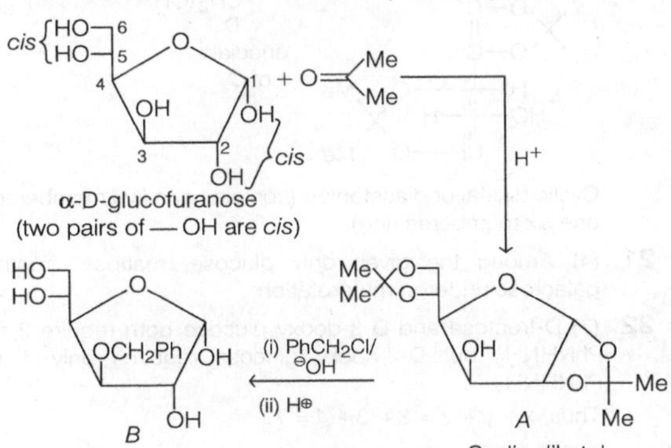

α-D-glucofuranose (two pairs of — OH are *cis*)

B *A* Cyclic diketal (Diacetonide)

(ii) Only this pair is *cis*

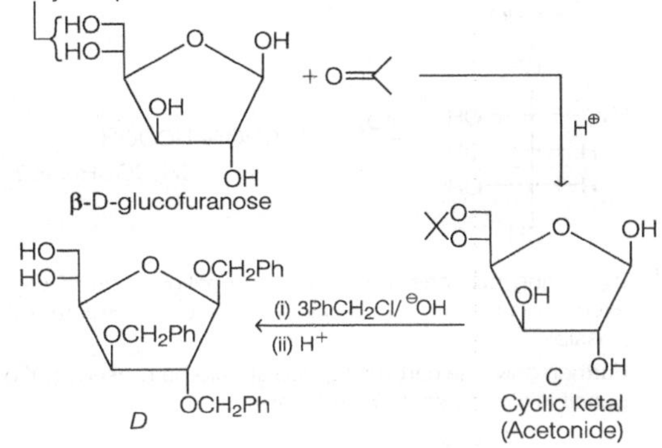

β-D-glucofuranose *D* *C* Cyclic ketal (Acetonide)

(iii)

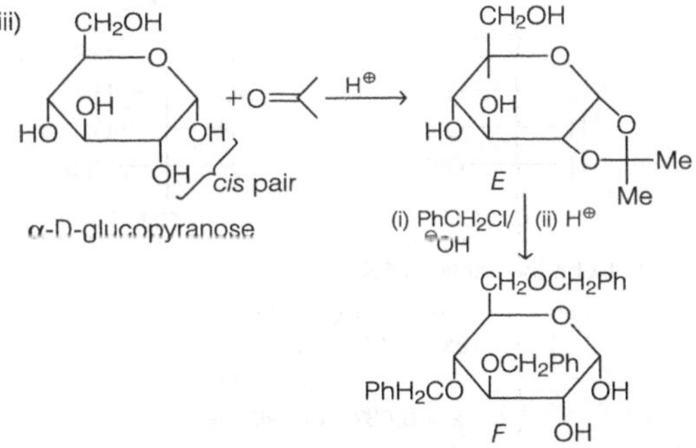

α-D-glucopyranose *E* *F*

(iv)

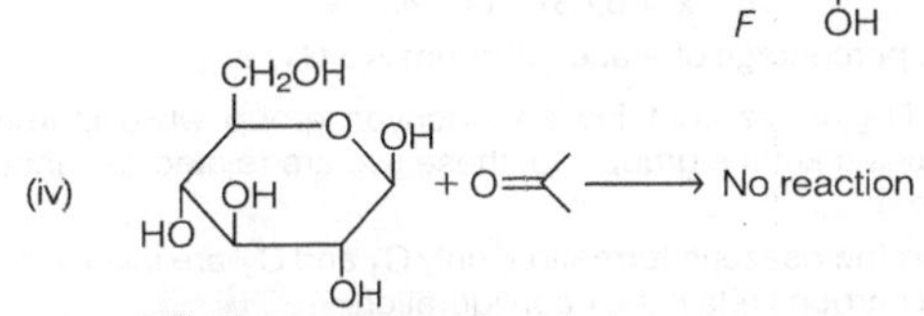

β-D-glucopyranose (no pair of *cis* — OH group)

6. (b) In pentacetate form, glucose does not react with any of the given reagents because of the absence of free aldehyde group.

7. (c) Tetraacetate means four —OH groups.
Oxidation with Br_2 -H_2O ⇒ a —CHO group.
Formation of *iso*-pentane ⇒ presence of a branch.
Thus, the possible structures for *A* are

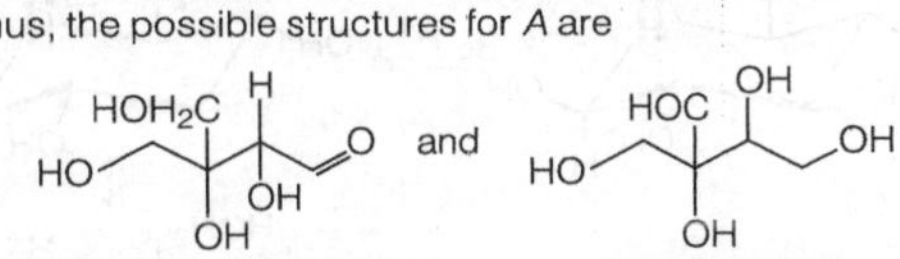

(Forms diphenyl hydrazone, i.e. osazone derivative) (It does not form diphenyl hydrazone)

Thus, phenyl hydrazine can be used to distinguish between them.

8. (a) For giving the same chiral alditol after reduction, the configuration of rest of the chain should be the same.

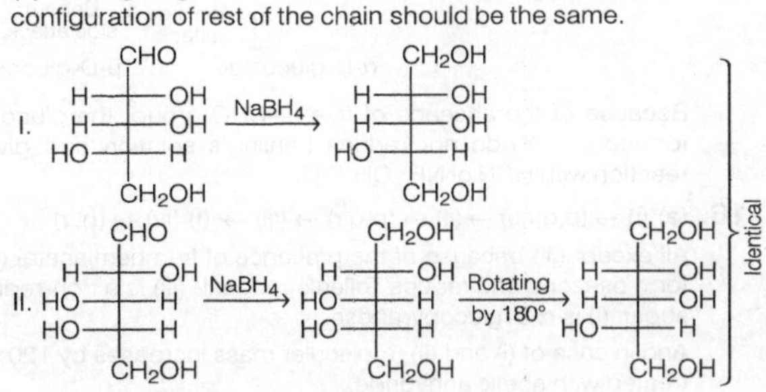

9. (a) At the end all the — OH are replaced by — OMe Although, the configuration remains the same.

10. (a) Its open chain structure is

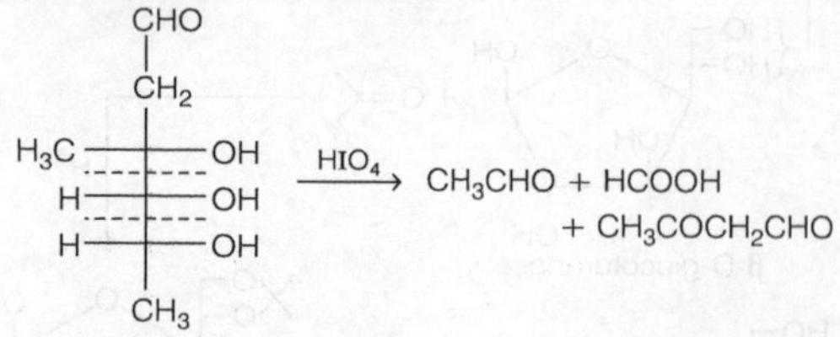

11. (b) During osazone formation, the configuration of all carbons except the C_1 and C_2 remain the same, so structure IV is not possible.

Further, only one carbonyl group should be present in *Y* or *Z*. So, possible structures for *Y* and *Z* are

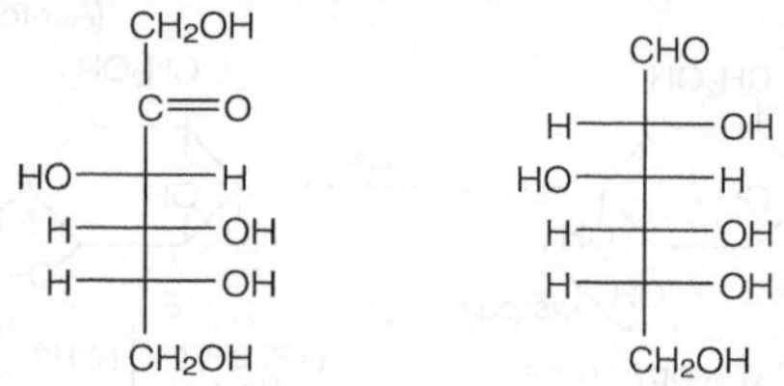

12. (b) Let *x* be the fraction of *A*.

$$19x + 112(1 - x) = 52.7$$
$$19x + 112 - 112x = 52.7$$
$$93x = 59.3$$
$$x = 0.637 \quad \text{or} \quad 63.7\%$$

Thus, percentage of *A* at equilibrium is 64%.

13. (c,d) D-glucose contains an aldehyde group while D-fructose contains a ketone group. So, these two are related as functional isomers.

During the osazone formation, only C_1 and C_2 are involved. Rest of the carbons retain their configuration.

14. (a,d) Epimers are the diastereomers that differ in configuration only at one chiral carbon. (i.e. they are non-superimposable non-mirror images).

It is not necessary that epimers rotate the plane polarised light in opposite direction.

15. (a,b,c) The ring structure of glucose is

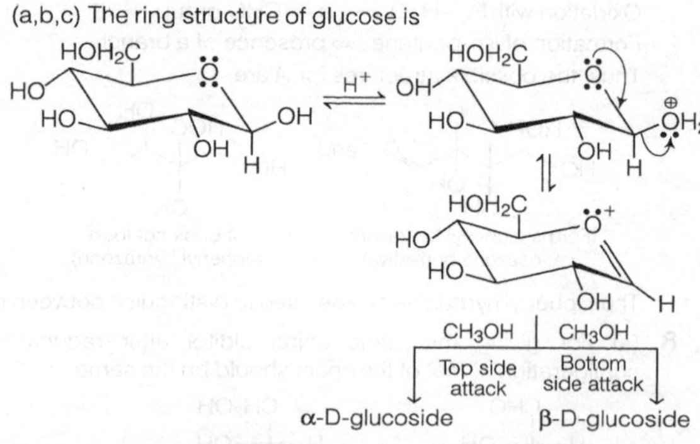

Because of the absence of free —CHO group, the glucosides formed (i.e. X) do not reduce Fehling's solution and give no reaction with HCN or NH_2OH.

16. (a) (i) → (p,q,r,s) → (ii) → (p,q,r) → (iii) → (t) (iv) → (p, r)

All except (iii) because of the presence of free hemiacetal group form osazone and reduce Tollen's reagent. (iii) is a non-reducing sugar. (i) is α-D-glucopyranose.

And in case of (i) and (ii), molecular mass increases by 120 when treated with acetic anhydride.

17. (a) (i) → (p,q,s), (ii) → (p,q), (iii) → (p,q,s), (iv) → (r)

Glucose, fructose and mannose, all because of the presence of free carbonyl group (—CHO group) reduce Fehling's solution. Except fructose (which is a ketone), the remaining two are oxidised by mild reagents like bromine water. All these show mutarotation in mild alkaline medium. Glucopyranoside is the acetal form.

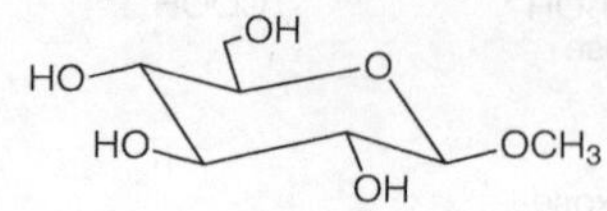

Only four —OH groups are available for reaction with anhydride in (I) so result in the formation of tetraacetate derivative.

18. (c) The anion formed by the removal of enolic H is stable because the negative charge is stabilised by resonance, i.e. by delocalisation of the negative charge to the O of the (C=O) through (C=C) as shown below:

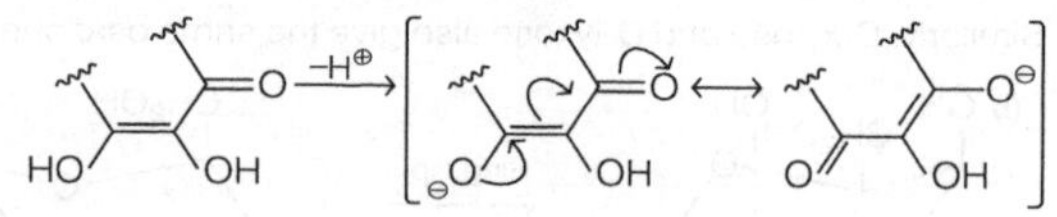

19. (c)

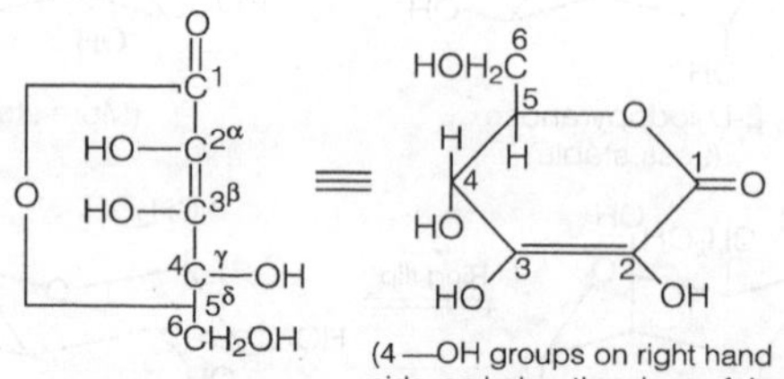

(4—OH groups on right hand side, so below the plane of ring)

20. (c,d)

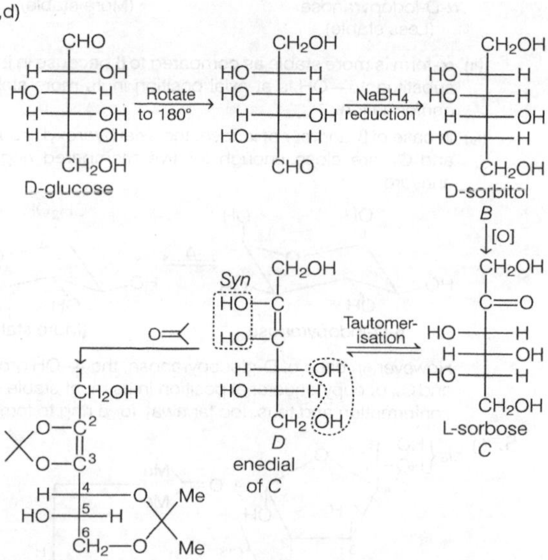

Cyclic diketal or diacetonide (contains one five membered and one six membered ring)

21. (4) Among the given only glucose, maltose, fructose and galactose undergo mutarotation.

22. (7) D-fructose and D-3-deoxy glucose both require 3 moles of $PhNHNH_2$ but D-2-deoxy glucose requires only 1 mole of $PhNHNH_2$.

Thus, $x + y + z = 3 + 3 + 1 = 7$.

DPP-3 Some Important Oligos and Polysaccharides

1. (i) (a) α (1 $\rightarrow$ 4), α (1 $\rightarrow$ 6) (b) β (1 $\rightarrow$ 4) (c) α(1 $\rightarrow$ 4), α(1 $\rightarrow$ 6) (d) β (1 $\rightarrow$ 4).

(ii)

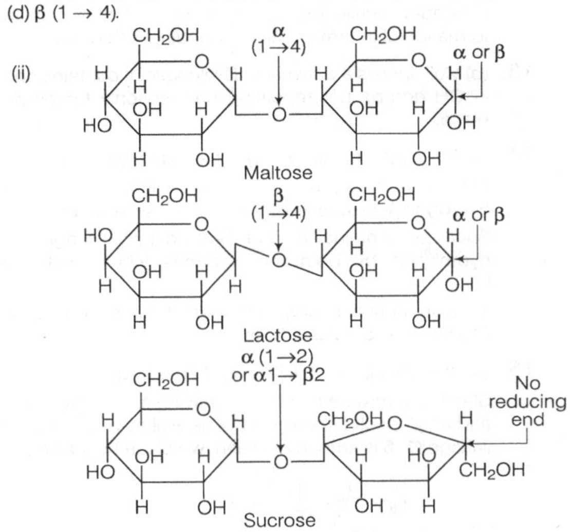

Maltose : α-D-glucopyranosyl- (1 $\rightarrow$ 4)-α-D-glucopyranoside.

Lactose : β-D-glucopyranosyl-(1 $\rightarrow$ 4) α-D-glucopyranoside.

Sucrose : α-D-glucopyranosyl-(1 $\rightarrow$ 2) β-D-fructofuranoside.

(iii) The β (1 $\rightarrow$ 4) linkage of cellulose allows the polymer to be linear while the α(1 $\rightarrow$ 4) linkage in glycogen (starch) causes the polymer to coil. Glycogen (and some starches) also has α(1 $\rightarrow$ 6) branch points which make the overall structure puffier and highly soluble in water unlike the tightly packed cellulose fibres.

2. *A* does not reduce Fehling's solution and does not mutarotate, which indicates the involvement of hemiacetalic —OH at C-1 of aldose and hemiketalic —OH group at C-2 of ketose in the formation of glycosidic linkage. Hydrolysis by maltase suggests the presence of α-linkage and by emulsin suggests presence of β-linkage.

Products obtained after the methylation and hydrolysis of *A* suggest (C – 1) OH (α) of aldose and C-2 (OH) β of ketose are involved in glycosidic linkage.

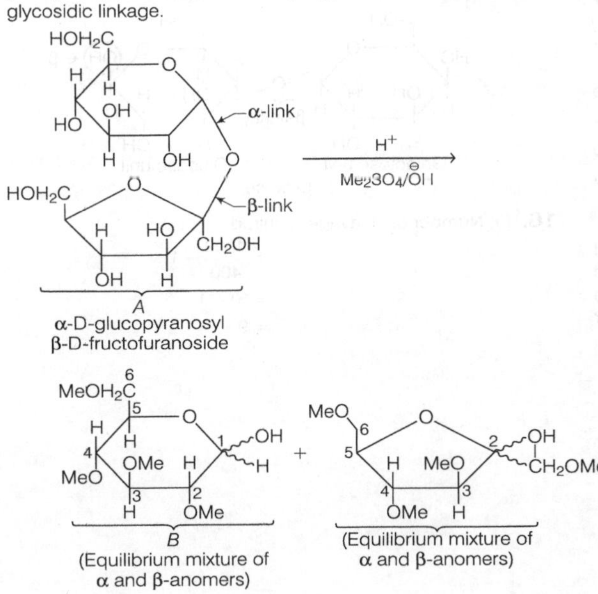

Aldose (glucose) is in pyranose form which shows the presence of six-membered ring.

Ketose (fructose) is in furanose form and suggests the presence of five membered ring.

All the observations suggest that *A* is sucrose.

3. Formation of mannose and galactose derivatives on treatment with $CH_3I/Ag_2O/H^+$ suggests the presence of these two units in the disaccharide.

Reduction of Tollen's reagent to silver shows the presence of atleast one hemiacetal group.

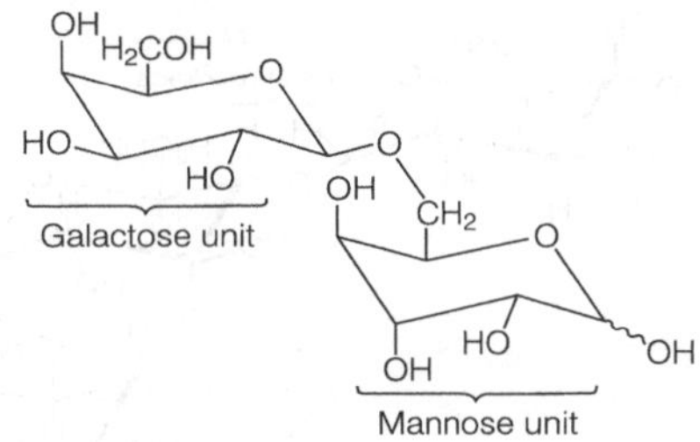

Hydrolysis by β-glycosidase indicates that the monomer units (i.e. mannose and galactose) are bonded together through β-glycosidic linkage. Thus, the structure of *X* is given above.

Here, Ag_2O gives ppt of AgI and hence, shifts the reaction in forward direction.

4. Formation of tetra-O-methyl glucopyranose and tri-O-methyl glucopyranose on treatment with $CH_3I/H_3O^{\oplus}$ suggests that the oxygen in tri-O-methylated ring is not a part of hemiacetal group and is not methylated, so it forms the acetal bond. Thus, C-6 oxygen is involved in the formation of glycosidic linkage.

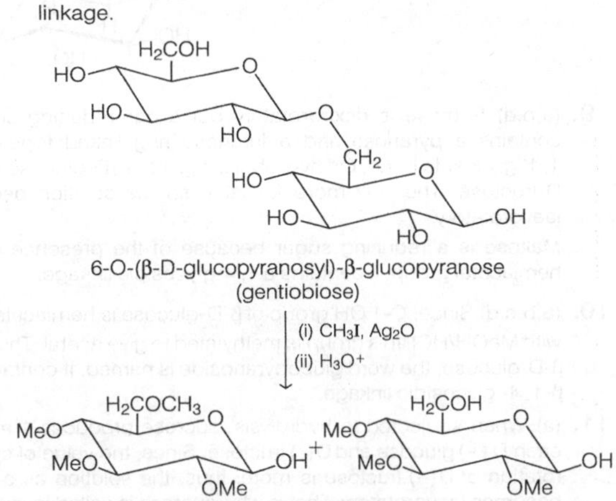

5. (a) Because of the absence of free hemiacetal group. it is a non-reducing sugar.

Here, the two units are joint together through α- (1, 2) glycosidic linkage.

6. (c) Aldose contains aldehyde group while ketose have ketone group. In the given molecule, both the units have one —CHO group, i.e. it contains two aldose and zero ketose.

Six membered -O- containing cyclic unit is called pyranose. So, no. of pyranose units = 2.

(i.e. furanose unit = 0)

It contains only one hemiacetal unit.

7. (c)

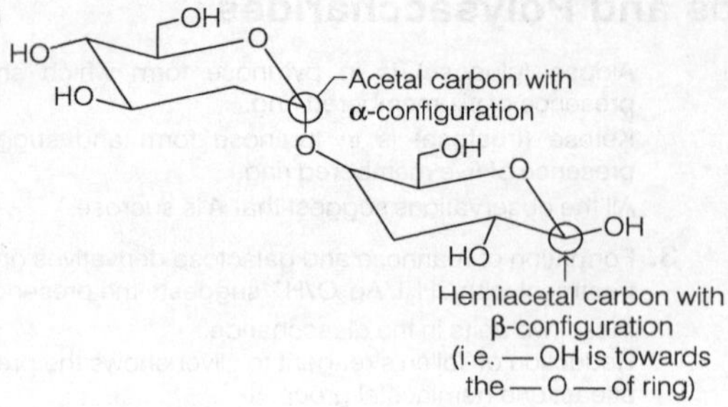

8. (a)

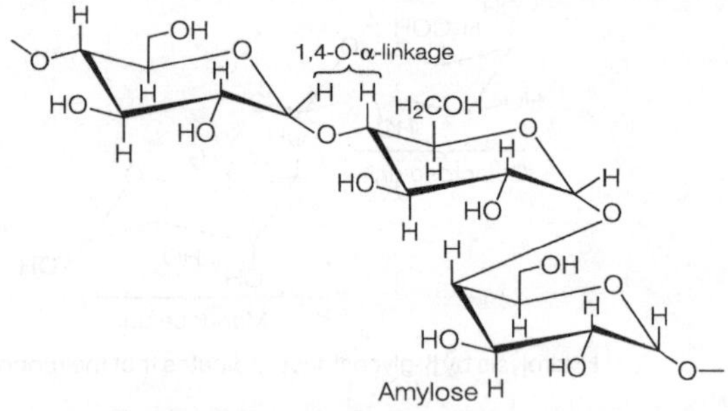

Amylose

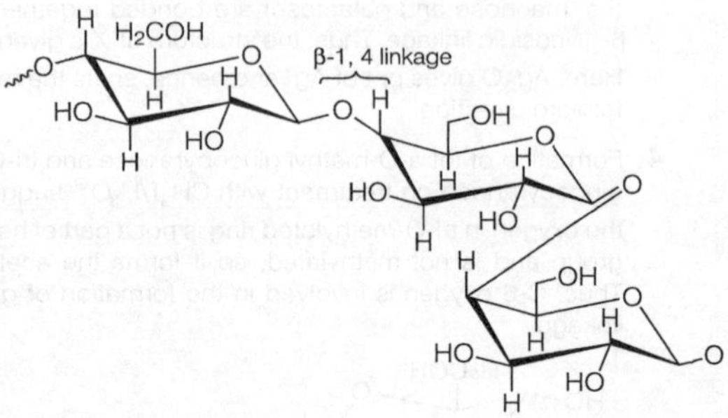

9. (a,b,d) Sucrose is dextrorotatory but a non-reducing sugar. It contains a pyranose and a furanose ring linked together by 1, 1′-glycoside bond. On hydrolysis, it gives (+) D-glucose and (–) D-fructose. The α is more for later, so the solution becomes laevorotatory.

Maltose is a reducing sugar because of the presence of free hemiacetal group. It contains a 1, 4-glycosidic linkage.

10. (a,b,c,d) Since, C-1 OH group of β-D-glucose is hemiacetalic, so with MeOH/HCl, this group is methylated to give acetal. Thus, with β-D-glucose, the word glucopyranoside is named. It contains a β-1, 4- glycosidic linkage.

11. (a) When subjected to hydrolysis, sucrose produces 1 mole of each D (+) glucose and D(–) fructose. Since, the value of specific rotation of D(–) fructose is more, thus, the solution as a whole becomes laevorotatory. That is why, sucrose is called invert sugar (as rotation changes from dextro to laevo).

12. (d) Cellulose is insoluble in water while starch is partially soluble in it. This is because hydroxy groups present on the repeating glucoside units of cellulose are linked through intramolecular H-bonding while the same in case of starch is free for the formation of intermolecular H-bonding with water.

13. (b) Cellulose is a polymer of D-glucose. It contains only three free —OH groups per repeating unit, so only trimethylation takes place.

14. (a) (i) → (p,q,r,s), (ii) → (p,r), (iii) → (q,s), (iv) → (p,r)

Maltase and amylase hydrolyse α-linkage. Emulsin and lactase hydrolyse β-linkage. Invertase hydrolyse sucrose.

Sucrose contains α and β-glycosidic linkages. It can be hydrolysed by both the enzymes which hydrolyse α- and β-linkage.

Maltose contains only α-linkage. Lactose contains β-linkage. Starch also contains α-linkage.

15. (a) (i) → (t), (ii) → (p), (iii) → (s), (iv) → (p,q)

Starch is a mixture of amylose (having α- 1, 4-glycosidic linkage) and amylopectin (having α-1, 4 as well as α-1, 6- glycosidic linkage (1, 6 is formed between two different chains).

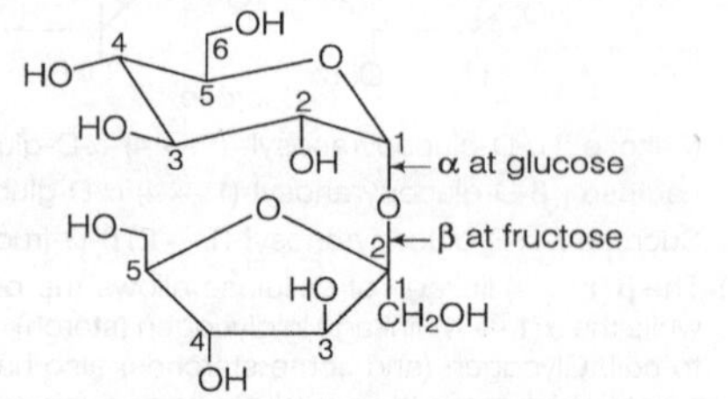

Sucrose

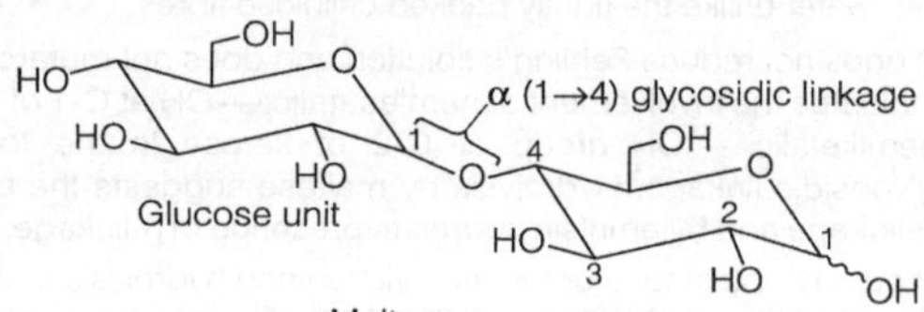

Maltose

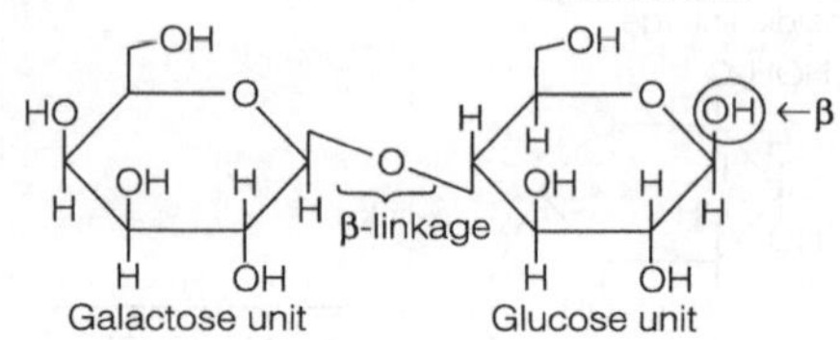

Lactose

16. (9) Number of cleavage required

$$= \frac{4000}{400} - 1$$

$$= 10 - 1$$

$$= 9$$

DPP-4 Classification and Physical Properties of Amino Acids

1. (i) (a) 2-amino ethanoic acid

(b) 2-amino propanoic acid

(c) 2-amino-3-hydroxy propanoic acid

(d) 2-amino-3-mercapto propanoic acid

(ii)

(a) 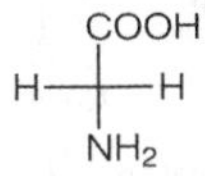 (not possible)

(b) 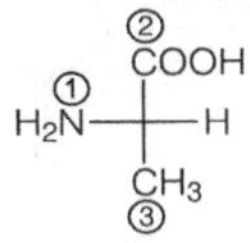 (L-alanine) Configuration-*S*

© 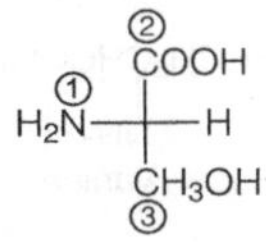 (L-serine) Configuration-*S*

(d) 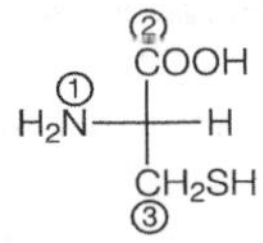 (L-cysteine) Configuration-*S*

(iii) Except glycine ($CH_2 NH_2 COOH$), all are chiral.

2. (i)

(a) 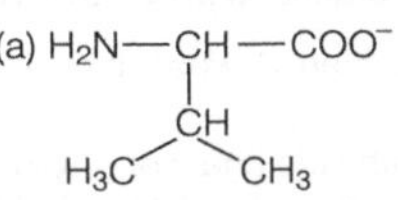 (b) 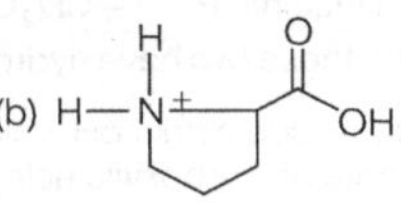

(c) 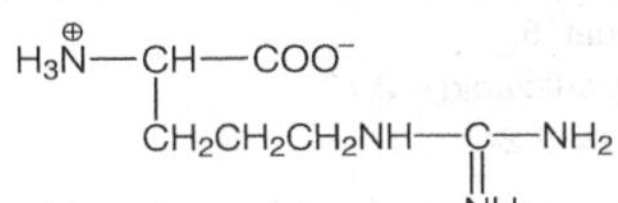

(d) 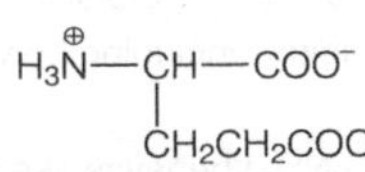

(e) At pH 6,

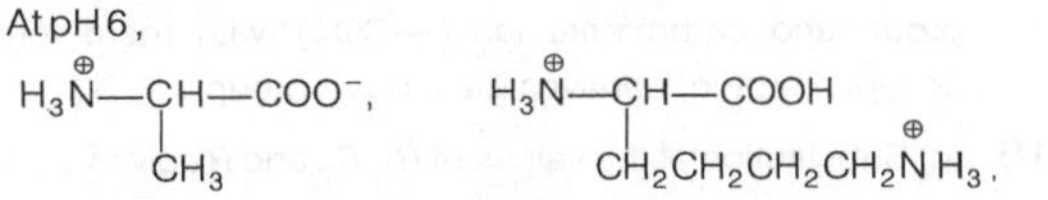

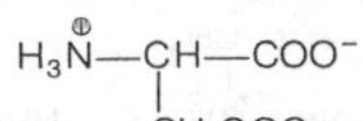

At pH 3,

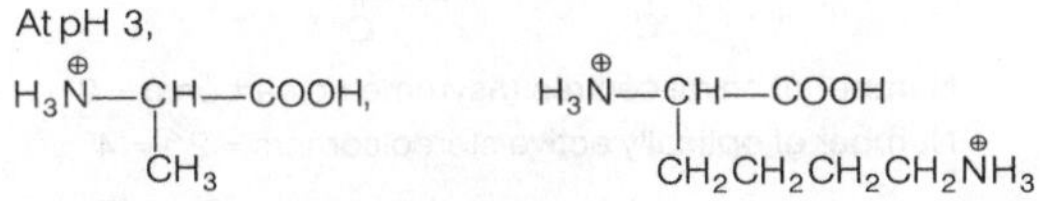

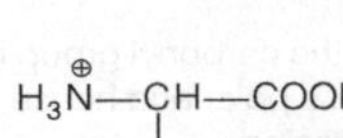

(ii) Physiological pH means at a pH of 7.3

(a) 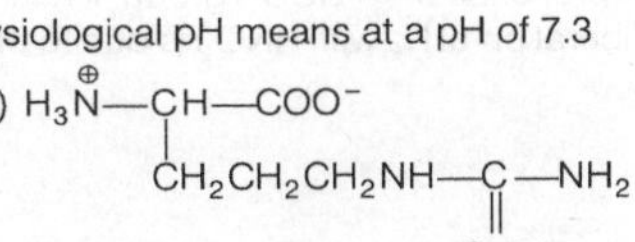

(b) 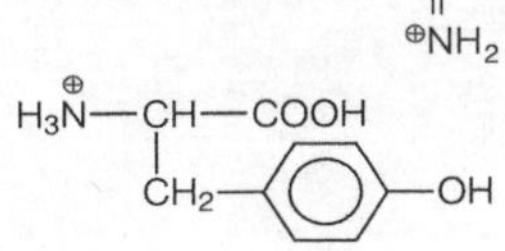

(c) 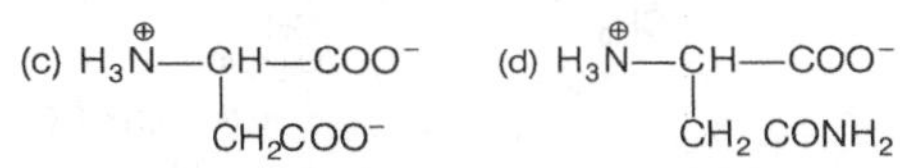

(e) 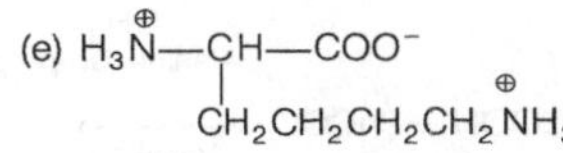

(iii) (a) In alkaline medium (i.e. *aq*. NaOH)

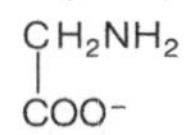

(b) In acidic medium (i.e. *aq*. HCl)

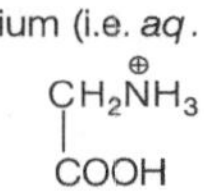

3. (i) (a) $pI = \frac{2.02 + 8.84}{2} = 5.43$

(b) $pI = \frac{9.04 + 12.48}{2} = 10.76$

(c) $pI = \frac{pK_{a_1} + pK_{a_2}}{2}$

$= \frac{(14 - 4.4) + 2.2}{2} = \frac{11.8}{2} = 5.9$

(d) $pI = \frac{2.21 + 9.15}{2} = 5.68$

(e) $pI = \frac{4 + 2}{2} = 3.0$

(ii) Methionine, as it has pI lower than that of glycine.

4. (i)

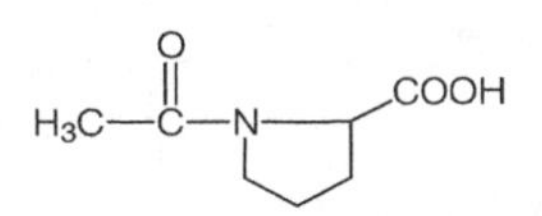

(ii) 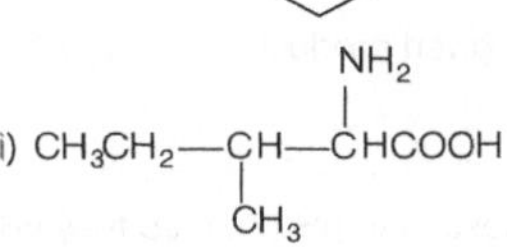

(iii) 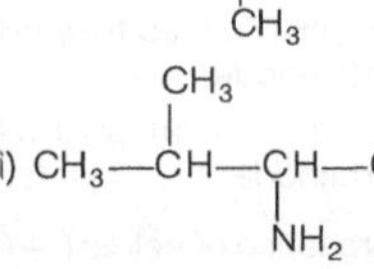 (involve HVZ reaction i.e. attack at α – C)

(iv) 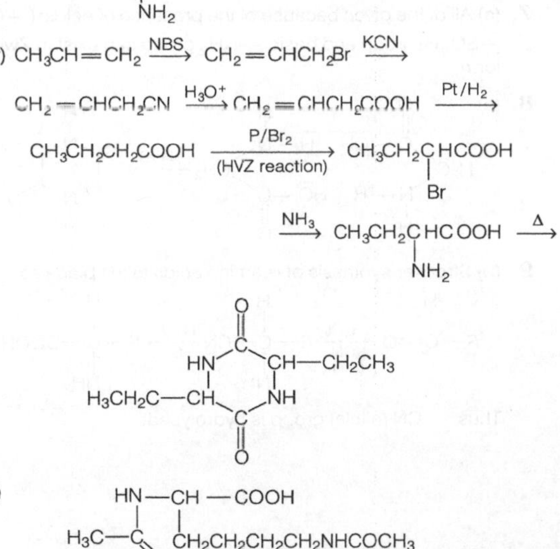

(v) 

5. (i) (a) $H_2N—\underset{CH_3}{\underset{|}{C}}H—COOH + CH_3\underset{\underset{(Excess)}{OH}}{\underset{|}{C}}H—CH_3 \xrightarrow{HCl} H_2N—\underset{CH_3}{\underset{|}{C}}H\ COOCH(CH_3)_2$

Alanine *iso*-propyl ester

(b) $H_2N—\underset{CH_3}{\underset{|}{C}}H—COOH + PhCOCl \xrightarrow[\text{of amine}]{\text{Acylation}} Ph—\overset{O}{\overset{||}{C}}—NH—\underset{CH_3}{\underset{|}{C}}H—COOH$

N-benzoylalanine

(c) $H_2N—\underset{CH_3}{\underset{|}{C}}H—COOH + (CH_3)_3CCOCl \longrightarrow (CH_3)_3C—\overset{O}{\overset{||}{C}}—NH—\underset{CH_3}{\underset{|}{C}}H—COOH$

t- butyloxy carboxylalanine

(ii) (a) $CH_3—\underset{NH_2}{\underset{|}{C}}H—CH_2COOH \xrightarrow{\Delta} CH_3CH{=}CHCOOH$

β- amino acid

(b)

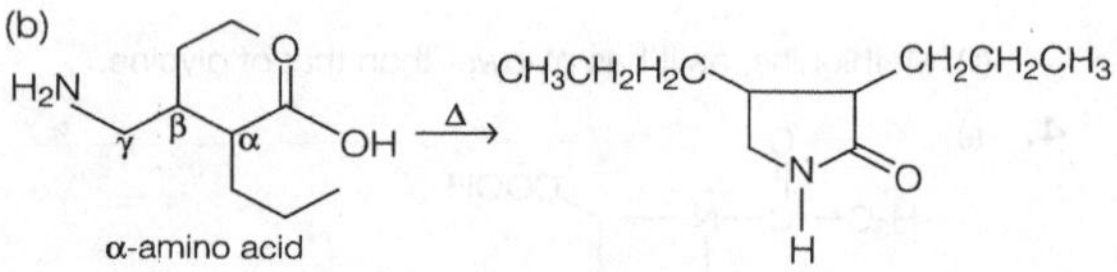

α-amino acid

(c) $C_3H_7\underset{NH_2}{\underset{|}{C}}H\ COOH \xrightarrow{\Delta}$ given product

α- amino acid

6. (d) Since, pI of A and B is less than pH (7.7), so they will exist in anionic form and will migrate towards anode.

Since, pI of C is greater than the pH of medium, so it will exist in cationic form and migrate towards cathode.

7. (a) All of the given because of the presence of acidic (—COOH or —SO_3H) group and basic —NH_2 group can exist in Zwitter ionic form.

8. (a)

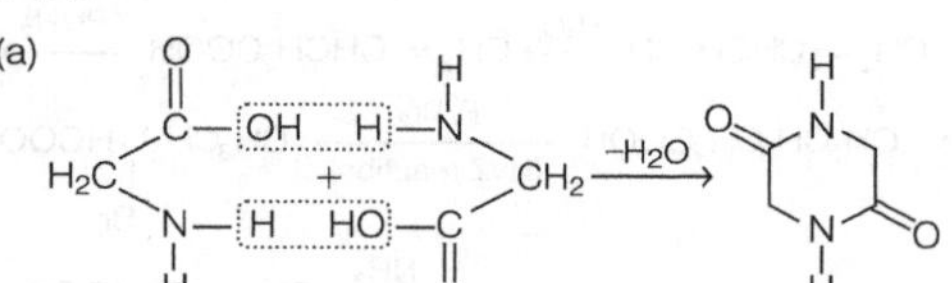

9. (b) Strecker synthesis of α-amino acids takes place as

$$R—\overset{H}{\overset{|}{C}}{=}O \xrightarrow[\text{KCN}\ (X/Y)]{NH_4Cl/} R—\underset{NH_2}{\underset{|}{\overset{H}{\overset{|}{C}}}}—CN \xrightarrow[(Z)]{H_3O^+} R—\underset{NH_2}{\underset{|}{\overset{H}{\overset{|}{C}}}}—COOH$$

Thus, —CN (nitrile) group is hydrolysed.

10. (b) Proline contains secondary amine group, so forms nitrosoamine with HNO_2.

Others because of the presence of primary amine group form unstable diazonium salt which decomposes to give nitrogen.

11. (a) Structure (a) is responsible for purple colour in ninhydrin test irrespective of the nature of α-amino acid.

12. (d) $X =$ serine, $H_3\overset{\oplus}{N}—\underset{CH_2OH}{\underset{|}{C}}HC\overset{\ominus}{O_2} \xrightarrow[{[H]}]{LiAlH_4} H_3\overset{\oplus}{N}\underset{CH_2OH}{\underset{|}{C}}HCH_2OH$

No chiral centre

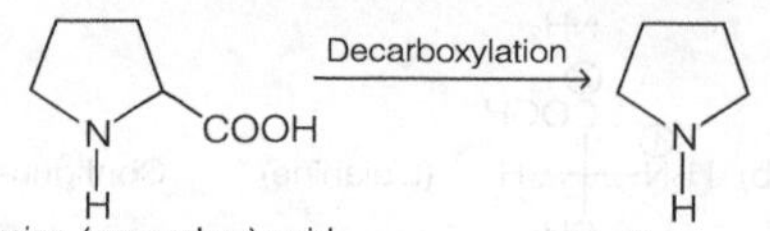

α-amino (secondary) acid $Y =$

$$CH_3\overset{O}{\overset{||}{C}}—\overset{O}{\overset{||}{C}}—OH \xrightarrow[\text{(ii) } NaBH_4]{\text{(i) } NH_3} CH_3\underset{NH_2}{\underset{|}{C}}HCOOH$$

$Z =$ Alanine

13. (b,c,d) In the general structure of amino acids, i.e.

$$H_3\overset{\oplus}{N}—\underset{R}{\underset{|}{C}}H—\overset{O}{\overset{||}{C}}—O^-$$

R is a non-polar group only in case of leucine and alanine [In leucine, R = —$CH_2CH(CH_3)_2$ and in alanine, R = —CH_3]

So, these two have hydrophobic interactions while others are not.

14. (b,c) Isoelectric pH value is less than 5 when the side chain contains carboxylic acid group.

Among the given, aspartic acid (with side chain of —CH_2COOH) have glutamic acid (with side chain — CH_2CH_2COOH) have isoelectric pH less than 5.

Isoelectric pH of aspartic acid = 2.76

and of glutamic acid = 3.24

15. (b) Because of the presence of acidic (—COOH) and basic (—NH_2) groups both, amino acids including glycine are amphoteric in nature.

In these acids ammonium ions (—$\overset{+}{N}H_3$) behaves like an acidic group and carboxylate ion (—CO_2^-) (with more tendency to accept a proton) behaves like a basic group.

16. (c) Substitution of the values of R_1, R_2 and R_3 gives

$$H_2N—CH_2—\underset{O}{\underset{||}{C}}—NH—\overset{CH_3}{\overset{|}{\overset{*}{C}}}H—\underset{O}{\underset{||}{C}}—NH—\overset{CH_2Ph}{\overset{|}{\overset{*}{C}}}H—COOH$$

Number of chiral centres (asymmetric carbons) = 2

Number of optically active stereoisomers = $2^2 = 4$

17. (c) In this compound, the carbonyl group is present in the form of amide, so becomes less polar and hence, does not react with 2, 4-dinitrophenyl hydrazine.

Because of the presence of —COOH group, it releases CO_2 with $NaHCO_3$ and liberation of N_2 with HNO_2 is due to the presence of —NH_2 group.

18. (a) (i) → (p,q) (ii) → (p,q) (iii) → (p,r) (iv) → (p,s)

(i) 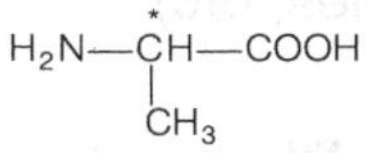

Chiral, one NH_2 group and one —COOH group, so neutral.

(ii) 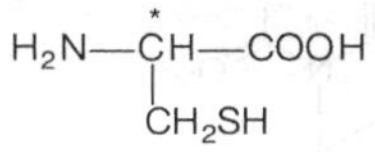

Chiral, one —NH_2 and one —COOH group, so neutral.

(iii) 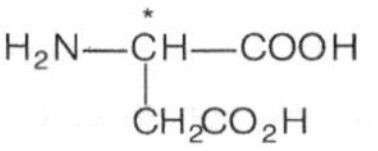

Chiral, 2— COOH groups and 1 —NH_2 group, so acidic.

(iv) 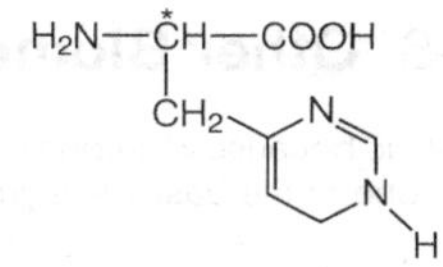

Chiral, more $\ddot{N}$ for protonation, so basic.

19. (3) Proline, tyrosine and alanine can be synthesised in the human body and hence, are considered as non-essential amino acids.

20. (5) Lone pair of sp^3- N of ring is a part of aromatic system, so not available for donation to H^+. On the other hand, lone pair of sp^2 - N of ring is not even in resonance and hence, more available on nitrogen for protonation. Thus, N(1) and N(4) get protonated easily.
So, $x + y = 1 + 4 = 5$.

DPP-5 **Peptides and Proteins**

1. (i) It is a pentapeptide.
   (ii) Between five molecules of amino acids, only 4 amide (peptide) linkages are possible.
   (iii) Glycine (C-terminal) and serine (N- terminal).
   (iv) Ser - Ala - Leu - Phe - Gly.

2. (d) The N-terminal of the peptide is towards left hand side, so

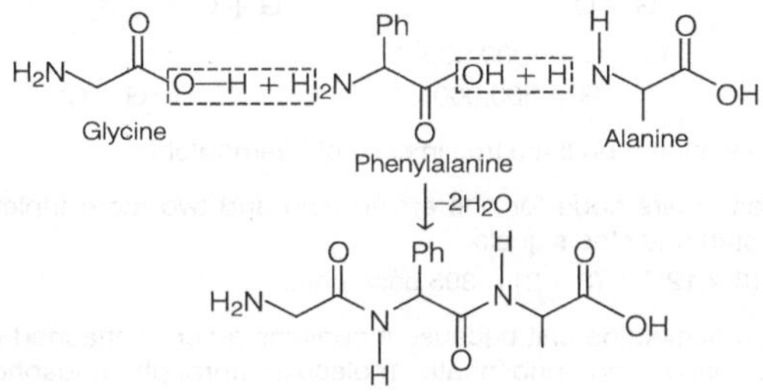

3. (c) Hydrogen bonding plays an important role in the secondary structure of proteins. These are formed between amide groups of two chains.

4. (a) Except hydrophobic effect, all the given are involved in the tertiary structure of proteins.

5. (c) Several weak electrostatic attraction exist between the amino acid units of protein. These interactions lower the free energy to a very great extent.

6. (a,b,c,d) Peptide linkage is infact an amide linkage which shows resonance and hence, the C—N bond acquires partial double bond character as

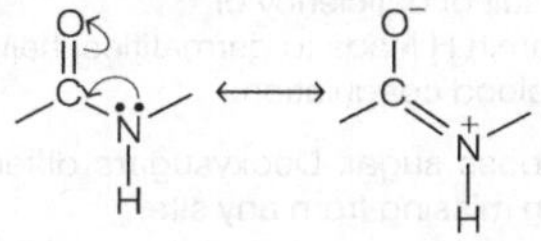

The above bond is hydrophilic (i.e. water loving) and the protein chains are associated through intermolecular H-bonding.

7. (a) Hydrolysis of the heptapeptide with carboxypeptidase gave an initial high concentration of alanine which indicates that alanine forms the C-terminal. In the paragraph, valine labelled at the α-amino group confirms valine as N-terminal.

8. (b) *B* on reaction with carboxypeptidase gives initial high concentration of glutamic acid, which shows that glutamic acid forms the C-terminal. So, the arrangement is Phe, Ala, Glu.

9. (c) From all the above information, the sequence of amino acids in heptapeptide is Val, Leu, Lys, Phe, Ala, Glu, Ala.

10. (b) By the definitions of the given terms, the correct match is
(i) → (q), (ii) → (p), (iii) → (s), (iv) → (r)

11. (d) (i) → (p), (ii) → (q), (iii) → (s), (iv) → (r)
   (i) Carboxypeptidase enzyme releases C-terminal amino acid, so 'Ala' is released.
   (ii) Leucine aminopeptidase enzyme releases N-terminal amino acid, so, 'Gly' is released.
   (iii) Hydrazinolysis releases C-terminal amino acid so 'his' is released.
   (iv) Enzyme leucine aminopeptidase releases N-terminal amino acid, so first 'Leu' is released. The dipeptide is treated with (Ph—N=C=S) (Edmann method) which releases N-terminal amino acid, so Val is released.

12. (4) Leucine, cysteine, tyrosine and glutamic acid total 4 amino acids form the given polypeptide chain.

13. (6) The following sequences are possible

Val-tyr-ala,	Tyr-ala-val
Val-ala-tyr,	Ala-tyr-val
Tyr-val-ala,	ala-val-tyr

14. (a) In proteins, the adjacent *R*-groups occupy alternate sides of the plane in order to have low conformational energy and high stability.

15. (a) Amide or peptide linkages are hydrolysed by the treatment of alkali.

16. (b) Because of resonance, a peptide bond has about 40% double bond character.
Further due to steric hindrance, *trans*-configuration is more stable so, α-carbons of adjacent amino acids are *trans* to each other.

DPP-6 **Other Biomolecules** (DNA, RNA, Vitamins, Hormones, etc)

1. (i) This is a DNA molecule because 2′ carbon is lacking a hydroxyl group and one of the bases is thymine not uracil.

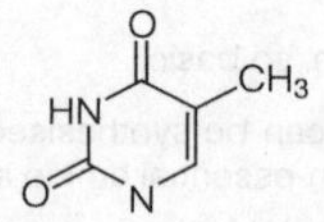

(ii) Arrangements I and II are possible because they maintain the antiparallel direction of the two strands forming hydrogen bonds. III is not possible because two strands forming hydrogen bonds are parallel to each other.

2. (i)

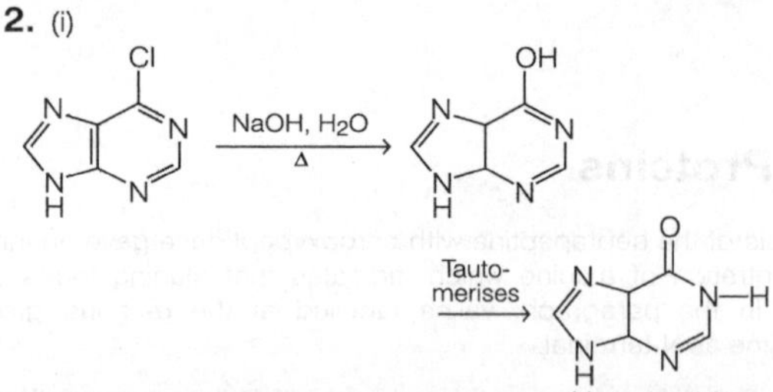

Hypoxanthine
A

(ii) In a DNA molecule, adenine (A) pairs with thymine (T) and cytosine (C) always pairs with guanine (G). Thus, if sequence of bases in one strand is.

A T G C T T C A (given)

∴ Sequence of bases in complementary strand

T A C G A A G T

(iii) A U A G A U G G A C C U

∵ Here, adenine pair with uracil (U), not with thymine.

3. (d) Enzymes act as catalyst, i.e. they catalyse the reactions occurring in human beings, so also known as biocatalysts. They increase the rate of forward and reverse reaction to the same extent, so have no effect on equilibrium constant.

4. (c) With the exception of a few enzymes, all the enzymes are proteins and hence, have amide (or peptide) linkages.

5. (c) Among the given, only insulin controls the burning process of biomolecules for the liberation of energy. Thus, X is insulin.

6. (a) Vitamins A, D, E and K are soluble in fat so absorbed from the intestine along with the fats.

7. (d) This is the messenger (*m*) RNA, which carries the code for the synthesis of proteins. All other given statements are true.

8. (b) Because of the presence of five membered ring, it is a furanose structure. Oxygen is missing from 2 position, so it is β-glycoside of D-deoxyribofuranose.

9. (c) The structure of DNA or RNA is as

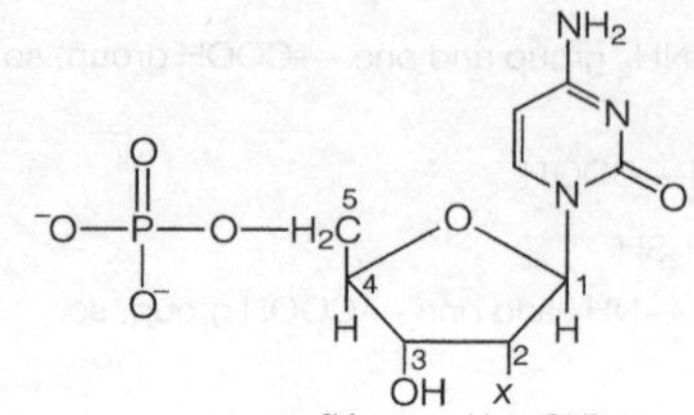

[Here, *x*=H or OH]

So, heterocyclic base is present at C_1 and phosphate ester linkage is present at C_5 of ribose or deoxyribose sugar.

10. (a) The structure of thymine is

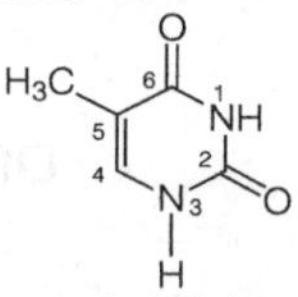

5-methyl uracil

11. (c) Hyperglycemia is the result of diabetes mellitus. All other given statements are true.

12. (b) In general, A = T and G = C

$$\therefore \quad \frac{A+T}{G+C} = 0.93 \quad \Rightarrow \quad \frac{2 \times 465{,}000}{G+C} = 0.93$$

$$G + C = 1000{,}000$$

$$G = 500{,}000 \qquad (\because G = C)$$

13. (d) All the given statements are true in case of haemoglobin.

14. (b) Three base pairs code for one amino acid and two more triplets are required for start and stop signals.

$$\therefore \quad (3 \times 129) + (3 \times 2) = 393 \text{ base pairs}$$

15. (a,c) This is a nucleotide unit because it contains a sugar attached with a heterocyclic base and phosphate molecule (through phosphoester linkage).

16. (a,b,c) With the exception of haemoglobin, all the given molecules being proteinous contain disulphide linkages.

17. (b) (i) → (p,r), (ii) → (q,r), (iii) → (s), (iv) → (t)

RNA (ribonucleic acid) has single stranded α-helix structure with adenine and guanine as purine bases. DNA (deoxyribonucleic acid) has double stranded α-helix structure with adenine and guanine as purine bases. *r*-RNA forms the site of protein synthesis while *t*-RNA acts as a carrier of amino acid.

18. (b) (i) → (s), (ii) → (p), (iii) → (q), (iv) → (r)

Beri-beri is a nervous disease caused by the deficiency of vitamin B. Pernicious anaemia is the result of deficiency of vitamin B_{12}. Deficiency of vitamin H leads to dermatities, hair loss, etc. Vitamin K is responsible for blood coagulation.

19. (b) DNA contains 2-deoxyribose sugar. Deoxysugars differ from the other sugars in having one O-atom missing from any site.

20. (c) Anaerobic reaction occurs in the absence of oxygen. However, ATP here also is the main source of energy.

Revisal Problems (JEE Main)

1. (d) Proteins contain several peptide linkages $\left(-\overset{\overset{O}{\|}}{C}-NH-\right)$ but at the terminals of the chain —COOH and —NH_2 groups are present.

2. (c) Since, D-mannose differ from D-glucose in the configuration of C_2 carbon, so the structure of β-D-mannopyranose is

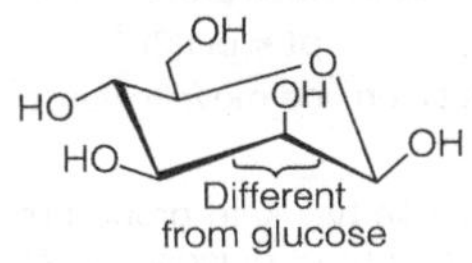

3. (c)

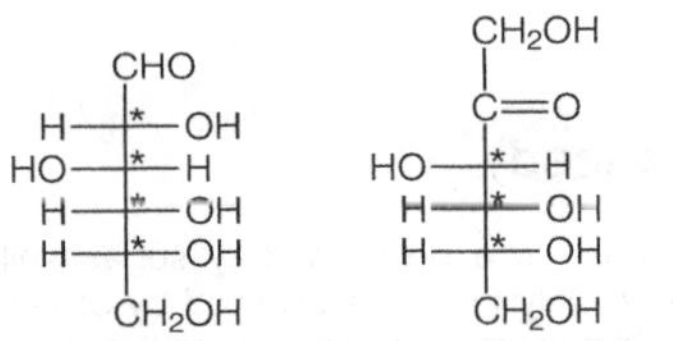

Glucose Chiral centre =4 | Fructose Chiral centre =3

(where, C^*= asymmetric carbon)

4. (a) In the α-helix structure, the *R* groups occupy alternate positions between the inner and outer side of the helix.

5. (d) Insulin is a globular protein.

6. (a) Enzyme acts as catalyst and increases the rate of biochemical reactions by decreasing activation energy.

7. (a) Chiral or achiral nature of monomer unit (including glycine) has no effect over the secondary or tertiary structure of proteins.

8. (b) α-D-glucopyranose and β-D-glucopyranose differ in the configuration of C_1 (anomeric carbon) so, these are the anomers, not epimers.

9. (c) In D-sugar, the —OH group of second last carbon occupy right hand side position.

10. (a) Reactant is β-D-glucopyranosyl, so the product formed will also be D-glucitol.
[$NaBH_4$ reduces —CHO group into —CH_2OH]

11. (a)

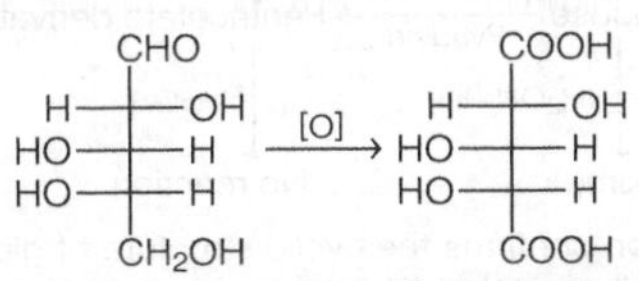

Optically active

(The dibasic acid formed by other given compounds is optically inactive because of the presence of plane of symmetry or symmetry elements)

12. (b)

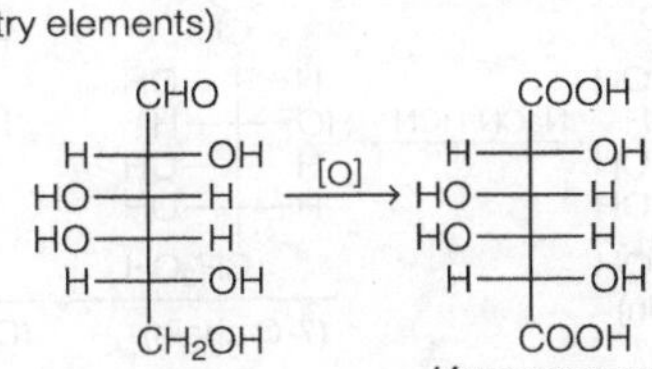

Meso compound
(so optically inactive)

13. (b)

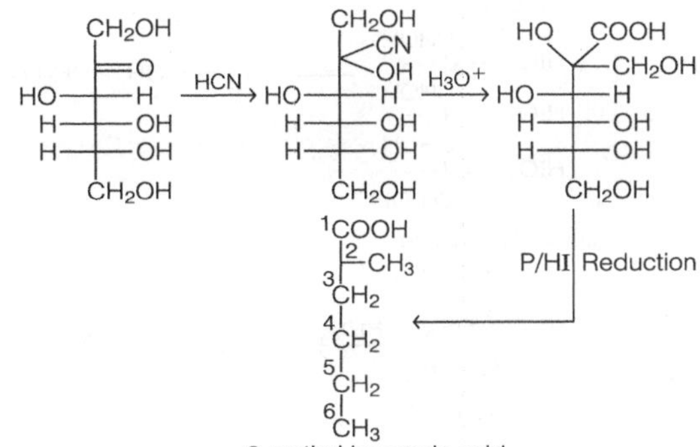

2-methyl hexanoic acid
A

14. (b) C-2 epimeric structures (i.e. structures differing only at the configuration of C-2) form same osazone derivatives when treated with excess of $PhNH \cdot NH_2$.

(I)

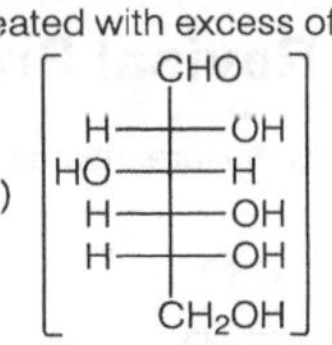

(II)

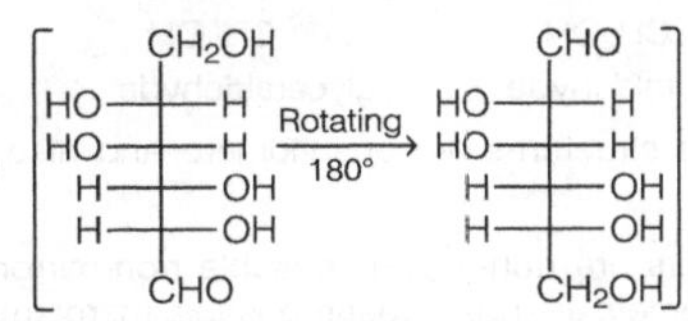

are C-2 epimers, so form same osazone derivative.

15. (a) In α-helix structure, H-bonds are parallel to the axis of helix.

16. (c)

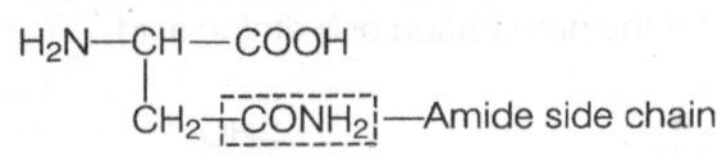

Asparagine

17. (d) Proline has a ring structure, so does not allow turning of a corner in protein.

18. (d) Adding SDS during electrophoresis of amino acids preserves the nature, structure and biological activity of the protein molecule.

19. (a) Since, pI > pH so, it exists in cationic form.

20. (a) $pI = \dfrac{pK_{a_1} + pK_{a_2}}{2}$

21. (c) Because of the presence of acetal bond, it is non-reducing in nature.

22. (d) Proline, it is a β-amino acid with no aromatic ring.

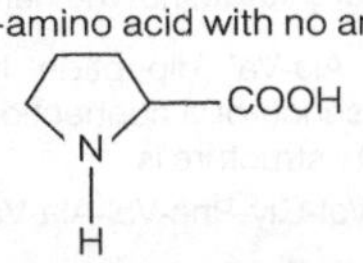

23. (c) It is a method used to increase the length of carbon chain.

24. (c) This is because one methoxy group is involved in the formation of acetal while the all others are ethers.

25. (a)

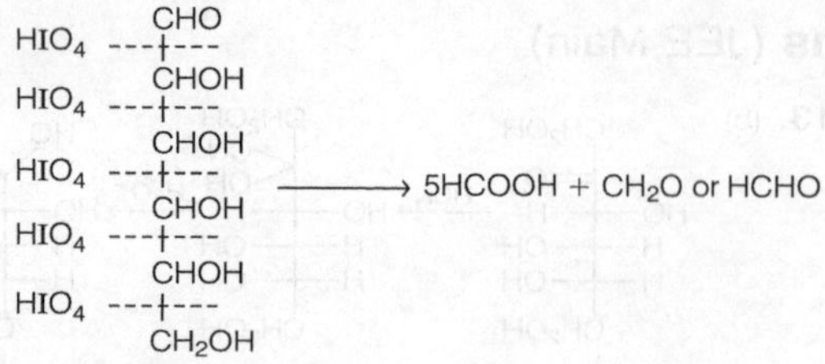

Ratio = 5/1 Glucose

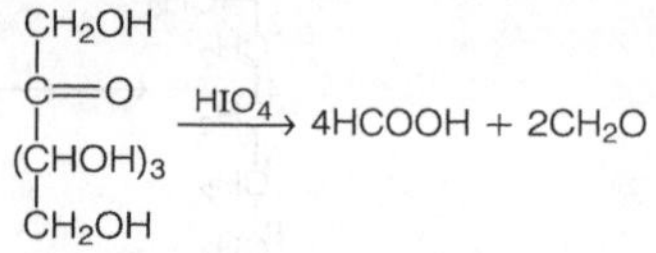

Fructose

So, ratio is 4/2.

26. (d)

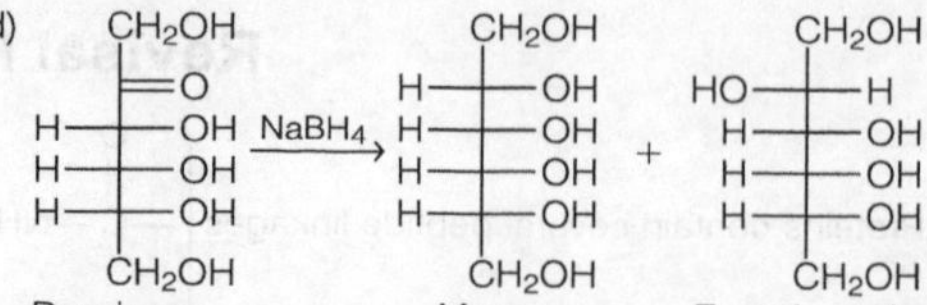

D-psicose — *Meso* B — Enantiomeric C

27. (d) Conjugated proteins produce amino acids and other biomolecules when hydrolysed.

28. (c) Aspartic acid has pI value lower than 5.

Since, pI < pH(6).

So, it will exist as anion and move towards anode under the given conditions.

29. (d) Although fructose has keto group but in alkaline medium, it converts into a mixture of glucose, mannose, etc., and hence, reduces the Tollen's reagent. It is also called fruit sugar.

Revisal Problems (JEE Advanced)

1. (d) Conformers of polypeptide backbone are related with its secondary structure.

2. (a)

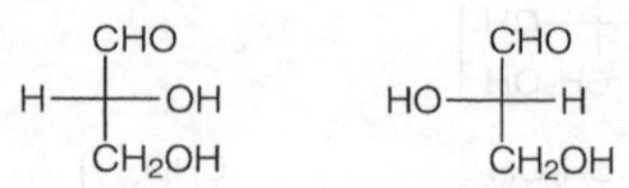

D-glyceraldehyde L-glyceraldehyde

Thus, these two structures are enantiomers and also the C-2 epimers.

3. (b) Diastereomers are non-superimposable non-mirror images whereas enantiomers are non-superimposable mirror images.

4. (c) Benzoic acid (C_6H_5COOH) because of the absence of labile α-H and formic acid (HCOOH) because of the absence of α-C are not used for the preparation of α-amino acid.

5. (c)

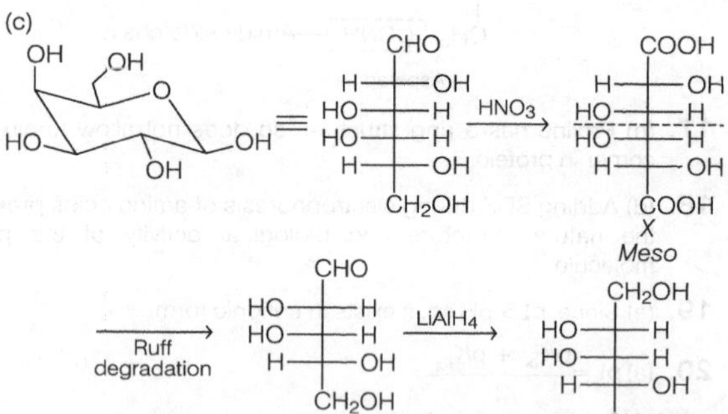

X *Meso*

Optically active

6. (a) It contains acetal linkage, not the hemiacetal linkage.

7. (a) Formation of Ala-Val dipeptide fragment on hydrolysis, suggested the presence of a connection between Ala-Val. Thus, the possible primary structure is

Val-Gly-Phe-Val-Ala-Val.

8. (b) Formation of triacetate shows the presence of only three —OH groups. Oxidation with Br_2-water shows the presence of a CHO group.

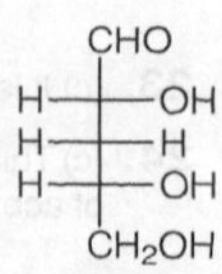

Reduction with HI to *n*-pentane, shows the presence of a straight chain of 5 carbons.

Formation of only 1 molecule of CH_2O and 1 molecule of HCOOH suggests that its structure is

9. (a,d) D-glucose is a dextrorotatory sugar while D-fructose is laevorotatory. Only former is oxidised by bromine water. Since, these two differ in first two carbons, so they form same osazone derivative with $PhNHNH_2$.

C-2 epimers

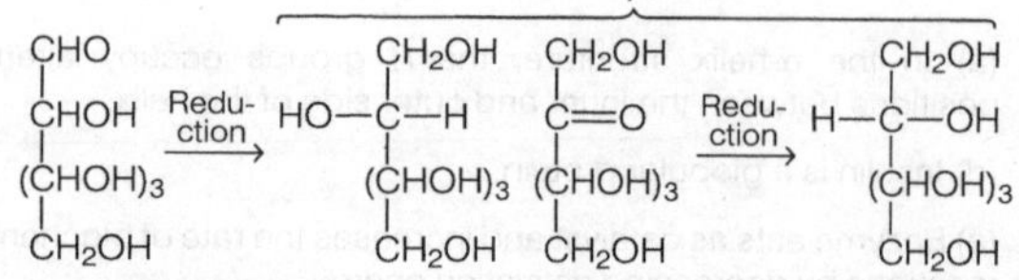

Glucose Fructose

10. (b,c) D-erythrose and D-threose are the smallest aldose that form cyclic hemiacetal.

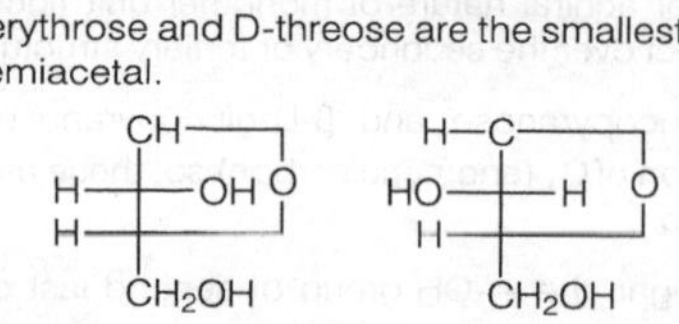

11. (a,b,c,d) All the given are non-covalent interactions and play a key role in maintaining the 2°, 3° and quarternary structures of proteins.

12. (a,d)

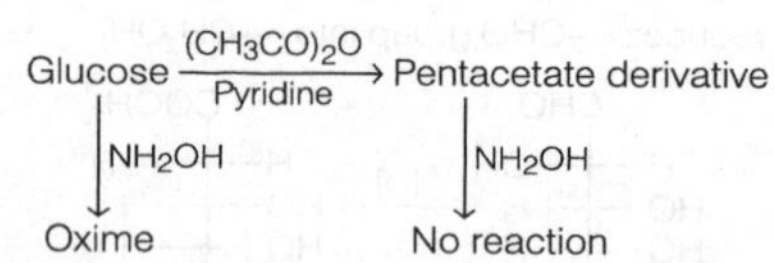

This reaction confirms the cyclic structure of glucose. Glycoside is functionally acetal which is obtained from hemiacetal (or cyclic) form.

13. (a,b,d)

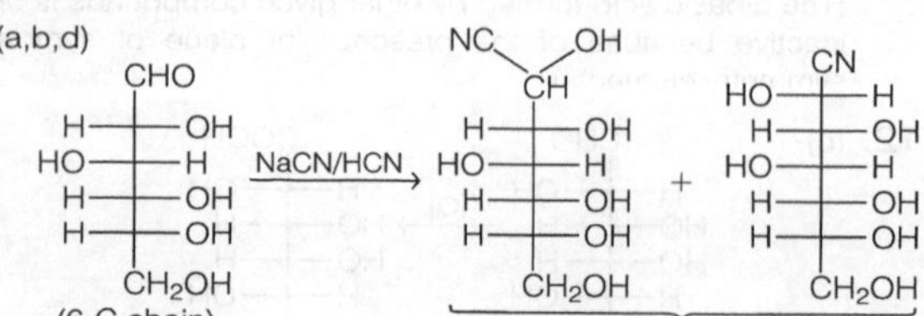

(6-C chain) (7-C chain) (C-2 epimers)

Due to different rates of the reactions in the formation of these two products, the epimers (diastereomers) are formed in different amounts.

14. (c, d) In sugars, D and L indicates only the relative configuration of —OH at the second last carbon. It has no relation with the dextro or laevorotatory nature.

This relation is also true in case of amino acids but here, we see the configuration of $—NH_2$ group at penultimate C-atom. In this case, serine is taken as a standard.

15. (a) In the aqueous solution, α and β-forms of glucose are in equilibrium with its open chain structure, so give the same gluconic acid, when treated with bromine water.

16. (c) α-helix has coiled ring of protein chain. Because of coiling, the structure becomes more ordered and hence, entropy decreases.

17. (c) Carboxypeptidase is an exopeptidase. It cleaves C-terminal bond of the amino acid chain (polypeptide chain).

Explanations for Q. Nos. (18-20)

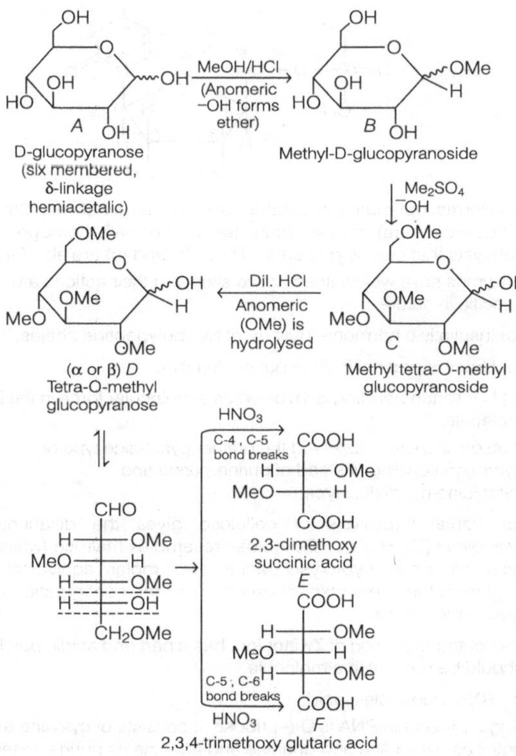

18. (c) **19.** (d) **20.** (a)

Explanations for Q. Nos. (21 and 22)

DNP method or Leucine aminopeptidase enzyme cleaves N-terminal amino acid.

Tetrapeptide (I) $\longrightarrow$ Alanine (N-terminal) + Tripeptide

Ala (N-terminal) $\longrightarrow$ Tripeptide

Hydrazinolysis or carboxy peptidase cleaves C-terminal amino acid.

Tripeptide (II) $\longrightarrow$ Dipeptide + glycine (C-terminal)

Ala (N-terminal) $\longrightarrow$ Dipeptide + Gly (C-terminal)

Dipeptide (III) now contains only two amino acids, valine and leucine. Edmann method releases N-terminal amino acid.

Dipeptide (III) $\longrightarrow$ Valine (N-terminal) + Leucine (C-terminal)

Tripeptide (II) $\longrightarrow$ Dipeptide (Val $\rightarrow$ Leu) + Gly (C-terminal)

$\longrightarrow$ Val $\longrightarrow$ Leu $\longrightarrow$ Gly

Tetrapeptide (I) $\longrightarrow$ Tripeptide (Va $\rightarrow$ Leu $\rightarrow$ Gly) + Alanine (N-terminal)

$\longrightarrow$ Ala $\rightarrow$ Val $\rightarrow$ Leu $\rightarrow$ Gly

21. (a) **22.** (d)

Explanations for Q. Nos. (23-25)

$$pK_a = -\log K_a$$

So, $$pK_{a_1} = -\log K_{a_1} = -\log(4.6 \times 10^{-3}) = 2.34$$

$$pK_{a_2} = -\log(2 \times 10^{-10}) = 9.69$$

$$\text{pI} = \frac{2.34 + 9.69}{2} = 6.015$$

$$\text{pH} = 10 > \text{pI}$$

At pH 10 (i.e. point 4) both the groups will be in the basic form i.e.

$CH_3—CH(NH_2)—COO^-$

23. (d) **24.** (c) **25.** (c)

26. (c) (i) $\rightarrow$ (p), (ii) $\rightarrow$ (p,q,r), (iii) $\rightarrow$ (q,r), (iv) $\rightarrow$ (r,s)

Glucose is oxidised by Br_2 water but fructose being a ketone does not.

Glucose or fructose being reducing sugar reduces Tollen's reagent or Fehling's solution but sucrose does not give these reactions due to the absence of free hemiacetal group.

Acetaldehyde forms iodoform with I_2/OH^- and reduces Fehling's solution but these reactions are not given by benzaldehyde.

27. (d) (i) $\rightarrow$ (p,r,s), (ii) $\rightarrow$ (q,r,s), (iii) $\rightarrow$ (p,r,s), (iv) $\rightarrow$ (q,r).

H_2NCH_2COOH Glycine

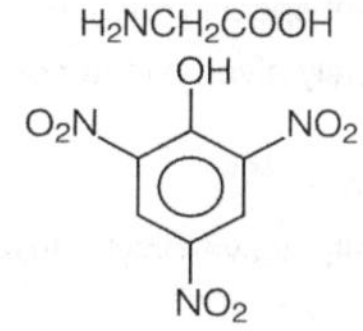

Picric acid

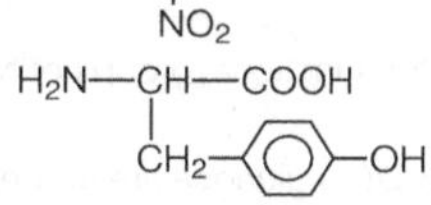

Tyrosine

$H_2N—CH(CH_2OH)—COOH$ Serine

28. (b) (i) $\rightarrow$ (q,r), (ii) $\rightarrow$ (s), (iii) $\rightarrow$ (p), (iv) $\rightarrow$ (q)

LDBVE rearrangement is given by glucose and fructose.

Ninhydrin test is a characteristic of proteins or amino acids, so given by alanine.

Sucrose undergoes inversion in configuration when hydrolysed.

Glucose is oxidised by bromine water as it contains a —CHO group and thus decolourises it.

29. (4) Glucose and fructose, Glucose and mannose, Ribose and arabinose, Mannose and fructose — These pairs form same osazone.

30. (2)

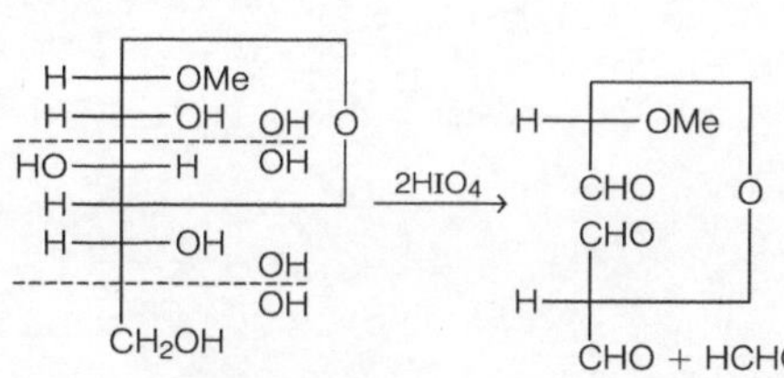

31. (2) Here, end product contains 2 moles α and β-D-glucose.

JEE Main & AIEEE Archive

1. (a) DNA contains four nitrogenous bases : adenine, guanine, cytosine, thymine. While quinoline is an alkaloid, hence, it is not present in DNA.

2. (a) $6CO_2 + 12NADPH + 18ATP \longrightarrow C_6H_{12}O_6 + 12NADP + 18ADP$

3. (b) Molisch's test is used for the detection of carbohydrates. Add two drops of alcoholic solution of α-naphthol to the carbohydrate solution under study. Add conc. H_2SO_4 slowly by the side of the test tube. A violet ring is formed at the junction of the two liquids, if carbohydrate is present.

4. (c) Glycine 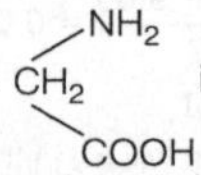 is α-amino acetic acid with no chiral carbon, thus optically inactive.

5. (b) In RNA molecule, sugar moeity is β-D-ribose in which — OH group is present at 2nd carbon. Whereas in DNA, it is β-D-2-deoxyribose in which — OH group is absent at 2nd carbon.

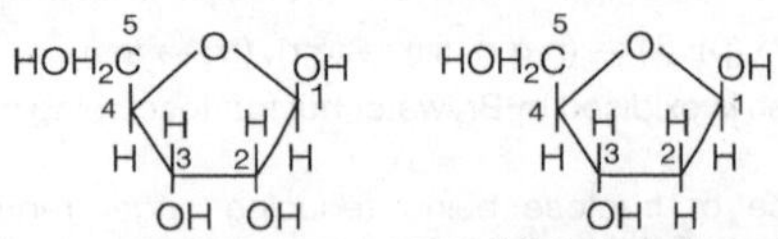

β-D-ribose β-D-2-deoxyribose

6. (d) A spontaneous change in the specific rotation of a solution of an optically active compound is called mutarotation.

Hemiacetal forms of α- and β-D- glucose are stable in solid state but in aqueous solution, there is opening of the cyclic structure which gives solution of constant specific rotation.

7. (a) Biuret test is characteristically given by the compound having $-\overset{\overset{\displaystyle O}{||}}{C}-NH-$ functional group.

8. (c) Carbohydrates are optically active polyhydroxy aldehyde or polyhydroxy ketones.

$-C=O$, $-OH$ are the functional groups of typical carbohydrate.

9. (c) α-D(+) glucose and β-D(+) glucose are anomers of glucose.

10. (a) Primary structure involves sequence of α-amino acids in a polypeptide chain.

Secondary structure involves α-helical and β-pleated sheet like structures.

11. (d) Anomers of glucose are cyclic diastereomers (epimers) differing in configuration of the hydroxyl group at C-1, existing in two forms α and β, respectively.

12. (c) In DNA, cytosine and thymine are pyrimidine bases.

13. (b) Synthesis of RNA/DNA from phosphoric acid, ribose and cytosine is given below. Thus, ester linkage formation and heterocyclic attachment place at C'_5 and C'_1 of sugar molecule.

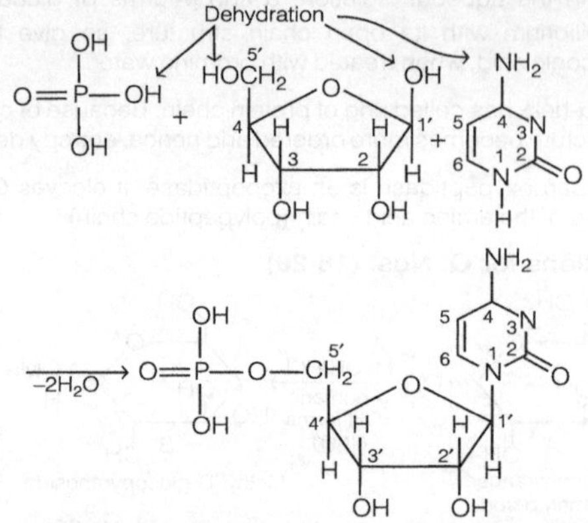

14. (d) Normal optimum temperature of enzymes is between 25°C to 40°C. Hence, (a) is false. Enzymes are normally homogeneous catalysts that can be poisoned. Thus, (b) and (c) are also false. Enzymes have well defined active sites and their actions are specific in nature.

15. (b) Insulin is a hormone, built up of two polypeptide chains.

16. (a) Uracil is present in RNA but not in DNA.

17. (c) Hydrogen bonding is involved as a molecular force in the DNA molecule.

Watson and Crick observed the purine- pyrimidine type of hydrogen bonding (instead of purine-purine and pyrimidine-pyrimidine type).

18. (c) Partial hydrolysis of cellulose gives the disaccharide cellobiose ($C_{12}H_{22}O_{11}$). Cellobiose resembles maltose (which on acid catalysed hydrolysis yields two molar equivalents of D-glucose) in every respect except one ; the configuration of its glycosidic linkage.

19. (c) For the formation of Zwitter ion, basic part and acidic part both should be present in a molecule.

20. (b) RNA, ribonucleic acid.

Sugar present in RNA is D (–) ribose. It consists of cytosine and uracil as pyrimidine and guanine and adenine as purine bases.

JEE Advanced & IIT JEE Archive

1. (4) Chemical reaction and product formed after hydrolysis of given peptide can be represented as

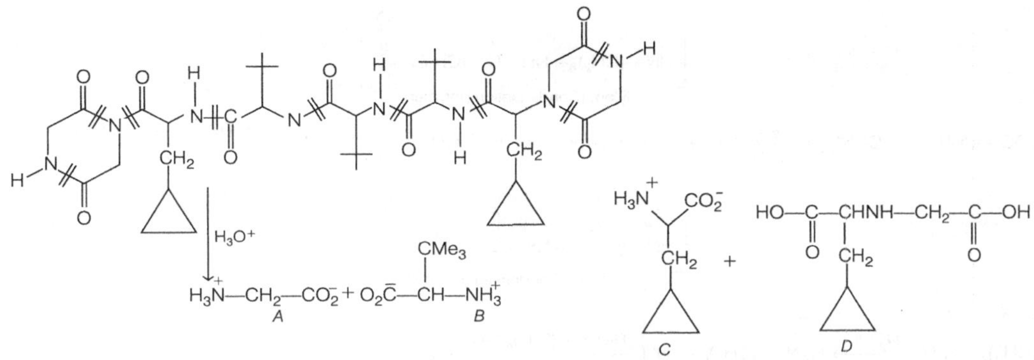

2. (4) A peptide linkage is hydrolysed to two free amino acids.

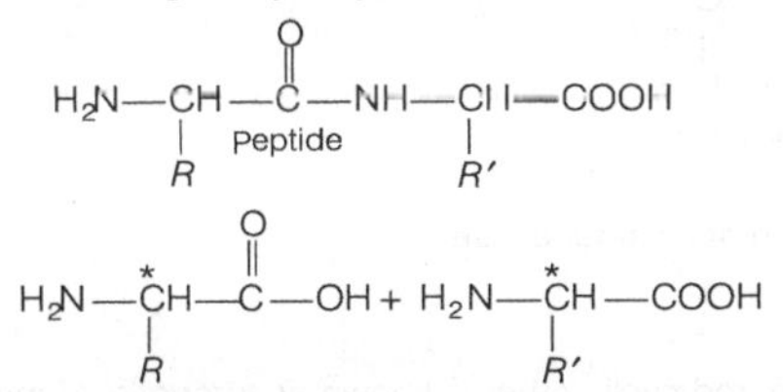

C^* is chiral carbon. Tetrapeptide has four amino acids joined by three peptide linkages.

—COOH group is on alanine part, thus it is at fixed C-terminal position in each combination.

Glycine is optically inactive thus, it cannot be on the N-terminal side.

Thus, possible combinations are

Phe-Gly-Val-Ala
Phe-Val-Gly-Ala
Val-Gly-Phe-Ala
Val-Phe-Gly-Ala

Thus, in all four combinations are possible.

3. (4) The amino acid remain completely in Zwitter ionic form at its isoelectric point. Amino acids with additional acidic group have their isoelectric pH less than 7.0 and increasing pH above isoelectric point makes them anionic. On the other hand, amino acids with additional basic group have their isoelectric pH greater than 7.0 and decreasing pH below isoelectric point (by adding acid solution) makes them cationic. The given peptide with following R_1 and R_2 are basic, will remain protonated (cationic) at pH = 7.0.

Peptide	R_1	R_2
IV	CH_2CONH_2	$(CH_2)_4NH_2$
VI	$(CH_2)_4NH_2$	$(CH_2)_4NH_2$
VIII	CH_2OH	$(CH_2)_4NH_2$
IX	$(CH_2)_4NH_2$	CH_3

Thus, 4 is the correct integer.

4. (8) The D-form of given sugar is

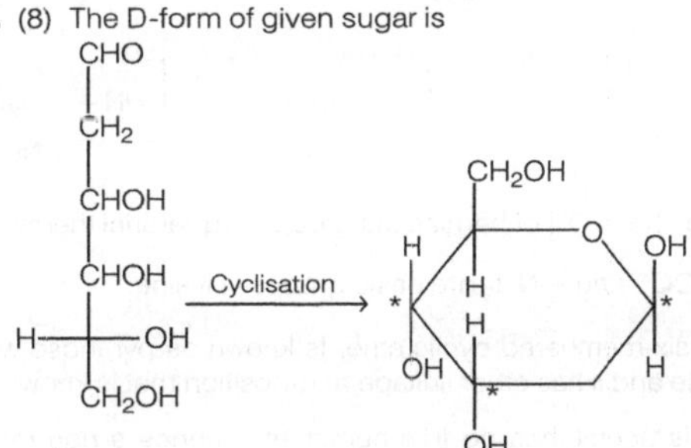

D- sugar D-pyranose

Configurations at the three chiral carbons (starred) can be changed maintaining D-configuration. Hence, the total number of stereoisomers of D-pyranose $= 2^3 = 8$

Thus, the correct integer is 8.

5. (6) A decapeptide has nine peptide (amide) linkages.
Therefore, on hydrolysis, it will absorb nine water molecules.
Hence, total mass of hydrolysis product

$$= 796 + 18 \times 9$$
$$= 958$$

Mass of glycine in hydrolysis product

$$= \frac{958 \times 47}{100} = 450$$

Number of glycine molecule in one molecule of decapeptide

$$= \frac{450}{75} = 6$$

6. (a,b,c,d) (a) When $X = COOCH_3$

$$CH_3OOC-(CH_2)_4-COOCH_3 \xrightarrow[\text{Heat}]{H_2/Ni} HOCH_2-(CH_2)_4-CH_2OH + 2CH_3OH \xrightarrow[\text{Heat}]{HOOC-(CH_2)_4-COOH}$$

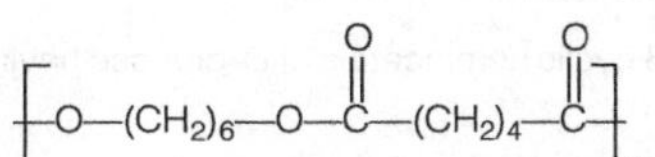

Ester, (Condensation polymer)

(b) When $X = CONH_2$

$$H_2NOC-(CH_2)_4-CONH_2 \xrightarrow[\text{Heat}]{H_2/Ni} H_2N-(CH_2)_6-NH_2 \xrightarrow[\text{Heat}]{HOOC-(CH_2)_4-COOH}$$

$$\left[-HN-(CH_2)_6-NH-\overset{O}{\overset{\|}{C}}-(CH_2)_4-\overset{O}{\overset{\|}{C}}-\right]_n$$

Nylon, (Condensation polymer)

(c) $$H_2NOC-(CH_2)_4-CONH_2 \xrightarrow[NaOH]{Br_2} \text{Hofmann's bromamide reaction } H_2N-(CH_2)_4-NH_2 \xrightarrow[\text{Heat}]{HOOC-(CH_2)_4-COOH}$$

$$\left[-HN-(CH_2)_4-NH-\overset{O}{\overset{\|}{C}}-(CH_2)_4-\overset{O}{\overset{\|}{C}}-\right]_n$$

Nylon (Condensation polymer)

(d) When $X = CN$

$$NC-(CH_2)_4-CN \xrightarrow[\text{Heat}]{H_2/Ni} H_2N-(CH_2)_6-NH_2 \xrightarrow[\text{Heat}]{HOOC-(CH_2)_4-COOH}$$

$$\left[-HN-(CH_2)_6-NH-\overset{O}{\overset{\|}{C}}-(CH_2)_4-\overset{O}{\overset{\|}{C}}-\right]_n$$

Nylon (Condensation polymer)

7. (b) Here, the —OH of hemiacetal group is equatorial therefore, it is a β-pyranose of an aldohexose.

8. (2) $-COO^-$ and $-NH_2$ are basic groups in lysine.

9. (a) The six-membered cyclic ether is known as pyranose while the five membered cyclic ether is known as furanose. Hence, ring (a) is a pyranose and it has ether linkage at α-position that is known as α-glycosidic linkage in carbohydrate chemistry.

10. (b,c) X is acetal, has no free hemiacetal, hence a non-reducing sugar while Y has a free hemiacetal group, it is reducing sugar. Also, glucosidic linkage of X is α while that of Y is β-linkage.

11. (a) Cellulose is biopolymer of β-D-glucopyranose as

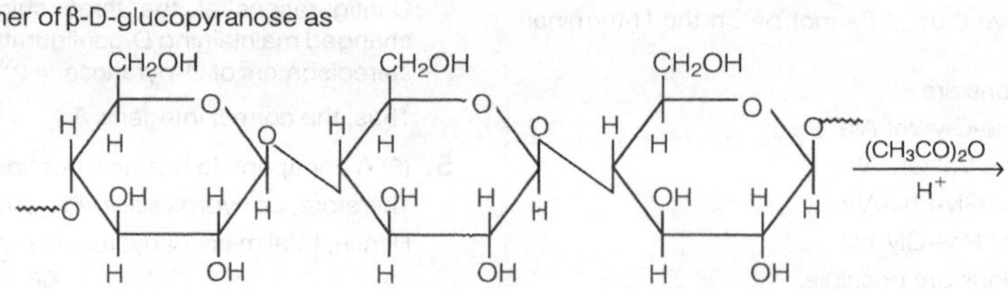

Cellulose

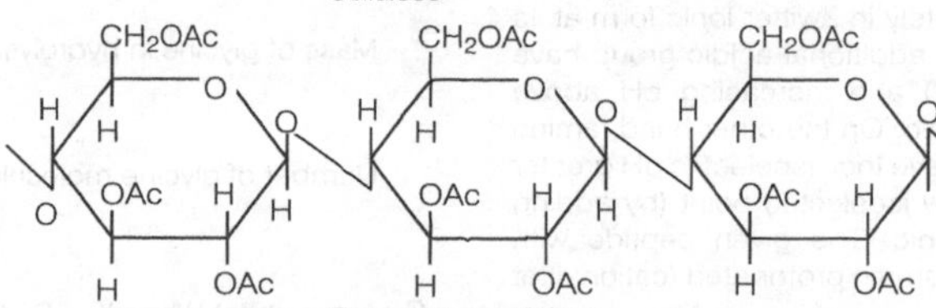

Triacetylated cellulose

12. (c) Statement I is correct : Presence of —CHO group in glucose is tested by Fehling's solution test where a reddish-brown precipitate of Cu_2O is formed.
Hence, Statement II is incorrect.

13. (b) α and β cyclic hemiacetals of D-glucose having difference in configuration at C-1 only are called anomers.

6. Polymers

DPP-1 General Introduction and Classification of Polymers

1. (i) Repeat unit is the unit, repetition of which again and again gives complete structure of the polymer. Thus, the repeat unit in the given cases are as follows:

(I) 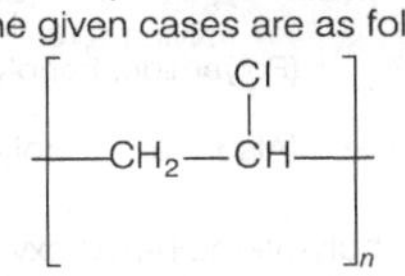 (II) 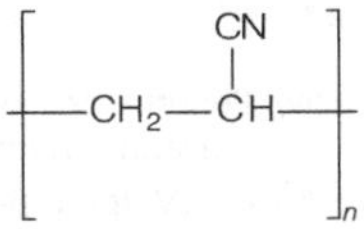

(III) 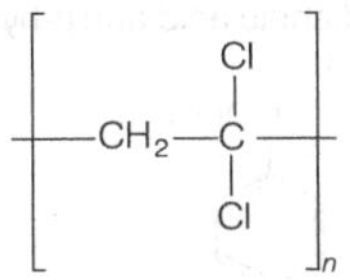

(ii) (I) 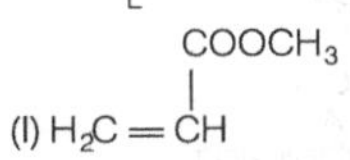 II. 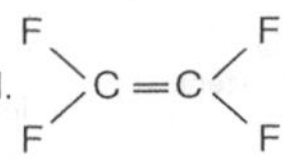

(III) HOOC—$(CH_2)_8$—COOH (decanoic or sebacic acid) + $H_2N(CH_2)_6NH_2$ (hexamethylenediamine)

(IV) 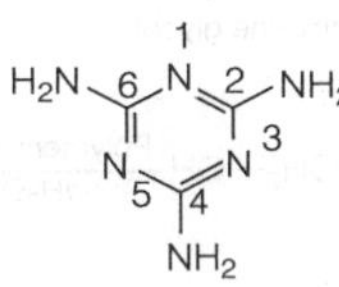 (2,4,6-triamino-1, 3, 5-triazine)

+ HCHO (formaldehyde)

2. (i) Because of the presence of stronger C—F bond and a highly regular polymeric chain, teflon is highly stable to heat.

(ii) Teflon has no chiral centres, so it has no stereochemical complications. Also, it has no hydrogens that can be abstracted from the interior of chain, so it is linear.

3. (i) (I) Its monomer unit

$$H_2N—\overset{6}{C}H_2—\overset{5}{C}H_2—\overset{4}{C}H_2—\overset{3}{C}H_2—\overset{2}{C}H_2—\overset{1}{C}H_2—NH_2$$

IUPAC name : (1,6-diaminohexane or hexan-1, 6-diamine)

$$\text{and } HO—\overset{O}{\overset{\|}{C^6}}—\overset{5}{C}H_2—\overset{4}{C}H_2—\overset{3}{C}H_2—\overset{2}{C}H_2—\overset{O}{\overset{\|}{\overset{1}{C}}}—OH$$

IUPAC name : Hexan-1, 6-dioic acid

(II) 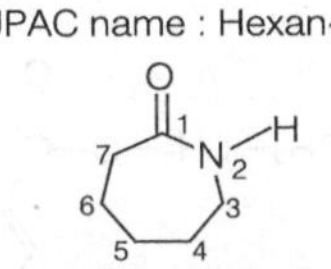

2-iminocyclohexanone

(III) $CF_2=CF_2$

1, 1, 2, 2-tetrafluoroethene

(ii) (I) Nylon-6 6 is a fibre, buna-S is an elastomer and polythene is a thermoplastic. Elastomers have the weakest whereas fibres have the maximum forces among the given.

Thus, the order of intermolecular forces is

Buna-S < polythene < nylon-6 6

(II) Neoprene (Elastomer) < PVC (Thermoplastic) < terylene (Fibre) < nylon-6. (Fibre)

Nylon-6 has more intermolecular forces due to the formation of H-bonds in between the chains.

4. (i) 1,1-dichloroethene produces a more crystalline polymer because it has no chiral centres and is more stereoregular.

(ii) Coordination polymer produces a less branched and more stereoregular polymer which is more crystalline.

5. (b) Cellulose acetate is obtained by chemical modification (acetylation followed by hydrolysis) of cellulose (the natural polymer) so, it is a semisynthetic polymer.

6. (c) Thermoplastics are the plastics which become soft on heating and hard on cooling, i.e. they can be moulded again and again, e.g. PVC, polythene, PMMA, polystyrene, polypropylene, teflon, etc.

Note *Melmac, bakelite and urea-formaldehyde are thermosetting polymers.*

7. (c) Structure of terylene is

(It is attached with the benzene nuclei.)

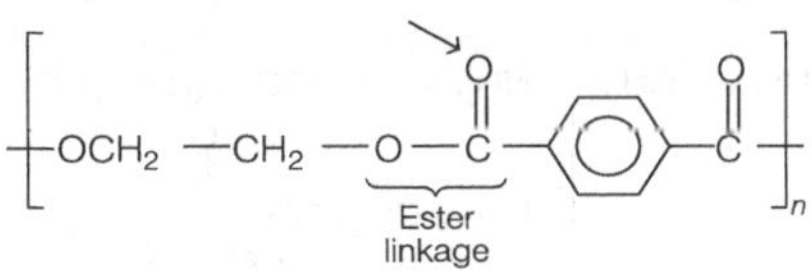

8. (a,b) Polymers essential for the existence of life are called biopolymers, e.g. nucleic acid, leather, etc.

9. (b,c) Polystyrene and PVC have forces intermediate of fibres and elastomers. Moreover, they can be remoulded. So, these are the examples of thermoplastics.

Bakelite and melmac because of the formation of stable cross-linked structure (on heating) are classified as thermosetting polymers.

10. (b,c) Polymers degraded by microorganisms are called biodegradable polymers, e.g. PHBV, nylon-2-nylon-6, etc.

11. (c) Since, intermolecular forces of attraction in nylon are H-bonding, so it is stronger than terylene (where forces of attraction are van der Waals' forces).

12. (c) On heating, bakelite gets converted into highly cross-linked structure.

Due to the formation of cross-links, it is hard and possesses high melting point.

13. (b) (i) → (p), (ii) → (s), (iii) → (q), (iv) → (r)

(i) 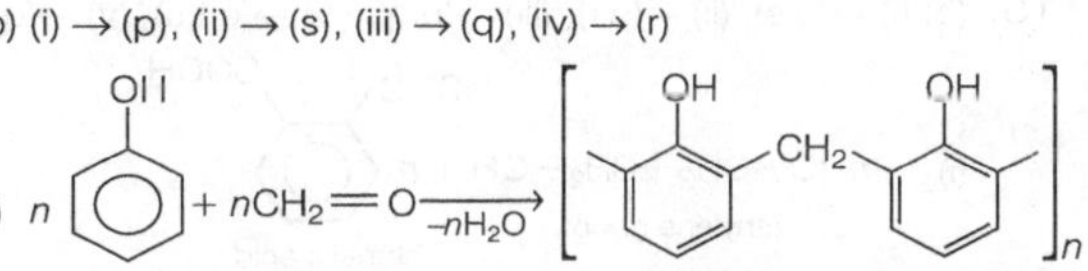

Resol or novolac

(ii) 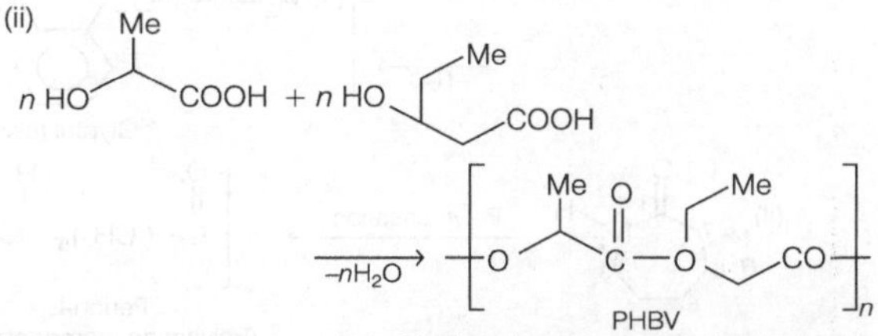

PHBV

Poly β-hydroxy butyrate-co-β-hydroxy valerate

(iii)

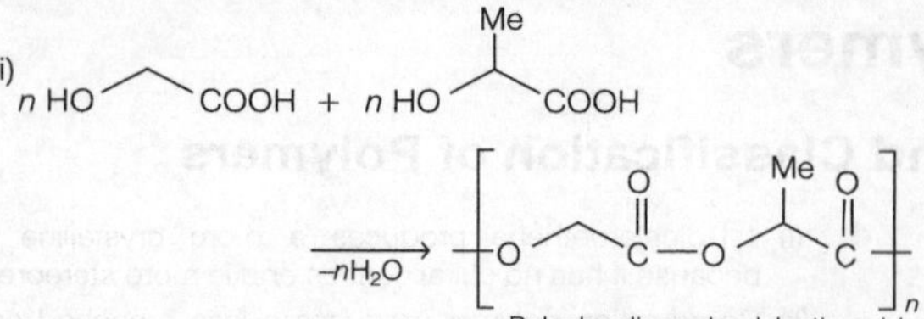

Polyglycolic and polylactic acid

(iv)

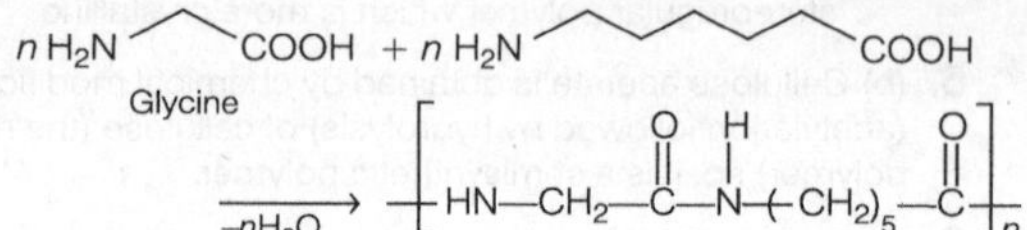

Glycine

Nylon-2-nylon-6

14. (a) (i) →(p,s), (ii) →(q,r), (iii) →(p,r), (iv) →(q,r)

Addition polymerisation involves no loss of small molecules which are lost in case of condensation polymerisation. Homopolymers contain only one kind of monomer units while in case of copolymer, more than one kind of monomer unit is involved.

Thus,

(i) $n\text{NH}_2(\text{—CH}_2)_6\text{—NH}_2 + n\ \text{HOOC—(CH}_2)_4\text{COOH}$

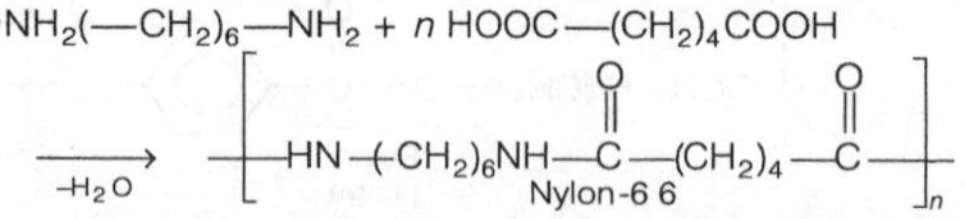

Nylon-6 6

So, it is a condensation copolymer.

(ii)

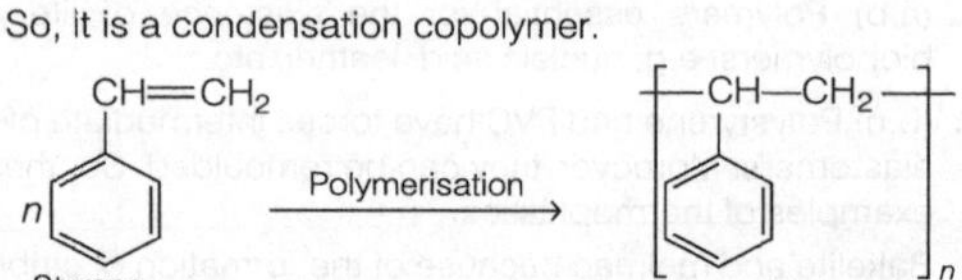

Styrene

Polystyrene or styrone
(Addition, homopolymer)

(iii)

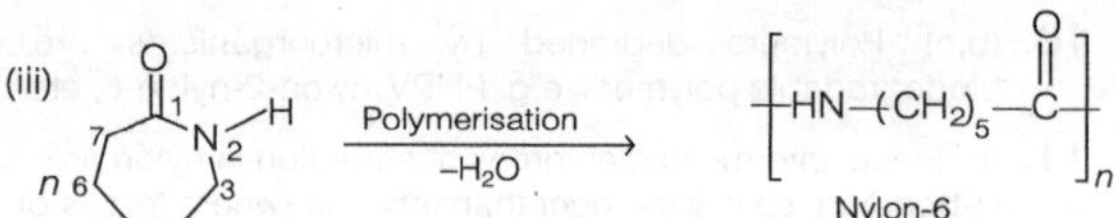

Nylon-6
(Condensation homopolymer)

(iv)

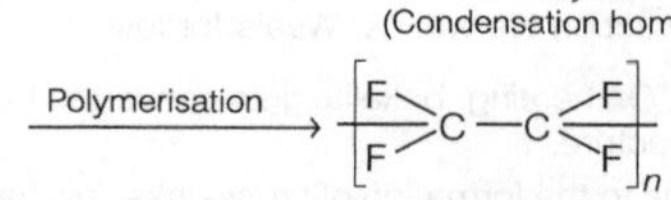

Teflon
(Addition homopolymer)

15. (b) (i) →(q,s), (ii) →(p,r), (iii) →(p,q), (iv) →(q,t,u), (v) →(q,s)

(i)

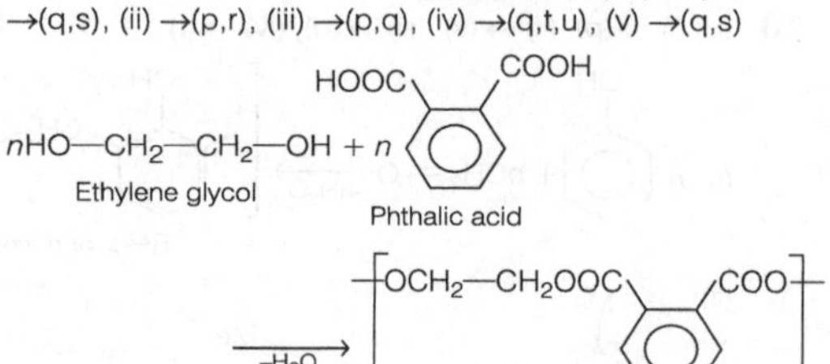

Ethylene glycol

Phthalic acid

Glyptal resin

(ii)

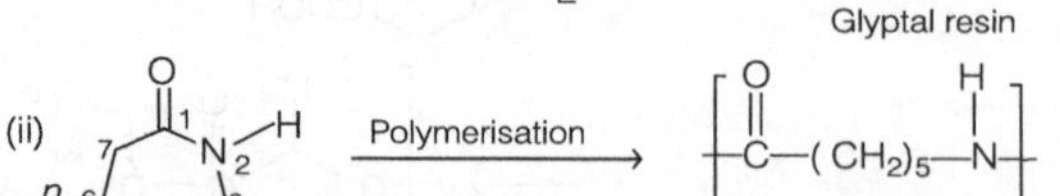

Perlon-L
(Polyamide, homopolymer)

(iii) $n\text{H}_2\text{N}\text{—}(\text{CH}_2)_6\text{NH}_2 + n\text{HOOC—}(\text{CH}_2)_8\text{COOH}$

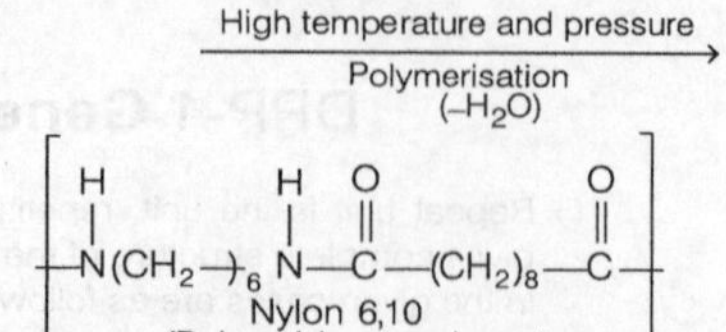

Nylon 6,10
(Polyamide, copolymer)

(iv) Perlon or polyurethane (Also a copolymer and is polycarbamate ester).

(v) PHBV (poly β-hydroxy butyrate-co-β-hydroxy valerate) is a polyester of β-hydroxy butanoic acid and β-hydroxy pentanoic acid (i.e. it is a copolymer).

16. (a)

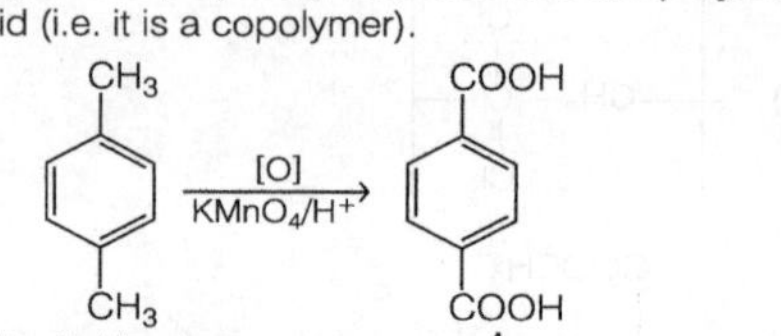

1, 4-dimethyl benzene

Terephthalic acid

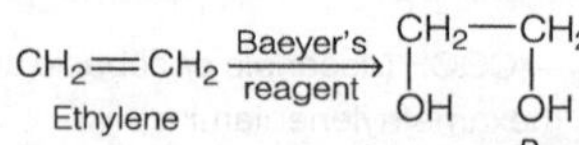

Ethylene

Ethylene glycol

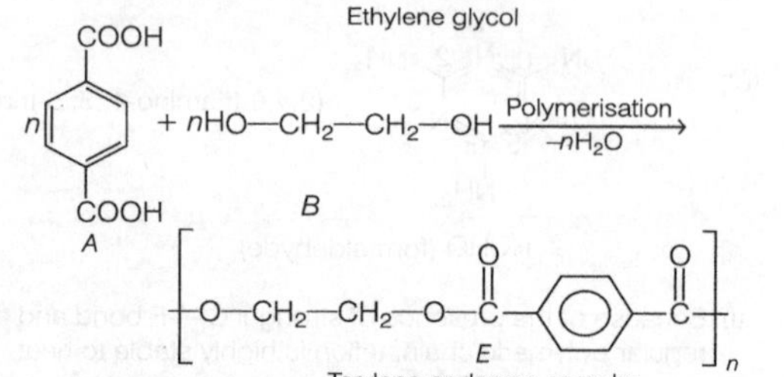

Terylene or dacron or mylar
(Polyester)

17. (c)

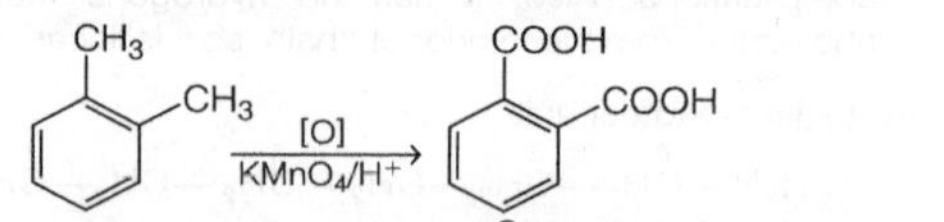

Phthalic acid

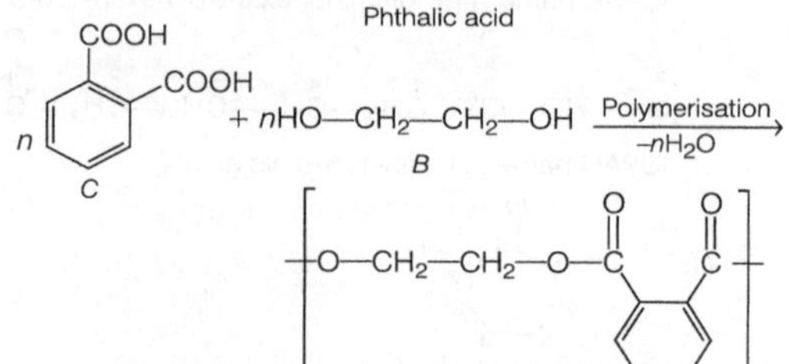

Glyptal or alkyl resin
(*F*, polyester)

18. (5) Nylon-6 6 dacron, glyptal, PHBV and perlon-U all are condensation copolymers.

19. (6) This condition is true in case of addition polymers and acrilon, vinyon, dynel, natural rubber, PCTFE (polychlorotrifluoro ethylene) and PMMA are addition polymers.

DPP-2 Mechanism of Polymerisation

1. (i) Free radical polymerisation takes place in unsaturated monomers and during this process, pi (π) -bonds are converted into stronger σ-bonds. Hence, energy is released, i.e. it is an exothermic process.

(ii) III is the most preferred way of polymerisation.

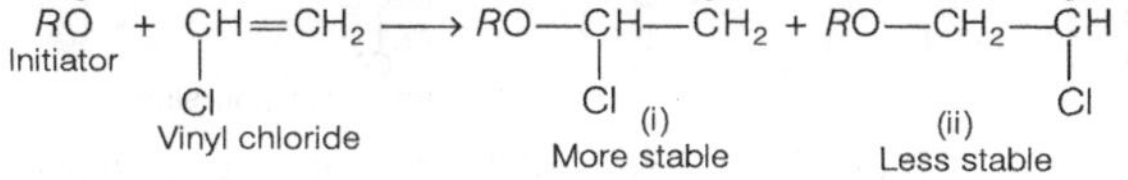

Propagation occurs through (i) as

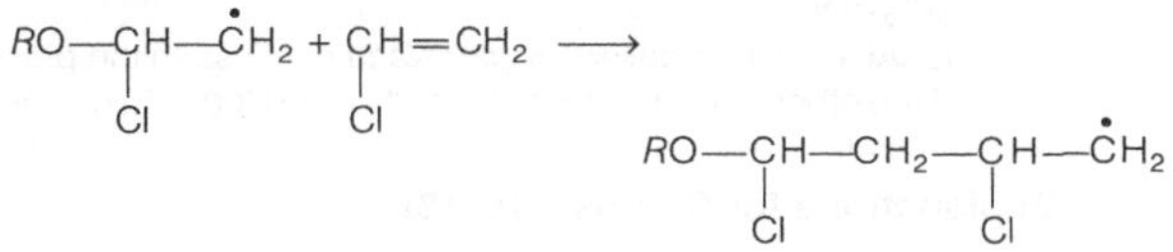

(Other way to this step willproduce radical at α-C to Cl, which is a less stable free radical as compared to the present one).

2. (i) In the cationic addition polymerisation, formation of an intermediate carbocation takes place. More the stability of carbocation formed, more is the reactivity of parent alkene towards cation addition polymerisation. In general, electron releasing groups (like Me) make the carbocation more stable, and hence, enhance the activity of alkene towards this reaction.

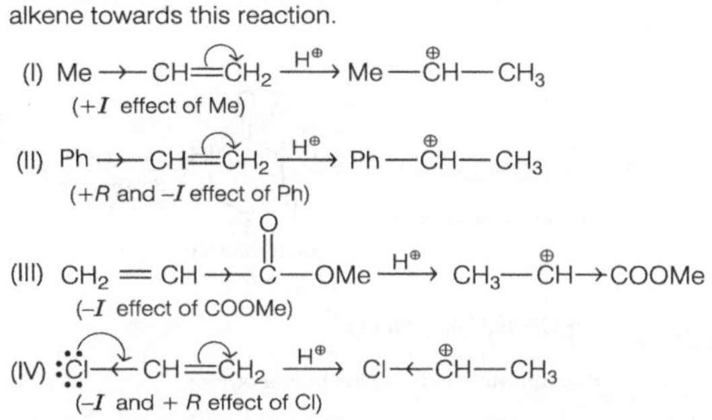

Thus, the decreasing order of reactivity towards cationic polymerisation is II > I > IV > III.

(ii) Carbanion intermediate is involved in case of anionic polymerisation and it is generally stabilised by the presence of electron withdrawing group like —X, —COOR, —CN, etc. Thus, electron withdrawing group, if present on the alkene makes it more reactive towards anionic polymerisation.

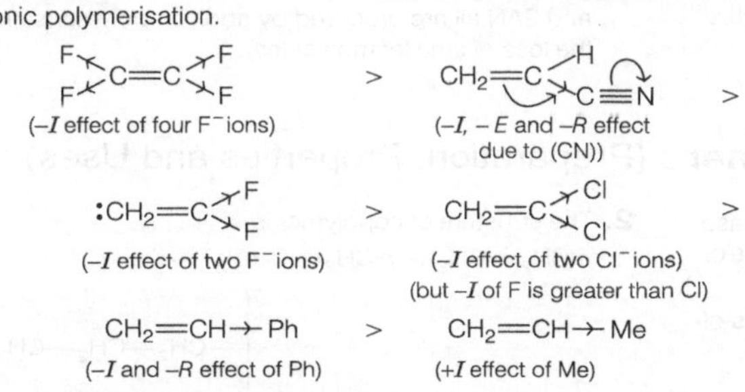

The decreasing order of reactivity towards anionic polymerisation is

VI > III > V> IV > II > I

3. (i) $CH_3—CH=CH—CH_3$, free radical polymerisation.

(ii) $HO—CH_2CH_2COOH$, condensation polymerisation.

(iii) $CH_2=CHOCOCH_3$, addition polymerisation.

(iv) Polymer in "(iii)" on hydrolysis gives this polymer.

4. The reaction takes place in the following manner.

Chain initiation step

$$R—O—O—R \xrightarrow{h\nu} 2RO^{\bullet}$$

Chain propagation step

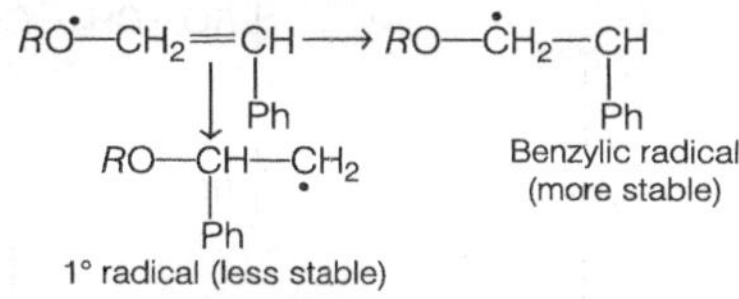

So, the reaction proceeds *via* more stable radical (I) rather than (II) to give product (I).

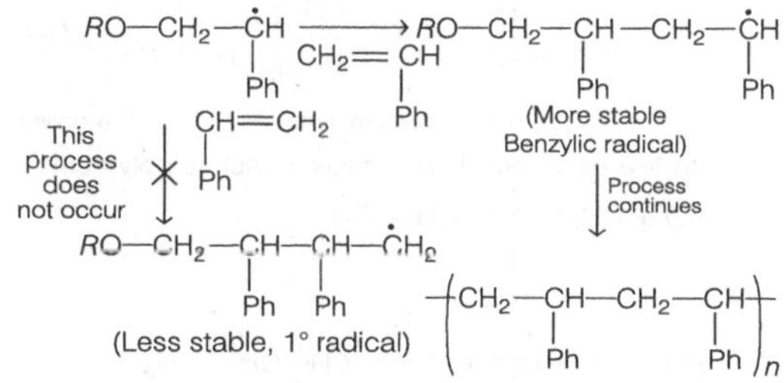

Hence, product (I) is formed.

5. (a) Benzoyl peroxide and azo*bis*-isobutyronitrile are the two commonly used free radical initiators.

$$C_6H_5—\overset{O}{\overset{\|}{C}}—O—O—\overset{O}{\overset{\|}{C}}—C_6H_5 \longrightarrow 2C_6H_5\overset{O}{\overset{\|}{C}}—\dot{O}.$$

6. (a) Structure of the isotactic polypropylene is as

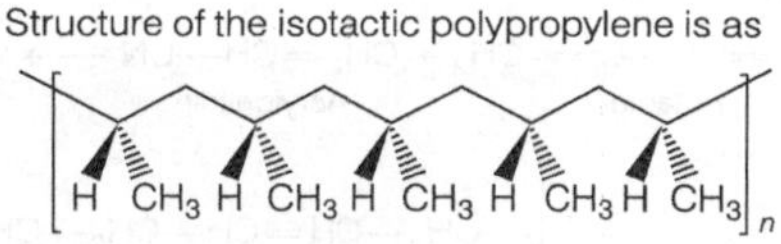

[Thus, it has high crystallinity, melting point and strength]

7. (b) Benzyl cation, anion and free radical all are stable species so, it undergoes radical, cationic and anionic polymerisation with equal ease.

8. (c) Chain branching always takes place when an alkene undergoes free radical polymerisation. This is due to the abstraction of some H-radicals from the growing polymer chain.

9. (b) PP (polypropylene) is prepared by cationic addition polymerisation of propylene ($CH_2=CH—CH_3$) as presence of electron donating —Me group stabilises the carbocation intermediate formed.

10. (b,c) The substance which lowers the melting point (softening point) of a polymer, thereby reducing the interparticle forces is called plasticiser. Generally, high boiling esters or haloalkanes are used for this purpose. Thus, di-*n*-butylphthalate and tricresyl phosphate are the examples of plasticisers.

11. (b,c) Compounds (olefins) in which two groups attached to the same carbon atom are different, give isotactic, syndiotactic or atactic forms when subjected to polymerisation.

12. (a,b,c) Buta-1, 3-diene can polymerise either by 1, 4-addition or by 1, 2-addition mechanism. 1,4-polymerisation gives unbranched polymer in the following manner.

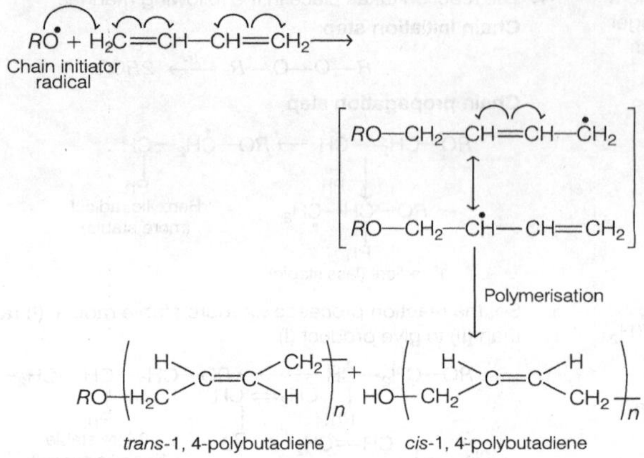

In 1, 2-polymerisation, it gives branched polymer.

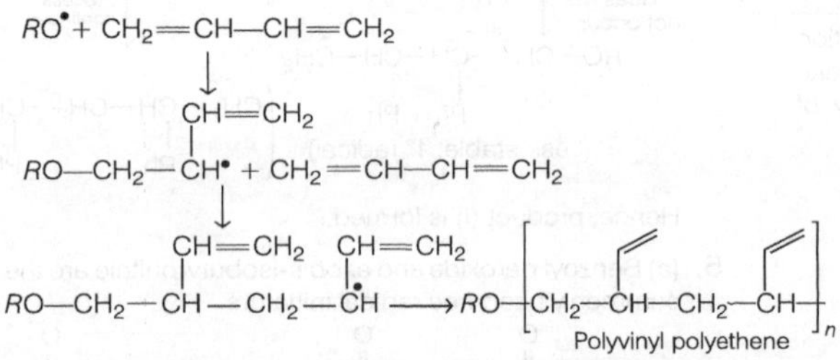

13. (d) Polybutadiene is formed by the free radical addition polymerisation.

$$\underset{\text{Butadiene}}{CH_2{=}CH{-}CH{=}CH_2} + \underset{\text{Acrylonitrile}}{CH_2{=}CH{-}CN} \longrightarrow$$

$$\left[-CH_2-CH{=}CH-CH_2-CH_2-\underset{CN}{\underset{|}{CH}}-\right]_n$$

Buna-N

14. (d) 2-methylbuta-1,3-diene (isoprene) is the monomer of gutta-percha. It is actually *trans*-polyisoprene unit.

Presence of electron releasing Me group makes it more reactive towards cationic polymerisation.

15. (c) (i)→(p), (ii)→(q), (iii)→(p), (iv)→(r,s).

Benzoquinone and phenols inhibit the growth of the chain due to the formation of less reactive, highly stable radical.

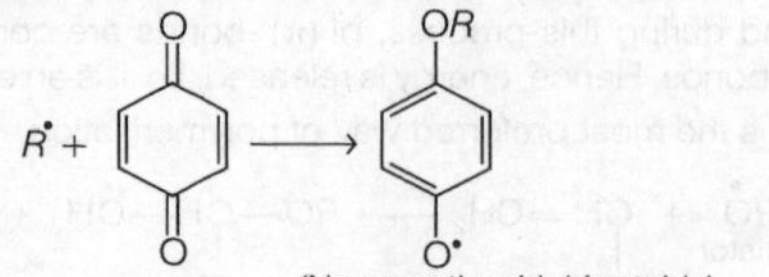

(Non-reactive, highly stable)

CCl_4 or CBr_4, when added, terminates the chain and are called chain transfer agents. These controls the molecular mass of the polymer.

$Et_3Al + TiCl_4$ is called *Ziegler-Natta* catalyst. In the presence of it, if the ethene polymerises, it results in the formation of high density polyethylene.

Explanations for Q. Nos. (16-18)

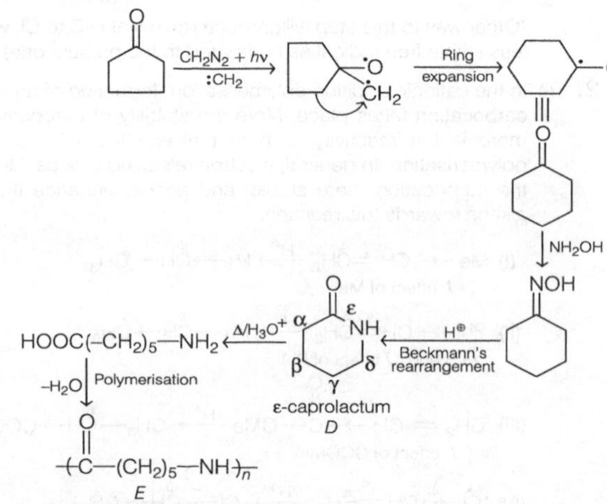

Step growth condensation homopolymer

16. (c) **17.** (c) **18.** (a)

19. (5) Among the given dacron, nylon-6 6, nylon-6, bakelite and kevlar (total five) are prepared from the monomers containing bifunctional group and hence, obtain by condensation polymerisation.

20. (7) Among the given orlon, PTFE, neoprene, buna-S, vinyon, ABS and SAN all are prepared by addition polymerisation (i.e. without the loss of smaller molecules).

DPP-3 **Some Important Polymers** (Preparation, Properties and Uses)

1. (i) Strength and melting point of nylons decreases with increase in the number of carbon atoms in the chain of either diamine or diacid.

(ii) Hole is developed due to the breaking or acidic hydrolysis of amide bond $\left(-\overset{O}{\overset{\|}{C}}-\underset{H}{\underset{|}{N}}-\right)$ present in the nylon stocking.

(iii) Due to irradiation with γ-rays, some of the C—H bonds undergo homolytic fission to give alkyl free radical as

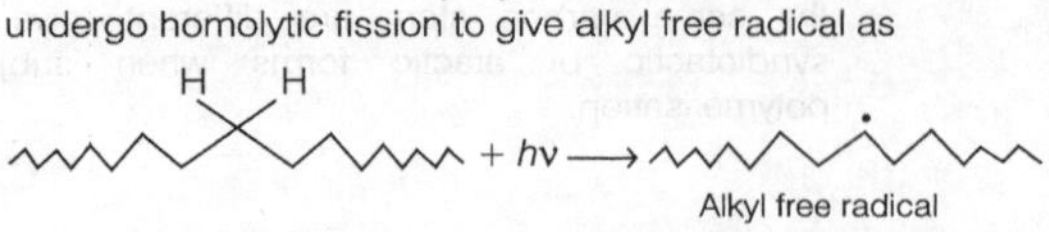

This free radical (alkyl radical) may combine with other similar radical leading to the formation of cross-linked structure.

2. The structure of copolymer is

$$n(CH_2{=}CH_2) + n(CH_2{=}CHCl) \longrightarrow \left[-CH_2-CH_2-CH_2-\overset{Cl}{\overset{|}{C}H}-\right]_n$$

Molecular mass of ethene $(CH_2{=}CH_2) = 24 + 4 = 28$

Molecular mass of vinyl chloride

$(CH_2{=}CHCl) = 24 + 3 + 35.5 = 62.5$

Empirical formula mass of copolymer $= 28 + 62.5 = 90.5$

Mass percentage of vinyl chloride in the copolymer

$$= \frac{62.5 \times 100}{90.5} = 69.06$$

3. (c) PCTFE and PTFE both have same carbon backbone.

$$-[-ClF-C-CF_2-]_n$$

Polymonochlorotrifluoroethylene (PCTFE)

$$-[-F_2C-CF_2-]_n$$

That is why, these two resemble in their properties.

4. (d) Polystyrene or styrone, a polymer of styrene, is used for making soft drink and baby feeding bottles.

5. (a) GRA stands for general purpose acrylonitrile rubber. It is obtained by the reaction of buta-1,3 diene and acrylonitrile in the presence of sodium metal.

$NC—CH=CH_2 + CH_2=CH—CH=CH_2$

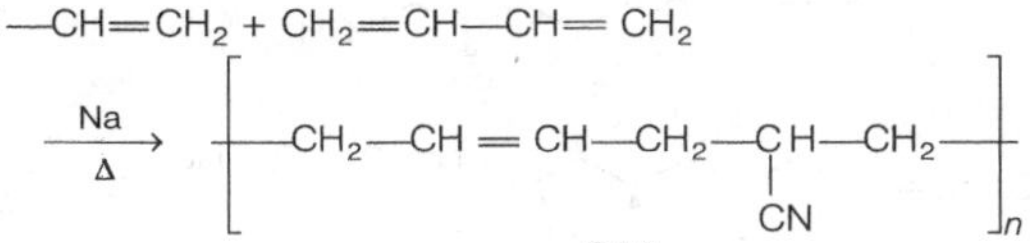

GRA

6. (d) A copolymer of acrylonitrile and vinyl chloride in equimolar quantities is called dynel or vinyon-N. It can be spun into fibres from acetone solution and used for making synthetic human hair wigs.

7. (d)

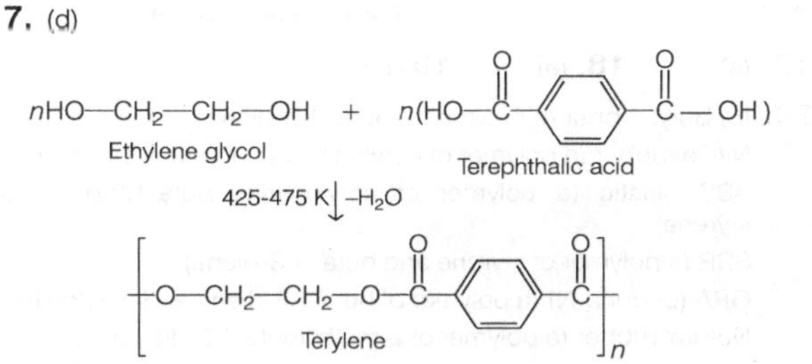

8. (d) $n\ CH_2=C(CH_3)—CH=CH_2 \xrightarrow{\text{Polymerisation}}$

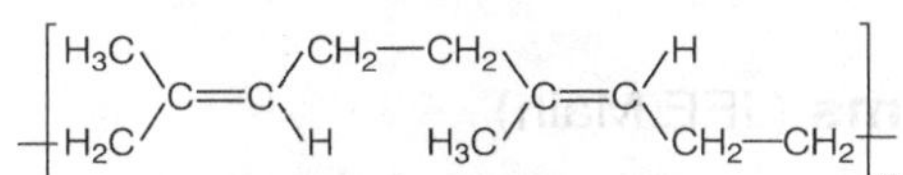

trans-polyisoprene or gutta-percha (soften on heating)

It is used for making underwater cables and golf balls.

9. (c) Since, formaldehyde ($CH_2=O$) and O=C(Cl)—C(H)=O are formed by the ozonolysis of one mole of monomer, so the structure of monomer should be

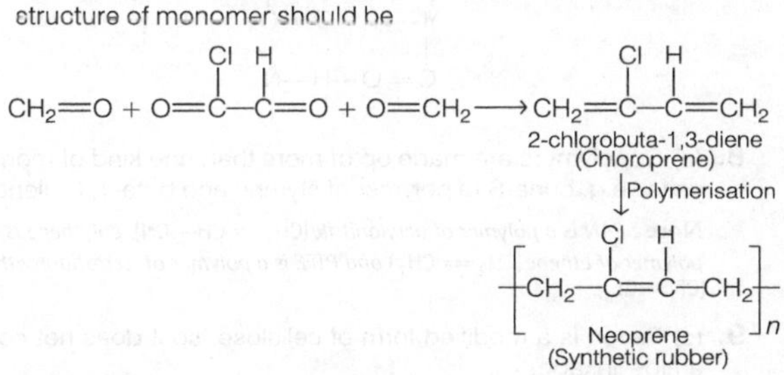

10. (a,c) In case of nylon, H-bonding exists between the polymeric chain while in terylene, there exists only weak dipole-dipole interactions. Hence, tensile strength of nylon fibres is much higher as compared to the terylene fibres.

11. (a,b,d) Natural rubber is a *cis*-polymer of 2-methyl buta-1,3-diene (i.e. isoprene unit).

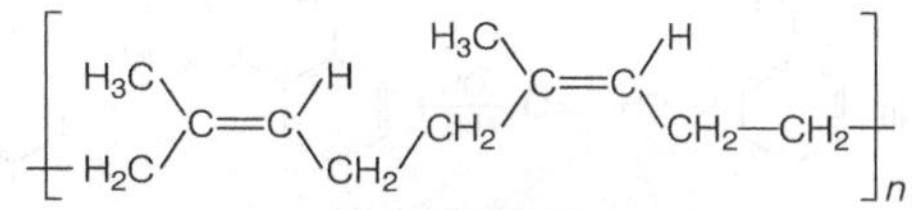

cis-polyisoprene (amorphous with coiled conformation)

Its stereochemistry is isotactic, i.e. not syndiotactic.

12. (a,c)

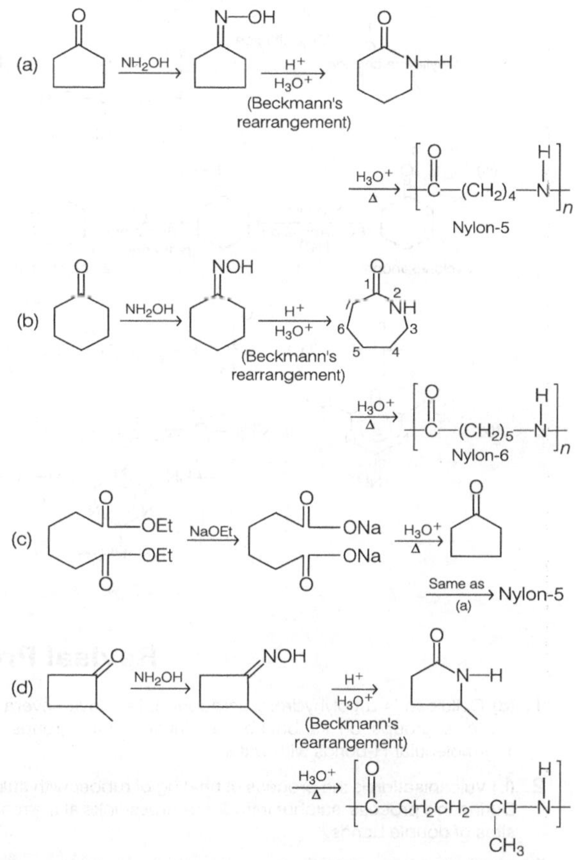

13. (a) PMMA (polymethylmethacrylate) is commercially called plexiglas because it is hard and transparent, with excellent transmission property and hence, used for making contact lens.

14. (a) PUF is prepared by mixing a little water with diol. Some of the diisocyanate reacts with water to give carbamic acid, which spontaneously loses carbon dioxide to give an aromatic diamine. The CO_2 thus evolved forms bubbles that are trapped within the bulk of polymer. When it solidified, a spongy product PUF is formed.

15. (b) (i)→(q), (ii)→(r), (iii)→(s), (iv)→(p)

Neoprene is a polymer of 2-chlorobuta-1,3-diene (i.e. chloroprene), gutta-percha is *trans*-polyisoprene, i.e. *trans*-poly 2-methyl buta-1,3-diene. Buna-S is styrene butadiene rubber.

Buna-N is acrylonitrile butadiene rubber.

16. (d) (i)→(q,s), (ii)→(p), (iii)→(r,s,t), (iv)→(s,t)

(i)

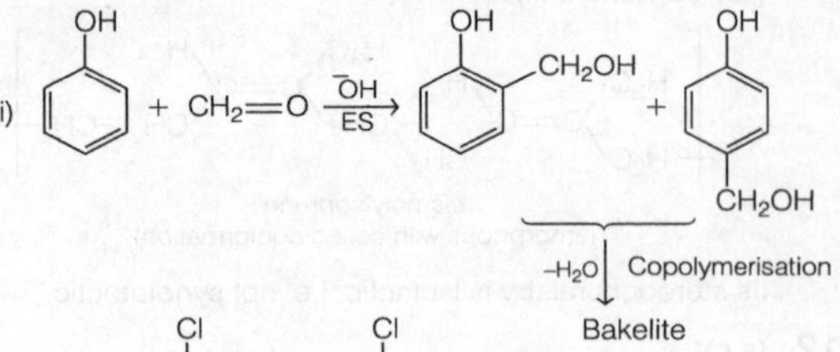

(ii)

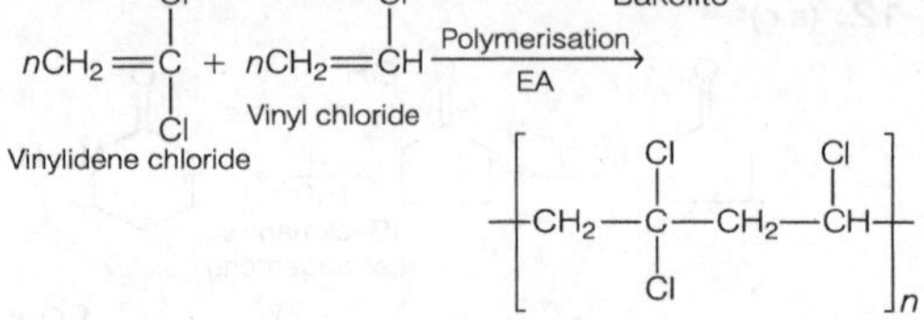

(iii)

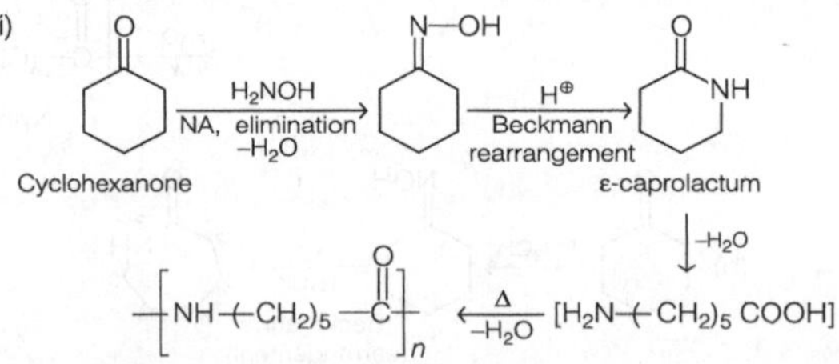

(iv)

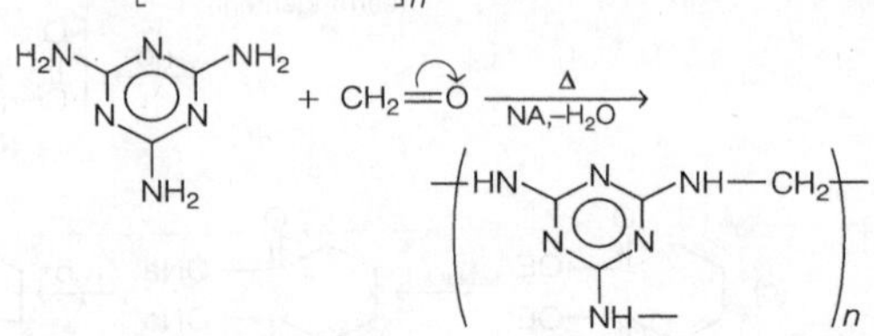

Explanations for Q. Nos. (17-19)

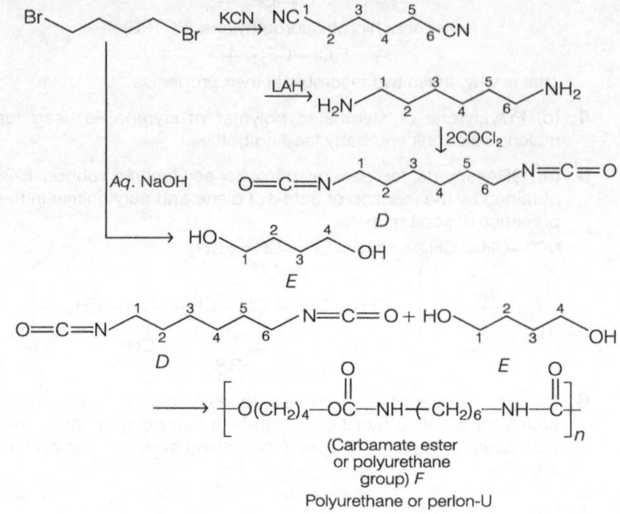

17. (a) **18.** (a) **19.** (c)

20. (5) Butyl rubber (a polymer of buta -1,3-diene)
Nitrile rubber (a polymer of acrylonitrile and buta-1,3-diene)
ABS plastic (a polymer of acrylonitrile, buta-1,3-diene and styrene)
SBR (a polymer of styrene and buta -1,3-diene)
GRA (or buna-N) (a polymer of buta-1,3-diene and acrylonitrile)
Natural rubber (a polymer of 2-methylbuta-1,3-diene)
Neoprene (a polymer of 2-chlorobuta-1,3-diene)

Revisal Problems (JEE Main)

1. (d) Cellulose is a polyhydroxy compound, i.e. have several free hydroxyl groups on its backbone. These —OH groups form intermolecular H-bonds with water.

2. (b) Vulcanisation is the process of heating of rubber with sulphur. During this process, sulphur introduces cross-links at the reactive sites of double bonds.

3. (d) Chain growth polymers are generally formed by unsaturated moieties.
Terylene is a polymer of terephthalic acid.

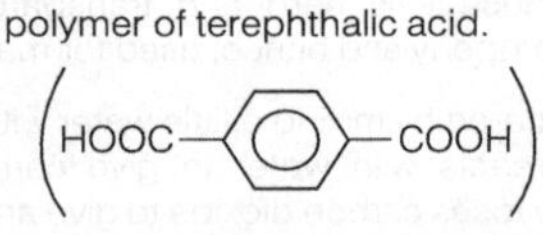

and ethylene glycol ($CH_2OH—CH_2OH$), and formed by the condensation reaction. So, it is a step growth polymer.

4. (d) Perlon-U because of the presence of stronger bond, it is not degraded by microorganisms, i.e. it is a non-biodegradable polymer.

5. (b) Generally, the amount of S varies from 4 to 10%., 5% is most common.

6. (d) In isotactic polyethenyl benzene, all its substituents are present on the same side of the polymer chain, i.e. a more dense structure is formed due to which it is highly crystalline with better mechanical strength.

7. (a) The linear chains of nylon-6 6 are binded together through H-bonding.

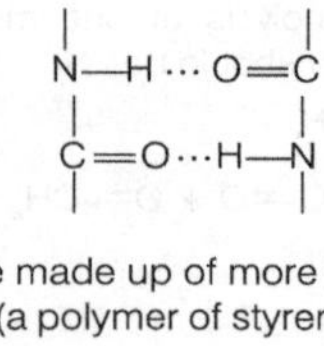

8. (a) Copolymers are made up of more than one kind of monomer units, e.g. buna-S (a polymer of styrene and buta-1, 3-diene).

Note : *PAN is a polymer of acrylonitrile* ($CH_2 = CH—CN$), *polythene is a polymer of ethene* ($CH_2 = CH_2$) *and PTFE is a polymer of tetrafluoroethylene* ($CF_2=CF_2$).

9. (c) Rayon is a modified form of cellulose, so it does not contain amide linkage.

10. (d) A typical *Ziegler-Natta* catalyst is a mixture of $TiCl_4$ and $(C_2H_5)_3Al$. It is used to have a control over stereochemistry of polymerisation process. In its presence, stereospecific polymer (addition) is obtain.

11. (d)

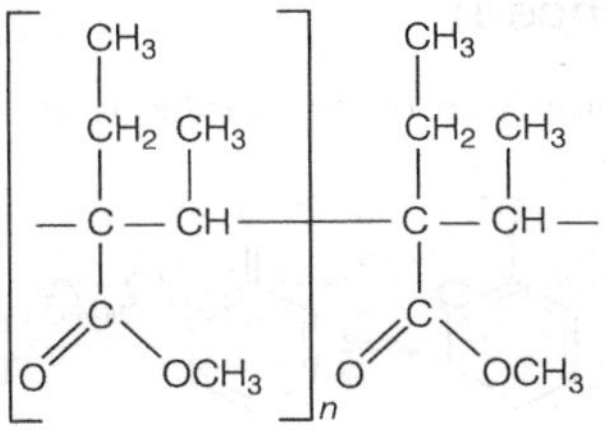

Repeating unit

So, monomer is $CH_3CH_2C(COOCH_3){=}CH{-}CH_3$

12. (d) PMMA is a polymer of methylmethacrylate, an unsaturated compound $\left(CH_2{=}C(CH_3)COOCH_3\right)$, so it is formed by addition or chain growth polymerisation.

13. (b) On hydrolysis, the given polymer gives

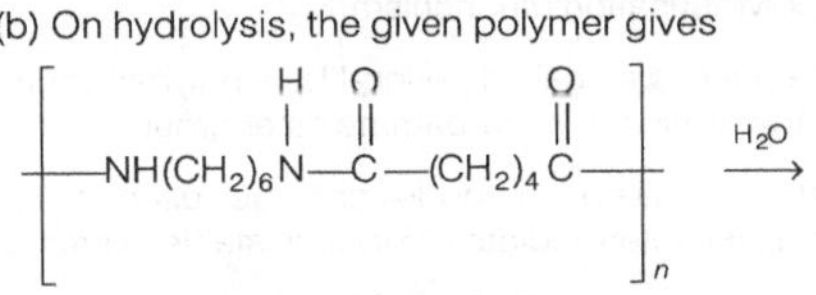

$NH_2(CH_2)_6NH_2 + HOOC(CH_2)_4COOH$

Since, two kinds of monomers are involved, so it is a copolymer.

14. (d) Because of the presence of electron withdrawing F-atoms, $CF_2{=}CF_2$ undergoes anionic addition polymerisation. Others because of the absence of electron releasing or electron withdrawing groups are prepared by free radical polymerisation.

15. (b) Cellulose is a condensation homopolymer of β-D-glucose units.

16. (c) Turbidity shows light scattering power of a solution.

17. (c) Step-growth or condensation polymers are formed when the monomer units contain functional groups.

e.g. $H_2N(CH_2)_6NH_2 + HOOC{-}(CH_2)_4{-}COOH \longrightarrow$

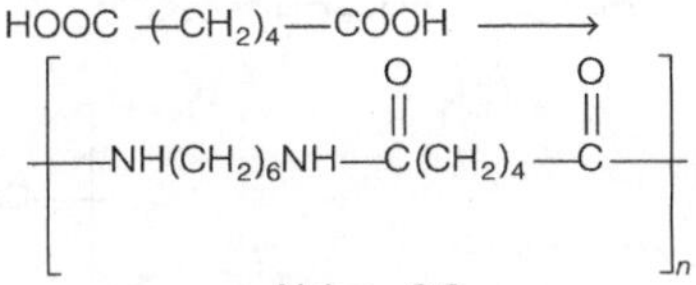

Nylon - 6 6

18. (b) Neoprene is a polymer of chloroprene (2-chlorobuta-1, 3-diene).

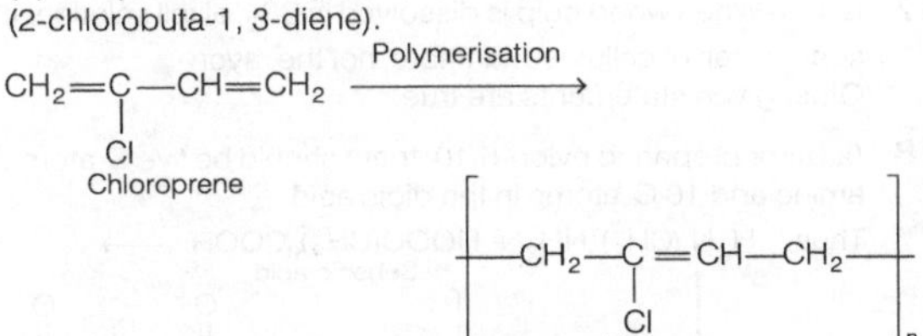

19. (b) Vulcanisation is a process of heating natural rubber with S. This brings about linking of polymer chains at various points of backbone through S—S linkage. This provides strength to the rubber.

20. (d) If P/F > 1, novolac is formed and if P/F < 1, bakelite is formed.

21. (d) Cellulose is a natural polymer of D-glucose and it has not yet been synthesised in laboratory or industry.

22. (a)

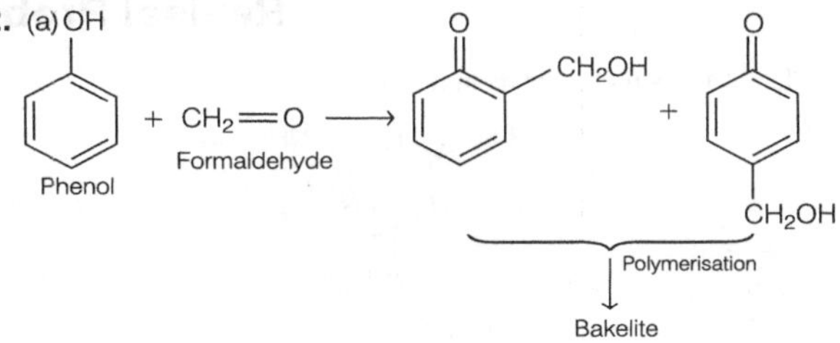

23. (d)

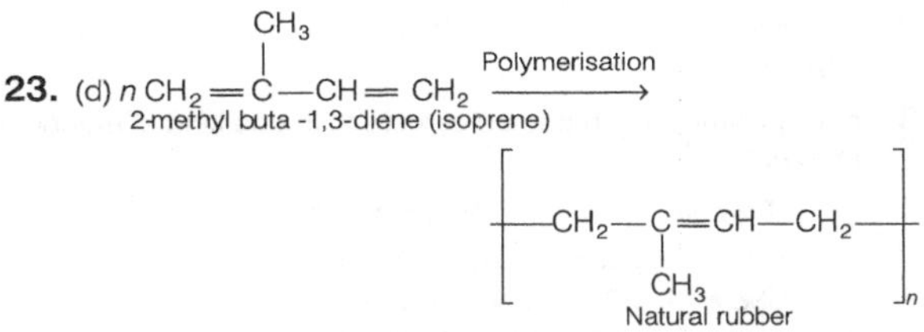

24. (c) Polyurethane are obtained by the reaction of diisocyanate and diol. The group present in it is carbamate ester or urethane group. These can be represented as

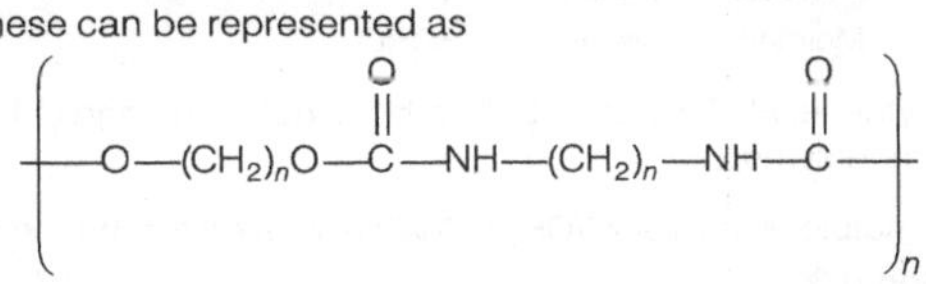

25. (a) SBR (styrene butadiene rubber) is a copolymer of 70% buta-1,3-diene and 30% styrene, (i.e. phenyl ethene, $C_6H_5CH{=}CH_2$).

26. (d) Because of the symmetry of $F_2C{=}CF_2$ and strength of C—F bond, branching does not take place and a linear chain teflon is always formed. High bond energy of C—F bond gives high thermal stability to polymer.

Large difference in the electronegativity of C and F makes the C—F bonds highly polar.

27. (d) Free radical reactions are generally initiated by the use of free radical initiator (either chemical initiator like benzyl peroxide or radiation). The polymerisation in case of olefinic monomers takes place with the evolution of heat, (i.e. a exothermic process) due to the conversion of π-bonds into σ-bond.

Propagation step result in increased chain length and hence, increased molar mass.

28. (a)

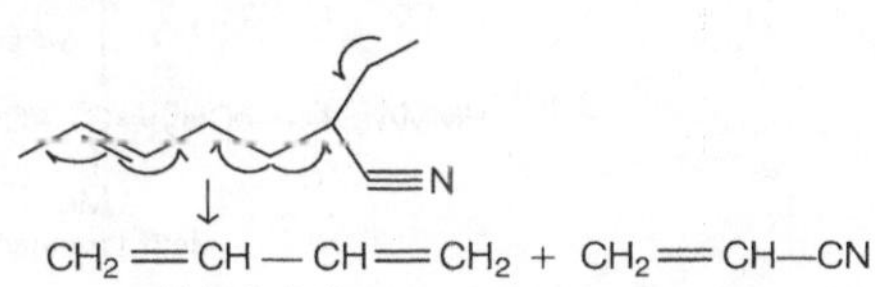

Because of the involvement of two monomer units, it is an addition copolymer.

29. (b) In nylon-6, 10, 10 represents the number of carbon atoms present in the acid. Thus, here Y is sebacic or decandioic acid, $HOOC(—CH_2)_8COOH$.

30. (c)

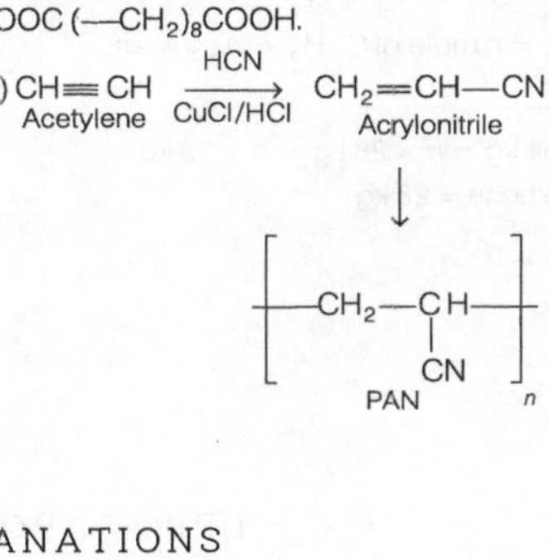

Revisal Problems (JEE Advanced)

1. (c) The structure of melamine is

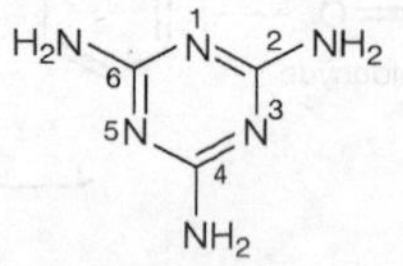

2, 4, 6-triamino-1, 3, 5-triazine

2. (b) Nylon-2-nylon-6 is an amide and also it is a biodegradable polymer.

3. (a) Copolymers are made up of more than one kind of monomer units. e.g.

Polymer	Monomer units
SBR	Styrene + buta-1,3-diene
Glyptal	Glycol + phthalic acid
Nylon-6 6	Adipic acid + hexamethylenediamine
Bakelite	Phenol + formaldehyde
Polyester	Acid + alcohol
Melmac	Melamine + HCHO

Polythene, PVC, nylon-6, butyl rubber, neoprene and teflon all are homopolymers.

4. (b) Adipic acid $[HOOC(CH_2)_4COOH]$ is used for the formation of nylon-6 6.

5. (a) $\{CH_2—CH(—COOH)\}_n \xrightarrow{NaOH} \{CH_2—CH(—COO^-Na^+)\}$

Thus, in X intermolecular forces are stronger than in the parent compound.

6. (a) $CHCl_3 \xrightarrow[HF]{SbF_3} CHF_2Cl \xrightarrow{1070\ K} F_2C{=}CF_2$

$X +$ HCl (Forms white fumes of NH_4Cl with NH_3)

7. (a)

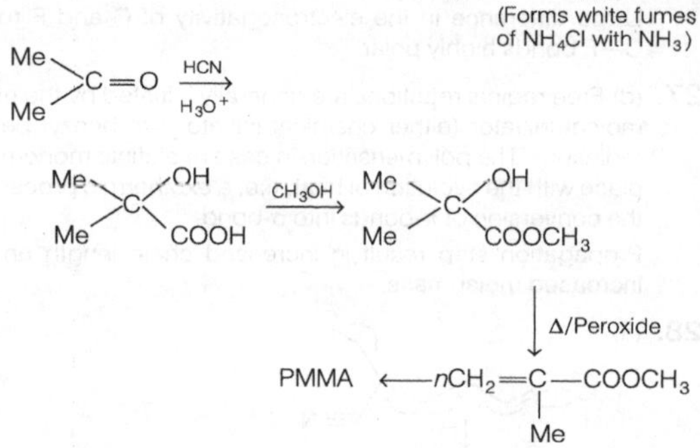

Methyl methacrylate

8. (c) Because of the presence of polar group, intermolecular forces of attraction are greater, hence it does not behave like an elastomer.

9. (b) If cross-links and branching is introduced, substance 'X' which is an amide (a thermoplastic) behaves like an elastomer.

10. (d) n mole of $CaC_2 = n$ mole of $C_2H_4 = n$ mole of

$$—\!(CH_2—CH_2)\!—$$

$$n \times 64\ kg = n \times 28\ kg = n \times 28\ kg$$

$\therefore$ Amount of polyethene = 28 kg

11. (d) First (initial) step of this reaction is aromatic electrophilic substitution.

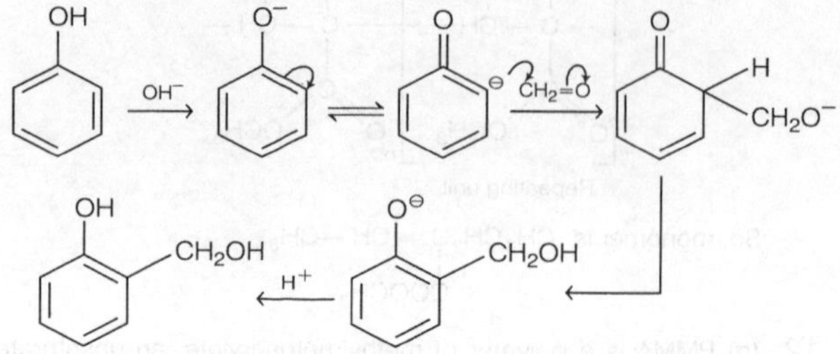

12. (a) On vulcanisation, S introduces cross-links at the reactive sites of double bonds and makes the rubber stiffened, more elastic, and resistant to oxidation and organic solvents.

13. (a,b,c) Polymer, monomer of which contains some electron withdrawing groups like —CN, —Cl, —F, etc., generally prepared by anionic polymerisation mechanism.

14. (a,b) Polyurethane, also called perlon-U is a polycarbamate ester because of the presence of carbamate ester group.

15. (a,b,c,d) All the given methods can be used to prepare polystyrene as the intermediate involved is stable in each case.

16. (a,c,d)

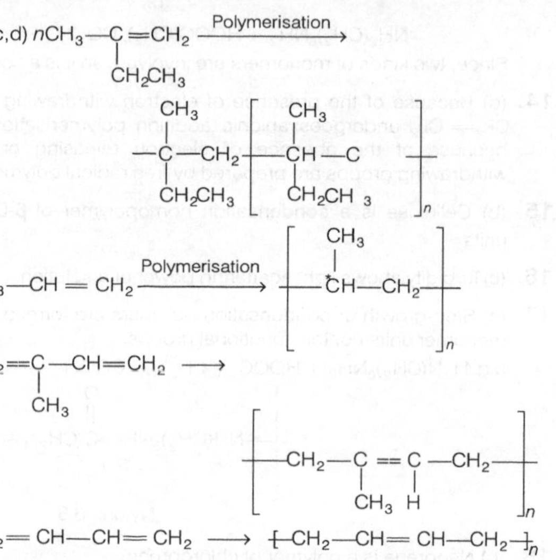

17. (a,c,d) When wood pulp is dissolved in CS_2, alkali solution, it gives sodium salt of cellulose xanthate, not the rayon. Other given statements are true.

18. (a,b) For preparing nylon-5,10, there should be five C-atoms in the amine and 10 C-atoms in the dioic acid.

Thus, $H_2N(CH_2)_5NH_2 + HOOC(CH_2)_8COOH \longrightarrow$
Sebacic acid

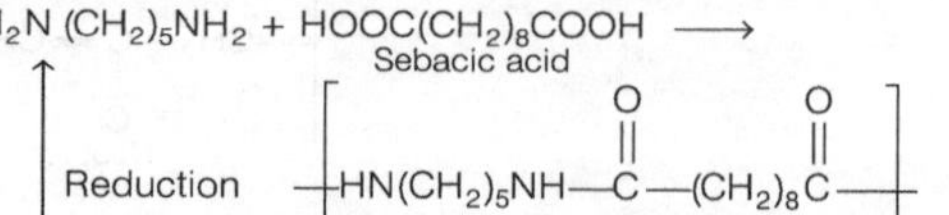

Nylon-5, 10

$CNCH_2CH_2CH_2CN$ (Pentan-1, 5-dinitrile)

Explanations for Q. Nos. (19-20)

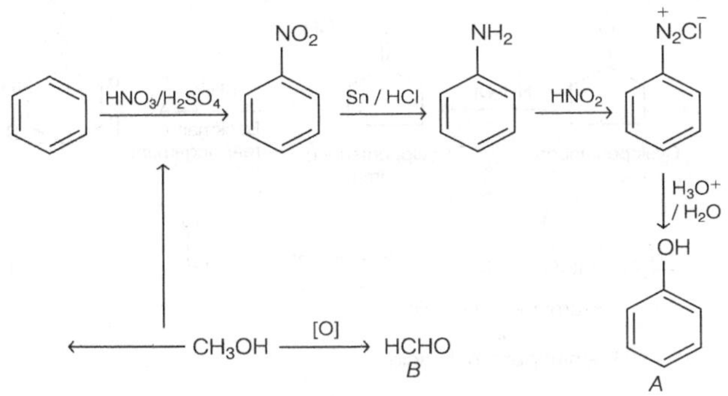

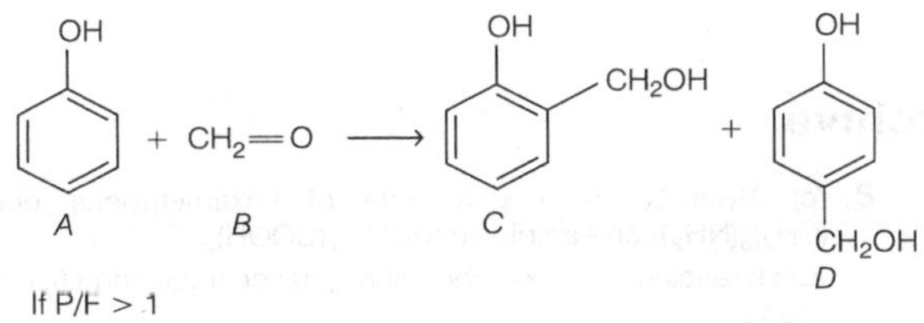

If P/F > 1

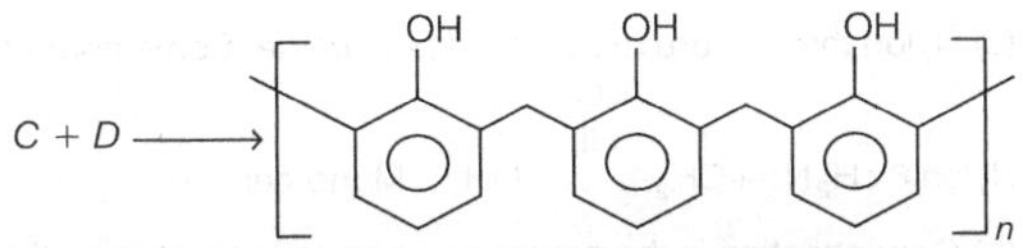

Linear polymer
(Resol or novolac)
E

If P/F < 1

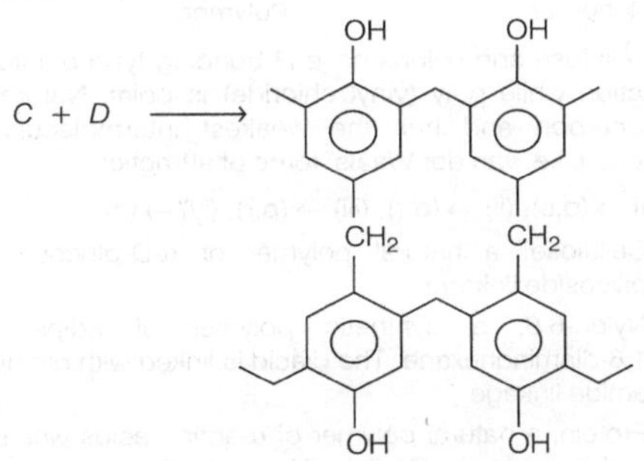

Cross-linked polymer
F
Bakelite
(a thermosetting polymer, used for making switches and plugs)

19. (b)

20. (b,d)

Explanations for Q. Nos. (21-22)

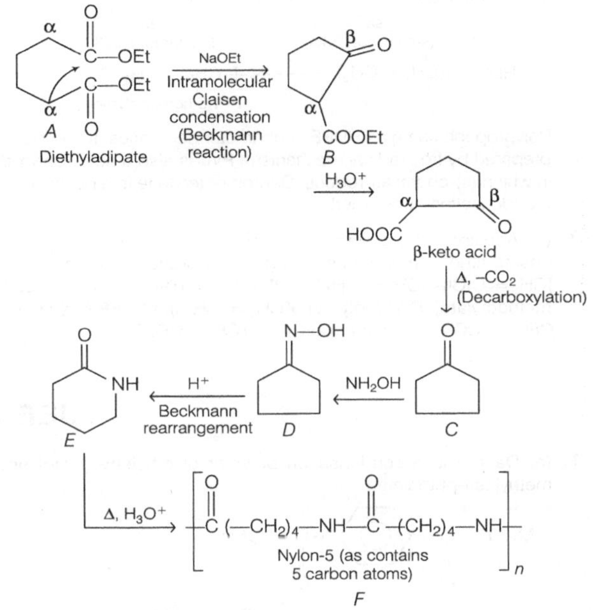

21. (b,c) **22.** (b)

23. (a) Rayon is a semisynthetic polymer which is obtained from cellulose through acetylation process. It is taken as a better choice as compared to cotton fabric.

24. (d) Polyvinyl alcohol cannot be prepared by polymerisation of vinyl alcohol since, it readily tautomerises into more stable keto form (which is acetaldehyde).

$$\underset{\text{(Very less stable)}}{CH_2{=}CHOH} \xrightarrow{\text{Tautomerisation}} \underset{\text{(Highly stable)}}{CH_3CHO}$$

25. (d) HDPE (high density polyethylene) is prepared by the polymerisation of ethene in the presence of *Ziegler-Natta* catalyst at 350 K. Ionic intermediates are involved here. It is an example of coordination polymerisation.

The polymer (HDPE) obtained by this method consists of linear chains.

26. (a) (i)→(q), (ii)→(p,s), (iii)→(p,s), (iv)→(r,s)

o-cresyl phosphate is a plasticiser which is added to such thermoplastic, which do not soften very much with rise in temperature.

Bakelite is a synthetic polymer. It is a thermosetting plastic.

Resol or novolac is also a thermosetting synthetic polymer. Cellulose, although is a natural polymer but in modified form acts as a semisynthetic polymer.

27. (a) (i)→(q,s), (ii)→(q,r,s), (iii)→(q,s), (iv)→(p)

$$\text{HDPE } n(\overset{sp^2\ (s\text{-character }33\%)}{CH_2{=}CH_2}) \longrightarrow \underset{\text{HDPE homopolymer}}{\overset{sp^3\ (s\text{-character }25\%)}{\left[CH_2-CH_2\right]_n}}$$

Polypropene same as HDPE, *s*-character decreases. It can be prepared by free radical mechanism. PVC is also a homopolymer in which (s)-character reduce. Dacron or terylene is a polyester (condensation copolymer).

28. (5) Addition homopolymers are made up of only one type of unsaturated monomers. e.g. neoprene [polymer of $[CH_2{=}C(Cl){-}CH = CH_2]$, PMMA (polymer of methyl methacrylate], PVC (polymer of $CH_2{=}CHCl$), PCTFE (polymer of $CFCl{=}CCl_2$) and teflon (polymer of $CF_2{=}CF_2$).

29. (5)

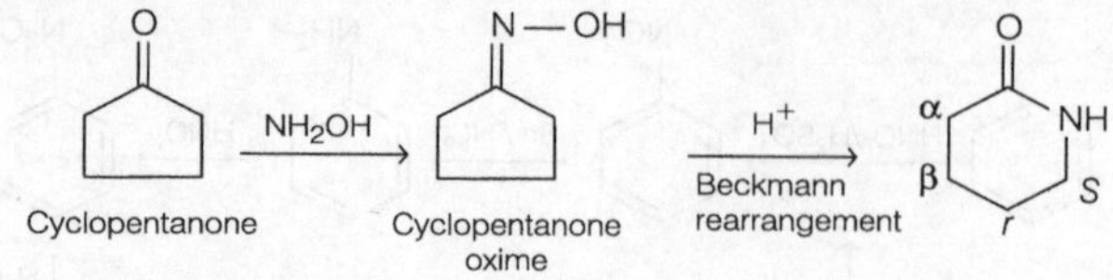

$$\xrightarrow[\Delta]{H_2O} HOOC(CH_2)_4NH_2 \xrightarrow[-nH_2O]{\text{Polymerisation}} \left[\overset{O}{\overset{\|}{C}}-(CH_2)_4-NH\right]_n$$

5-aminovaleric acid
or
5-aminopentanoic acid

JEE Archive

1. (a) Dacron is a condensation polymer of ethylene glycol and methyl terephthalate.

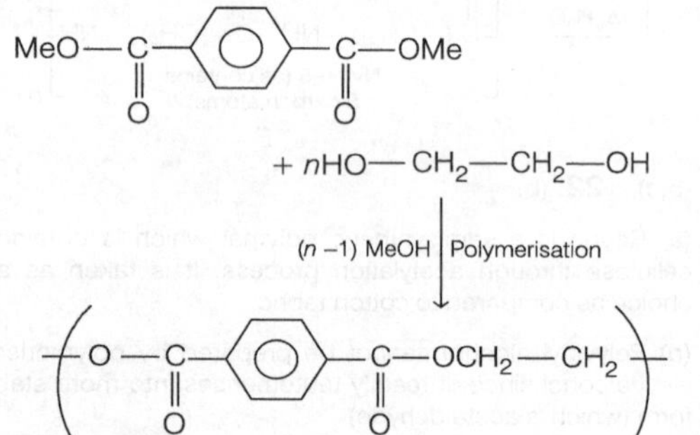

Dacron

2. (c) Electron-deficient species (Lewis acid) is used as an initiator for cationic polymerisation.

3. (c) Bakelite is a thermosetting polymer formed by the condensation reaction of phenol with HCHO in the presence of conc. H_2SO_4.

It is thus, a cross-linked polymer in which condensation takes place at *o*- and *p*–positions.

4. (b) Nylon-6 6 contains strong intermolecular forces like hydrogen bonds that are formed between $-\overset{O}{\overset{\|}{C}}-NH$ group of successive chains.

5. (c) Buna-N is actually abbreviated form, where Bu represents 1,3-butadiene, na represents Na, (sodium) and N represents nitrile (acrylonitrile).

Thus, buna-N is a copolymer of 1,3-butadiene and acrylonitrile usually polymerise in the presence of sodium.

6. (d) Bakelite is obtained from phenol by reacting with HCHO in the acidic or alkaline medium.

7. (c) Teflon is $\left[CF_2-CF_2\right]_n$.

8. (c) Nylon-66 is a polyamide of hexamethylene diamine $(CH_2)_6(NH_2)_2$ and adipic acid $(CH_2)_4(COOH)_2$.

(Each reactant has six carbon chain, hence trade code 6, 6 is used.)

9. (c) Nylon threads are made up of polyamide. Some common are

Nylon 6 : $H_2N\left(CH_2\right)_5\overset{O}{\overset{\|}{C}}-OH$; Monomer

10. (b) Condensation is the process of aggregation of more than one molecule without loosing any atom or group (sometimes smaller group or atoms H_2O, *R*—OH, etc are released).

$$\underset{\text{Monomer}}{nCH_2{=}CH_2} \longrightarrow \underset{\text{Polymer}}{-CH_2-(CH_2-CH_2)_n-CH_2-}$$

11. (d) Cellulose and nylons have H-bonding type of intermolecular attraction while poly (vinyl chloride) is polar. Natural rubber is hydrocarbon and has the weakest intermolecular force of attraction, i.e. van der Waals' force of attraction.

12. (b) (i) → (p,s), (ii) → (q,r), (iii) → (p,r), (iv) → (s)

(i) Cellulose, a natural polymer of α-D-glucose, linked by glycoside linkage.

(ii) Nylon-6 6, a synthetic polymer of adipic acid and 1,6-diaminohexane. The diacid is linked with diamine through amide linkage.

(iii) Protein, a natural polymer of α-amino acids where individual amino acid units are linked by amide linkage.

(iv) Sucrose, has glycoside linkage, a disaccharide.

13. (i) $$\underset{\text{Isoprene}}{H_2C{=}\overset{CH_3}{\overset{|}{C}}-CH{=}CH_2} \xrightarrow[Zn\text{-}H_2O]{O_3} 2HCHO + CH_3-\overset{O}{\overset{\|}{C}}-CHO$$

(ii) Isoprene ⟶

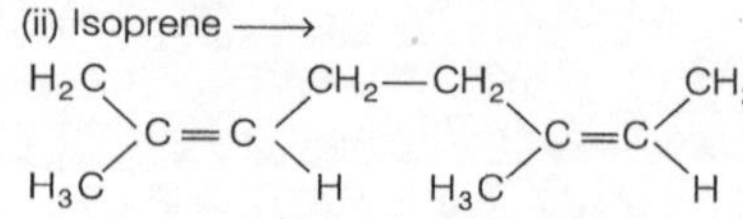

Natural rubber

Last 3 Years' Questions
JEE Main & Advanced
(2016-2018)

JEE Main

2016

1. Which of the following statements about low density polythene is false? **[Polymers]**

(a) It is a poor conductor of electricity
(b) Its synthesis required dioxygen or a peroxide initiator as a catalyst
(c) It is used in the manufacture of buckets, dustbins etc.
(d) Its synthesis requires high pressure

2. Thiol group is present in **[Biomolecules]**

(a) cystine
(b) cysteine
(c) methionine
(d) cytosine

2017

3. Which of the following compounds will give significant amount of *meta*-product during mononitration reaction? **[Aromatic Nitrocompounds, Anilines, Diazonium Salts]**

(a) 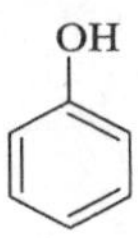

(b) 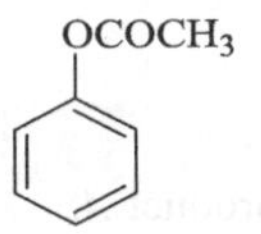

(c) 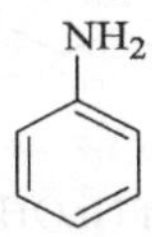

(d) 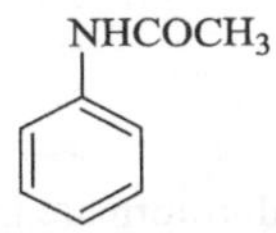

4. The formation of which of the following polymers involves hydrolysis reaction? **[Polymers]**

(a) Nylon-6
(b) Bakelite
(c) Nylon-6, 6
(d) Terylene

5. Which of the following compounds will behave as a reducing sugar in an aqueous KOH solution? **[Biomolecules]**

(a) 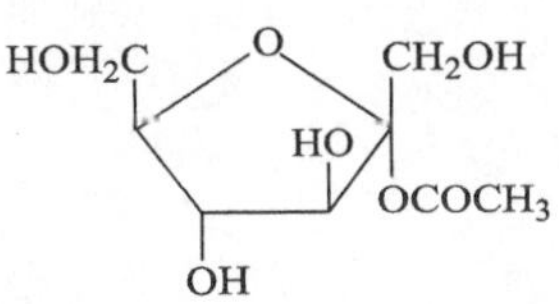

(b) 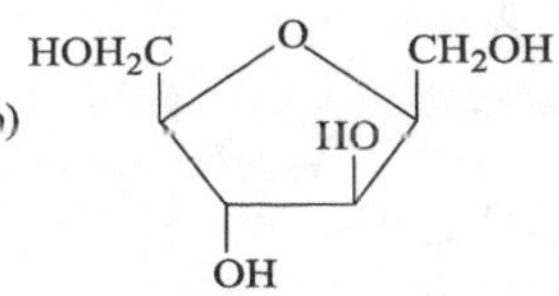

(c) 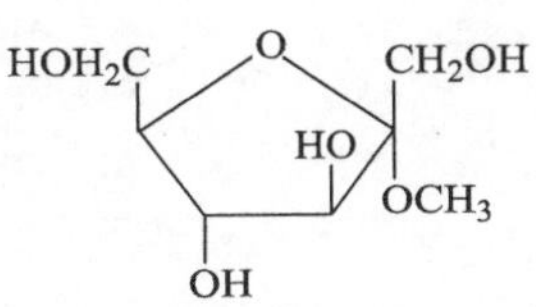

(d) 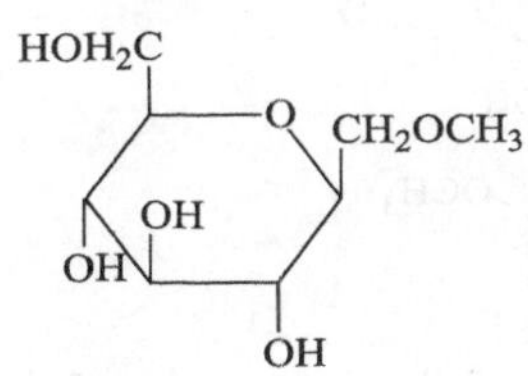

2018

6. Glucose on prolonged heating with HI gives **[Bimolecules]**

(a) *n*-hexane
(b) 1-hexene
(c) Hexanoic acid
(d) 6-iodohexanal

7. Phenol on treatment with CO_2 in the presence of NaOH followed by acidification produces compound X as the major product. X on treatment with $(CH_3CO)_2O$ in the presence of catalytic amount of H_2SO_4 produces: **[Aryl Halides and Phenols]**

(a) 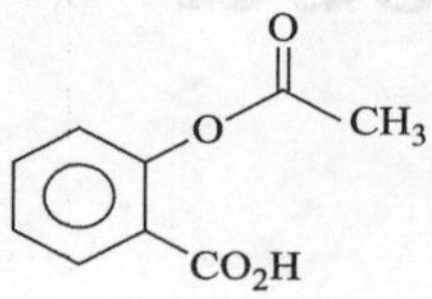

(b) 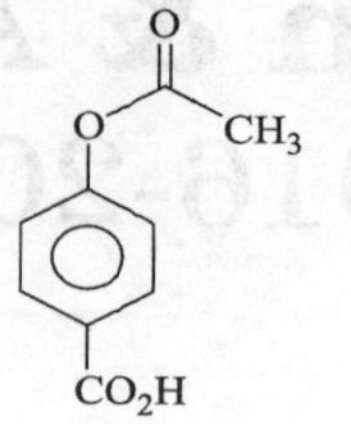

(c) 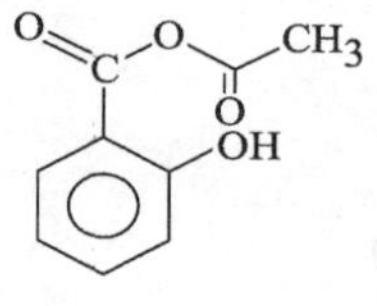

(d) 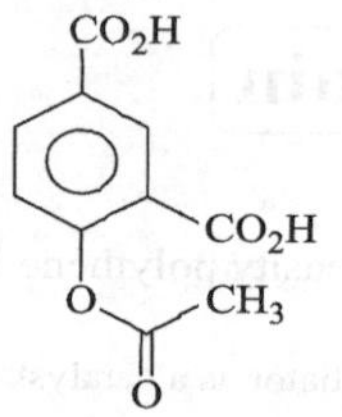

8. The predominant form of histamine present in human blood is (pK_a, Histidine = 6.0) **[Biomolecules]**

(a) 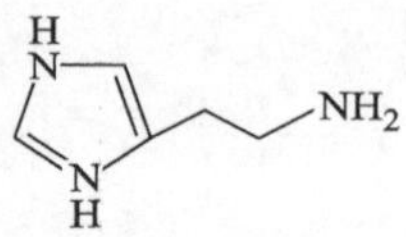

(b) 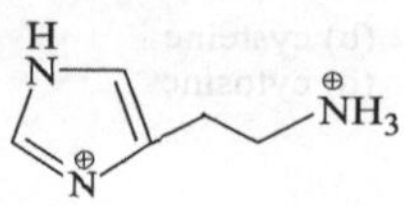

(c) 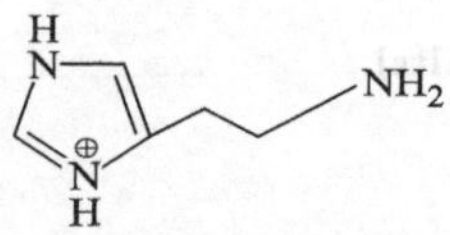

(d) 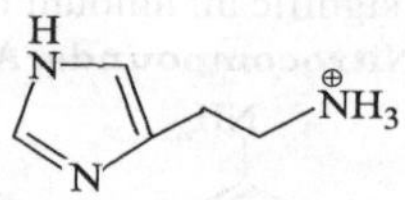

9. Phenol reacts with methyl chloroformate in the presence of NaOH to form product A. A reacts with Br_2 to form product B. A and B are respectively **[Aryl Halides and Phenols]**

(a) 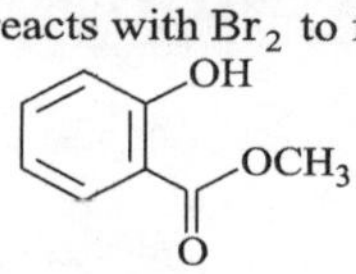 and 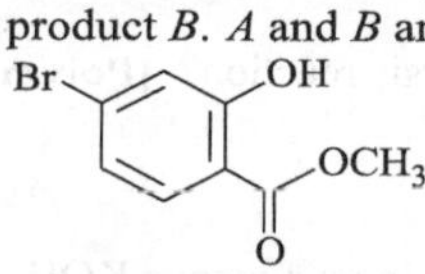

(b) 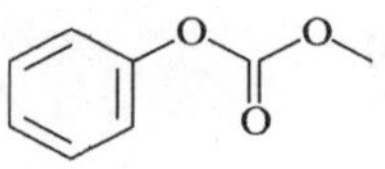 and 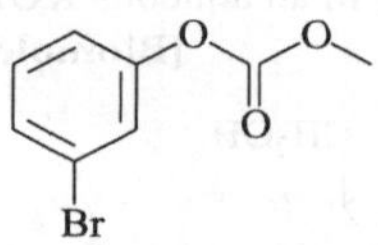

(c) 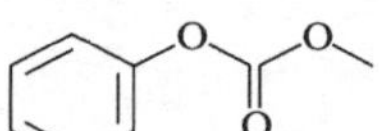 and 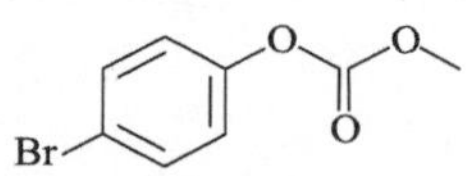

(d) 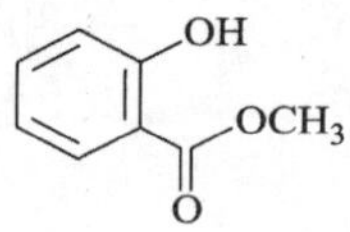 and 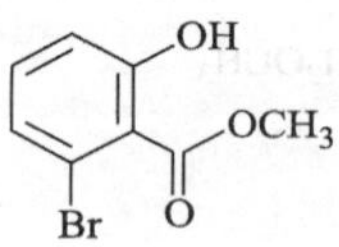

10. The increasing order of basicity of the following compounds is

I.  II. 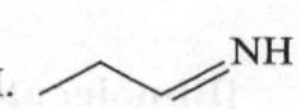 III. 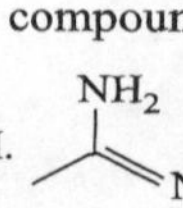 IV. 

[Aliphatic Nitrogen Containing Compounds]

(a) (I) < (II) < (III) < (IV)
(b) (II) < (I) < (III) < (IV)
(c) (II) < (I) < (IV) < (III)
(d) (IV) < (II) < (I) < (III)

JEE Advanced

2016

1. On complete hydrogenation, natural rubber produces **[Polymers, Type 1]**

(a) ethylene-propylene copolymer (b) vulcanised rubber

(c) polypropylene (d) polybutylene

2. The product(s) of the following reaction sequence is (are)

[Aromatic Nitro Compounds, Anilines and Diazonium Salt, Type 1]

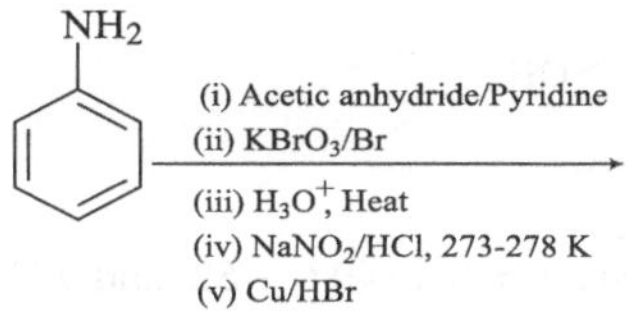

(a) 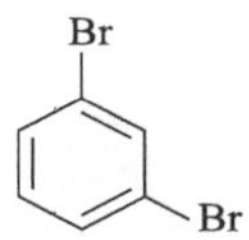

(b) 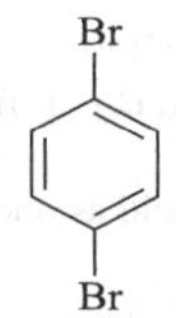

(c) 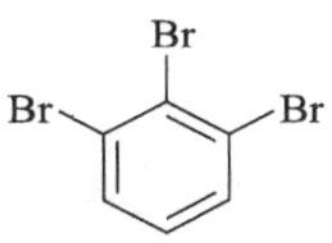

(d) 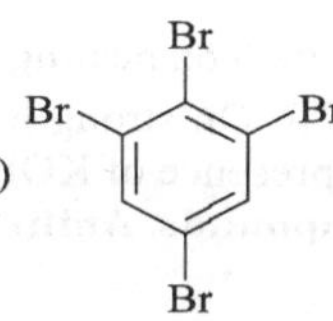

3. The correct statement(s) about the following reaction sequence is (are)

$$\text{Cumene } (C_9H_{12}) \xrightarrow[\text{(ii) } H_3O^+]{\text{(i) } O_2} P \xrightarrow{CHCl_3/NaOH} Q(\text{major}) + R(\text{minor}),\ Q \xrightarrow[PhCH_2Br]{NaOH} S$$

[Aromatic Aldehyde, Ketones and Acids, Type 2]

(a) R is steam volatile

(b) Q gives dark violet colouration with 1% aqueous $FeCl_3$ solution

(c) S gives yellow precipitate with 2, 4-dinitrophenylhydrazine

(d) S gives dark violet colouration with 1% aqueous $FeCl_3$ solution

4. The correct order of acidity for the following compounds is

[Aromatic Aldehydes, Ketones and Acids, Type 2]

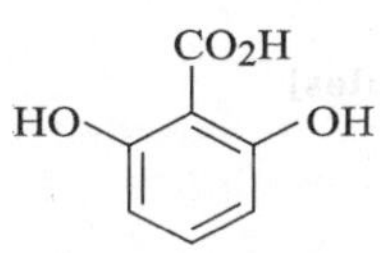

I

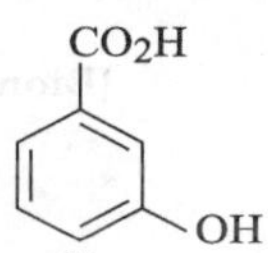

II

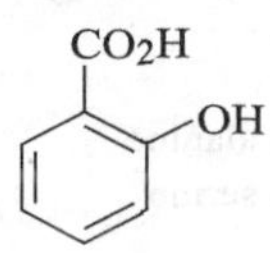

III

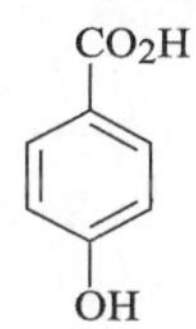

IV

(a) I > II > III > IV (b) III > I > II > IV

(c) III > IV > II > I (d) I > III > IV > II

5. The major product of the following reaction sequence is

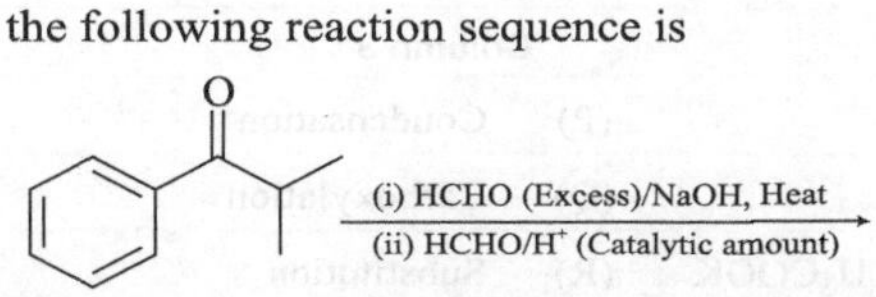

[Aromatic Aldehyde, Ketone and Acids, Type 1]

Note ***Type-1*** *Only one correct option,* ***Type-2*** *More then one correct option,* ***Type-3*** *Single digit Answer,* ***Type-4*** *Numerical value based*

(a) 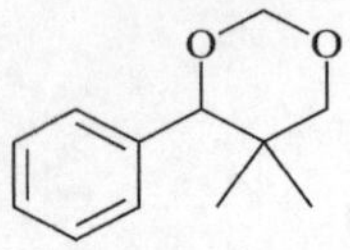

(b) 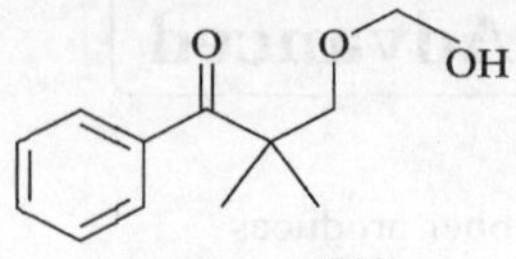

(c) 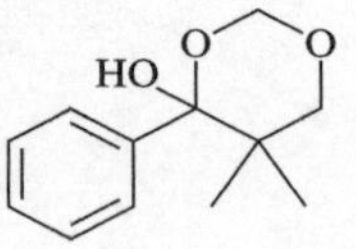

(d) 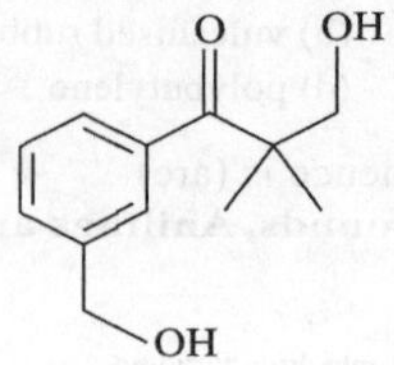

6. For 'invert sugar', the correct statement(s) is (are)
(Given: specific rotations of (+) - sucrose, (+) - maltose, L-(−)-glucose and L-(+)-fructose in aqueous solution are + 66°, +140°, − 52° and 92°, respectively) **[Biomolecules, Type 2]**
(a) Invert sugar is prepared by acid catalysed hydrolysis of maltose
(b) Invert sugar is an equimolar mixture of D-(+) -glucose and D-(−)- fructose
(c) Specific rotation of invert sugar is − 20°
(d) On reaction with Br_2 water, invert sugar forms saccharic acid as one of the products

Paragraph (Q. Nos. 7-8)

Treatment of compound *O* with $KMnO_4$ / H^+ gave *P*, which on heating with ammonia gave *Q*. The compound *Q* on treatment with Br_2 / NaOH produced *R*. On strong heating, *Q* gave *S*, which on further treatment with ethyl 2-bromopropanoate in the presence of KOH followed by acidification, gave a compound *T*. **[Aromatic Nitrocompounds, Anilines, and Diazonium Salts]**

7. The compound *R* is

(a) 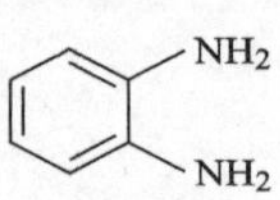

(b) 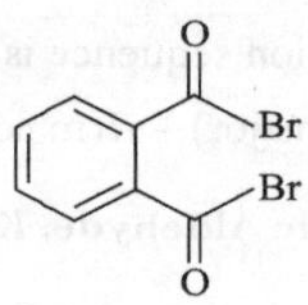

(c) 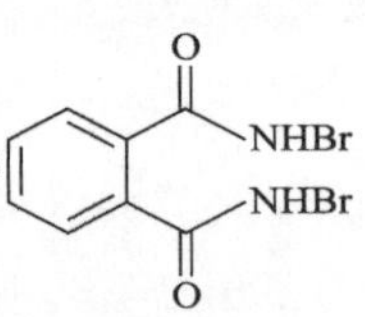

(d) 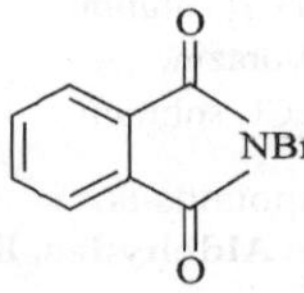

8. The compound *T* is **[Biomolecules]**
(a) glycine
(b) alanine
(c) valine
(d) serine

2017

Answer **Q. 9, Q. 10 and Q. 11** by appropriately matching the information given in the three columns of the following table.

Column 1, 2 and 3 contain starting materials, reaction conditions, and type of reactions, respectively.

Column 1		Column 2		Column 3	
(I)	Toluene	(i)	NaOH/ Br_2	(P)	Condensation
(II)	Acetophenone	(ii)	$Br_2/h\nu$	(Q)	Carboxylation
(III)	Benzaldehyde	(iii)	$(CH_3CO)_2O$/ CH_3COOK	(R)	Substitution
(IV)	Phenol	(iv)	NaOH/CO_2	(S)	Haloform

[Aromatic Aldehydes, Ketones and Acids]

9. The only **CORRECT** combination in which the reaction proceeds through radical mechanism is

(a) (IV) (i) (Q) (b) (III) (ii) (P)
(c) (II) (iii) (R) (d) (I) (ii) (R)

10. For the synthesis of benzoic acid, the only CORRECT combination is

(a) (II) (i) (S) (b) (I) (iv) (Q)
(c) (IV) (ii) (P) (d) (III) (iv) (R)

11. The only **CORRECT** combination that gives two different carboxylic acids is

(a) (IV) (iii) (Q) (b) (II) (iv) (R)
(c) (I) (i) (S) (d) (III) (iii) (P)

12. The major product of the following reaction is

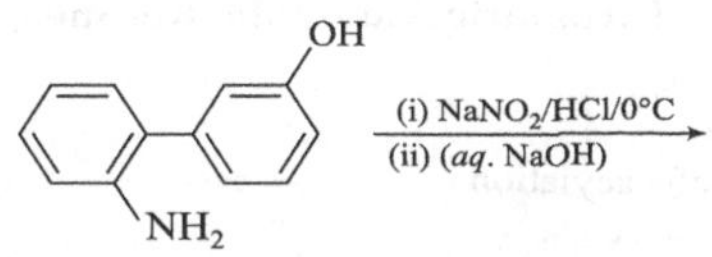

[Aromatic Nitrocompounds, Aniline and Diazonium Salts, Type 1]

(a) 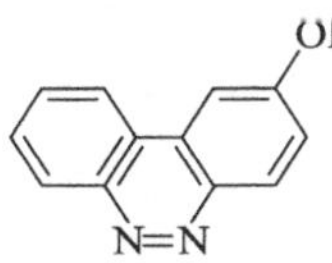

(b) 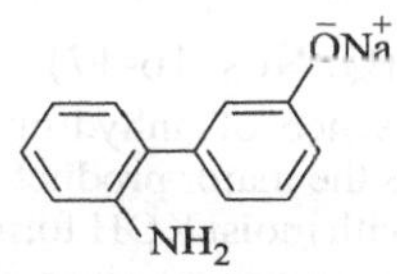

(c) 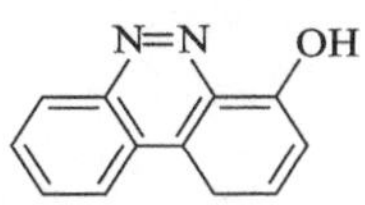

(d) 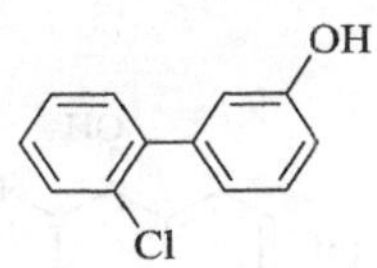

13. Compound P and R upon ozonolysis produce Q and S, respectively. The molecular formula of Q and S is C_8H_8O. Q undergoes Cannizzaro reaction but not haloform reaction, whereas S undergoes haloform reaction but not Cannizzaro reaction.

(i) $P \xrightarrow[\text{(ii) Zn/H}_2\text{O}]{\text{(i) O}_3/\text{CH}_2\text{Cl}_2} \underset{(C_8H_8O)}{Q}$ (ii) $R \xrightarrow[\text{(ii) Zn/H}_2\text{O}]{\text{(i) O}_3/\text{CH}_2\text{Cl}_2} \underset{(C_8H_8O)}{S}$

The option(s) with suitable combination of P and R, respectively, is(are)

[Aromatic Aldehyde, Ketones and Acids, Type 2]

(a) 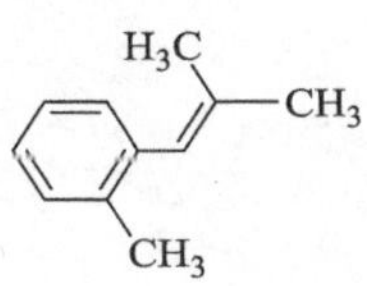

and 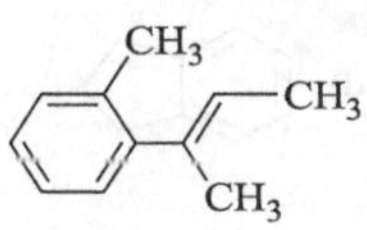

(b) 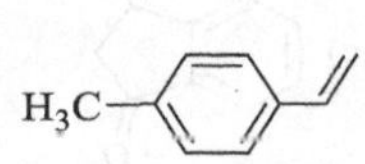

and 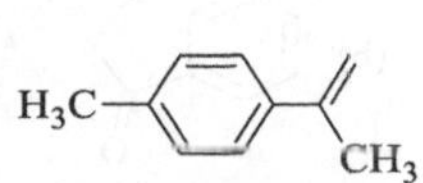

(c) 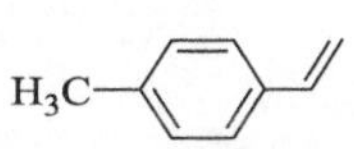

and 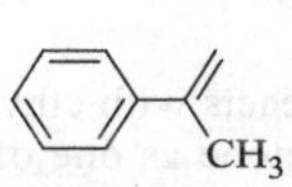

(d) 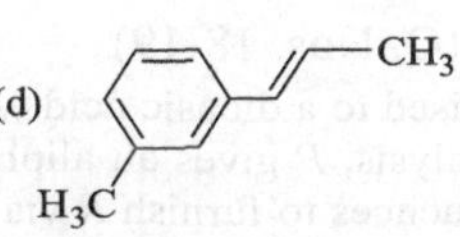

and 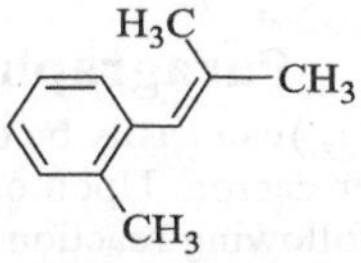

Paragraph 2 (Q. Nos. 14-15)

The reaction of compound P with CH_3MgBr (excess) in $(C_2H_5)_2O$ followed by addition of H_2O gives Q. The compound Q on treatment with H_2SO_4 at 0° C gives R. The reaction of R with CH_3COCl in the presence of anhydrous $AlCl_3$ in CH_2Cl_2 followed by treatment with H_2O produces compound S. [Et in compound P is ethyl group]

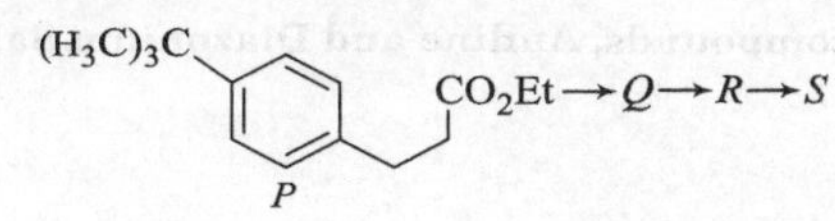

[Aromatic Aldehyde, Ketones, Acids]

14. The product S is

(a) 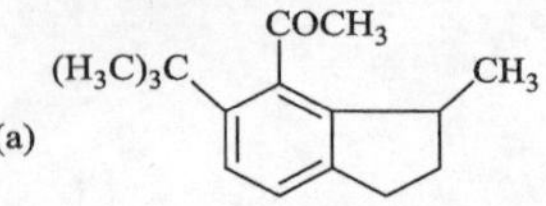

(b) 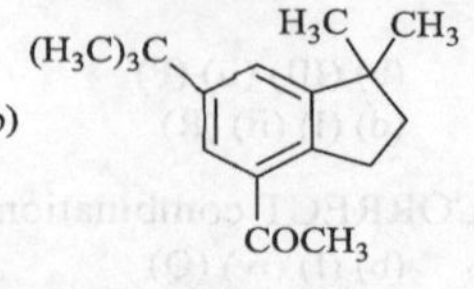

(c) 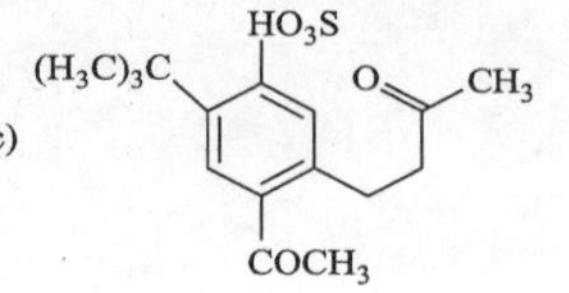

(d) 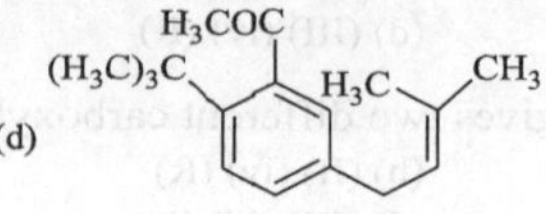

15. The reactions, Q to R and R to S, are **[Aromatic Aldehyde, Ketones, Acids]**

(a) Aromatic sulfonation and Friedel-Crafts acylation
(b) Friedel-Crafts alkylation and Friedel-Crafts acylation
(c) Friedel-Crafts alkylation, dehydration and Friedel-Crafts acylation
(d) Dehydration and Friedel-Crafts acylation

2018

Paragraph X (Q. Nos. 16-17)

Treatment of benzene with CO/HCl in the presence of anhydrous $AlCl_3$ / $CuCl$ followed by reaction with $Ac_2O/NaOAc$ gives compound X as the major product. Compound X upon reaction with Br_2 / Na_2CO_3 followed by heating at 473 K with moist KOH furnishes Y as the major product. Reaction of X with H_2 / Pd - C, followed by H_3PO_4 treatment gives Z as the major product.

16. The compound Y is **[Aldehydes Ketones]**

(a) 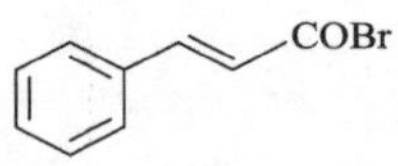

(b) 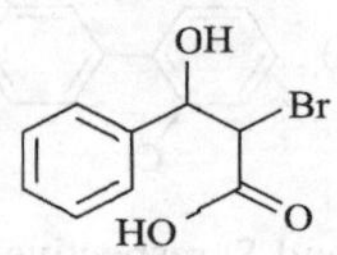

(c) 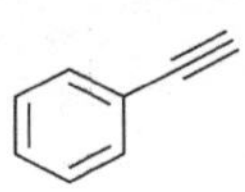

(d) 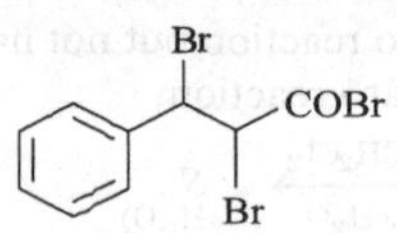

17. The compound Z is

(a) 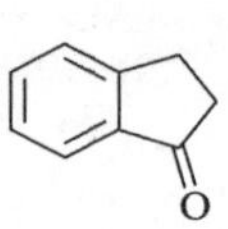
(b) 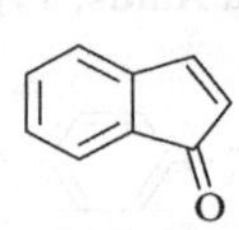
(c) 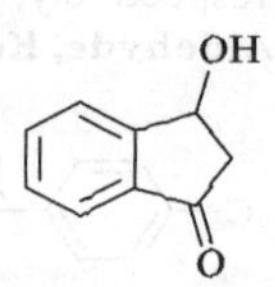
(d) 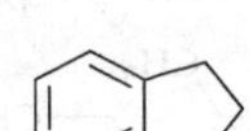

Paragraph A (Q. Nos. 18-19)

An organic acid P ($C_{11}H_{12}O_2$) can easily be oxidised to a dibasic acid which reacts with ethylene glycol to produce a polymer dacron. Upon ozonolysis, P gives an aliphatic ketone as one of the products. P undergoes the following reaction sequences to furnish R via Q. The compound P also undergoes another set of reactions to produce S.

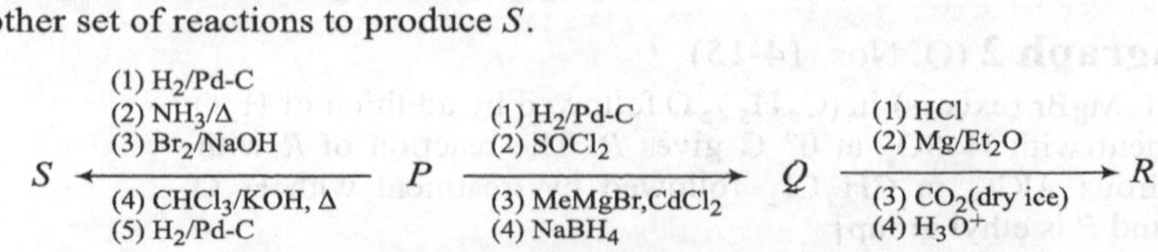

[Aromatic Nitrocompounds, Aniline and Diazonium Salts]

18. The Compound *R* is

(a) 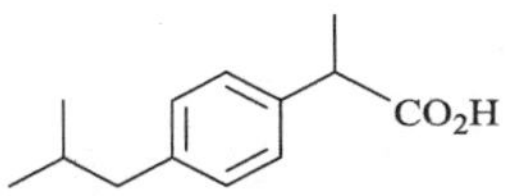

(b) 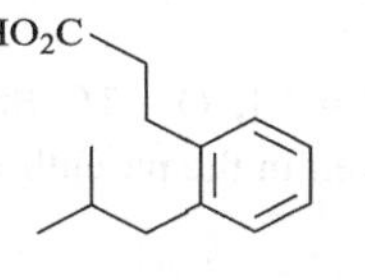

(c) 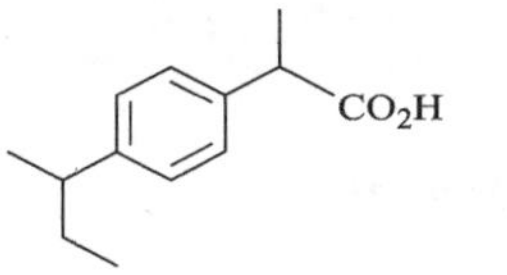

(d) 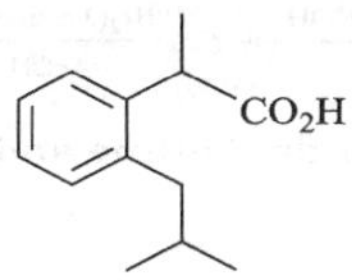

19. The compound *S* is

(a) 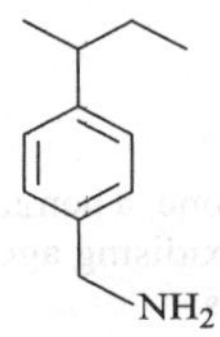

(b) 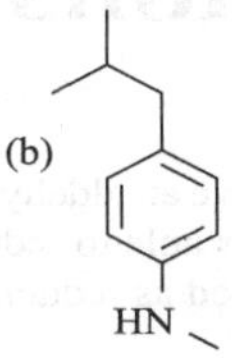

(c) 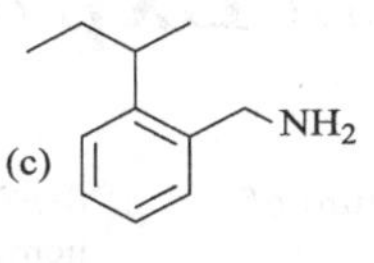

(d) 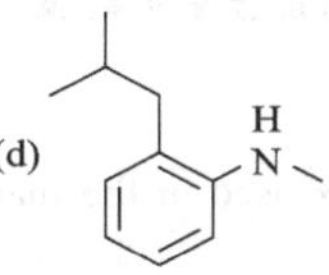

20. Aniline reacts with mixed acid (conc. HNO_3 and conc. H_2SO_4) at 288 K to give P (51%), Q (47%) and R (2%). The major product(s) of the following sequence is (are)

$$R \xrightarrow[\substack{(3)\ H_3O^+ \\ (4)\ NaNO_2,\ HCl/273\text{-}278\ K \\ (5)\ EtOH,\ \Delta}]{\substack{(1)\ Ac_2O,\ pyridine \\ (2)\ Br_2,\ CH_3CO_2H}} S \xrightarrow[\substack{(3)\ NaNO_2,\ HCl/273\text{-}278\ K \\ (4)\ H_3PO_2}]{\substack{(1)\ Sn/HCl \\ (2)\ Br_2/H_2O\ (excess)}} \text{Major product(s)}$$

[Aromatic Nitrocompounds, Anilines and Diazonium Salts, Type 2]

(a) 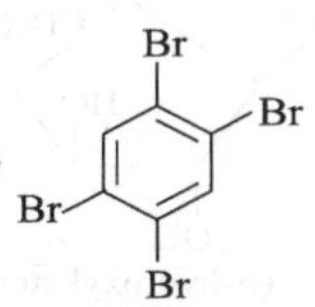

(b) 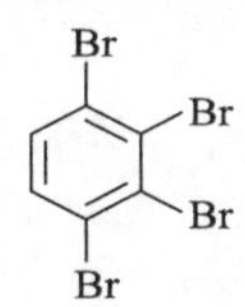

(c) 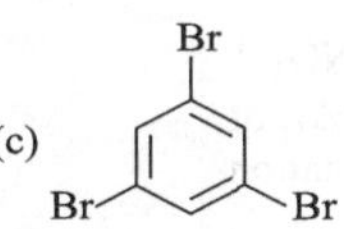

(d) 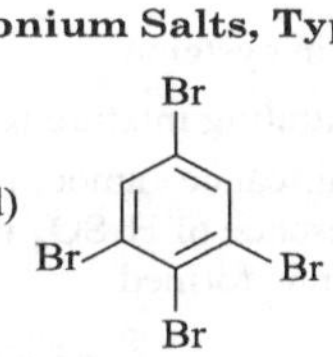

21. The Fischer presentation of D-glucose is given below.

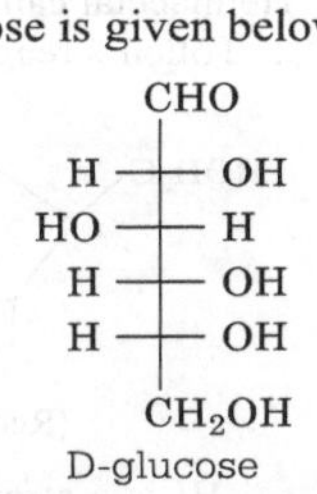

D-glucose

The correct structure(s) of β-L-glucopyranose is (are) **[Biomolecules, Type 2]**

(a) 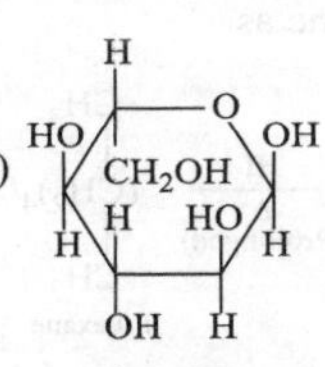

(b) 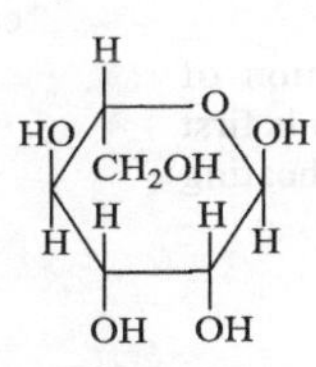

(c) 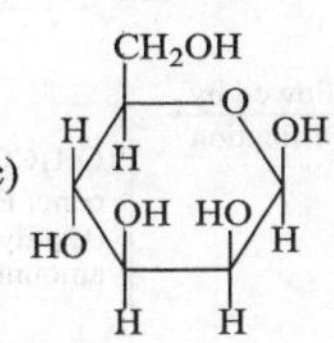

(d) 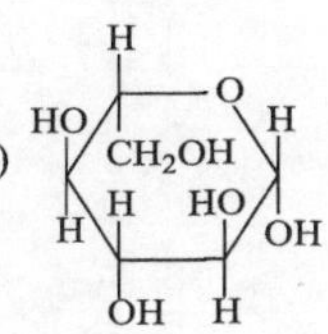

22. In the following reaction sequence, the amount of D (in gram) formed from 10 moles of acetophenone is ....
(Atomic weights in g mol^{-1} : H = 1, C = 12, N = 14, O = 16, Br = 80. The yield (%) corresponding to the product in each step is given in the parenthesis)

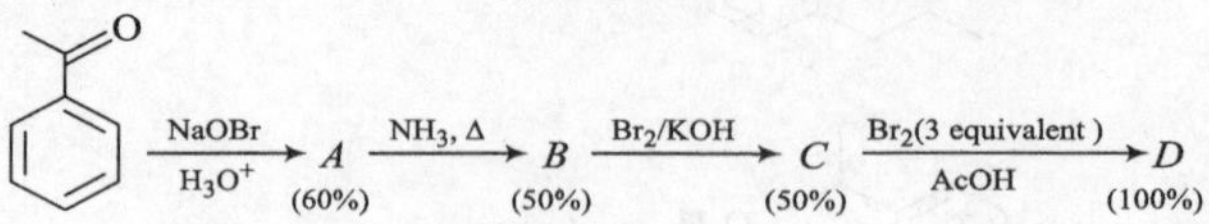

[Aromatic Nitrocompounds, Aniline and Diazonium Salts, Type 4]

Answer *with* Explanations

JEE Main

1. *(c)* High density polythene is used in the manufacture of buckets, dustbins etc.
2. *(b)*

Methionine

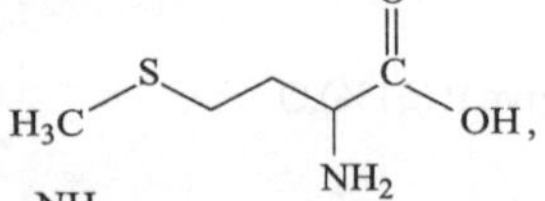

Cytosine

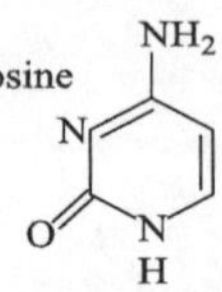

Thiol group (SH) is present in cysteine.

3. *(c)* Aniline in presence of nitrating mixture (conc. HNO_3 + conc. H_2SO_4) gives significance amount ($\approx$ 47%) of meta-product because in presence of H_2SO_4 its protonation takes place and anilinium ion is formed.

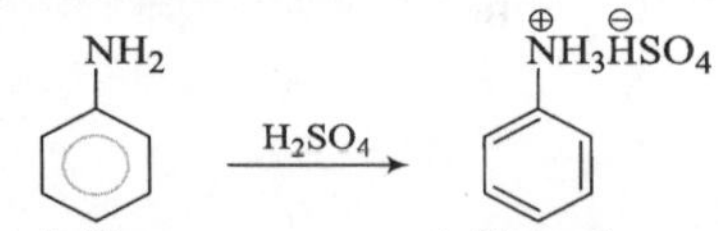

Aniline Anilinium ion

Here, anilinium ion is strongly deactivating group and *meta*-directing in nature. So, it gives *meta*-nitration product.

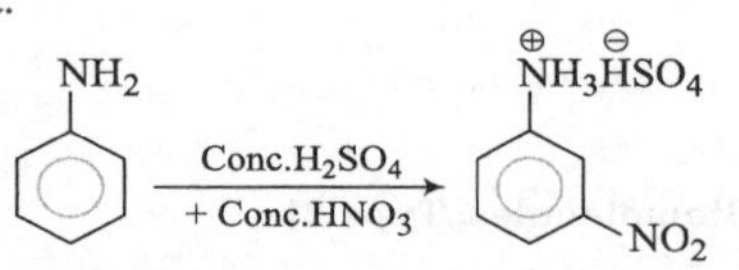

4. *(a)* Nylon-6 or perlon is prepared by polymerisation of amino caproic acid at high temperature. Caprolactam is first hydrolysed with water to form amino acid which on heating undergoes polymerisation to give nylon-6.

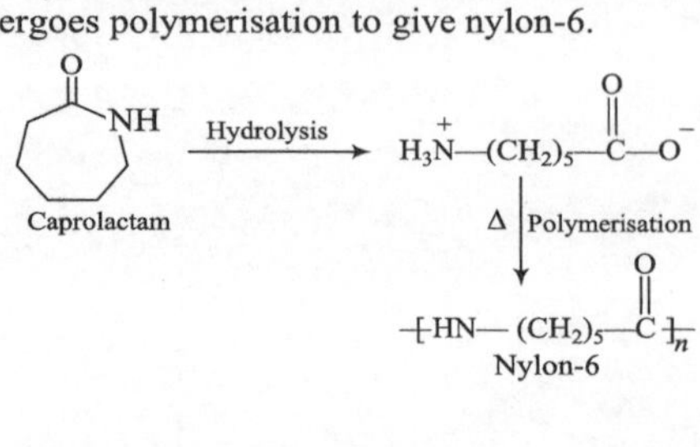

5. *(a)* Sugars that have an aldehyde, a ketone, a hemiacetal or a hemiketal group is able to reduce an oxidising agent. These sugars are classified as reducing sugars.

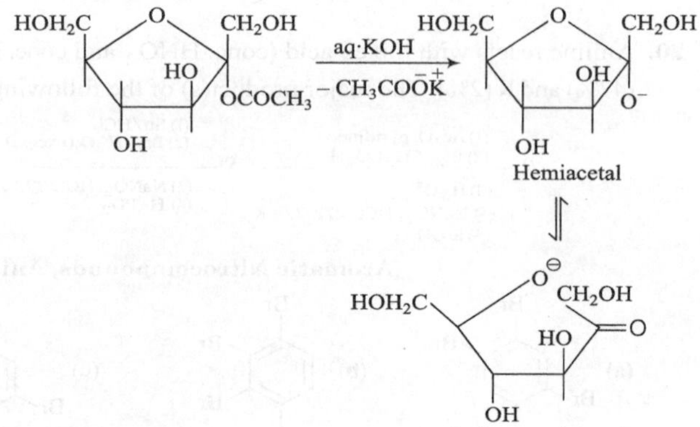

Hemiacetal can be easily reduced by oxidising agent such as Tollen's reagent.

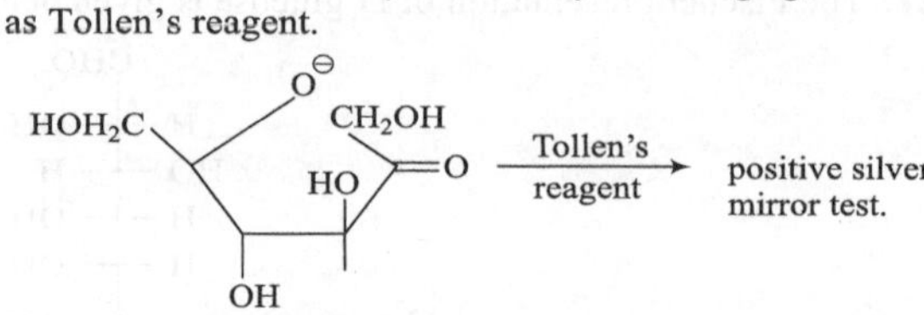

(Reducing sugar)

6. *(a)* HI is a strong reducing agent. It reduces both primary and secondary alcoholic groups of glucose along with the carbonyl group to produce *n*-hexane as

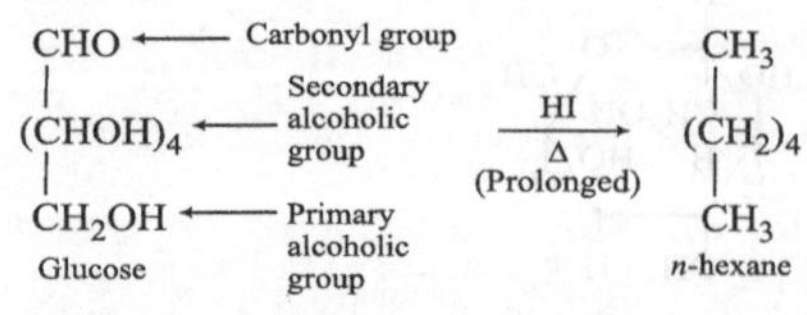

7. *(a)*

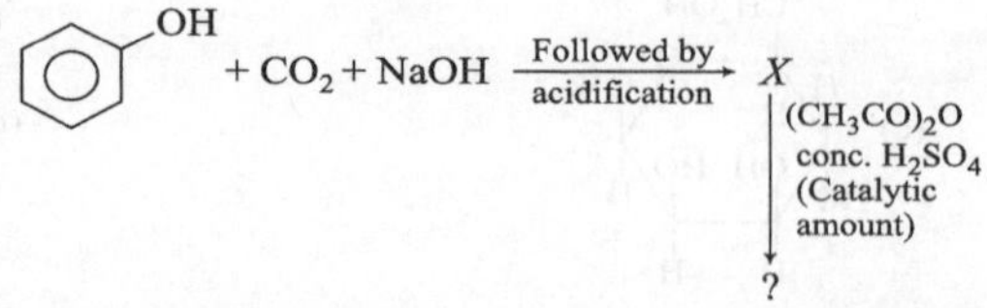

The very first reaction in the above road map looks like Kolbe's reaction which results to salicylic acid as

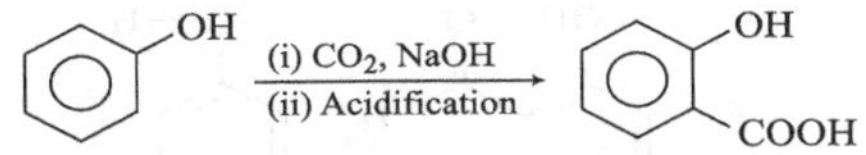

Salicylic acid
(*X*)

The salicylic acid with acetic anhydride $[(CH_3CO)_2O]$ in the presence of catalytic amount of conc. H_2SO_4 undergoes acylation to produce aspirin as

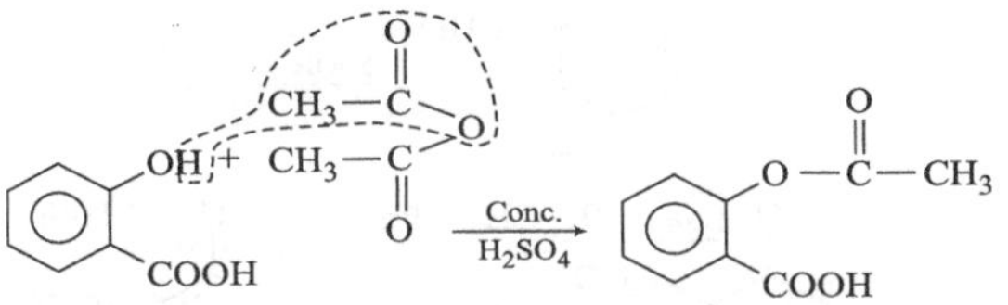

Acetyl salicylic acid (Aspirin)

$+ CH_3COOH$

Aspirin is a non-narcotic analgesic (Pain killer).

8. (*d*) Our blood is slightly basic in nature with pH range from 7.35-7.4.

The structure of histamine is given below:

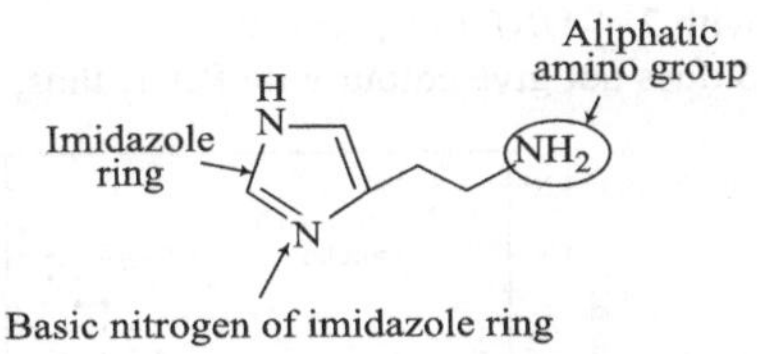

It is produced by decarboxylation of histidine having following structure. It is clearly visible from the above structure that histamine has two basic centres namely aliphatic amino group and basic nitrogen of imidazole ring.

The aliphatic amino group has pK_a around 9.4. In blood with pH around 7.4 the aliphatic amino group of histamine become protonated to give a single charged cation as shown below

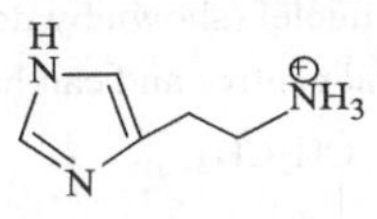

9. (*c*) Given

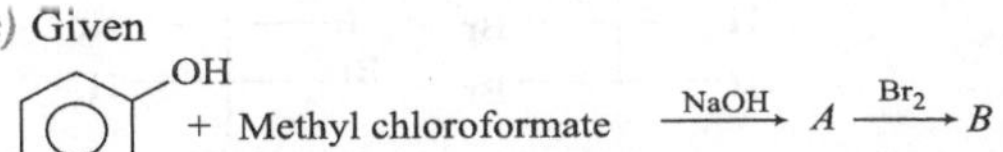

In the above road map, first reaction appears as acid base reaction followed by S_NAE (Nucleophilic substitution through Addition and Elimination). Both the steps are shown below

(i) **Acid base reaction**

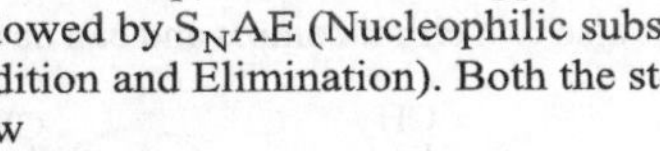

(ii) **S_NAE**

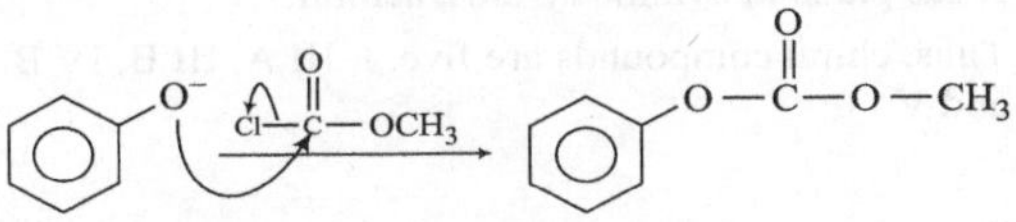

In the product of S_NAE the attached group is *ortho* and *para*-directing due to following cross conjugation

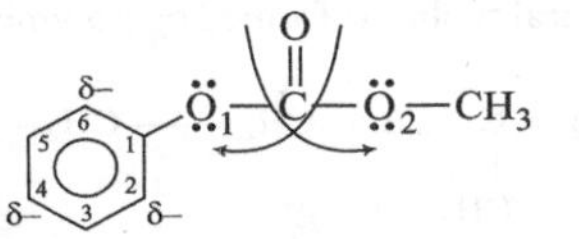

Cross conjugation due to which lone pair of oxygen 1 will be easily available to ring resulting to higher electron density at 2, 4,6 position with respect to group. However from the stability point of view *ortho* positions are not preferred by substituents as group —O—C(=O)—O—CH$_3$ is bulky.

Hence, on further bromination of S_NAE product *para* bromo derivative will be the preferred product i.e.

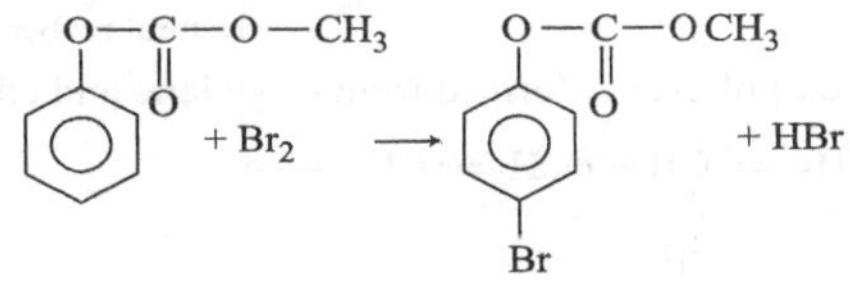

10. (*c*) **Key Idea** *Among the given compounds the basic nature depends upon their tendency to donate electron pair.*

Among the given compounds in 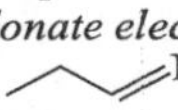, Nitrogen is *sp*^{2}-hybridised. This marginally increases the electronegativity of nitrogen which in turn decreases the electron donation tendency of nitrogen. Thus making compound least basic.

Among the rest 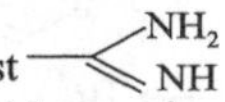 is totally different from others as in this compound lone pair of one nitrogen are in conjugation with π bond i.e. As a result of this conjugation the cation formed after protonation becomes resonance stabilised

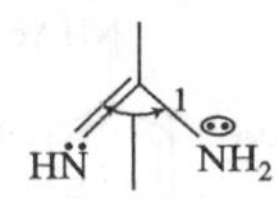

In conjugation

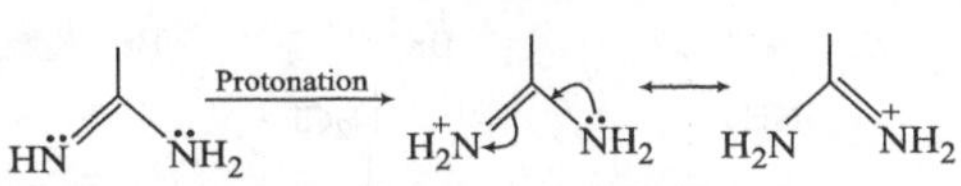

Equivalent resonance

This equivalent resonance in cation makes 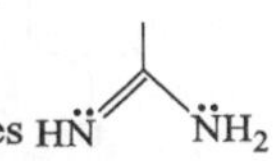 most basic among all.

Categorisation is very simple between rest two as 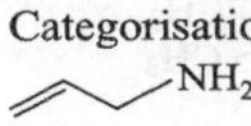 (primary amine) is less basic than 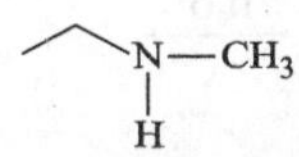 (secondary amine)

Hence, the correct order is

(II) < (I) < (IV) < (III) i.e. option (*c*) is correct.

JEE Advanced

1. *(a)* Natural rubber is formed by polymerisation of isoprene.

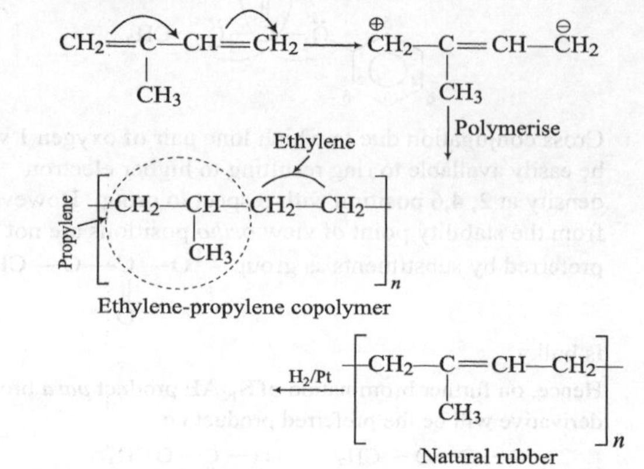

This co-polymer is formed from propylene and ethylene.

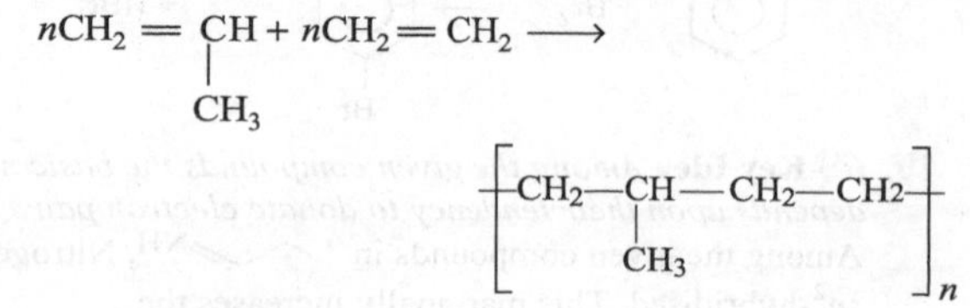

2. *(b)*

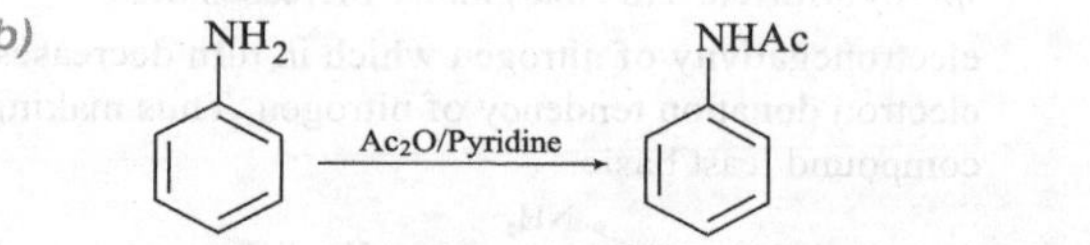

$KBrO_3 + HBr \longrightarrow Br_2$

Ac is CH_3CO (acetyl), it protects $-NH_2$ group from being oxidised.

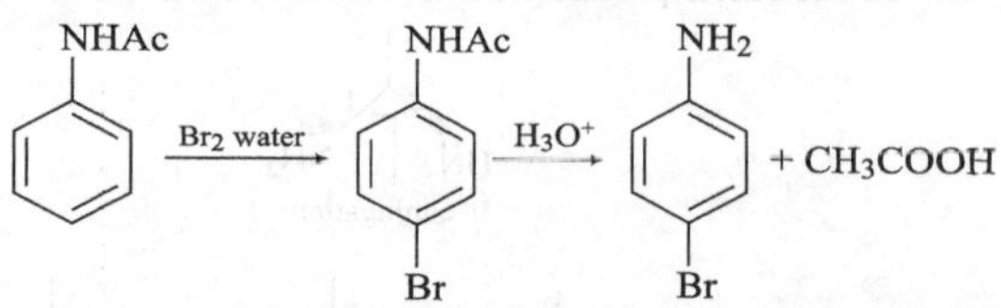

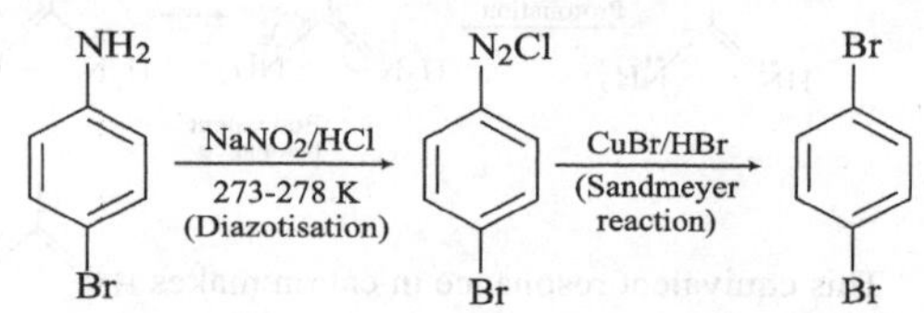

3. *(b, c)*

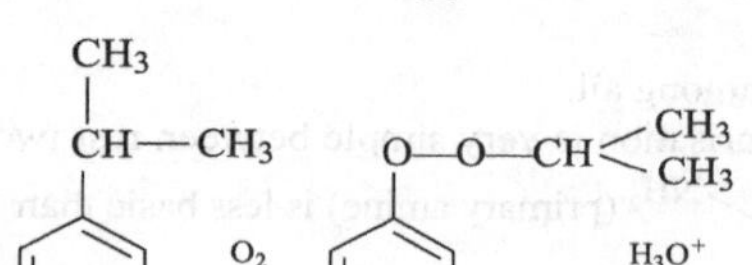

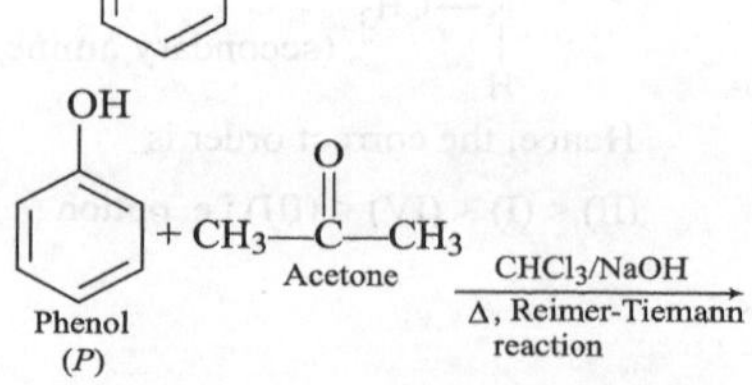

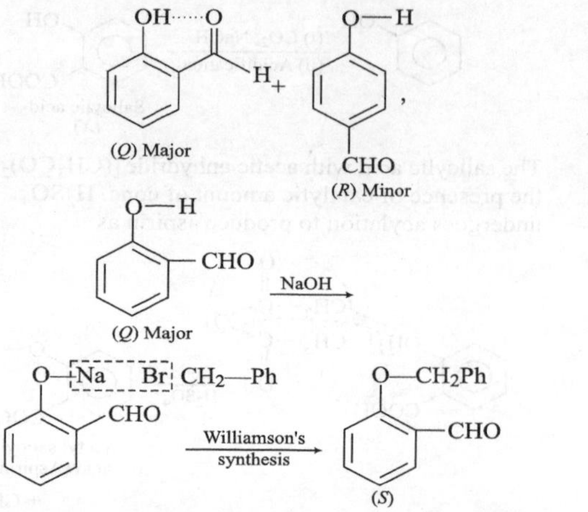

(a) R is not steam volatile, but Q is steam volatile thus, incorrect.

(b) Q has enolic group thus, gives violet colour with 1% aqueous $FeCl_3$ solution thus, correct.

(c) S has carbonyl group hence, gives yellow precipitate with 2, 4-DNP thus, correct.

(d) S does not give colour with $FeCl_3$ thus, incorrect.

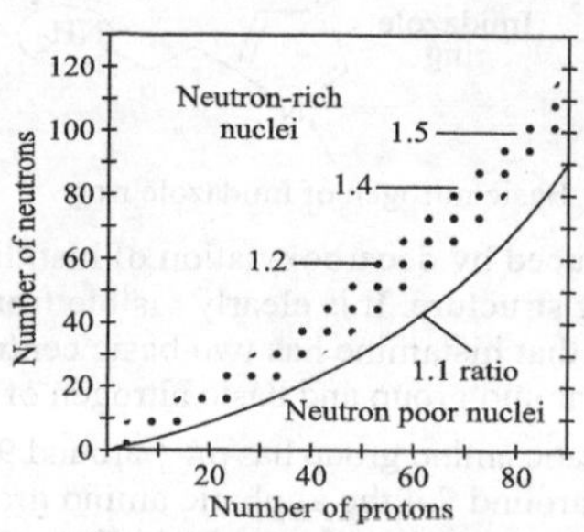

Plot of the number of neutrons against the number of protons in stable nuclei (shown by dots).

(III) has two chiral centres and can have two structures.

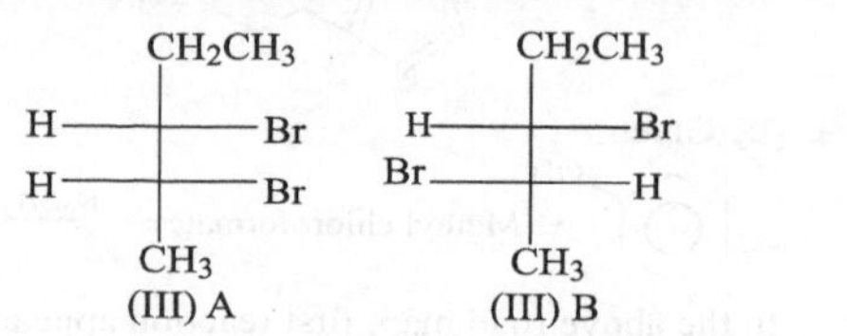

(IV) has also two chiral centres and can have two structures.

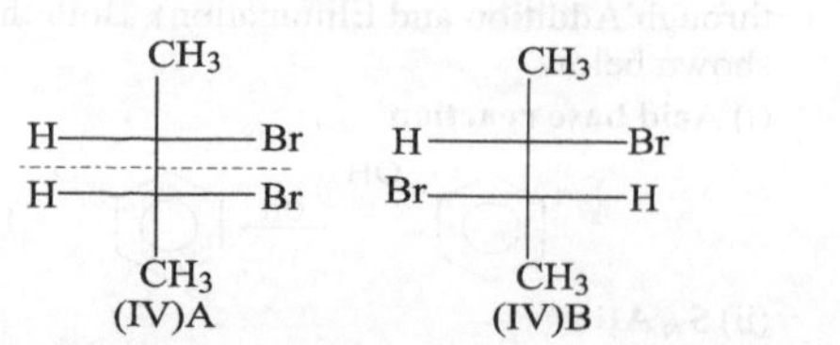

It has plane of symmetry thus, achiral.

Thus, chiral compounds are five. I, III A, III B, IV B and V.

4. *(a)* — OH group displays both kinds of effect; an electron withdrawing acid- strengthening inductive effect from the *meta*-position and an electron-releasing acid weakening resonance effect from the *para*-position (at this position, resonance effect overweighs the inductive effect). Thus, III > IV.

o-hydroxybenzoic acid (II) is far stronger than the corresponding *meta* and *para* isomers as the carboxylate ion is stabilised by intramolecular H-bonding.

2,6-dihydroxybenzoic acid (I) forms carboxylate ion which is further stabilised by intramolecular H-bonding, Thus, correct order is

I > II > III > IV

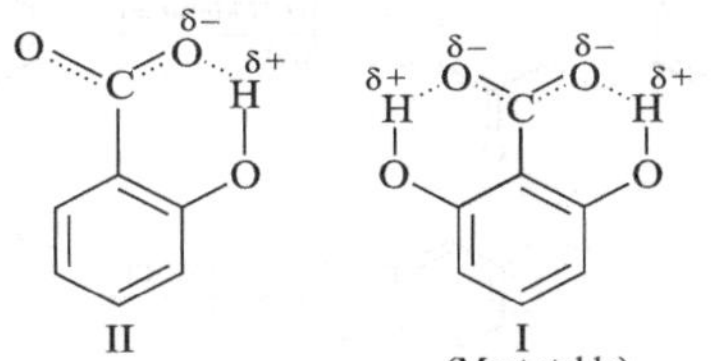

II I (Most stable)

5. *(a)*

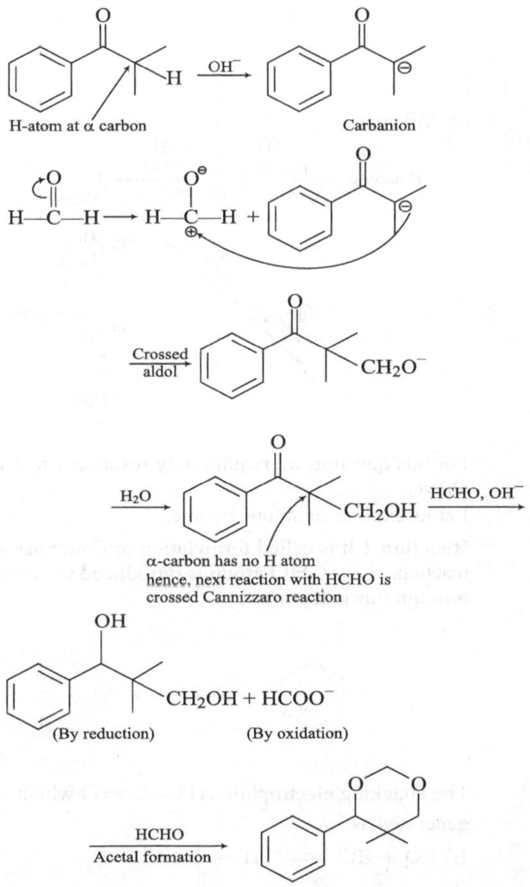

6. *(b, c)* If there is inversion of specific rotation from (+) to (−), then invert sugar is formed.

(a) $\underset{\substack{(+)\text{Maltose}\\140^\circ}}{C_{12}H_{22}O_{11}} + H_2O \longrightarrow \underset{\substack{D(+)\\52^\circ}}{\text{Glucose}}$

(b) $\underset{\substack{(+)\text{Sucrose}\\+66^\circ}}{C_{12}H_{22}O_{11}} + H_2O \longrightarrow \underset{\substack{D(+)\\52^\circ}}{\text{Glucose}} + \underset{\substack{L(-)\\-92^\circ}}{\text{Fructose}}$

-40° for 2 moles mixture
-20° for 1 mole mixture

There is formation of invert sugar. Thus, correct.

(c) Specific rotation of invert sugar is -20° per mole. Thus, correct.

(d) Br_2 water is a weak oxidising agent. It oxidises —CHO to —COOH. —CH_2OH group is not affected.

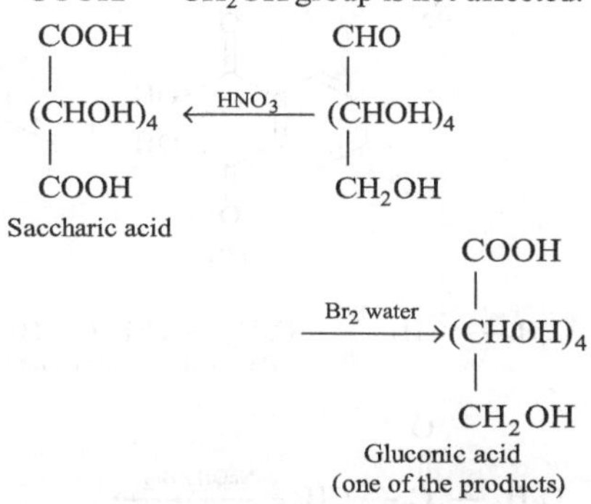

HNO_3 (a strong oxidising agent) oxidises invert sugar to saccharic acid. Thus, incorrect.

7. *(a)*

8. *(b)* Explanation

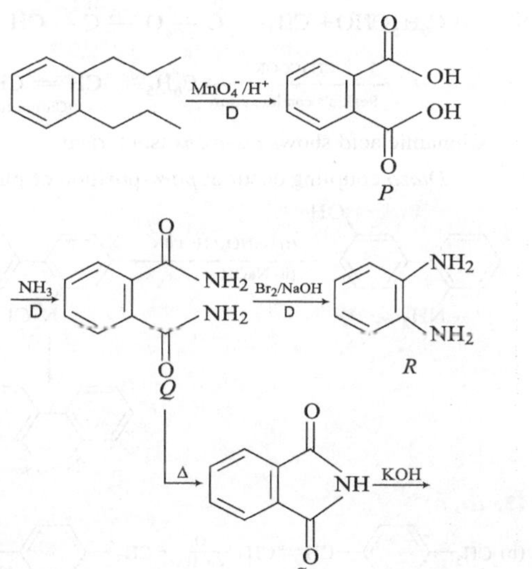

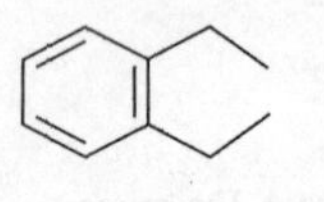

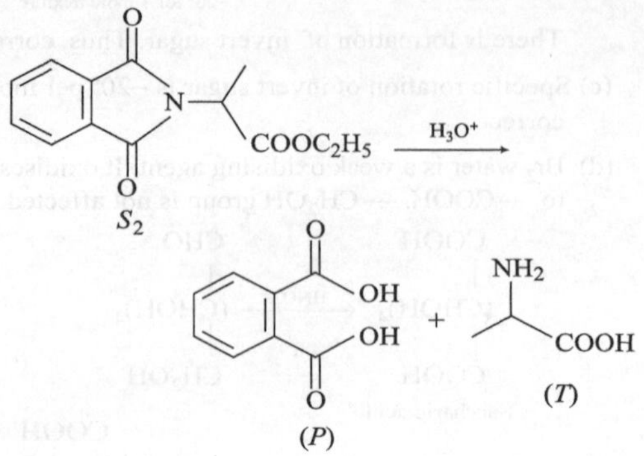

9. *(a)* $C_6H_5-CH_3 \xrightarrow[hv]{Br_2} C_6H_5-CH_2Br + HBr$
(Free radical bromination)

10. *(d)* $C_6H_5-\overset{\overset{O}{||}}{C}-CH_3 \xrightarrow[(i)]{NaOH/Br_2}$ (II)

$C_6H_5COONa + CHBr_3(P)$
Bromoform

11. *(b)* $C_6H_5CHO + CH_3-\overset{\overset{O}{||}}{C}-O-\overset{\overset{O}{||}}{C}-CH_3$

$\xrightarrow[\text{Perkin's condensation}]{CH_3COOK} C_6H_5-CH=CH-COOH$
(Cinnamic acid)

Cinnamic acid shows *cis-trans* isomerism.

12. *(c)* *Diazo* coupling occur at *para*-position of phenol.

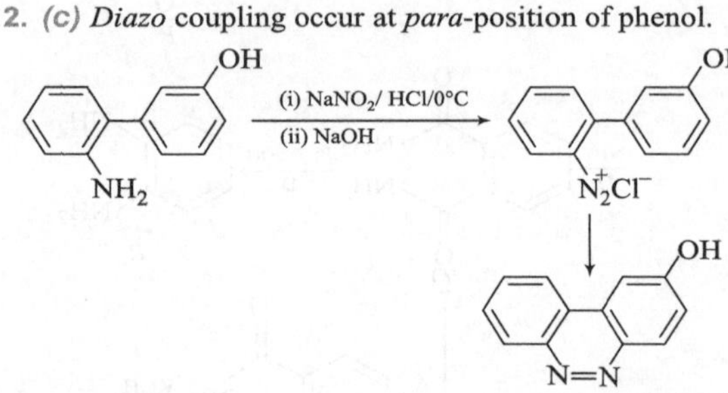

13. *(b, c)*

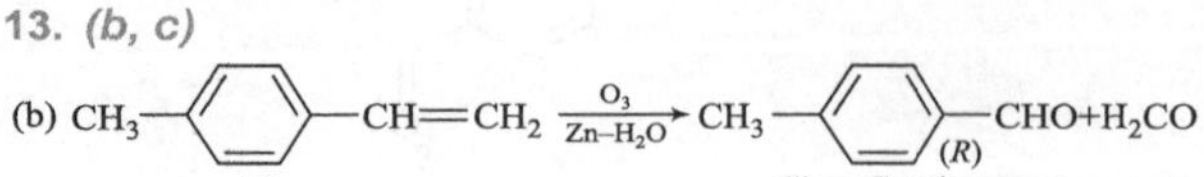

Gives Cannizzaro but not haloform reaction

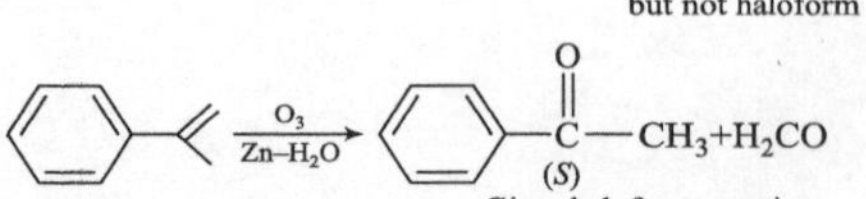

Gives haloform reaction but not Cannizzaro reaction

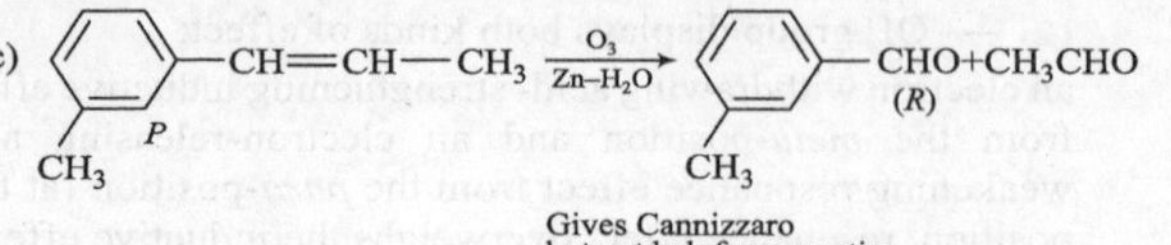

Gives Cannizzaro but not haloform reaction

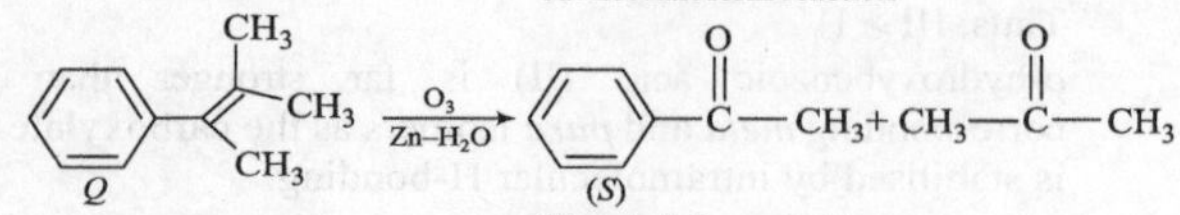

Gives haloform reaction but not Cannizzaro reaction

Paragraph-2

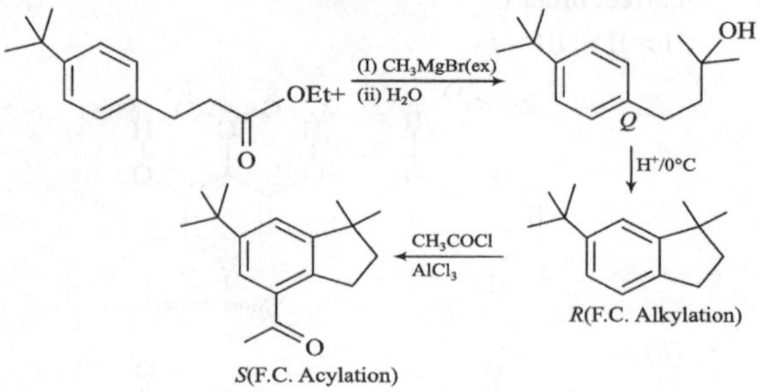

14. *(a)*

15. *(c)*

16. *(c)* Given,

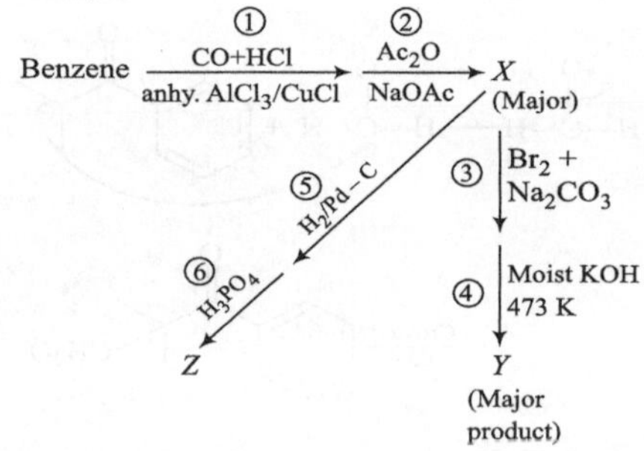

For this question we require only reaction 1 to 4 written above.

Let us explore them one by one.

Reaction 1 It is called formylation or Gatterman Koch reaction. $A-CHO$ group is introduced to benzene ring through this reaction as

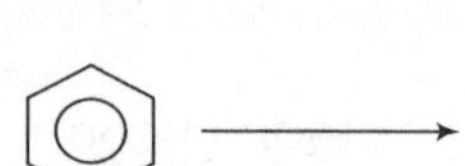

The attacking electrophile is $H-\overset{+}{C}=O$ which is generated as

(i) $CO + HCl \rightleftharpoons H-\underset{\underset{O}{||}}{C}-Cl$

(ii) $H-\underset{\underset{O}{||}}{C}-Cl + AlCl_3 \rightleftharpoons H-\underset{\underset{O}{||}}{\overset{+}{C}} + AlCl_4^-$

Reaction 2 It is Perkin condensation which results in α, β unsaturated acid as

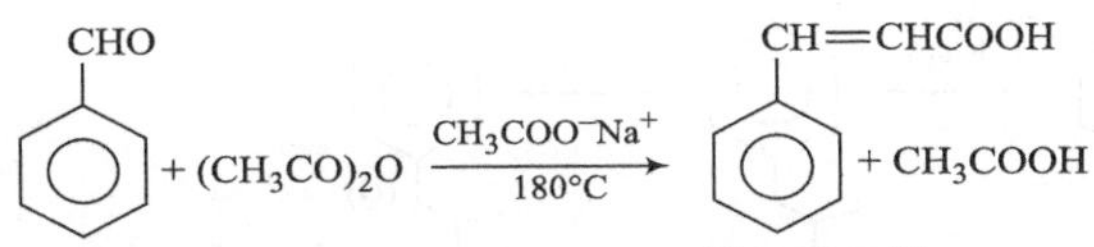

Cinnamic acid

Note Besides $CH_3COO^-Na^+$, quinoline, pyridine, Na_2CO_3, triethylamine can also be used as bases in this reaction.

Reaction 3 It is simple addition of bromine to unsaturated acid formed through reaction 2.

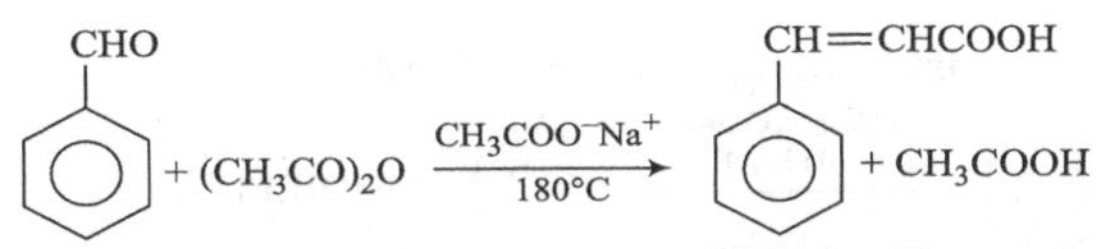

Cinnamic acid

Na_2CO_3 works as a base in the reaction to trap H^+ to be released by — COOH of the reactant. 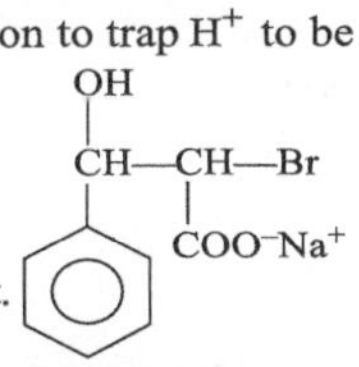 is also formed in the reaction as the minor product.

Reaction 4 It is decarboxylation and dehydrohalogenation of product produced by reaction 3 as

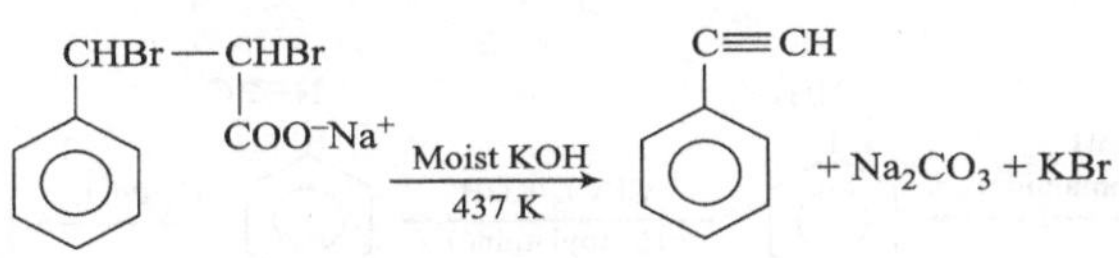

Hence, Y is 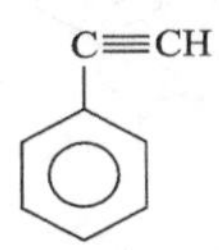 i.e., (c) is the correct answer.

17. *(a)* **Reaction 5** The Perkin condensation product X is

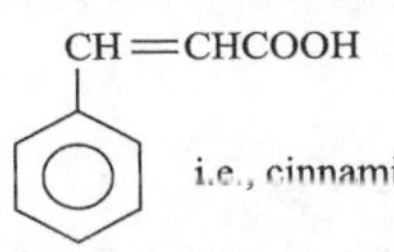

i.e., cinnamic acid (Ref. Q. 15)

This compound on hydrogenation with H_2 in the presence of Pd activated with charcoal (Pd – C) gives

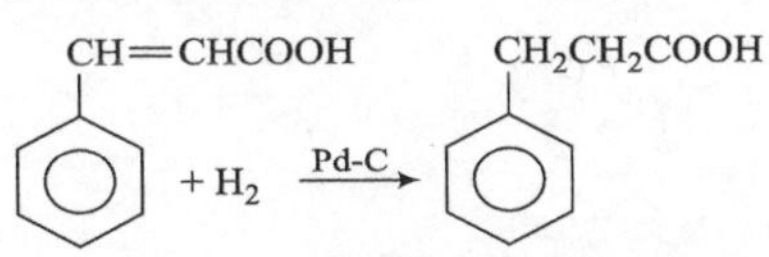

3-Phenyl propanoic acid

Reaction 6 The product of reaction 5 on heating with H_3PO_4 dehydrates to give

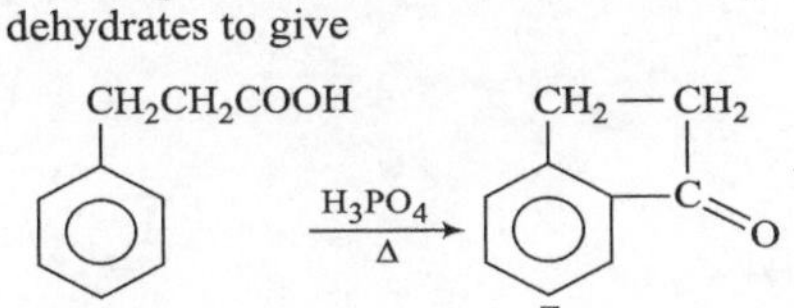

Z

18. *(a)* Given,

(i) $C_{11}H_{12}O_2 \xrightarrow{\text{Oxidation}}$ Dibasic acid (An organic acid 'P')
This indicates the presence of alkyl or alkenyl branch in P along with –COOH group.

(ii) Dibasic acid produced by oxidation of P

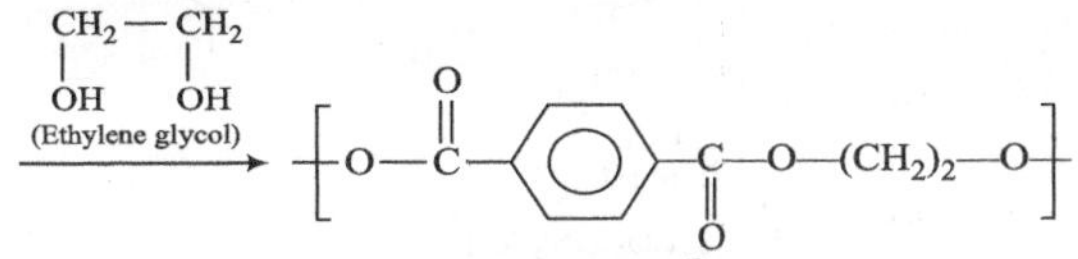

Dacron

This indicates presence of benzene ring in P; as concluded from the structure of dacron given below.

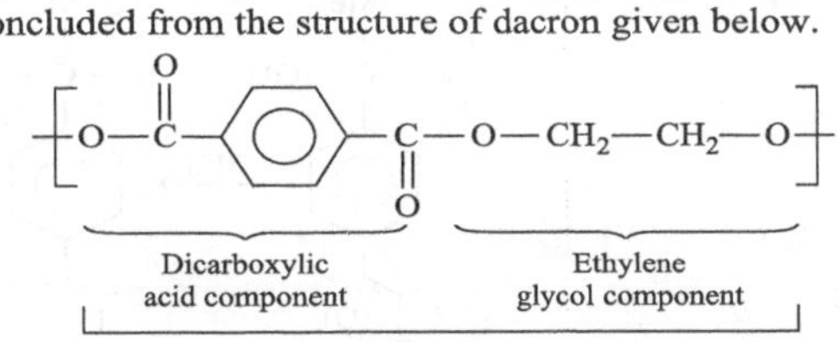

Dacron

(iii) $P \xrightarrow{\text{Ozonolysis}}$ Aliphatic ketone + other oxidised products.
This reaction confirms the presence of multiple bonded branch i.e., alkenyl group in P.

Thus P can be

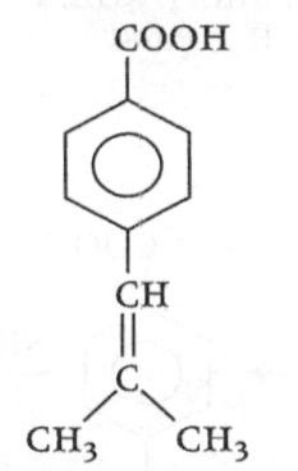

Attachment of —COO group in dacron also confirm the *para* position of branch with respect to —COOH group in P.

IUPAC name : 4-(2-methyl) prop-l-enyl benzoic acid

Now look for the reactions

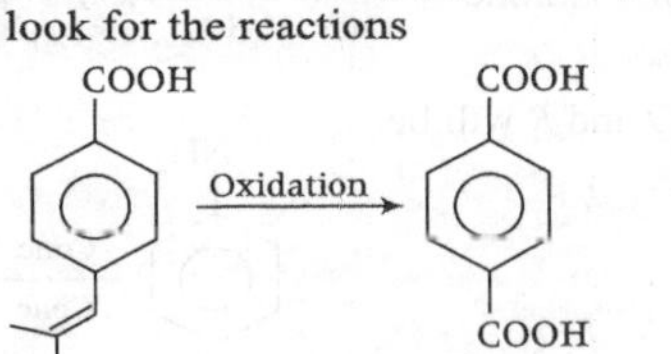

(P) Terephthalic acid

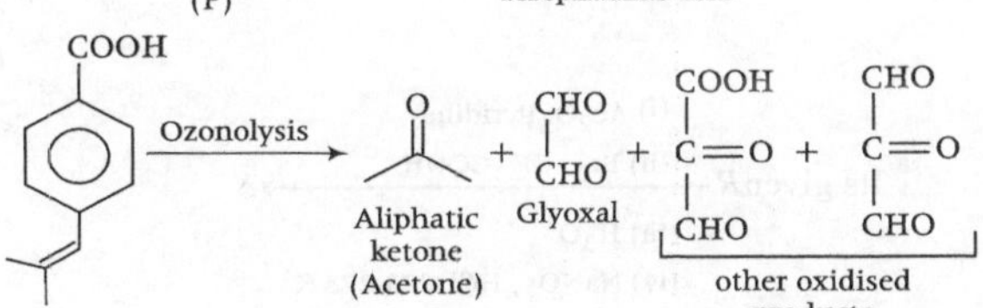

(P)

$$\text{Given, } P \xrightarrow[\text{3. MeMgBr, CdCl}_2\text{ 4. NaBH}_4]{\text{1. H}_2\text{/Pd–C 2. SOCl}_2} Q$$

So,

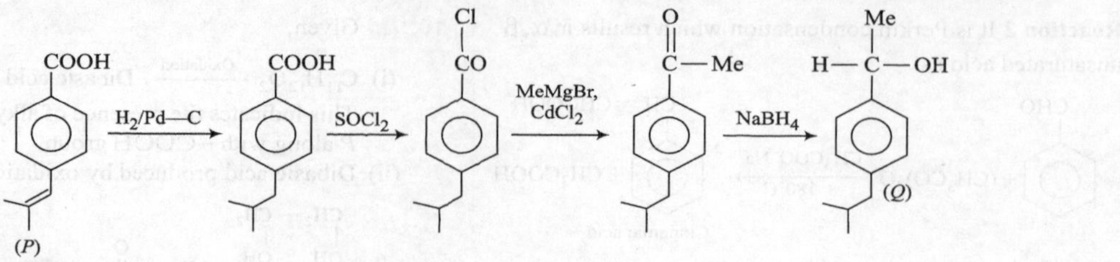

Further, $Q \xrightarrow[\text{3. } CO_2 \text{ (Dry ice)} \quad \text{4. } H_3O^+]{\text{1. HCl} \quad \text{2. } Mg/Et_2O} R$

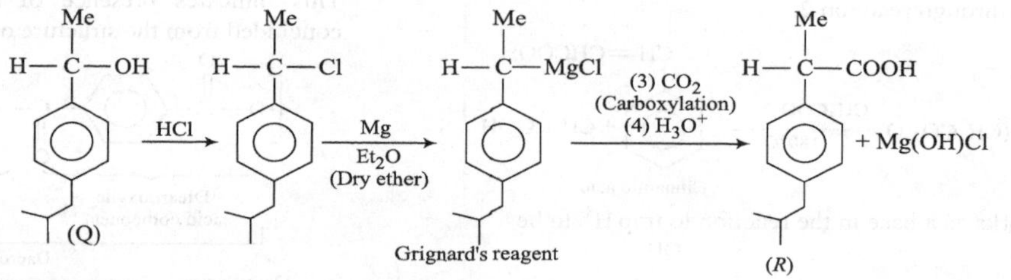

19. *(b)* Given (In connection with Q. 17)

$P \xrightarrow[\text{4. } CHCl_3/KOH,\Delta \quad \text{5. } H_2/\text{Pd–C}]{\text{1. } H_2/\text{Pd–C} \quad \text{2. } NH_3/\Delta \quad \text{3. } Br_2/NaOH} S$

So,

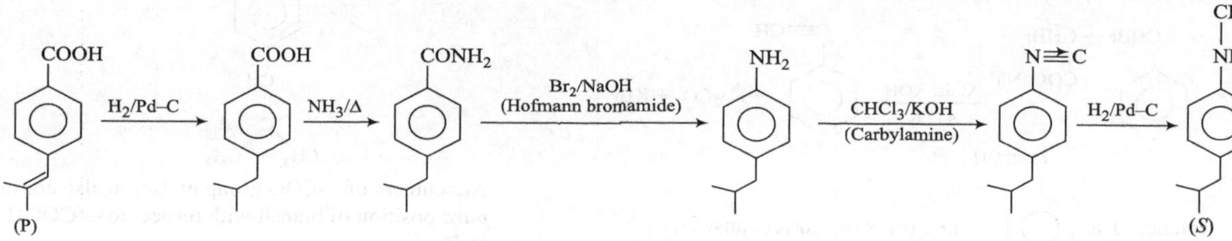

20. *(d)* Given, Aniline $\xrightarrow[\text{Conc. } H_2SO_4]{\text{Conc. } HNO_3 +} P\ (51\%) + Q\ (47\%) + R\ (2\%)$

Then P, Q and R will be

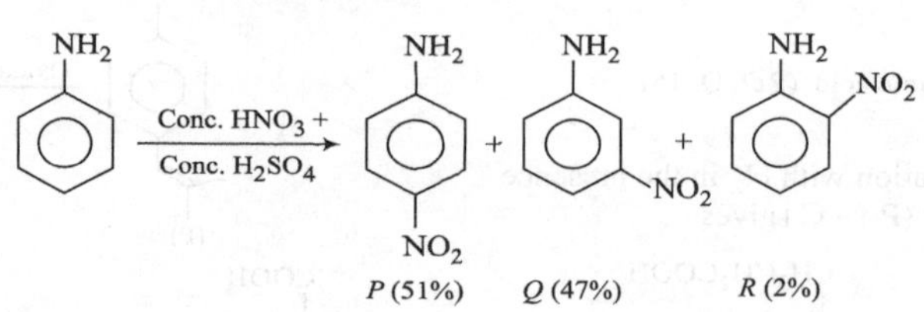

Its given $R \xrightarrow[\text{(iii) } H_3O^+ \quad \text{(iv) } NaNO_2, \text{ HCl/273-278 K} \quad \text{(v) EtOH, } \Delta]{\text{(i) } Ac_2O, \text{ Pyridine} \quad \text{(ii) } Br_2, CH_3COOH} S$

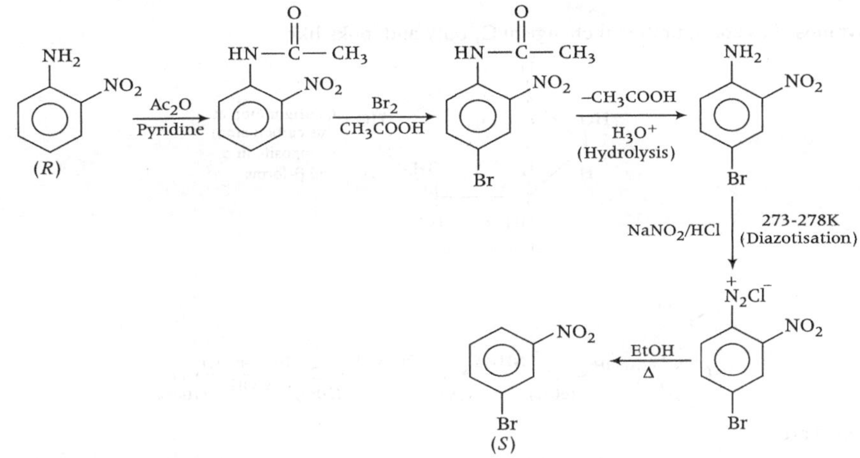

Now from S to major products its given.

$$S \xrightarrow[\text{(iii) } NaNO_2,\ HCl/273\text{-}278K \quad \text{(iv) } H_3PO_2]{\text{(i) } Sn/HCl \quad \text{(ii) } Br_2/H_2O \text{ (Excess)}} \text{Major product}$$

So,

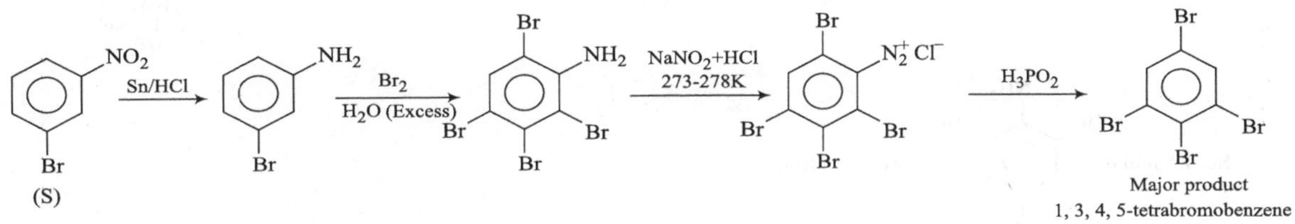

Hence, only (d) is the correct answer.

21. *(d)*

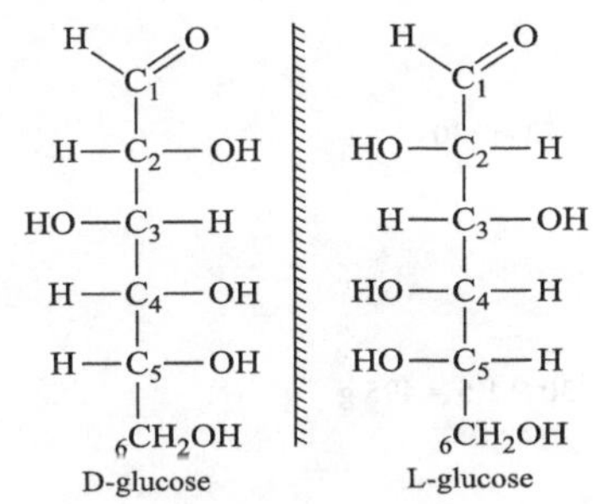

So, β-L glucopyranose is formed as

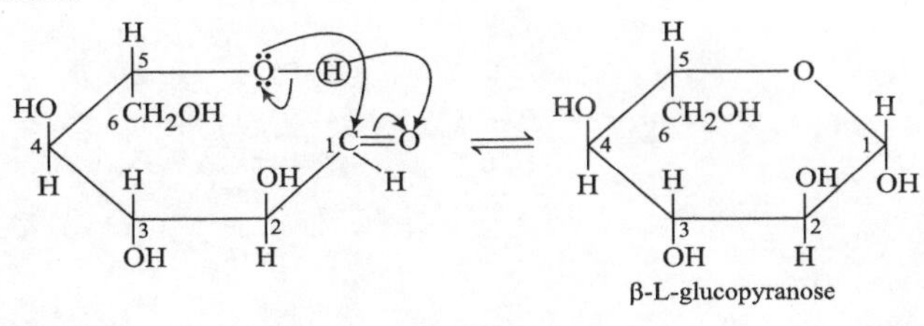

The α-L-glucopyranose has configurational change at C_1 only and looks like

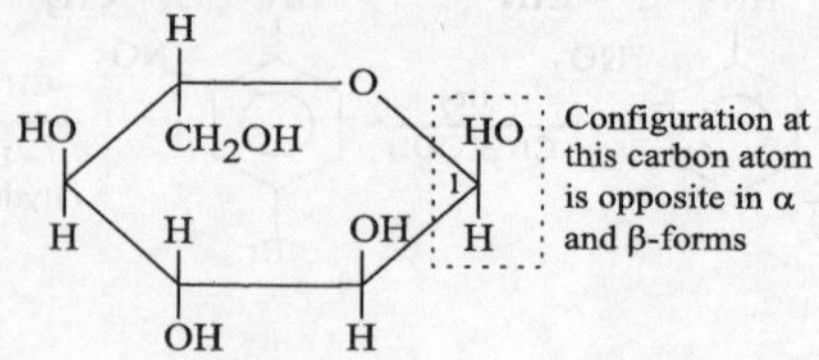

α-L-glucopyranose

22. *(495 g)* Given,

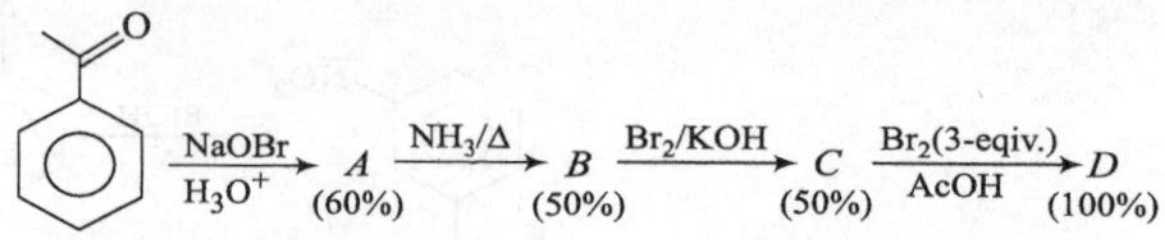

The products formed are

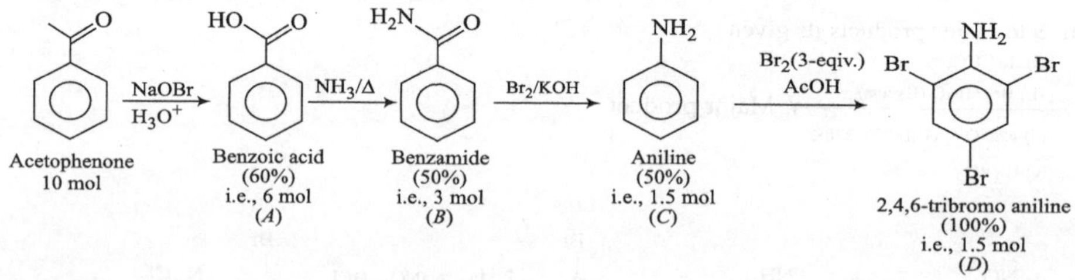

So, 1.5 mol of 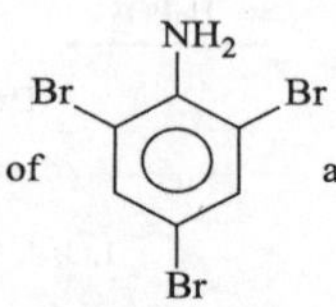

are produced from

10 moles of acetophenone.

Molar mass of 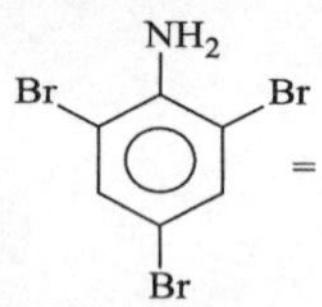

$= 240 + 14 + 4 + 72 = 330$

Hence, amount of 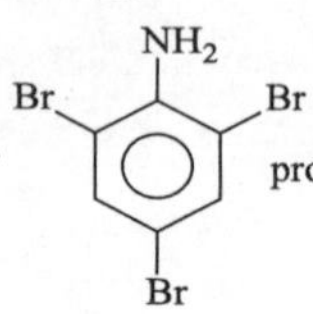

produced is $330 \times 1.5 = 495$ g